MW00844959

Principles of Water Treatment

Principles of Water Treatment

Kerry J. Howe, Ph.D., P.E., BCEE
Associate Professor of Civil Engineering
University of New Mexico

David W. Hand, Ph.D., BCEEM
Professor of Civil and Environmental Engineering
Michigan Technological University

John C. Crittenden, Ph.D., P.E., BCEE, NAE
Hightower Chair and Georgia Research Alliance Eminent Scholar
Director of the Brook Byers Institute for Sustainable Systems
Georgia Institute of Technology

R. Rhodes Trussell, Ph.D., P.E., BCEE, NAE
Principal
Trussell Technologies, Inc.

George Tchobanoglous, Ph.D., P.E., NAE
Professor Emeritus of Civil and Environmental Engineering
University of California at Davis

Cover Design: Michael Rutkowski
Cover Photographs: Main photograph courtesy of George Tchobanoglous; top photographs courtesy of MWH file photographs.
Cover photo is the Vineyard Surface Water Treatment Plant, owned by the Sacramento County Water Agency.

This book is printed on acid-free paper. ♾

Published by John Wiley & Sons, Inc., Hoboken, New Jersey
Published simultaneously in Canada

Library of Congress Cataloging-in-Publication Data:

Principles of water treatment / Kerry J. Howe, David W. Hand, John C. Crittenden, R. Rhodes Trussell, George Tchobanoglous.
pages cm
Includes index.
ISBN 978-0-470-40538-3 (hardback); ISBN 978-1-118-30167-8 (ebk); ISBN 978-1-118-30168-5 (ebk); ISBN 978-1-118-30967-4 (ebk); ISBN 978-1-118-30969-8 (ebk); ISBN 978-1-118-30970-4 (ebk)
1. Water–Purification. I. Howe, Kerry J. II. Hand, David W. III. Crittenden, John C. (John Charles), 1949- IV. Trussell, R. Rhodes V. Tchobanoglous, George.
TD430.W3752 2012
628.1′62—dc23

2012017207

Printed in the United States of America
SKY10028468_072621

About the Authors

Dr. Kerry J. Howe is an associate professor in the Department of Civil Engineering at the University of New Mexico. His career in water treatment spans both consulting and academia. He has a B.S. degree in civil and environmental engineering from the University of Wisconsin-Madison, an M.S. degree in environmental health engineering from the University of Texas at Austin, and a Ph.D. degree in environmental engineering from the University of Illinois at Urbana-Champaign. After a stint at CH2M-Hill, he worked for over 10 years at MWH, Inc., where he was involved in the planning, design, and construction of water and wastewater treatment facilities up to 380 ML/d (100 mgd) in capacity. He has experience with conventional surface water treatment and other treatment technologies such as membrane treatment, ozonation, and packed-tower aeration. At the University of New Mexico, his teaching and research focuses on membrane processes and desalination, physicochemical treatment processes, water quality, sustainability, and engineering design. Dr. Howe is a registered professional engineer in Wisconsin and New Mexico and a Board Certified Environmental Engineer by the American Academy of Environmental Engineers.

Dr. David W. Hand is a professor of civil and environmental engineering at the Michigan Technological University. He received his B.S. degree in engineering at Michigan Technological University, an M.S. degree in civil engineering at Michigan Technological University, and a Ph.D. in engineering from Michigan Technological University. His teaching and research focuses on water and wastewater treatment engineering with emphasis on physicochemical treatment processes. He has authored and co-authored over 130 technical publications including six textbooks, two patents, and eight copyrighted software programs. He received the ASCE Rudolf Hering Medal, an outstanding teaching award and publication award from the Association of Environmental Engineering and Science Professors, and a publication award from American Water Works Association. He is a Board Certified Environmental Engineering Member of the American Academy of Environmental Engineers.

Dr. John C. Crittenden is a professor in the School of Civil and Environmental Engineering at the Georgia Institute of Technology and the director of the Brook Byers Institute for Sustainable Systems. In this position, he leads the creation of an integrated initiative in Sustainable Urban Systems. He is a Georgia Research Alliance (GRA) Eminent Scholar in Sustainable Systems and occupies the Hightower Chair for Sustainable Technologies. Dr. Crittenden is an accomplished expert in sustainability, pollution prevention, physicochemical treatment processes, nanotechnology, air and water treatment, mass transfer, numerical methods, and modeling of air, wastewater, and water treatment processes. He has received multiple awards for his research in the treatment and removal of hazardous materials from drinking water and groundwater. He has four copyrighted software products and three patents in the areas of pollution prevention, stripping, ion exchange, advanced oxidation/catalysis, adsorption and groundwater transport. The American Institute of Chemical Engineers (AIChE) Centennial Celebration Committee named Dr. Crittenden as one of the top 100 Chemical Engineers of the Modern Era at their 100th annual meeting in 2008. He is a member of the National Academy of Engineering.

Dr. R. Rhodes Trussell is a registered Civil and Corrosion Engineer in the State of California with 40 years of water treatment experience. He has a B.S., M.S., and Ph.D. in environmental engineering from the University of California at Berkeley. He founded Trussell Technologies, Inc., a consulting firm specializing in the application of science to engineering, after working for 33 years for MWH, Inc. He has authored more than 200 publications, including several chapters in all three editions of *MWH's Water Treatment: Principles and Design*. Dr. Trussell has served as Chair of the EPA Science Advisory Board's Committee on Drinking Water, serves on the Membership Committee for the National Academy of Engineering, and as Chair of the Water Science and Technology Board for the National Academies. For the International Water Association, Dr. Trussell serves as a member of the Scientific and Technical Council, the Editorial Board, and on the Program Committee. In 2010, Dr. Trussell was awarded the prestigious A.P. Black Award from the American Water Works Association.

Dr. George Tchobanoglous is a professor emeritus of environmental engineering in the Department of Civil and Environmental Engineering at the University of California at Davis. He received a B.S. degree in civil engineering from the University of the Pacific, an M.S. degree in sanitary engineering from the University of California at Berkeley, and a Ph.D. in environmental engineering from Stanford University. His principal research interests are in the areas of wastewater treatment, wastewater filtration, UV disinfection, wastewater reclamation and reuse, solid waste management, and wastewater management for small systems. He has authored or coauthored over 500 technical publications, including

22 textbooks and 8 reference works. Professor Tchobanoglous serves nationally and internationally as a consultant to both governmental agencies and private concerns. An active member of numerous professional societies, he is a past president of the Association of Environmental Engineering and Science Professors. He is a registered civil engineer in California and a member of the National Academy of Engineering.

Contents

Preface

Without water, life cannot exist. Thus, securing an adequate supply of fresh, clean water is essential to the health of humankind and the functioning of modern society. Water is also known as the universal solvent—it is capable of dissolving a vast number of natural and synthetic chemicals. Increasing population and the contamination of water with municipal, agricultural, and industrial wastes has led to a deterioration of water quality and nearly all sources of water require some form of treatment before potable use. This textbook is designed to serve as an introduction to the field of water treatment and the processes that are used to make water safe to drink.

The authors of this book have collaborated on two books that are intertwined with each other, both published by John Wiley and Sons, Inc. The other book, *MWH's Water Treatment: Principles and Design,* 3rd ed. (Crittenden et al., 2012), was the source for a significant portion of the material in this book. The focus of this present book is on principles of water treatment; it is suitable as a textbook for both undergraduate and graduate courses. The other book is an expanded edition, nearly triple the length of this one, that provides more comprehensive coverage of the field of drinking water treatment and is suitable as both a textbook and a reference for practicing professionals. The unit process chapters of *MWH's Water Treatment: Principles and Design* contain a detailed analysis of the principles of treatment processes as well as in-depth material on design. *MWH's Water Treatment: Principles and Design* also provides extensive chapters on the physical, chemical, and microbiological quality of water, removal of selected contaminants, internal corrosion of water conduits, and case studies that are not included in this book. Students who use this textbook in a class on water treatment and go on to a career in design of water treatment facilities are encouraged to consult *MWH's Water Treatment: Principles and Design* on topics that were beyond the scope of this textbook.

Acknowledgments

The authors gratefully acknowledge the people who assisted with the preparation of this book. Particular credit goes to Dr. Harold Leverenz of the University of California at Davis, who adapted most of the figures for this textbook after preparing them for the companion book, *MWH's Water Treatment: Principles and Design*, 3rd Ed. Figures for several chapters were prepared by Mr. James Howe of Rice University. Mr. Daniel Birdsell and Ms. Lana Mitchell of the University of New Mexico reviewed and checked the chapters, including the figure, table, and equation numbers, the math in example problems, and the references at the ends of the chapters. Ms. Lana Mitchell also helped prepare the solutions manual for the homework problems. Dr. Sangam Tiwari of Trussell Technologies assisted with the writing of Chap. 2, Dr. Daisuke Minakata of Georgia Tech assisted with the writing of Chaps. 10 and 12, and Dr. Zhongming Lu of Georgia Tech assisted with the writing of Chap. 10.

Several chapters were reviewed by external reviewers and their comments helped improve the quality of this book. The reviewers included:

Ms. Elaine W. Howe, Trussell Technologies Inc.

Dr. Jaehong Kim, Georgia Institute of Technology

Dr. David A. Ladner, Clemson University

Dr. Qilin Li, Rice University

Dr. Edward D. Schroeder, University of California-Davis

Dr. John E. Tobiason, University of Massachusetts-Amherst

We gratefully acknowledge the support and help of the Wiley staff, particularly Bob Hilbert, James Harper, Robert Argentieri, and Daniel Magers.

Kerry J. Howe
David W. Hand
John C. Crittenden
R. Rhodes Trussell
George Tchobanoglous

1 Introduction

Securing and maintaining an adequate supply of water has been one of the essential factors in the development of human settlements. The earliest communities were primarily concerned with the quantity of water available. Increasing population, however, has exerted more pressure on limited high-quality surface sources, and contamination of water with municipal, agricultural, and industrial wastes has led to a deterioration of water quality in many other sources. At the same time, water quality regulations have become more rigorous, analytical capabilities for detecting contaminants have become more sensitive, and the public has become more discriminating about water quality. Thus, the quality of a water source cannot be overlooked in water supply development. In fact, most sources of water require some form of treatment before potable use.

Water treatment can be defined as the processing of water to achieve a water quality that meets specified goals or standards set by the end user or a community through its regulatory agencies. Goals and standards can include the requirements of regulatory agencies, additional requirements set by a local community, and requirements associated with specific industrial processes.

The primary focus of this book is the principles of water treatment for the production of potable or drinking water on a municipal level. Water treatment, however, encompasses a much wider range of problems and ultimate uses, including home treatment units and facilities for industrial water treatment with a wide variety of water quality requirements that depend on the specific industry. Water treatment processes are also applicable to remediation of contaminated groundwater and other water sources and wastewater treatment when the treated wastewater is to be recycled for new uses. The issues and processes covered in this book are relevant to all of these applications.

This book thoroughly covers the fundamental principles that govern the design and operation of water treatment processes. Following this introduction, the next three chapters provide background information that is necessary to understand the scope and complexity of treatment processes. Chapter 2 describes the relationship between water quality and public health, introduces the types of constituents that are present in various water supplies, and outlines some of the challenges faced by water treatment professionals. Chapter 3 introduces how the physicochemical properties of constituents in water and other factors guide the selection of treatment processes. Chapter 4 introduces the core principles necessary for understanding treatment processes, such as chemical equilibrium and kinetics, mass balance analysis, reactor analysis, and mass transfer. Chapters 5 through 13 are the heart of the book, presenting in-depth material on each of the principal unit processes traditionally used in municipal water treatment. Chapter 14 presents material on the processing of treatment residuals, a subject that can have a significant impact on the design and operation of treatment facilities.

1-1 The Importance of Principles

From the 1850s to about the 1950s, water treatment facilities were frequently designed by experienced engineers who drew upon previous successful design practices. Improvements were made by incremental changes from one plant to the next. Treatment processes were often treated as a "black box," and detailed understanding of the scientific principles governing the process was not essential in completing a successful design. In recent years, however, significant changes have taken place in the water treatment industry that require engineers to have a greater understanding of fundamental principles underlying treatment processes. Some of these changes include increasing contamination of water supplies, increasing rate of technological development, and increasing sophistication of treatment facilities.

Early treatment practices were primarily focused on the aesthetic quality of water and prevention of contamination by pathogenic organisms. These treatment goals were relatively clear-cut compared to today's requirements. Since about the 1950s, tens of thousands of chemicals have been developed for a wide variety of purposes—about 3300 chemicals are produced in quantities greater than 454,000 kg/yr (1,000,000 lb/yr) in the United States. Some chemicals have leaked into water supplies and have carcinogenic or other negative health impacts on humans. Many water supplies are now impacted by discharges from wastewater treatment plants and urban storm sewers. Engineers may be required to identify and design treatment strategies for chemicals for which no previous experience is available. As will be demonstrated in Chap. 3, treatment processes depend on well-established physicochemical principles. If the scientific principles are understood, it is possible to identify candidate processes based on the expected interaction between the properties of the contaminants and the capabilities of the processes. For instance, by knowing the volatility and hydrophobicity of a synthetic organic chemical, it is possible to predict whether air stripping or adsorption onto activated carbon is a more suitable treatment strategy.

Technology has been accelerating the pace at which treatment equipment is being developed. Engineers are faced with situations in which equipment vendors and manufacturers have developed new or innovative processes, and the engineer is assigned the task of recommending to a client whether or not the equipment should be evaluated as a viable option. Potable water is a necessary part of modern society, properly working processes are a matter of public health, and consumers expect to have water available continuously. Practical knowledge of previous successful design practices may not be sufficient for predicting whether new equipment will work. Understanding the scientific principles that govern treatment processes gives the engineer a basis for evaluating process innovations.

Treatment plants have gotten more complex. Sometimes facilities fail to work properly and the engineer is called in to identify factors that are preventing the plant from working or to recommend strategies to improve performance. Often, the difference between effective and ineffective performance is the result of scientific principles—a coagulant dose too low to destabilize particles, a change in water density because of a change in temperature, treatment being attempted outside the effective pH range. In these instances, scientific principles can guide the decision-making process regarding why a process is not working and what changes to operation would fix the problem.

As a result, the range of knowledge and experience needed to design water treatment facilities is extensive and cannot be learned in a single semester in college; today's design engineers need both knowledge about the fundamental principles of processes and practical design experience. This book provides a solid foundation in the former; other books focus more on the latter, such as books by Kawamura (2000) and AWWA and ASCE (2004). In addition, a companion book written by the authors, *MWH's Water Treatment Principles and Design,* 3rd ed. (Crittenden et al., 2012), covers both principles and design. While the coverage of that book is broad, it is nearly triple the length of this book and is difficult to cover in detail in a single engineering course. This book takes a focused approach on principles of water treatment and does so with the perspective of applying principles during design and operation so that it will serve as a useful introduction into the field of water treatment.

1-2 The Importance of Sustainability

Another concept in this book is that sustainability and energy consumption should be considered in selecting treatment processes, designing them, and operating them. There are several reasons for this approach. First, the withdrawal, conveyance, treatment, and distribution of potable water—and subsequent collection, treatment, and discharge of domestic wastewater—is one of the most energy-intensive industries in the United States. Only the primary metal and chemical industries use more energy. A focus on sustainability and energy considerations will help the water treatment industry develop ways to be more efficient while conserving resources.

Water demand has grown in urban areas and adequate supplies of locally available, high-quality water are increasingly scarce. Simultaneously, the ability to detect contaminants has become more sophisticated, negative health effects of some constituents have become more evident, regulations have become more stringent, and consumer expectations of high-quality water have become more strident. The growing trend toward use of poor-quality water sources, coupled with these other effects, has stimulated a trend toward more advanced treatment that requires more energy and resources. Increasing energy and resource use will contribute to greater pollution and environmental degradation; incorporating sustainability and energy consumption into process and design practices will offset that trend and allow higher levels of water treatment without the negative impacts.

Ultimately, the most important reason to consider sustainability in water treatment plant design is an issue of leadership. Environmental engineering professionals—the engineers who design water treatment facilities—ought to be more knowledgeable about environmental considerations than the general public and should demonstrate to other professions that successful design can be achieved when the environmental impacts are taken into account. The section on sustainability and energy considerations at the end of each of the process chapters in this book is a small start in that direction.

References

AWWA and ASCE. (2004) *Water Treatment Plant Design,* 4th ed., McGraw-Hill, New York.

Crittenden. J. C., Trussell, R. R., Hand, D. W., Howe, K. J., and Tchobanoglous, G. (2012) *MWH's Water Treatment: Principles and Design*, 3rd ed., Wiley, Hoboken, NJ.

Kawamura, S. (2000) *Integrated Design and Operation of Water Treatment Facilities,* Wiley, New York.

2 Water Quality and Public Health

The primary purpose of municipal water treatment is to protect public health. Water can contain a wide array of constituents that can make people ill and has a unique ability to rapidly transmit disease to large numbers of people. The purpose of this chapter is to introduce the relationship between water quality and public health and identify the major sources of contaminants in water supplies. The basic features of drinking water regulations in the United States are introduced. The chapter ends with a description of some of the challenges, competing issues, and compromises that water treatment engineers must balance to successfully design a water treatment system.

2-1 Relationship between Water Quality and Public Health

History of Waterborne Disease

Prior to the middle of the nineteenth century, it was commonly believed that diseases such as cholera and typhoid fever were primarily transmitted by breathing miasma, vapors emanating from a decaying victim and drifting through the night. Serious engagement in treatment of public drinking water supplies began to develop in the last half of the nineteenth century after Dr. John Snow identified the connection between contamination of drinking water and waterborne disease. Snow's discovery was later supported by the advocacy of the germ theory of disease by the French scientist Louis Pasteur in the 1860s and the discovery of important microbial *pathogens* (microorganisms capable of causing disease) by the German scientist Robert Koch. These developments led to the understanding that gastrointestinal disease spreads when the pathogens in the feces of infected human beings are transported into the food and water of healthy individuals—exposure via the so-called *fecal-to-oral route*. As a result, a number of strategies were developed to break the connection between drinking water systems and systems for disposal of human waste. These strategies included the use of water sources that are not exposed to sewage contamination, the use of water treatment on contaminated supplies, the use of continuously pressurized water systems that ensure that safe water, once it is obtained, could be delivered to the consumer without exposure to further contamination, and the use of bacterial indices of human fecal contamination.

Continuous chlorination of drinking water as a means for bacteriological control was introduced at the beginning of the twentieth century. In the next four decades, the focus was on the implementation of conventional water treatment and chlorine disinfection of surface water supplies. By 1940, the vast majority of water supplies in developed countries had "complete treatment" and was considered microbiologically safe. The success of filtration and disinfection practices lead to the virtual elimination of the most deadly waterborne diseases in developed countries, particularly typhoid fever and cholera, as depicted on Fig. 2-1 (CDC, 2011).

In 1974, however, both in the United States and in Europe, it was discovered that chlorine, the chemical most commonly used for disinfection, reacted with the natural organic matter in the water to produce synthetic organic chemicals, particularly chloroform. Since that time, decades of research have shown that chlorine produces a large number of disinfection by-products (DBPs), and that alternate chemical disinfectants produce DBPs of their own. The challenge to protect the public from waterborne diseases continues as engineers balance disinfection and the formation of treatment by-products.

In the 1970s and 1980s, it became apparent that some waterborne diseases spread by means other than from one human to another via the fecal-to-oral route. First among these are *zoonotic diseases*, diseases that humans can contract via the fecal-to-oral route from the feces of other animals. Examples of zoonotic pathogens are *Giardia lamblia* and *Cryptosporidium parvum*. Second are diseases caused by *opportunistic pathogens* that make their home in aquatic environments but will infect humans when the opportunity arises. Examples of opportunistic pathogens are *Legionella pneumophila, Aeromonas hydrophilia, Mycobacterium avium* complex, and *Pseudomonas aeruginosa*. An opportunistic pathogen is a microorganism

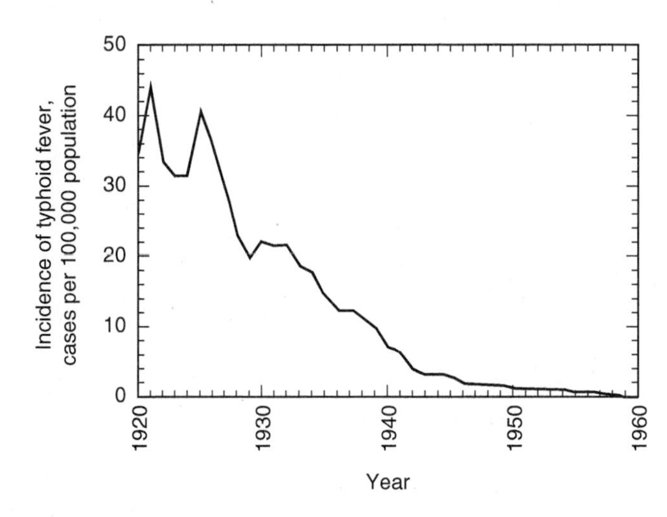

Figure 2-1
Decline in the incidence of typhoid fever in the United States due to the provision of higher quality drinking water and other sanitation and hygiene practice improvements. [Data from CDC (2011).]

that is not ordinarily able to overcome the natural defenses of a healthy human host. Under certain circumstances, however, such organisms are able to cause infection resulting in serious damage to the host. There are two circumstances when opportunistic pathogens are more successful: (a) when the immune response of the host has been compromised [e.g., persons with human immunodeficiency virus (HIV), persons on drugs that suppress the immune system, the very elderly] or (b) when the host is exposed to such high levels of the organism in question that the infection becomes overwhelming before the body can develop a suitable immune response. As a result of the possible presence of zoonotic pathogens, finding a water supply free of sewage contamination does not assure the absence of pathogens and does not obviate the need for water treatment. Also, understanding the role of opportunistic pathogens makes it clear that purifying water and transporting it under pressure does not provide complete protection, and growth of opportunistic pathogens must also be controlled in distribution systems and in water system appurtenances.

Role of Water in Transmitting Disease

A unique aspect of water as a vehicle for transmitting disease is that a contaminated water supply can rapidly expose a large number of people. When food is contaminated with a pathogen, tens to hundreds of persons are commonly infected. If a large, centralized food-packaging facility is involved, thousands might be infected. However, when drinking water is contaminated with a pathogen, typically hundreds of people are infected and occasionally hundreds of thousands are infected. For example, it is estimated that 500,000 people became ill from contaminated drinking water in the 1993 Milwaukee *Cryptosporidium* incident (MacKenzie et al., 1994).

The principal mechanisms for the transmission of *enteric* (gastrointestinal) diseases are shown on Fig. 2-2. Suppose that, while infecting an adult,

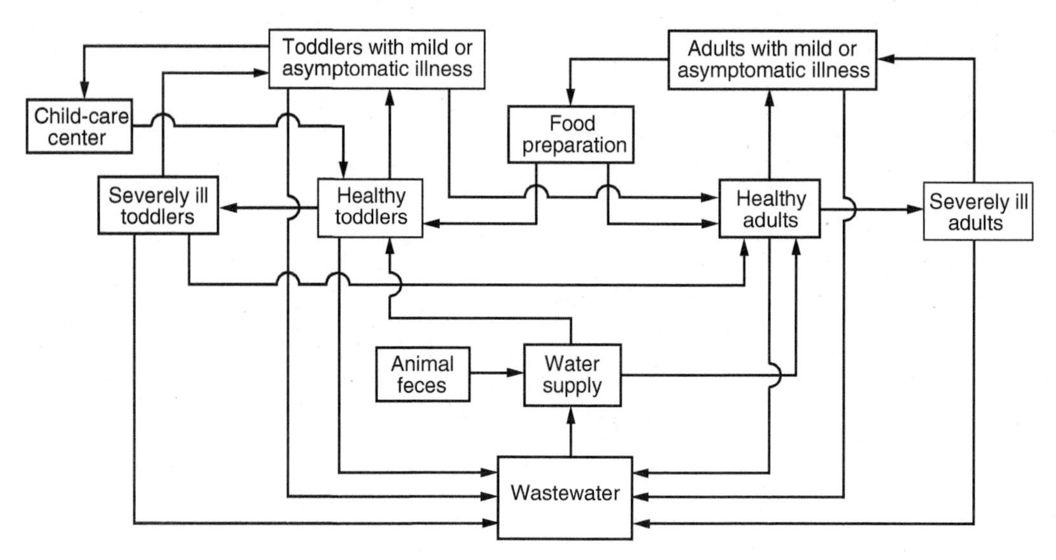

Figure 2-2
Schematic of routes of transmission for enteric disease.

a pathogen causes a severe, debilitating enteric disease that immobilizes and seriously injures the infected person. The route of transmission can be analyzed using Fig. 2-2. If an adult with severe illness is too debilitated to prepare food, the organism cannot get into the food supply. However, the organism does get in the sewer even if the sick person cannot get out of bed. Once in the sewer, the organism is then transported to the wastewater treatment plant. If the organism is not removed or inactivated at the wastewater treatment plant, it enters the receiving watercourse. If that watercourse serves as a water supply and water treatment does not remove or inactivate the organism, both healthy toddlers and adults who drink the water are exposed and may get infected. Thus, the entire population drinking the water supply is potentially exposed to the disease-causing agent. Under these conditions, an organism can successfully reproduce even if it causes a severe disease from which the host rarely recovers. According to some historical accounts, the classic form of Asiatic cholera that appeared in the middle of the nineteenth century behaved in this way. The route of transmission can be interrupted by removing or inactivating the organism from the water either at the wastewater treatment plant or at the drinking water treatment plant.

Figure 2-2 can also be used to consider the spread of the disease via the food route. Adults with mild symptoms of the disease, if they do not use adequate hygiene, may contaminate food when they prepare it. Both toddlers and adults who eat the contaminated food may then get infected. Some of those who get infected will be asymptomatic; others may exhibit mild symptoms. Infected adults may again prepare and contaminate food, and some infected toddlers will go to child-care centers. Toddlers in child-care centers will expose other toddlers. Adult caregivers can also expose themselves while handling the sick toddlers.

Debilitating diseases are less likely to spread this way because seriously ill adults are unlikely to be preparing food for others and seriously ill children are unlikely to go to child-care centers. Furthermore, the drinking water has no connection to this route of communication so treating the drinking water will not stop it. The value of a water treatment intervention is much greater where severe, debilitating disease is concerned.

Enteric organisms that cause seriously debilitating disease can be nearly eliminated through water treatment because they depend on this route of exposure for survival. When enteric organisms cause mild disease or asymptomatic infections, water treatment can prevent the largest scale epidemic events but the disease remains in the community. This is because mildly ill or asymptomatic carriers will spread the disease via food preparation and in child-care centers.

2-2 Source Waters for Municipal Drinking Water Systems

Designing on effective water treatment plant is a complex process because of the wide variety of undesireable constituents that can be in the source water. Even waters thought of as "pristine" might contain some constituents that should be removed. The specific constituents in water, the relative concentrations of those constituents, and other water quality parameters that affect treatment depend heavily on local conditions of geology, climate, and human activity. Thus, treatment processes must be tailored to the specific source water. The specific treatment challenges, however, are heavily influenced by the type of source water, which can include groundwater, lakes and reservoirs, rivers, seawater, and wastewater impaired waters. Each type of source will require different treatment processes and present different challenges to the water treatment engineer. Constituents can enter the water supply through several pathways, as depicted on Fig. 2-3. Potential types of contamination and general characteristics of each type of source are described in the following sections.

Groundwater

Groundwater is water that exists in the pore spaces between sand, gravel, and rocks in the earth and can be brought to the surface using wells. About 35 percent of people served by public water systems in the United States are supplied with groundwater; nearly all the rest are supplied with fresh surface water. Undesirable constituents in groundwater can be either naturally occurring or *anthropogenic* (of human origin). The natural constituents result from dissolution caused by long-term contact between the water and the rocks and minerals. Some natural constituents that might need to be removed by water treatment include:

- Iron and manganese: Depending on local conditions, groundwater can be aerobic (in the presence of oxygen gas) or anaerobic (in

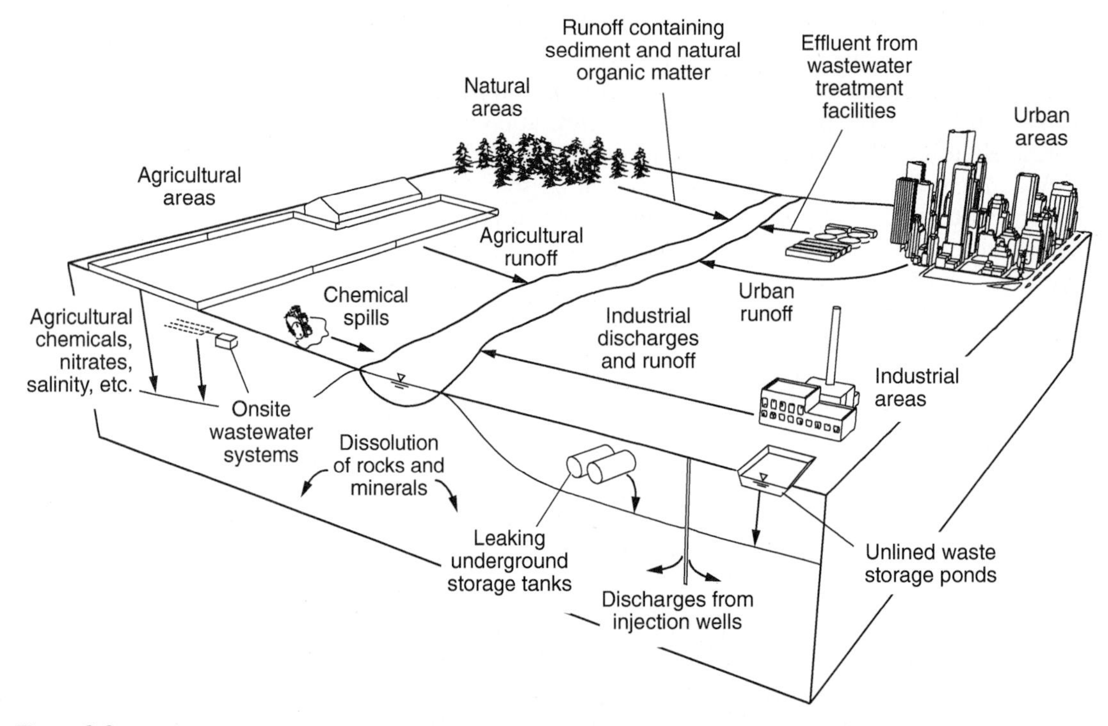

Figure 2-3
Sources of naturally occurring constituents and contaminants in drinking water supplies.

the absence of oxygen-containing electron acceptors). In anaerobic conditions, iron- and manganese-containing minerals are relatively soluble and can dissolve into the water. When the water is aerated and/or chlorinated, the iron and manganese react to form insoluble species that precipitate and cause rust- and black-colored stains on laundry and plumbing fixtures.

- Hardness: *Hardness* is a characteristic of water caused by the presence of calcium and magnesium, which are abundant in the Earth's crust. Hard water does not cause negative health impacts, but it reacts with soap to form a white precipitate (soap scum), leaves water spots on surfaces, and forms precipitates in water heaters, tea pots, heat exchangers, boilers valves, and pipes, clogging them and/or reducing their efficiency.
- Trace inorganics: Minerals can contain many trace elements, including arsenic, barium, chromium, fluoride, selenium, and species that exhibit radioactivity such as radium, radon, and uranium. Many trace inorganics exhibit toxicity, carcinogenicity, or other adverse health effects, if concentrations are too high.

- Salinity: Brackish groundwater with low to moderate salinity, ranging from about 1000 to 5000 mg/L total dissolved solids (TDS), is relatively common. *Brackish water* is too salty for potable, industrial, or agricultural applications (the United States secondary drinking water standard for TDS is 500 mg/L). Interest in desalinating these sources has increased in areas short on freshwater, such as Florida, Texas, and the Southwest region of the United States.
- Natural organic matter: Most groundwaters have low concentrations of natural organic matter (NOM), but some locations have shallow groundwater that is hydraulically connected with swampy areas. The Biscayne Aquifer in southeast Florida is an example of this type of water source. These waters are highly colored (like weak ice tea), which is not only undesirable aesthetically but can react with chlorine during disinfection to form disinfection by-products that may be carcinogenic.

In addition to these natural constituents, groundwater can contain a variety of anthropogenic contaminants. The potential number of anthropogenic contaminants is vast. In the United States, about 70,000 chemicals are used commercially and about 3300 are considered by the U.S. Environmental Protection Agency (EPA) to be high-volume production chemicals [i.e., are produced at a level greater than or equal to 454,000 kg/yr (1,000,000 lb/yr)]. Anthropogenic contributions to groundwater can come from the following sources:

- Leaking underground storage tanks: Gas stations store gasoline in underground tanks, which can corrode, leak, and contaminate groundwater. Benzene, toluene, ethylbenzene, and xylene (BTEX) are constituents in gasoline that must be removed from groundwater to make it potable, and methyl *t*-butyl ether (MTBE) is a gasoline additive that is particularly difficult to remove with conventional water treatment processes.
- Leaking residential septic systems: Improperly constructed septic systems can leak nitrate, household chemicals, and other contaminants into the water supply.
- Industrial contamination: Past practices of discharging chemical wastes on the ground, in landfills, in open pits, or into waste disposal wells have contaminated water supplies with many kinds of industrial chemicals. Industrial solvents like trichloroethene (TCE) and tetrachloroethene (PCE) are particularly common contaminants in groundwater. Inadvertent chemical spills also lead to contamination.
- Agricultural contamination: During irrigation, plants uptake some water but excess water can percolate downward and reach the underlying groundwater table. Pesticides, herbicides, and fertilizers applied to the land can travel down with the water and contaminate the groundwater. In addition, irrigation water will contain some dissolved

salts and the plants can selectively uptake water, leaving the excess water with higher salinity that can contaminate the groundwater with excess salts.

In aquifers where groundwater withdrawals exceed rates of recharge, seawater migrates inland. This process, called *saltwater intrusion*, can result in high concentrations of TDS (mainly sodium and chloride) at potable water supply wells. Coastal areas in Florida and California have been affected by saltwater intrusion. The only long-term solution is to balance supply and demand, but saltwater intrusion can be slowed or reversed by injection of water between the supply wells and the ocean, as shown on Fig. 2-4. Such saltwater intrusion barriers typically consist of a network of wells arrayed parallel to the shoreline to form a hydrostatic barrier. In several cases, including four saltwater intrusion barriers in southern California, highly polished reclaimed water has been used to create the groundwater barrier.

Despite the potential for many constituents to be in groundwater, an advantage of this type of water supply is that the quality tends to be consistent over time with little or no seasonal variation. Changes due to migration of contaminants tend to happen slowly. Groundwater withdrawn from properly constructed wells is free from pathogenic organisms and does not need to be filtered. A disadvantage, however, is that the quality of the water is not known until the well has actually been drilled and pumped long enough to exert its full zone of influence for some time. While general water quality can often be predicted from the local geology, there have been many cases of wells drilled to different depths or a few hundred meters apart that contain significantly different concentrations of trace constituents such as arsenic, which then affects treatment requirements. The lack of reliable information on the specifics of water quality prior to installing a well complicates the treatment selection and design process in some locations.

It is important to realize that not all groundwaters will exhibit all of these problems. *Confined aquifers* (isolated from the surface by a zone of lower permeability) can be less susceptible to anthropogenic contamination, depending on where the recharge zone is. Depending on local geology and human activity, many groundwaters might be relatively pure and have essentially no treatment requirements, others might have excessive iron and manganese or high hardness, still others might have contamination from septic tanks or fertilizers, and some will have a combination of these problems. The treatment required will be different in each case, leading to the reality that treatment practices must be tailored to the individual water supply.

Rivers

The water in rivers often has less mineral content than groundwater but can dissolve natural materials during overland flow after rain or during interaction with groundwater. Surface waters can contain floating and suspended material like sediment, leaves, branches, algae, and other plants or animals that wash into the water during overland flow or live in the

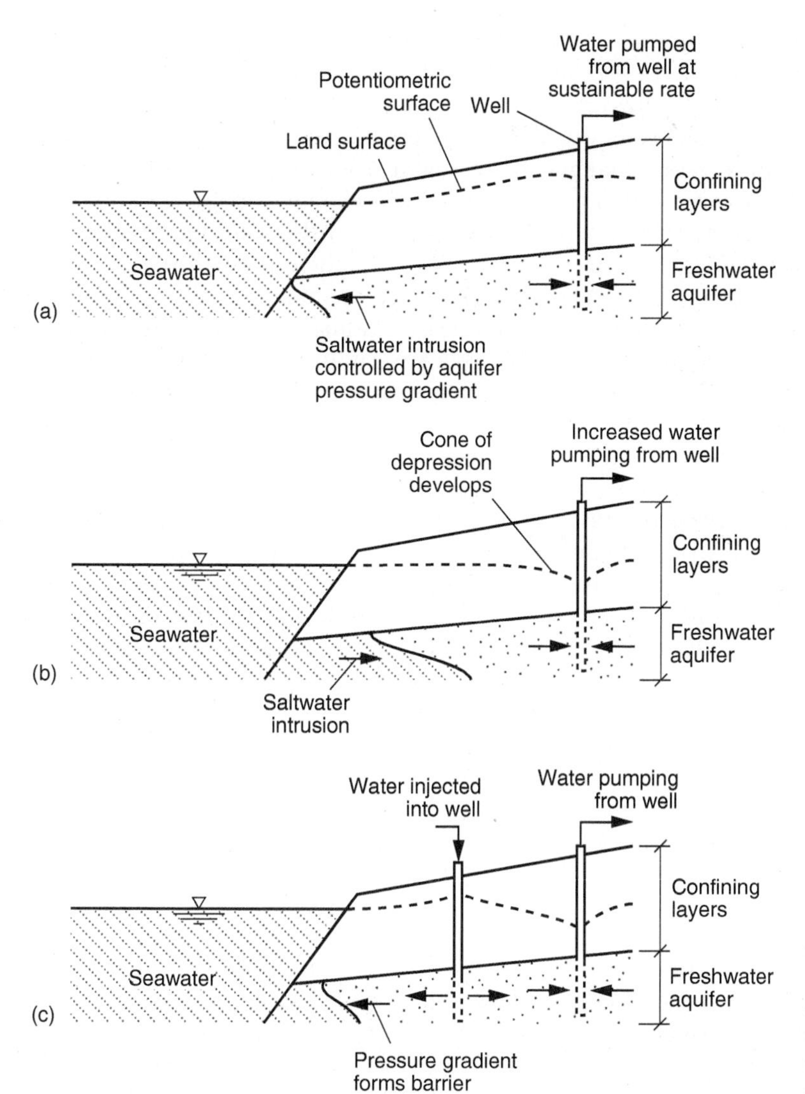

Figure 2-4
Saltwater intrusion into a groundwater supply: (a) natural hydrologic condition, (b) saltwater intrusion caused by depression of the water table and reversal of the hydraulic gradient by pumping, and (c) prevention of saltwater intrusion by creation of a hydrostatic barrier.

water itself. The key element that distinguishes all surface waters from groundwater is the potential for the presence of pathogenic bacteria and other microorganisms that must be eliminated to make water safe to drink. The necessity of removing pathogenic organisms makes surface water treatment dramatically different from groundwater treatment; nearly all surface water treatment plants have filtration systems designed to physically remove microorganisms and engineered disinfectant contact basins to disinfect the water. In contrast, treatment facilities for groundwater have processes focused on removing dissolved contaminants.

Large storm events in the watershed can have a significant impact on water quality in rivers. A rainfall event can lead to rapid increases in turbidity and simultaneous changes in the temperature, pH, alkalinity, dissolved oxygen, and other water quality parameters. These changes in water quality often require rapid changes in treatment operation to successfully treat the water. Rivers in which water quality can change rapidly are known as ''flashy'' rivers. Turbidity in the Rio Grande in the southwestern United States can change from less than 100 NTU (nephelometric turbidity units) to greater than 10,000 NTU in a matter of hours after a storm event. In addition, surface waters are susceptible to seasonal changes in water quality; in temperate climates, surface waters are warm in the summer and cold in the winter and many other water quality parameters can change seasonally as well.

The presence of plants and animals living in the watershed contributes to NOM in river water supplies. *Natural organic matter* is the term used to describe the complex matrix of organic chemicals originating from natural biological activity, including secretions from the metabolic activity of algae, protozoa, microorganisms, and higher life forms; decay of organic matter by bacteria; and excretions from fish or other aquatic organisms. The bodies and cellular material of aquatic plants and animals contribute to NOM. Natural organic matter can be washed into a watercourse from land originating from many of the same biological activities but undergoing different reactions due to the presence of soil and different organisms. Surface water generally contains more NOM than groundwater and is more likely to require treatment to remove NOM prior to disinfection than are groundwater sources. The amount NOM and the chemical by-products it forms when reacting with chlorine often influences the choices for disinfection.

Surface water can be susceptible to exposure to anthropogenic contamination, particularly if wastewater treatment facilities, industrial plants discharges, or farms that use fertilizers and pesticides are located upstream of the water treatment facility intake. Some utilities are successful in limiting access to their watershed; for instance, Portland, Oregon, has been able to remain exempt from filtration requirements in their Bull Run supply because of their ability to protect the high quality of their source water.

In general, naturally occurring inorganics such as arsenic and selenium are less of a concern in surface water than in groundwater because of the shorter time for exposure to minerals. Hardness in an exception; surface water can be fairly hard in regions with large deposits of limestone and other calcium-bearing minerals. Many treatment facilities using the Missouri, Mississippi, and Ohio Rivers in the central United States practice lime softening to reduce hardness.

An advantage of surface waters is that water quality is easier to measure and predict before the intake structure for the treatment facility is built.

Historical water quality data can be obtained from agencies such as the U.S. Geological Survey or other water utilities located upstream or downstream of the proposed intake. Sampling can be conducted for a period of time before facilities are built. Water quality can be similar to measurements upstream or downstream provided that flows from tributaries, runoff, and point source discharges are taken into account, in contrast to groundwaters where the water quality from one well may not necessarily be similar to nearby wells.

Lakes and Reservoirs

Lakes and reservoirs share many water quality characteristics with rivers. Significant similarities include the presence of bacteria and other microorganisms, the potential for anthropogenic contamination, and typically higher NOM concentrations than groundwater. The differences between rivers and lakes are related to factors affected by water velocity. The low velocity in lakes allows sediment to drop out. Lakes typically have much lower and more consistent turbidity than rivers, which makes treatment easier. Alkalinity, pH, and other parameters are also more consistent over time.

Lakes and reservoirs can be so quiescent that they become thermally stratified during certain times of the year, as shown on Fig. 2-5. In the summer, a layer of warm water forms at the surface (epilimnion) and does not mix with colder water at lower depths (hypolimnion). The warm water and sunlight at the surface can lead to algae blooms that contribute to taste and odor problems in the water. The lack of exchange between the upper and lower layers allows the hypolimnion to become depleted in oxygen and the anaerobic conditions allow iron and manganese to dissolve from sediments on the lake bottom. Water withdrawn through intakes located at lower depths in the lake will need treatment for iron and manganese during portions of the year. In the fall, the surface layer

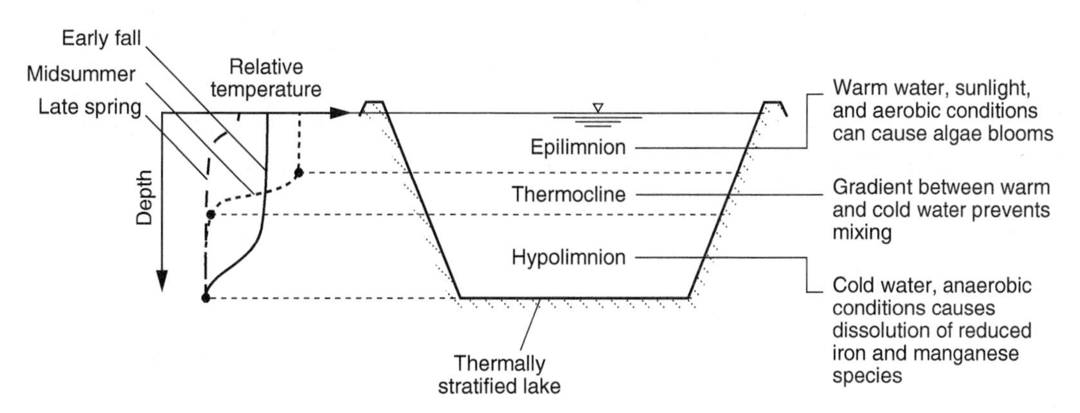

Figure 2-5
Stratification of a lake.

can cool to below the temperature of the deeper water, at which time the more dense water at the surface sinks to the bottom of the lake, causing the water in the entire lake to *turn over*. Lake turnover can be a relatively rapid event that changes the water quality at the location of the intake, requiring changes in treatment practices.

Seawater

Declining availability of freshwater sources may portend an increase in the use of ocean water or seawater as a water supply. About 97.5 percent of the Earth's water is in the oceans and about 75 percent of the world's population lives in coastal areas. The salinity of the ocean ranges from about 34,000 to 38,000 mg/L as TDS, nearly two orders of magnitude higher than that of potable water. Tampa Bay, Florida, is an example of a community using seawater for its water supply. The challenges for using seawater as a source for potable water are primarily related to removing the salinity, but individual species such as bromide and boron can complicate the treatment processes. High levels of these parameters lead to a wide range of effects, including impacts on health, aesthetics, and the suitability of the water for purposes such as irrigation. Also the low hardness and alkalinity and relatively high chloride content of desalted seawater present special corrosion control challenges.

Wastewater-Impaired Waters

Communities generally discharge their treated municipal wastewater into rivers, which then become the water supply for downstream communities. It is not uncommon for the treated wastewater to be a significant portion of the flow of a river; the Trinity River system between Dallas and Houston in Texas, for example, contains significant amounts of treated wastewater. Significant increases in population density in regions with limited water resources have prompted interest in treated wastewater as a potential water supply, which in its most comprehensive form would be known as direct potable reuse. Regulations currently restrict direct potable reuse because potential health impacts resulting from long-term, low-level exposure to chemicals and mixtures of chemicals present in wastewater effluent have yet to be fully elucidated. However, the large contribution of wastewater in some rivers results in de facto water reuse that raises the same issues. De facto water reuse also increases the potential for pathogenic organisms to be in the source water and the potential for household chemicals, pharmaceuticals, and personal care products in the water supply has been a concern among the public in recent years. A systematic analysis of the contribution of municipal wastewater effluent to potable water supplies has not been made in the United States for over 30 years. The lack of such data impedes efforts to identify health impacts of de facto water reuse, and additional research is needed regarding the appropriate level of treatment for rivers with large contributions of treatment municipal wastewater.

2-3 Regulations of Water Treatment in the United States

In the United States, regulations (or rules) are developed by regulatory agencies to implement statutes, which are enacted by Congress and are legally enforceable. Standards are the portion of a rule that defines the allowable amount of a constituent in water. As analytical techniques for measuring constituents in water have gotten more sophisticated and knowledge of how human health is impacted by these constituents has grown, standards and regulations have become more stringent, meaning more constituents are regulated and at lower concentrations. Drinking water standards and regulations are designed to protect human health and are often so comprehensive that the design treatment process is dictated by these mandates.

U.S. Public Health Service

The United States began regulating drinking water quality in the early 1900s. The first drinking water quality regulations were developed by the U.S. Public Health Service (U.S. PHS) and established bacteriological quality standards for water supplied to the public by interstate carriers. After the initial emphasis on controlling waterborne bacteria, new parameters were regulated to limit exposure to contaminants that cause acute health effects, such as arsenic, or that adversely affect the aesthetic quality of the water. The U.S. PHS continued to set drinking water regulations over the next 50 years, expanding into minerals, metals, radionuclides, and organics. By the 1940s, with minor modifications, all 50 states adopted the U.S. PHS standards.

U.S. Environmental Protection Agency

Due to growing public concern with environmental issues, on July 9, 1970, President Nixon sent an executive reorganization plan to Congress with the goal of consolidating all federal environmental regulatory activities into one agency. The U.S. Environmental Protection Agency (U.S. EPA) was created on December 2, 1970, with the mandate to protect public health and the environment, which included drinking water quality.

Safe Drinking Water Act

The *Safe Drinking Water Act* (SDWA) was passed by Congress and signed into law by President Ford on December 16, 1974 (Public Law 93-523). Following the passage of the SDWA, the principal responsibility for setting water quality standards shifted from state and local agencies to the federal government. The SDWA gave the federal government, through the U.S. EPA, the authority to set standards and regulations for drinking water quality delivered by community (public) water suppliers. The SDWA created the framework for developing drinking water quality regulations by defining specific steps and timetables that were to be taken to establish the National Interim Primary Drinking Water Regulations (NIPDWR), National Primary Drinking Water Regulations (NPDWR), and National

Secondary Drinking Water Regulations (NSDWR). Although the U.S. EPA sets national regulations, the SDWA gives states the opportunity to obtain primary enforcement responsibility (primacy). States with primacy must develop their own drinking water standards, which must be at least as stringent as the U.S. EPA standards. Almost all states have applied for and have been granted primacy.

SDWA Amendments and Updates

The SDWA has been reauthorized and amended since its original passage in 1974. The most significant changes were made when the SDWA was reauthorized on June 16, 1986 (Public Law 99-339), and when it was amended in 1996 (Public Law 104-182). The amendments of 1986 were driven by public and congressional concern over the slow process of establishing NPDWRs. The amendments enacted in 1996 included an emphasis on the use of sound science and risk-based standard setting.

Since the inception of the SDWA, the number of regulated contaminants has increased dramatically and continues to grow, as shown on Fig. 2-6.

Current Updating Process for Drinking Water Contaminants

Our ability to identify the presence of contaminants at increasingly lower levels, and the fact that it continues to outstrip our ability to understand their consequences, presents a significant challenge to drinking water regulators. In an attempt to address this disparity, the U.S. EPA regularly updates the list of constituents within the NPDWR. There are two avenues that ensure that the regulated contaminants are kept up to date, as illustrated on Fig. 2-7.

The first strategy is a regular review and revision of the existing regulations, which occurs once every 6 years. The second strategy is the

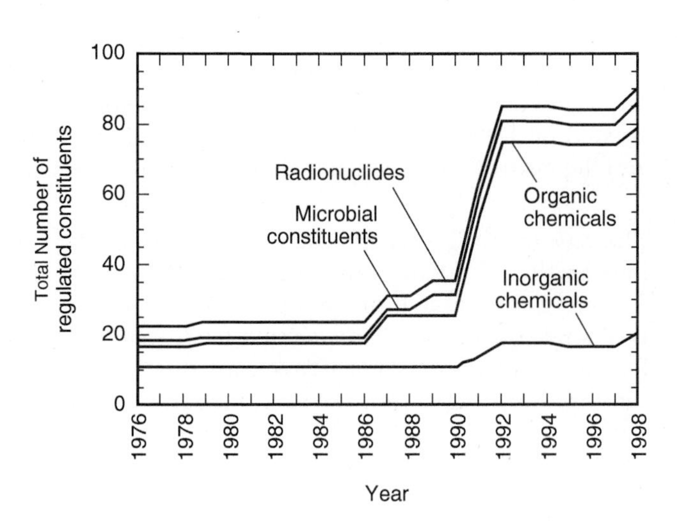

Figure 2-6
Growth in the number of regulated constituents since the inception of the SDWA (U.S. EPA, 1999).

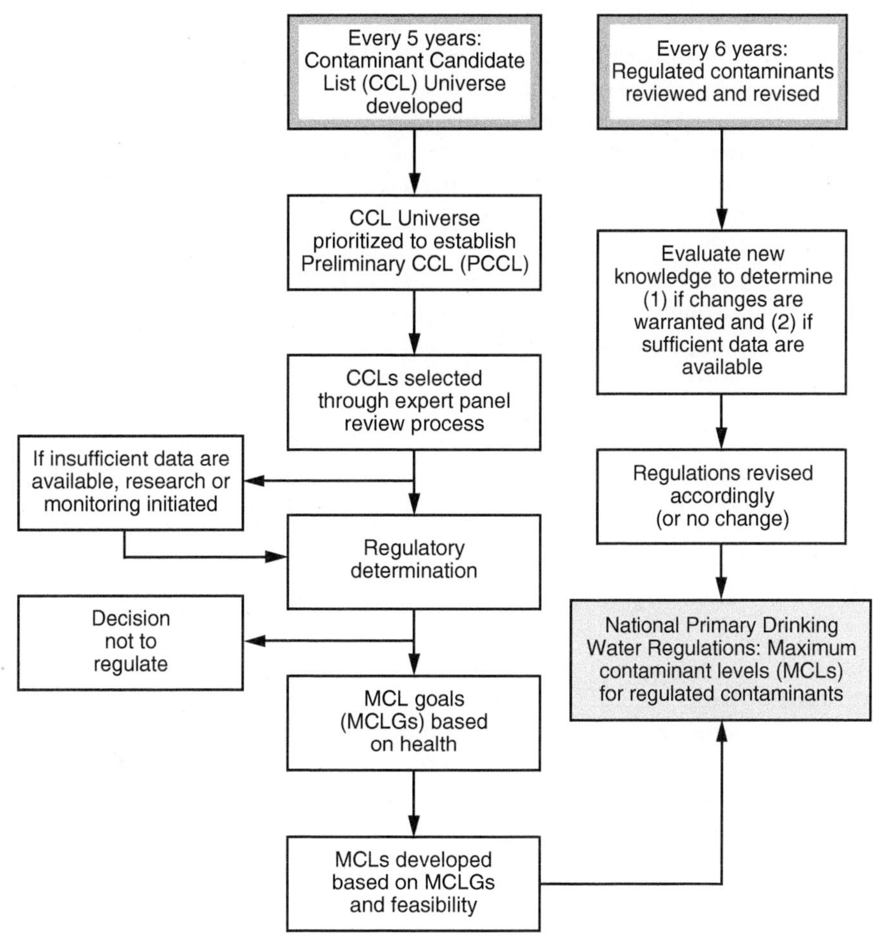

Figure 2-7
Illustration of current protocol to maintain regulated contaminant list.

identification and evaluation of potential water contaminants that may deserve regulation, through a process that centers round the generation and review of the *Contaminant Candidate List* (CCL). This list is regenerated every 5 years. CCL1 was announced in 1998; CCL2 was announced in 2005; CCL3 was announced in 2009; and CCL4 is due in 2014. The current process for developing the CCL is based on advice from the National Research Council (NRC, 2001). The initial step in developing the list involves the establishment of a broad spectrum list of potential drinking water contaminants (called the Universe of Chemicals). This step is followed by a screening step to narrow the universe to those contaminants that deserve further assessment based on their potential prevalence in drinking water and impact on human health. This list is called the pre-CCL (PCLL). Next, an expert panel is charged with the task to select, from the PCCL, the contaminants for which sufficient information is available to make a regulatory determination as well as those warranting additional research and monitoring to bridge the gaps necessary so that a regulatory determination can be made. In order to gather occurrence information to support the CCL approach, the U.S. EPA also maintains an unregulated contaminant monitoring regulation (UCMR). The list of compounds to be monitored through this regulation is updated each time the CCL is updated. Finally, if a given compound from the CCL is elected to be regulated, a *maximum contaminant level goal* (MCLG) is established. An MCLG is a nonenforceable concentration of a drinking water contaminant set at the level at which no known or anticipated adverse effects on human health occur and that allows an adequate safety margin. The MCLG, along with information on treatment and limits of analytical detection, is then used as guidance for the establishment of a *maximum contaminant level* (MCL). An MCL is an enforceable standard set as close as feasible to the MCLG taking cost and technology into consideration.

Acute versus Chronic Exposure

Regulations and treatment practices are both influenced by a contaminant's health effect. Contaminants in drinking water can have effects that are acute or chronic. As these terms are used here, they refer to the time of exposure that is normally required to cause the identified health effect. A contaminant is said to have acute effects when health effects can result from a brief exposure. The infections that result from exposure to pathogens are acute. A contaminant is said to have chronic effects when health effects are normally associated with long-term exposure. Carcinogens almost always have chronic effects.

Acute contaminants often have instantaneous maximums for indicators that cannot be exceeded, whereas chronic contaminants are more appropriately regulated on the basis of long-term averages. Where the design of treatment processes is concerned, whether a contaminant is acute or chronic can affect the type of multiple barriers that might be appropriate. For example, for contaminants of all types, multiple barriers can be used to expand the variety of contaminants the process train can effectively address (i.e., robustness), but when an acute contaminant must be addressed, it is especially important to use multiple barriers to improve the degree to which the process train can be relied upon to remove it (i.e., reliability). These principles are illustrated on Fig. 2-8 (Olivieri et al., 1999). In designing a treatment system, reliability is paramount for a treatment scheme that is intended to reduce acute health risks, however, robustness is sufficient for a treatment system intended to reduce chronic health risks. The prevailing challenge is addressing those constituents that engender chronic consequences.

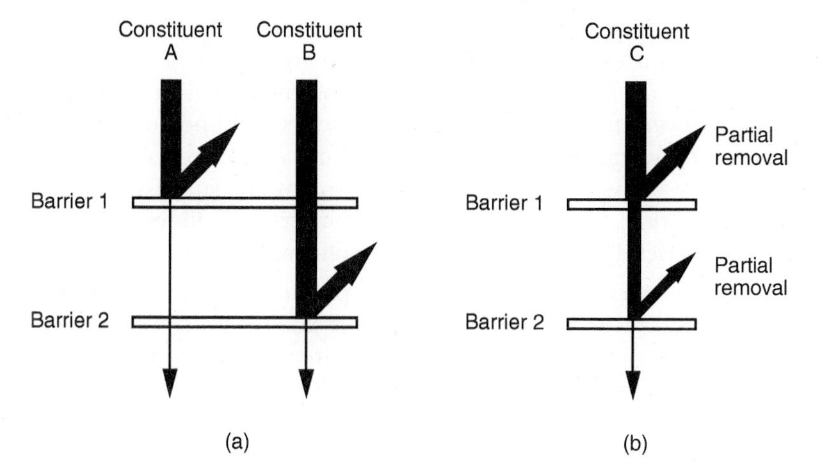

Figure 2-8
Depiction of multiple barriers to achieve robustness and reliability.

2-4 Evolving Trends and Challenges in Drinking Water Treatment

Engineers have been involved in the planning, design, and construction of municipal water treatment systems for about 200 years. The last 30 or 40 years, however, have been a time of dramatic changes in the interrelationship between water quality and public health because of increases in scientific understanding and growing human impact on water sources. As a result, the modern water treatment engineer faces an increasingly complex array of challenges, competing issues, and compromises that must be balanced to successfully design a water treatment system. The overall impact of these complexities is a need for engineers to have a solid grasp on the scientific and fundamental principles underlying water treatment processes, rather than designing solely from the perspective of applying previously successful practices. Some of these complexities faced by water treatment engineers include:

- Since Dr. Snow identified the Broad Street well as the source of a cholera epidemic in 1854, water has been recognized as an important vehicle for transmitting disease by carrying fecal matter from sick people to healthy people. As a result, water quality management for many years was focused on disrupting this fecal-to-oral route; minimizing contamination of water supplies (through wastewater treatment) or protecting watersheds were important factors. With the recognition in the 1970s and 1980s that *G. lamblia* and *C. parvum* do not only follow the fecal-to-oral route but are also present in the natural environment, it was realized that merely disrupting the fecal-to-oral route is insufficient. Modern water quality management practices must protect against and provide treatment for a wider array of potential sources of microbial contamination.

- In the early 1970s, it was discovered that chlorination of water containing natural organic matter causes the formation of potentially carcinogenic disinfection by-products. Thus, the benefits of chlorination for preventing acute illness are in conflict with the potential for chlorination to cause chronic illness. When it comes to disinfectants, more is not always better. Modern water treatment must balance the need to provide disinfection to prevent waterborne illness with the need to restrict disinfection to minimize chronic health effects.
- A layperson's view of water quality and public health might be that water with no measurable contaminants is safe to drink and that the goal of water treatment is to remove all measurable contaminants. That view is unrealistic. Improvements in analytical equipment over the last 30 to 40 years have made it possible to measure constituents in water at exceedingly low concentrations. The result is that anthropogenic chemicals can be detected in most water sources. Polychlorinated biphenyls (PCBs) and other anthropogenic chemicals have been detected in remote high mountain lakes in the Pyrenees and Alps because of atmospheric deposition.
- The mere presence of constituents, however, does not imply negative health impacts. People have different sensitivities to chemicals; when exposed to the same concentration of the same chemical, one person might be affected and another might not. Lowering the allowed concentration in water decreases the fraction of the population who may be affected by a contaminant. The challenge is to find the appropriate concentration that reduces the probability of harm to an acceptable level; for instance, a probability of less than one in a million. Unfortunately, human response to anthropogenic and natural chemicals is exceedingly complex and identifying the "correct" concentration that is protective of human health can be difficult. Chemicals may have a threshold level below which they have no negative health effect, or may even be beneficial to health at low concentrations. For instance, at high concentrations selenium, copper, and chromium are harmful (EPA has MCLs for these contaminants), but at low concentrations they are essential minerals (they are present in multivitamins). At some threshold level, achieving increasingly lower concentrations in water may have considerable costs but no public health benefit. Modern analytical instruments are able to detect the presence of some chemicals at concentrations substantially lower than that at which they have a measurable impact on human health. A challenge in future water treatment practice is balancing the level of treatment with actual health benefits.
- Water is treated to exacting standards in central water treatment facilities and then delivered to the community through underground pipes—pipes that in some cases are decades old, full of deposits, corroded, or leaking. In addition, the quality of water sitting stagnant

in storage tanks and home plumbing fixtures naturally degrades as it comes into equilibrium with adjacent materials. It is now possible to achieve considerably better water quality at the discharge of a water treatment plant than what actually arrives at the kitchen faucet. Water treatment practices must consider the impact of water distribution on water quality and balance the objectives at the plant effluent with the objectives at the point of use.

- Water treatment plants supply water that is used for drinking, cooking, bathing, cleaning clothes, flushing toilets, watering lawns, industrial applications, and other uses. Only 3 to 4 percent of the water delivered to a residence is actually destined for human consumption but all water is treated to the same high level. Future water management practices must balance the level of water quality achieved with the actual use of the water, potentially supplying drinking water separately from water for other uses.
- Many communities are experiencing shortages of locally available high-quality water sources. Options for additional water supply include greater use of local impaired water, such as treated wastewater effluent, or transporting better quality water tens or hundreds of miles through pipes and aqueducts. Neither option has clear advantages over the other. Both may involve greater expenditure of energy and resources than previous water treatment projects, with commensurate negative impacts on the environment or human health. Future water treatment practices must evaluate water treatment strategies from a holistic perspective that considers all benefits and impacts to the community, environment, and society.

The issues introduced in this chapter make it clear that water treatment engineering continues to evolve. At the same time, the public's expectations for water quality have never been higher. An integration of past strategies and progressive tactics are essential as new challenges continue to surface and the fundamental mission expands.

2-5 Summary and Study Guide

After studying this chapter, you should be able to:

1. Define the following terms and phrases and describe the significance of each in the context of water quality and public health:

anthropogenic	hardness	opportunistic pathogen
brackish water	lake turnover	Safe Drinking Water Act
confined aquifer	MCL	saltwater intrusion
Contaminant Candidate List	MCLG	zoonotic disease
enteric disease	natural organic matter	
fecal-to-oral route	pathogen	

2. Explain the role of water treatment in virtually eliminating deadly waterborne diseases such as cholera and typhoid fever in developed countries.
3. Explain why outbreaks of debilitating diseases that transmit via the fecal-to-oral route can be effectively prevented by water treatment but mild diseases cannot be prevented that way.
4. Describe the types of constituents that can be present in groundwater and the pathways for these constituents to enter groundwater.
5. Describe the types of constituents that can be present in surface water and the pathways for these constituents to enter surface water.
6. Describe the differences in water quality between groundwater and surface water.
7. Describe the differences in water quality between rivers and lakes.
8. Describe what causes saltwater intrusion and how it can be prevented.
9. Explain the current process for updating drinking water quality regulations in the United States.
10. Explain the difference between acute and chronic exposure to contaminants in drinking water.
11. Describe some of the evolving trends and challenges in drinking water treatment.

References

CDC (2011) Available at: http://www.cdc.gov/Features/DrinkingWater/graph.html, accessed on Aug. 10, 2011.

U.S. EPA (1999) *25 Years of the Safe Drinking Water Act: History and Trends*, EPA816-R-99-007, 5, Cincinnati, OH.

MacKenzie, W., Hoxie, N., Proctor, M., Gradus, M., Blair, K., Peterson, D., Kazmierczak, J., Addiss, D., Fox, K., Rose, J., and Davis, J. (1994) "A Massive Outbreak in Milwaukee of *Cryptosporidium* Infection Transmitted through the Public Water Supply," *N. Engl. J. Med.*, **331**, 161–167.

NRC (2001) *Classifying Drinking Water Contaminants for Regulatory Consideration*, National Academy Press, Washington, DC.

Olivieri, A., Eisenberg, D., Soller, J., Eisenberg, J., Cooper, R., Tchobanoglous, G. Trussell, R., and Gagliardo, P. (1999) "Estimation of Pathogen Removal in an Advanced Water Treatment Facility Using Monte Carlo Simulation", *Water Sci. Tech.*, **40**, 4–5, 223–233.

3 Process Selection

Chapters 5 through 13 in this book introduce individual separation processes that are widely used in municipal drinking water treatment. Each process is effective for some contaminants but not others. Some contaminants are treatable by several different processes. An important question for the water treatment engineer is how to select the processes that should be used for a particular situation. This chapter is devoted to the principles that are used to answer that question.

An individual process is known throughout environmental engineering and chemical engineering literature as a *unit process*. Water treatment plants rarely contain a single unit process; instead, they typically have a series of processes. Multiple processes may be needed when different processes are needed for different contaminants. In addition, sometimes processes are effective only when used in concert with another; that is, two processes individually may be useless for removing a compound, but together they may be effective if the first process preconditions the compound so that the second process can remove it. A series of unit processes is called a *process train* or *treatment train*. The treatment train for a typical conventional surface water treatment plant is shown on Fig. 3-1.

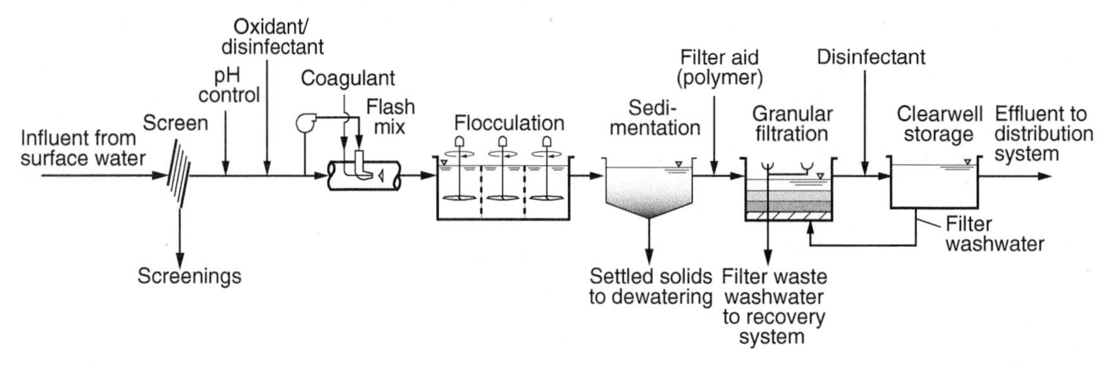

Figure 3-1
Typical treatment train for a surface water treatment plant.

The central ingredient in process selection—the relationship between the properties of constituents and the capabilities of separation processes—is the first topic of this chapter. The following two sections address additional considerations in process selection such as cost, reliability, and energy consumption. The final section describes the design process and the steps involved in selecting the process train.

3-1 Process Selection Based on Contaminant Properties

The source water for a treatment facility can contain a wide variety of constituents that may be undesirable in potable water. Section 2-2 described the constituents that can be in water, their sources, and the general differences between groundwater and surface water. The specific constituents in water, the relative concentrations of those constituents, and other water quality parameters that affect treatment depend heavily on local conditions of geology, climate, and human activity. Thus, water treatment facility processes must be tailored for the specific situation.

Removing a constituent from water is done by exploiting differences between that constituent and water; that is, if every physical, chemical, and biological property of a constituent were identical to those of water molecules, removal would be impossible. If, however, some property is different and a process is able to exploit that difference, removal is possible. The primary properties of interest include size, density, charge, solubility, volatility, polarity, hydrophobicity, boiling point, chemical reactivity, and biodegradability.

Classes of compounds tend to have similar *physicochemical* properties (the collective physical and chemical properties of a substance). For instance, inorganic constituents frequently (but not always) are nonvolatile, nonbiodegradable, and charged. Table 3-1 indicates general trends of

Table 3-1
General trends of physicochemical properties of some classes of constituents in water

	Microorganisms	Inorganics	Synthetic Organics	Natural Organics	Radionuclides
Examples of compounds	Viruses, bacteria, protozoa	Na, Cl, Fe, Mn, As, Pb, Cu, NO_3^-	Pesticides, solvents, pharmaceuticals	Products of decaying plants and animals	Ra, U, radioactive inorganic chemicals
Size	Particles (0.0025–10 μm)	Small molecules (low MW)	Molecules (usually low MW)	Large molecules (high MW)	Small molecules
Density	Close to that of water	Varies (as a precipitate). Does not apply if dissolved.	Varies (as a liquid phase). Does not apply if dissolved.	Does not apply (is dissolved).	Does not apply (is dissolved).
Charge	Some negative surface charge	Positive or negative	Usually no	Negative charge	Varies
Soluble	No	Varies	Varies	Yes	Varies
Volatile	No	No	Varies	No	No (except radon)
Polar	N/A	Yes	Varies	Yes	No
Hydrophobic	No	No	Usually yes	No	No
Boiling point	N/A	Very high	Varies	Very high	Very high
Chemically reactive	Yes	Yes	Yes	Yes	No
Biodegradable	Yes	No	Usually yes	Usually no	No

properties of various constituents in water. This table is a starting point for deciding what processes might be appropriate for removing a particular constituent. The next step would be a more detailed investigation of the properties of the specific constituents in the particular source water of interest.

The second essential element for process selection is the ability of a unit process to capitalize on differences in the properties of constituents. Each unit process relies on one or more key properties. For instance, air stripping relies on the difference in volatility between a constituent and water. The more volatile a chemical is, the easier it is to remove from water. The properties exploited by each unit process covered in this book are listed in Table 3-2. Comparing Tables 3-1 and 3-2 reveals the processes that might be most appropriate for specific constituents. Air stripping and adsorption are the most common processes for removing organic contaminants, although reverse osmosis and advanced oxidation can also be effective. Granular filtration or membrane filtration, sometimes preceded by coagulation, flocculation, and sedimentation, are the best processes for removing particles and microorganisms. Common processes for removing inorganic constituents include coagulation followed by filtration (contaminants are co-precipitated and/or adsorbed onto particles after addition of a chemical), oxidation followed by filtration (solubility decreases with a change in the oxidation state), lime softening (co-precipitation and/or adsorption after addition of lime), adsorption onto activated alumina, ion exchange, and reverse osmosis. Process selection can only proceed when the properties of the constituents and the principles of the unit processes are understood.

The properties shown in Table 3-1 are general trends; specific chemicals may have different properties. For instance, ammonia and silica, both inorganic chemicals, are volatile and uncharged, respectively, at ambient

Table 3-2
Properties exploited by unit processes and the constituents in water for which each is commonly used

Process	Chapter	Properties Exploited	Most Common Target Constituents
Adsorption	10	Polarity, hydrophobicity	Dissolved organics
Air stripping	11	Volatility	Dissolved organics
Disinfection	13	Chemical reactivity	Microorganisms
Granular filtration	7	Adhesive molecular forces	Particles
Ion exchange	10	Charge	Dissolved inorganics
Membrane filtration	8	Size	Particles
Oxidation	12	Chemical reactivity	Dissolved organics and inorganics
Precipitation	5	Solubility	Dissolved inorganics
Reverse osmosis	9	Size, charge, polarity	Dissolved inorganics
Sedimentation	6	Density, size	Particles

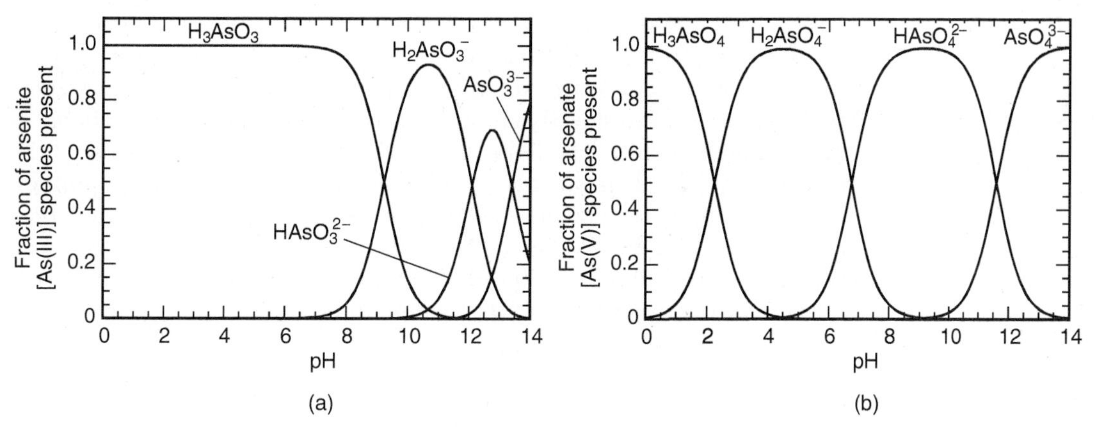

Figure 3-2
Speciation of arsenic as a function of the pH of the water: (a) arsenite [As(III)] and (b) arsenate [As(V)].

pH conditions. Properties may depend on the speciation of the chemical, which in turn depends on solution chemistry. As an example, arsenic commonly exists in water in the III (arsenite) or V (arsenate) oxidation states. Arsenic is not present as As^{3+} or As^{5+} ions, but forms triprotic weak acids in water; As(III) forms H_3AsO_3 and As(V) forms H_3AsO_4. As weak acids, arsenic species dissociate to form charged species. The speciation of arsenic as a function of pH and oxidation state is displayed on Fig. 3-2. As shown on this figure, neutral H_3AsO_3 is the predominant arsenite species below pH 9.2 and negatively charged $H_2AsO_3^-$ is predominant between pH values of 9.2 and 12.1. For arsenate species, $H_2AsO_4^-$ is the predominant species between pH values of 2.2 and 6.8, and $HAsO_4^{2-}$ is predominant between pH values of 6.8 and 11.6. Charged inorganics tend to be easier to remove from water than neutral ones. Thus, removal of arsenic might involve addition of an oxidant to convert As(III) to As(V), followed by addition of an acid or base to change the pH and convert arsenic to the desired species, followed by a separation process that exploits charge as a removal mechanism. In fact, ion exchange relies on charge and is one process that can remove arsenic from water.

The overall message is that the effectiveness of various unit processes at removing specific contaminants is founded in well-established scientific principles. It should be possible to predict the effectiveness of a process for any contaminant if the properties are understood. When process performance is not as expected, it is often because some aspect of the physicochemical properties (such as pH dependence) has been overlooked, that properties are not well understood, that various properties of a chemical may have contradictory effects, or that other constituents in the water compete or interfere with treatment for the desired chemical (i.e., sulfate interferes with removal of arsenic by ion exchange). A final consideration is kinetics (see Chap. 4). In some cases, the rate of a reaction may be as important as the equilibrium condition. For instance, physicochemical properties may indicate that a precipitation or oxidation reaction may take place, but the rate of reaction could be so slow that it does not occur within the time available in the treatment facility. The importance of chemistry—the chemistry of the constituents, the chemistry of the processes, and the chemical composition of the water—should be evident to students who are studying water treatment. Basic concepts of chemical equilibrium and kinetics are introduced in Chap. 4, but additional knowledge and understanding of chemistry is necessary to be an effective water treatment engineer.

3-2 Other Considerations in Process Selection

While constituent properties and process capabilities are the cornerstone of process selection, other factors must be considered. Some important considerations are removal efficiency, reliability, flexibility, a successful operating history, utility experience, and cost.

Removal Efficiency

The objective of treatment processes is to remove contaminants. Removal can be determined for bulk water quality measures (e.g., turbidity, total dissolved solids) or for individual constituents of interest (e.g., perchlorate, *Cryptosporidium oocysts*). The fraction of a constituent removed by a process can be calculated with the equation

$$R = \left(1 - \frac{C_e}{C_i}\right) \tag{3-1}$$

where R = removal expressed as a fraction, dimensionless
C_e = effluent concentration, mg/L
C_i = influent concentration, mg/L

In general, Eq. 3-1 is used where the removal efficiency for a given constituent is three orders of magnitude or less (i.e., 99.9%). For some constituents, such as microorganisms and trace organics, and some processes, such as membrane filtration, the concentration in the effluent is typically three or more orders of magnitude less than the influent concentration. For these situations, the removal is expressed in terms of base 10 *log removal value* (LRV) as given by the equation

$$\text{LRV} = \log(C_i) - \log(C_e) = \log\left(\frac{C_i}{C_e}\right) \tag{3-2}$$

The log removal notation is used routinely to express the removals achieved with membrane filtration (Chap. 8) and for disinfection (Chap. 13).

Calculation of removal and log removal value is demonstrated in Example 3-1.

Example 3-1 Calculation of removal and log removal value

During testing of a prototype membrane filter, bacteriophage concentrations of 10^7 mL^{-1} and 13 mL^{-1} were measured in the influent and effluent, respectively. Calculate the removal and log removal value.

Solution

1. Calculate removal using Eq. 3-1:

$$R = 1 - \frac{C_e}{C_i} = 1 - \frac{13\ \text{mL}^{-1}}{10^7\ \text{mL}^{-1}} = 0.9999987$$

2. Calculate the log removal value using Eq. 3-2:

$$\text{LRV} = \log\left(\frac{C_i}{C_e}\right) = \log\left(\frac{10^7\ \text{mL}^{-1}}{13\ \text{mL}^{-1}}\right) = 5.89$$

Comment

Note that seven significant digits are necessary to express removal adequately in arithmetic units, but only three significant digits are necessary to express log removal value for this example. Also note that LRV = 5 corresponds to 99.999 percent and LRV = 6 corresponds to 99.9999 percent removal (i.e., the log removal value equals the "number of 9's").

Reliability

Reliability has at least two meanings with respect to water treatment. First, process reliability indicates a process's ability to continuously meet the treatment objective. Some processes are very reliable and are able to meet treatment objectives despite changes in raw-water quality or operating parameters. Other processes are more sensitive to changes. Reliable processes are always preferred but are particularly important for contaminants such as pathogens that can cause acute health effects.

Second, a process must have mechanical and hydraulic reliability. Readily accessible potable water is a necessary part of modern society and is provided to customers as a utility such as gas or electricity. Consumers expect to have water available continuously. Processes that require very little oversight or maintenance, have few moving parts, or operate by gravity tend to be more reliable than processes with many complex components.

An example of the difference between process reliability and hydraulic reliability is the difference between granular filtration and membrane filtration. Granular filters operate by gravity and are hydraulically very reliable but require operator attention and proper pretreatment. Rapid changes in raw-water quality can lead to poor effluent water quality if operators do not respond correctly. In contrast, the removal efficiency of membrane filters is independent of raw-water quality; the effluent always meets treatment goals when they are working correctly. Changes in raw-water quality, however, can lead to membrane fouling and decreased passage of water through the filters. Thus, although water quality would be acceptable, the quantity of water produced may be insufficient.

Multiple-Barrier Concept

The reliability of a treatment train can be increased by providing multiple barriers for the same contaminant in series. Multiple barriers provide a factor of safety in the event one process fails even for a short period of time. Multiple-barrier reliability is particularly important for pathogens because acute effects can result from short-term exposure. The multiple-barrier approach is more than just redundancy. Multiple barriers will increase the reliability of the system even if the overall removal capability is not significantly different. A thought experiment that illustrates the increased reliability of multiple barriers is presented in Example 3-2.

Example 3-2 Effect of multiple barriers on reliability

Consider two treatment train alternatives. Train 1 has one unit process that reduces the target contaminant by six orders of magnitude (a 6 log reduction) when operating normally. Train 2 has three unit processes in series, each of which reduces the target contaminant by two orders of magnitude (a 2 log reduction in each step) when operating normally.

For the purpose of this analysis, assume that each unit process fails about 1 percent of the time and that when it fails it achieves half the removal that it normally achieves. With this information, estimate (a) the overall removal for trains 1 and 2 when all the unit processes are operating normally and (b) the frequency (in days per year) of various levels of treatment for each train assuming that process failures occur randomly.

Solution

1. Overall removal during normal operation:
   a. Train 1. Normal operation = 6 log removal.
   b. Train 2. Normal operation = 2 + 2 + 2 = 6 log removal.
2. Frequency of various levels of removal:
   a. Train 1:
      i. Provides 6 log removal 99 percent of time = 0.99 × 365 d = 361.35 d.
      ii. Provides 3 log removal 1 percent of time = 0.01 × 365 d = 3.65 d.

b. Train 2:
   i. Provides 6 log removal when all three processes are operating normally = 0.99 × 0.99 × 0.99 × 365 d = 354.16 d.
   ii. Provides 5 log removal when two processes are operating normally and one is in failure mode = 0.99 × 0.99 × 0.01 × 3 (failure mode combinations) × 365 d = 10.73 d.
   iii. Provides 4 log removal when one process is operating normally and two are in failure mode = 0.99 × 0.01 × 0.01 × 3 (failure mode combinations) × 365 d = 0.11 d = 2.6 h.
   iv. Provides 3 log removal when all three processes are in failure mode = 0.01 × 0.01 × 0.01 × 365 d = 0.00037 d = 32 s.

3. The results of this analysis are displayed in the following table:

Log Removal	Time of Operation During Typical Year, d	
	Train 1	Train 2
6	361.35	354.16
5		10.73
4		0.11
3	3.65	0.00037
Total	365.0	365.0

Comment

These results show that multiple barriers (train 2) are more robust. If the regulatory treatment requirement is for 4-log removal, train 2 reduces the time during which the customer is exposed to removal below this level by 10,000-fold, from 3.65 d per year to 0.00037 d per year (32 s). The use of multiple barriers in treatment provides reduced exposure to the risks that are associated with process failure.

Flexibility

Flexibility is an important consideration in process selection. Processes and process trains need to accommodate changes in raw-water quality. For instance, some types of sedimentation facilities can produce consistent effluent quality in spite of rapid changes in influent quality, whereas other types cannot accommodate rapid changes in influent water quality. Regulations for water treatment have changed frequently over the past several decades and will undoubtedly do so in the future as additional research on new contaminants and processes becomes known. Processes and process trains should have the flexibility to accommodate changes in regulations so that utilities are not forced to upgrade or replace processes every time a new regulation is passed. Additional processes can be added to a process train if space and hydraulic capacity has been made available in the original design. Processes also need the flexibility to accommodate an increase in capacity as the water demand in the community increases over time.

Successful Operating History

Some treatment processes have been used successfully for more than 100 years. Newer processes can offer advantages such as improved process performance, less waste production, easier operation, less maintenance, or lower cost. Equipment manufacturers sometimes develop new or updated processes that offer distinct advantages over existing equipment options. Other times, perceived benefits are nothing more than marketing claims to improve equipment sales. New equipment and processes must be considered cautiously—public health depends on a properly working treatment facility. Thus, a successful operating history in other applications should be considered during process selection, and newer processes should be considered only when the water treatment engineer can validate the claims of superior performance. One objective of this book is for water treatment engineers to understand the principles of unit processes. With that knowledge, an engineer can more reliably assess the claims of a manufacturer selling a new product. New products must still follow scientific principles.

Utility Experience

The unit processes in a treatment facility must be within the ability of the utility to properly operate and maintain them. Small water utilities often do not have the resources to hire and pay experienced operators who can be dedicated to proper operation of complex processes. In those cases, simple and automated processes may be more appropriate.

Cost

Cost must be a consideration in process selection. Since potable water is a utility provided at a municipal scale, costs must be affordable by the public. Both construction and operating costs are important, and many times the operating cost of a process will be a more significant factor than the construction cost.

3-3 Sustainability and Energy Considerations

Society has recently become concerned with climate change and other issues related to sustainability. While *sustainability* can mean different things to different people, a commonly accepted definition, from the 1987 Brundtland Commission report *Our Common Future*, is "the ability of a society to meets their needs without compromising the ability of future generations to meet their own needs" (WCOED, 1987, p. 24). In other words, our society should not consume resources at such a rate that they would be unavailable in the future nor degrade the environment to such an extent that it would be unusable in the future.

Sustainability is particularly relevant to the water industry because water use has a large impact on the environment. The Electric Power Research Institute (EPRI) (2009) reports that the water and wastewater industry in the United States used 123.45 billion kWh of electricity in 2000. This value was about 3.4 percent of all end-use electricity in the United States, making the water and wastewater industry sector the third largest consumer of electricity, behind only the chemical and primary metals industries.

Individual consumers can reduce their environmental impact by considering gas mileage as one factor when choosing one car over another or choosing compact fluorescent lightbulbs over incandescent ones. The water treatment industry can make similar choices. While water quality, physicochemical properties, and treatment mechanisms are clearly important in process selection, sustainability should be a consideration when two or more processes may be effective at meeting a treatment goal. Sustainability should also be a consideration when setting process design criteria; small changes in design criteria can have significant impacts on the energy consumed over the lifetime of a treatment plant.

Life-Cycle Assessment

While sustainability is a broad and general term, standard procedures are available for quantifying potential environmental impacts from a product, process, or service. The approach is called *life-cycle assessment* (LCA) and is codified in the International Standard Organization (ISO) 14040 series standards. An LCA is a cradle-to-grave analysis, examining the total environmental impact of a product through every step of its life, from raw material acquisition, manufacturing, distribution, use by consumers, and ultimate disposal.

An LCA has four components. First, the goals and scope of the assessment are defined, followed by an inventory assessment in which the relevant inputs and outputs to the system are quantified. Next, the potential environmental impacts associated with those inputs and outputs are calculated. Finally, the results are interpreted and opportunities to reduce the environmental impact are identified.

Society is currently focused on climate change. However, climate change is just one of many potential impacts on the environment. LCAs may consider a number of potential environmental impact categories, such as

- Global warming
- Stratospheric ozone depletion
- Acidification potential
- Eutrophication potential
- Photochemical smog formation
- Terrestrial toxicity
- Aquatic toxicity
- Human health
- Resource depletion

These impacts cannot easily be compared to each other (i.e., how do you compare two design options when one might have a greater impact on global warming and the other might have a greater impact on human health?). Thus, overall environmental indicators have been developed that weight and normalize impacts on a common scale so that a single final score can be reported. Eco-Indicator 99 and Eco-Points 97 are two examples of overall environmental indicators used in LCA.

A full LCA is a data-intensive and laborious activity. Thus, many LCAs are streamlined by limiting the scope, for instance, by neglecting components that are expected to have minimal impact. In addition, software packages are available to assist with the collection and interpretation of LCA data. SimaPro by Pré Consultants (2011) and GaBi by PE International (2011) are two commonly used software packages.

Life-Cycle Assessment of Water Treatment Facilities

A number of LCAs have been conducted of individual unit processes and full water treatment plant trains over the past 10 to 15 years. Analyses have considered the impacts of constructing the facility, operating the treatment facility, and decommissioning the plant after its useful life. The construction phase considers acquisition of materials needed to build the plant, such as concrete and steel, and the impact associated with the actual construction process. A conclusion from the existing LCA literature of water treatment processes is that construction is usually a minor component, typically 5 to 20 percent, of the overall environmental impact. Similarly, decommissioning of a treatment facility at the end of its useful life has a very small impact, less than 1 percent of the total (Vince et al., 2008). Operation typically has the largest environmental impact.

Three potential sources of environmental impact from water treatment plant operation are energy consumption, obtaining chemicals and other consumable materials, and waste production. Of these, energy consumption has generally been found to have the largest single impact; in some processes, such as reverse osmosis, energy use during operation accounts for more than 80 percent of all environmental impacts over all plant life stages.

Pumping is a major source of energy consumption in water treatment. Energy consumed during pumping depends on the flow rate and the pressure:

$$P_W = \frac{Q_F P}{e} \tag{3-3}$$

where P_W = power, W (or rate of energy consumption, kWh/d)
Q_F = feed water flow rate, m^3/d or ML/d
P = pressure, Pa
e = efficiency

Pumps are rated in units of either pressure or head, but the two are related:

$$P = \rho g h \tag{3-4}$$

where ρ = density of fluid, kg/m^3
g = gravitational constant, 9.81 m/s^2
h = head, m

Specific energy consumption is the energy consumed per unit volume of water produced and can be calculated from

$$E = \frac{P_W}{Q_P} \quad (3\text{-}5)$$

where E = specific energy consumption, kWh/m^3
Q_P = product water flow rate, m^3/d or ML/d

In a single pump, the feed flow rate and product flow rate are the same. In many treatment processes, however, a portion of the feed water is used within the process (e.g., for backwashing during granular filtration) or becomes a waste stream (sludge withdrawal in sedimentation or concentrate from reverse osmosis). In these processes, the fraction of product water produced by the process is called the recovery:

$$r = \frac{Q_P}{Q_F} \quad (3\text{-}6)$$

where r = is the recovery.

Recovery can have an important impact on specific energy consumption. Examples of specific energy consumption calculations are shown in Example 3-3.

Example 3-3 Specific energy consumption during pumping

Calculate the specific energy consumption of the following scenarios: (a) a reverse osmosis (RO) system designed to produce 19,000 m^3/d (5 mgd) at 80 percent recovery. The RO feed pumps operate at 16 bar (232 psi) and 87 percent efficiency, and (b) distribution pumps operating at 3785 m^3/d (1 mgd), 90 m head (295 ft), and 85 percent efficiency.

Solution

Part 1

1. Calculate the feed water flow using Eq. 3-6:

$$Q_F = \frac{Q_P}{r} = \frac{19{,}000 \text{ m}^3/\text{d}}{0.80} = 23{,}750 \text{ m}^3/\text{d}$$

2. Calculate the pump power using Eq. 3-3. Note: 1 bar = 100 kPa = 10^5 N/m^2:

$$P_W = \frac{Q_F P}{e} = \frac{(23{,}750\ \text{m}^3/\text{d})(16 \times 10^5\ \text{N/m}^2)}{0.87(86{,}400\ \text{s/d})}$$

$$= 5.06 \times 10^5\ \text{N} \cdot \text{m/s} = 506\ \text{kW}$$

3. Calculate specific energy consumption using Eq. 3-5:

$$E = \frac{P_W}{Q_P} = \frac{506\ \text{kW}(24\ \text{h/d})}{19{,}000\ \text{m}^3/\text{d}} = 0.64\ \text{kWh/m}^3$$

Part 2

1. Calculate the pressure produced by the pump using Eq. 3-4. Note: 1 N = 1 kg · m/s^2:

$$P = \rho g h = (1000\ \text{kg/m}^3)(9.81\ \text{m/s}^2)(90\ \text{m}) = 8.83 \times 10^5\ \text{N/m}^2$$

2. Calculate specific energy consumption. Note that $Q_F = Q_P$ so the flow cancels out if Eq. 3-3 is substituted into 3-5. Also note that 1 N · m = 1 J = 1 W · s, so 1 kWh = 3.6 × 10^6 N · m:

$$E = \frac{P}{e} = \frac{8.83 \times 10^5\ \text{N/m}^2}{0.85}\left(\frac{1\ \text{kWh}}{3.6 \times 10^6\ \text{N} \cdot \text{m}}\right) = 0.29\ \text{kWh/m}^3$$

Because of the overall significance of energy consumption in life-cycle impacts, energy consumption can be used as an overall environmental indicator. Studies have found the average overall energy consumption at typical surface water treatment plants, including raw-water pumping, treatment processes, and distribution pumping, is between 0.37 and 0.50 kWh/m^3. Energy use will vary significantly depending on raw and distribution pumping requirements and on the unit processes in the plant. Figure 3-3 summarizes specific energy consumption data for various water treatment processes. This table is a starting point for considering sustainability when evaluating alternative unit processes. Additional energy and sustainability considerations are addressed in the individual unit process chapters throughout this book.

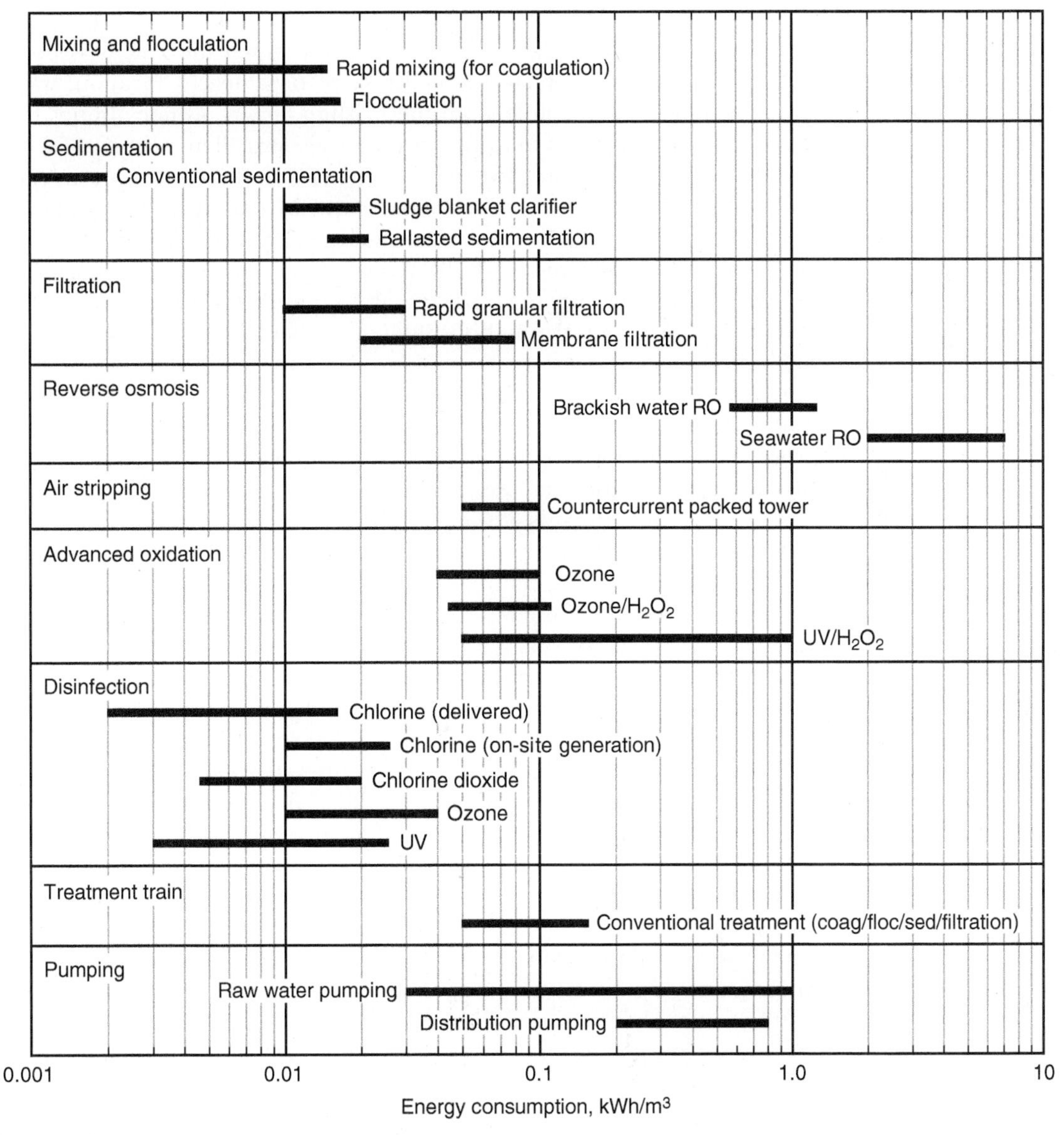

Figure 3-3
Electricity consumption by common water treatment processes (data obtained from Elliott et al., 2003; Vince et al., 2008; EPRI, 2009; Veerapani et al., 2011; and authors' experience).

3-4 Design and Selection of Process Trains

The treatment train selection process starts with at least three key pieces of information: (1) the source water quality, (2) the desired finished-water quality, and (3) the quantity of water needed (the capacity of the facility).

Source water quality may be available from several sources. First, historical data may be available. If the utility for which the facility is being constructed has another facility at the same or a nearby location, water quality data will be available from the existing facility. Other utilities that withdraw water upstream or downstream are also excellent sources of water quality data. Finally, state and federal agencies may have long-term sampling programs that have collected water quality data from the proposed source water. For instance, the U.S. Geological Survey (USGS) National Water Information System (NWIS) (USGS, 2011) is a compilation of the results from millions of water quality analyses sampled at many surface water and groundwater sites throughout the United States, free and available on the Internet.

In addition to historical data, it may be beneficial to conduct a directed study to collect additional water quality data at the actual site of the proposed intake or well. Sampling may be conducted as part of a pilot study (discussed later in this chapter), to gain information about specific, new, or unregulated contaminants, if the historical record is not sufficient to make process decisions or to support permitting or regulatory requirements.

The primary factors affecting the selection of finished-water quality goals are the intended use of the water and the regulatory parameters governing that use. The primary focus of this book is municipal drinking water. Guidelines or regulations for drinking water are set at a national or state level. Sources for drinking water quality guidance in several nations are shown in Table 3-3.

States or member nations can set limits on water quality more stringent than the guidance sources listed in Table 3-3. Utilities also sometimes set drinking water target levels lower than regulated limits. Lower limits established during design are useful because they provide a factor of safety for variability during operation. They may also instill public confidence in the utility. Unregulated parameters may also be part of the finished-water quality goals when the source water has unique sources of contamination, when future regulation of unregulated parameters is anticipated, or to instill additional public confidence in the water supply and the utility.

Armed with raw-water quality data and finished-water quality goals, water treatment engineers can begin to select the treatment train. Sources

Table 3-3
National and international guidelines for drinking water quality

Country or Region	Guidance or Regulatory Document	Reference
United States	Safe Drinking Water Act	EPA (2011)
Canada	Guidelines for Drinking Water Quality	Health Canada (2011)
European Union	Drinking Water Directive	Europa (2011)
International	Guidelines for Drinking Water Quality	WHO (2011)

of information that are useful during process selection include the following:

- **Textbooks, design guides, and reference materials:** Numerous textbooks are available with detailed information on the design of unit processes and the contaminants for which they are effective. Examples include *MWH's Water Treatment: Principles and Design* (Crittenden et al., 2012), and *Integrated Design and Operatic of Water Treatment Facilities* (Kawamura, 2000). A popular reference for treatment plant design used in the mid and eastern United States is known as the Ten State Standards (Great Lakes–Upper Mississippi River Board, 2007). The U.S. EPA and the American Water Works Association (AWWA) have published manuals and reports on treatment processes.
- **Regulatory guidance:** For contaminants regulated in the Safe Drinking Water Act, the U.S. EPA has designated certain processes as *best available technology* (BAT), which are processes that EPA certifies as being the most effective for removing a contaminant. For some contaminants, the U.S. EPA identifies *treatment techniques,* which are specific processes and associated requirements that are required in order to meet the regulations.
- **Engineering experience:** Experience acquired through treatment of the same or similar source waters provides an excellent guide in selecting the treatment process scheme. Experience may come from other engineers within the organization, from the utility, or other utilities in the region.
- **Recent research:** For contaminants that are not currently regulated, treatment information can often be found in recent scientific literature such as journals and conference proceedings.
- **Laboratory (bench) testing:** Bench testing involves transporting a small quality of source water to an offsite location for analysis. Testing is typically done in batch reactors (see Chap. 4), compared to the continuous-flow reactors common in pilot testing or full-scale facilities. Bench testing can be used to determine chemical doses needed to achieve treatment or to verify that specific chemical reactions will take place as expected.
- **Pilot testing:** *Pilot plants* are small-scale versions of actual treatment processes. The scale is typically small enough to fit on a trailer or in a small shed, but large enough that they must be located at the site of the source water because it would be impractical to transport the water to a distant location. Pilot studies are appropriate when the applicability of a process for a given situation is unknown but the potential benefits of using the process are significant. They are necessary when the hydraulics of a process is as important as the chemistry in achieving effective treatment; the relationship between reactor hydraulics and effluent concentrations of reactors is presented in Chap. 4. Pilot tests

are particularly important for testing new or innovative processes or when processes might be designed with high loading rates. They can be used to establish the suitability of the process in the treatment of specific water under specific environmental conditions, verify performance claims by manufacturers, optimize or document process performance, satisfy regulatory agency requirements, and generate the necessary data on which to base a full-scale design.

3-5 Summary and Study Guide

After studying this chapter, you should be able to:

1. Define the following terms and phrases and describe the significance of each in the context of process selection and water treatment:

best available technology	pilot plant	sustainability
life-cycle assessment	process train	treatment technique
log removal value	specific energy consumption	treatment train
multibarrier concept		unit process
physicochemical		

2. Sketch a typical treatment train for a surface water treatment plant.
3. List the major classes of constituents in natural waters and identify some compounds within each major class.
4. Describe the physicochemical properties that can be used to separate constituents from water.
5. Describe common separation processes, the physicochemical properties that each will exploit to accomplish treatment, and the types of constituents that each can effectively remove from water.
6. Propose unit processes that might be effective for removing a contaminant, if given physicochemical properties of the contaminant.
7. Explain some reasons why a process might not remove a constituent from water as predicted solely from known properties.
8. Identify important considerations in process selection in addition to contaminant properties.
9. Calculate the removal efficiency and log removal value of a compound.
10. Explain how multiple barriers improve the reliability of a treatment train.
11. Explain why sustainability should be considered in process selection.
12. Describe the objective of a life-cycle assessment and the general steps in conducting one.
13. Calculate specific energy consumed during pumping.

14. Describ... ation an engineer can use to go about iden-tifying/sele... t process that could be used to eliminate a particular c...

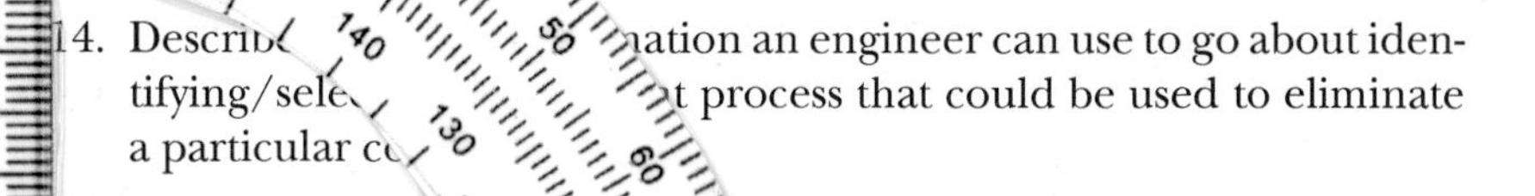

omework Problems

1 Calculate rejection a... al value for the following filtration process (to be selecte... or). Use the number of significant figures necessary to c... trate the removal being obtained.

	A		C	D	E
nfluent concentration (#/mL)	10...		7.1 × 10^5	1.65 × 10^7	2.8 × 10^6
ffluent concentration (#/mL)			0.16	65	96

2 You work for ... onmental engineering consulting firm and a pote... lled and said that a new contaminant has recently ... n their water supply. She wants your firm to id... s might be able to remove the contaminant. ... t (to be selected by your instructor), suggest ... ght be used and explain how you arrived at your

...icivirus

c. 17-α ethynyl estradiol

d. *Mycobacterium avium*

e. *Naegleria fowleri*

f. perchlorate

g. plutonium-239

h. *Salmonella enterica*

i. strontium

j. 1,1,1,2-tetrachloroethane

k. vanadium

l. vinclozolin

3-3 Pick a city in the United States that is of interest to you. Any city is acceptable with one limitation: The water utility must use a surface water for at least part of its water supply. Read the Consumer Confidence Report (often called a Water Quality Report) provided by the water utility. The water quality report is typically 2 to 8 pages long and provides specific information dictated by EPA regulations. These reports are often posted on the utility's website as pdf files. A list of websites for some larger utilities is available at http://www.epa.gov/safewater/dwinfo.htm. If the utility is not shown on the EPA website, try finding the utility website or contact the utility department directly to get a copy. Answer the following questions:

a. Describe the source water for the utility.

b. Describe the treatment provided by the utility. Draw a schematic of the treatment train (don't just cut and paste from the report or the Internet), identify the chemicals added, and describe the purpose of each unit process and each chemical added. In some cases, it may be necessary to obtain information beyond what is provided in the Consumer Confidence Report (utilities often have additional details about the treatment train on their websites).

c. If you were a consumer in this community, would you be concerned about the water quality based on the information provided in the report?

d. Develop a list of questions about this utility and its treatment practices that you hope to be able to answer after studying this book.

e. Provide a copy of the utility's water quality report with your assignment.

3-4 Calculate the specific energy consumption by the following process or system (to be selected by the instructor):

a. A raw-water supply pump operating at 8200 m^3/d, 10 m head, and 85 percent efficiency.

b. The feed water pumps for a seawater reverse osmosis system generate a flow of 9 ML/d at 75 bar, operating at 86 percent efficiency. The RO system operates at 55 percent recovery.

c. A granular media filter that generates 2.7 m of head loss as the water passes through the filter.

d. Filters at a treatment plant are in backwash mode for 15 min each day and filtration mode for the rest of the day. During filtration, the filters produce 1200 m^3/h of filtered water. The backwash pump operates at 12 m of head and 80 percent efficiency, and pumps a total of 1500 m^3 of water each time it operates. Specific energy is the energy of the backwash pump per volume of filtered water.

e. A hydraulic pump rapid mix system draws a side stream from the process flow and reinjects it at higher pressure to create turbulence that facilitates chemical mixing. The side-stream pump operates at 2700 L/min, 150 kPa pressure, and 85 percent efficiency. The main process flow is 190,000 m^3/d. Specific energy is the energy of the side-stream pump per volume of main process flow.

References

Crittenden, J. C., Trussell, R. R., Hand, D. W., Howe, K. J, and Tchobanoglous, G. (2012) *MWH's Water Treatment: Principles and Design,* 3rd ed., Wiley, Hoboken, NJ.

Elliott, T., Zeier, B., Xagoraraki, I., and Harrington, G. W. (2003) *Energy Use at Wisconsin's Drinking Water Facilities.* Report Number 222-1, Energy Center of Wisconsin, Madison, WI.

EPA (2011) Safe Drinking Water Act Home. Available at: http://www.epa.gov/safewater/sdwa/; accessed on Aug. 20, 2011.

EPRI (2009) *Program on Technology Innovation: Electric Efficiency through Water Supply Technologies - A Roadmap,* Technical Report 1019360, Electric Power Research Institute, Pala Alto, CA.

Europa (2011) Environment—Water—Drinking Water. Available at: http://ec.europa.eu/environment/water/water-drink/index_en.html; accessed on Aug. 20, 2011.

Great Lakes–Upper Mississippi River Board (2007) 10 States Standards—Recommended Standards for Water Works. Available at: http://10statesstandards.com/waterstandards.html; accessed on Aug. 20, 2011.

Health Canada (2011) Drinking Water—Water Quality. Available at: http://www.hc-sc.gc.ca/ewh-semt/water-eau/drink-potab/guide/index-eng.php; accessed on Aug. 20, 2011.

Kawamura, S. (2000) *Integrated Design and Operation of Water Treatment Facilities,* Wiley, Hoboken, NJ.

PE International (2011) GaBi—Life Cycle Assessment (LCE/LCA) Software System. Available at: http://www.gabi-software.com/; accessed on Aug. 20, 2011.

Pré Consultants (2011) SimaPro LCA Software. Available at: http://www.pre.nl/simapro/; accessed on Aug. 20, 2011.

USGS (2011) USGS Water Data for the Nation. Available at: http://waterdata.usgs.gov/nwis; accessed on Aug. 20, 2011.

Veerapaneni, S., Klayman, B., Wang, S., and Bond, R. (2011) *Desalination Facility Design and Operation for Maximum Efficiency,* Water Research Foundation, Denver, CO.

Vince, F., Aoustin, E., Bréant, P., and Marechal, F. (2008) "LCA Tool for the Environmental Evaluation of Potable Water Production," *Desalination,* **220**, 1–3, 37–56.

WCOED (World Commission on Environment and Development) (1987) *Our Common Future.* WCOED Document A/42/427.

WHO (2011) Available at: http://www.who.int/water_sanitation_health/dwq/guidelines/en/; accessed on Aug. 20, 2011.

4 Fundamental Principles of Environmental Engineering

A number of principles are essential to the development and understanding of water treatment processes; these same principles are important throughout the environmental engineering profession. These principles include the equilibrium and kinetics of chemical reactions (Secs. 4-1 through 4-4), mass balance analysis (Sec. 4-5), reactor analysis (Secs. 4-6 through 4-12), and mass transfer (Secs. 4-13 through 4-17). Each of these topics is complex. The environmental engineering curriculum typically contains an entire course on water chemistry, and it is not uncommon in chemical engineering curriculum to have a course on reactor analysis and a course on mass transfer. This chapter is more focused and contains material in sufficient detail to understand the principles of the water treatment processes discussed in Chaps. 5 through 13.

4-1 Units of Expression for Chemical Concentrations

The quantity of components (i.e., species, solutes, or particles) present in various media (water, air, and solids) can be expressed in a variety of ways. Some common methods for expressing quantity and concentration are as follows:

1. *Mass concentration* is expressed as units of mass of a component per volume of solution. Many constituents are present in water in mg/L or μg/L concentrations. [note: Details of the SI system of measurement are available in Thompson and Taylor, (2008)]. In air, units of $\mu g/m^3$ are common. Parts per million (ppm) is often used as equivalent for mg/L in water because the density of water is about 1 kg/L. Dividing a mass concentration of 1 mg/L of solute by the density of water yields

$$\frac{1\ \text{mg/L}}{(1\ \text{kg/L})(10^6\ \text{mg/kg})} = \frac{1\ \text{mg solute}}{10^6\ \text{mg solution}} = 1\ \text{part per million, ppm}$$

In general, ppm, parts per billion (ppb), and parts per trillion (ppt) should be avoided as replacements for mg/L, μg/L, and ng/L because they are only equivalent when the solution has a density of 1 kg/L.

2. *Molar concentration* or *molarity* is units of amount of solute per volume of solution. A *mole* is an amount of something (like a dozen is an amount), equal to 6.022×10^{23} (Avogadro's number). Molar concentration (mol/L or M) is more unambiguous than mass concentration and is preferred, particularly when working with chemical stoichiometry or when the basis for the mass is not clear. Molar concentration is often designated by square brackets, [A], and molar concentrations can be converted to mass concentrations if the molecular weight is known:

$$[A](MW) = C_A \qquad (4\text{-}1)$$

where $[A]$ = molar concentration of component A, mol/L
MW = molecular weight of component A, g/mol
C_A = mass concentration of component A, g/L

3. *Mole fraction* or *mass fraction* is the ratio of the amount or mass of a given component to the total amount or mass of all components:

$$X_A = \frac{n_A}{\sum_{i=1}^{N} n_i} \quad (4\text{-}2)$$

$$C_A = \frac{m_A}{\sum_{i=1}^{N} m_i} \quad (4\text{-}3)$$

where X_A = mole fraction of component A
n_A, n_i = amounts of component A and component i, mol
C_A = mass fraction of component A
m_A, m_i = mass of component A and component i, kg
N = number of components

Percent by amount or mass can be calculated by multiplying the mole fraction or mass fraction by 100, respectively. A solute that is present at 1 percent by mass or has a mass fraction of 0.01 has a mass concentration of 10,000 mg/L if the solution density is 1 kg/L, so mass fractions and percent are most suitable for concentrated solutions.

4. *Mass concentration as "X"* is a common method of expressing concentration in environmental engineering because water quality parameters are often composed of multiple constituents. For example, nitrogen can be present in water as NH_3, NH_4^+, NO_3^-, or NO_2^-, each of which has a different molecular weight. A change of pH or oxidation state can change which species is present, leading to a change in the mass concentration of nitrogen species even though the total amount of nitrogen in the water has not changed. Thus, the concentration of nitrogen can be expressed as mg/L of N, where the MW of N is used to calculate the mass concentration, rather than the MW of the particular nitrogen species present. The concentration of hardness, alkalinity, and individual species like silica and arsenic are frequently expressed in this form. An example of this type of concentration is illustrated in Example 4-1.
5. *Normality (N)* or *equivalents/volume (eq/L)* is used to express concentration in specific cases related to ionic species in water,

acid/base chemistry, and oxidation/reduction chemistry. Normality is defined as

$$N = \frac{m_A}{(EW)\,V} \tag{4-4}$$

where N = normality of component A, eq/L
m_A = mass of component A, g
EW = equivalent weight of component A, g/eq
V = volume of solution, L

The equivalent weight is expressed as

$$EW = \frac{MW}{z} \tag{4-5}$$

where z is the equivalents per mole of the component. For ionic species in water, z is equal to the valence; for oxidation–reduction reactions, z is equal to the number of electrons transferred; and for acid/base reactions, z is equal to the number of replaceable hydrogen atoms or their equivalent. For example, for hydrochloric and sulfuric acids:

- HCl: 1 M = 1 N because 1 M HCl releases 1 M H^+ ions; the valence of Cl^- is 1; therefore HCl has 1 eq/mol.
- H_2SO_4: 1 M = 2 N because 1 M H_2SO_4 releases 2 M H^+ ions; the valence of SO_4^{2-} is 2; therefore H_2SO_4 has 2 eq/mol.

An example of normality is illustrated in Example 4-1.

6. *Log molar concentrations* are used because concentrations often vary by many orders of magnitude, making logarithms convenient. For instance, if $[C] = 2 \times 10^{-5}$ mol/L, then $\log[C] = -4.7$ and $[C] = 10^{-4.7}$ mol/L.
7. The *p notation* is another convenient way of expressing the low concentrations of chemical species that are often found in natural waters. The p operand is defined as the negative of the base-10 log of the value:

$$pC = -\log(C) \tag{4-6}$$

where C = is the concentration of a constituent in solution (in mol/L).

The reporting of the hydrogen ion concentration as pH is a familiar example of p notation. The pH of a solution is defined as

$$pH = -\log[H^+] \tag{4-7}$$

The p notation can be used for any value, not just concentrations. Equilibrium constants, introduced later in this section, are often expressed using p notation.

The conversion of concentrations between various sets of units is demonstrated in Example 4-1.

Example 4-1 Converting between units of concentration

Calculate the concentration of 0.85 mM solution of calcium in units of mg/L, meq/L, mg/L as $CaCO_3$, log molar concentration, and p notation.

Solution

1. Determine the concentration of Ca in mg/L. (Note: mM = millimole/liter.) From a periodic table (App. D), the MW of Ca = 40 g/mol = 40 mg/mmol:

$$0.85 \text{ mmol/L} = (0.85 \text{ mmol/L})(40 \text{ mg/mmol}) = 34 \text{ mg/L}$$

2. Determine the concentration of Ca in meq/L. (Note: meq = milliequivalents.) Calcium ion are divalent and have a charge of +2:

$$0.85 \text{ mmol/L} = (0.85 \text{ mmol/L})(2 \text{ meq/mmol}) = 1.7 \text{ meq/L}$$

3. Determine the concentration of Ca in mg/L as $CaCO_3$. (Note, hardness is a bulk parameter of water that consists of the concentrations of Ca and Mg in water, but is expressed as mg/L as $CaCO_3$). The MW of $CaCO_3$ = 100 g/mol.

$$0.85 \text{ mmol/L} = (0.85 \text{ mmol/L})(100 \text{ mg/mmol}) = 85 \text{ mg/L as } CaCO_3$$

4. Determine the concentration of Ca in log molar concentration and p notation.

$$\log(0.85 \times 10^{-3} \text{ mol/L}) = -3.07$$

$$[\text{Ca}] = 10^{-3.07} \text{ M}$$

$$\text{pCa} = 3.07$$

4-2 Chemical Equilibrium

Chemical reactions are used in water treatment to change the physical, chemical, and biological nature of water to accomplish water quality objectives. The reactions of acids and bases, precipitation of solids, complexation of metals, and oxidation of reduced species are all important reactions used in water treatment. An understanding of chemical reaction stoichiometry, equilibrium, and kinetics is needed to develop mathematical expressions that can be used to describe the rate at which these reactions proceed. Stoichiometry and equilibrium are discussed in this section, and chemical kinetics are introduced in the next section.

Environmental engineering curriculum generally includes an entire course or more devoted to water chemistry, and many books have been written on the subject. Students are urged to consult one of the books on the topic. Water chemistry textbooks (Benjamin, 2002; Pankow, 1991; Sawyer et al., 2003; Snoeyink and Jenkins, 1980; Stumm and Morgan, 1996) may be reviewed for more complete treatment of these concepts and other principles of water chemistry.

Chemical reactions used for water treatment are described using chemical equations. Chemical reactions are shown with reactants on the left side and products on the right side; in the following reaction, reactants A and B react to form products C and D:

$$A + B \rightarrow C + D \tag{4-8}$$

Symbols commonly used in chemical equations are described in Table 4-1. Reactions can be thought of as reversible or irreversible. Irreversible reactions consume reactants and form products until one of the reactants is totally consumed. Oxidation–reduction reactions are considered to be irreversible. Reversible reactions are those that proceed until an equilibrium condition is reached; at this equilibrium, both reactants and products may be present. Acid–base, precipitation, and complexation reactions are reversible. Reversible reactions can proceed in either direction. For example, in Eq. 4-8 the reactants A and B react to form products C and D, whereas in Eq. 4-9 the reactants C and D react to form products A and B:

$$C + D \rightarrow A + B \tag{4-9}$$

The reactions presented in Eqs. 4-8 and 4-9 can be combined as follows:

$$A + B \rightleftarrows C + D \tag{4-10}$$

For example, the reaction between bicarbonate (HCO_3^-) and carbonate (CO_3^{2-}) can be written as:

$$HCO_3^- \rightleftarrows H^+ + CO_3^{2-} \tag{4-11}$$

At equilibrium, both bicarbonate and carbonate can be present in solution, and the relative concentration of each will depend on the solution pH (which defines the H^+ concentration). This equilibrium is exactly the same regardless of whether the solution was created by adding bicarbonate and allowing carbonate to form, or by adding carbonate and allowing bicarbonate to form. Note that although the reaction can proceed in either direction, by definition the species on the left are called reactants and the species on the right side are called products.

Table 4-1
Symbols used in chemical equations

Symbol	Description	Comments
$\rightarrow$	Irreversible reaction	Single arrow points from the reactants to the products, e.g., $A + B \rightarrow C$.
$\rightleftarrows$	Reversible reaction	Double arrows used to show that the reaction can proceed in the forward or reverse direction, depending on the solution characteristics.
[ ]	Brackets	Concentration of a chemical species in standard units (mol/L for aqueous phase).
{ }	Braces	Activity of a chemical constituent.
(s)	Solid phase	Designates a chemical species present in solid phase, e.g., calcium carbonate, $CaCO_3(s)$.
(l)	Liquid phase	Designates a chemical species present in liquid phase, e.g., liquid benzene, C_6H_6 (l).
(aq)	Aqueous (dissolved)	Designates a chemical species dissolved in water, e.g., ammonia in water, $NH_3(aq)$.
(g)	Gas	Designates a chemical species present in gas phase, e.g., chlorine gas, $Cl_2(g)$.
$\xrightarrow{x}$	Catalysis	Chemical species, represented by x, catalyzes reaction, e.g, cobalt (Co) is the catalyst in the reaction $SO_3{}^{2-} + 0.5O_2 \xrightarrow{Co} SO_4{}^{2-}$.
$\uparrow$	Volatilization	Arrow directed up following a component is used to show volatilization of given component, e.g., $CO_3{}^{2-} + 2H^+ \rightleftarrows CO_2(g) \uparrow + H_2O$.
$\downarrow$	Precipitation	Arrow directed down following a component is used to show precipitation of given component, e.g., $Ca^{2+} + CO_3{}^{2-} \rightleftarrows CaCO_3(s) \downarrow$.

Source: Adapted from Benefield et al. (1982).

Reaction Stoichiometry

The relationship between the relative amount of each reactant needed to produce an amount of each product is called the reaction stoichiometry. The general equation for a chemical reaction that describes the relative amounts of reactants A and B needed to form products C and D can be written:

$$aA + bB \rightleftarrows cC + dD \tag{4-12}$$

where a, b, c, d = stoichiometric coefficients of species A, B, C, D, respectively, unitless

Stoichiometry is based on an understanding of chemical formulas, conservation of mass, and atomic masses. For example, ferrous iron (Fe^{2+}) is an undesirable constituent in drinking water because it can impart an unpleasant taste and cause rust-colored stains on plumbing fixtures. One method for removing Fe^{2+} from water is to oxidize it with oxygen to produce insoluble ferric hydroxide $[Fe(OH)_3]$ according to the following reaction:

$$4Fe^{2+} + O_2 + 10H_2O \rightarrow 4Fe(OH)_3 + 8H^+ \qquad (4\text{-}13)$$

As shown in Eq. 4-13, 1 mol of O_2 is capable of oxidizing 4 mol of Fe^{2+}; in doing so, it will form 4 mol of $Fe(OH)_3$ and 8 mol of H^+. Using reaction stoichiometry and the molecular weight of the chemical species, it is possible to calculate the mass of reactants and products participating in a reaction, as shown in Example 4-2.

Example 4-2 Using reaction stoichiometry to calculate the mass of reactants and products

A groundwater used as a drinking water supply contains 2.6 mg/L of Fe^{2+}. Calculate the amount of O_2 that will be needed to oxidize it and the amount of $Fe(OH)_3$ that will be produced. Assume that the reaction proceeds to completion.

Solution

1. Determine the molecular weight of each species involved in the reaction from the atomic weights of the constituent elements. From a periodic table (App. D), the atomic weights of the elements are: H = 1 g/mol, O = 16 g/mol, and Fe = 55.8 g/mol. The molecular weights of the species then are

$$\text{MW of } Fe^{2+} = 55.8 \text{ g/mol}$$

$$\text{MW of } O_2 = (2)16.0 \text{ g/mol} = 32.0 \text{ g/mol}$$

$$\text{MW of } Fe(OH)_3 = 55.8 + (3)(1.0) + (3)(16.0) = 106.8 \text{ g/mol}$$

2. Calculate the concentration of oxygen required to oxidize the iron.

$$2.6 \text{ mg/L } Fe^{2+} \left(\frac{1 \text{ mol } Fe^{2+}}{55.8 \text{ g } Fe^{2+}}\right)\left(\frac{1 \text{ mol } O_2}{4 \text{ mol } Fe^{2+}}\right)\left(\frac{32.0 \text{ g } O_2}{\text{mol } O_2}\right)$$

$$= 0.37 \text{ mg/L } O_2$$

Therefore, 0.37 mg/L of O_2 is capable of oxidizing 2.6 mg/L of Fe^{2+}.

3. Calculate the concentration $Fe(OH)_3$ that will be produced.

$$2.6 \text{ mg/L } Fe^{2+} \left(\frac{1 \text{ mol } Fe^{2+}}{55.8 \text{ g } Fe^{2+}}\right)\left(\frac{4 \text{ mol } Fe(OH)_3}{4 \text{ mol } Fe^{2+}}\right)\left(\frac{106.8 \text{ g } Fe(OH)_3}{\text{mol } Fe(OH)_3}\right)$$

$$= 4.98 \text{ mg/L } Fe(OH)_3$$

Therefore, 4.98 mg/L of $Fe(OH)_3$ will be produced when 2.6 mg/L of Fe^{2+} is oxidized.

Comment

Note that the amount of each element (Fe, H, and O) is conserved in the chemical reaction on a molar basis, but that the mass of O_2 consumed and $Fe(OH)_3$ produced depends on the stoichiometry of the reaction and the molecular weights of the species. On a mass basis, more $Fe(OH)_3$ is produced than Fe^{2+} is consumed. The amount of residuals produced is an important consideration during water treatment.

Concentration and Activity

The discussion above noted that reactants and products can both be present when a reversible reaction reaches equilibrium. The ability of a species to participate in chemical reactions depends on its chemical *activity*. At equilibrium, the amounts of reactants and products present will depend on the activity of each species according to an equilibrium relationship that will be defined later in this section. The activity of a species is related to its concentration by an activity coefficient:

$$\{A\} = \gamma[A] \tag{4-14}$$

where $\{A\}$ = activity of species A
γ = activity coefficient for species A
$[A]$ = concentration of species A

In a formal sense, the activity and activity coefficient of a species are both unitless while the concentration has units. This apparent contradiction results from the common practice of expressing Eq. 4-14 as shown above, whereas a more rigorous formulation of the equation is (Benjamin, 2002)

$$\{A\} = \gamma\frac{[A]}{[A]_{ST}} \tag{4-15}$$

where $[A]_{ST}$ = is the standard concentration of species A in the reference state.

The standard concentration of the species in the reference state has a value of 1 and the same units as the real system. The standard concentration with appropriate units is 1 mol/L for solutes in solution, 1 bar for gases, and 1 mol/mol (mole fraction) for solids, solvents, and miscible liquids. Thus, the denominator in Eq. 4-16 has the effect of canceling the units of concentration without changing the values, leading to the normal practice of expressing the relationship between activity and concentration with Eq. 4-15.

In most water treatment applications, the deviation between activity and concentration expressed by the activity coefficient is most relevant for ionic species. The activity coefficient for ionic species depends on the overall ionic content of the solution, which is characterized by the ionic strength. The ionic strength is calculated using the equation

$$I = \frac{1}{2} \sum_i C_i z_i^2 \tag{4-16}$$

where I = ionic strength of solution, mol/L
C_i = concentration of species i, mol/L
z_i = charge (valence) on species i, unitless

The ionic strength may also be estimated from the total dissolved solids concentration or the conductivity of the solution using the following empirical correlations (Snoeyink and Jenkins, 1980):

$$I = (2.5 \times 10^{-5})\ (\text{TDS}) \tag{4-17}$$

$$I = (1.6 \times 10^{-5})\ (\text{EC}) \tag{4-18}$$

where TDS = total dissolved solids, mg/L
EC = electrical conductivity, μS/cm

Freshwater is typically considered to be water with a TDS concentration of less than 1000 mg/L. Based on Eq. 4-17, water with TDS = 1000 mg/L has an ionic strength of 0.025 M.

A number of relationships have been developed for calculating the activity coefficient. For solutions up to $I \leq 0.5$ M, the activity coefficient can be calculated with the following expression, known as the Davies equation:

$$\log(\gamma) = -Az^2 \left(\frac{\sqrt{I}}{1 + \sqrt{I}} - 0.3I \right) \tag{4-19}$$

where A = constant (for water at 25°C, A = 0.50).

Calculation of activity coefficient with the Davies equation is demonstrated in Example 4-3.

Example 4-3 Calculating activity coefficients

For water with an ionic strength of 5 mM (corresponding to TDS of about 200 mg/L), calculate the activity coefficients of Na^+ and Ca^{2+} at 25°C.

Solution

1. Calculate the activity coefficients for Na^+ using Eq. 4-19. Note 5 mM = 0.005 M.

$$\log(\gamma_{Na^+}) = -0.50(1)^2\left[\frac{\sqrt{0.005}}{1+\sqrt{0.005}} - 0.3(0.005)\right] = -0.0323$$

$$\gamma_{Na^+} = 10^{-0.0323} = 0.93$$

2. Calculate the activity coefficient for Ca^{2+} using Eq. 4-19.

$$\log(\gamma_{Ca^{2+}}) = -0.50(2)^2\left[\frac{\sqrt{0.005}}{1+\sqrt{0.005}} - 0.3(0.005)\right] = -0.129$$

$$\gamma_{Ca^{2+}} = 10^{-0.129} = 0.74$$

Comment

The charge on the species has a large influence on the value of the activity coefficient.

As can be demonstrated by calculations similar to Example 4-3, the activity coefficient is between about 0.9 and 1.0 for monovalent species and between about 0.6 and 1.0 for divalent species in typical drinking water applications (TDS $\leq$ 500 mg/L). The deviation between activity and concentration can have a significant impact on the rate (kinetics) and fate (equilibrium condition) of chemical reactions, particularly as the charge of the species and the ionic strength of the solution increases. Calculating activity coefficients, however, increases the computational requirements associated with chemical calculations. In many cases, complexities such as unknown species in solution, unknown reaction mechanisms, competing reactions, the accuracy of rate constants and equilibrium constants, and the application of factors of safety during design reduce the value of calculating the activity coefficients when evaluating chemical systems. Thus, it is common to ignore the application of activity coefficients in many water treatment applications when freshwater systems are being considered, and activity {A} is assumed to be equal to concentration [A]. Nevertheless, the activity coefficients should be calculated when evaluating chemical equilibrium and kinetics for improved accuracy.

Equilibrium Constants

When chemical reactions come to a state of equilibrium, the numerical value of the ratio of the activity of the products over the activity of the reactants all raised to the power of the corresponding stoichiometric coefficients is known as the equilibrium constant (K). For the reaction shown in Eq. 4-12, the equilibrium constant is written as

$$K = \frac{\{C\}^c\{D\}^d}{\{A\}^a\{B\}^b} \tag{4-20}$$

where K = equilibrium constant
$\{\ \}$ = activity of species
a, b, c, d = stoichiometric coefficients of species A, B, C, D, respectively

The units corresponding to the activity of each species in Eq. 4-20 are the units of the standard concentration of the species in the reference state; as noted earlier these are mol/L for solutes in solution, partial pressure in bars for gases, and mole fractions for solids, solvents, and miscible liquids. Thus, the activity of a pure solid or liquid in an equilibrium expression is simply equal to a mole fraction of 1.0, that is, $\{A\} = 1$. Species for which $\{A\} = 1$ can be taken out of the equilibrium expression; thus, equilibrium expressions are always written without including the activity of solids or water (because $\{H_2O\} = 1$).

Equilibrium constants are frequently reported using p notation (see Eq. 4-6) and reported as pK values, which are defined as

$$pK = -\log(K) \tag{4-21}$$

Substituting Eq. 4-14 into Eq. 4-20 to incorporate activity coefficients, the equation for the equilibrium constant can be written

$$K = \frac{(\gamma_C[C])^c(\gamma_D[D])^d}{(\gamma_A[A])^a(\gamma_B[B])^b} = \frac{\gamma_C^c\gamma_D^d}{\gamma_A^a\gamma_B^b}\frac{[C]^c[D]^d}{[A]^a[B]^b} \tag{4-22}$$

In some cases, a reactant and product of the reaction will have the same valence and same stoichiometric coefficient, in which case a set of activity coefficient values in the numerator and denominator of Eq. 4-22 will cancel each other. Furthermore, as was noted above, it is common to ignore activity coefficients when evaluating the chemistry of relatively dilute solutions. If activity coefficients cancel each other or are ignored, Eq. 4-22 reduces to

$$K = \frac{[C]^c[D]^d}{[A]^a[B]^b} \tag{4-23}$$

The form of the equilibrium constant shown in Eq. 4-23, in which activity coefficients have been ignored, will be used extensively throughout this book to focus attention on the underlying principles of specific chemical reactions rather than the mechanics of calculating activity coefficients. Spreadsheets or chemical speciation software can facilitate the calculation of activities for use in Eq. 4-20. The use of equilibrium constants to determine the concentrations of constituents at equilibrium is demonstrated in Example 4-4.

Example 4-4 Calculating species concentrations using equilibrium constants

Sodium hypochlorite (NaOCl, aka bleach) is added to water as a disinfectant. Upon addition, it immediately dissociates according to the following reaction:

$$\text{NaOCl} \rightarrow \text{Na}^+ + \text{OCl}^-$$

The hypochlorite then participates in the following reversible acid–base reaction:

$$\text{HOCl} \rightleftarrows \text{H}^+ + \text{OCl}^- \qquad \text{p}K_a = 7.6$$

The strength of hypochlorite as a disinfectant depends on which species is present; thus, it is important to know how much is present as HOCl and how much as OCl^- at equilibrium. If 2 mg/L of NaOCl is added, determine how much is present as each species at pH 7.0.

Solution

1. Calculate the molar concentration of NaOCl added (see Example 4-1). From a periodic table (App. D), the MW of NaOCl can be determined to be 74.5 g/mol:

$$\text{Total OCl}^- = [\text{NaOCl}] = \frac{2 \text{ mg/L}}{(74.5 \text{ g/mol})(10^3 \text{ mg/g})} = 2.68 \times 10^{-5} \text{ M}$$

2. Write the equilibrium relationship for the equation provided in the problem statement. From Eq. 4-21, K is calculated as the antilog of the negative of the pK_a value:

$$K_a = \frac{[\text{H}^+][\text{OCl}^-]}{[\text{HOCl}]} = 10^{-7.6}$$

3. Determine the ratio of $[OCl^-]$ to [HOCl] at pH = 7.0. From Eq. 4-7, the hydrogen ion concentration $[H^+]$ at pH 7.0 is equal to $10^{-7.0}$ M and the relationship can be written as

$$\frac{[\text{OCl}^-]}{[\text{HOCl}]} = \frac{K_a}{[\text{H}^+]} = \frac{10^{-7.6}}{10^{-7.0}} = 10^{-0.6} = 0.25$$

4. At pH = 7.0, 25 percent of the total hypochlorite added is present as OCl^- and the rest is present as HOCl. Thus, the concentration of each is

$$[\text{OCl}^-] = 0.25(2.68 \times 10^{-5} \text{ M}) = 6.71 \times 10^{-6} \text{ M} = 6.71 \ \mu\text{M}$$

$$[\text{HOCl}] = 0.75(2.68 \times 10^{-5} \text{ M}) = 2.01 \times 10^{-5} \text{ M} = 20.1 \ \mu\text{M}$$

Comment

The ratio shown in step 3 indicates that OCl^- is the predominant form of hypochlorite at pH values above the pK_a value and HOCl is the predominant form at pH values below the pK_a value. HOCl is the stronger disinfectant, so disinfection with NaOCl is more effective at pH values below 7.6.

Temperature Dependence of Equilibrium Constants

Equilibrium constants are dependent on the temperature at which the reaction occurs. Reference books typically list equilibrium constants at the standard temperature of 25°C. The equilibrium constant at a different temperature can be calculated from the equilibrium constant at 25°C using the van't Hoff relationship:

$$\ln \frac{K_{T2}}{K_{T1}} = \frac{\Delta H^\circ}{R}\left(\frac{1}{T_1} - \frac{1}{T_2}\right) \tag{4-24}$$

where K_{T1}, K_{T2} = equilibrium constants at temperatures T_1 and T_2
ΔH° = standard enthalpy of the reaction, J/mol
R = universal gas constant, J/mol·K
T_1, T_2 = temperatures of known and unknown equilibrium constants, K

4-3 Chemical Kinetics

Chemical kinetics is the study of the rate at which chemical reactions take place, that is, the speed at which reactants are consumed and products are formed. The rate is not constant but normally depends on the chemical activity of the reacting species. Generally, the higher the concentration (and, therefore, the activity) of the reacting species, the faster the reaction will occur. Mechanistically, the reason for this trend is that reactions result from the collision of molecules; the more molecules present, the more often they come in contact with each other and the faster the reaction proceeds.

The rate of a reaction is expressed as the amount of reactants consumed or products generated by the reaction per unit of volume and per unit time. In equation form, this can be expressed as

$$r_A = \frac{n}{Vt} \tag{4-25}$$

where r_A = reaction rate, mol/L·s
n = amount of reactant consumed or product generated, mol
V = volume of reactor, L
t = time, s

The reaction rate will have a negative value for reactants that are being consumed and a positive value for products that are being generated.

Reaction rates are often expressed as a change in concentration over time, but the concentration of species depends on other factors in a reactor. In a reactor with no inputs, outputs, or other reactions, the rate of a reaction will indeed be equal to the change in concentration over time, that is, $r_A = dC/dt$. In other systems, reactants continually enter a reactor and a reaction consumes them at the same rate that they are entering; thus the concentration of reactants in the reactor is constant even though a reaction is taking place. These and other types of reactors will be introduced later in this chapter.

Rate Equations and Reaction Order

The dependence of reaction rates on the activity of the chemical species present leads to the development of rate equations to describe the relationship between the reacting species and the reaction rate. The simplest form is for that of an irreversible elementary reaction. An elementary reaction is a reaction in which the species react directly to form products in a single reaction step and with a single transition state. In this case, the collision of reactant molecules leads directly to the formation of product molecules. The kinetics of such a reaction are such that the rate will be directly proportional to the activity of each reactant participating in the reaction. A general reaction for an irreversible elementary reaction can be written as

$$a\mathrm{A} + b\mathrm{B} \rightarrow \text{products} \tag{4-26}$$

The rate equation for the reaction in Eq. 4-25 is

$$r_A = -k\{\mathrm{A}\}^m\{\mathrm{B}\}^n \tag{4-27}$$

where k = reaction rate constant, units vary (see below)
m, n = reaction order constants, unitless

The concentration dependence of the reaction rate is accounted for in the reactant exponents m and n and is known as the reaction order. For Eq. 4-27, the reaction order is m for species A and n for species B, and the overall reaction order is $m+n$. The reaction order is typically a small positive integer.

Two common forms of rate equations are first- and second-order reactions. First-order reactions depend on the activity of only one species and have the rate equation

$$r_A = -k\{\mathrm{A}\} \tag{4-28}$$

where k = first-order reaction rate constant, s^{-1}

Second-order reactions depend on collisions of two molecules of the same species or on collisions between molecules of two different species.

The rate equations corresponding to these two situations, respectively, are

$$r_A = -k\{A\}^2 \tag{4-29}$$

$$r_A = -k\{A\}\{B\} \tag{4-30}$$

where k = second-order reaction rate constant, L/mol·s

As evident from the above equations, the units of the reaction rate constant depend on the form and reaction order of the rate equation. This dependence is because the reaction rate always has units of mol/L·s and the number of species in the equation (with units mol/L) varies.

A reaction rate can be determined for each species in a reaction. The reaction rate for each species in a reaction will be related to the others based on the stoichiometry of the reaction. Considering the general reversible reaction presented in Eq. 4-12,

$$aA + bB \rightleftarrows cC + dD \quad \text{(Eq. 4-12)}$$

The relative reaction rates would be related by

$$\frac{r_C}{c} = \frac{r_D}{d} = \frac{-r_A}{a} = \frac{-r_B}{b} \tag{4-31}$$

Empirical Reaction Rate Expressions

In some cases, a reaction does not follow straightforward first- or second-order reaction rates and more complex equations are necessary to describe it. For instance, in biological systems, the rate at which organic materials (known as substrate) are consumed depends on the concentrations of both the substrate and the microorganisms (known as biomass). Increasing the concentration of the biomass can increase the rate of the reaction, but increasing the substrate may or may not increase the reaction rate. The rate of substrate consumption in a biological system is often described using saturation-type kinetics known as the Monod equation:

$$r_s = -\frac{\mu XS}{K_s + S} \tag{4-32}$$

where r_s = rate of substrate consumption in a biological system, mg/L·s
μ = maximum specific substrate utilization rate, s^{-1}
X = biomass concentration, mg/L
S = substrate concentration, mg/L
K_s = half-saturation constant, substrate concentration at which the reaction rate is half of the maximum, mg/L

Many other forms of reaction rates exist.

Effect of Temperature on Rate Constants

Reaction rates are dependent on the temperature at which the reaction occurs. The dependence of the reaction rate constant on temperature can be described with the Arrhenius equation:

$$k = Ae^{-E_a/RT} \tag{4-33}$$

where A = frequency factor, same units as k
E_a = activation energy, J/mol
R = universal gas constant, J/mol·K
T = temperature, K

The parameter A is constant; solving Eq. 4-33 for A for two values of temperature and setting the equations equal to each other and rearranging yields

$$\ln\left(\frac{k_{T2}}{k_{T1}}\right) = \frac{E_a}{R}\left(\frac{1}{T_1} - \frac{1}{T_2}\right) \tag{4-34}$$

Determining Reaction Rate Constants

The rate at which reactions occur is usually determined experimentally by measuring the concentration of either a reactant or a product as the reaction proceeds in a batch reactor. Details of the measurement of reaction rates is described in more detail in Sec. 4-7 and demonstrated in Example 4-6.

4-4 Reactions Used in Water Treatment

The major chemical reactions used in water treatment processes include (1) acid–base reactions, (2) precipitation–dissolution reactions, and (3) oxidation–reduction reactions. These types of reactions are introduced in this section.

Acid–Base Reactions

Acid–base reactions involve the transfer of a hydrogen ion, or proton, between two species. The hydrogen ion is the species that contributes acid character to water; thus, the transfer of a proton changes the pH of a solution. pH has a significant effect on many treatment processes and is one of the most important water quality parameters. Alkalinity, which is the buffering capacity of water, is also affected by acid–base reactions. Acid–base reactions are written

$$HA \rightleftarrows H^+ + A^- \tag{4-35}$$

where HA = acid species
H^+ = hydrogen ion (hydrated proton, i.e., H_3O^+)
A^- = conjugate base species

Some species can lose more than one proton. For instance, the carbonate system is one of the most important acid–base systems in natural waters and loses two protons according to the following reactions:

$$H_2CO_3 \rightleftarrows H^+ + HCO_3{}^- \tag{4-36}$$

$$HCO_3^- \rightleftarrows H^+ + CO_3{}^{2-} \tag{4-37}$$

Acid–base reactions are very fast (reaching equilibrium in less than a second), reversible reactions. The acid species and the conjugate base can exist simultaneously, depending on the pH of the solution. The equilibrium constant for an acid–base reaction is known as the acid dissociation constant, K_a:

$$K_a = \frac{\{H^+\}\{A^-\}}{\{HA\}} = \frac{\gamma_{H^+}[H^+]\gamma_{A^-}[A^-]}{\gamma_{HA}\,[HA]} \tag{4-38}$$

where K_a = acid dissociation constant

Acid dissociation constants are frequently expressed as pK_a values (see Eq. 4-21). Acids with pK_a values below 2 are called strong acids and are completely dissociated at environmentally relevant pH values. Weak acids have pK_a values greater than 2 and the degree of dissociation depends on the solution pH.

An important relationship for analyzing acid–base reactions is the total concentration of acid and conjugate base species in solution. For a diprotic (i.e., containing two protons) acid:

$$C_{T,A} = [H_2A] + [HA^-] + [A^{2-}] \tag{4-39}$$

where $C_{T,A}$ = total concentration of species A, mol/L

Because of the importance of pH in many treatment processes and acids and conjugate bases can exist simultaneously in solution, it is important to know the concentration of each species present at a given pH value. Algebraic manipulation of the equilibrium constant equations and Eq. 4-39 yields a convenient convention known as α notation, in which the concentration of each species is expressed as a fraction of the total conjugate base. For the carbonate system:

$$[H_2CO_3] = \alpha_0 C_T \tag{4-40}$$

$$[HCO_3^-] = \alpha_1 C_T \tag{4-41}$$

$$[CO_3^{2-}] = \alpha_2 C_T \tag{4-42}$$

where α_i = fraction of total acid–base species present as species i, starting with the most protonated species as $i = 0$
C_T = total concentration of carbonate species in solution, mol/L

Rearranging the equilibrium constant and total concentration equations yields equations for the α values. For any diprotic acid–base system:

$$\alpha_0 = \frac{[H^+]^2}{[H^+]^2 + [H^+]K_1 + K_1K_2} \quad (4\text{-}43)$$

$$\alpha_1 = \frac{[H^+]K_1}{[H^+]^2 + [H^+]K_1 + K_1K_2} \quad (4\text{-}44)$$

$$\alpha_2 = \frac{K_1K_2}{[H^+]^2 + [H^+]K_1 + K_1K_2} \quad (4\text{-}45)$$

where K_1, K_2 = equilibrium constants for the first and second dissociations of the diprotic acid, respectively.

For a monoprotic acid–base system:

$$\alpha_0 = \frac{[H^+]}{[H^+] + K_1} \quad (4\text{-}46)$$

$$\alpha_1 = \frac{K_1}{[H^+] + K_1} \quad (4\text{-}47)$$

Precipitation–Dissolution Reactions

In water treatment processes, dissolved contaminants can be removed by causing them to precipitate and the removing the solids from water. Also, chemicals can be purchased as solids and then dissolved into the water; thus, both precipitation and dissolution reactions are important. The equilibrium constant between a solid and its ions in solution is known as the solubility product. A solubility product is always written with the solid phase as the reactant. The precipitation–dissolution reaction for gypsum is

$$CaSO_4 \cdot 2H_2O \rightleftarrows Ca^{2+} + SO_4^{2-} + 2H_2O \quad (4\text{-}48)$$

And the corresponding solubility product is

$$K_{sp} = \{Ca^{2+}\}\{SO_4^{2-}\} = \gamma_{Ca^{2+}}\gamma_{SO4^{2-}}[Ca^{2+}][SO_4^{2-}] \quad (4\text{-}49)$$

where K_{sp} = solubility product

It is important to recognize that the equation for the solubility product does not contain terms for the solid phase or water. This is because pure solids and liquids have an activity of 1.0 (see Sec. 4-2).

Precipitation–dissolution reactions are reversible but are not as fast as acid–base reactions.

Oxidation–Reduction Reactions

Reactions that involve the transfer of electrons between two chemical species are known as oxidation–reduction, or redox, reactions. In water treatment, disinfection and chemical oxidation are common redox reactions. In a redox reaction, one species is reduced (gains electrons) and one species is oxidized (loses electrons). Redox reactions are typically reported as half reactions to show the number of electrons transferred. Thus, to obtain a complete oxidation–reduction reaction, an oxidation half reaction and a reduction half reaction must be combined. The general expression of a half reaction for the reduction of a species is as follows:

$$\text{Ox}_\text{A} + ne^- \rightarrow \text{Red}_\text{A} \tag{4-50}$$

where Ox_A = oxidized species A
n = number of electrons transferred
e^- = electron
Red_A = reduced species A

Oxidized species A is called an oxidant (or electron acceptor) because it oxidizes another species as it is reduced during this reaction. The half reaction for the oxidation of a species may be expressed as

$$\text{Red}_\text{B} \rightarrow \text{Ox}_\text{B} + ne^- \tag{4-51}$$

where Ox_B = oxidized species B
Red_B = reduced species B

Reduced species B is called a reductant (or electron donor) because it reduces another species as it is oxidized. The two half reactions may be combined to obtain the following overall oxidation–reduction reaction:

$$\text{Ox}_\text{A} + \text{Red}_\text{B} \rightarrow \text{Ox}_\text{B} + \text{Red}_\text{A} \tag{4-52}$$

Redox reactions are irreversible reactions that proceed until one of the reactants is totally consumed. The reactions can be kinetically limited so that the reactants can exist in contact with each other indefinitely without reacting, but then react rapidly when enough activation energy is applied.

4-5 Mass Balance Analysis

The quantitative analysis of many problems in environmental engineering begins with an accounting of all materials that enter, leave, accumulate in, or are transformed within the boundaries of a system. The basis for this accounting procedure, known as a *mass balance analysis*, is the law of conservation of mass, which states that matter can neither be created or destroyed. Matter can, however, be transferred from one phase to another (such as from water to air) or participate in chemical transformations that may lead to the appearance or disappearance of individual chemical species. This law allows matter to be accounted for as it flows through or is transformed within a system.

Before starting a mass balance analysis, it is important to determine what problem is being investigated and what answer is needed. For instance, a mass balance analysis can be used to track the movement and fate of matter (often contaminants) in the environment or to develop the governing equations of many water treatment processes. Governing equations can either give an effluent concentration as a function of time or a single value of concentration for steady-state processes. It is important to identify the component to be tracked through the system. For instance, oxygen can be pumped into a basin to encourage the growth of microorganisms that will consume a contaminant (substrate) in the feed water. A mass balance could be written around the oxygen, the substrate, or the microorganisms, and the resulting equation may be different in each case. When a single reaction is taking place within the system, the concentrations of the other components can often be determined from the stoichiometry of the reaction.

The basic steps for performing a mass balance analysis are as follows:

1. Draw a picture of the system.
2. Identify the control volume, for example, the boundaries of the system of interest.
3. Identify all inflows, outflows, or transformations to the components in the system.
4. Write the mass balance equation and identify simplifying assumptions, such as whether the system is at steady state or whether any reactions are occurring. Clearly identify the answer you are looking for, whether the solution should be dependent on time, and the like.
5. Solve the equations.
6. Do a ''reality check,'' verifying items such as (a) are the units correct?, (b) is the time dependence of the final equation correct?, (c) are the assumptions valid?, and (d) is the magnitude of the answers reasonable?

Control Volumes and System Boundaries

The first steps of a mass balance analysis are to draw a picture and determine the space in which the law of conservation of mass will be applied. A definition sketch of a control volume for a completely mixed reactor with inflow and outflow is shown on Fig. 4-1. The region where the mass balance takes place is known as the *control volume*, and the edges of

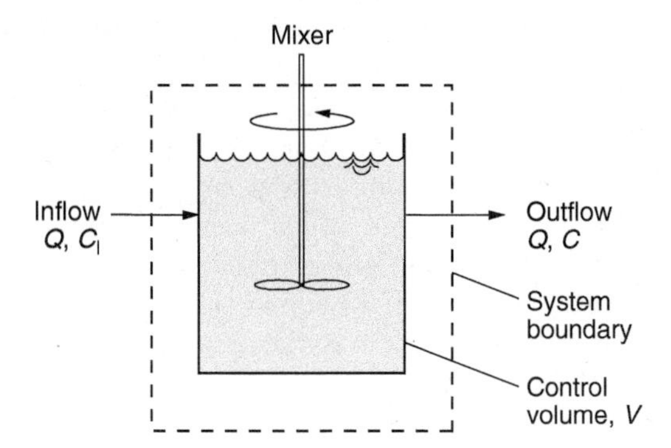

Figure 4-1
Definition sketch for a mass balance analysis of a completely mixed flow reactor (CMFR).

this volume are called the *system boundary*. The control volume can be any region of space and may or may not have physical boundaries. The control volume may be fixed, moving, changing size, contain only a single phase, or contain multiple phases of matter. Two key principles in choosing a control volume are that it should be easy to visualize or mathematically describe the control volume, and the mass flux (in and out) across the boundaries should be easily determined. A third common constraint is that if reactions are taking place with the system or if integration is needed to solve the equations, the control volume must be chosen so that conditions within it are uniform (i.e., intensive properties such as temperature, pressure, and concentration are constant throughout the control volume).

Two common types of control volumes are the "reactor" and the "differential element." Reactors can be any contained volume, including lakes, basins, and tanks that contain treatment processes at treatment plants. This type of control volume typically has discrete inputs and outputs that enter the system through pipes or other defined points. Reactors are sometimes known as black boxes and this type of mass balance analysis is sometimes called a box model. A differential element control volume is small segment of a reactor, typically without physical boundaries. The inputs and outputs are fluxes across the boundaries instead of discrete flows. Differential element control volumes can be helpful in developing governing equations for treatment processes.

Fundamental Mass Balance Equation

The principle of the conservation of mass within a system can be stated as a rate equation with the following form:

$$\begin{bmatrix} \text{rate of accumulation} \\ \text{of component A} \\ \text{in control volume} \end{bmatrix} = \begin{bmatrix} \text{rate of mass flow} \\ \text{of A into} \\ \text{the control volume} \end{bmatrix} - \begin{bmatrix} \text{rate of mass flow} \\ \text{of A out of} \\ \text{the control volume} \end{bmatrix} + \begin{bmatrix} \text{rate of transformation} \\ \text{of A due to reactions} \\ \text{in the control volume} \end{bmatrix} \tag{4-53}$$

The units for every term in Eq. 4-53 are mass/time. The terms in Eq 4-53 are simplified in this book to be

$$[\text{accum}] = [\text{mass in}] - [\text{mass out}] + [\text{rxn}] \tag{4-54}$$

These terms are described in more detail in the following sections.

Accumulation Term

When inputs, outputs, and reactions are not perfectly balanced, the mass of the constituent of interest within the control volume will change over time, which is expressed as dM/dt. Frequently, the property of interest for a contaminant is the concentration, and mass is related to concentration simply by $M = CV$, where C is the concentration and V is the volume of the system. Applying the product rule from calculus yields

$$[\text{accum}] = \frac{dM}{dt} = \frac{d(CV)}{dt} = V\frac{dC}{dt} + C\frac{dV}{dt} \tag{4-55}$$

where $[\text{accum}]$ = accumulation term in mass balance analysis, mg/s
M = mass of constituent within the control volume, mg
C = concentration of constituent within the control volume, mg/L
V = volume of the control volume, L

In nearly all mass balance analyses, the control volume is chosen so that the volume of the system is not changing. In this case, $dV/dt = 0$, so

$$[\text{accum}] = V\frac{dC}{dt} \tag{4-56}$$

Equation 4-56 is the standard form for the accumulation term when the mass within the system is changing. To evaluate this integral, the concentration must be the same at every point throughout the control volume; otherwise a different control volume must be chosen. When the state of the system (including the mass) is not changing, the system is said to be at steady state. At steady state

$$[\text{accum}] = 0 \tag{4-57}$$

Input and Output Terms

Mass can be transported across system boundaries by bulk fluid flow (advection) or separately from fluid movement (via processes such as molecular diffusion). When the fluid enters or leaves the control volume as a discrete, measurable flow (such as through a stream or pipe), the rate at which mass enters or leaves the control volume by advection can be written

$$[\text{mass in}] \text{ or } [\text{mass out}] = QC \tag{4-58}$$

where [mass in], [mass out] = mass balance input and output terms, mg/s
Q = flow rate of fluid entering or leaving the control volume, L/s

In many mass balance and mass transfer operations, matter or fluid enters or leaves the control volume by flow across regions of the overall system boundary instead of through discrete entrances or exits such as pipes. The flow of material through a unit of area per unit time is called the mass flux:

$$J = \frac{m}{At} \tag{4-59}$$

where J = mass flux of a constituent through an area, mg/m$^2 \cdot$ s
m = mass of the constituent, mg
A = area perpendicular to the direction of flux, m^2
t = time, s

Because flux is defined per unit area, mass flow is the product of the flux and the area:

$$[\text{mass in}] \text{ or } [\text{mass out}] = JA \tag{4-60}$$

As will be seen later in this chapter, surface area is a key parameter for the rate of mass transfer and, hence, the efficiency of a separation process that relies on mass transfer. Equation 4-60 describes the flux through a control volume regardless of whether the fluid is moving. In the absence of fluid movement, mass can enter or leave a control volume in response to a concentration gradient (see Sec. 4-14), which is known as a diffusive flux. When the fluid is moving and the constituent is transported with the fluid, the flux is known as the convective flux. Flux due to fluid movement is equal to the product of the velocity of the fluid and concentration of the constituent ($J = vC$), thus

$$[\text{mass in}] \text{ or } [\text{mass out}] = vAC \tag{4-61}$$

where v = fluid velocity, m/s

It should be noted that flux can be reported in other units. In some cases (principally membrane processes), the material moving across the interface is measured in units of volume instead of mass, and the corresponding flux is called a volumetric flux instead of a mass flux. An example of units for volumetric flux is L/m$^2 \cdot$ s. Other situations are best described with molar units (mol/m$^2 \cdot$ s). Molar fluxes can be converted to mass fluxes by multiplying by the molecular weight.

Reaction Term

Reaction rates were introduced in Sec. 4-3, and it was noted that the reaction rate r is equal to the change in concentration of a constituent over time due to chemical reactions. The change in mass due to chemical reactions can be determined simply by multiplying the reaction rate by the volume of the control volume:

$$[\text{rxn}] = Vr \tag{4-62}$$

The sign on terms in the mass balance analysis equation depends on whether constituents are entering or leaving the control volume; an increase in mass in the control volume is considered the positive direction. Thus, a decay reaction (consumption of reactants) will have a negative sign and a generation reaction (production of products) will have a positive sign in the mass balance equation; similarly, input terms are positive, output terms are negative, and the accumulation term will be positive if mass is increasing in the control volume and negative if mass is decreasing.

A mass balance analysis of a separation process is presented in Example 4-5. Numerous additional examples of the mass balance analysis are presented later in this chapter.

Example 4-5 Mass balance analysis of a separation process

A well that flows at 45 m^3/h is contaminated with trichloroethylene (TCE) at a concentration of 400 $\mu g/L$. A treatment device is installed that removes most of the TCE from the water, leaving a constant concentration of 5.0 $\mu g/L$ in the effluent stream. The effluent flow rate is 97 percent of the influent flow rate, and the remaining water goes to a waste stream. Calculate the TCE concentration in the waste stream.

Solution

1. The control volume will be the treatment device, and the inflows and outflows are identified in the problem statement. A diagram of the system with all relevant information labeled is shown below, using subscripts *in* = influent, *ef* = effluent, and *w* = waste stream.

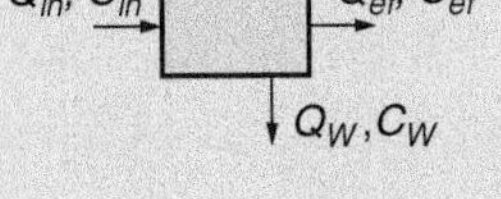

Q_{in}, C_{in} → [ ] → Q_{ef}, C_{ef}
↓ Q_W, C_W

2. The problem statement describes a continuous process; that is, there is no mention of any time dependence to the input parameters or requested effluent concentration, and it can be assumed that the

treatment device achieves that level of treatment continuously. Furthermore, the problem statement describes only influent and effluent streams to the treatment device; thus, it can be assumed that no reactions are occurring (e.g., [rxn] = 0). Furthermore, no mass accumulates in the treatment device; the system is at steady state and [accum] = 0.

3. Two values are unknown (Q_w and C_w). Thus, two equations will be necessary. These equations will be (a) a mass balance on the TCE and (b) a mass balance on the water. Addressing the TCE first, writing the mass balance equation and applying the simplifying assumptions from step 2 yields

$$\text{[}\cancel{\text{accum}}\text{]} = \text{[mass in]} - \text{[mass out]} + \text{[}\cancel{\text{rxn}}\text{]}$$

$$0 = Q_{in}C_{in} - Q_{ef}C_{ef} - Q_wC_w$$

4. Rearranging terms yields

$$C_w = \frac{Q_{in}}{Q_w}(C_{in} - C_{ef})$$

5. If water is an incompressible fluid with density ρ and the presence of the TCE does not affect the density of the solution, a mass balance on the solution can be written:

$$Q_{in}\rho = Q_{ef}\rho + Q_w\rho$$

Dividing every term by the density and solving yields

$$Q_w = Q_{in} - Q_{ef} = 45\ \text{m}^3/\text{h} - (0.97)(45\ \text{m}^3/\text{h}) = 1.35\ \text{m}^3/\text{h}$$

6. Substituting the values in the problem statement and step 5 into the equation from step 4 and solving yields

$$C_w = \frac{45\ \text{m}^3/\text{h}}{1.35\ \text{m}^3/\text{h}}(400 - 5.0\ \mu\text{g/L}) = 13{,}200\ \mu\text{g/L}$$

Comments

1. The equation shown in step 5 is common in mass balance analyses. In environmental engineering applications, contaminants are normally so dilute that they have no effect on the density of the solution and the steps of writing the equation with density and then canceling it from every term are skipped.
2. The solution is a single value for the waste concentration because the system is at steady state. A non–steady-state system has time

dependence and the solution would be an equation that is a function of time.

3. The waste concentration is much higher than the influent concentration. This is to be expected because the TCE from the influent has been concentrated into a much smaller volume.

4-6 Introduction to Reactors and Reactor Analysis

In the environment, many contaminants in water are removed gradually by naturally occurring physical, chemical, and biological processes. In engineered systems, the same processes are carried out in vessels, basins, and tanks known as *reactors*. The rate at which such processes occur depends on the constituents involved and conditions in the reactor, including temperature and hydraulic (mixing) characteristics.

The goals of reactor analysis are to understand the conditions within a reactor and to use that understanding to develop models and equations that describe the hydraulic conditions or chemical concentrations within the reactor or the concentration of reactants and products leaving a reactor. The equations can then be used to design reactors for use in water treatment processes.

To analyze a reactor situation and develop equations, four key elements of the analysis must be defined: (1) type of reactor, (2) time dependence, (3) reaction characteristics, and (4) input characteristics (for flow reactors). These elements and key options for each element are represented on the concept map shown on Fig. 4-2. The selection of options for each element depends on the objectives of the reactor analysis. For instance, equations that describe the hydraulic characteristics of a reactor are developed by a non–steady-state analysis of a conservative chemical in a reactor with a step or pulse input, whereas equations for the effluent concentration of a contaminant in a full-scale treatment process are developed by steady-state analysis of a reactive chemical with a continuous input into the reactor. The elements of reactor analysis and options for each element are described in the following sections.

Types of Reactors

Reactors can be divided into ideal reactors and real (or nonideal) reactors. The category of ideal reactors can then be subdivided into batch reactors and flow reactors, and ideal flow reactors can subsequently be divided into plug flow reactors (PFRs) and completely mixed flow reactors (CMFRs).

Ideal reactors are characterized by specific assumptions, such as instantaneous, perfect homogeneity throughout the entire reactor or an absolute

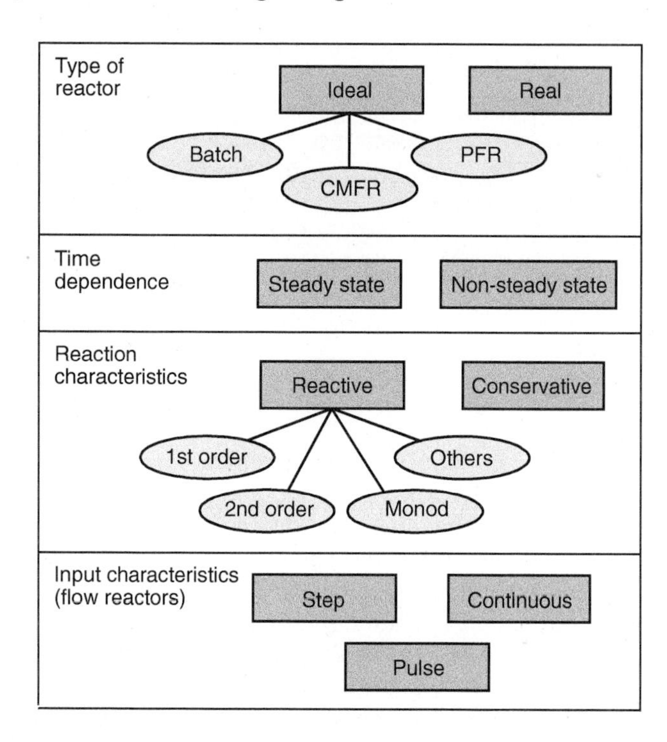

Figure 4-2
Concept map for reactor analysis.

lack of diffusion and dispersion. Real reactors are, simply, those that do not achieve the ideal assumptions and tend to have more complex hydraulic and mixing conditions. Ideal conditions can be approached in small, laboratory-scale devices and some well-engineered pilot-scale equipment but are impossible at the size of many full-scale water treatment processes.

BATCH REACTORS

Ideal *batch reactors* are characterized by intermittent operation with no flow in or out (see Fig. 4-3a). Reactants are mixed together and the reaction is allowed to proceed over time. The principle assumptions of an ideal batch reactor are (1) no reactants or products enter or leave the container during the analysis period (e.g., there are no input or output terms in a mass balance analysis), (2) complete mixing occurs instantaneously and uniformly so that concentration, temperature, density, and other variables are uniform throughout the reactor, and (3) the reaction proceeds at the same rate everywhere in the reactor. A beaker on a laboratory countertop is an example of a batch reactor. Batch reactors are used widely in the production of small-volume, specialty chemicals in the chemical processing industries, in laboratory-scale investigations, and in the preparation of chemical feed solutions at water treatment facilities but are not commonly used for large-volume water treatment processes.

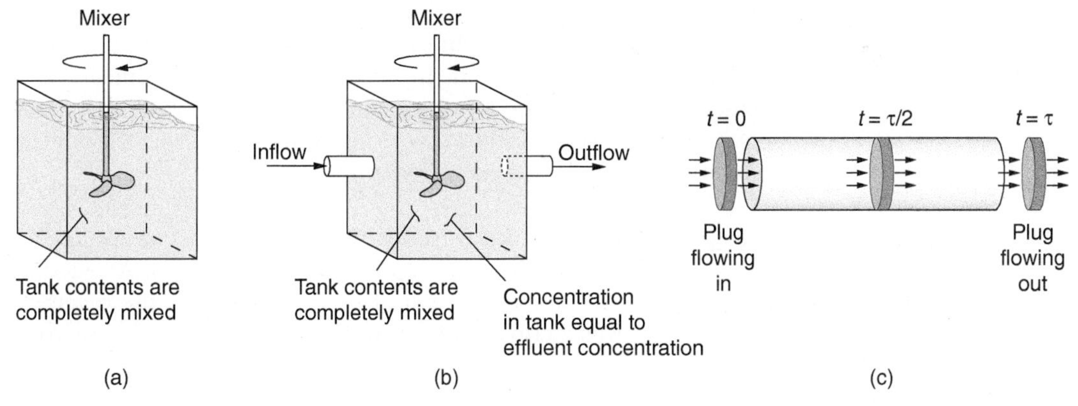

Figure 4-3
Diagrams of three ideal reactors: (a) batch reactor, (b) completely mixed flow reactor, and (c) plug flow reactor.

CONTINUOUS-FLOW REACTORS

Ideal *continuous-flow* reactors operate on a continuous basis with flow into and out of the reactor. Typically the reactor is a basin or tank with process water continuously flowing in through a pipe or gate at one end and flowing out over a weir or through a pipe at the other end. Reactants (disinfectants, coagulants, oxidants, etc.) are mixed into the process water immediately prior to the water entering the reactor and the reaction is allowed to take place while the water is in the reactor. Large volumes of water can be processed efficiently in this manner.

The two types of ideal continuous-flow reactors are completely mixed flow reactors and plug flow reactors.

CONTINUOUSLY MIXED FLOW REACTORS

When process water and reactants flow into a CMFR, they are instantaneously and completely mixed with the contents of the reactor. Thus, some assumptions for a CMFR are similar to a batch reactor, specifically that the concentration, temperature, and other variables are uniform throughout the reactor, and the reaction proceeds at the same rate everywhere in the reactor. Because of this uniformity within the reactor, the effluent concentration must be the same as the concentration within the reactor, regardless of where the effluent is located. Some older texts and literature refer to a CMFR as a continuously stirred tank reactor (CSTR). A conceptual sketch of a CMFR is shown on Fig. 4-3b.

PLUG FLOW REACTORS

A plug flow reactor is an ideal reactor in which water passes through without mixing with the water in front of or behind it. The plug flow concept can be thought of as flow consisting of a series of elements (or plugs) with the same diameter as the reactor. Each time a new element is introduced in one end of the reactor, an element of the same size must exit the other end. A conceptual sketch of a PFR is shown on Fig. 4-3c. No mixing of contents occurs between one element and the next. In a PFR, the concentrations of the reactants will decrease as a function of position as the water passes through the reactor, which is different from a CMFR, in which the contents of the reactor are identical everywhere.

Time Dependence

The concept of steady state was introduced in Sec. 4-5 and its applicability to reactor analysis is the same as for mass balance analysis. A reactor at steady state is one in which conditions at each point within the reactor do not change with time. Many water treatment processes use flow-through reactors, and the design process typically assumes that the flow through the reactor and the concentrations of all reactants in the influent are constant over time; thus, the steady-state assumption is widely used for developing models and equations that describe water treatment processes. A non-steady-state analysis is used to develop equations that describe the hydraulic properties of a reactor or to develop equations that describe how long a flow reactor takes to respond to a change in input conditions before steady-state conditions are reestablished.

An assumption of whether or not a reactor is at steady state must be made to conduct a reactor analysis, and it is necessary to check the final equations for consistency with the assumption. If non-steady-state was assumed, time should be a variable in the final equations; conversely, if the analysis was conducted assuming steady state, time-dependent variables should not be present in the final equations.

Reaction Characteristics

Obviously, the purpose of reactors is to provide a container in which reactions that accomplish treatment can take place. To use reactor analysis principles to develop design equations, it is necessary to know the reaction rate for the reactions of interest. As noted in Sec. 4-3, common reactions include first-order, second-order, and Monod-type reactions, but other types can also be used in reactor analysis.

Some analyses are conducted with conservative (or nonreactive) constituents. While it may seem that conservative chemicals would be of little interest in reactor analysis, in fact they provide a mechanism for understanding the hydraulic characteristics of a reactor. Since conservative constituents do not react, they flow with the water and stay in a reactor as long as the water stays in the reactor. Thus, a curve of effluent concentration of a conservative constituent reveals the residence time distribution of the water in the reactor. Conservative constituents are commonly called *tracers*, and tests to determine the residence time distribution of a reactor are called tracer tests.

Input Characteristics for Flow Reactors

The flow into a reactor can be constant or change over time. In normal full-scale treatment processes, it is desirable to have a constant flow of water through a reactor because it leads to stable and predictable operating conditions. A constant flow of water, coupled with constant concentrations of reactants in the influent, leads to steady-state conditions in the reactor.

It is sometimes necessary to analyze the situation that occurs when a change in the input to a reactor has occurred, which leads to a non-steady-state situation. Normally, the analysis considers an instantaneous change that occurs immediately prior to the start of the analysis period. A common non-steady-state situation is a tracer test. Two types of changes to the reactor input techniques are considered for tracer tests:

- *Pulse input*: At the beginning of the testing period (i.e., time = 0), a known mass of tracer is added to the reactor influent instantaneously (i.e., added as a pulse or slug) and then flows through the reactor. Measurement of the effluent concentration continues until the pulse has completely passed through the reactor.
- *Step input*: At time = 0, a feed pump is turned on and feeds a tracer into the reactor influent. The concentration of the tracer in the influent stays constant over the duration of the test. Measurement of the effluent concentration continues until it is the same as the new influent concentration.

The next six sections of this chapter use the concepts of reactor analysis to describe the performance of reactors, including the hydraulic characteristics of both ideal and real reactors, and the concentration of reactants in decay reactions in the effluent flow from both ideal and real reactors.

4-7 Reactions in Batch Reactors

When a batch reactor is used as a vessel for a chemical reaction, the primary interest is how the concentrations of reactants and products change over time. The main objective in many environmental engineering applications is to remove contaminants; thus, reactants are the contaminants and reagents added to degrade the contaminants, and the information that is needed is how much time is required for the reactant concentration to be reduced to some acceptable level. The mass balance analysis presented in Sec. 4-5 can be used to develop an equation for the concentration of a reactant in a batch reactor as a function of time. Batch reactors have no inputs or outputs, so applying the accumulation and reaction terms from Sec. 4-5 for a constant-volume reactor yields

$$[\text{accum}] = [\cancel{\text{mass in}}] - [\cancel{\text{mass out}}] + [\text{rxn}]$$

$$V\frac{dC}{dt} = Vr \tag{4-63}$$

where V = reactor volume, L
C = concentration of reactant, mg/L

t = time, s
r = reaction rate, mg/L·s

Equation 4-63 can be simplified to

$$\frac{dC}{dt} = r \tag{4-64}$$

The reaction rate equation can be substituted for r and Eq. 4-64 can be integrated to yield an equation for C as a function of t. For instance, a first-order decay reaction has a reaction rate equation $r = -kC$. Substituting this relation into Eq. 4-64 yields

$$\frac{dC}{dt} = -kC \tag{4-65}$$

Rearranging and setting up an integration of both sides yields

$$\int_{C_0}^{C} \frac{dC}{C} = -k \int_{0}^{t} dt \tag{4-66}$$

where C_0 = initial reactant concentration, mol/L

Integration yields

$$C = C_0 e^{-kt} \tag{4-67}$$

Equation 4-67 describes the concentration of a reactant in a batch reactor as a function of time for a first-order decay reaction. A similar mass balance analysis for a second-order decay reaction (i.e., $r = -kC^2$) results in the following:

$$\frac{1}{C} = \frac{1}{C_0} + kt \tag{4-68}$$

A common use of batch reactors in laboratories is to determine the reaction equation and rate constant for a chemical reaction. The kinetic information determined in a batch reactor can be used to design other types of reactors and full-scale treatment facilities. If the reaction order is not known, it is not possible to determine *a priori* whether Eq. 4-67, 4-68, or some other equation describes the concentration in the reactor. The approach to analyzing experimental kinetic data is to develop a linearized form (i.e., an equation of the form $y = mx + b$) for each possible rate equation, plot concentration versus time data in these various forms, and observe which formulation of the data provides the best fit of a straight line. The linear form of the first-order batch reactor equation is developed by taking the natural logarithm of both sides of Eq. 4-67, which produces the following relationship:

$$\ln(C) = \ln(C_0) - kt \tag{4-69}$$

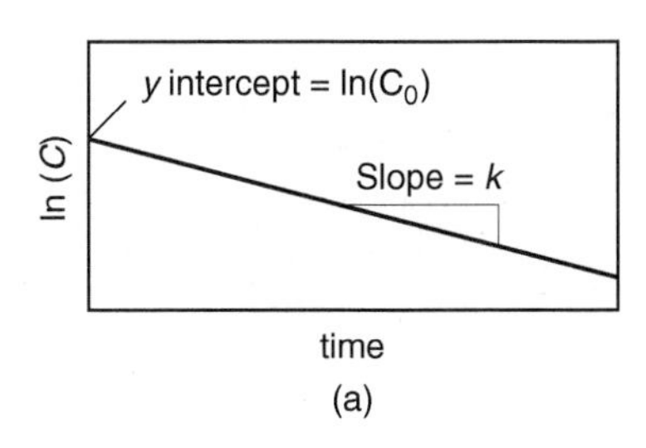

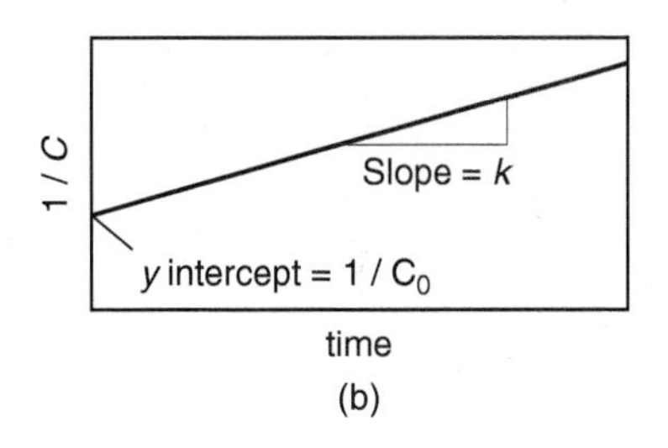

Figure 4-4
Linearized form of concentration of (a) first- and (b) second-order reactions in a batch reactor.

For a first-order reaction, a plot of $\ln(C)$ as a function of t will result in a linear relationship. Such a plot is illustrated on Fig. 4-4a. The slope of the line in the plot is equal to the first-order rate constant k and the intercept is equal to $\ln(C_0)$.

Similarly, straightforward graphical solutions can be demonstrated for second-order reactions (Fig. 4-4b). The use of these equations to determine the reaction rate equation and rate constant is demonstrated in Example 4-6.

Example 4-6 Determining the reaction order and rate constant for decomposition of ozone

In laboratory experiments, ozone was added to a beaker (batch reactor) of water and the concentration of ozone remaining was measured periodically. The initial concentration of ozone, C_0, was 5 mg/L for all experiments. The concentration of ozone remaining in the water at various times is presented in the following table:

Time, min	O_3 conc, mg/L
0	5.00
1	4.25
5.5	2.10
9	1.10

Determine the reaction order and reaction rate constant for the decomposition of ozone in water, considering first- and second-order reactions.

Solution

1. Calculate $\ln(C)$ and $1/C$ for plotting as a function of time. The values are tabulated below.

Time, min	C	$\ln(C)$	$1/C$
0	5.0	1.61	0.20
1.0	4.25	1.45	0.24
5.5	2.1	0.74	0.48
9.0	1.1	0.095	0.91

2. Plot $\ln(C)$ and $1/C$ as a function of time. The graphs shown below are tabulated below. For a first-order reaction, a plot of $\ln(C)$ vs. t is shown in panel (a) below. For a second-order reaction, a plot of $1/C$ vs. t is shown in panel (b).

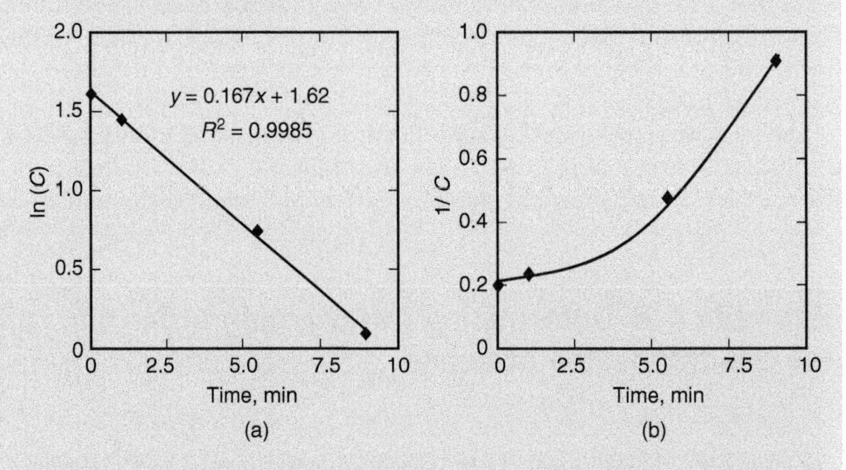

3. Because the plot constructed in panel (a) results in a linear relationship, ozone decomposition in water can be described using first-order kinetics.
4. The reaction rate constant is determined by finding the slope of the best-fit line for the data. As shown in panel (a) above, the first-order reaction rate constant for the decomposition of ozone in water is 0.167 min^{-1}.

4-8 Hydraulic Characteristics of Ideal Flow Reactors

Section 4-9 will demonstrate that the hydraulic characteristics of a flow-through reactor can influence the outcome of reactions. Thus, it is essential to be able to measure and describe the hydraulic characteristics. The hydraulics of a reactor can be determined using non-steady-state reactor analysis by injecting a tracer into the reactor influent using either a pulse or step input and then observing its concentration in the reactor's effluent over time. Both input methods yield the exact same information about the reactor hydraulics. Tracer studies are discussed in greater depth in Sec. 4-10.

Completely Mixed Flow Reactor

When a pulse input is introduced into a CMFR, the effluent tracer concentration instantly reaches a maximum as the tracer is uniformly distributed throughout the reactor. As clean water (containing no tracer) continues to enter the reactor after time = 0, the tracer gradually dissipates in an exponential manner as the tracer material leaves the effluent. The exponential shape of the tracer curve can be demonstrated using a mass balance analysis of a CMFR. The mass balance equation contains an accumulation term because the concentration of tracer in the reactor will be changing over time (not at steady state), but no input term (for pulse input, no tracer enters the reactor after $t = 0$) or reaction term (the tracer is a nonreactive chemical). The mass balance is written

$$[\text{accum}] = [\cancel{\text{mass in}}] - [\text{mass out}] + [\cancel{\text{rxn}}]$$

$$V\frac{dC}{dt} = -QC \tag{4-70}$$

where C = effluent concentration of tracer at time t, mg/L
t = time since pulse of tracer was added to reactor, s

Algebraically rearranging Eq. 4-70 yields

$$\frac{dC}{C} = -\frac{Q}{V}\,dt \tag{4-71}$$

At $t = 0^+$ (time immediately after tracer is added), the tracer pulse has entered the reactor and is uniformly dispersed within the CMFR. Consequently, Eq. 4-71 may be integrated:

$$\int_{C_0}^{C} \frac{dC}{C} = -\frac{Q}{V}\int_0^t dt \tag{4-72}$$

where C_0 = initial mass of tracer added divided by volume of reactor, M/V, mg/L
M = mass of tracer added, mg

The hypothetical time that water stays in a reactor is related to the volume and flow rate:

$$\tau = \frac{V}{Q} \tag{4-73}$$

where τ = hydraulic residence time, s
V = reactor volume, m^3
Q = flow rate through the reactor, m^3/s

After substitution of τ and integration of Eq. 4-72, the following expression is obtained:

$$C = C_0 e^{-t/\tau} \tag{4-74}$$

Equation 4-74 is the equation for the tracer curve for a pulse input in a CMFR as shown on Fig. 4-5a. The equation demonstrates that the effluent concentration from a CMFR will be $C = C_0$ at $t = 0$, $C = 0$ at infinite time, and decay exponentially between those extremes. A similar mass balance analysis can be used to generate the equation for the tracer curve for a step input in a CMFR. The formulation of the mass balance equation can be written

$$[\text{accum}] = [\text{mass in}] - [\text{mass out}] + [\cancel{\text{rxn}}]$$

$$V\frac{dC}{dt} = QC_I - QC \tag{4-75}$$

where C_i = influent tracer concentration mg/L.

After rearranging, integrating, and substituting Eq. 4-75, the equation describing the effluent concentration from a CMFR following a step input is

$$C = C_I(1 - e^{-t/\tau}) \tag{4-76}$$

Equation 4-76 is shown on Fig. 4-5b.

Plug Flow Reactor

The tracer curves that result from the addition of a pulse input and step input to a PFR are more straightforward and are illustrated on panels (c) and (d) of Fig. 4-5, respectively. In both cases, the effluent concentration curve has exactly the same shape as the influent but delayed by a time equal to the reactor's hydraulic residence time, τ. The reason the influent and effluent curves are identical is because PFRs have no mixing or dispersion in the axial direction and every drop of water and molecule of tracer takes the same amount of time to pass through the reactor. Thus, there is no opportunity for any molecule of tracer to come out sooner or later than the hydraulic residence time.

A primary conclusion from this analysis is that CMFRs and PFRs have dramatically different tracer effluent curves. Since the tracer describes the effluent concentration of a nonreactive chemical, the residence time of the tracer also reflects the residence time of the water in the reactor; in other words, these curves describe the hydraulic characteristics of the reactors. When reactive chemicals are present in a PFR, every reactant molecule will have the exact same amount of time in the reactor. In

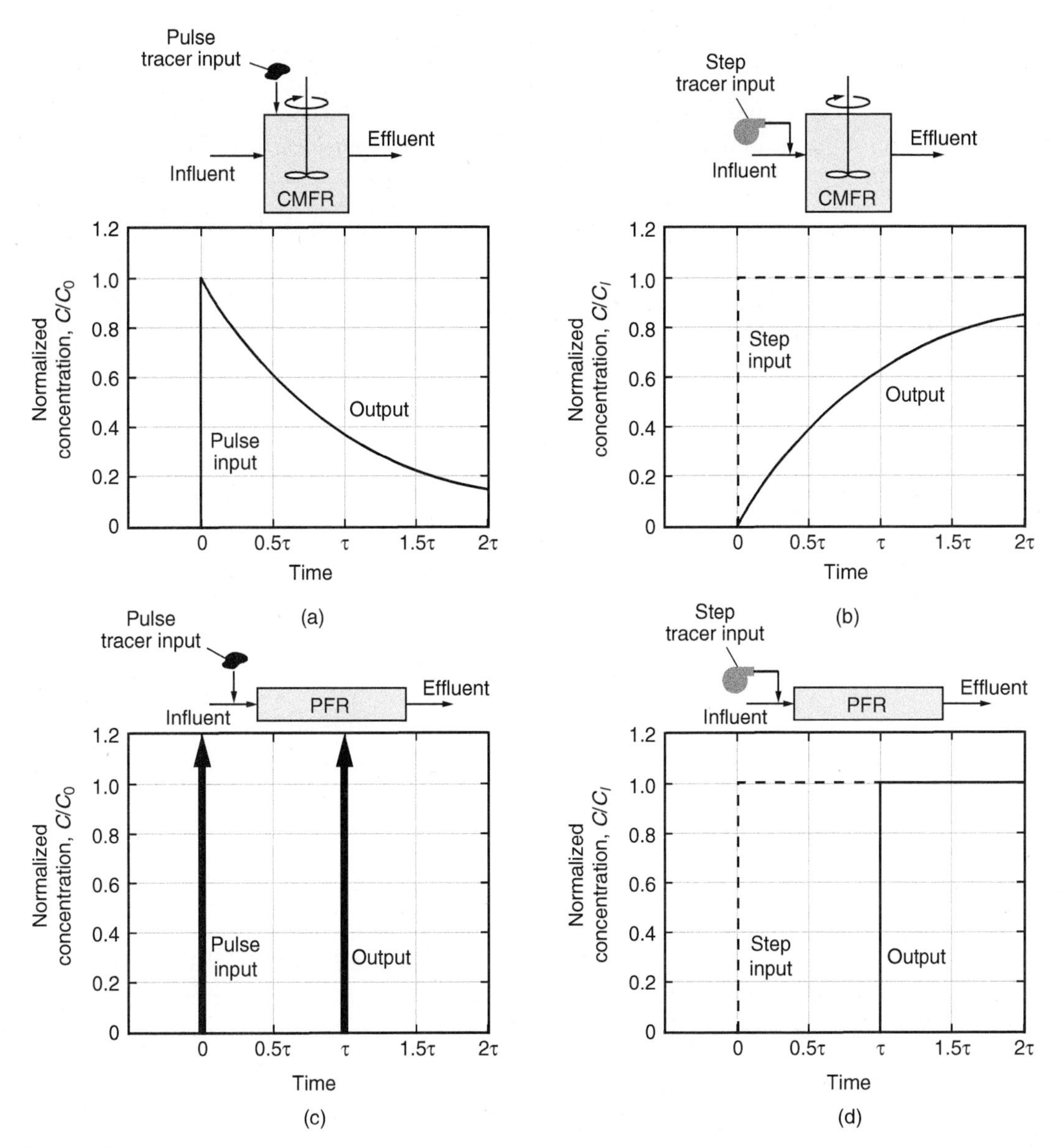

Figure 4-5
Tracer curves from ideal reactors: (a) CMFR with pulse input, (b) CMFR with step input, (c) PFR with pulse input, and (d) PFR with step input.

a CMFR, however, the droplets of water have different residence times. Because of complete mixing, some droplets of water are transported immediately to the vicinity of the reactor effluent and leave the reactor after a short period of time, while other droplets can reside in the reactor much longer. Thus, reactant molecules in a CFMR will spend a range of times in the reactor and have different amounts of time to react. This difference in residence time leads to a difference in reaction time available, which then leads to a difference in the extent to which chemicals react within the reactor, as will be demonstrated in the next section.

4-9 Reactions in Ideal Flow Reactors

In normal environmental engineering practice, flow reactors are designed to treat a constant flow and achieve a particular level of treatment on a continuous basis (e.g., an ozone contactor may be designed to achieve 95 percent removal of a contaminant continuously). The most common design question thus is, for a given application, how big should the reactor be to achieve a particular effluent concentration of the reactant of interest. Once again, the mass balance analysis presented earlier can be used to develop the equations. Design of flow-through reactors is typically based on steady-state conditions. This section develops and presents the equations for the effluent concentration from a single PFR and single CMFR under steady-state conditions.

Completely Mixed Flow Reactor

For a steady-state mass balance analysis of a CMFR, the accumulation term is zero. Mathematically, this mass balance may be written as

$$[\text{\sout{accum}}] = [\text{mass in}] - [\text{mass out}] + [\text{rxn}]$$

$$0 = QC_I - QC + Vr \tag{4-77}$$

where Q = flow rate, L/s
C_I = influent concentration, mg/L
C = effluent concentration, mg/L
V = reactor volume, L
r = reaction rate in reactor at effluent concentration C, mg/L·s

Use of Eq. 4-77 to develop an equation for the effluent concentration depends on the form of the reaction rate equation. For a first-order decay reaction ($r = -kC$), the effluent concentration can be developed as follows:

$$QC_I - QC - VkC = 0 \tag{4-78}$$

Substituting Eq. 4-73 ($\tau = V/Q$) and rearranging yields

$$C_I - (1 + k\tau)C = 0 \tag{4-79}$$

$$C = \frac{C_I}{1 + k\tau} \tag{4-80}$$

Use of Eq. 4-80 to calculate the steady-state effluent concentration from a CMFR will be demonstrated in Example 4-8. For a second-order decay reaction $r = -kC^2$), Eq. 4-77 becomes

$$QC_I - QC - VkC^2 = 0 \tag{4-81}$$

Substituting Eq. 4-73 and rearranging yields a quadratic equation:

$$k\tau C^2 + C - C_I = 0 \tag{4-82}$$

One of the roots of the quadratic equation will necessarily be a negative number, so the effluent concentration of a second-order decay reaction from a CFMR must be the other root and is equal to

$$C = \frac{-1 + \sqrt{1 + 4k\tau C_I}}{2k\tau} \tag{4-83}$$

By rearranging Eq. 4-77, the volume and the hydraulic residence time of the reactor can be estimated if flow rate, influent concentration, effluent concentration (treatment objective), and reaction kinetics are known, as follows:

$$V = \frac{Q(C_I - C)}{-r} \tag{4-84}$$

$$\tau = \frac{C_I - C}{-r} \tag{4-85}$$

The use of Eq. 4-85 to determine the hydraulic residence time needed to achieve a specific effluent concentration is demonstrated in Example 4-7.

Example 4-7 Hydraulic residence time in a CMFR

A CMFR has an influent concentration of 200 mg/L and a first-order reaction rate constant of 4 min^{-1}. Assuming steady-state conditions, calculate the required hydraulic residence time for an effluent concentration of 10 mg/L.

Solution

Determine the hydraulic residence time using Eq. 4-85:

$$\tau = \frac{C_I - C}{kC} = \frac{200\ \text{mg/L} - 10\ \text{mg/L}}{(4\ \text{min}^{-1})(10\ \text{mg/L})} = 4.75\ \text{min}$$

Plug Flow Reactor

In a PFR, reactions occur as the water passes from the influent to the effluent end of the reactor. Because there is no mixing in the axial direction, the concentrations of the reactants and products change along the length of the reactor. Thus, the whole reactor cannot be used as the control volume in a mass balance analysis because it violates the criterion that the conditions be constant throughout the control volume. The only way to meet this criterion is to choose a control volume that is so small that there is essentially no change in concentration in the axial direction, that is, a differential element with a length Δx, as shown in Fig. 4-6. Although the concentration changes axially, when the flow and reaction proceed at constant rates, the concentration at each point in the reactor is a constant and the system is at steady state. A steady-state mass balance on this differential element can be written

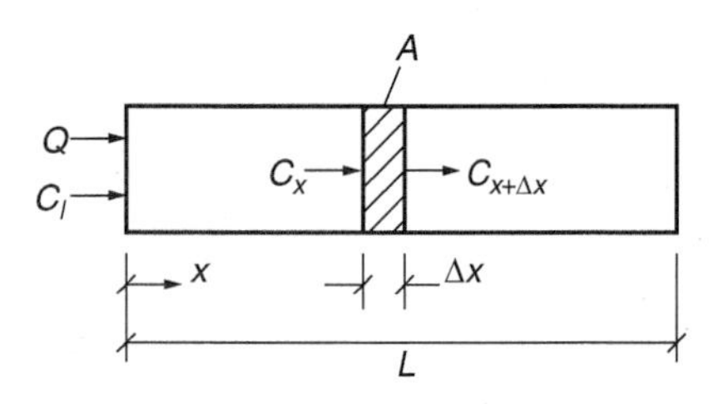

Figure 4-6
Definition sketch for a mass balance analysis of a differential element in a plug flow reactor.

[~~accum~~] = [mass in] − [mass out] + [rxn]

$$0 - QC_x - QC_{x+\Delta x} + Vr \tag{4-86}$$

The volume of differential element is $V = A\Delta x$. Substituting this into Eq. 4-86 and rearranging yields

$$\left(\frac{Q}{A}\right)\frac{C_{x+\Delta x} - C_x}{\Delta x} = r \tag{4-87}$$

Using calculus, the term on the left side of Eq. 4-87 can be recognized as the derivative dC/dx by taking the limit as $\Delta x \to 0$, as shown as

$$\lim_{\Delta x \to 0} \frac{C_{x+\Delta x} - C_x}{\Delta x} = \frac{dC}{dx} \tag{4-88}$$

Substituting Eq. 4-88 into Eq. 4-87 yields

$$\left(\frac{Q}{A}\right)\frac{dC}{dx} = r \tag{4-89}$$

Equation 4-89 can be used to develop equations for the effluent concentration from a PFR. For instance, for a first-order decay reaction ($r = -kC$), substitution and integration yields

$$\left(\frac{Q}{A}\right)\frac{dC}{dx} = -kC \tag{4-90}$$

$$\int_{C_I}^{C} \frac{dC}{C} = -\frac{kA}{Q}\int_0^x dx \tag{4-91}$$

$$\ln(C) - \ln(C_I) = -\frac{kAx}{Q} \tag{4-92}$$

Equation 4-92 can be used to plot the concentration profile of a reactant in a PFR as a function of the axial position in the reactor. The effluent concentration is the concentration when $x = L$. Noting that $AL = V$ and $\tau = V/Q$, a final rearrangement of Eq. 4-92 yields the effluent concentration from a PFR:

$$C = C_I e^{-k\tau} \tag{4-93}$$

An inspection of Eqs. 4-67 and 4-93 reveals that they are essentially identical except $t = \tau$ in the equation for a PFR. The effluent concertration from a PFR when a second-order reaction is taking place will sinilarly be identical to Eq. 4-68. Thus, a PFR can be visualized as a reactor in which discrete packets of water enter the reactor (as if each packet were a batch reactor) and travel through the reactor for a time equal to τ. The effluent of a PFR with residence time τ has the same concentration as a batch reactor at time t.

An identical analysis can be performed for other reaction rates (second-order reaction, Monod reaction, etc.) using Eq. 4-86 as a starting point. A comparison of the effluent concentration from a CMFR and a PFR of the same size is demonstrated in Example 4-8.

Example 4-8 Steady-state effluent concentrations from a CMFR and a PFR

A groundwater supply has soluble iron (Fe^{2+}) concentration of 7.5 mg/L, which is in excess of what is desired for a potable supply. The Fe^{2+} is to be removed by aeration with oxygen (O_2) followed by precipitation and filtration. O_2 reacts with Fe^{2+} in a first-order reaction with a rate constant of 0.168 min^{-1}. Find the concentration of Fe^{2+} in the effluent of a flow-through reactor with a hydraulic residence time of 15 min.

Solution

1. Calculate the effluent concentration from a CMFR using Eq. 4-80:

$$C = \frac{C_I}{1 + k\tau} = \frac{7.5 \text{ mg/L}}{1 + (0.168 \text{ min}^{-1})(15 \text{ min})} = 2.13 \text{ mg/L}$$

2. Calculate the effluent concentration from a PFR using Eq. 4-93:

$$C = C_I e^{-k\tau} = (7.5 \text{ mg/L}) e^{-(0.168 \text{ min}^{-1})(15 \text{ min})} = 0.6 \text{ mg/L}$$

Comment

Even though the reactors have the same residence time and the same reaction is taking place, the effluent concentration from a PFR is significantly lower than that from a CMFR. The results demonstrate the importance of mixing in reactors where chemical reactions are taking place.

The results in Example 4-8 demonstrate that a PFR is more efficient than a CMFR for consuming the reactants in a first-order reaction. The reason for this behavior is that the rate of a first-order reaction depends on the concentration of the reactant. In the CMFR, the reaction proceeds at a rate governed by the same concentration everywhere in the reactor, corresponding to the effluent concentration. At the inlet of a PFR, the reaction proceeds at a rate governed by the influent concentration, which is much higher than the effluent concentration from a CMFR. Although the rate of reaction in a PFR declines as the fluid moves through the reactor, the high rate of reaction in the early portion of the reactor results in an overall greater extent of reaction than in a CMFR.

This difference between CMFRs and PFRs has important implications in environmental engineering. The CMFR is an ideal reactor in which the contents are perfectly mixed and the PFR is an ideal reactor in which no mixing occurs; thus, these ideal reactors occupy two ends of a spectrum of the extent of actual mixing in a real reactor. The hydraulics of real reactors are considered in the next section.

4-10 Measuring the Hydraulic Characteristics of Flow Reactors with Tracer Tests

CMFRs and PFRs are relatively easy to achieve for small or moderately sized reactors. Small reactors can be designed so that the effects of dispersion are negligible or with large mixers that can achieve near-perfect mixing. Water treatment facilities, however, can include very large continuous-flow reactors. Sedimentation basins and chlorine contact chambers can be tens or hundreds of meters long. Because of their large size, virtually all water treatment processes take place in turbulent flow. The mixing that results from the shearing forces between fluid layers and by the random fluid motion of turbulence is known as dispersion. In large reactors, it may not be possible to install a mixer capable of achieving perfect mixing or to avoid the effects of dispersion and currents caused by wind, temperature, and density differences or other forms of nonideal flow.

The previous section demonstrated that hydraulic characteristics of reactors affect the extent to which reactions occur and that a PFR can achieve a lower effluent concentration than a CMFR if they are the identical size and all other conditions are the same. Thus, the nonideality associated with large reactors has important implications for the treatment of drinking water because it affects the effluent concentrations of contaminants from treatment systems. A lack of appreciation for the effects of nonideality leads to poor design of treatment facilities, which then leads to reduced treatment performance.

The best way to determine the hydraulic characteristics and quantify the amount of dispersion in a real reactor is to measure it with a tracer test. The results of the tracer test are used to generate the residence time distribution (RTD) for the reactor, which can then be used to calculate the actual performance (effluent concentrations of reactants such as contaminants and oxidants). Tracer tests are also used to quantify dispersion and determine the contact time for disinfection regulations.

A tracer test is conducted by injecting a conservative chemical (tracer) at the reactor inlet using a step or pulse input and measuring the concentration at the reactor outlet over time. The tracer concentration may be measured using a spectrophotometer if a dye is used, and a conductivity meter or specific ion (e.g., fluoride or lithium) measurements if salts are used.

Analysis of Tracer Data

The raw data from a tracer test are values of tracer concentration exiting the reactor as a function of time. Concentration plotted as a function of time is known as the *C* curve. The *C* curve from a pulse input tracer test of a real reactor is shown on Fig. 4-7a. Tracer data is analyzed to determine the mean and variance and generate two additional curves, known as the exit age distribution (*E* curve) and cumulative exit age distribution (*F* curve). From those results, additional parameters that characterize the extent of dispersion can be generated.

MEAN AND VARIANCE

The mean residence time, $\bar{t}$, is the average amount of time that water stays in the reactor as determined by the tracer test. The mean is the first moment of the area under a curve of effluent concentration versus

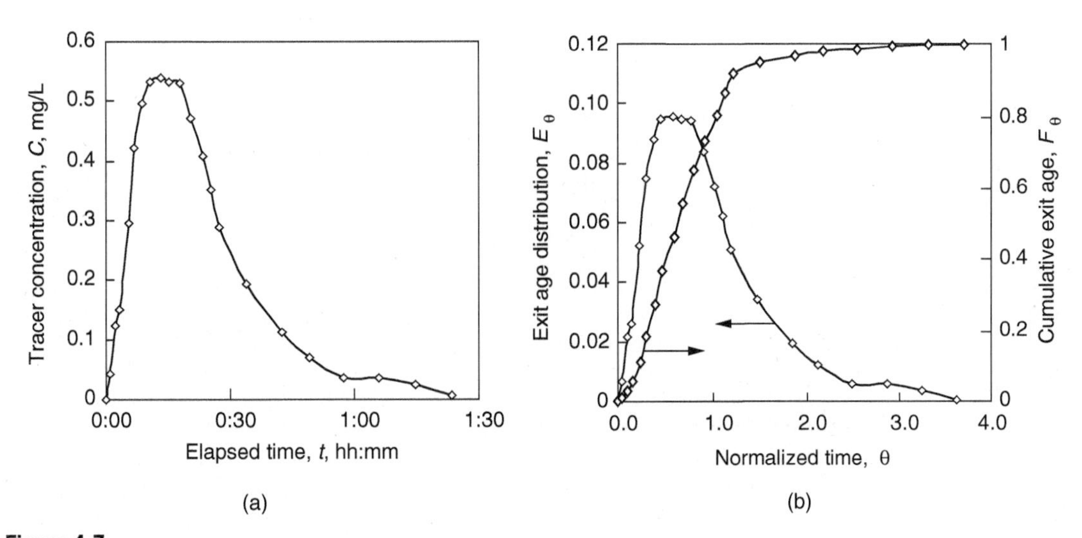

Figure 4-7
Results of tracer test from three CMFRs in series: (a) concentration *C* as function of time and (b) exit age distribution *E* and cumulative exit age distribution *F*.

time. If concentration is described by a continuous function, the mean is determined from

$$\bar{t} = \frac{\int_0^\infty Ct\,dt}{\int_0^\infty C\,dt} \tag{4-94}$$

where $\bar{t}$ = mean residence time of tracer in reactor, min or s
C = concentration exiting reactor at time t, mg/L
t = time since addition of tracer pulse to reactor's entrance, min or s

Of course, the tracer test produces discrete data instead of a continuous function, and it is therefore necessary to estimate the area under the C curve using numerical integration. Several methods can be used, including the rectangular or Simpson rules, but the trapezoidal rule is simple and provides acceptable accuracy. The trapezoidal rule can be written

$$\bar{t} \cong \frac{\sum_{k=1}^{N} C_k t_k \Delta t_k}{\sum_{k=1}^{N} C_k \Delta t_k} \tag{4-95}$$

where $C_k = \frac{1}{2}(C_{i-1} + C_i)$
$t_k = \frac{1}{2}(t_{i-1} + t_i)$
$\Delta t_k = t_i - t_{i-1}$
C_i = effluent concentration from tracer test at time i, mg/L
t_i = time at which concentration measurement was taken, min
N = number of time and concentration measurements

Ideally, the mean residence time $\bar{t}$ is equal to the hydraulic residence time τ, but this is generally not the case. A principal cause of this deviation is the presence of dead spaces in the reactor (spaces that do not mix well with the remainder of the contents) where the volume is not used.

The variance σ_t^2, the second moment of the area under the C curve, is used to determine the spread of the tracer curve using the following equation:

$$\sigma_t^2 = \frac{\int_0^\infty C(t-\bar{t})^2\,dt}{\int_0^\infty C\,dt} \cong \frac{\sum_{k=1}^{N} C_k (t_k - \bar{t})^2 \Delta t_k}{\sum_{k=1}^{N} C_k \Delta t_k} \tag{4-96}$$

where σ_t^2 = variance with respect to t, min^2

MASS OF TRACER RECOVERED

The C curve can be used to determine the amount of tracer recovered during the test. The mass of tracer recovered in each trapezoid of the C curve is equal to the product of the flow, concentration, and time, such that

$$M_T = \sum_{i=1}^{N} M_k = Q \sum_{k=1}^{N} C_k \Delta t_k \tag{4-97}$$

where M_T = mass of tracer recovered in the reactor effluent, mg

The mass recovered in Eq. 4-97 can be compared to the mass of tracer injected into the reactor. A well-conducted tracer test should recover 95 percent or more of the tracer.

EXIT AGE DISTRIBUTION (*E* CURVE)

As discussed in previous sections, different elements of fluid can take different amounts of time to get through the reactor. The exit age distribution, E_θ, is the residence time distribution (RTD) of the fluid in the reactor. The RTD is strictly a function of the flow characteristics of the reactor and is independent of how the tracer test was conducted. The RTD is a probability function and the area under the *E* curve is dimensionless and has a value equal to 1. The area under the curve up to a particular time t represents the fraction of water that had a residence time less than or equal to that time.

The manipulation of the raw data to generate the *E* curve depends on whether pulse or step input was used. If the test was done as a pulse input, the *E* curve can be generated by normalizing the time and concentration values of the *C* curve so that the area under the curve is 1.

Time is normalized by dividing all original values of time by the mean residence time determined in Eq. 4-95.

$$\theta_i = \frac{t_i}{\bar{t}} \tag{4-98}$$

where θ_i = normalized time, dimensionless

Because τ may not take dead spaces in the reactor into account, it is important that $\bar{t}$, not τ, be used in normalizing tracer curves.

The area under the *C* curve was determined in the denominator of Eq. 4-95. Because the x axis was scaled by dividing by $\bar{t}$, multiplying values on the y axis by $\bar{t}$ and dividing by the total area under the *C* curve will yield a new curve with an area of 1. Thus, values for the *E* curve are determined from

$$E_{\theta i} = \frac{C_i \bar{t}}{\sum_{k=1}^{N} C_k \Delta t_k} \tag{4-99}$$

where E_{θ_i} = value of the *E* curve at time i, dimensionless

The exit age distribution for the tracer curve presented on Fig. 4-7a is shown on Fig. 4-7b.

CUMULATIVE EXIT AGE DISTRIBUTION (*F* CURVE)

Although the *E* curve is the residence time distribution function for the reactor, it can be difficult to use to determine specific residence time values. For instance, t_{10} is the time at which 10 percent of the water has had a residence time less than or equal to that time. To determine t_{10} from the E curve, it is necessary to find the point on the x axis that divides the area with 10 percent of the area to the left and 90 percent to the right.

This analysis can be facilitated by generating the cumulative exit age distribution F_θ, or F curve. The F curve is the integral of the E curve:

$$F_\theta = \int_0^\theta E_\theta \, d\theta \tag{4-100}$$

Using the discrete data from the tracer test, the values for the F curve can be calculated from

$$F_{\theta i} = \frac{\sum_{k=1}^{i} C_k \Delta t_k}{\sum_{k=1}^{N} C_k \Delta t_k} \tag{4-101}$$

The cumulative exit age distribution for the tracer curve presented on Fig. 4-7a is shown on Fig. 4-7c. The cumulative area under the E curve is shown on the y axis, so it is possible to read specific residence time values directly from the F curve. The analysis of tracer data is demonstrated in Example 4-9.

Example 4-9 Analysis of tracer data to determine mean, variance, and *E* and *F* curves

A pulse tracer test with 5.0 kg of dye was conducted on a flow-through reactor operating at a flow rate of 3.17 ML/d. The reactor has a volume of 175 m^3. The results are reported in the table below. Calculate the mean and variance, the mass of dye recovered, and plot the tracer curve, E curve, and *F* curve.

Time, min	C, mg/L	Time, min	C, mg/L	Time, min	C, mg/L
0	0	65	38	90	31
20	1	68	58	100	15
40	4	71	63	110	6
55	15	75	64	130	1
62	28	80	58	140	0

Solution

Tracer data is best analyzed in a spreadsheet. The setup for the spreadsheet is as follows:

1. Enter the given time and concentration in the first two columns.

2. In the next three columns, calculate the mean time t_k, mean concentration C_k, and time interval Δt for each trapezoid using the definitions after Eq. 4-95. For the second data point:

$$\text{Col. 3:} \quad t_k = \tfrac{1}{2}(t_{i-1} + t_i) = \tfrac{1}{2}(0 + 20) = 10 \text{ min}$$

$$\text{Col. 4:} \quad C_k = \tfrac{1}{2}(C_{i-1} + C_i) = \tfrac{1}{2}(0 + 1) = 0.5 \text{ mg/L}$$

$$\text{Col. 5:} \quad \Delta t_k = t_i - t_{i-1} = 20 - 0 = 20 \text{ min}$$

3. In the next two columns, calculate the product terms in the numerator and denominator of Eq. 4-95. For the second data point:

$$\text{Col. 6:} \quad C_k t_k \Delta t_k = (0.5 \text{ mg/L})(10 \text{ min})(20 \text{ min}) = 100 \text{ mg} \cdot \text{min}^2/\text{L}$$

$$\text{Col. 7:} \quad C_k \Delta t_k = (0.5 \text{ mg/L})(20 \text{ min}) = 10 \text{ mg} \cdot \text{min/L}$$

4. Complete the calculations in steps 2 and 3 for all remaining data and calculate the sum of columns 6 and 7 at the bottom of the column. Then calculate the mean using Eq. 4-95:

$$\bar{t} = \frac{\sum_{k=1}^{N} C_k t_k \Delta t_k}{\sum_{k=1}^{N} C_k \Delta t_k} = \frac{167{,}604 \text{ mg} \cdot \text{min}^2/\text{L}}{2192 \text{ mg} \cdot \text{min/L}} = 76.48 \text{ min}$$

5. Calculate the product terms in the numerator of Eq. 4-96 in column 8. For the second data point:

$$C_k(t_k - \bar{t})^2 \Delta t_k = (0.5 \text{ mg/L})(10 - 76.48 \text{ min})^2(20 \text{ min})$$

$$= 44{,}195 \text{ mg} \cdot \text{min}^3/\text{L}$$

6. Complete the calculation for all remaining data and then calculate the summation at the bottom of the column. Then calculate the variance using Eq. 4-96:

$$\sigma_t^2 = \frac{\sum_{k=1}^{N} C_k(t_k - \bar{t})^2 \Delta t_k}{\sum_{k=1}^{N} C_k \Delta t_k} = \frac{710{,}103 \text{ mg} \cdot \text{min}^3/\text{L}}{2192 \text{ mg} \cdot \text{min/L}} = 324.0 \text{ min}^2$$

7. Calculate the mass of dye recovered using Eq. 4-97:

$$M_T = Q \sum_{k=1}^{N} C_k \Delta t_k = \frac{(3.17 \text{ ML/d})(2192 \text{ mg} \cdot \text{min/L})(10^6 \text{ L/ML})}{(1440 \text{ min/d})(10^6 \text{ mg/kg})}$$

$$= 4.83 \text{ kg}$$

8. In the next two columns, calculate θ_i and E_{θ_i} for each value of t_i using Eqs. 4-98 and 4-99. For the second row:

$$\theta_i = \frac{t_i}{\bar{t}} = \frac{20 \text{ min}}{76.48 \text{ min}} = 0.262$$

$$E_{\theta i} = \frac{C_i \bar{t}}{\sum_{k=1}^{N} C_k \Delta t_k} = \frac{(1 \text{ mg/L})(76.48 \text{ min})}{2192 \text{ mg} \cdot \text{min/L}} = 0.0349$$

9. Values for the F curve are determined by calculating a running sum of the values $C_k \Delta t_k$ up to that point on the curve, divided by the total area under the curve. The first two values for the F curve are

$$F_{\theta 1} = \frac{C_1 \Delta t_1}{\sum_{k=1}^{N} C_k \Delta t_k} = \frac{(0.5 \text{ mg/L})(20 \text{ min})}{2192 \text{ mg} \cdot \text{min/L}} = 0.00456$$

$$F_{\theta 2} = F_{\theta 1} + \frac{C_2 \Delta t_2}{\sum_{k=1}^{N} C_k \Delta t_k} = 0.00456 + \frac{(2.5 \text{ mg/L})(20 \text{ min})}{2192 \text{ mg} \cdot \text{min/L}}$$

$$= 0.02738$$

The data is tabulated below.

(1)	(2)	(3)	(4)	(5)	(6)	(7)	(8)	(9)	(10)	(11)
t_i	C_i	t_k	C_k	Δt_k	$C_k t_k \Delta t_k$	$C_k \Delta t_k$	$C_k(t_k - \bar{t}_k)^2 \Delta t_k$	θ_i	E_{θ_i}	F_{θ_i}
0	0	0								
20	1	10	0.5	20	100	10	44,195	0.262	0.0349	0.00456
40	4	30	2.5	20	1,500	50	108,016	0.523	0.1396	0.02738
55	15	47.5	9.5	15	6,769	143	119,671	0.719	0.5235	0.09240
62	28	58.5	21.5	7	8,804	151	48,650	0.811	0.9771	0.16108
65	38	63.5	33.0	3	6,287	99	16,678	0.850	1.3261	0.20625
68	58	66.5	48.0	3	9,576	144	14,340	0.889	2.0241	0.27196
71	63	69.5	60.5	3	12,614	182	8,841	0.928	2.1986	0.35478
75	64	73	63.5	4	18,542	254	3,075	0.981	2.2335	0.47068
80	58	77.5	61.0	5	23,638	305	318	1.046	2.0241	0.60986
90	31	85	44.5	10	37,825	445	32,309	1.177	1.0818	0.81291
100	15	95	23.0	10	21,850	230	78,894	1.308	0.5235	0.91786
110	6	105	10.5	10	11,025	105	85,411	1.438	0.2094	0.96578
130	1	120	3.5	20	8,400	70	132,584	1.700	0.0349	0.99772
140	0	135	0.5	10	675	5	17,123	1.831	0.0000	1.00000
				Sums:	167,604	2,192	710,103			

The *C*, *E*, and *F* curves are shown below.

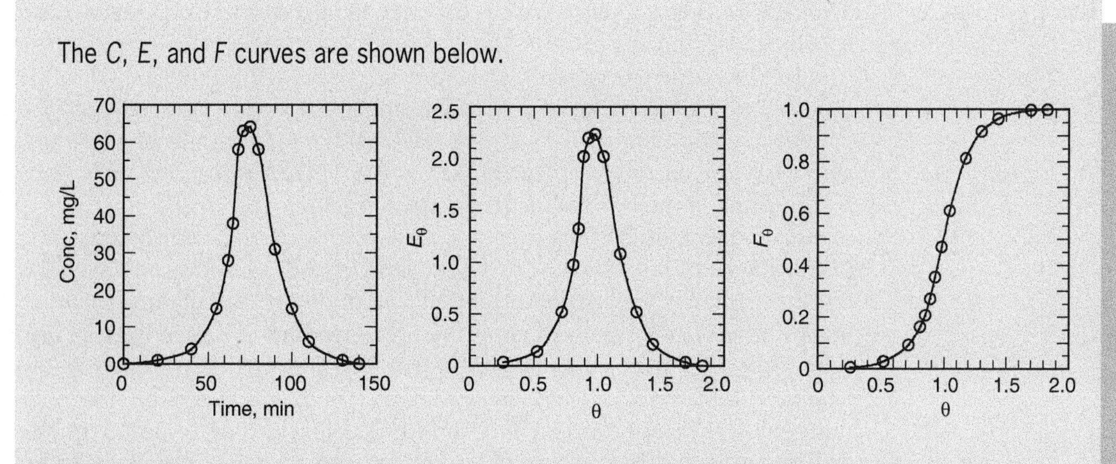

Comments

The mass of dye recovered was 4.83 kg, or about 97 percent of the dye that was added, indicating a successful tracer test. The mean, variance, and *E* and *F* curves can be used for additional analysis of tracer data, as demonstrated in the next sections.

4-11 Describing the Hydraulic Performance of Real Flow Reactors

The preceding section demonstrates that tracer tests can produce useful information about the RTD of a real reactor. However, the *E* and *F* curves are graphical representations of the residence time distribution, and it is preferable to have a single parameter (or a small number of parameters) that describe the RTD. The variance could be used to describe the extent of dispersion in a real reactor, but it is hard to relate to the physical meaning of the variance. As a result, a number of parameters and models have been developed that describe the RTD. Two single-parameter models that are typically used to describe the RTD of a real reactor are (1) the t_{10}/τ ratio and (2) the tanks-in-series (TIS) model. The t_{10}/τ ratio is used by regulatory agencies to assess the level of dispersion in a reactor without using complicated models. The TIS model provides a clear conceptual image of how the RTD of a real reactor fits on the spectrum of mixing or dispersion ranging from no dispersion (represented by a PFR) to perfect mixing (represented by a CMFR). The t_{10}/τ ratio and TIS model are described in this section. More sophisticated parameters and models also exist. In particular, the axial dispersion model or dispersed-flow model (DFM) and the segregated-flow model (SFM) are described in the companion reference book to this textbook (Crittenden, et al., 2012).

The t_{10}/τ Ratio

The time for a particular fraction of water to pass through the reactor can be obtained from the F curve. For instance, t_{10} is the time it takes for the first 10 percent of the water to pass through the reactor and is equal to the time for which the F value is 0.1. The more dispersion in a reactor, the greater deviation between t_{10} and τ. A CMFR has a t_{10}/τ ratio of about 0.1, whereas a PFR has a t_{10}/τ ratio of exactly 1.0. Thus, this simple ratio can be an indicator of the level of dispersion in a reactor.

Regulatory authorities often regulate reactor design using simplified performance criteria such as the t_{10}/τ ratio. As will be shown in Chap. 13, the disinfectant concentration and contact time are equally important in inactivating microorganisms, such that the product of concentration and time (called Ct) is the basis for disinfection regulations in the United States. However, because the water in a real reactor has a range of residence times, different microorganisms in the reactor will be exposed to the disinfectant for different amounts of time, and some will be exposed for less than τ. Thus, disinfection would be inadequate if regulations specified the τ of a reactor as the appropriate amount of time for disinfection. The time used in Ct regulations in the United States is the t_{10} of the disinfectant contactor from a tracer test. A reactor with low dispersion as expressed by a high t_{10}/τ ratio can achieve similar disinfection effectiveness with less volume than a reactor with high dispersion, as demonstrated in Example 4-10.

Example 4-10 Volume required for disinfection in reactors with low and high dispersion

To achieve a certain level of disinfection, a treatment facility must achieve a Ct value of 56.4 mg·min/L, where C is the concentration of chlorine and t is t_{10} from a tracer test. Assume the plant has a flow rate of 38 ML/d and the acceptable chlorine concentration is 1.1 mg/L. Determine the required chlorine contactor volume if the contactor is designed with (a) low dispersion and has $t_{10}/\tau = 0.65$ and (b) high dispersion and has $t_{10}/\tau = 0.4$.

Solution

1. Determine the required t_{10}:

$$t_{10} = \frac{Ct \text{ value}}{C} = \frac{56.4 \text{ mg} \cdot \text{min/L}}{1.1 \text{ mg/L}} = 51 \text{ min}$$

2. Determine the hydraulic residence time and volume of the low dispersion contactor:

$$\tau = \frac{t_{10}}{0.65} = \frac{51\text{ min}}{0.65} = 78.5\text{ min}$$

$$V = Q\tau = \frac{(38\text{ ML/d})(78.5\text{ min})(10^3\text{ m}^3/\text{ML})}{1440\text{ min/d}} = 2070\text{ m}^3$$

3. Determine the hydraulic residence time and volume of the high dispersion contactor:

$$\tau = \frac{t_{10}}{0.4} = \frac{51\text{ min}}{0.4} = 128\text{ min}$$

$$V = Q\tau = \frac{(38\text{ ML/d})(128\text{ min})(10^3\text{ m}^3/\text{ML})}{1440\text{ min/d}} = 3380\text{ m}^3$$

Comments

The high dispersion contactor must be over 60 percent larger than the low dispersion contactor to achieve the same regulatory disinfection level.

The Tanks-in-Series Model

A PFR is modeled by performing a mass balance on a differential element in the reactor and integrating over the length of the reactor. The differential element can be considered to be a CMFR, so conceptually a PFR has performance comparable to an infinite series of CMFRs. Since the PFR and CMFR represent two ends of a spectrum on the degree of mixing (CMFR is perfect mixing, PFR is no mixing), it follows that a discrete number of CMFR tanks in series might represent a degree of mixing between the extremes represented by the PFR and CMFR. The TIS model is built on this foundation. The number of tanks, n, is a single parameter that can be used to approximate the performance of a real tank.

DERIVATION OF THE TIS MODEL

The TIS model is developed by assuming that the residence time distribution of a real reactor can be compared to a series of CMFRs, and the number of tanks in the series will represent the degree of dispersion in the real reactor. A real reactor with a high degree of mixing or dispersion (nearing the performance of a CMFR) will be represented by a low number of tanks, and a reactor with very low dispersion (nearing plug flow) will have a high number of tanks.

The model is developed by performing a mass balance on each CMFR in the series. Several adjustments to the mass balance need to be made to account for the difference in the volume of the real reactor and the volume of each CMFR in the series. The total volume of all the CMFRs in the series is set equal to the volume of the real reactor, such that

$$V_R = \frac{V}{n} \tag{4-102}$$

where V_R = volume of each individual CMFR, m^3
V = volume of the entire reactor, m^3
n = number of CMFR in the series

With a pulse input, the initial concentration is $C_0 = M/V$ if the tracer is mixed into the entire volume of a single reactor, but in this analysis the tracer initially mixes only into the first CMFR, that is,

$$C_{0,R} = \frac{M}{V_R} \tag{4-103}$$

where $C_{0,R}$ = initial concentration of tracer in the first CMFR, mg/L
C_0 = initial concentration of tracer if mixed into entire reactor, mg/L
M = mass of tracer used in the tracer test

Substituting Eq. 4-103 into Eq. 4-102 yields

$$C_{0,R} = \frac{M}{V_R} = \frac{nM}{V} = nC_0 \tag{4-104}$$

Similarly, the hydraulic residence time of each CMFR can be related to the overall hydraulic residence time:

$$\tau_R = \frac{V_R}{Q} = \frac{V}{nQ} = \frac{\tau}{n} \tag{4-105}$$

where τ_R = hydraulic residence time of each CMFR, min
τ = hydraulic residence time of the entire reactor, min
Q = flow rate, m^3/min

Using these definitions, the effluent from the first CMFR (which is the influent to the second CMFR) can be written

$$C_1 = C_{0,R}e^{-t/\tau_R} = nC_0e^{-nt/\tau} \tag{4-106}$$

where C_1 = effluent concentration of a tracer from the first CMFR in a series mg/L

The effluent concentration from the first CMFR is described by Eq. 4-106. If Eq. 4-106 is used as the influent in a mass balance on a second CMFR in the series, the resulting equation is

[accum] = [mass in] − [mass out] + [~~rxn~~]

$$V_R \frac{dC_2}{dt} = QC_1 - QC_2 \tag{4-107}$$

$$V_R \frac{dC_2}{dt} = QnC_0 e^{-nt/\tau} - QC_2 \tag{4-108}$$

After rearranging and integrating:

$$C_2 = (nC_0)\left(\frac{nt}{\tau}\right) e^{-nt/\tau} \tag{4-109}$$

where C_2 = effluent concentration of a tracer from the second CMFR mg/L

Continuing this analysis for a third CMFR in a series yields

$$C_3 = \left(\frac{nC_0}{2}\right)\left(\frac{nt}{\tau}\right)^2 e^{-nt/\tau} \tag{4-110}$$

where C_3 = effluent concentration of a tracer from the third CMFR, mg/L

Continuing the analysis for additional CMFRs eventually yields a general equation for any number of CMFRs:

$$C_n = C_0 \frac{n^n}{(n-1)!}\left(\frac{t}{\tau}\right)^{n-1} e^{-nt/\tau} \tag{4-111}$$

where C_n = effluent concentration of a tracer from a series of n CMFRs, mg/L

Equation 4-111 plotted for several values of n is shown on Fig. 4-8. An examination of this equation and figure indicates that a single parameter, n, describes the amount of dispersion or mixing in a real reactor. A value of $n = 1$ is equivalent to a CMFR and a value of $n = \infty$ is equivalent to a PFR. It is important to note that this model was developed with a mass balance on a CMFR, then extended to more and more CMFRs in series, so that the model has a fundamental basis.

Data from a tracer test can be used to determine the number of tanks in the TIS model that approximates the residence time distribution of the real reactor. The number of tanks is calculated from

$$n = \frac{\bar{t}^2}{\sigma^2} + 1 \tag{4-112}$$

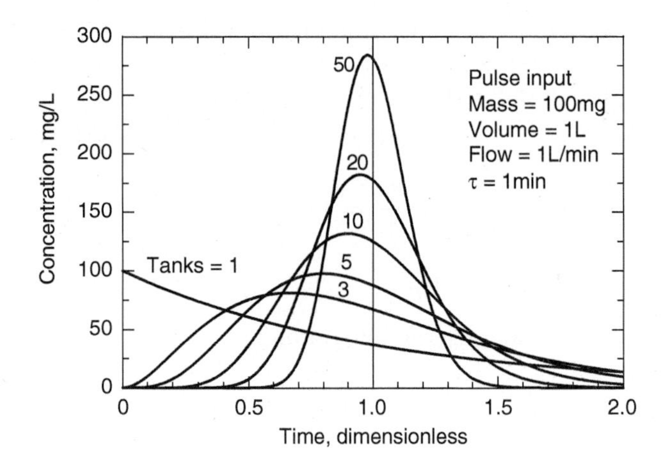

Figure 4-8
Effluent concentration of a tracer from the tanks-in-series model.

where n = number of tanks in the TIS model
$\bar{t}$ = mean residence time from tracer data, min
σ^2 = variance from tracer data, min^2

Models generally work best when the real conditions are reasonably close to the assumptions. In this case, the model development started with a mass balance on a CMFR, thus reactors that are closer to a CMFR (i.e., lots of dispersion) will generally fit better to the TIS model than reactors with very little dispersion.

A comparison between the t_{10}/τ value from a tracer test and the number of tanks in the TIS model is shown in Table 4-2. As expected, as the number of tanks increases, the t_{10}/τ value increases, since higher values of both parameters indicate less dispersion.

Table 4-2
Comparison between $\mathbf{t_{10}/\tau}$ and number of tanks in the TIS model

n	t_{10}/τ
1	0.10
3	0.35
6	0.5
10	0.6
20	0.7
50	0.8
Infinity	1.0

Example 4-11 Determining t_{10}/τ and number of tanks in the TIS model from tracer data

Determine the t_{10}/τ ratio and the number of tanks for the TIS model from the tracer test in Example 4-9.

Solution

1. The t_{10}/τ ratio is calculated by determining the values of t_{10} and τ.
   a. The value of t_{10} is determined from the F curve in Example 4-9. The F curve has a value of 0.1 at $\theta = 0.73$. The mean residence time was determined to be 76.48 min; thus, by rearranging Eq. 4-98, t_{10} is

$$t_{10} = \theta_{10}\bar{t} = (0.73)(76.48 \text{ min}) = 55.8 \text{ min}$$

   b. The hydraulic residence time τ is calculated from Eq. 4-73. The volume and flow rate through the reactor were given in the problem statement in Example 4-9.

$$\tau = \frac{V}{Q} = \frac{(175 \text{ m}^3)(1440 \text{ min/d})}{(3.17 \text{ ML/d})(10^3 \text{ m}^3/\text{ML})} = 79.5 \text{ min}$$

   c. The t_{10}/τ ratio is $t_{10}/\tau = 55.8 \text{ min} / 79.5 \text{ min} = 0.70$
2. The number of tanks for the TIS model is calculated from Eq. 4-112 using the mean and variance from Example 4-9.

$$n = \frac{\bar{t}^2}{\sigma^2} + 1 = \frac{(76.48 \text{ min})^2}{324.0 \text{ min}^2} + 1 = 19 \text{ tanks}$$

The relation between the t_{10}/τ ratio and the number of tanks for the TIS model compares favorably with the values given in Table 4-2.

4-12 Reactions in Real Flow Reactors

The modeling of chemical reactions occurring in ideal reactors was introduced in Secs. 4-7 and 4-9. However, the nonideal nature of the hydraulics of real reactors, as described in Secs. 4-10 and 4-11, affects the actual performance. Therefore, it is necessary to describe the performance of reactors in terms of the nonideal nature of reactor hydraulics. When a tracer curve is not available, the TIS model may be used with appropriate kinetic expressions to model reactor performance.

The reactor performance for the TIS model can be estimated from mass balances for a number of tanks in series. Using the same mathematical strategy that was used in Sec. 4-11, by performing a mass balance analysis on each CMFR in the series and using the effluent concentration equation from one CMFR as the influent concentration for the next CMFR, a general equation can be developed. The form of the equation will depend on the reaction rate term used in the mass balance analysis. For a first-order reaction, the following expression is obtained:

$$C = C_I \frac{1}{(1 + k\tau/n)^n} \tag{4-113}$$

where C = effluent concentration of a reactant with first-order kinetics from a series of n CMFRs, mg/L
C_I = influent concentration of reactant, mg/L
k = first-order reaction rate constant, min^{-1}
τ = hydraulic residence time of the entire reactor, min
n = number of tanks in the TIS model

Example 4-12 Effluent concentration from a real reactor

The reactor evaluated in Examples 4-9 and 4-11 is to be used to degrade a contaminant with an oxidant. The influent concentration of the contaminant is 100 μg/L. Experiments in a batch reactor have determined that the reaction between the contaminant and oxidant is a first-order reaction with a rate constant of 0.063 min^{-1}. Calculate the effluent concentration from the reactor, and compare it to the effluent concentration from a CMFR and a PFR with identical hydraulic residence times.

Solution

In Example 4-11, it was determined that the reactor hydraulic residence time was $\tau = 79.5$ min and the number of tanks for the TIS model was $n = 19$ tanks.

1. Calculate the effluent concentration from the real reactor using Eq. 4-113 for the TIS model:

$$C = C_I \frac{1}{(1 + k\tau/n)^n}$$

$$= \frac{100\ \mu\text{g/L}}{[1 + (0.063\ \text{min}^{-1})(79.5\ \text{min})/19]^{19}}$$

$$= 1.17\ \mu\text{g/L}$$

2. Calculate the effluent concentration from a CMFR and a PFR using Eqs. 4-80 and 4-93, respectively:

$$\text{CMFR:}\quad C = \frac{C_I}{1 + k\tau} = \frac{100\ \mu\text{g/L}}{1 + (0.063\ \text{min}^{-1})(79.5\ \text{min})} = 16.6\ \mu\text{g/L}$$

$$\text{PFR:}\quad C = C_I e^{-k\tau} = (100\ \mu\text{g/L})e^{-(0.063\ \text{min}^{-1})(79.5\ \text{min})} = 0.67\ \mu\text{g/L}$$

Comment

As demonstrated earlier in this chapter (see Example 4-8), mixing and dispersion is important in reactor performance for degrading contaminants. The real reactor performance is between that of a CMFR and PFR because the degree of dispersion is between that of a CFMR (perfectly mixed) and a PFR (no mixing). In addition, the performance is closer to that of a PFR than to a CMFR because the number of tanks in series is large, which is indicative of a low amount of dispersion.

4-13 Introduction to Mass Transfer

Several water treatment processes involve the transfer of material from one phase to another, such as from liquid to gas in air stripping, or liquid to solid in adsorption. In these processes, the contaminant removal efficiency, the rate of separation, and/or the size of the equipment can be governed by the rate of mass transfer.

The next few sections of this chapter introduce important concepts about the rate of movement of matter from one location to another, particularly from one phase to another. Consider a contaminant removal process that relies on an instantaneous reaction at a surface. Since the reaction is instantaneous, the rate at which the contaminant is degraded is controlled not by the rate of the reaction but by the rate at which the reactants can be transported to the surface. Such a process is called "mass transfer limited."

Mass transfer is a complex topic. Books have been written about the topic, and the chemical engineering curriculum at many universities includes an entire course on mass transfer. This book focuses on key principles that are relevant to environmental engineering and water treatment processes. Topics discussed in this chapter include an introduction to mass transfer, molecular diffusion and diffusion coefficients, models and correlations for mass transfer coefficients, operating diagrams, and mass transfer across a gas–liquid interface.

Mass transfer occurs in response to a driving force. Forces that can move matter include gravity, magnetism, electrical potential, pressure, and others.

In each case, the flux of material is proportional to the driving force. In environmental engineering, the driving force of interest is a concentration gradient or, in more general terms, a gradient in chemical potential or Gibbs energy. When a concentration gradient is present between two phases in contact with each other or between two locations within a single phase, matter will flow from the region of higher concentration to the region of lower concentration at a rate that is proportional to the difference between the two concentrations, as given by the following equation:

$$J = k(\Delta C) \tag{4-114}$$

where J = mass flux of a solute, $\mathrm{g/m^2 \cdot s}$
k = mass transfer coefficient, m/s
ΔC = concentration gradient of the solute, mg/L

Equation 4-114 has only two components (the mass transfer coefficient and the concentration gradient), and while this equation seems simple, it has profound implications for many treatment processes. The bulk of the rest of this chapter is devoted to the examination of variations of this equation. The next three sections are devoted to development of the mass transfer coefficient and models that describe mass transfer. Following that, Sec. 4-17 will explore how operating diagrams can be used to describe the concentration gradient.

4-14 Molecular Diffusion

Molecular diffusion is a form of mass transfer in which solute molecules or particles flow from a region of higher concentration to a region of lower concentration solely due to the kinetic energy of the solution molecules, that is, when no external forces are present to cause fluid movement. Molecular diffusion is a fundamental concept in mass transfer and an understanding of molecular diffusion is a necessary part of an understanding of mass transfer. Key concepts of molecular diffusion include Brownian motion and Fick's first law.

Brownian Motion

Brownian motion is the random motion of a particle or solute molecule due to the internal energy of the molecules in the fluid. As a result of this internal thermal energy, all molecules are in constant motion. A solute molecule or small particle suspended in a gas or liquid phase will be bombarded on all sides by the movement of the surrounding gas or liquid molecules. The random collisions cause unequal forces that cause the solute molecule to move in random directions. The random motion caused by these collisions is called Brownian motion after Robert Brown, who described it in 1827.

Although Brownian motion of individual molecules or particles is completely random, it causes bulk matter to flow from regions of high concentration to regions of low concentration. Consider a beaker containing water in which one drop of a blue dye has been placed. Molecules, both water molecules and dye molecules, are randomly moving in all directions. An imaginary boundary in the solution, as shown on Fig. 4-9, has a greater concentration of dye molecules on one side than the other. In response to completely random movement, the rate at which dye molecules cross the boundary in each direction is proportional to the number of dye molecules on each side; that is, the more dye molecules present, the more that can randomly cross the boundary from that direction. The net result is a bulk movement from concentrated regions to dilute ones. Net movement of dye molecules across any particular interface ceases when the concentration is the same on both sides. In this way, molecular diffusion stops (although Brownian motion continues) when the dye is uniformly distributed throughout the beaker, that is, the concentration is the same everywhere. At this point the solution in the beaker has reached equilibrium.

Fick's First Law

With Brownian motion as a foundation, molecular diffusion can be described by Fick's first law:

$$J_A = -D_{AB}\frac{dC_A}{dz} \tag{4-115}$$

where J_A = mass flux of component A due to diffusion, $mg/m^2 \cdot s$
D_{AB} = diffusion coefficient of component A in solvent B, m^2/s
C_A = concentration of component A, mg/L
z = distance in direction of concentration gradient, m

The term dC_A/dz is the concentration gradient, that is, the change in concentration per unit change in distance. The negative sign in Fick's

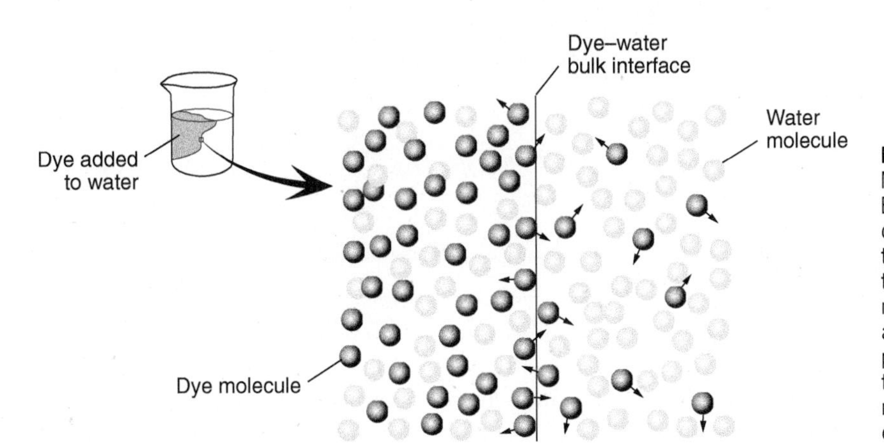

Figure 4-9
Mechanism by which Brownian motion leads to diffusion. In this diagram, the left side has about 4 times as many dye molecules, consequently about 4 times as many pass the interface from left to right compared to the number passing in the other direction.

first law arises because material flows from regions of high concentration to low concentration; thus, positive flux is in the direction of a negative concentration gradient. The diffusion coefficient describes the proportionality between a measured concentration gradient and the measured flux of material.

Diffusion in the Presence of Fluid Flow

Strictly speaking, Fick's first law describes the flux with respect to the centroid of the diffusing mass of solute. In other words, Fick's first law describes the rate of diffusion from a relative point of view; if the fluid is moving, the mass transfer due to diffusion is superimposed on top of, or in addition to, mass transfer due to the movement of the fluid.

The mass flow of component A due strictly to advection (in the absence of diffusion) may be written as

$$M_A = QC_A \tag{4-116}$$

where M_A = mass flow of solute A due to advection, mg/s
Q = flow rate of fluid, m^3/s

In terms of flux, the mass flow is divided by the perpendicular area:

$$J_A = \frac{QC_A}{A} = v(C_A) \tag{4-117}$$

where J_A = mass flux of component A due to advection, $mg/m^2 \cdot s$
A = cross-sectional area perpendicular to direction of flow, m^2
v = fluid velocity in direction of concentration gradient, where $v = Q/A$

Consequently, when matter is being transported by both fluid flow and diffusion, Eqs. 4-60, 4-115, 4-116, and 4-117 can be combined to define the net mass flow and mass flux as follows:

$$M_A = QC_A - D_{AB}\frac{dC_A}{dz}A \tag{4-118}$$

and

$$J_A = v(C_A) - D_{AB}\frac{dC_A}{dz} \tag{4-119}$$

4-15 Diffusion Coefficients

The diffusion coefficient is an essential parameter for calculating the rate of mass transfer in a wide variety of situations. Diffusion coefficients can be obtained from (1) laboratory measurements, (2) reference books or published literature, and (3) models and empirical correlations.

Diffusion coefficients can be determined experimentally in the laboratory and procedures for doing so are available in the literature (Robinson and Stokes, 1959; Malik and Hayduk, 1968). Measured diffusion coefficients of some common solutes found in water treatment are presented in Table 4-3. Diffusion coefficients for other constituents are available in the literature and reference books, such as Robinson and Stokes (1959), Marrero and Mason (1972), Poling et al. (2001), and CRC (2003).

Measured values of diffusion coefficients are not readily available for many compounds of interest. In addition, diffusion varies with temperature, and coefficients in reference books are often not at the temperature desired for the process application. In these cases, it is possible to estimate the diffusion coefficient based on chemical properties and structure using various models and empirical correlations. For each class of compound, a variety of calculation methods are available (Lyman et al., 1990; Poling et al., 2001). Some common correlations are described in this section. Use of these correlations is the most common way of estimating diffusion coefficients for many applications.

Liquid-Phase Diffusion Coefficients for Large Molecules and Particles

Based on the principle that diffusion is caused by Brownian motion, and Brownian motion is caused by collisions with the solvent molecules, it ought to be possible to derive a theoretical value for the diffusion coefficient from the kinetic theory of matter. Albert Einstein derived this relationship in papers published in 1905 and 1908. The derivation is beyond the scope of this book. Relating the mean square distance traveled by a molecule (or particle) during diffusion to the diffusion coefficient defined by Fick's law, and then determining the mean square distance traveled by a solute

Table 4-3
Measured values of molecular diffusion coefficients in water (at 25°C, unless noted otherwise)

Constituent	D_L, m^2/s	Constituent	D_L, m^2/s
Neutral species		Strong electrolytes (0.001 M)	
Acetic acid	1.29×10^{-9}	$BaCl_2$	1.32×10^{-9}
Acetone	1.28×10^{-9}	$CaCl_2$	1.25×10^{-9}
Benzene (20°C)	1.02×10^{-9}	KCl	1.96×10^{-9}
Ethanol	1.24×10^{-9}	KNO_3	1.90×10^{-9}
Ethylbenzene (20°C)	0.81×10^{-9}	NaCl	1.58×10^{-9}
Methane	1.49×10^{-9}	Na_2SO_4	1.18×10^{-9}
Sucrose	0.52×10^{-9}	$MgCl_2$	1.19×10^{-9}
Toluene (20°C)	0.85×10^{-9}	$MgSO_4$	0.77×10^{-9}
Vinyl chloride	1.34×10^{-9}	$SrCl_2$	1.27×10^{-9}

Sources: Robinson and Stokes (1959), Poling et al. (2001), and CRC (2003).

molecule as a result of collisions with solvent molecules, results in a relationship knows as the Stokes–Einstein equation:

$$D_L = \frac{k_b T}{3\pi\mu_L d} \tag{4-120}$$

where D_L = liquid-phase diffusion coefficient, m^2/s
k_b = Boltzmann's constant, 1.381×10^{-23} J/K ($kg \cdot m^2/s^2 \cdot K$)
T = absolute temperature, K (273.15 +°C)
μ_L = viscosity of water, kg/m·s
d = diameter of solute molecule or particle, m

Equation 4-120 predicts that diffusion increases with temperature and decreases with viscosity and molecular size, which have been observed experimentally. Equation 4-120 was derived from the kinetic theory of gases and does not strictly apply to liquids. Nonetheless, Eq. 4-120 can be used to obtain a good prediction of the liquid diffusion coefficient for large spherical molecules [molecular weight (MW) > 1000 daltons (Da)] and particles. The Stokes–Einstein equation has been compared to experimental data for globular proteins and other large molecules and found to be accurate within about 15 percent in many cases.

Liquid-Phase Diffusion Coefficients for Small Neutral Molecules

The diffusivities of small uncharged molecules (such as synthetic organic chemicals) in water can be calculated using the Hayduk–Laudie correlation, which is an empirical equation given by

$$D_L = \frac{13.26 \times 10^{-9}}{(\mu_L)^{1.14}(V_b)^{0.589}} \tag{4-121}$$

where D_L = liquid-phase diffusion coefficient of solute, m^2/s
μ_L = viscosity of water, cP (1 cP = 10^{-3} kg/m · s)
V_b = molar volume of solute at normal boiling point, cm^3/mol

Because the Hayduk–Laudie correlation was developed as a regression of experimental data and is not dimensionally consistent, it is important to use the units given for the equation. The molar volume is the volume occupied by one mole of a substance and is equal to the molecular weight divided by the density. One method for estimating the molar volume at the normal boiling point is the LeBas (1915) method. In this method, contributions of various functional groups are added together (with deductions for certain ring structures) using the group contributions listed in Table 4-4. Calculation of the diffusion coefficient of a small neutral molecule using the Hayduk–Laudie correlation is illustrated in Example 4-13.

Example 4-13 Calculating diffusion coefficients for small neutral molecules in water with the Hayduk–Laudie correlation

Estimate the liquid-phase diffusion coefficient of vinyl chloride at 25°C and compare it to the measured value reported in Table 4-3.

Solution

1. Estimate the molar volume at the boiling point using the contributions listed in Table 4-4. The chemical formula for vinyl chloride is C_2H_3Cl. The contribution of each atom to the molar volume is

$$2C = 2(14.8) = 29.6 \text{ cm}^3/\text{mol}$$

$$3H = 3(3.7) = 11.1 \text{ cm}^3/\text{mol}$$

$$Cl = (21.6) = 21.6 \text{ cm}^3/\text{mol}$$

The molar volume is determined by adding the contributions of each atom:

$$V_b = 29.6 + 11.1 + 21.6 = 62.3 \text{ cm}^3/\text{mol}$$

2. Calculate the diffusion coefficient using Eq. 4-121. The viscosity of water is available in App. C and must be converted to units of centipoise (cP). At 25°C, the viscosity of water is 0.89×10^{-3} kg/m · s = 0.89 cP:

$$D_L = \frac{13.26 \times 10^{-9}}{(0.89 \text{ cP})^{1.14} (62.3 \text{ cm}^3/\text{mol})^{0.589}} = 1.33 \times 10^{-9} \text{ m}^2/\text{s}$$

3. Compare the calculated value to the measured value in Table 4-3:

$$\frac{1.34 \times 10^{-9} - 1.33 \times 10^{-9}}{1.34 \times 10^{-9}} \times 100 = 1\% \text{ error}$$

Comment

The value estimated with the Hayduk–Laudie correlation is within 1 percent of the measured value for vinyl chloride. This result is common; the Hayduk–Laudie correlation is within 10 to 15 percent of measured values for many compounds (of course, values measured by different researchers with different methods also vary). As a result of this level of accuracy, it is common to estimate liquid-phase diffusion coefficients with the Hayduk–Laudie correlation rather than obtaining measured values for the species of interest.

Table 4-4
Atomic volumes for use in computing molar volumes at normal boiling point with the LeBas method

Substituent or Functional Group	Atomic Volume, cm^3/mol
Bromine	27.0
Carbon	14.8
Chlorine	
Terminal as in R–Cl	21.6
Medial as in R–CHCl–R′	24.6
Hydrogen	
In organic compound	3.7
In hydrogen molecule	7.15
Nitrogen	
Non-amine substitutions	15.6
In primary amines, R-NH_2	10.5
In secondary amines, R-NH-R′	12.0
Oxygen	
Double bond, aldehydes RCOH or ketones RCOR′	7.4
Single bond, methyl esters CH_3COOR	9.1
Single bond, methyl ethers CH_3OR	9.9
Single bond, higher ethers RCOOR′ and esters ROR′	11.0
In carboxylic acids, RCOOH	12.0
In union with S, P, or N	8.3
Phosphorus	27.0
Sulfur	25.6
Water	18.8
Ring deductions	
3-member, as in ethylene oxide C_2H_5O	−6.0
4-member, as in cyclobutane C_4H_8	−8.5
5-member, as in furan C_4H_4O	−11.5
6-member, as in benzene C_6H_6	−15
Naphthalene ring, $C_{10}H_8$	−30
Anthracene ring, $C_{14}H_{10}$	−47.5

Source: Adapted from LeBas (1915).

Liquid-Phase Diffusion Coefficients for Electrolytes

Electroneutrality requires that positive and negative ions migrate together, so diffusion coefficients are calculated for electrolytes (solutions of charged ions) instead of being calculated for each ion individually. As an example, the values of diffusion coefficients in Table 4-3 demonstrate that sodium and magnesium each diffuse faster when the counterion is chloride than when it is sulfate. In the absence of an electric field, diffusion of ions will generate an electric current in a solution. Conversely, the current through a unit area that results from applying an electric field for a given electrolyte concentration is known as the equivalent conductance. Thus, liquid-phase diffusion coefficients of electrolytes in the absence of an electric field are related to the equivalent conductance and can be calculated using the Nernst–Haskell equation:

$$D_L^\circ = \frac{RT}{(100\ \text{cm/m})^2 F^2}\left(\frac{1/n^+ + 1/n^-}{1/\lambda_+^\circ + 1/\lambda_-^\circ}\right) \tag{4-122}$$

where D_L° = liquid-phase diffusion coefficient at infinite dilution, m^2/s

R = universal gas constant, 8.314 J/mol·K

T = absolute temperature, K (273.15 +°C)

n^+, n^- = cation and anion valence, eq/mol

F = Faraday's constant, 96,500 C/eq

$\lambda_+^\circ, \lambda_-^\circ$ = limiting cation and anion ionic conductance, $S \cdot cm^2/eq$ or $C^2 \cdot cm^2/(J \cdot s \cdot eq)$

Values for limiting ionic conductance at 25°C are tabulated in Table 4-5. Values at other temperatures are available in reference books such as Robinson and Stokes (1959). Calculation of the diffusion coefficient of electrolytes with the Nernst–Haskell equation is shown in Example 4-14.

Example 4-14 Calculating diffusion coefficients for electrolytes in water with the Nernst–Haskell Equation

Estimate the diffusion coefficient of $MgCl_2$ in a dilute aqueous solution at 25°C and compare it to the measured value in Table 4-3.

Solution

1. From Table 4-5, the limiting ionic conductances are 53.0 $S \cdot cm^2/eq$ for Mg^{2+} and 76.4 $S \cdot cm^2/eq$ for Cl^-.
2. Calculate the diffusion coefficient at infinite dilution using Eq. 4-122: Note from footnote a in Table 4-5 that 1 S = 1 $C^2/J \cdot s$.

$$D_L^\circ = \frac{(8.314\ \text{J/mol}\cdot\text{K})(298\text{K})}{(100\ \text{cm/m})^2(96{,}500\ \text{C/eq})^2} \times \left[\frac{\left(\frac{1\ \text{C}^2/\text{J}\cdot\text{s}}{1\ \text{S}}\right)\left(\frac{1}{2\ \text{eq/mol}} + \frac{1}{1\ \text{eq/mol}}\right)}{\left(\frac{1}{53.0\ \text{S}\cdot\text{cm}^2/\text{eq}} + \frac{1}{76.4\ \text{S}\cdot\text{cm}^2/\text{eq}}\right)}\right]$$

$$= 1.25 \times 10^{-9}\ \text{m}^2/\text{s}$$

3. Compare the calculated diffusion coefficient to the measured value reported in Table 4-3:

$$\frac{1.25 \times 10^{-9} - 1.19 \times 10^{-9}}{1.19 \times 10^{-9}} \times 100 = 5\% \text{ error}$$

Comment

The value calculated with the Nernst–Haskell equation is the diffusion coefficient in an infinitely dilute solution, and the measured value in Table 4-3 is for a 0.001-M solution, but the values are within 5 percent of each other.

Gas-Phase Diffusion Coefficients for Organic Compounds

The diffusion coefficient of an organic compound in the gas phase can be calculated using a variety of correlations (Lyman et al., 1990). The Wilke–Lee correlation is appropriate for a wide variety of organic compounds and is

$$D_G = \frac{\left(1.084 - 0.249\sqrt{1/M_A + 1/M_B}\right)(T^{1.5})\sqrt{1/M_A + 1/M_B}}{P\left[\frac{1}{2}(r_A + r_B)\right]^2 f(k_b T/\varepsilon_{AB})(100 \text{ cm/m})^2} \tag{4-123}$$

Table 4-5
Limiting ionic conductances in water at 25°C [S · cm²/eq or (C² · cm²)/(J · s · eq)] [a]

Cation	Formula	λ_+°	Anion	Formula	λ_-°
Hydrogen	H^+	349.8	Hydroxide	OH^-	199.1
Lithium	Li^+	38.6	Fluoride	F^-	55.4
Sodium	Na^+	50.1	Chloride	Cl^-	76.4
Potassium	K^+	73.5	Bromide	Br^-	78.1
Rubidium	Rb^+	77.8	Iodide	I^-	76.8
Cesium	Cs^+	77.2	Bicarbonate	HCO_3^-	44.5
Ammonium	NH_4^+	73.5	Nitrate	NO_3^-	71.5
Silver	Ag^+	61.9	Perchlorate	ClO_4^-	67.3
Magnesium	Mg^{2+}	53.0	Bromate	BrO_3^-	55.7
Calcium	Ca^{2+}	59.5	Formate	$HCOO^-$	54.5
Strontium	Sr^{2+}	59.4	Acetate	CH_3COO^-	40.9
Barium	Ba^{2+}	63.6	Chloroacetate	$ClCH_2COO^-$	42.2
Copper	Cu^{2+}	53.6	Propionate	$CH_3CH_2COO^-$	35.8
Zinc	Zn^{2+}	52.8	Benzoate	$C_6H_5COO^-$	32.3
Lead	Pb^{2+}	69.5	Carbonate	CO_3^{2-}	69.3
Lanthanum	La^{3+}	69.7	Sulfate	SO_4^{2-}	80.0

[a] The siemen (S) is the SI derived unit for electrical conductance, 1 S = 1 A/V. Since 1 A = 1 C/s and 1 V = 1 J/C, then 1 S = 1 C²/J·s.
Source: Robinson and Stokes (1959).

where D_G = gas-phase diffusion coefficient of organic compound A in stagnant gas B, m^2/s
T = absolute temperature, K (273.15 +°C)
M_A, M_B = molecular weights of A and B, respectively, Da or g/mol
P = absolute pressure, N/m^2
r_A, r_B = molecular separation at collision for diffusing organic component A and stagnant gas B, nm
$f(k_bT/\varepsilon_{AB})$ = collision function

The collision function is related to the energy of molecular attraction and is calculated from the following equations:

$$f\left(\frac{k_bT}{\varepsilon_{AB}}\right) = 10^{\xi} \tag{4-124}$$

$$\xi = \left(\begin{array}{c} -0.14329 - 0.48343\,(\mathrm{ee}) + 0.1939\,(\mathrm{ee})^2 + 0.13612\,(\mathrm{ee})^3 \\ -0.20578\,(\mathrm{ee})^4 + 0.083899\,(\mathrm{ee})^5 - 0.011491\,(\mathrm{ee})^6 \end{array}\right) \tag{4-125}$$

$$\mathrm{ee} = \log\left(\frac{k_bT}{\varepsilon_{AB}}\right) \tag{4-126}$$

where k_b = Boltzmann constant, $1.381 \times 10^{-16}\ g \cdot cm^2/s^2 \cdot K$
T = absolute temperature, K (273.15 +°C)
ε_{AB} = energy of molecular attraction, equal to $\sqrt{\varepsilon_A \varepsilon_B}$, ergs (1 erg = 10^{-7} J)

The energy of molecular attraction is calculated by determining values of ε_A/k_b and ε_B/k_b and substituting them into the expression below:

$$\frac{\varepsilon_{AB}}{k_b} = \sqrt{\left(\frac{\varepsilon_A}{k_b}\right)\left(\frac{\varepsilon_B}{k_b}\right)} \tag{4-127}$$

where $\varepsilon_A, \varepsilon_B$ = energy of molecular attraction for component A and stagnant gas B, ergs (1 erg = 10^{-7} J)

When the stagnant gas B is air, the diffusion coefficient of a substance can be calculated by assuming that air behaves like a single substance with respect to molecular collisions. The value of ε_A/k_b for air is 78.6 and the value for the diffusing component is calculated from

$$\frac{\varepsilon_A}{k_b} = 1.21\,T_b \tag{4-128}$$

where T_b = normal boiling point of component A, K

The last parameter needed for calculating the gas-phase diffusion coefficient is the molecular separation at collision. The molecular separation for air is $r_B = 0.3711$ nm and the molecular separation for the diffusing component is calculated from

$$r_A = 1.18 V_b^{1/3} \quad \text{(in nm for } V_b \text{ in L/mol)} \tag{4-129}$$

where V_b = molar volume of component A at normal boiling point, L/mol

Calculation of gas-phase diffusion coefficients using the Wilke–Lee correlation is demonstrated in Example 4-15.

Example 4-15 Calculating gas-phase diffusion coefficients with the Wilke–Lee correlation

Calculate the gas-phase diffusion coefficient of trichloroethene (TCE) in air at 20°C at 1 bar.

Solution

1. In Eq. 4-123, the subscript A refers to TCE and B refers to air. The MW of air is 29 g/mol. Necessary parameters for TCE, available in reference books, are MW = 131.39 g/mol and $T_b = 360$ K. V_b is determined from the LeBas method (see Example 4-13) and is $V_b = 98.1\ \text{cm}^3/\text{mol} = 0.0981$ L/mol.
2. Calculate ε_{AB}/k_b with Eq. 4-127, by first calculating ε_A/k_b with Eq. 4-128 and using $\varepsilon_B/k_b = 78.6$:

$$\frac{\varepsilon_A}{k_b} = 1.21 T_b = 1.21(360\text{K}) = 435.6$$

$$\frac{\varepsilon_{AB}}{k_b} = \sqrt{\left(\frac{\varepsilon_A}{k_b}\right)\left(\frac{\varepsilon_B}{k_b}\right)} = \sqrt{(435.6)(78.6)} = 185$$

3. Calculate the collision function $f(k_B T/\varepsilon_{AB})$:
   a. Calculate $k_b T/\varepsilon_{AB}$. Note that 20°C = 293 K.

$$\frac{k_b T}{\varepsilon_{AB}} = \frac{T}{\varepsilon_{AB}/k_b} = \frac{293}{185} = 1.58$$

   b. Calculate ee using Eq. 4-126:

$$\text{ee} = \log\left(\frac{k_b T}{\varepsilon_{AB}}\right) = \log(1.58) = 0.200$$

c. Calculate ξ using Eq. 4-125:

$$\xi = \left\{ \begin{array}{l} -0.14329 - [0.48343(0.200)] + [0.1939(0.200)^2] \\ +[0.13612(0.200)^3] \\ -[0.20578(0.200)^4] + [0.083899(0.200)^5] \\ -[0.011491(0.200)^6] \end{array} \right\}$$

$$= -0.231$$

d. Calculate $f(k_b T/\varepsilon_{AB})$ using Eq. 4-124:

$$f\left(\frac{k_b T}{\varepsilon_{AB}}\right) = 10^{\xi} = 10^{-0.231} = 0.587$$

4. Calculate the values for r_A and r_B.$r_B = 0.3711$ nm and r_A is calculated using Eq. 4-129.

$$r_A = 1.18(V_b)^{1/3} = 1.18(0.0981)^{1/3} = 0.544 \text{ nm}$$

5. Calculate the gas-phase diffusion coefficient of TCE in air using Eq. 4-123. Note that pressure must be converted into the correct units, 1 bar = 10^5 N/m^2.

$$D_G = \frac{(1.084 - 0.249\sqrt{1/131.39 + 1/29})(293)^{1.5}\sqrt{1/131.39 + 1/29}}{(10^5)\left[\frac{1}{2}(0.3711 + 0.544)\right]^2 (0.587)(100)^2}$$

$$= 8.65 \times 10^{-6} \text{ m}^2/\text{s}$$

4-16 Models and Correlations for Mass Transfer at an Interface

In many common treatment processes, such as air stripping, adsorption, ion exchange, and reverse osmosis, mass transfer occurs at an interface. The interface is the boundary between the phase containing the solute or contaminant (typically the water) and the extracting phase (e.g., air or activated carbon). An understanding of mass transfer at an interface is essential to understanding the principles of these processes. Common models used to describe the mass transfer include (1) the film model, (2) the two-film model, and (3) the boundary layer model. These models are described in more detail in this section.

Surface Area Available for Mass Transfer

When mass transfer occurs at an interface, the concentration gradient is given by the concentrations in the bulk solution and at the interface, as shown on Fig. 4-10 and in following expression:

$$J = k_f\,(C_b - C_s) \tag{4-130}$$

where J = mass flux of solute A to interface, $\text{mg/m}^2 \cdot \text{s}$
k_f = mass transfer coefficient, m/s
C_s = concentration of solute A at interface, mg/L
C_b = concentration of solute A in bulk solution, mg/L

The mass transfer coefficient depends on the diffusion coefficient and the mass transfer boundary layer thickness δ, as shown on Fig. 4-10. As shown on Fig. 4-10, the direction of flux depends on the direction of the concentration gradient.

To calculate the mass flow rate, the flux must be multiplied by the surface area (see Eq. 4-60). It is common to express the area of the interface between phases as a function of the contactor volume (e.g., the surface area of carbon grains is expressed as a function of the volume of the carbon bed). Thus, the mass flow rate is given by the expression

$$M = JA = k_f\,a\,(C_b - C_s)\,V \tag{4-131}$$

where M = mass flow of solute A, mg/s
a = specific surface area, A/V, surface area available for mass transfer per unit volume of the contactor, m^2/m^3
V = contactor volume, m^3

The specific area is an important concept. For a given contactor volume, the mass transfer rate can increase linearly with an increase in specific area. Thus, designing a mass transfer device with a high specific area can result in a high rate of mass transfer in a small contactor. Mass transfer devices are often designed to have the highest possible specific area within the limitations imposed by hydraulic considerations. Increases in specific area

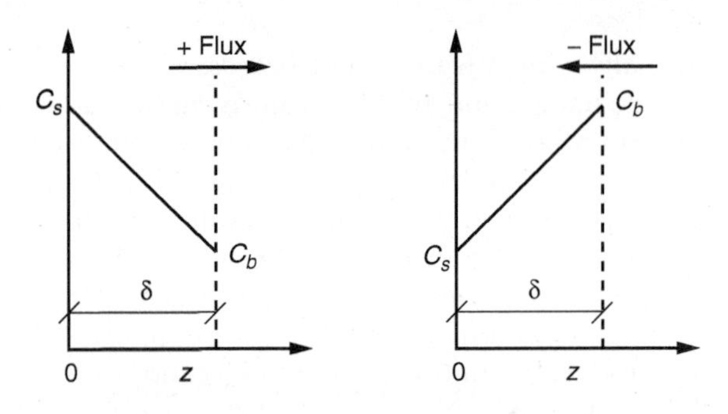

Figure 4-10
Hypothetical fluxes at interface at steady state.

often come at the expense of higher headloss. For example, in a packed bed of activated carbon it would be advantageous to use small carbon grains to increase the specific area, but the pressure drop would become too large and the cost of pumping water through the contactor would be high. In addition, the contactor would have to withstand the increased pressure. The relationship between grain size and specific area is demonstrated in Example 4-16.

Example 4-16 Calculating area available for mass transfer

Determine the specific area for the transport of a solute to granular activated carbon (GAC) particles in a carbon adsorber. The porosity (ε, fraction of void volume) of the carbon bed is 0.45 and the GAC particle diameter is $d_p = 1$ mm. Assume the surface of the GAC is like that of a smooth sphere.

Solution

$$a = \left(\frac{\text{surface area of particle}}{\text{volume of particle}}\right)\left(\frac{\text{volume of particles}}{\text{volume of contactor}}\right)$$

$$= \left(\frac{\pi d_p^2}{\left(\frac{1}{6}\right)\pi d_p^3}\right)(1-\varepsilon) = \frac{6(1-\varepsilon)}{d_p} = \frac{6(1-0.45)}{0.001\text{ m}} = 3300\text{ m}^2/\text{m}^3$$

Comment

The grain diameter is in the denominator so decreasing the size will increase the specific area for the same amount of GAC in the contactor (decreasing the grain size to 0.1 mm would increase the specific area to 33,000 m^2/m^3, which would increase the rate of mass transfer by a factor of 10 for the same size contactor if diffusion from the bulk solution to the particle surface is the limiting rate). This action, however, would increase the headloss and make it more difficult to pass water through the contactor.

Film Model

The film model is the most straightforward of the models that explain mass transfer at an interface. The system is considered to be composed of a well-mixed bulk solution (either gas or liquid), a stagnant film layer, and an interface to another phase, as shown on Fig. 4-10. As a result of the solution being well-mixed, solutes are transported continually to the edge of the stagnant film layer, and no concentration gradients exist in the bulk solution. Mass transfer in the film layer occurs when the concentration at the interface to the other phase is different than the concentration in the bulk solution, causing a concentration gradient across the film layer. Because this layer is quiescent, the sole mechanism for transport across this layer is molecular diffusion. In the simplest case, processes that occur at the actual interface (such as a chemical reaction or adsorption to the surface) occur much faster than the rate of diffusion and, as a result, the rate of mass transfer is described by Fick's first law for diffusion across the film layer:

$$J = -D_f \frac{dC}{dz} = -\frac{D_f}{\delta}(C_s - C_b) = k_f (C_b - C_s) \tag{4-132}$$

where J = mass flux of solute A to the interface, $mg/m^2 \cdot s$
D_f = fluid-phase diffusion coefficient of solute A, m^2/s
k_f = fluid-phase mass transfer coefficient of solute A, m/s
δ = film thickness, as shown on Fig. 4-10, m
C_b = concentration of solute A in bulk solution, mg/L
C_s = concentration of solute A at the interface, mg/L
z = distance in direction of mass transfer (or in direction of decreasing concentration gradient), m

In the film model, the mass transfer coefficient is explicitly related to the film thickness, as shown in the expressions

$$k_f = \frac{D_f}{\delta} \tag{4-133}$$

The theoretical stagnant film thickness will vary from 10 to 100 μm for liquids and from 0.1 to 1 cm for stagnant gases. Unfortunately, there is no way to calculate the stagnant film thickness based on fluid mixing; consequently, the film model cannot be used to calculate the local mass transfer coefficient. Nevertheless, the film model is used frequently to develop a conceptual view of mass transfer across an interface and to illustrate the importance of diffusion in controlling the rate of mass transfer.

Two-Film Model

When liquid is in contact with a gas, a stagnant film can form on both sides of the interface (on the liquid side and on the gas side). The two-film model extends the film model to describe mass transfer in this situation. The two situations where mass transfer occurs between air and water at steady state are shown on Fig. 4-11. The situation for stripping where mass is transferred from the water to the air is shown on Fig. 4-11a, and the situation for absorption in which mass is transferred from the air to the water is shown on Fig. 4-11b. The following discussion describes the mechanisms and assumptions of the two-film model from the perspective of stripping, but it should be noted that the model is essentially identical for both cases, and the only difference is that mass is transferred in the opposite direction.

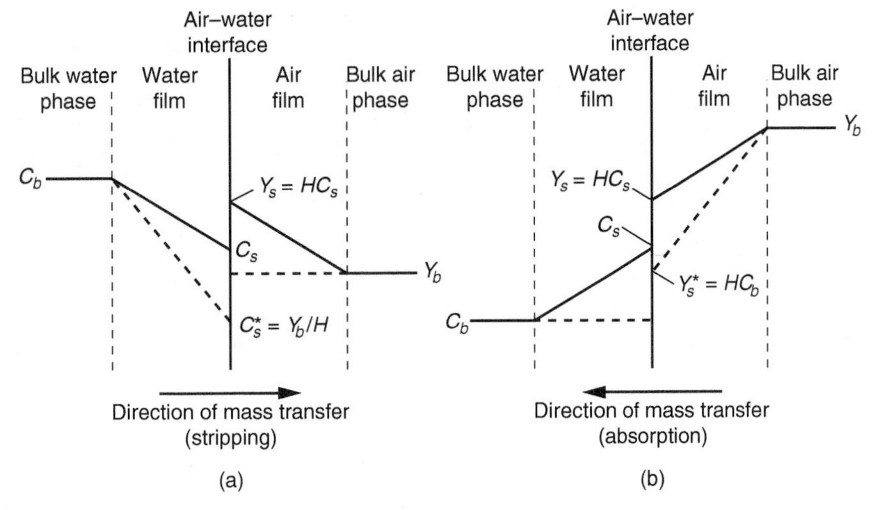

Figure 4-11
Two-film model: mass transfer driving gradients that occur for (a) stripping and (b) absorption.

CONDITIONS IN THE STAGNANT LAYERS

Figure 4-11a presents conditions for addressing the stripping of a volatile component A from water. As shown on Fig. 4-11a, the concentration of A in the bulk water, C_s, is larger than the concentration of A at the air–water interface, C_b. Consequently, A diffuses from the bulk solution to the air–water interface. The concentration gradient, $C_s - C_b$, is the driving force for stripping in the liquid phase. The discontinuity in concentrations at the air–water interface is because A partitions into air at a different concentration based on equilibrium, as described below. Similarly, the concentration of A in the air at the air–water interface, Y_s, is larger than the concentration of A in the bulk air, Y_b, and it diffuses from the air–water interface to the bulk air. The concentration gradient, $Y_s - Y_b$, is the driving force for stripping in the gas phase.

CONDITIONS AT THE INTERFACE

Local equilibrium occurs at the air–water interface because random molecular movement (on a local scale of nanometers in water and thousands of nanometers on the air side) causes constituent A to dissolve in the aqueous phase and volatilize into the air more rapidly than diffusion to or away from the air–water interface. Accordingly, the concentrations at the actual interface are in equilibrium and Henry's law can be used to relate Y_s to C_s (see Chap. 11):

$$Y_s = HC_s \tag{4-134}$$

where Y_s = gas-phase concentration of A at air–water interface, mg/L
H = Henry's law constant, L of water/L of air, dimensionless
C_s = liquid-phase concentration of A at air–water interface, mg/L

For a dilute solution where no accumulation occurs at the surface, the flux of A through the gas-phase film must be equal to the flux through the liquid-phase film. Thus

$$J = k_L (C_b - C_s) = k_G (Y_s - Y_b) \tag{4-135}$$

where J = flux of A across air–water interface, $mg/m^2 \cdot s$
k_L, k_G = local liquid-phase and gas-phase mass transfer coefficients, respectively, m/s
C_b, C_s = liquid-phase concentration of A in bulk solution and at the air–water interface, respectively, mg/L
Y_s, Y_b = gas-phase concentration of A at air–water interface and in the bulk solution, respectively, mg/L

Both k_L and k_G are sometimes referred to as local mass transfer coefficients for the liquid and gas phases because they describe mass transfer occurring only in their particular phase.

OVERALL MASS TRANSFER RELATIONSHIP

The flux across the interface cannot be calculated directly from Eq. 4-135 because the interfacial concentrations Y_s and C_s are not known and cannot be measured easily. Consequently, it is necessary to define another flux equation in terms of hypothetical concentrations that are easy to determine. If it is hypothesized that all the resistance to mass transfer is on the liquid side, then there is no concentration gradient on the gas side and a hypothetical concentration, C_s^*, can be defined as shown on Fig. 4-11a:

$$Y_b = HC_s^* \tag{4-136}$$

where C_s^* = liquid-phase concentration of A that is in equilibrium with bulk air concentration, mg/L

With all resistance to mass transfer on the liquid side, it is now possible to envision the rate of mass transfer being dependent on the concentration gradient between the bulk solution and the hypothetical concentration C_s^* using an overall mass transfer coefficient K_L, as shown in the equation

$$J = K_L \left(C_b - C_s^*\right) \tag{4-137}$$

where J = mass flux of A across air–water interface, $mg/m^2 \cdot s$
K_L = overall mass transfer coefficient, m/s

Since no mass accumulates at the interface, the hypothetical, gas-side, and liquid-side mass fluxes given in Eqs. 4-135 and 4-137 must all be equal to one another:

$$J = k_L\left(C_b - C_s\right) = k_G\left(Y_s - Y_b\right) = K_L\left(C_b - C_s^*\right) \tag{4-138}$$

Equation 4-138 relates K_L to k_L and k_G and accounts for mass transfer resistances on both the gas and liquid sides of the interface. The individual expressions in Eq. 4-138 can be rearranged as follows:

$$\frac{J}{k_L} = C_b - C_s \tag{4-139}$$

$$\frac{J}{k_G} = Y_s - Y_b \tag{4-140}$$

$$\frac{J}{K_L} = C_b - C_s^* \tag{4-141}$$

The overall mass transfer coefficient can be related to the local mass transfer coefficients starting with the relationship

$$C_b - C_s^* = \left(C_b - C_s\right) + \left(C_s - C_s^*\right) \tag{4-142}$$

Substituting Eqs. 4-134 and 4-136 into Eq. 4-140, and then substituting Eqs. 4-139, 4-140, and 4-141 into Eq. 4-142 yields

$$\frac{J}{K_L} = \frac{J}{k_L} + \frac{J}{Hk_G} \tag{4-143}$$

Or simply

$$\frac{1}{K_L} = \frac{1}{k_L} + \frac{1}{Hk_G} \tag{4-144}$$

Thus, according to the two-film model, the mass flux across the interface can be calculated using the expression

$$J = K_L\left(C_b - \frac{Y_b}{H}\right) \tag{4-145}$$

Equation 4-145 is convenient to use because the driving force for stripping $(C_b - Y_b/H)$ involves concentrations that are easy to measure. The overall mass transfer coefficient can be estimated from the local mass transfer coefficients, and the local mass transfer coefficients can be determined from correlations.

APPLICATION OF THE TWO-FILM MODEL

Equipment for aeration and stripping processes often define the interfacial area on a volumetric basis; that is, a particular contactor has a certain amount of interfacial surface area per unit volume of contactor. In this case, mass transfer across the gas-liquid interface is described using the specific surface area $(a = A/V)$ and contactor volume. The overall mass transfer coefficient (K_L) and specific area (a) are then combined into a single parameter ($K_L a$) as follows:

$$M = K_L a \left(C_b - \frac{Y_b}{H} \right) V \tag{4-146}$$

where M = mass flow of A, mg/s
a = specific surface area, area of interface per unit volume of contactor, m^{-1}
$K_L a$ = overall liquid-side mass transfer coefficient, s^{-1}
V = volume of contactor, m^3

The specific area is then incorporated into the expression relating the overall mass transfer coefficient to the local mass transfer coefficients:

$$\frac{1}{K_L a} = \frac{1}{k_L a} + \frac{1}{H k_G a} \tag{4-147}$$

Boundary Layer Models

The film model is somewhat simplistic in assuming that a bulk fluid can be completely mixed but that a completely stagnant film layer forms adjacent to a surface. Boundary layer models attempt a more realistic analysis. Consider a situation when fluid flows parallel to a solid surface, such as when water flows through a pipe. In this situation, a velocity gradient forms because the fluid velocity is assumed to be zero at the surface (no slip condition) but greater than zero away from the surface. Students who have studied fluid mechanics will be familiar with the parabolic velocity profile that develops during laminar flow in a pipe. In a larger pipe with turbulent flow, most of the fluid will be traveling at the same net average velocity, with regions of lower velocity near the pipe wall. This region of lower velocity near the surface is known as the velocity boundary layer.

Simultaneously, material in the bulk solution can adsorb to the surface or material on the surface can dissolve or leach into solution. Adsorption or leaching of material at the surface causes a concentration gradient to form between the concentration at the surface and the concentration in the bulk solution. The concentration gradient then leads to mass transfer to (for adsorption) or from (for leaching) the surface. The region of the concentration gradient is known as the concentration boundary layer. The limit of the concentration gradient is not necessarily the same as the velocity gradient, but the two will be related. A conceptual view of velocity and concentration boundary layers forming adjacent to a flat plate in turbulent flow is shown on Fig. 4-12. The relationship between the concentration and velocity gradients depends on conditions of the fluid flow. As fluid velocity increases, the velocity boundary layer will become thinner, leading to an increase in the slope of the concentration gradient and an increase in the rate of mass transfer.

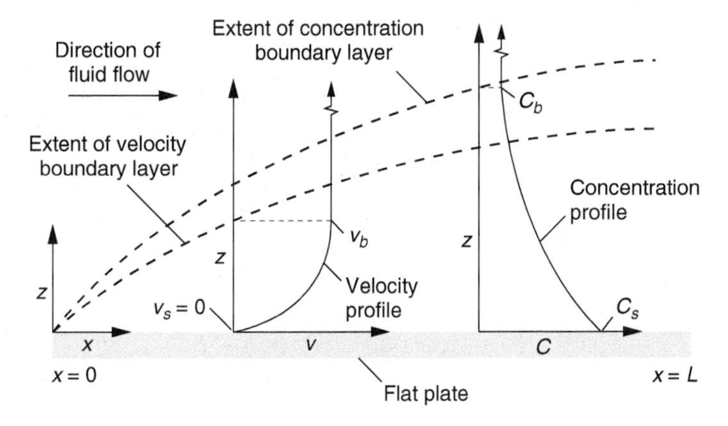

Figure 4-12
Boundary layer model diagram showing velocity and concentration profiles for laminar flow across flat plate.

The Sherwood number is a dimensionless parameter group that describes the relationship between the mass transfer coefficient and the diffusion coefficient:

$$\mathrm{Sh} = \frac{k_f L}{D_f} \tag{4-148}$$

where Sh = Sherwood number, dimensionless
k_f = fluid-phase mass transfer coefficient, m/s
L = characteristic length scale, m
D_f = fluid-phase diffusion coefficient, m^2/s

The fluid can be either a gas or a liquid. For a given length scale, a higher Sherwood number indicates that mass transfer is faster compared to the mass transfer that would occur by pure molecular diffusion. For instance, in the film model presented earlier the characteristic length scale is the stagnant film layer thickness (δ), mass transfer occurs only by molecular diffusion, and Sh = 1.

When fluid is flowing, the Sherwood number depends on the values of the Schmidt and Reynolds numbers according to the following general relationship:

$$\mathrm{Sh} = A + B\,\mathrm{Re}^c\,\mathrm{Sc}^d \tag{4-149}$$

where A, B, c, d = coefficients that depend on the specific system, unitless
Sc = Schmidt number, dimensionless
Re = Reynolds number, dimensionless

The coefficients in Eq. 4-149 (i.e., A, B, c, d) depend on the geometry (e.g., sphere, cylinder, plate) and flow regime (e.g., laminar, transition, or turbulent) of the particular system. Many investigators have developed the theoretical bases for Eq. 4-149 for various geometries and flow regimes and have also developed mass transfer correlations by fitting data to Eq. 4-149 for specific situations (e.g., laminar or turbulent flow past a flat plate, through a pipe, through a packed bed, bubbles rising through a water column, etc.). These correlations are a particularly powerful concept in mass transfer.

The Schmidt number describes the importance of viscous versus diffusive forces in contributing to mass transfer, and the Reynolds number describes the importance of viscous versus inertial forces in fluid flow. These dimensionless parameter groups are defined as

$$\mathrm{Sc} = \frac{\nu}{D_f} = \frac{\mu}{\rho D_f} \tag{4-150}$$

$$\mathrm{Re} = \frac{vL}{\nu} = \frac{\rho v L}{\mu} \tag{4-151}$$

where ν = kinematic viscosity, equal to μ/ρ, m^2/s
μ = absolute viscosity, $kg/m \cdot s$
ρ = fluid density, kg/m^3
v = superficial fluid velocity (outside the boundary layer), m/s

The characteristic length in the equations for the Sherwood number (Eq. 4-148) and Reynolds number (Eq. 4-151) depends on the geometry of the system. For flow through pipes, L is taken as the diameter of the pipe, and for flow though packed beds or around particles, L is taken as the diameter of the particle. An example of a correlation developed from the boundary layer model is the Gilliland correlation, which describes mass transfer due to turbulent flow through pipes:

$$\mathrm{Sh} = 0.023\mathrm{Re}^{0.83}\,\mathrm{Sc}^{0.33} \tag{4-152}$$

The Gilliland correlation uses the pipe diameter as the length scale and is appropriate for turbulent flow when $\mathrm{Re} > 2100$ and $0.6 < \mathrm{Sc} < 3000$.

It is sometimes necessary to take additional factors into account in the length scale. For instance, the porosity of the bed and shape of the granular media are important in the Sherwood and Reynolds numbers for use in the Gnielinski correlation, which describes mass transfer in packed beds of granular material. The Gnielinski correlation is

$$\mathrm{Sh} = 2 + 0.644\mathrm{Re}^{1/2}\,\mathrm{Sc}^{1/3} \tag{4-153}$$

Where the Sherwood and Reynolds numbers are defined as

$$\mathrm{Sh} = \frac{k_f d_p}{[1 + 1.5(1 - \varepsilon)]\,D_f} \tag{4-154}$$

$$\mathrm{Re} = \frac{\rho \phi d_p v}{\varepsilon \mu} \tag{4-155}$$

where k_f = fluid-phase mass transfer coefficient, m/s
d_p = media grain diameter, kg/m^3
ε = bed porosity (void fraction), dimensionless
D_f = fluid-phase diffusion coefficient, m^2/s
ϕ = sphericity, equal to ratio of surface area of equivalent-volume sphere to actual surface area of particle, dimensionless

The Gnielinski correlation is suitable when $0.7 < Sc < 10^4$, $Re < 2 \times 10^4$, $0.26 < \varepsilon < 0.93$, and $Pe = Re \times Sc > 500$. The use of a correlation based on the boundary layer model to calculate a mass transfer coefficient is demonstrated in Example 4-17.

Example 4-17 Application of a correlation to determine a mass transfer coefficient

A resort in the mountains has a good water source; however, the water is extremely soft (no hardness) and acidic, which makes cleaning and bathing difficult. One solution is to pass the low-pH water through a packed bed containing crushed limestone ($CaCO_3$). Determine the film transfer coefficient for limestone media. Given: The media diameter d_p is 1.0 cm, the bed porosity ε is 0.43, the particle sphericity ϕ is 0.8, the temperature is 20°C, and the superficial velocity v through the bed is 12 m/h.

Solution

Determine the mass transfer coefficient k_f for limestone particles using the Gnielinski correlation in Eq. 4-153.

1. Calculate the diffusion coefficient for aqueous calcium carbonate using Eq. 4-122 (see Example 4-14). From Table 4-5, the limiting conductances are 59.5 $S \cdot cm^2/eq$ for Ca^{2+} and 69.3 $S \cdot cm^2/eq$ for CO_3^{2-}.

$$D_L^\circ = \frac{(8.314\ \text{J/mol}\cdot\text{K})(298\ \text{K})}{(100\ \text{cm/m})^2(96{,}500\ \text{C/eq})^2}$$

$$\times \left[\frac{\left(\frac{1\ \text{C}^2/\text{J}\cdot\text{s}}{1\ \text{S}}\right)\left(\frac{1}{2\ \text{eq/mol}} + \frac{1}{2\ \text{eq/mol}}\right)}{\left(\frac{1}{59.5\ \text{S}\cdot\text{cm}^2/\text{eq}} + \frac{1}{69.3\ \text{S}\cdot\text{cm}^2/\text{eq}}\right)} \right]$$

$$= 8.52 \times 10^{-10}\ \text{m}^2/\text{s}$$

2. Calculate Re from Eq. 4-155. From App. C, $\rho = 998.2\ \text{kg/m}^3$ and $\mu = 1.002 \times 10^{-3}\ \text{kg/m} \cdot \text{s}$ at 20°C.

$$\text{Re} = \frac{\rho \phi d_p v}{\varepsilon \mu}$$

$$= \frac{(998.2\ \text{kg/m}^3)(0.8)(1.0\ \text{cm})(1\ \text{m}/100\ \text{cm})(12\ \text{m/h})(1\ \text{h}/3600\ \text{s})}{(0.43)(1.002 \times 10^{-3}\ \text{kg/m} \cdot \text{s})}$$

$$= 61.8$$

3. Calculate Sc using Eq. 4-150:

$$\text{Sc} = \frac{\mu}{\rho D_L} = \frac{(1.002 \times 10^{-3}\ \text{kg/m} \cdot \text{s})}{(998.2\ \text{kg/m}^3)(8.52 \times 10^{-10}\ \text{m}^2/\text{s})} = 1180$$

4. Calculate Sh using Eq. 4-153:

$$\text{Sh} = 2 + 0.644\,\text{Re}^{1/2}\,\text{Sc}^{1/3} = 2 + 0.644\,(61.8)^{1/2}\,(1180)^{1/3}$$

$$= 55.5$$

5. Calculate k_f using Eq. 4-154:

$$k_f = \frac{[1 + 1.5\,(1 - \varepsilon)]\,D_L\,\text{Sh}}{d_p}$$

$$= \frac{[1 + 1.5(1 - 0.43)](8.52 \times 10^{-10}\ \text{m}^2/\text{s})(55.5)}{(1\ \text{cm})(1\ \text{m}/100\ \text{cm})}$$

$$= 8.76 \times 10^{-6}\ \text{m/s}$$

4-17 Evaluating the Concentration Gradient with Operating Diagrams

The last sections have dealt with development of theory and correlations needed to determine mass transfer coefficients. This section explores the other half of the primary mass transfer equation (Eq. 4-114), the concentration gradient. The concentration gradient and the impact it has on mass transfer can be evaluated graphically. Graphical analysis of concentration gradients depends on the type of contacting equipment. The major types of contacting equipment are described next, followed by a discussion of operating diagrams, also known as McCabe–Thiele diagrams.

Development of Operating Diagrams

The impact of the concentration gradient on the rate of mass transfer between two phases can be evaluated graphically using a concept called operating diagrams, or McCabe–Thiele diagrams (McCabe and Thiele,

1925). Operating diagrams are drawn by plotting the solute concentration in the extracting phase (e.g., air for gas transfer, activated carbon for adsorption) as a function of the solute concentration in the aqueous phase. The operating diagram consists of two lines: (1) an equilibrium line and (2) an operating line. Operating diagrams can be used to determine the minimum amount of the extracting phase needed for treatment and to examine graphically the trade-off between the size of the mass transfer contacting device and the quantity of extracting phase needed [e.g., air–water ratio for stripping or powdered activated carbon (PAC) required for adsorption].

EQUILIBRIUM LINE

The equilibrium line is derived from two-phase equilibrium relationships and gives the solute concentration in the extracting phase that exists when the extracting and aqueous phases are in equilibrium with each other. Examples of two-phase equilibrium relationships are Henry's law for air stripping and the Freundlich isotherm for adsorption. Equilibrium relationships were introduced in Sec. 4-2., and additional details on Freundlich isotherms and Henry's Law will be provided in Chaps. 10 and 11, respectively.

OPERATING LINE

The operating line is derived from a mass balance on the contacting device, relating the solute concentration in each phase initially to the solute concentration in each phase after contact has begun. An example using a batch reactor, in which PAC is added to a vessel containing a solution of water and an organic solute, is shown on Fig. 4-13. Initially, there is no solute adsorbed onto the PAC. The mass balance for this system is as follows:

$$\begin{bmatrix} \text{mass initially} \\ \text{present in solution} \end{bmatrix} = \begin{bmatrix} \text{mass} \\ \text{adsorbed} \end{bmatrix} + \begin{bmatrix} \text{mass remaining in} \\ \text{solution after adsorption} \end{bmatrix} \tag{4-156}$$

$$VC_0 = Mq + VC \tag{4-157}$$

where V = volume of liquid in vessel, L
C_0 = initial concentration of solute in vessel, mg/L
M = mass of carbon, g
q = concentration of solute adsorbed to the activated carbon at any time, mg/g
C = concentration of the solute in the water after adsorption, mg/L

Equation 4-157 can be rearranged as follows:

$$q = \frac{V}{M}(C_0 - C) \tag{4-158}$$

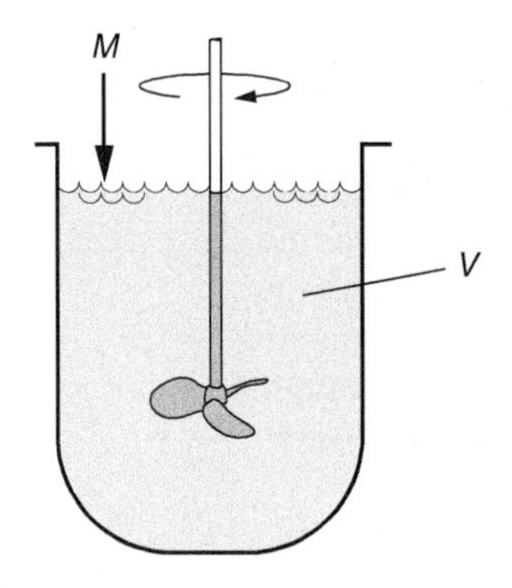

Figure 4-13
Batch contactor for powdered activated carbon.

The operating line, which is the solute concentration in the extracting phase as a function of the concentration in the aqueous phase at any point in time after contact has started, is defined by Eq. 4-158. When the PAC is first added to the vessel, there is no solute on the PAC. As time proceeds, the solute becomes adsorbed onto the PAC, and q and C at a particular time are related to one another by the operating line. It should be noted that although adsorption in a batch reactor proceeds toward equilibrium over the passage of time, the operating line does not identify the time progression of the process, but only relates the dependent variables q and C.

The operating diagram for the relationship described in Eq. 4-158 is shown on Fig. 4-14. Equation 4-158 is the equation of a straight line with a slope of $-V/M$, and several operating lines with different values for V/M have been shown. The equilibrium line is shown on Fig. 4-14 as a dashed line.

DRIVING FORCE

The driving force for mass transfer, as shown on Fig. 4-14, is the difference between the actual solute concentration in solution and the concentration in solution that would be in equilibrium with the extracting phase. Initially, the solute is entirely in the aqueous phase, and the solute is transferred rapidly to the PAC. As time progresses, the concentration on the PAC increases and the concentration in the aqueous phase decreases, which slows the rate of mass transfer. After a very long time, the solute concentration in the water is in equilibrium with the concentration on the PAC, and bulk mass transfer ceases. Thus, the concentration gradient, or driving force, is defined as the difference between the actual and equilibrium concentration C_e in the aqueous phase.

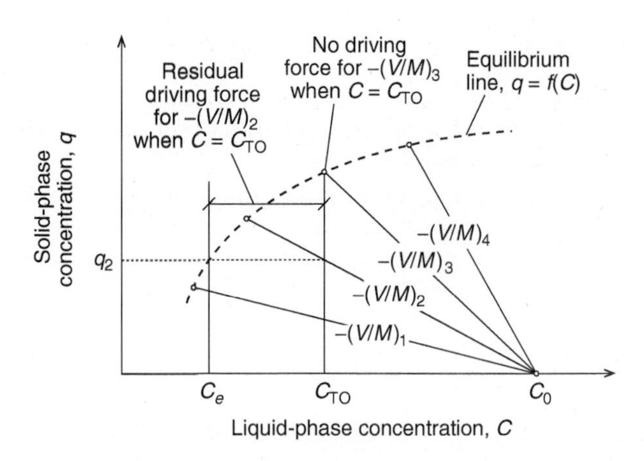

Figure 4-14
Operating lines for a constant initial concentration C_0 and different adsorbent doses, V/M (equilibrium line is also plotted for reference).

Because the equilibrium concentration is identified by the equilibrium line and the actual concentration (determined by mass balance) is identified by the operating line, the horizontal distance between these lines describes the concentration gradient. Equilibrium occurs and mass transfer ceases when the operating line and equilibrium line intersect.

Analysis Using Operating Diagrams

The operating diagram can be used to determine the minimum amount of extracting phase required for treatment, which is an initial indicator of the feasibility of a process. For example, if millions of tons of activated carbon are required to treat a given water, then adsorption with activated carbon is not a feasible treatment option and no further analysis is necessary. If a separation process appears to be feasible based on the amount of extracting phase, then more detailed design and economic calculations are warranted.

An operating line analysis for an adsorption process is shown on Fig. 4-14. For a given volume of water, the quantity of PAC required can be defined by the V/M ratio, with greater values of V/M (greater slope of the operating line) corresponding to smaller amounts of PAC. If the treatment objective is the concentration shown as C_{TO} on Fig. 4-14, the minimum amount of PAC required can be determined from the operating line with the slope of $(V/M)_3$, which is the operating line that intersects the equilibrium line at the value of C_{TO}. Operating lines with greater slope, shown as $(V/M)_4$, intersect the equilibrium line at a concentration higher than C_{TO} and therefore would be unable to meet the treatment objective.

The operating diagram also qualitatively demonstrates the trade-off between the quantity of the extracting phase and the size of the contacting device. For the operating line identified as $(V/M)_3$, the driving force (horizontal distance between the equilibrium and operating lines) becomes infinitesimally small as equilibrium is approached. The small driving force results in a slow rate of mass transfer, requiring an exceedingly long time to reach the treatment objective. In a flow-through system treating a specified water flow rate, a long time corresponds to a long residence time and hence a very large contactor. The operating lines labeled as $(V/M)_1$ and $(V/M)_2$ have lower slopes, which correspond to greater quantities of carbon, but have larger concentration gradients when the actual concentration (operating line) reaches the treatment objective, resulting in shorter contact times. Thus, for the operating lines shown, the line labeled $(V/M)_1$ would use the most carbon but have the smallest contactor, the line labeled $(V/M)_2$ would have an intermediate carbon usage rate and contactor size, the line labeled $(V/M)_3$ would use the minimum amount of carbon but have a large (theoretically, infinitely large) contactor, and the line labeled $(V/M)_4$ would be unable to meet the treatment objective.

An equation similar to Eq. 4-157 can be developed for co-current continuous plug flow operation. If a quantity of PAC per time, M_r, is added to water with a flow rate, Q, the mass balance analysis can be written

$$[\text{accum}] = [\text{mass in}] - [\text{mass out}] + [\text{rxn}]$$

$$0 = QC_I - QC - M_r q \tag{4-159}$$

$$q = \frac{Q}{M_r}(C_I - C) \tag{4-160}$$

where Q = flow rate, L/s
C_I = influent solute concentration, mg/L
M_r = PAC feed rate, mass added per time, g/s
C = effluent solute concentration, mg/L
q = concentration of solute adsorbed to the activated carbon, mg/g

The PAC dose in the plug flow system, M_r/Q, is identical to the PAC dose in the batch reactor, M/V, and Eqs. 4-158 and 4-160 are essentially identical.

An example calculation of the minimum amount of extracting phase required for treatment is presented for PAC in Example 4-18.

Example 4-18 Minimum amount of PAC required to achieve given level of treatment

Many adsorption equilibrium lines, as discussed in Chap. 10, can be described by the Freundlich isotherm:

$$q = KC^{1/n}$$

where

q = equilibrium concentration of solute in solid phase, mg/g
K = Freundlich capacity factor, $(\text{mg/g})(\text{L/mg})^{1/n}$
C = equilibrium concentration of solute in aqueous phase, mg/L
$1/n$ = Freundlich intensity factor, dimensionless

Calculate the minimum dose of PAC that is required for the removal of geosmin, an odor-producing compound. The initial concentration is 50 ng/L, and the treatment objective is 5 ng/L. The K and $1/n$ values for geosmin are 200 $(\text{mg/g})(\text{L/mg})^{1/n}$ and 0.39, respectively. A reasonable PAC dose would be less than 10 to 20 mg/L. Is the process feasible and should more detailed studies be conducted?

Solution

The lowest PAC dose occurs when the PAC is used to capacity, which is when the concentration on the PAC would be in equilibrium with the treatment objective and the operating line (Eq. 4-160) intersects the equilibrium line at the treatment objective. The intersection of the equilibrium and operating lines is determined by equating the equilibrium equation with the operating line and solving for the minimum dose:

$$\left(\frac{M_r}{Q}\right)_{\min} = \frac{C_I - C}{KC^{1/n}} = \frac{\left(50 \times 10^{-6} - 5 \times 10^{-6}\right)\ \text{mg/L}}{(200)\,(5 \times 10^{-6})^{0.39}\ \text{mg/g}}$$

$$= 2.63 \times 10^{-5}\ \text{g/L} = 0.0263\ \text{mg/L}$$

Comment

A dose of 0.0263 mg/L is within the acceptable range, and additional tests that simulate water plant conditions (jar tests) can be planned. The tests would be needed because the presence of natural organic matter (NOM) will reduce the adsorption capacity. Further, the computed value is the minimum dose of PAC, which yields an exceedingly small driving force as equilibrium is approached, resulting in an extremely low rate of mass transfer and an unreasonably large PAC contactor.

4-18 Summary and Study Guide

After studying this chapter, you should be able to:

1. Define the following terms and phrases and describe the significance of each in the context of environmental engineering

activity	film model	reactor
activity coefficient	first-order reaction	residence time distribution
Arrhenius equation	Hayduk–Laudie correlation	Reynolds number
batch reactor	hydraulic retention time	Schmidt number
boundary layer model	mass balance analysis	Sherwood number
Brownian motion	mass transfer coefficient	steady state
completely mixed tank reactor	molarity	stoichiometry
concentration	molecular diffusion	t_{10}/τ ratio
concentration gradient	molecular weight	tanks-in-series model
conservative species	Nernst–Haskell equation	tracer
control volume	operating diagram	tracer test
diffusion coefficient	plug flow reactor	two-film model
equilibrium constant	p notation	van't Hoff equation
E curve	reaction order	Wilke–Lee correlation
F curve	reaction rate constant	
Fick's first law	reactive species	

2. Convert concentrations into different units, that is, mg/L to mol/L to percent by weight, and from molarity to normality.
3. Calculate species concentrations using equilibrium constants.
4. Describe the importance of the mass balance to environmental engineering.
5. Describe the conditions necessary for something to be a good control volume for mass balances and what constitutes a steady-state system.
6. Analyze an environmental or engineered system and determine how to apply a mass balance, including definition of the control volume, inputs, outputs, and reactions, determination of the appropriate assumptions, and development of the governing equation. Solve the mass balance analysis equations, including integration of the fundamental equation if the system is not at steady state and with or without reactions.
7. Calculate the change in concentration in a batch reactor over time due to chemical reactions and demonstrate how this data can be used to determine reaction kinetics.
8. Describe the characteristics of a batch reactor, a PFR, and a CMFR.
9. Develop an equation for and calculate the influent or effluent concentrations, volume, or flow rate from PFRs and CMFRs under steady or non–steady-state conditions, with or without reactions.
10. Explain why a PFR will have a lower effluent concentration than a CMFR if both are the same size and treating the same contaminant at the same flow rate.
11. Evaluate tracer test data, generating *C*, *E*, and *F* curves, and determine model parameters such as mean detention time and variance.
12. Assess whether a reactor exhibits poor or good mixing based on tracer test data using the t_{10}/τ value or tanks-in-series model.
13. Describe the significance of mass transfer in physical–chemical treatment processes.
14. Explain the concept of molecular diffusion and how the random motion of molecules can lead to mass transfer in a defined direction.
15. Calculate diffusion coefficients and mass transfer coefficients.
16. Identify variables that influence the rate of mass transfer, and predict changes in the rate of mass transfer when process conditions are changed.
17. Calculate the rate of mass transfer, given concentrations and other pertinent information about a system.
18. Explain the relationship between the concentration gradient and operating diagrams.

Homework Problems

4-1 Using the principles of stoichiometry, (a) balance the reaction for the coagulation of water with 50 mg/L of ferric sulfate, $Fe_2(SO_4)_3 \cdot 9H_2O$, shown below, (b) calculate the amount of $Fe(OH)_3$ precipitate formed in mg/L, and (c) calculate the amount of alkalinity consumed in meq/L if the alkalinity consumed is equal to the sulfate (SO_4^{2-}) generated:

$$Fe_2(SO_4)_3 \cdot 9H_2O \rightleftarrows Fe(OH)_3 + H^+ + SO_4{}^{2-} + H_2O$$

4-2 Using information obtained from your local water utility, compute the ionic strength of your drinking water. In addition, estimate the TDS concentration and electrical conductivity (EC) of the water. If available, measure the TDS and/or EC of the water and compare to the computed values.

4-3 Plot the activity coefficients of Na^+, Ca^{2+}, and Al^{3+} for ionic strengths from 0.001 M (very fresh water) to 0.5 M (seawater). Determine the ionic strength and TDS at which the activity coefficient corrections become important (activity coefficient less than 0.95) for monovalent, divalent, and trivalent ions.

4-4 Un-ionized ammonia (NH_3) is toxic to fish at low concentrations. The dissociation of ammonia in water has an equilibrium constant of $pK_a = 9.25$ and is described by the reaction

$$NH_4^+ \rightleftarrows NH_3 + H^+$$

Calculate and plot the concentrations of NH_3 and $NH_4{}^+$ at pH values between 6 and 10 if the total ammonia concentration ($NH_3 + NH_4{}^+$) is 1 mg/L as N.

4-5 A scrubber is used to remove sulfur dioxide (SO_2) from the flue gas from a coal-fired power plant. The scrubber works by spraying high-pH water downward through a tower while the flue gas passes upward, transferring the SO_2 from the gas to the water. The influent flue gas enters the tower at a rate of 50,000 m^3/h and contains 645 mg/m^3 of SO_2. The scrubber must reduce the SO_2 in the exhaust flue gas by 90 percent to meet emission requirements. The maximum possible concentration of SO_2 in the water is 820 mg/L. Calculate the required water flow rate to meet emission requirements. Assume there is no SO_2 in the influent water and the air and water flow rates do not change in the tower.

4-6 A rancher needs to provide water for his cattle, but the only water source is a brackish well that has a total dissolved solids concentration (TDS) of 4800 mg/L. The cattle need 400 L/d of water with TDS $<$ 1600 mg/L. The rancher has purchased a solar still that operates at 37 percent recovery of water (distillate) and 96 percent removal of dissolved solids. The rancher wants to recycle the blowdown from the still to a 20-m^3 feed tank to maximize his freshwater recovery and minimize the waste that has to be hauled off, but the still cannot operate effectively above 52,000 mg/L TDS in the blowdown because of scaling problems. The system will operate as shown in the following diagram.

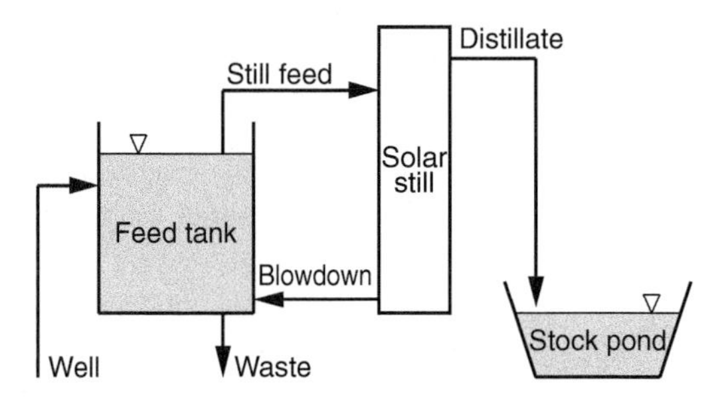

a. Prepare a table showing the flow rate and concentration of TDS in the (i) well, (ii) still feed, (iii) distillate, (iv) blowdown, and (v) waste. Explain all assumptions you make.

b. Propose a modification (i.e., using the existing equipment) that would decrease the waste that has to be hauled off, and determine how much reduction in waste flow this modification would achieve.

4-7 The following time and concentration data were measured in a batch reactor. For the specified data set (to be selected by instructor), determine the reaction order that yields the best fit and estimate the rate constant for the reaction.

	Concentration, mg/L				
Time, min	A	B	C	D	E
0	40.0	1.18	120.0	120.0	20.0
1	31.5	1.11	36.1	51.0	9.52
2	21.5	1.06	21.5	24.0	6.38
3	17.9	1.00	16.3	8.7	4.27
4	12.2	0.93	11.5	4.1	3.96
5	10.1	0.92	9.3	1.8	3.11
6	6.84	0.81	7.8	0.55	2.65
7	5.25	0.76	6.9	0.35	2.25
8	4.30	0.73	5.9	0.096	2.15
9	2.95	0.66	5.4	0.052	1.97
10	2.42	0.59	4.9	0.022	1.70

4-8 Calculate the hydraulic residence time and volume of a PFR required to achieve the given effluent concentration for the reaction given below (to be selected by the instructor).

a. $Q = 38$ ML/d, $C_I = 100$ μg/L, $C_E = 2.0$ μg/L, first-order decay reaction, $k = 0.375$ min^{-1}.

b. $Q = 190$ ML/d, $C_I = 15$ mg/L, $C_E = 1.8$ mg/L, first-order decay reaction, $k = 0.057$ min^{-1}.

c. $Q = 5000$ m^3/d, $C_I = 55$ μg/L, $C_E = 21$ μg/L, first-order decay reaction, $k = 0.0086$ s^{-1}.

d. $Q = 5000$ m^3/d, $C_I = 55$ μg/L, $C_E = 21$ μg/L, second-order decay reaction, $k = 0.0075$ L/mg·min.

e. $Q = 3.30$ m^3/s, $C_I = 1.25$ mg/L, $C_E = 0.045$ mg/L, second-order decay reaction, $k = 0.0936$ L/mg·s.

4-9 For the system given in Problem 4-8 (to be selected by the instructor), calculate the hydraulic residence time and volume if the reactor is a CMFR.

4-10 For the given problem below (to be selected by the instructor), calculate the effluent concentration from the following reactor or system of reactors:

a. A CMFR with a volume of 125 m^3, treating a flow rate of 20 ML/d that has an influent concentration of 100 μg/L of a contaminant that degrades as a second-order reaction with a rate constant of 0.51 L/mg·min.

b. A PFR with a volume of 50 m^3, treating a flow rate of 15.2 ML/d that has an influent concentration of 60 mg/L of a contaminant that degrades as a first-order reaction with a rate constant of 0.426 min^{-1}.

c. A laboratory CMFR with a volume of 4 L, treating a flow rate of 350 mL/min that has an influent concentration of 1.0 g/L of a contaminant that degrades as a first-order reaction with a rate constant of 0.0817 s^{-1}.

d. A pipeline 2 m in diameter and 100 m long (which behaves as a PFR), treating a flow rate of 380 ML/d that has an influent concentration of 80 mg/L of a contaminant that degrades as a second-order reaction with a rate constant of 0.36 L/mg·min.

e. A series of two reactors consisting of a PFR followed by a CMFR (analogous to a pipeline followed by a storage tank) treating a flow rate of 500 m^3/d that has an influent concentration of 250 μg/L of a contaminant that degrades as a first-order reaction with a rate constant of 0.01 s^{-1}. The PFR has a volume of 2 m^3 and the CMFR has a volume of 4 m^3.

f. A real reactor with hydraulic performance equivalent to 9 tanks in series according to the TIS model, with a volume of 125 m^3, treating a flow rate of 20 ML/d that has an influent concentration of 100 μg/L of a contaminant that degrades as a first-order reaction with a rate constant of 0.51 min^{-1}.

4-11 Compare the size of a CMFR and a PFR to achieve 50 percent removal of a contaminant, given a flow rate of 10^4 m^3/d and a first-order rate constant of -0.4 h^{-1}. Repeat for 99 percent removal. Comment on the relative efficiency of each type of reactor and the situations where each type of reactor may be useful.

4-12 The following concentration data expressed in mg/L were obtained from tracer studies conducted on five different reactors. For a given reactor (to be selected by the instructor), plot the tracer curve, the E curve, and the F curve, and determine the hydraulic residence time, mean residence time, variance of the residence time distribution and the mass and percent of the dye recovered.

	Reactor A	Reactor B	Reactor C	Reactor D	Reactor E
Reactor volume (m^3)	4000	4200	1450	304.7	682
Plant flow rate (ML/d)	70	100	25	4.16	12.5
Mass of dye (kg)	39.1	60	10	2.23	3.75

Time, min	Reactor A	Reactor B	Reactor C	Reactor D	Reactor E
0	0	0	0	0	0
10	0	2	0	0	0
20	1	5.4	0	0	0
30	2	8.4	0.1	0	0
40	5.1	11.4	0.2	0	0
50	8.9	13	0.5	0	2
60	11.2	12.1	6.3	0	6.2
70	10.5	9.3	15.2	4.5	13
80	9.2	7.2	18.1	9	10.4
90	8	5.2	8.5	14.1	5.1
100	6.5	3.6	3.2	15.6	2.8
110	5	2.5	1.8	12.9	1.1
120	3.5	1.4	1.2	9.2	0.5
130	2	0.9	0.8	5.3	0.4
140	1.4	0.4	0.6	2.3	0.1
150	0.8	0.1	0.3	1.1	0
160	0.4	0	0.2	0.8	0
170	0.2	0	0.2	0.5	0
180	0	0	0.1	0.2	0
190	0	0	0	0.1	0
200	0	0	0	0	0

4-13 Using the tracer data for the reactor in Problem 4-12 (to be selected by instructor), determine the t_{10}/τ value and the equivalent number of tanks for the TIS model for the selected reactor.

4-14 Using the tracer data for the reactor in Problem 4-12 (to be selected by instructor) and the equivalent number of tanks from Problem 4-13, calculate the expected effluent concentration using the TIS model assuming a first-order reaction rate constant $k = 0.085\ \text{min}^{-1}$ and influent concentration of 1 mg/L.

4-15 For an ideal reactor with the same hydraulic residence time as the reactor in Problem 4-12 (to be selected by instructor), calculate the expected effluent concentration assuming a first-order reaction rate constant $k = 0.085\ \text{min}^{-1}$ if the influent concentration is 1 mg/L and the reactor is (a) a PFR and (b) a CMFR. Compare your answers to the result from Problem 4-14.

4-16 Calculate the diffusion coefficient for the following compound in water at 20°C (to be selected by instructor):

a. Trichloroethylene (TCE)

b. Trichloromethane

c. Toluene

d. Sodium bicarbonate

e. Sodium sulfate

f. Barium chloride

4-17 Calculate the diffusion coefficient for the following compound in air at 20°C (to be selected by instructor).

a. Tetrachloroethene

b. Benzene

c. Vinyl chloride

4-18 A 150-mm ID potable water distribution pipe has water flowing at a velocity of 1.52 m/s. The water entering the pipe contains 1.1 mg/L of chlorine. The walls of the pipe are covered with an aggressive biofilm that completely consumes the chlorine (i.e., the chlorine concentration at the pipe wall is zero). Calculate the chlorine consumption rate by a 1.0 km length of pipe, using the Gilliland correlation for the mass transfer coefficient. The water temperature is 25°C. For the purposes of this calculation, assume that the 1.1 mg/L chlorine concentration is maintained constant in the bulk water through the entire length of pipe. Based on your calculated results, is this assumption reasonable? (When is an assumption reasonable? What makes a good assumption, anyway?) If the assumption is not reasonable, how would you have to modify your approach to solve the problem correctly?

4-19 Raw water that has an influent pH of 2.8 is to be fed to a packed bed of crushed limestone ($CaCO_3$) to raise the pH and add hardness (as Ca^{2+}). The temperature is 25°C, the bed porosity is 0.5, and the particle sphericity is 0.75. Calculate the mass transfer coefficient for limestone media for 0.5-, 1.5-, 2-, or 3-cm limestone particles (particle size to be specified by instructor). The flow rate is 800 L/min and the superficial velocity is 10 m/h.

References

Benefield, L. D., Judkins, J. F., and Weand, B. L. (1982) *Process Chemistry for Water and Wastewater Treatment*, Prentice-Hall, Englewood Cliffs, NJ.

Benjamin, M. M. (2002) *Water Chemistry*, McGraw-Hill, New York.

CRC (2003) *CRC Handbook of Chemistry and Physics*, 84th ed., CRC Press, Boca Raton, FL.

Crittenden, J. C., Trussell, R. R., Hand, D. W., Howe, K. J., and Tchobanoglous, G. (2012) *MWH's Water Treatment: Principles and Design*, 3rd ed., Wiley, Hoboken, NJ.

LeBas, G. (1915) *The Molecular Volumes of Liquid Chemical Compounds*, Longmans, London.

Lyman, W. J., Reehl, W. F., and Rosenblatt, D. H. (1990) *Handbook of Chemical Property Estimation Methods: Environmental Behavior of Organic Compounds*, American Chemical Society, Washington, DC.

McCabe, W. L., and Thiele, E. W. (1925) "Graphical Design of Fractionating Columns," *Ind. Eng. Chem.*, **17**, 6, 605–611.

Malik, V. K., and Hayduk, W. (1968) "A Steady-State Capillary Cell Method for Measuring Gas–Liquid Diffusion Coefficients," *Canadian J. Chem. Eng.*, **46**, 6, 462–466.

Marrero, T. R., and Mason, E. A. (1972) "Gaseous Diffusion Coefficients," *J. Phys. Chem. Ref. Data*, **1**, 1, 3–118.

Pankow, J. F. (1991) *Aquatic Chemistry Concepts*, Lewis, Chelesa, MI.

Poling, B. E., Prausnitz, J. M., and O'Connell, J. P. (2001) *The Properties of Liquids and Gases*, 5th ed., McGraw-Hill, New York.

Robinson, R. A., and Stokes, R. H. (1959) *Electrolyte Solutions: The Measurement and Interpretation of Conductance, Chemical Potential and Diffusion in Solutions of Simple Electrolytes*, 2nd ed., Butterworths, London.

Sawyer, C. N., McCarty, P. L., and Parkin, G. F. (2003) *Chemistry for Environmental Engineering*, 5th ed., McGraw-Hill, New York.

Snoeyink, V. L., and Jenkins, D. (1980) *Water Chemistry*, Wiley, New York.

Stumm, W., and Morgan, J. J. (1996) *Aquatic Chemistry: Chemical Equilibria and Rates in Natural Waters*, 3rd ed., Wiley, New York.

Thompson, A. and Taylor, B. N. (2008) Guide for the Use of the International System of Units (SI), NIST Special Publication 811, 2008 Edition. Available at: http://physics.nist.gov/cuu/Units/index.html.

5 Coagulation and Flocculation

Natural surface waters contain inorganic and organic particles. Inorganic particles, including clay, silt, and mineral oxides, typically enter surface water by natural erosion processes. Organic particles may include viruses, bacteria, algae, protozoan cysts and oocysts, as well as detritus litter that have fallen into the water source. In addition, surface waters will contain very fine colloidal and dissolved organic constituents such as humic acids, a product of decay and leaching of organic debris. Particulate and dissolved organic matter is often identified as natural organic matter (NOM).

Removal of particles is required because they can (1) reduce the clarity of water to unacceptable levels (i.e., cause turbidity) as well as impart color to water (aesthetic reasons), (2) be infectious agents (e.g., viruses, bacteria, and protozoa), and (3) have toxic compounds adsorbed to their external surfaces. The removal of dissolved NOM is of importance because many of the constituents that comprise dissolved NOM are precursors to the formation of disinfection by-products (see Chap. 13) when chlorine is used for disinfection. NOM can also impart color to the water.

The most common method used to remove particulate matter and a portion of the dissolved NOM from surface waters is by sedimentation and/or filtration following the conditioning of the water by coagulation and flocculation, the subject of this chapter. Thus the purpose of this chapter is to present the chemical and physical basis for the phenomena occurring in the coagulation and flocculation processes. Specific topics addressed in this chapter include the role of coagulation and flocculation processes in water treatment, the basis for stability of particles in water, principles and design of coagulation processes, and principles and design of flocculation processes.

5-1 Role of Coagulation and Flocculation in Water Treatment

The importance of the coagulation and flocculation processes in water treatment can be appreciated by reviewing the process flow diagram illustrated on Fig. 5-1. As used in this book, *coagulation* involves the addition of a chemical coagulant or coagulants for the purpose of conditioning the suspended, colloidal, and dissolved matter for subsequent processing by flocculation or to create conditions that will allow for the subsequent removal of particulate and dissolved matter. *Flocculation* is the aggregation of destabilized particles (particles from which the electrical surface charge has been reduced) and precipitation products formed by the addition of coagulants into larger particles known as flocculant particles or, more commonly, "floc." The aggregated floc can then be removed by gravity

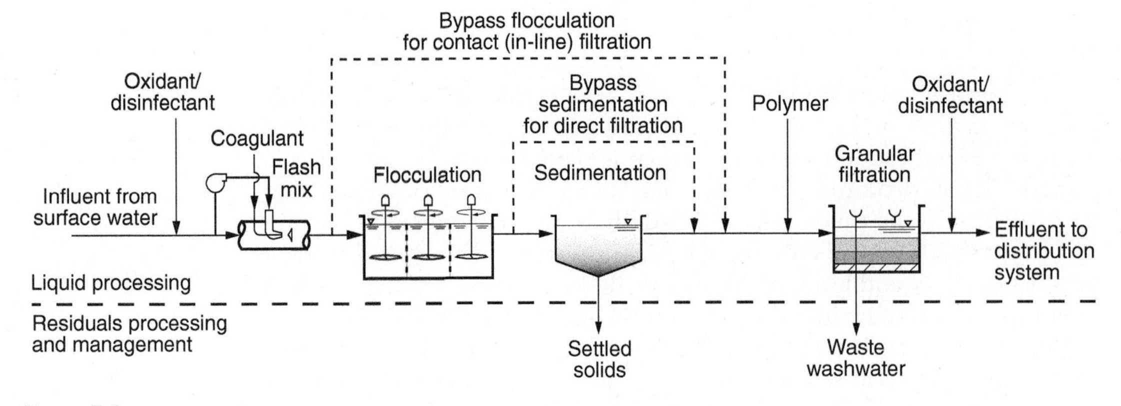

Figure 5-1
Typical water treatment process flow diagram employing coagulation (chemical mixing) with conventional treatment, direct filtration, or contact filtration.

sedimentation and/or filtration. An overview of the coagulation and flocculation processes is provided below.

Coagulation Process

The objective of the coagulation process depends on the water source and the nature of the suspended, colloidal, and dissolved organic constituents. Coagulation by the addition of chemicals such as alum and iron salts and/or organic polymers can involve:

1. destabilization of small suspended and colloidal particulate matter
2. adsorption and/or reaction of portions of the colloidal and dissolved NOM to particles
3. creation of flocculant precipitates that sweep through the water enmeshing small suspended, colloidal, and dissolved material as they settle

Coagulants such as aluminum sulfate (alum), ferric chloride, and ferric sulfate hydrolyze rapidly when mixed with the water to be treated. As these chemicals hydrolyze, they form insoluble precipitates that destabilize particles by adsorbing to the surface of the particles and neutralizing the charge (thus reducing the repulsive forces). Natural or synthetic organic polyelectrolytes (polymers with multiple charged functional groups) are also used for particle destabilization. Because of the many competing reactions, the theory of chemical coagulation is complex. Thus, the simplified reactions presented in this and other textbooks to describe the various coagulation processes can only be considered approximations, as the reactions may not necessarily proceed exactly as indicated.

Flocculation Process

The purpose of flocculation is to produce particles, by means of aggregation, that can be removed by subsequent particle separation procedures such as gravity sedimentation and/or filtration. Two general types of flocculation can be identified: (1) microflocculation (also known as perikinetic flocculation) in which particle aggregation is brought about by the random thermal motion of fluid molecules (known as Brownian motion, see Sec. 4-14) and (2) macroflocculation (also known as orthokinetic flocculation) in which particle aggregation is brought about by inducing velocity gradients and gentle mixing in the fluid containing the particles. Mixing for flocculation generally lasts for 20 to 40 min. Another form of macroflocculation is brought about by differential settling in which large particles overtake small particles to form larger particles. The aggregated particles form large masses of loosely bound particles known as floc, and this floc is sufficiently large that it will settle relatively rapidly or be easier to remove from water by filtration.

Practical Design Issues

When it comes to the design of coagulation and flocculation facilities, engineers must consider four process issues: (1) the type and concentration of coagulants and flocculant aids, (2) the mixing intensity and the method used to disperse chemicals into the water for destabilization, (3) the mixing intensity and time for flocculation, and (4) the selection of the liquid–solid separation process (e.g., sedimentation and filtration). With the exception of sedimentation (considered in Chap. 6), and filtration (considered in Chaps. 7 and 8), these subjects are addressed in the subsequent sections of this chapter.

5-2 Stability of Particles in Water

Particles in water may, for practical purposes, be classified as suspended and colloidal, according to particle size, where colloidal particles are those that are smaller than about 1 μm. Small suspended and colloidal particles and dissolved constituents will not settle in a reasonable period of time. Particles that won't settle are stable particles and chemicals must be used to help remove them. To appreciate the role of chemical coagulants, it is important to understand particle–water interactions and the electrical properties of particles in water. These subjects along with the nature of particle stability and the compression of the electrical double layer are considered in this section.

Particle–Solvent Interactions

Particles in natural water can be classified as hydrophobic (water repelling) and hydrophilic (water attracting). Hydrophobic particles have a well-defined interface between the water and solid phases and have a low affinity for water molecules. In addition, hydrophobic particles are thermodynamically unstable and will aggregate over time.

Hydrophilic particles such as clays, metal oxides, proteins, or humic acids have polar or ionized surface functional groups. Many inorganic particles in natural waters, including hydrated metal oxides (iron or aluminum oxides), silica (SiO_2), and asbestos fibers, are hydrophilic because water molecules will bind to the polar or ionized surface functional groups (Stumm and Morgan, 1996). Many organic particles are also hydrophilic and include a wide diversity of biocolloids (humic acids, viruses) and suspended living or dead microorganisms (bacteria, protozoa, algae). Because biocolloids can adsorb on the surfaces of inorganic particles, the particles in natural waters often exhibit heterogeneous surface properties.

Electrical Properties of Particles

The principal electrical property of fine particles in water is surface charge, which contributes to relative stability, causing particles to remain in suspension without aggregating for long periods of time. Given sufficient time, colloids and fine particles will flocculate and settle, but this process is not economically feasible because it is very slow. A review of the causes of particle stability will provide an understanding of the techniques that can be used to destabilize particles, which are discussed in the following section.

ORIGIN OF PARTICLE SURFACE CHARGE

Most particles have complex surface chemistry and surface charges may arise from several sources. Surface charge arises in four principal ways, as discussed below (Stumm and Morgan, 1996).

Isomorphous Replacement (Crystal Imperfections)

Under geological conditions, metals in metal oxide minerals can be replaced by metal atoms with lower valence, and this will impart a negative charge to the crystal material. An example where an aluminum atom replaced a silicon atom in a silica particle is shown on Fig. 5-2. This process, known as isomorphous replacement, produces negative charges on the surface of clay particles.

Structural Imperfections

In clay and similar mineral particles, imperfections that occur in the formation of the crystal and broken bonds on the crystal edge can lead to the development of surface charges.

Preferential Adsorption of Specific Ions

Particles adsorb NOM, and these large macromolecules typically have a negative charge because they contain carboxylic acid groups:

$$\text{R–COOH} \rightleftarrows \text{R–COO}^- + \text{H}^+ \left(\text{p}K_a = 4 \text{ to } 5\right) \tag{5-1}$$

Consequently, particle surfaces that have adsorbed NOM will be negatively charged for pH values greater than ~5.

Ionization of Inorganic Surface Functional Groups

Many mineral surfaces contain surface functional groups (e.g., hydroxyl) and their charge depends on pH. For example, silica has hydroxyl groups on its exterior surface, and these can accept or donate protons as shown here:

$$\begin{array}{ccccc} \text{Si–OH}_2^+ & \rightleftarrows & \text{Si–OH} + \text{H}^+ & \rightleftarrows & \text{Si–O}^- + 2\text{H}^+ \\ \text{pH} \ll 2 & & \text{pH} = 2 & & \text{pH} \gg 2 \end{array} \tag{5-2}$$

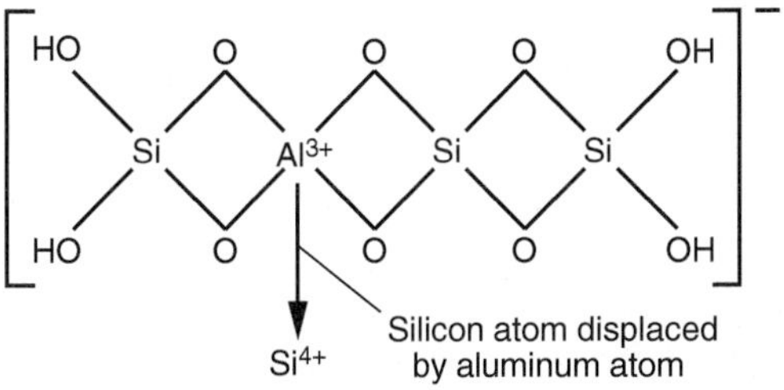

Figure 5-2
Charge acquisition through isomorphous substitution of Al for Si. Since the silicon has a charge of 4 and the aluminum has a charge of 3, the replacement with an aluminum atom leave the crystal with less positive charge.

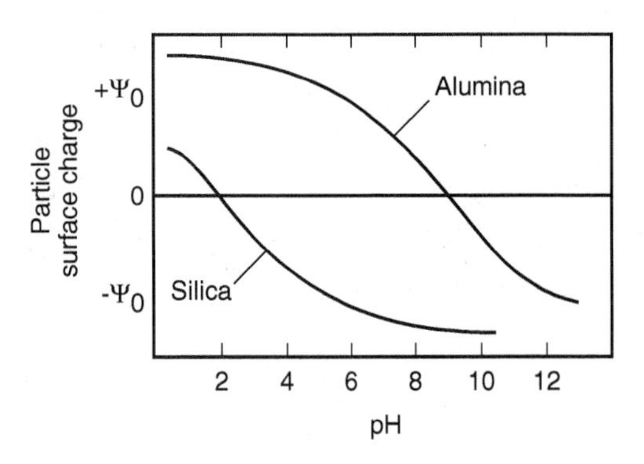

Figure 5-3
Variation in particle charge with pH.

The pH corresponding to a surface charge of zero is defined as the *zero point of charge* (ZPC). Above the ZPC the surface charge will be negative (anionic), and below the ZPC the charge will be positive (cationic). The zero point of charge, as shown on Fig. 5-3, for silica is at pH 2, whereas the zero point of charge for alumina is about pH 9. The ZPC for other particles that commonly occur in water are listed in Table 5-1. Many of the measurements reported in Table 5-1 are in low-ionic-strength waters (i.e., distilled water); consequently, the reported pH_{ZPC} values are higher than are observed in natural waters.

ELECTRICAL DOUBLE LAYER

In natural waters, the processes described above nearly always result in a negative surface charge on particles. Negatively charged particles accumulate positive counterions on and near the particle's surface to satisfy electroneutrality. As shown on Fig. 5-4, a layer of cations will bind tightly to the surface of a negatively charged particle to form a fixed adsorption layer. This adsorbed layer of cations, bound to the particle surface by electrostatic and adsorption forces, is about 0.5 nm thick and is known as the *Helmholtz*

Table 5-1
Surface characteristics of inorganic and organic particles commonly found in natural waters

Type of Particle	Zero Point of Charge, pH_{ZPC}
Inorganic	
$Al(OH)_3$ (amorphous)	7.5–8.5
Al_2O_3	9.1
CuO_3	9.5
$Fe(OH)_3$ (amorphous)	8.5
MgO	12.4
MnO_2	2–4.5
SiO_2	2–3.5
Clays	
Kaolinite	3.3–4.6
Montmorillonite	2.5
Organic	
Algae	3–5
Bacteria	2–4
Humic acid	3

Source: Adapted from Parks (1967) and Stumm and Morgan (1996).

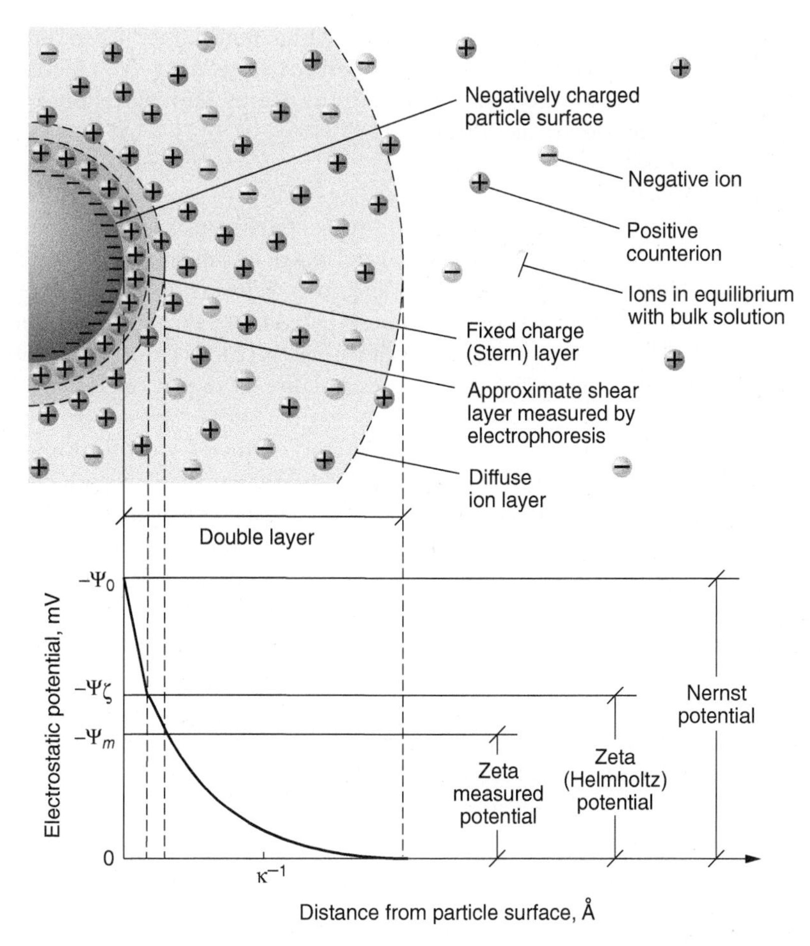

Figure 5-4
Structure of the electrical double layer. It should be noted that the potential measured at the shear plane is known as the zeta potential. The shear plane typically occurs in the diffuse layer.

layer (also known as the *Stern layer* after Stern, who proposed the model shown on Fig. 5-4.). Beyond the Helmholtz layer, a net negative charge and electric field is present that attracts an excess of cations (over the bulk solution concentration) and repels anions, neither of which are in a fixed position. These cations and anions move about under the influence of diffusion (caused by collisions with solvent molecules), and the excess concentration of cations extends out into solution until all the surface charge and electric potential is eliminated and electroneutrality is satisfied.

The layer of cations and anions that extends from the Helmholtz layer to the bulk solution where the charge is zero and electroneutrality is satisfied is known as the diffuse layer. Taken together the adsorbed (Helmholtz) and diffuse layer are known as the *electric double layer* (EDL). Depending on the solution characteristics, the EDL can extend up to 30 nm into the solution. Techniques have been developed for measuring the electrical properties of particles and particle systems and they have been presented in detail (Crittenden et al., 2012).

ZETA POTENTIAL

When a charged particle is subjected to an electric field between two electrodes, a negatively charged particle will migrate toward the positive electrode, as shown on Fig. 5-5, and vice versa. This movement is termed *electrophoresis*. It should be noted that when a particle moves in an electrical field some portion of the water near the surface of the particle moves with it, which gives rise to the shear plane, as shown on Fig. 5-4. Typically, as shown on Fig. 5-4, the actual shear plane lies in the diffuse layer to the right of the theoretical fixed shear plane defined by the Helmholtz layer. The electrical potential between the actual shear plane and the bulk solution is called the *zeta potential*. Zeta potential can be measured to give an indication of particle stability; particles tend to be stable when the zeta potential is above 20 mV and unstable when the zeta potential is below that value.

Particle Stability

The stability of particles in natural waters depends on a balance between the repulsive electrostatic force of the particles and the attractive forces known as the van der Waals forces. Since particles in water have a net negative surface charge, the principal mechanism controlling particle stability is electrostatic repulsion.

Van der Waals forces originate from magnetic and electronic resonance that occurs when two particles approach one another. This resonance is

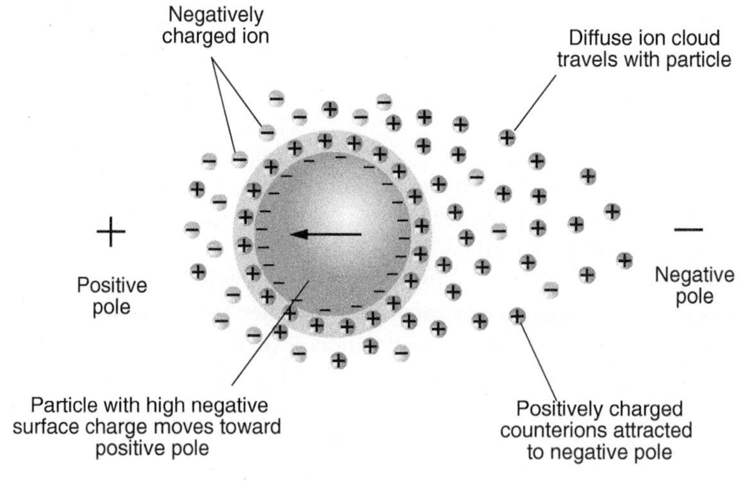

Figure 5-5
Schematic illustration of electrophoresis in which a charged particle moves in an electrical field, dragging with it a cloud of ions.

caused by electrons in atoms on the particle surface, which develop a strong attractive force between the particles when these electrons orient themselves in such a way as to induce synergistic electric and magnetic fields. Van der Waals attractive forces ($<\sim 20$ kJ/mol) are strong enough to overcome electrostatic repulsion, but they are unable to do so because electrostatic repulsive forces and the EDL extend further into solution than do the van der Waals forces. As a result, the electrostatic repulsion represents an energy barrier that must be overcome for particles to be destabilized.

Particle–particle interactions are extremely important in bringing about aggregation by means of Brownian motion. The theory of particle–particle interaction is based on the interaction of the repulsive and attractive forces on two charged particles as they are brought closer and closer together. The theory, first worked out by Derjaguin, later improved upon together with Landau, and subsequently extended by Verwey and Overbeek, is known as the DLVO theory after the scientists who developed it.

A conceptual diagram of the DLVO model is provided on Fig. 5-6 in which the interaction between two particles represented by flat plates with similar charge is illustrated. As shown on Fig. 5-6, the two principal forces involved are the forces of repulsion due to the electrical properties of the charged plates and the van der Waals forces of attraction. Two cases are illustrated on Fig. 5-6 with respect to the forces of repulsion. In the first case, the repulsive force extends $4/\kappa$ into solution where κ is the double-layer thickness. In the second case, the extent of the repulsive force is reduced considerably and the repulsive forces only extend about $2/\kappa$ into solution. The net total energy shown by the solid lines on Fig. 5-6 is the difference between the forces of repulsion and attraction.

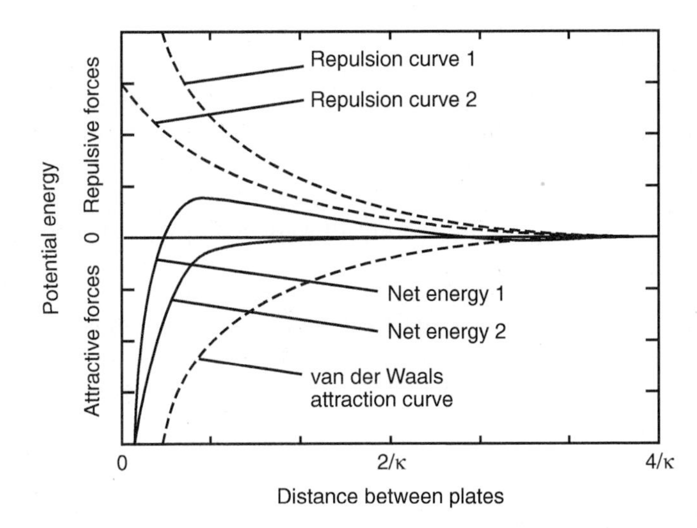

Figure 5-6
Attractive and repulsive potential energy that result when two particles are brought together.

For case 1, the forces of attraction will predominate at very short and long distances. The net energy curve for condition 1 contains a repulsive maximum that must be overcome if the particles are to be held together by the van der Waals force of attraction. Although floc particles can form at long distances as shown by the net energy curve for case 1, the net force holding these particles together is weak and the floc particles that are formed can be ruptured easily. In case 2, there is no energy barrier to overcome and particles can contact each other relatively easily.

Destabilization

If the repulsive energy barrier were not present, the attractive van der Waals forces would cause any particles that came near each other to stick to each other. The process of flocculation introduced at the beginning of the chapter provides the mechanisms for particles to come near each other. Particles will collide with each other due to the random movement of particles caused by Brownian motion (microflocculation) or due to gentle mixing of the water (macroflocculation). Particles that stick to each other form aggregations of larger particles, and these larger particles will settle relatively rapidly or be easier to remove by filtration. Eliminating the repulsive forces, then, is an essential step in the removal of colloids from water. Reducing or eliminating the repulsive forces so that particles have the opportunity to stick to each other is known as *destabilization.*

One method of destabilizing particles is to compress the double layer so that it does not reach as far from the particle surface. As noted earlier, the double layer forms to counteract the negative surface charge of particles and satisfy electroneutrality. If more ions are in solution or if the ions have greater charge (divalent or trivalent instead of monovalent), then electroneutrality can be satisfied in a shorter distance. The DVLO theory mentioned earlier accurately predicts that an increase in ionic strength or ion valence can compress the EDL thickness sufficiently to allow van der Waals forces to extend further than the EDL, resulting in destabilized particles that will flocculate. The effect of ionic strength explains why particles are stable in freshwater (low ionic strength, EDL extends beyond van der Waals forces) and flocculate rapidly in saltwater (high ionic strength, compressed EDL), such that a river that flows into the sea will drop sediment close to its mouth even though the turbulence of wave action should keep particles suspended.

Unfortunately, reducing the thickness of the EDL by adding salt to increase the ionic strength is not a practical method for destabilizing particles in drinking water treatment because the required ionic strengths are greater than are considered acceptable in potable water. Coagulating chemicals must be added to destabilize particles, as described in the following section.

5-3 Principles of Coagulation

The electrical properties of particles were considered in the previous section. Coagulation, as described in Sec. 5-1, is the process used to destabilize the particles found in waters so that they may be removed by subsequent separation processes. The purpose of this section is to introduce the principal coagulation mechanisms responsible for particle destabilization and removal. Coagulation practice including the principal chemicals used for coagulation in water treatment and jar testing is presented and discussed in Sec. 5-4.

Mechanisms that can be exploited to achieve particle destabilization include (1) compression of the electrical double layer, (2) adsorption and charge neutralization, (3) adsorption and interparticle bridging, and (4) enmeshment in a precipitate, or "sweep floc." While these mechanisms are discussed separately here, destabilization strategies often exploit several mechanisms simultaneously. It should also be noted that compression of the electrical double layer, discussed in the previous section, is considered a coagulation mechanism but is not discussed here because increasing the ionic strength is not practical in water treatment.

Adsorption and Charge Neutralization

Particles can be destabilized by adsorption of oppositely charged ions or polymers. Most particles in natural waters are negatively charged (clays, humic acids, bacteria) in the neutral pH range (pH 6 to 8); consequently, positively charged hydrolyzed metal salts, prehydrolyzed metal salts, and cationic organic polymers can be used to destabilize particles through neutralizing the charge on the particle surface. If the particle surface has no net charge, the EDL will not exist and van der Waals forces can cause particles to stick together.

Adsorption and Interparticle Bridging

Polymer bridging is complex and has not been adequately described analytically. Schematically, polymer chains adsorb on particle surfaces at one or more sites along the polymer chain as a result of (1) coulombic (charge–charge) interactions, (2) dipole interaction, (3) hydrogen bonding, and (4) van der Waals forces of attraction (Hunter, 2001). The rest of the polymer may remain extended into the solution and adsorb on available surface sites of other particles, thus creating a "bridge" between particle surfaces that results in a larger particle that can settle more efficiently. Polymer bridging is an adsorption phenomenon; consequently, the optimum coagulant dose will generally be proportional to the concentration of particles present. Adsorption and interparticle bridging occur with nonionic polymers and high-molecular-weight (MW 10^5 to 10^7 g/mol), low-surface-charge polymers. High-molecular-weight cationic polymers have a high charge density to neutralize surface charge.

Precipitation and Enmeshment

When high enough dosages are used, aluminum and iron form insoluble precipitates and particles become entrapped in the amorphous precipitates. This type of destabilization has been described as *precipitation and enmeshment* or *sweep floc*. Although the molecular events leading to sweep floc have not been defined clearly, the steps for iron and aluminum salts at lower coagulant dosages are as follows: (1) hydrolysis and polymerization of metal ions, (2) adsorption of hydrolysis products at the particle surface interface, and (3) charge neutralization. At high dosages, it is likely that nucleation of the precipitate occurs on the surface of particles, leading to the growth of an amorphous precipitate with the entrapment of particles in this amorphous structure. This mechanism predominates in water treatment applications where pH values are generally maintained between pH 6 and 8, and aluminum or iron salts are used at concentrations exceeding saturation with respect to the amorphous metal hydroxide solid that is formed.

5-4 Coagulation Practice

Selection of the type and dose of coagulant depends on the characteristics of the coagulant, the concentration and type of particles, concentration and characteristics of NOM, water temperature, and water quality. Presently, the interdependence of these five parameters is only understood qualitatively, and prediction of the optimum coagulant combination from characteristics of the particles and the water quality is not yet possible. The purpose of this section is to introduce coagulation practice, including the types of inorganic and organic coagulants and coagulant aids used, and alternative techniques used to reduce coagulant dosages.

Inorganic Metallic Coagulants

The principal inorganic coagulants used in water treatment are sulfide or chloride salts of aluminum and ferric ions and prehydrolyzed salts of these metals. These hydrolyzable metal cations are readily available in both liquid and solid (dry) form. In the United States, the predominant water treatment coagulant is aluminum sulfate, or "alum," sold in a hydrated form as $Al_2(SO_4)_3 \cdot xH_2O$, where x is usually about 14. The action, solubility, and application of these coagulants are considered in the following discussion.

COMPLEXATION AND DEPROTONATION OF ALUMINUM AND IRON SALTS

When ferric or aluminum ions are added to water, a number of parallel and sequential reactions occur. Initially, when a salt of Al(III) and Fe(III) is added to water, it will dissociate to yield trivalent Al^{3+} and Fe^{3+} ions, as given below:

$$Al_2(SO_4)_3 \rightleftarrows 2Al^{3+} + 3SO_4^{2-} \tag{5-3}$$

$$FeCl_3 \rightleftarrows Fe^{3+} + 3Cl^- \tag{5-4}$$

The trivalent ions of Al^{3+} and Fe^{3+} then hydrate to form the aquometal complexes $Al(H_2O)_6^{3+}$ and $Fe(H_2O)_6^{3+}$, as shown on the left-hand side of Eq. 5-5. As shown, the metal ion has a coordination number of 6 and six water molecules orient themselves around the metal ion. The complexed water molecules then often lose protons as shown in the following reaction:

$$\begin{bmatrix} H_2O & & OH_2 \\ H_2O{-}Me{-}OH_2 & & \\ H_2O & & OH_2 \end{bmatrix}^{3+} \rightarrow \begin{bmatrix} H_2O & & OH \\ H_2O{-}Me{-}OH_2 & & \\ H_2O & & OH_2 \end{bmatrix}^{2+} + H^+ \tag{5-5}$$

These aquometal complexes then successively lose additional protons to form a variety of soluble mononuclear [$Al(H_2O)_5(OH)^{2+}$, $Al(H_2O)_4(OH)_2^+$, $Al(H_2O)_3(OH)_3^0$] and polynuclear [$Al_{18}(OH)_{20}^{4+}$, $(Al_8(OH)_{20}{\cdot}28H_2O)^{4+}$] species. Similarly, iron forms a variety of soluble species, including mononuclear species such as $Fe(H_2O)_5(OH)^{2+}$ [or just $Fe(OH)^{2+}$] and $Fe(H_2O)_4(OH)_2^+$ [or just $Fe(OH)_2^+$]. It should be noted that all of these mononuclear and polynuclear species can interact with the particles in water, depending on the characteristics of the water and the number of particles. Unfortunately, it is difficult to control and know which mononuclear and polynuclear species are operative. As will be discussed later, this uncertainty has led to the development of prehydrolized metal salt coagulants.

SOLUBILITY OF ALUMINUM AND FERRIC HYDROXIDE

The solubility of the various alum [Al(III)] and iron [Fe(III)] species are illustrated on Figs. 5-7a and 5-7b, respectively, in which the log molar concentrations have been plotted versus pH. The equilibrium diagrams shown on Figs. 5-7a and 5-7b were created using equilibrium constants for the major hydrolysis reactions that have been estimated after approximately 1 h of reaction time (upper limit of coagulation/flocculation detention times). Aluminum hydroxide and ferric hydroxide are precipitated within the shaded region, and polynuclear and polymeric species are formed outside of the shaded region at higher and lower pH values. It should also be noted that the structure of the precipitated iron is far more compact and inert as compared to the amorphous nature of precipitated aluminum.

In most water treatment applications for removal of turbidity, disinfection by-product precursors (NOM), and color, the pH during coagulation ranges between 6 and 8. The lower limit is imposed to prevent accelerated corrosion rates that occur at pH values below pH 6. The darker shaded areas corresponding to sweep coagulation region shown on Figs. 5-7a and 5-7b

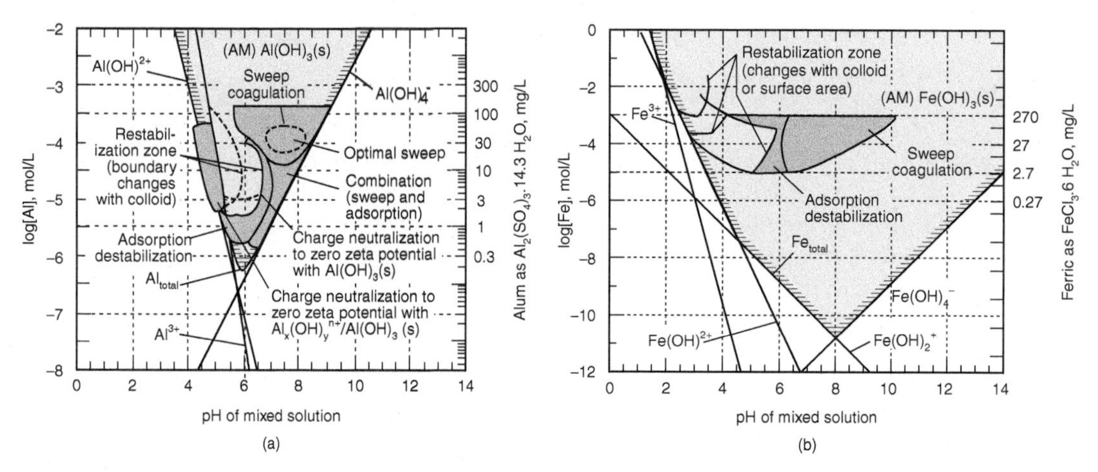

Figure 5-7
Solubility diagram for (a) Al(III) and (b) Fe(III) at 25°C. Only the mononuclear species have been plotted. The metal species are assumed to be in equilibrium with the amorphous precipitated solid phase. Typical operating ranges for coagulants: (a) alum and (b) iron. [Adapted from Amirtharajah and Mills (1982).]

correspond to the operating pH and dosage ranges that are normally used in water treatment when alum and iron are used in the sweep floc mode of operation. For example, the operating region for aluminum hydroxide precipitation is in a pH range of 7.0 to 8.0 and an alum dose from 20 to 60 mg/L. The minimum alum solubility occurs at a pH of about 6.2 at 25°C. The importance of pH in controlling the concentration of soluble metal species that will pass through the treatment process is illustrated on Figs. 5-7a and 5-7b. Comparing the solubility of alum and ferric species, the ferric species are more insoluble than aluminum species and are also insoluble over a wider pH range. Thus, ferric ion is often the coagulant of choice to aid destabilization in the lime-softening process, which is carried out at higher pH values (pH 9).

STOICHIOMETRY OF METAL ION COAGULANTS

When alum is added to water and aluminium hydroxide precipitates, the overall reaction is

$$Al_2(SO_4)_3{\cdot}14H_2O \rightarrow 2Al(OH)_3(s) + 6H^+ + 3SO_4^{2-} + 8H_2O \qquad (5\text{-}6)$$

After $Al(OH)_3$ precipitates, the species left in water is the same as if H_2SO_4 had been added to the water. Thus, adding alum is like adding a strong acid. A strong acid will lower the pH and consume alkalinity. The change in pH depends on the initial alkalinity. Alkalinity is the acid-neutralizing capacity of water and is consumed on an equivalent basis; that is, 1 meq/L of alum will consume 1 meq/L of alkalinity. If the natural alkalinity of the water is not sufficient to buffer the pH, it may be necessary to add alkalinity to the water to keep the pH from dropping too low. Alkalinity can be added in the form of caustic soda (NaOH), lime [$Ca(OH)_2$], or soda ash (Na_2CO_3). In many water plants, caustic soda is often used because it is easy to handle and the required dosage is relatively small. The reaction for alum with caustic soda is

$$Al_2(SO_4)_3 \cdot 14H_2O + 6NaOH \rightarrow 2Al(OH)_3(s) + 3Na_2SO_4 + 14H_2O \tag{5-7}$$

The corresponding reaction for lime is given by the expression

$$Al_2(SO_4)_3 \cdot 14H_2O + 3Ca(OH)_2 \rightarrow 2Al(OH)_3(s) + 3CaSO_4 + 14H_2O \tag{5-8}$$

Similarly, the overall precipitation reactions for ferric sulfate and ferric chloride are as follows.

Ferric sulfate:

$$Fe_2(SO_4)_3 \cdot 9H_2O \rightarrow 2Fe(OH)_3(s) + 6H^+ + 3SO_4^{2-} + 3H_2O \tag{5-9}$$

Ferric chloride:

$$FeCl_3 \cdot 6H_2O \rightarrow Fe(OH)_3(s) + 3H^+ + 3Cl^- + 3H_2O \tag{5-10}$$

The application of the above equations is illustrated in Example 5-1.

Example 5-1 Calculation of coagulant doses, alkalinity consumption, and precipitate formation

A chemical supplier reports the concentration of stock alum chemical as 8.37 percent as Al_2O_3 with a specific gravity of 1.32. For the stock chemical, calculate (a) the molarity of Al^{3+} and (b) the alum concentration if reported as g/L $Al_2(SO_4)_3 \cdot 14H_2O$. Also, for a 30-mg/L alum dose applied to a treatment plant with a capacity of 43.2 ML/d (0.5 m^3/s), calculate (c) the chemical feed rate in L/min, (d) the alkalinity consumed (expressed as mg/L as $CaCO_3$), and (e) the amount of precipitate produced in mg/L and kg/d.

Solution

1. Calculate the formula weights (FW) for Al_2O_3, $Al_2(SO_4)_3 \cdot 14H_2O$, $Al(OH)_3$, and NaOH, given molecular weights: Al = 27, O = 16, H = 1, S = 32 g/mol.

   FW: $Al_2O_3 = 2(27) + 3(16) = 102$ g/mol

   FW: $Al_2(SO_4)_3 \cdot 14H_2O = 2(27) + 3(32) + 26(16) + 28(1) = 594$ g/mol

   FW: $Al(OH)_3 = 27 + 3(16) + 3(1) = 78$ g/mol

2. Calculate the molar concentration of Al^{3+} in the stock alum chemical.
   a. Calculate the density of stock chemical.

$$\rho_{stock} = 1.32\,(1\ \text{kg/L}) = 1.32\ \text{kg/L}$$

   b. Calculate the concentration of alum in the stock chemical as mg/L Al_2O_3.

$$C_{stock} = 0.0837\,(1.32\ \text{kg/L})\left(10^3\ \text{g/kg}\right) = 110.5\ \text{g/L}\ Al_2O_3$$

   c. Calculate the molar concentration of Al^{3+} in the stock alum chemical.

$$\left[Al^{3+}\right] = (110.5\ \text{g/L}\ Al_2O_3)\left(\frac{\text{mol}\ Al_2O_3}{102\ \text{g}\ Al_2O_3}\right)\left(\frac{2\ \text{mol}\ Al^{3+}}{\text{mol}\ Al_2O_3}\right)$$

$$= 2.17\ \text{mol/L}$$

3. Calculate the stock alum concentration if reported as g/L $Al_2(SO_4)_3{\cdot}14H_2O$.

$$C_{stock} = (110.5\ \text{g/L}\ Al_2O_3)\left(\frac{594\ \text{g/mol alum}}{102\ \text{g/mol}\ Al_2O_3}\right)$$

$$= 643.5\ \text{g/L}\ Al_2(SO_4)_3{\cdot}14H_2O$$

4. Calculate the chemical feed rate. Note 43.2 ML/d = 43,200 m^3/d.

   By mass balance: $C_{stock}Q_{feed} = C_{process}Q_{process}$

$$Q_{feed} = \frac{C_{process}Q_{process}}{C_{feed}} = \frac{(30\ \text{mg/L})(43{,}200\ \text{m}^3\text{/d})(10^3\ \text{L/m}^3)}{(643.5\ \text{g/L})\,(10^3\ \text{mg/g})\,(1440\ \text{min/d})}$$

$$= 1.40\ \text{L/min}$$

5. Calculate the alkalinity consumed using Eq. 5-6:
   Note that alkalinity is commonly expressed as mg/L as $CaCO_3$. $CaCO_3$ has a molecular weight of 100 g/mol and 2 equivalents per mole. Thus the equivalent weight is 50 g/eq or 50 mg/meq.

$$\text{Alk} = \left(30\ \text{mg/L alum}\right)\left(\frac{1\ \text{mmol alum}}{594\ \text{mg alum}}\right)\left(\frac{3\ \text{mmol}\ SO_4^{2-}}{\text{mmol alum}}\right)\left(\frac{2\ \text{meq}\ SO_4^{2-}}{\text{mmol}\ SO_4^{2-}}\right)$$

$$\times\left(\frac{1\ \text{meq alk}}{\text{meq}\ SO_4^{2-}}\right)\left(\frac{50\ \text{mg}\ CaCO_3}{\text{meq alk}}\right) = 15\ \text{mg/L as}\ CaCO_3$$

6. Calculate the precipitate formed using Eq. 5-6:

$$[Al(OH)_3] = (30 \text{ mg/L alum})\left(\frac{1 \text{ mmol alum}}{594 \text{ mg alum}}\right)\left(\frac{2 \text{ mmol Al(OH)}_3}{\text{mmol alum}}\right)$$

$$\times \left(\frac{78 \text{ mg Al(OH)}_3}{\text{mmol Al(OH)}_3}\right) = 7.88 \text{ mg/L Al(OH)}_3$$

$$[Al(OH)_3] = \frac{(7.88 \text{ mg/L})(43{,}200 \text{ m}^3/\text{d})(10^3 \text{ L/m}^3)}{10^6 \text{ mg/kg}} = 340 \text{ kg/d}$$

Comment

The sludge produced by coagulation has two components: the precipitate formed by the reactions shown above and the particles from the raw water. Example 14-2 in Chap. 14 addresses the additional solids generated from the raw-water turbidity.

PREHYDROLYZED METAL SALTS

Since it is difficult to control the Al and Fe metal species formed, especially at low dosages, this has led to the use of prehydrolyzed metal salts. Prehydrolyzed metal salts are prepared by reacting alum or ferric with various salts (e.g., chloride, sulfate) and water and hydroxide under controlled mixing conditions. For example, the commercial prehydrolyzed alum salts, commonly known as PACl, have the following overall formula: $Al_a(OH)_b(Cl)_c(SO_4)_d$. Although many formulations do not contain any sulfate, the presence of sulfate ions helps to stabilize the aluminum polymers and keep them from precipitating. These polymers can be more effective than those formed by simply adding aluminum salts to solution because the larger cationic polymers can be formed by increasing the hydroxide-to-aluminum ratio ($R = OH/Al$), which can lead to enhanced charge neutralization. Another benefit is that, as the polymer becomes larger, it becomes more crystalline, compact, and dense. However, as the value of R increases, the polymers become less stable and may begin to precipitate, which can cause a problem in the storage of PACl.

Several advantages of preformed aluminum metal salts include the following: (1) lower dosages may be required for effective coagulation (on the basis of Al^{3+}) for cases where NOM does not dictate the coagulant dosage at neutral or slightly acidic conditions, (2) flocs tend to be tougher and denser (although flocculation aids are still necessary in many cases), and (3) the performance of prehydrolyzed alum salts is less temperature dependent as compared to unmodified alum salts.

IMPACT OF WATER QUALITY

Coagulation can be affected by pH, alkalinity, temperature, other ions in the water, and NOM. The optimal pH range corresponds to the region of minimum solubility and is about 5.5 to 7.5 for alum and from 5 to 8.5 for ferric salts. Since alum and ferric salts are strong acids, the alkalinity of the water is important in coagulation. If the alkalinity is too low, addition of a high alum dose may drive the pH lower than the optimal range, and a base may need to be added. PACl may be advantageous in low-alkalinity waters because the acidity has been partially neutralized and the pH will not drop as much. Alternatively, water with high alkalinity may require acid in addition the coagulant to lower the pH into the optimal range.

Temperature affects the solubility constants for the precipitation reactions and may have an impact on the amount of metal hydroxide that forms if coagulation is not near the pH of minimum solubility. The most important effect of temperature, however, is that the floc that forms can be close to the density of water. Cold water is more dense than warm water, and when the water temperature is near 4°C, the density of alum floc is very close to the density of water and the floc may not settle well. Ferric hydroxide floc is more dense than alum floc and can be a better option in locations with very cold water.

As with all cations in water, hydrolysis products of aluminum and iron react with various ligands (e.g., SO_4^{2-}, NOM, F^-, PO_4^{3-}) forming both soluble and insoluble products that will influence the quantity or dose of the coagulant required to achieve a desired level of particle destabilization. Thus, the optimum dose of a coagulant depends on the particular water chemistry and the types of particles.

Natural organic matter is the term used to describe the complex matrix of organics originating from natural sources that are present in all water bodies. It has been observed that NOM reacts or binds with metal ion coagulants, and some evidence suggests that the coagulant dosages at many, if not most, operating plants are determined by the NOM–metal ion interactions and not particle–metal ion interactions (O'Melia et al., 1999). Qualitatively, as pH increases, humic substances become more ionized because the carboxyl groups lose protons, and the positive charge on metal coagulants will decrease. Consequently, higher coagulant dosages will be required at higher pH values. At neutral pH, the amount of positively charged Al species decreases with increasing temperature and a higher alum dosage may be required.

Typical Dosages

A typical dosage of alum ranges from 10 to 150 mg/L, depending on raw-water quality and turbidity. Typical dosages of ferric sulfate [$Fe_2(SO_4)_3 \cdot 9H_2O$] and ferric chloride ($FeCl_3 \cdot 6H_2O$) range from 10 to 250 mg/L and 5 to 150 mg/L, respectively, depending on raw-water quality and turbidity. Ferric chloride is more commonly used than ferric sulfate and comes as a liquid.

Importance of Initial Mixing with Metal Salts
The rapid initial mixing of the metal salts in water treatment is extremely important. The sequence of reactions shown on Fig. 5-8 occurs rather rapidly. For example, at a pH of 4, half of the Al^{3+} hydrolyzes to $Al(OH)^{2+}$ within 10^{-5} s. Hahn and Stumm (1968), studying the coagulation of silica dispersions with Al(III), reported that the time required for the formation of mono- and polynuclear hydroxide species was on the order of 10^{-3} s, and the time of formation for the polymer species was on the order of 10^{-2} s.

Clearly, based on the literature and actual field evaluations, the instantaneous rapid and intense mixing of metal salts is of critical importance, especially where the metal salts are to be used as coagulants to lower the surface charge of the colloidal particles. It should be noted that, although achieving extremely low mixing times in large treatment plants is often difficult, low mixing times can be achieved by using multiple mixers. Recommend mixing times of less than 1 s. For sweep flow coagulation with aluminum hydroxide precipitate, short times are not as critical because the precipitate forms slower in the range of 1 to 7 (Amirtharajah and Mills, 1982).

Organic Polymers

Organic polymers are long-chain molecules consisting of repeating chemical units with a structure designed to provide distinctive physicochemical properties. The chemical units usually have an ionic functional group that imparts an electrical charge to the polymer chain. Hence, organic polymers

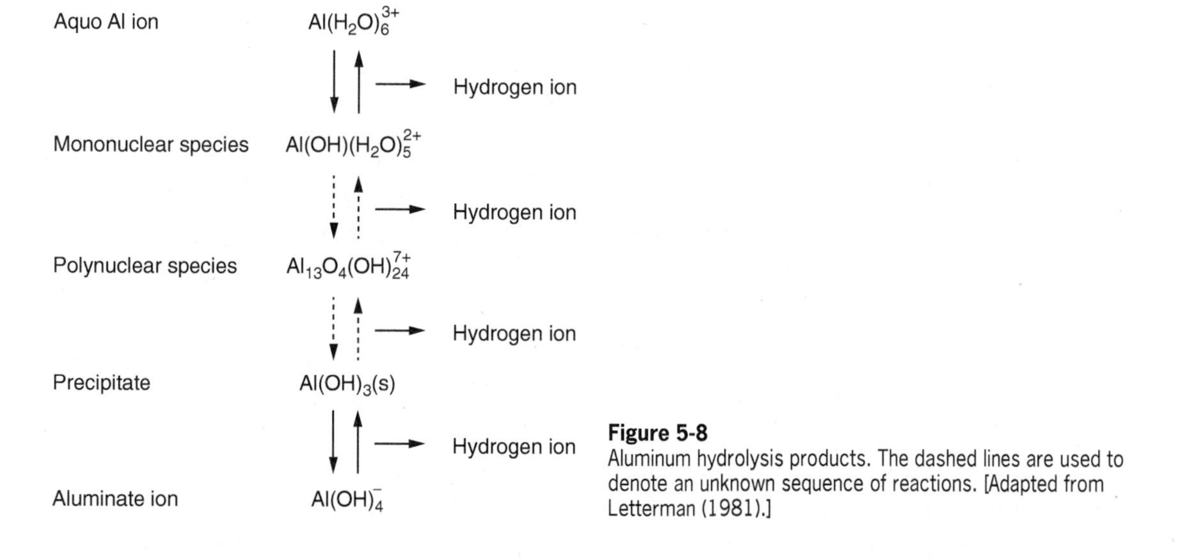

Figure 5-8
Aluminum hydrolysis products. The dashed lines are used to denote an unknown sequence of reactions. [Adapted from Letterman (1981).]

are often termed *polyelectrolytes.* Organic polymers have two principal uses in water treatment: (1) as a coagulant for the destabilization of particles and (2) as a filter aid to promote the formation of larger and more shear-resistant flocs. While destabilization occurs primarily through charge neutralization, nonionic and anionic polymers can be used to form a bridge between particles. Organic polymers are not generally used as primary coagulants and are often used after the particles have been destabilized to some degree with metal coagulants. Table 5-2 summarizes some organic polymers commonly used in water treatment.

It is common to use cationic organic polymers and metallic ion coagulants together. The main advantage of the combined usage is that the

Table 5-2
Typical organic coagulants used in water treatment

Type	Charge	Molecular Weight, g/mol	Common Applications	Typical Examples [a]
Cationic	Positive	$10^4 - 10^6$	Coagulant aid, primary coagulant, sludge conditioning	Epichlorohydrin dimethylamine (epi-DMA) 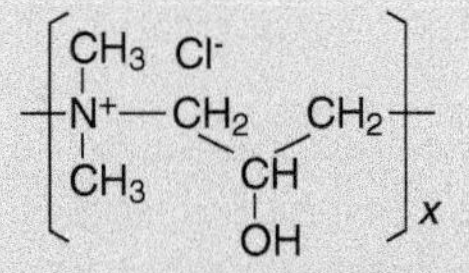 $[-N^+(CH_3)_2Cl^- - CH_2 - CH(OH) - CH_2-]_x$ Polydiallyldimethyl ammonium chloride (poly-DADMAC) 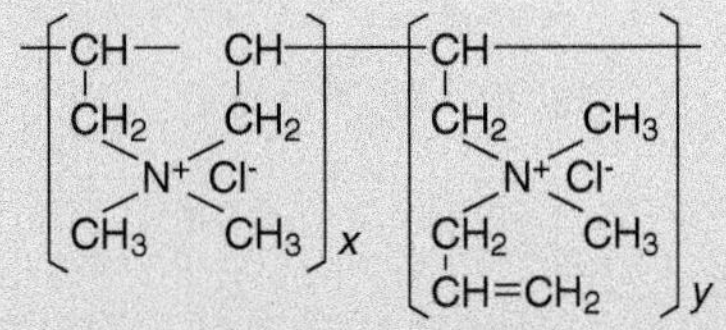 $[-CH-CH-]_x$ with CH_2, CH_2, N^+ Cl^-, CH_3, CH_3; $[-CH-]_y$ with CH_2, CH_3, N^+ Cl^-, CH_2, CH_3, $CH{=}CH_2$
Anionic	Negative	$10^4 - 10^7$	Coagulant aid, filter aid, flocculant aid, sludge conditioning	Hydrolyzed polyacrylamides 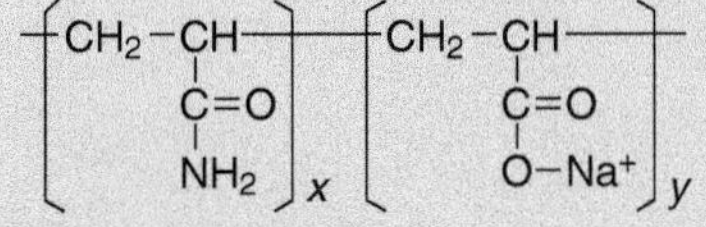 $[-CH_2-CH(C{=}O)(NH_2)-]_x[-CH_2-CH(C{=}O)(O^-Na^+)-]_y$
Nonionic	Neutral	$10^5 - 10^7$	Coagulant aid, filter aid, filter conditioning	Polyacrylamides 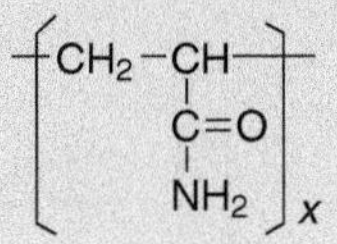 $[-CH_2-CH(C{=}O)(NH_2)-]_x$

[a]Number of monomer molecules in polymer designated by x and y.

dosage of metallic ion coagulants can be reduced by 40 to 80 percent. The lower metallic ion coagulant dosage in turn reduces sludge and alkalinity consumption. With lower alkalinity consumption, the pH will not be depressed as much, which can improve metallic ion coagulation.

Polymer Dosages

Polymer selection is empirical. The typical dosage rates for DADMAC and epi-DMA are on the order of 1 to 10 mg/L. Low dosages of high-molecular-weight nonionic polymers (0.005 to 0.05 mg/L) are often added before granular filtration and to the backwash water to improve filter performance. Incorrect dosing can cause mudball formation in the filters, which are not usually broken apart during normal backwashing operations.

Uncharged and negatively charged organic polymers are used as flocculant aids as opposed to primary coagulants. The main advantage of using flocculant aids is that a stronger floc is formed. Flocculant aids are added after the coagulants are added and the particles are already destabilized.

Enhanced Coagulation

Disinfection by-products (DBPs) are formed as a result of chemical reactions between chlorine and NOM. While trihalomethanes (THMs) and haloacetic acids (HAAs) are the primary DBPs that form during chlorination, the DBP regulations in the United States recognize that MCLs for specific DBPs may not address the total risk associated with adding chlorine to water containing NOM. Consequently, the regulations include a treatment technique that requires the removal of NOM prior to disinfection under certain conditions. The process of performing coagulation for the purpose of achieving specified removal of DBP precursors (NOM) is known as *enhanced coagulation*. The total organic carbon (TOC, a measure of the amount of NOM) removal requirements range from 15 to 50 percent removal depending on the raw-water TOC and alkalinity at the specific site. Specific requirements associated with enhanced coagulation are described in the Stage 1 D/DBP Rule and the *Enhanced Coagulation Guidance Manual* (U.S. EPA, 1999).

The dose required to achieve enhanced coagulation is typically higher than the dose for turbidity removal. When enhanced coagulation is required, the coagulant demand of the NOM and the required degree of TOC removal, not turbidity, will usually dictate the coagulant dosage. Of the metal salts and prehydrolyzed metal salts, the most effective for the removal of NOM is typically iron, followed by alum and PACl. Additional details for enhanced coagulation can be found in the *Enhanced Coagulation Guidance Manual* (U.S. EPA, 1999) and Crittenden et al. (2012).

Figure 5-7 demonstrated that the solubility of coagulants is dependent on pH; the minimum solubility of alum precipitate is around a pH of 6.3 at 25°C. As a result, the optimum NOM removal with alum is at a pH ranging from 5.5 to 6.5, depending on the water temperature and total dissolved solids (TDS) concentration. Removal of NOM with alum can also occur at higher pH values, but higher alum doses are required to meet the same NOM removal that can be achieved at optimum pH. In instances of high-pH conditions at the point of coagulation, acid addition to lower the pH can help improve NOM removal.

The coagulant dose and time required for particle destabilization and effective NOM removal depends on water temperature and the type of particles; consequently, jar tests have to be conducted. The important factors that need to be evaluated in jars and full-scale implementation are floc strength, size, and settling rate.

Jar Testing for Coagulant Evaluation

Because of the many competing reactions and mechanisms that are operative in the coagulation process, the selection of coagulants and dosage is typically determined empirically using bench-scale and pilot-scale studies. The standard bench-scale testing procedure for determining coagulant doses and types is the "jar test" procedure. The jar testing permits rapid evaluation of a range of coagulant types and doses. A modern jar test apparatus is shown on Fig. 5-9. As shown on Fig. 5-9, the apparatus consists of six batch reactors, each equipped with a paddle mixer. Square-shaped jars are used to avoid vortex flow, which can occur if circular beakers are used.

The purpose of the jar test is to simulate, to the extent possible, the expected or desired conditions in the coagulation–flocculation facilities. Standard jar testing procedures are available in AWWA (2011). Generally

Figure 5-9
Jar test apparatus. Note use of square containers to limit the formation of vortex flow in which the particles rotate in the same position relative to each other.

the test consists of a rapid-mix phase (high mixing intensity) with simple batch addition of the coagulant or coagulants followed by a slow-mix period to simulate flocculation. Flocs are allowed to settle and samples are taken from the supernatant. These parameters should be measured as part of the jar test routine: (1) turbidity or suspended solids removal; (2) NOM removal as measured by dissolved organic carbon (DOC) or a surrogate measure of dissolved NOM, such as ultraviolet(UV) at 254 nm; (3) residual dissolved coagulant concentrations of Fe or Al coagulants; (4) pH, and (5) alkalinity. If direct filtration is to be used, the filterability should be evaluated using a filterability test. The filterability is evaluated by filtering the mixed suspension through a 5- or 8-μm laboratory filter to simulate a granular medium filter.

The results of a series of jar tests to determine the optimum alum dose and pH for turbidity removal for given water are summarized on Fig. 5-10. As shown on Fig. 5-10, the optimum alum dose and pH would be approximately 8 mg/L and 7, respectively, because the turbidity is minimized under these conditions. However, it must be emphasized that the raw-water particle

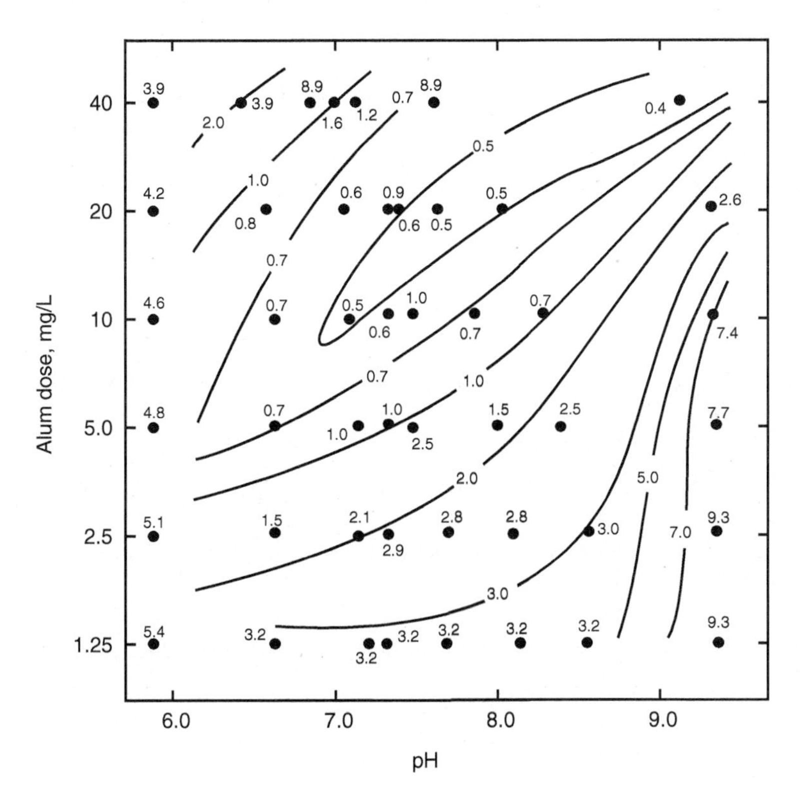

Figure 5-10
Turbidity topogram as function of pH and alum dosage. Points shown on the plot represent turbidity values and the isopleths represent constant turbidity at the value denoted on the isopleth. [Adapted from Trussell, (1978).]

concentration and NOM vary with water quality, and thus the optimum coagulant dosage also changes as the water quality changes.

5-5 Principles of Mixing for Coagulation and Flocculation

Mixing is an important part of coagulation and flocculation processes. Coagulation is performed in a rapid-mix unit that is designed to bring together the coagulant and particles in an efficient manner. Flocculation is performed in basins that are designed to bring the particles in contact with one another and form larger particles that can be removed by gravity separation or filtration. Mixing is a complex topic and significant amounts of research have been devoted to understanding, describing, and modeling turbulence caused by mixing. A detailed investigation of mixing is beyond the scope of this book, and other books such as Crittenden et al. (2012) devote more attention to the relationship between mixing and coagulation and flocculation.

On a macroscopic level, mixing is brought about when fluid at one point in the basin is moving at a different velocity than fluid at an adjacent point, a concept known as the velocity gradient. Under turbulent-flow conditions, the velocity gradient is not well defined and varies both in time and space throughout a coagulation or flocculation basin. Camp and Stein (1943) proposed that the velocity gradient averaged over the entire basin, the global root-mean-square (RMS) velocity gradient, might serve as a useful design parameter for flocculation facilities. Furthermore, they developed a relatively simple equation that related the RMS velocity gradient to the power input to the mixing facility by relating power to the forces on a fluid element.

Consider the fluid element illustrated on Fig. 5-11 and the forces acting on it. The shear stress in the x–y plane, τ_{xy}, is due to the velocity gradient in the z direction and the force exerted on it is given by the expression

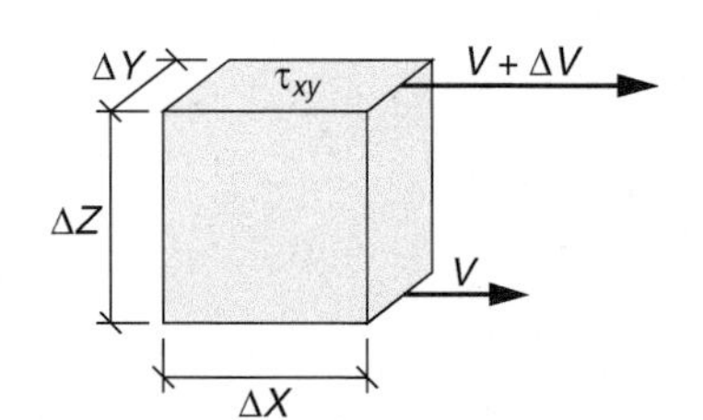

Figure 5-11
Schematic of forces acting on fluid element in flocculator.

$$\text{Force} = \tau_{xy}\,\Delta x \Delta y = \mu \frac{dv}{dz} \Delta x \Delta y \tag{5-11}$$

where μ = dynamic viscosity of water, $N{\cdot}s/m^2$

The product of force and velocity is power, so, using the velocity increment due to the shear stress in the fluid element, the power per unit volume can be written as

$$\frac{P}{V} = \frac{\text{Force} \times \text{velocity}}{\Delta x \Delta y \Delta z} = \frac{\left[\mu\,(dv/dz)\,\Delta x \Delta y\right]\left[(dv/dz)\,\Delta z\right]}{\Delta x \Delta y \Delta z} = \mu\left(\frac{dv}{dz}\right)^2 \tag{5-12}$$

where P/V = power dissipated in selected fluid element, $J/m^3 \cdot s$

Rearranging (Eq. 5-12) and defining the RMS velocity gradient (dv/dz) as $\overline{G}$,

$$\overline{G} = \sqrt{\frac{P}{\mu V}} \tag{5-13}$$

where $\overline{G}$ = global RMS velocity gradient (energy input rate), s^{-1}
P = power of mixing input to vessel, J/s or W
V = volume of mixing vessel, m^3

The Camp–Stein RMS velocity gradient $\overline{G}$ has since become a widely adopted standard used by engineers for assessing energy input in all kinds of mixing processes, including rapid–mix and particularly flocculation units. For rapid-mix units, very high velocity gradients are required (e.g., $\overline{G} = 600 - 5000\ s^{-1}$), whereas flocculation requires a velocity gradient high enough to contact the particles to allow them to flocculate without settling out of solution and yet low enough to prevent particles from falling apart due to shear forces caused by mixing (e.g., $\overline{G} = 20 - 50\ s^{-1}$).

5-6 Rapid-Mix Practice

Common rapid-mixing devices used for the addition of coagulants to the process water are summarized in Table 5-3 and some are displayed on Fig. 5-12. The most common devices include mechanical and hydraulic mixers, in-line static and mechanical mixers, and pressurized water jets. For situations where rapid-mix units are designed for coagulation by adsorption/destabilization, engineers will use $\overline{G}$ values in the range of 3000 to 5000 s^{-1}. For rapid mix units designed for sweep coagulation reactions, engineers will use $\overline{G}$ values in the range of 600 to 1000 s^{-1}.

Table 5-3
Types of common coagulant mixing devices and their typical application

Mixing Device	Application in Coagulant Addition
Mechanical mixers in stirred tanks	Sweep coagulation
Hydraulic mixers	Sweep coagulation
In-line mechanical mixers	Adsorption/destabilization
In-line static mixers	Adsorption/destabilization
Diffusion by pressurized water jets/pumps	Adsorption/destabilization

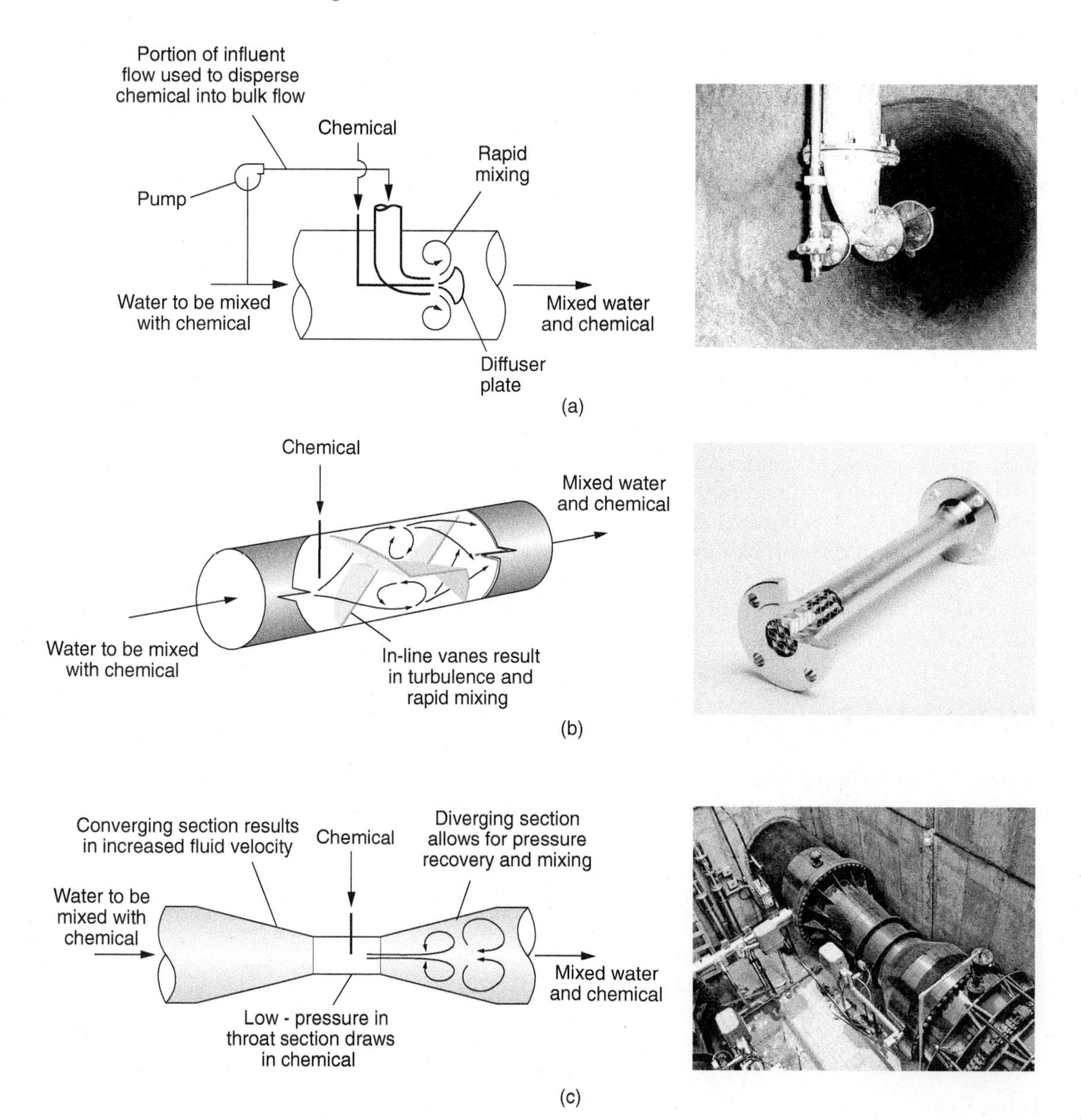

Figure 5-12
Illustrations of mixing approaches used in water treatment: (a) pumped flash mixing, (b) in-line static mixer, and (c) in-line venturi mixer.

For design of large-scale rapid-mix systems empirical methods are used for their design. Manufacturers typically supply the design information that can be used to size the system and determine the power requirements. An example design of an in-line static mixer is presented in Crittenden et al. (2012).

5-7 Principles of Flocculation

Flocculation theories have evolved from the following observations: (1) small particles undergo random Brownian motion due to collisions with fluid molecules resulting in particle–particle collisions and (2) stirring water containing particles creates velocity gradients that bring about particle collisions. These interactions are known as microscale (perikinetic) and macroscale (orthokinetic) flocculation, respectively. Another form of flocculation occurs due to differential settling in which large particles settling in a quiescent basin overtake small particles to form larger particles. In flocculation basins, however, the mixing provided to encourage macroscale flocculation is sufficient to keep particles from settling; thus, differential settling is not an effective mechanism where active mixing is occurring. A thorough discussion of the theory of flocculation is discussed in Crittenden et al. (2012). A brief discussion of the mechanisms and modeling of flocculation processes is presented in this section.

Rate of Particle Collision

The fundamental problem in mathematical modeling of the flocculation process is predicting the change of the particle size distribution as a function of time for a given set of chemical and hydrodynamic conditions. Any general kinetic model must account for changes in the number of particles found in all size classes. Particles of size d_i collide with particles of size d_j, forming particles of size d_k when collisions are successful. At the same time, aggregates of size d_k may break up into smaller aggregates due to hydrodynamic shearing forces.

The overall particle collision rate is a function of the rate of macroscale flocculation (r_M), rate of microscale flocculation (r_μ), and rate of differential settling flocculation (r_{DS}) between particles i and j.

The rate of particle attachments r_{ij} is a function of the particle concentrations and a collision frequency function β_{ij}:

$$r_{ij} = \alpha\beta_{ij}n_i n_j \tag{5-14}$$

where r_{ij} = rate of attachment between i and j particles
α = collision efficiency factor (attachments per collision)
β_{ij} = collision frequency function for particles of i and j size classes
n_i = concentration of i particles
n_j = concentration of j particles

The collision efficiency factor α, defined as the ratio of collisions that result in attachment to total collisions, has a range of values $0 \leq \alpha \leq 1$. The collision efficiency factor depends on the effectiveness of destabilization; for example, if particles have been destabilized completely, then $\alpha = 1$. Solution of mass balances on flocculation reactors that use Eq. 5-14 require the use of appropriate values of β to predict the change in the size distribution of the suspension as aggregation occurs.

Collision Frequency Function

The collision frequency function β_{ij} depends on the size of the particles, the flocculation transport mechanism, and the efficiency of particle collisions. The overall collision frequency function is a function of the individual mechanisms of flocculation as follows:

$$\beta_{ij} = \beta_M + \beta_\mu + \beta_{DS} \tag{5-15}$$

where β_{ij} = overall collision frequency between particles i and j
β_M = macroscale collision frequency, $= r_M/\alpha n_i n_j$
β_μ = microscale collision frequency, $= r_\mu/\alpha n_i n_j$
β_{DS} = differential settling collision frequency, $= r_{DS}/\alpha n_i n_j$

As noted above, differential settling is not effective when mixing is provided to encourage macroscale flocculation, so the overall collision frequency in a flocculation basin can be considered to be the sum of the microscale and macroscale collision frequencies.

Various models have been developed to describe the overall rate of flocculation ranging from simple models of spherical particles in a linear flow field to more complex models involving nonlinear flow fields and fractal geometry. The following discussion is a simplistic view of flocculation kinetics (spherical particles in a linear flow field) and is intended only to provide insight into the flocculation mechanisms that are most significant for various size particles.

MACROSCALE COLLISION FREQUENCY

Consider particles i and j with diameters d_i and d_j, respectively, suspended in and moving in fluid streamlines in the x direction with water subjected to a velocity gradient dv_x/dz, as shown on Fig. 5-13. When the distance between the centers of the particles, R_{ij}, becomes equal to $(d_i + d_j)/2$, a collision will occur. When fluid flow is laminar and steady, the velocity gradients are well defined, as shown on Fig. 5-13. The velocity gradient on Fig. 5-13 is proportional to the shear stress on the fluid elements because it is a Newtonian fluid. Given a uniform velocity gradient, the rate of flocculation can be determined from geometric considerations, as illustrated below.

The rate of macroscale flocculation in a system of unequal size (heterodisperse) particles subjected to uniform mixing may be derived using

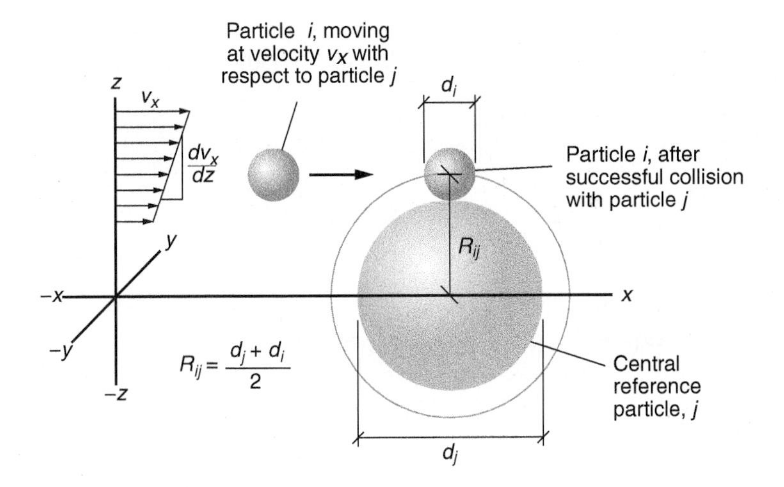

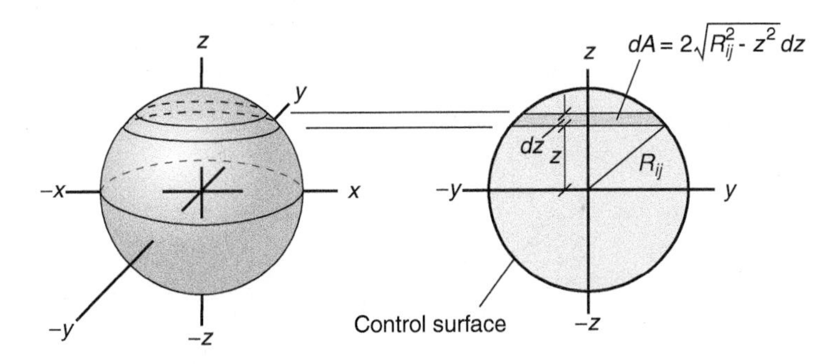

Figure 5-13
Definition sketch for analysis of the flocculation process.

the relationships shown on Fig. 5-13. The flow rate of fluid into an area element dA of the control surface is given by the following expression (Swift and Friedlander, 1964):

$$dq = \text{(velocity) (differential area)} = \left(z\frac{dv_x}{dz}\right)\left(2\sqrt{R_{ij}^2 - z^2}\,dz\right) \tag{5-16}$$

where dq = differential flow of fluid through area element dA, m^3/s
q = fluid flow rate through particle area projected onto y–z plane, m^3/s
z = vertical direction, m
dv_x/dz = velocity gradient in x-direction, $\overline{G}$, s^{-1}
R_{ij} = distance between centers of particles i and j, m

In a heterogeneous solution, the flow rate of particles through the control area may be expressed as the product of the i and j particle concentrations (n_i and n_j, respectively) and the differential flow of fluid through the control surface. Assuming that the velocity gradient is constant,

$$\text{Particle flow through control surface} = 2n_i n_j \int_{z=0}^{z=R_{ij}} dq$$

$$= 4n_i n_j \left(\frac{dv_x}{dz}\right) \int_0^{R_{ij}} z\sqrt{R_{ij}^2 - z^2}\, dz \tag{5-17}$$

Recalling from calculus that $\int x\sqrt{a^2 - x^2}\, dx = -\frac{1}{3}\left(a^2 - x^2\right)^{3/2} + c$, the integrated form of Eq. 5-17 is given by the expression

$$\text{Particle flow} = 4n_i n_j \left(\frac{dv_x}{dz}\right)\left[-\frac{1}{3}\left(R_{ij}^2 - z^2\right)^{3/2}\right]_0^{R_{ij}}$$

$$= \frac{4}{3}\left(\frac{dv_x}{dz}\right) R_{ij}^3 n_i n_j \tag{5-18}$$

The rate of flocculation is equal to the flow rate of particles times the collision efficiency α (i.e., the fraction of collisions that result in attachment):

$$r_{ij} = \frac{4}{3}\left(\frac{dv_x}{dz}\right) R_{ij}^3 n_i n_j \alpha \tag{5-19}$$

where r_{ij} = rate of collision between i and j particles (rate of flocculation)

Substituting the term $(d_i + d_j)/2$ for R_{ij} (see Fig. 5-13) results in the following expression for the rate of flocculation, by macroscale mechanisms, between i- and j-sized particles:

$$r_M = \frac{1}{6}\left(\frac{dv_x}{dz}\right)\left(d_i + d_j\right)^3 n_i n_j \alpha \tag{5-20}$$

where r_M = rate of macroscale flocculation

Under turbulent-flow conditions, the velocity gradient is not well defined and varies both in time and space throughout the flocculation basin. When averaged over the entire basin, the velocity gradient is known as the root-mean-square (RMS) velocity gradient and is given the symbol $\overline{G}$ (see Eq. 5-13).

Thus for unequal-sized (heterodisperse) particles the collision frequency function for the macroscale flocculation rate β_M can be determined from Eq. 5-20 by defining the velocity gradient as $\overline{G}$ (Eq. 5-13) and using the relationship given in the nomenclature for Eq. 5-15, resulting in

$$\beta_M = \frac{1}{6}\overline{G}\left(d_i + d_j\right)^3 \tag{5-21}$$

where $\overline{G}$ = *RMS* velocity gradient, s^{-1}

MICROSCALE COLLISION FREQUENCY

The flux of j-size particles to the surface of a single i-size particle by diffusion is given by the expression

$$J_A = -D_{lj}\left(\frac{\partial n_j}{\partial r}\right)_{r=d_i/2} = \frac{-2D_{lj}n_j}{d_i} \tag{5-22}$$

where J_A = flux of particles, m · number of particles/s
D_{lj} = liquid-phase diffusion coefficient for particle j to particle i, m^2/s

Thus, the flocculation rate $r_{\mu,j}$ is given by the expression

$$r_{\mu,j} = \text{sphere surface area} \times \text{flux} = \left(\pi d_i^2\right)\left(\frac{2D_{lj}n_j}{d_i}\right) = 2\pi d_i D_{lj} n_j \tag{5-23}$$

Substituting the Stokes–Einstein equation $D_{lj} = 2kT/6\pi\mu d_j$ (see Eq. 4-120) into Eq. 5-21 and incorporating the collision efficiency factor α and the number of particles, n_i, an expression for the rate of flocculation, $r_{\mu,ji}$ of all j-size particles diffusing to the surface of all i-size particles can be obtained:

$$r_{\mu,ji} = 2\pi d_i\, D_{lj}\, n_j n_i\, \alpha = 2\pi d_i\left(\frac{2kT}{6\pi\mu d_j}\right)\alpha n_i n_j = \frac{2}{3}\alpha\left(\frac{kT}{\mu}\right)\left(\frac{d_i}{d_j}\right) n_i n_j \tag{5-24}$$

where k = Boltzmann's constant, 1.3807×10^{-23} J/K
T = absolute temperature, K (273 + °C)
μ = dynamic viscosity of water, N·s/m^2

Generalizing to all possible combinations of i and j to form a particle of size k, the overall rate of r_μ is given by

$$r_\mu = \underbrace{\tfrac{2}{3}\alpha\left(\tfrac{kT}{\mu}\right)\left(\tfrac{d_i}{d_j}\right) n_i n_j}_{\substack{j \text{ diffusing to } i \\ \text{(different sizes)}}} + \underbrace{\tfrac{2}{3}\alpha\left(\tfrac{kT}{\mu}\right)\left(\tfrac{d_i}{d_j}\right) n_i n_j}_{\substack{i \text{ diffusing to } j \\ \text{(different sizes)}}} + \underbrace{\tfrac{2}{3}\alpha\left(\tfrac{kT}{\mu}\right)\left(\tfrac{d_i}{d_j}\right) n_i\,(2n_j)}_{\substack{i, j \text{ diffusing toward} \\ \text{each (other equal size)}}} \tag{5-25}$$

Grouping terms and simplifying the rate expression in Eq. 5-25 result in the expression

$$r_\mu = \frac{2}{3}\alpha\left(\frac{kT}{\mu}\right) n_i n_j\left(\frac{1}{d_i} + \frac{1}{d_j}\right)(d_i + d_j) \tag{5-26}$$

The collision function for microscale flocculation of heterodisperse particles can now be written as

$$\beta_\mu = \left(\frac{2kT}{3\mu}\right)\left(\frac{1}{d_i} + \frac{1}{d_j}\right)(d_i + d_j) \tag{5-27}$$

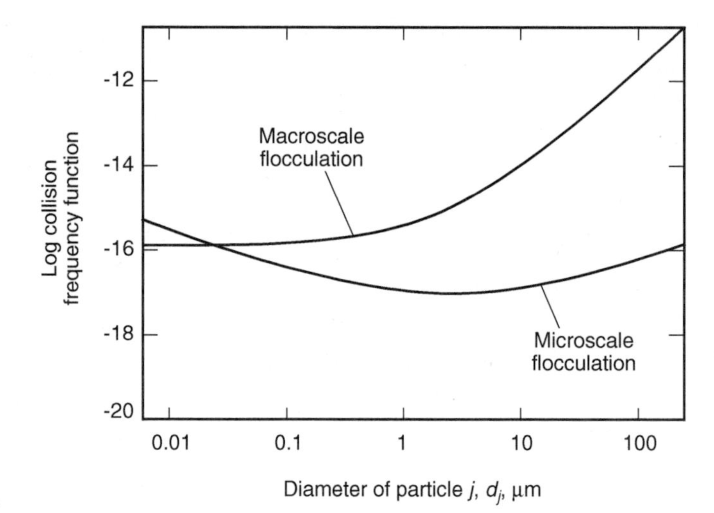

Figure 5-14
Collision frequency functions for macroflocculation (orthokinetic flocculation) and microflocculation (perikinetic flocculation).

Comparison of Collision Frequency Functions

The collision frequency functions for macroscale and microscale flocculation are given by Eqs. 5-21 and 5-27, respectively. The collision frequency function may be plotted for a given system to assess the relative effect of each type of flocculation mechanism. A plot of the collision frequency functions is presented on Fig. 5-14 for a system containing particles d_i of size 2.0 μm and particles d_j with sizes ranging from 0.01 to 100 μm. The curves shown on Fig. 5-14 are for a $\overline{G}$ value of 100 s^{-1} and water temperature of 15°C. As shown on Fig 5-14, microscale flocculation is the dominant mechanism for particles of size less than 0.1 μm and the macroscale mechanism is dominant for larger particles. Thus, the mixing provided in a flocculation basin is primarily focused on aggregating particles larger than 0.1 μm. Smaller particles are aggregated primarily by microscale flocculation, and the purpose of designing for a particular hydraulic detention time in a flocculation basin is to provide sufficient time for this process to occur.

5-8 Flocculation Practice

The purpose of flocculation basins is to provide gentle mixing that promotes macroflocculation for a sufficient amount of time to allow both microflocculation and macroflocculation to take place. Today's flocculation installations can be divided into two groups: mechanical and hydraulic. In mechanical flocculation horizontal paddles and vertical turbines have become the most common configurations for the prime mover, although new innovations continue to be developed. No particular arrangement dominates in hydraulic flocculation. Occasionally designers have used agitation with air or pumped water jets to create the velocity gradients for flocculation, but these efforts have met with limited success.

Some views of these three most common approaches to flocculation are given on Fig. 5-15. Information on how these approaches compare to each other with respect to a number of design and operational issues is presented in Table 5-4. All three of these approaches have been used successfully in numerous operations, and design details for a number of variations of each

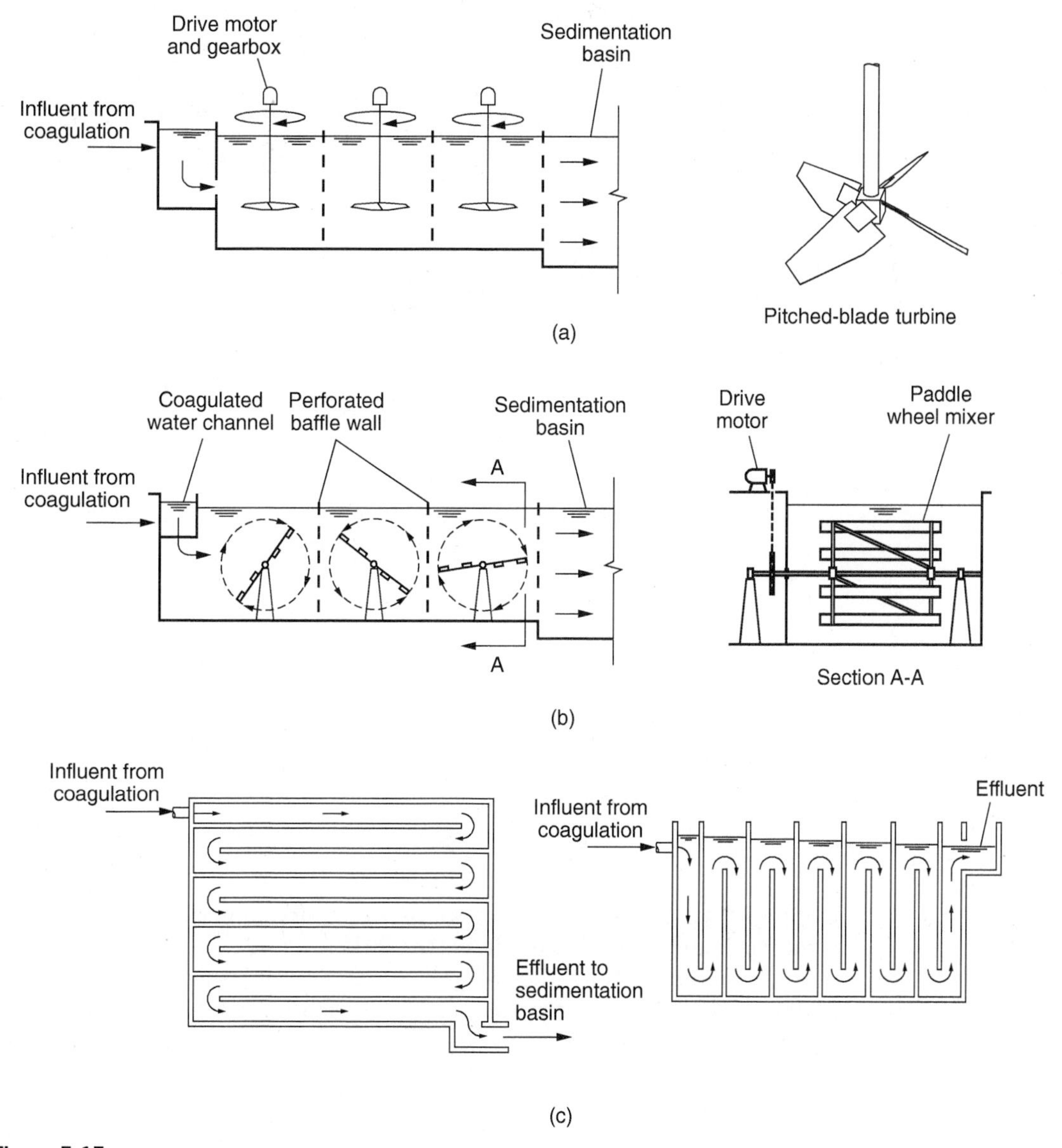

Figure 5-15
Common types of flocculation systems: (a) vertical-shaft turbine flocculation system, (b) horizontal paddle wheel flocculation system, and (c) hydraulic flocculation systems.

Table 5-4
Comparison of basic approaches to flocculation

Process Issue	Horizontal Shaft with Paddles	Vertical-Shaft Turbines	Hydraulic Flocculation
Type of floc produced	Large and fluffy	Small to medium, dense	Very large and fluffy
Head loss	None	None	0.05–0.15 m
Operational flexibility	Good, limited to low $\overline{G}$	Excellent	Moderate to poor
Capital cost	Moderate to high	Moderate	Low to moderate
Construction difficulty	Moderate	Easy to moderate	Easy to difficult
Maintenance effort	Moderate	Low to moderate	Low to moderate
Compartmentalization	Moderate compartmentalization	Excellent compartmentalization	Excellent compartmentalization, some designs nearly plug flow
Advantages	❑ Generally produces large floc ❑ Reliable ❑ No head loss ❑ One shaft for several mixers	❑ Flocculators can be maintained or replaced without basin shutdown ❑ No head loss ❑ Very flexible, reliable ❑ Highest energy input potential	❑ Simple and effective ❑ Easy, low-cost maintenance ❑ No moving parts ❑ Can produce very large flocs
Disadvantages	❑ Compartmentalization more difficult ❑ Replacement and some maintenance requires shutdown of basin ❑ Shaft breakage on startup because of high initial torque	❑ Difficult to specify proper impellers and reliable gear drives in competitive bidding process	❑ Little flexibility

of them can be found in other sources (e.g., Kawamura, 2000; Letterman and Yiacoumi, 2011; Crittenden et al., 2012).

The choice among these three alternatives is usually driven by personal preference, by downstream processes, and by the level of operational expertise available. Horizontal-shaft paddles and vertical turbines are both common in conventional treatment (includes sedimentation). Vertical turbines tend to dominate in direct filtration (no sedimentation) where horizontal-shaft paddles are rarely used. Hydraulic flocculation is usually employed with conventional treatment, although it has also been successfully used for direct filtration. Hydraulic flocculation is particularly popular in locations with poor access to resources and trained personnel for maintenance and operation, but it also plays an important role in some developed countries, particularly Japan. In recent years, vertical turbine flocculators have gained in popularity as impeller designs have improved and as design engineers learn how to specify them properly. One special attraction of vertical turbines is that these flocculators can be replaced or maintained while the process is operating.

Combining hydraulic and mechanical flocculation sometimes allows the water utility to capitalize on the strengths of both approaches. Using such a combination, the number of mechanical flocculators is reduced, reducing the capital and maintenance costs and increasing the reliability. In such combinations, Kawamura (2000) recommends that mechanical flocculators be located at the end of the process to keep the floc in suspension during low-flow conditions. The Houston East plant in Houston, Texas [570 ML/d (150 mgd)], and the Mohawk Water Treatment Plant in Tulsa, Oklahoma [380 ML/d (100 mgd)], both utilize this design and achieve excellent settled water turbidity and operate effectively during low-flow conditions by isolating some of the treatment trains (Kawamura, 2000).

The basic design criteria for mechanical flocculators are the Camp–Stein RMS velocity gradient $\overline{G}$ and the hydraulic detention time τ. Requirements of hydraulic detention time depend more on the downstream process than on the means of flocculation. Somewhat shorter flocculation times are often used for direct filtration (10 to 20 min) than for conventional treatment (20 to 30 min). Longer flocculation times are also required in colder climates. Representative design parameters for horizontal-shaft paddles and vertical turbines are shown in Table 5-5. Flocculation times depend on how the particles are going to be removed in the subsequent processes.

As floc grows larger, it becomes more susceptible to the shear imparted to the water by the impeller blades. The velocity gradient that provides optimal

Table 5-5
Typical design criteria for horizontal-shaft paddles and vertical-shaft turbines

Design Parameter	Unit	Horizontal Shaft with Paddles	Vertical-Shaft Turbines
Velocity gradient, $\overline{G}$	s^{-1}	20–50	10–80
Tip speed, maximum	m/s	1	2–3
Rotational speed	rev/min	1–5	10–30
Compartment dimensions (plan)			
Width	m	3–6	6–30
Length	m	3–6	3–5
Number of compartments	No.	2–6	4–6
Variable-speed drives	—	Usually	Usually

opportunities for collisions between particles early in the flocculation process might be excessive and tear large floc apart toward the end of the flocculation process. Thus, it is common to design flocculation facilities with multiple stages (typically two to four) with decreasing velocity gradient in each successive stage. This design strategy is known as *tapered flocculation.* For instance, a vertical turbine flocculation system might be designed with a $\overline{G}$ values of 80, 60, 40, and 20 s^{-1} in four successive stages. In addition, it is sometimes necessary to vary the flocculation energy from one season to the next because the viscosity of water is dependent on temperature and viscosity is a key term in the equation for $\overline{G}$. As a result, variable-speed drives are usually provided.

The size and shape of a flocculation basin are generally determined by the type of flocculator selected and the type of sedimentation process employed downstream. If mechanical flocculators are paired with rectangular, horizontal-flow sedimentation basins, the width and depth of the flocculation basins should match the width and depth of the sedimentation basins. Similar dimensions enhance constructability and reduce overall project costs.

Diffuser walls are often used to divide flocculation basins into separate compartments, to place a hydraulic division between flocculation and sedimentation basins, as well as in other situations where an even velocity profile is required and backmixing is undesirable.

Vertical Turbine Flocculators

Vertical-shaft turbine flocculators are impellers attached to a vertical shaft that is rotated by an electric motor through a speed reducer. The impellers used for mixing can be placed in two broad classifications: (1) radial flow impellers and (2) axial flow impellers. Examples of the two types of impellers and the differences between their performances are illustrated on Fig. 5-16. The radial impeller directs flow outward from the impeller blades in a horizontal direction, through centrifugal force, with a velocity profile that peaks at the center of the blades. The axial impeller directs the flow parallel to the vertical shaft. The circulation pattern in the mixing tank is also substantially different for these two types of impellers. Two circulation loops are generated from radial flow mixers: one above the impeller and one below. Axial flow impellers, on the other hand, create one circulation pattern from the bottom of the tank to the top and back through the impeller again.

Axial flow impellers can be configured in two ways: to pump downward or to pump upward. Down pumping is usually employed in flocculation because it helps keep the particles in the tank in suspension. The axial impeller shown on Fig. 5-16 is arranged to pump downward.

IMPELLER DESIGN CRITERIA

Important design considerations for vertical turbine impellers are the displacement capacity (the rate at which the impeller pumps water), the

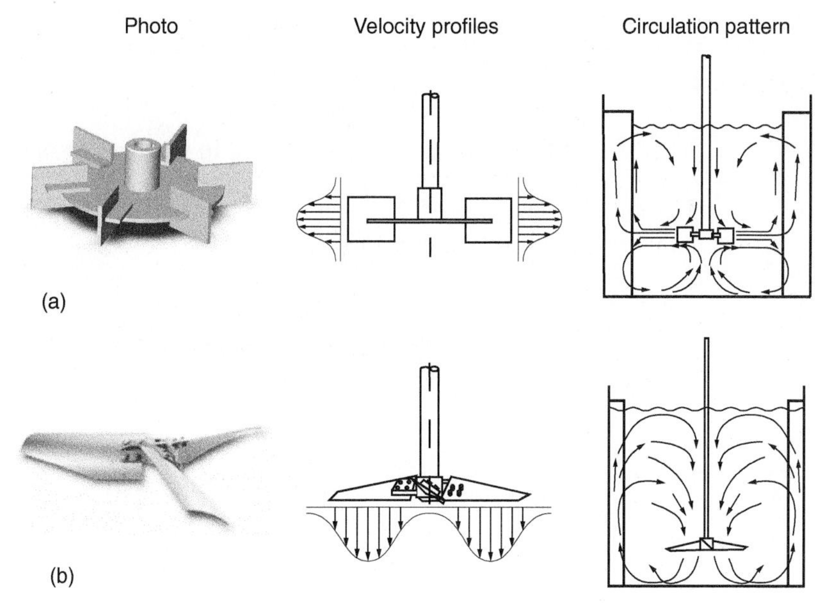

Figure 5-16
Comparison of (a) radial and (b) axial flow mixers with respect to shape, velocity profiles, and circulation patterns. [Adapted from Oldshue and Trussell (1991).]

power consumption, and the pumping head. Together, these determine much about the nature of the flow in the impeller's operating environment.

To evaluate the impeller's performance, it is important to know the nature of the flow in the mixing tank, specifically if the flow is laminar or turbulent as determined by the Reynolds number. Virtually all flocculation impellers operate in the turbulent-flow regime. The Reynolds number for a vertical turbine flocculator is given by the expression

$$\mathrm{Re} = \frac{D^2 N \rho}{\mu} \tag{5-28}$$

where Re = Reynolds number, dimensionless
D = diameter of impeller, m
N = impeller's rotational speed, s^{-1}
ρ = density of water, kg/m^3
μ = dynamic viscosity of water, kg/m·s

For the vertical turbines used in flocculation, full turbulence is developed at Reynolds numbers of 10,000 and greater.

Example 5-2 Estimating Reynolds number of vertical turbine flocculator

A vertical turbine 1.6 m in diameter is used to mix the contents of a flocculation tank 4 m in diameter. The turbine rotates at a speed of 20 rev/min. The absolute viscosity of the water is 1.31×10^{-3} kg/m·s. Determine if turbulent conditions are present.

Solution

1. Determine the Reynolds number using Eq. 5-28:

$$\text{Re} = \frac{D^2 N\rho}{\mu} = \frac{(1.6\text{ m})^2(20\text{ min}^{-1})(998\text{ kg/m}^3)}{(60\text{ s/min})(1.31 \times 10^{-3}\text{ kg/m·s})} = 6.5 \times 10^5$$

2. Because the computed value of R is greater than 10^4, the flow regime is turbulent.

Three parameters that are important to the design of mixing devices are the power number, the pumping number, and the head number. These have the following form:

$$N_p = \frac{P}{\rho N^3 D^5} \quad (5\text{-}29)$$

$$N_Q = \frac{Q}{ND^3} \quad (5\text{-}30)$$

$$N_H = \frac{\Delta H g}{(ND)^2} \quad (5\text{-}31)$$

where P = power required, J/s (W)
N_p = power number, dimensionless
D = diameter of impeller, m
N_Q = pumping number, dimensionless
N_H = head number, dimensionless
ρ = fluid density, kg/m^3
N = rotational speed, rev/min
Q = flow rate imparted by impeller, m^3/s
ΔH = head impeller imparts to impeller flow, m
g = acceleration due to gravity, 9.81 m/s^2

The power number is the most straightforward of these numbers to determine. All that is required is a torque meter on the shaft of the mixer and a tachometer to measure its rate of rotation. As a consequence, power numbers are available for most commercial impellers. The availability of power numbers is convenient because it is the power number and the rotational speed that determine the nominal Camp–Stein RMS velocity gradient $\overline{G}$ for the basin.

In general, as the pumping number increases, the circulation pattern becomes prevalent. As the head number increases for a given pumping number, more turbulence, shear, and mixing occur. In addition, if the pumping number and head number are available, they can be used to determine whether a particular impeller mixer is suitable for the mixing tank. For example, the circulation time (the volume of the flocculation chamber divide by the impeller flow rate) is related to the mixing time required to achieve completely mixed conditions. However, pumping numbers and head numbers are substantially more difficult to measure and, as a result, they are not as readily available as the power number.

IMPACT OF IMPELLER SHAPE

Several types of impellers used in water treatment along with their typical uses are displayed in Table 5-6. When impellers on vertical shafts

Table 5-6
Power and pumping numbers for common impellers

Impeller Type	Photograph	Power Number	Pumping Number	Application
Flat-bladed turbine (FBT)		3.6	0.9	Blending, maintaining suspensions, flocculation
Pitched-blade turbine (45° PBT)		1.26	0.75	Blending, maintaining suspensions, flocculation
Pitched-blade turbine with camber (hydrofoil, 3 blades)		0.2–0.3	0.45–0.55	Blending, maintaining suspensions, flocculation
Cast foil with proplets		0.23	0.59	Blending viscous liquids
Rushton turbine (6 blades)		4.5–5.5	0.72	Gas–liquid dispersion, solids suspension, flocculation
Propeller (pitch of 1:1)		0.32–0.36	0.4	Blending viscous liquids

Figure 5-17
Trailing vortex behind 45° pitched-blade turbine in turbulent flow. [From Shäfer et al. (1998).]

were first used for flocculation, some radial flow turbines were used, particularly Rushton turbines and flat-bladed turbines. As these impellers move through the water, however, they create substantial trailing vortices (Van't Reit et al., 1976). Vortices represent anisotropic turbulence that contributes significantly to floc breakup. Long pitch blade turbines subsequently became more popular, but, as illustrated on Fig. 5-17, even these produce substantial trailing vortices (Shäfer et al., 1998). Today hydrofoils, or pitched-blade turbines with cambered blades (blades with an upper surface shaped like an airplane wing) are the impellers of choice. Properly designed, flocculators using these devices can form large floc similar to that formed by more traditional horizontal paddle flocculators.

OTHER DESIGN CONSIDERATIONS

In addition to the choice of the impeller itself, the following design parameters should be carefully scrutinized: (1) the ratio of the blade diameter to equivalent tank diameter should be greater than 0.35, preferably between 0.4 and 0.5, and (2) the velocity profile caused by the mixing blade should have a maximum of 2.5 m/s (8 ft/s) in the first stage and less than 0.6 m/s (2 ft/s) in the last stage of the flocculator. Design criteria are summarized in Table 5-7.

Depth and Shape of Flocculation Chamber

The depth and shape of the flocculation chamber can be important. Most mixing tests are conducted in square tanks with the impeller held at two-thirds of the depth of the tank. The more closely the full-scale design emulates those conditions, the more likely it is that the full-scale performance will replicate the manufacturer's test data. As a result, when

Table 5-7
Key design criteria for vertical-turbine flocculator

Parameter	Range	Definition Sketch
Impeller	Hydrofoil or 45° pitched-blade turbine (PBT), hydrofoil preferred	
D/T_e^a	0.3–0.6, 0.4–0.5 preferred	
H/T_e	0.9–1.1	
C/H	0.5–0.33	
N	10–30 rev/min	
Tip speed	2–3 m/s	

[a] $T_e = \sqrt{4A_{plan}/\pi}$.

vertical turbine impellers are used, it is wise to stick to a nearly cubical shape flocculation chamber and to locate the impeller at approximately two-thirds of the chamber's water depth.

Example 5-3 Design of vertical turbine flocculator

Vertical turbines are to be used for flocculation in a water treatment plant with a design flow rate of 75 ML/d (20 mgd) and design temperature of 10°C. Flocculation is to be designed with four parallel trains and each train is to be made of four compartments in series. The total detention time in flocculation is to be 20 min. Determine the following design features for the first compartment in each flocculation train:

1. The dimensions of the compartment
2. The diameter of the impeller (assume a turbine having three pitched blades with camber, a foil)
3. The water power required to achieve a $\overline{G}$ of 80 s^{-1} (the power that must be input to the water through the impeller shaft)
4. The maximum rotational speed
5. The pumping capacity of the impeller and circulation time in the tank

At 10°C, the absolute viscosity of water is 1.31×10^{-3} kg/m·s and the density of water is 999.7 kg/m^3. The circulation time is the volume of the flocculation chamber divided by the impeller pumping rate.

Solution

1. Determine the dimensions of the compartment:

$$\text{Volume} = \frac{(75\ \text{ML/d})(1000\ \text{m}^3/\text{ML})(20\ \text{min})}{(1440\ \text{min/d})(4\ \text{trains})(4\ \text{stages/train})} = 65.1\ \text{m}^3$$

Assume a perfect cube of length L:

$$L = \sqrt[3]{65.1\ \text{m}^3} = 4.0\ \text{m} \quad (13.2\ \text{ft})$$

2. Determine the diameter of the impeller. Based on Table 5-7, choose an impeller diameter of $0.45T_e$:

$$T_e = \sqrt{\frac{4 \times A_{\text{plan}}}{\pi}}$$

Assume $A_{\text{plan}} = 4.0\ \text{m} \times 4.0\ \text{m} = 16\ \text{m}^2$:

$$T_e = 4.51\ \text{m}$$

$$D = 0.45 \times 4.51\ \text{m} = 2.03\ \text{m}$$

Choose $D = 2$ m.

3. Determine the power input to the water: The water power is determined by the requirement for $\overline{G} = 80\ \text{s}^{-1}$. Rearranging Eq. 5-13,

$$\begin{aligned} P &= \overline{G}^2 \mu V \\ &= (80\ \text{s}^{-1})^2(1.31 \times 10^{-3}\ \text{kg/m·s})(65.1\ \text{m}^3) = 546\ \text{kg·m}^2/\text{s}^3 \\ &= 546\ \text{J/s} \end{aligned}$$

4. Determine the maximum rotational speed: From Table 5-6, for a three-bladed foil, N_p values of 0.2 to 0.3, use 0.25. Rearranging Eq. 5-29,

$$\begin{aligned} N &= \sqrt[3]{\frac{P}{N_p \rho D^5}} \\ &= \sqrt[3]{\frac{546\ \text{J/s}}{(0.25)(999.7\ \text{kg/m}^3)(2\ \text{m})^5}} = 0.409\ \text{s}^{-1} \\ &= (0.409\ \text{s}^{-1})(60\ \text{s/min}) = 24.5\ \text{min}^{-1} \quad (\text{rev/min}) \end{aligned}$$

Note: N is within the operating range of 10 to 30 rev/min recommended in Table 5-7.

5. Determine the pumping capacity and circulation time:
   a. Pumping capacity: From Table 5-6, $N_Q \sim 0.5$. Rearranging Eq. 5-30,

$$Q = N_Q N D^3$$
$$= (0.5)(0.409\ \text{s}^{-1})(2\ \text{m})^3 = 1.64\ \text{m}^3/\text{s}$$

   b. Circulation time:

$$t_c = \frac{V}{Q} = \frac{65.1\ \text{m}^3}{1.64\ \text{m}^3/\text{s}} = 39.8\ \text{s}$$

The circulation time is a little less than 1 min.

Horizontal Paddle Wheel Flocculators

Horizontal-shaft paddle wheel flocculators are often employed if conventional treatment is used and a high degree of solids removal by sedimentation is required (see Fig. 5-18). However, they require more maintenance and expense, mainly because bearings and packings are typically submerged. An advantage of horizontal-shaft flocculators is that one shaft flocculates a larger basin volume, but with that advantage comes the liability that a significant amount of the mixing capacity is lost when one drive is out of commission. When these units first start rotating, a tremendous torque is suddenly applied. Consequently, most failures occur

(a)

(b)

(c)

Figure 5-18
Views of paddle flocculators: (a) horizontal paddle wheel arrangement and (b) and (c) vertical paddle arrangements. (Courtesy AMWELL. A Division of McNish Corp.)

during startup, especially if the unit is started at maximum rotational speed. Consequently, these mixers should be started at the lowest speed possible to minimize the initial torque.

The power input to the water by horizontal paddles may be estimated from the expression

$$P = \frac{C_D A_P \rho v_R^3}{2} \tag{5-32}$$

where C_D = drag coefficient on paddle (for turbulent flow), unitless
A_P = projected area of paddle, m^2
ρ = fluid density, kg/m^3
v_R = velocity of paddle relative to fluid, m/s

Here, v_R is usually assumed to be 70 to 80 percent of the paddle speed without tank baffles. With tank baffles, 100 percent of the paddle speed is approached. The Reynolds number for a paddle flocculator is

$$\text{Re} = \frac{D_{pw}^2 N \rho}{\mu} \tag{5-33}$$

where D_{pw} = diameter of paddle wheel, m

Criteria that are useful for the design of paddle wheel flocculators are summarized in Table 5-8. Two things can be done to increase or decrease

Table 5-8
Design criteria for paddle wheel flocculator

Parameter	Unit	Value
Diameter of wheel	m	3–4
Paddle board section	mm	100 × 150
Paddle board length	m	2–3.5
$A_{paddle\ boards}$/tank section area	%	<20
C_D (see Eq. 5-32) (for Re > 1000)	L/W = 1	C_D = 1.16
	L/W = 5	C_D = 1.20
	L/W = 20	C_D = 1.5
	L/W ≫ 20	C_D = 1.90
Paddle tip speed	m/s	Strong floc, 4
	m/s	Weak floc, 2
Spacing between paddle wheels on same shaft	m	1
Clearance from basin walls	m	0.7
Minimum basin depth	m	1 m greater than diameter of paddle wheel
Minimum clearance between stages	m	1

the $\overline{G}$ that is produced by a paddle wheel: (1) change the number of paddle boards or (2) change the rotational speed. It is difficult to achieve 50 to 60 s^{-1} with paddle wheel flocculators. Typical values of $\overline{G}$ for paddle wheel flocculators are 20 to 50 s^{-1}.

Example 5-4 Design of horizontal paddle wheel flocculator

Horizontal-shaft paddle wheel flocculators are to be used for flocculation in a water plant with a design flow rate of 150 ML/d (40 mgd) and water temperature of 10°C. Flocculation is to be designed with two parallel trains and each train is to be made of five stages of flocculation (compartments) in series. The total detention time for flocculation is to be 20 min. The paddle wheel flocculators to be used will have the design shown below:

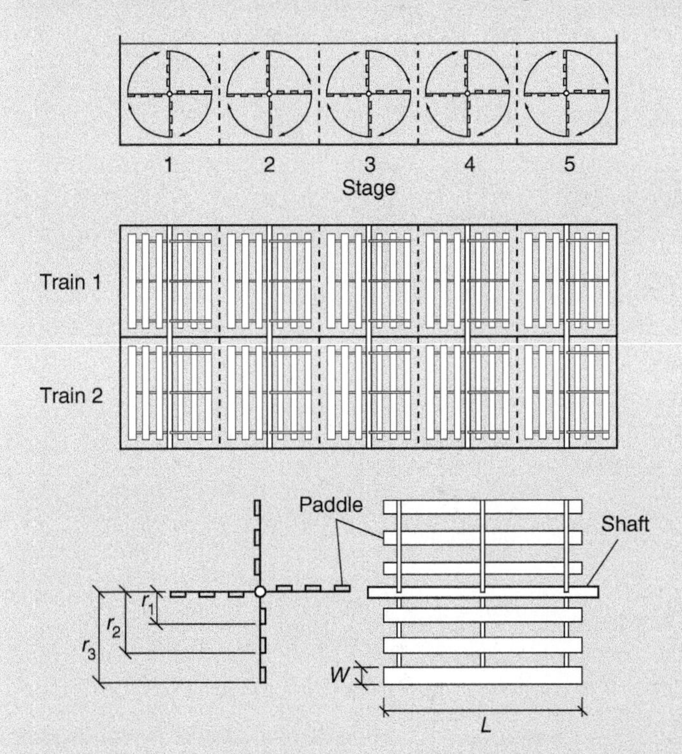

Two paddle wheels will be on the shaft in each compartment. The paddle wheel design should include three paddle boards per arm with leading edges located at 0.67, 1.33, and 2.0 m from the shaft centerline.

The width of the paddle boards is 0.15 m. Determine the following design features for the second compartment in each flocculation train:

1. The dimensions of the compartment in the stage (including the number of paddle wheels and their length)
2. The water power input required to achieve a $\overline{G}$ value of 40 s^{-1}
3. The rotational speed of the paddle shaft

Solution

1. Determine the physical features of the flocculation basins:
   a. The dimensions of the compartment in the first stage are as follows:

$$\text{Basin depth} = (2\text{ m})(2) + 1\text{ m} = 5\text{ m}$$

$$\text{Volume} = \frac{(150\text{ ML/d})(1000\text{ m}^3/\text{ML})(20\text{ min})}{(1440\text{ min/d})(2\text{ trains})(5\text{ stages/train})} = 208.3\text{ m}^3/\text{stage}$$

$$\text{Basin area (plan)} = \frac{208.3\text{ m}^3}{5\text{ m}} = 41.7\text{ m}^2$$

$$\text{Minimum length of stage} = 4\text{ m} + 2(0.5\text{ m}) = 5\text{ m}$$

$$\text{Nominal width} = \frac{41.7\text{ m}^2}{5\text{ m}} = 8.33\text{ m} \quad \text{(perpendicular to flow)}$$

   b. Determine paddle configuration: Two paddle wheel assemblies are needed. Clearance is needed at each end of each paddle and between the paddles.

$$\text{Required clearance} = 2(0.7\text{ m}) + 1\text{ m} = 2.4\text{ m}$$

$$\text{Length of both paddles} = 8.33\text{ m} - 2.4\text{ m} = 5.93\text{ m}$$

$$\text{Length of each paddle} = \frac{5.93\text{ m}}{2} = 2.97\text{ m}$$

   c. Summary:
      Compartment:
      Depth = 5 m
      Length = 5 m
      Width = 8.33 m
      Paddle wheel assemblies:
      Number = 2
      Length of paddles = 2.97 m

2. Determine the water power input required to achieve a $\overline{G}$ value of 40 s^{-1} using Eq 5-13:

$$P = \overline{G}^2 \mu V$$
$$= (40\ s^{-1})^2(1.31 \times 10^{-3} kg/m{\cdot}s)(208.3\ m^3) = 436.7\ kg{\cdot}m^2/s^3$$
$$= 436.7\ J/s$$

3. Determine the power required by the paddles by rearranging Eq. 5-32 and noting that the areas and shapes of the first, second, and third boards are the same; therefore

$$P = \frac{1}{2}\rho C_D A_P \left(v^r_{r,\text{inside paddles}} + v^r_{r,\text{middle paddles}} + v^r_{r,\text{outside paddles}}\right)$$

a. Determine the areas of the boards at each position (inside, middle, and outside):

$$A_P = (2\ \text{wheel})(4\ \text{boards/wheel})(0.15\ \text{m})(2.97\ \text{m}) = 3.56\ \text{m}^2$$

b. Check the length-to-width ratio and select the drag coefficient C_D:

$$\text{Paddle } L/W = 2.97/0.15 = 19.8$$

$$C_D \sim 1.5 \quad \text{(from Table 5-8)}$$

c. Develop parameters needed to determine the paddle power requirements:

$$\text{Velocity of paddles} = v_r = \frac{r2\pi N(0.75)}{60\ \text{s/min}}$$

where r = distance to centerline of paddle from center of rotation
N = shaft rotational speed, rev/min
0.75 = relative velocity of paddle with respect to fluid
$r_{\text{inside}} = r_1 = 0.67 - 0.15/2 = 0.595$ m
$r_{\text{middle}} = r_2 = 1.33 - 0.15/2 = 1.255$ m
$r_{\text{outside}} = r_3 = 2.0 - 0.15/2 = 1.925$ m

d. Substitute known values in the paddle power equation:

$$P = \frac{\rho C_D A_P}{2}\left(v^r_{r,\text{inside paddles}} + v^r_{r,\text{middle paddles}} + v^r_{r,\text{outside paddles}}\right)$$
$$= \frac{\rho C_D A_P}{2}\left[\frac{2\pi N(0.75)}{60\ \text{s/min}}\right]^3 (r_1^3 + r_2^3 + r_3^3)$$

$$= \left[\frac{(999.7\ \text{kg/m}^3)(1.5)(3.56\ \text{m}^3)}{2}\right]\left[\frac{2\pi N(0.75)}{60\ \text{s/min}}\right]^3$$

$$\times [(0.595)^3 + (1.255)^3 + (1.925)^3]$$

$$= (2669.2)(4.85 \times 10^{-4} N^3)(9.321)$$

4. Equate the required power determined in step 2 to meet the $\overline{G}$ value to the power required by the paddles as determined in step 3 above and solve for N:

$$436.7 = (2664.7)(4.85 \times 10^{-4} N^3)(9.321)$$

$$N = \sqrt[3]{\frac{436.7}{(2669.2)(4.85 \times 10^{-4})(9.321)}} = 3.31\ \text{rev/min}$$

5-9 Energy and Sustainability Considerations

The environmental impacts associated with the coagulation and flocculation processes is related to the production and transport of coagulant chemicals to the plant site and the energy associated with the mixing devices. Barrios et al. (2008) performed a life cycle assessment on the operational stage of two water treatment plants that supply Amsterdam with potable water. That analysis found that coagulation contributed about 23 percent of the environmental impact of the operational stage and most of that was due to chemical production. The plants used ferric chloride as a coagulant, and the analysis determined that switching to alum would reduce the environmental impact for the coagulation process by 3.8 percent, and switching to ferric sulfate would reduce it by 15.5 percent. Vince et al. (2008) also found that ferric chloride had more significant environmental impacts than alum; in the case of ozone layer depletion, 1 kg of ferric chloride production had an equivalent impact to 35 kg of alum production. Thus, the selection of the coagulant has an effect on the environmental impacts from a water treatment plant.

Since the goal of coagulation and flocculation processes is to treat particles in water so they can be removed by downstream sedimentation and/or filtration processes, engineers may want to consider upstream pretreatment steps such as bank filtration of surface waters to remove particles prior to treatment. Upstream particle reduction may reduce the required chemical doses.

The energy consumed in the rapid-mix and flocculation processes are relatively low compared to raw-and finished-water pumping at the plant. The energy requirements are related to the mixing requirements (velocity gradient or $\overline{G}$ value) and the duration of mixing. For instance, a pumped diffusion flash mix system with $\overline{G} = 1200\ \mathrm{s}^{-1}$ and $\tau = 1$ s has a theoretical energy requirement of 0.0004 kWh/m^3. Efficiency of energy transfer to the water will increase the energy requirements; a recent design with similar mixing characteristics was calculated to consume 0.0014 kWh/m^3. For flocculation, similar energy for mixing is required regardless of whether the energy powers an external motor (in the case of mechanical mixers) or is consumed as head loss through the process (in the case of hydraulic flocculation). For a flocculation process with $\overline{G} = 50\ \mathrm{s}^{-1}$ and $\tau = 30$ min, the theoretical energy requirement is 0.0013 kWh/m^3. As with the rapid-mix systems, the efficiency of transferring energy to the water will increase the energy consumption.

Energy consumption can be reduced by proper design of pumps and mixers to operate at their best efficiency points. Because the energy requirements of the coagulation and flocculation processes is relatively low, however, there is little to be gained by attempting to reduce the energy consumption by reducing the mixing energy. Mixing energy is an important aspect of the coagulation and flocculation processes, and minimizing energy by providing less mixing may be counterproductive by resulting in increased coagulant use.

5-10 Summary and Study Guide

After studying this chapter, you should be able to:

1. Define the following terms and phrases and describe the significance of each in context of coagulation and flocculation and water treatment:

Brownian motion	hydraulic flocculator	stable particle
coagulant aid	jar testing	suspended particle
coagulant	macroflocculation	sweep floc
charge neutralization	mechanical flocculator	tapered flocculation
destabilization	microflocculation	velocity gradient
electrical double layer	orthokinetic flocculation	zeta potential
enhanced coagulation	perikinetic flocculation	
floc		

2. Explain the role of coagulation and flocculation and explain how they fit into the overall process train for a surface water treatment plant.

3. Describe the origin of surface charge on particles and identify the surface charge (positive or negative) that exists on nearly all natural particles in water.
4. Draw a diagram of the electric double layer around particles and describe how the EDL contributes to particle stability.
5. Describe the mechanisms for particle destabilization.
6. Describe the chemicals used as coagulants and the primary mechanisms that coagulants use to destabilize particles.
7. For a given coagulant dose, calculate coagulant feed rate, the alkalinity consumed, and the amount of dry solids produced.
8. Describe the impact that temperature, concentration of NOM, pH, and other water quality parameters have on coagulation.
9. Explain what enhanced coagulation is.
10. Describe the jar test procedure for how to select the proper coagulant and dose to get effective coagulation.
11. Calculate the power required to achieve a specific value of velocity gradient (mixing energy) for a coagulation or flocculation process.
12. Describe the two primary mechanisms for flocculation and identify which mechanism has the most impact on particles of various sizes.
13. Calculate the collision frequency functions for flocculation.
14. Describe the advantages and disadvantages of common types of flocculators.
15. Evaluate an existing coagulation/flocculation process and explain how each of the following affect the process: coagulant dose, alkalinity, pH, temperature, rapid mixing energy, flocculation energy, and hydraulic residence time of the flocculation basin.
16. Design a flash mix process and/or a flocculation process (basic design criteria, mixing energy, mixer horsepower, and basin dimensions).
17. Describe the factors in the coagulation and flocculation processes that have the largest effects on sustainability and energy consumption and measures that can be considered to minimize the environmental impact of these processes.

Homework Problems

5-1 For the problem below (to be selected by the instructor), calculate (1) the molar concentration of metal ion in the stock coagulant solution, (2) the coagulant feed rate, (3) the alkalinity consumed expressed in units of mg/L as $CaCO_3$, and (4) the amount of precipitate produced.

Coagulant	Coagulant dose (mg/L)	Plant capacity (ML/d)	Stock coagulant chemical: Concentration	Stock coagulant chemical: Specific gravity
A alum	40	125	49% as $Al_2(SO_4)_3 \cdot 14H_2O$	1.33
B ferric chloride	25	75	41% as $FeCl_3$	1.42
C ferric sulfate	25	75	50% as $Fe_2(SO_4)_3$	1.45
D alum	45	20	8.3% as Al_2O_3	1.33
E alum	50	300	8.3% as Al_2O_3	1.33

5-2 An in-line mechanical mixer is used as the rapid mixing device for coagulant addition in a plant treating a flow rate of 40 ML/d. Calculate the power the mixer must dissipate into the water if the mixer is to achieve a velocity gradient of 1200 s^{-1} at a temperature of 15°C, if the hydraulic detention time in the mixing zone is 2 s. If the mixer is 65 percent efficient (that is, 65 percent of the input energy to the mixer is dissipated into the water), what is the required power of the mixer?

5-3 Assume that addition of ferric sulfate to surface water causes ferric hydroxide to precipitate as uniform spherical particles with an initial diameter of 0.5 μm. For a velocity gradient $\overline{G}$ of 60 s^{-1} and temperature of 20°C, calculate the collision frequency functions for flocculation between these particles and viruses (diameter = 25 nm) due to microscale flocculation and macroscale flocculation. Do the same for Cryptosporidium oocysts (diameter = 5 μm). Which mechanism predominates for the flocculation of each type of particle with ferric sulfate?

5-4 Graph the microscale and macroscale collision frequency functions for flocculation between a 1 μm particle and a second particle that ranges in size from 0.01 μm to 10 μm, for a velocity gradient of 50 s^{-1} and temperature of 10°C. What size of particle will have the slowest rate of collisions with 1 μm particles?

5-5 A flocculator is designed to achieve a velocity gradient $\overline{G} = 80$ s^{-1} at 25°C. Assuming that the flocculator operates that the same power, determine $\overline{G}$ imparted to the water at 5°C. What impact will this have on the flocculation process?

5-6 For the flocculation system below (to be selected by instructor), calculate the (1) Reynolds number and (2) amount of power that must be applied to the shaft to rotate it, (3) $\overline{G}$ value, (4) flow rate imparted by the impeller, and (5) tank turnover time.

a. A six-bladed Rushton turbine 2.0 m in diameter rotating at 25 rev/min in a tank 4 m square and 4 m deep, water temperature = 10°C.

b. A 45° pitch-bladed turbine 1.8 m in diameter rotating at 15 rev/min in a tank 3.8 m square and 3.8 m deep, water temperature = 15°C.

c. A 3-bladed hydrofoil 2.2 m in diameter rotating at 18 rev/min in a tank 4 m square and 4 m deep, water temperature = 5°C.

d. A 3-bladed hydrofoil 2.0 m in diameter rotating at 20 rev/min in a tank 4 m square and 4 m deep, water temperature = 25°C.

5-7 The flocculation process in one treatment train at a surface water treatment plant is to be designed to treat a flow rate of 12 ML/d with a hydraulic detention time of 30 minutes. The flocculation basin should be divided into 4 equally sized stages with successively lower velocity gradients in each stage (that is, tapered flocculation). The velocity gradients in stages 1, 2, 3, and 4 should be 80, 65, 50, and 35 s^{-1}, respectively. The water temperature is 15°C. Calculate the volume of the flocculation basin, the volume of each stage, and the power required for the flocculator in each stage.

5-8 Develop a design for the first stage of the flocculation basin in Problem 5-7 using a vertical shaft flocculator. Determine the (1) length, width, and depth of the stage, (2) diameter of the impeller using a 3-bladed hydrofoil, (3) the rotational speed, (4) flow rate imparted by the impeller, and (5) the circulation time in the tank.

5-9 What is the largest paddle wheel that meets the design criteria in Table 5-8? How many paddle boards may be used on such a wheel?

5-10 Design a flocculation compartment for a horizontal-shaft flocculator with two paddles like that in Example 5-8. The water temperature is 10°C. How fast must the paddle wheel rotate in that compartment to generate a $\overline{G}$ value 30 s^{-1}?

References

Amirtharajah, A., and Mills, K. M. (1982) "Rapid-Mix Design for Mechanisms of Alum Coagulation," *J. AWWA*, **74**, 4, 210–216.

AWWA (2011) *Operational Control of Coagulation and Filtration Processes, Manual of Water Supply Practices M37,* 3rd ed., American Water Works Assocation, Denver. CO.

Barrios, R., Siebel, M., van der Helm, A., Bosklopper, K., and Gijzen, H. (2008) "Environmental and Financial Life Cycle Impact Assessment of Drinking Water Production at Waternet," *J. Cleaner Prod.*, **16**, 4, 471–476.

Camp, T. R., and Stein, P. C. (1943) "Velocity Gradients and Hydraulic Work in Fluid Motion," *J. Boston Soc. Civil Eng.*, **30**, 203–221.

Crittenden, J. C., Trussell, R. R., Hand, D. W., Howe, K. J., and Tchobanoglous, G. (2012) *MWH's Water Treatment: Principles and Design*, 3rd ed., Wiley, Hoboken, NJ.

Hahn, H. H., and Stumm, W. (1968) "Kinetics of Coagulation with Hydrolyzed Al(III)," *J. Colloidal Interface Sci.*, **28**, 1, 134–144.

Hunter, R. J. (2001) *Foundations of Colloid Science,* Vols. 1 and 2, Oxford University Press, Oxford, UK.

Kawamura, S. (2000) *Integrated Design and Operation of Water Treatment Facilities,* 2nd ed., Wiley-Interscience, New York.

Letterman, R. D. (1981) Theoretical Principles of Flocculation, paper presented at Seminar Proceedings, AWWA Sunday Seminar Series, AWWA Annual Conference, St. Louis, June.

Letterman, R. D., and Yiacoumi, S. (2011) Coagulation and Flocculation, Chapter 8, in J.K. Edzwald *Water Quality and Treatment, A Handbook on Drinking Water,* 6th ed., American Water Works Association, Denver, CO.

Oldshue, J. Y., and Trussell, R. R. (1991) Design of Impellers for Mixing, in A. Amirtharajah, M. M. Clark, and R. R. Trussell (eds.), *Mixing in Coagulation and Flocculation,* AWWARF, Denver CO.

O'Melia, C. R., Becker, W. C., and Au, K. K. (1999) "Removal of Humic Substances by Coagulation," *Water Sci. Technol.*, **40**, 47–54.

Parks, G. A. (1967) Aqueous Surface Chemistry of Oxides and Complex Oxide Minerals; Isolectric Point and Zero Point of Charge, in *Equilibrium Concepts in Natural Water Systems, Advances in Chemistry Series,* No. 67, American Chemical Society, Washington, DC.

Shäfer, M., Yianneskis, M., Wächter, P., and Durst, F. (1998) "Trailing Vortices around a 45° Pitched Blade Impeller," *AIChE J.*, **44**, 1233–1246.

Stumm, W., and Morgan, J. J. (1996) *Aquatic Chemistry,* 3rd ed., Wiley, New York.

Swift, D. L., and Friedlander, S. K. (1964) "The Coagulation of Hydrosols by Brownian Motion and Laminar Shear Flow," *J. Colloid Sci.*, **19**, 621–647.

Trussell, R. R. (1978) Chapter 3, Predesign Studies, in R. L. Sanks (ed), *Water Treatment Plant Design,* Ann Arbor Science, Ann Arbor, MI.

U.S. EPA (1999) *Enhanced Coagulation and Enhanced Precipitative Softening Guidance Manual.* U.S. EPA, Washington, DC.

Van't Reit, K., Bruijn, W., and Smith, J. (1976) "Real and Pseudo Turbulence in the Discharge Stream from a Rushton Turbine," *Chem. Eng. Sci.*, **31**, 407–412.

Vince, F., Aoustin, E., Bréant, P., and Marechal, F. (2008) "LCA Tool for the Environmental Evaluation of Potable Water Production," *Desalination*, **220**, 1-3, 37–56.

6 Sedimentation

Drinking water sources include primarily freshwater (e.g., surface and groundwater). As discussed in Chap. 2, surface waters contain naturally suspended materials that can be observed in the water as cloudiness or turbidity. If turbid waters are placed into a large quiescent basin and left over time, the suspended material can settle to the bottom of the basin. Particles settle out of solution because they are large enough to settle out by gravitational forces. This process is called *sedimentation*.

Most raw surface waters contain mineral and organic particles. Mineral particles usually have densities ranging from 2000 to 3000 kg/m^3 and will settle out readily by gravity, whereas organic particles with densities of 1010 to 1100 kg/m^3 may require a long time to settle by gravity. Depending on their density, suspended particles larger than 1 μm can be removed by sedimentation. Sedimentation may be employed at the beginning of a water plant to remove mineral particles from highly turbid waters, called presedimentation, or after coagulation and flocculation processes, which is referred to as conventional sedimentation.

The simplest form of sedimentation basin is a large, open structure where the water can flow through quiescently. As water flows through the basin, particles settle to the bottom, where they form a sludge layer that is pushed by mechanical scrapers to a collection trough and removed. The clarified water flows over a weir at the far end of the basin. An example of a rectangular horizontal-flow sedimentation basin is shown on Fig. 6-1a. These basins take a lot of space, so various means have been developed to accelerate the settling process. Figure 6-1b shows an example of tube settlers, which accelerate settling by minimizing the distance particles have to fall before they are removed. Other options for accelerating the sedimentation process are to direct the water to flow up through a fluidized sludge blanket or buoyant media, where collisions with other particles causes aggregation (the larger aggregated particles will settle faster) or to add additives to the water that attract particles and make them heavier.

The removal of suspended matter from water at low cost and low energy consumption is conceptually simple but often involves complications that render proper sedimentation basin design a challenge for many engineers. The performance of a sedimentation basin for a given raw-water quality can be understood with the help of particle-settling theories. When supplemented with the understanding of the practical aspects of sedimentation basin design, sedimentation basins can be designed to perform reliably and consistently.

Particle suspensions are separated into four classifications based on their concentration and morphology, as shown on Fig. 6-2. Type I particles are discrete and do not interfere with one another during settling because the concentration is low and they do not flocculate. Type I suspensions are found in grit chambers, presedimentation basins for sand removal prior to coagulation, and settling of sand particles during backwashing of rapid sand filters. Type II suspensions consist of particles that can adhere

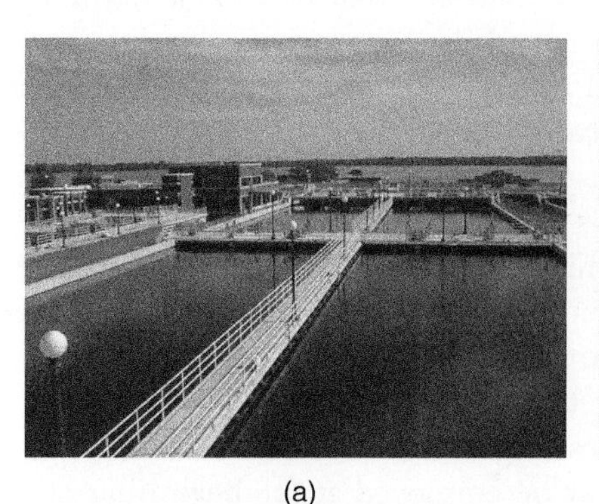

(a)

(b)

Figure 6-1
Photos of (a) rectangular horizontal-flow sedimentation basin and (b) tube settlers in a sedimentation basin.

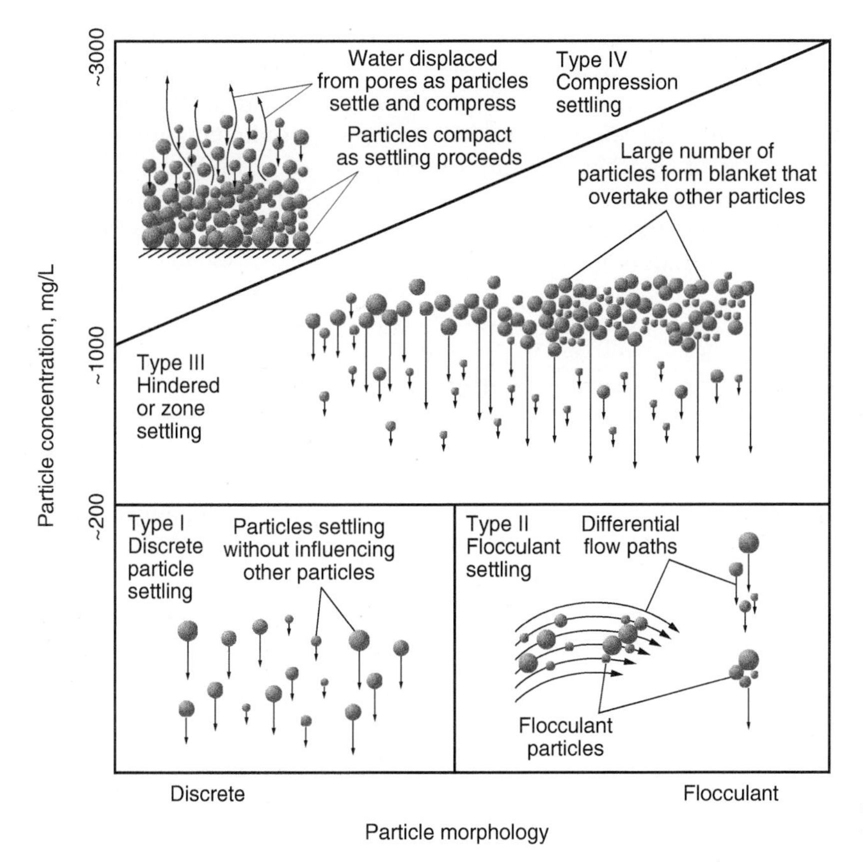

Figure 6-2
Relationship between settling type, concentration, and flocculant nature of particles.

to each other if they bump into each other (i.e., they are capable of flocculating). As particles aggregate and grow in size, they can settle faster. Type II suspensions are found when settling follows coagulation and in most conventional sedimentation basins.

At concentrations higher than Type I and II suspensions, hindered, or Type III, settling occurs. In hindered settling, a blanket of particles is formed. The blanket traps particles below it as it settles; consequently, a clear interface is found above the blanket. The settling velocity of the blanket depends on the suspended solids concentration, with the blanket velocity decreasing with increasing concentration. Type III suspensions are found in thickeners (sludge disposal) and the bottom of some sedimentation basins (e.g., lime-softening sedimentation).

At much higher concentrations than are found in Type III settling, the suspension begins to consolidate slowly. This type of settling or consolidation is known as Type IV settling or compression settling. For Type IV suspensions, the particles may not really settle, and a more correct visualization of what is occurring is that water flows or drains out of a mat of particles very slowly. Type IV suspensions are found in dewatering operations, and once they are dewatered, the suspension may become a paste or cake.

The discussion topics in this chapter include (1) principles of discrete particle settling, (2) discrete particle settling in sedimentation basins, (3) principles of flocculant settling, (4) principles of hindered settling, (5) conventional sedimentation basin design, (6) alternative sedimentation processes, (7) physical factors affecting sedimentation, and (8) energy and sustainability considerations.

6-1 Principles of Discrete (Type I) Particle Settling

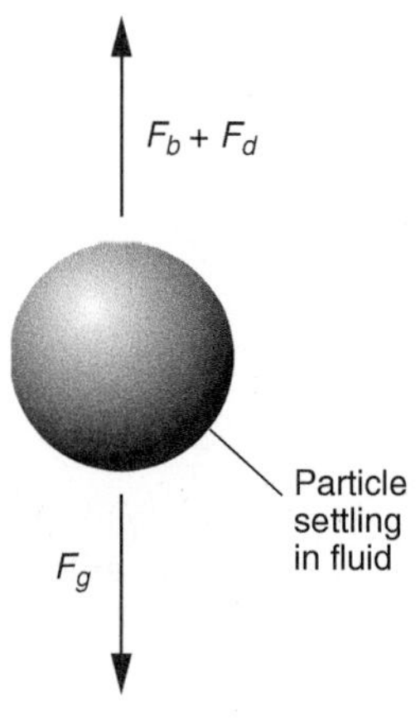

Figure 6-3
Forces acting on a settling particle.

In a dilute suspension, individual particles settle based on their size and density, and do not interact with each other. Settling only occurs if the vertical movement overcomes the random movement of particles. This section develops the equations that describe particle-settling velocity.

A particle settling vertically in a fluid is influenced by gravitational, buoyant, and drag forces. The vertical forces acting on the particle are shown in the free-body diagram on Fig. 6-3 and the force balance is given by the expression

$$\sum F = F_g - F_b - F_d \tag{6-1}$$

where F_g = gravitational force, N
F_b = buoyant force, N
F_d = drag force, N

The force balance is written in the direction of a positive gravitational force. A positive settling velocity means that the particle settles and a negative settling velocity means the particle rises. The gravitational and buoyant forces are given by $F = ma$, as follows:

$$F_g = ma = \rho_p V_p g \tag{6-2}$$

$$F_b = ma = \rho_w V_p g \tag{6-3}$$

where m = mass, kg
a = acceleration, m/s^2
ρ_p = density of particle, kg/m^3
ρ_w = density of water, kg/m^3
V_p = volume of particle, m^3
g = acceleration due to gravity, 9.81 m/s^2

In 1647, Isaac Newton proposed that the drag force could be described by the expression

$$F_d = C_d \rho_w A_p \frac{v_s^2}{2} \tag{6-4}$$

where C_d = drag coefficient, unitless
A_p = projected area of the particle in direction of flow, m^2
v_s = settling velocity of the particle, m/s

If the particles are spherical, the volume and projected area are given by the following expressions:

$$V_p = \frac{\pi}{6} d_p^3 \tag{6-5}$$

$$A_p = \frac{\pi}{4} d_p^2 \tag{6-6}$$

where d_p = particle diameter, m

If a particle starts at rest, it will accelerate due to an imbalance in forces. As the particle velocity increases, the buoyant force remains constant while the drag force increases until the vertical forces are balanced (i.e., $\Sigma F = 0$). At that time, the particle reaches a constant velocity known as terminal settling velocity where the drag force plus the buoyant force equals the gravitational force. For conditions typical in water treatment, the period of initial acceleration is extremely short (< 0.02 s) and not relevant in sedimentation basin design. Substituting Eqs. 6-2 to 6-6 into Eq. 6-1 and setting $\Sigma F = 0$, the following expression for terminal settling velocity is obtained:

$$v_s = \sqrt{\frac{4g(\rho_p - \rho_w)\, d_p}{3 C_d \rho_w}} \tag{6-7}$$

The drag coefficient shown in Eq. 6-4 generally cannot be predicted theoretically. Drag coefficients are determined experimentally by measuring settling velocity in laboratory experiments and then calculating the drag coefficient using Eq. 6-7. Analysis of experimental data reveals that the drag coefficient depends on the Reynolds number, where the Reynolds number is defined as

$$\text{Re} = \frac{\rho_w v_s d_p}{\mu} = \frac{v_s d_p}{\nu} \tag{6-8}$$

where Re = Reynolds number, dimensionless
μ = dynamic viscosity, $N \cdot s/m^2$ or $kg/m \cdot s$
ν = kinematic viscosity, m^2/s

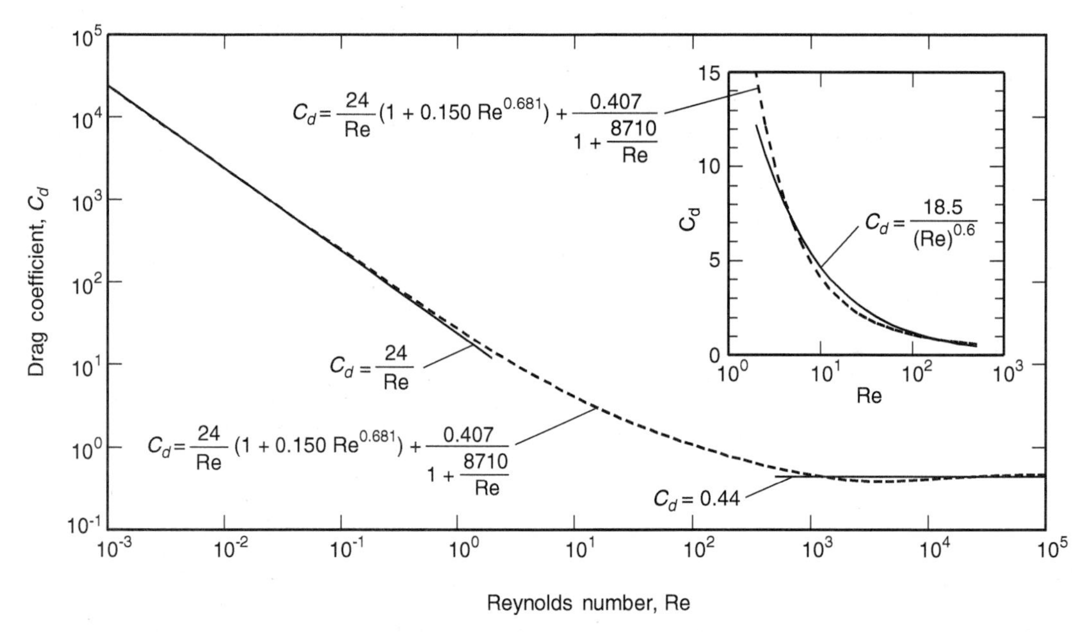

Figure 6-4
Newton's coefficient of drag for varying magnitudes of Reynolds numbers.

The drag coefficient for spheres as a function of Reynolds number is shown on Fig. 6-4. At low Reynolds numbers (laminar region), viscous forces control the drag force and the drag coefficient is larger because momentum is transferred farther into the fluid. As the Reynolds number increases, inertial forces become more significant. In the turbulent regime, inertial forces of displaced fluid control the drag force (the particle basically punches a hole in the fluid equal to the size of the projected area) and the drag coefficient becomes a constant.

Numerous researchers have collected experimental settling velocity data and developed empirical correlations for the drag coefficient. For Re values less than 2×10^5 the following correlation can be used to calculate C_d (Brown and Lawler, 2003):

$$C_d = \frac{24}{Re}(1 + 0.150\,Re^{0.681}) + \frac{0.407}{1 + 8710/Re} \tag{6-9}$$

While this correlation results in a single equation for drag coefficient that covers a wide range of Reynolds numbers, it cannot be substituted into Eq. 6-7 and easily manipulated to produce an equation for settling velocity. Thus, it is useful to develop simpler correlations that are reasonably accurate over smaller ranges of Reynolds numbers. For spherical particles, the drag coefficient C_d can be approximated by the following expressions, depending on the magnitude of the Reynolds number (Clark, 1996):

$$C_d = \frac{24}{\text{Re}} \quad \text{for Re} < 2 \quad \text{(laminar flow)} \tag{6-10}$$

$$C_d = \frac{18.5}{\text{Re}^{0.6}} \quad \text{for } 2 \leq \text{Re} \leq 500 \quad \text{(transition flow)} \tag{6-11}$$

$$C_d = 0.44 \quad \text{for } 500 < \text{Re} \leq 2 \times 10^5 \quad \text{(turbulent flow)} \tag{6-12}$$

For comparison purposes, drag coefficients calculated using Eqs. 6-9 to 6-12 are shown on Fig. 6-4. Equation 6-9 should be considered for rigorous laboratory studies, but in full-scale systems, confounding factors such as heterogeneities in particle size and geometry and currents in fluid flow reduce the usefulness of a highly accurate equation for drag coefficient. Consequently, Eqs. 6-10 to 6-12 are sufficient for full-scale design.

The equations for drag coefficients can be substituted into Eq. 6-7 to develop equations for settling velocity as a function of flow regime. In water treatment, particle settling generally occurs in the laminar and transition flow regimes. For laminar and transition flow, respectively, the equation becomes

$$v_s = \frac{g(\rho_p - \rho_w)\,d_p^2}{18\mu} \quad \text{(laminar flow)} \tag{6-13}$$

$$v_s = \left[\frac{g(\rho_p - \rho_w)\,d_p^{1.6}}{13.9\rho_w^{0.4}\mu^{0.6}}\right]^{1/1.4} \quad \text{(transition flow)} \tag{6-14}$$

Equation 6-13, for spherical particles and laminar flow, is commonly referred to as Stokes' law.

Equation 6-13 or 6-14 is used to calculate the settling velocity depending on whether the particle is in laminar or transition flow. However, the flow regime depends on the settling velocity, so it is not possible to predict a priori which equation applies. It is necessary to calculate the settling velocity using one of the equations, then calculate the flow regime based on the resultant settling velocity, and recalculate the settling velocity with the other equation if necessary. For sand particles (density = 2650 kg/m^3) and a temperature of 20°C, Stokes' law is valid for particles up to 0.13 mm (0.005 in.) in diameter, and Eq. 6-14 is valid for particles up to 1.7 mm (0.067 in.) in diameter. Calculating terminal settling velocity is demonstrated in Example 6-1.

Example 6-1 Calculating terminal settling velocity

Calculate the terminal settling velocity for sand in water at 10°C having particle diameters of 50 and 190 μm and a density of 2650 kg/m^3.

Solution

1. Calculate the settling velocity and Reynolds number for the 50-μm sand particles.
   a. Since the settling velocity is unknown, the Reynolds number is also unknown. First, calculate settling velocity using Eq. 6-13 (Stokes' law). From Table C-1 in App C $\mu = 1.307$ kg/m · s and $\rho_w = 999.7$ kg/m^3 at 10°C:

$$v_s = \frac{(9.81\ \text{m/s}^2)(2650 - 999.7\ \text{kg/m}^3)(50 \times 10^{-6}\ \text{m})^2}{18(1.307 \times 10^{-3}\ \text{kg/m} \cdot \text{s}}$$

$$= 0.00172\ \text{m/s}$$

   b. Check the Reynolds number using Eq. 6-8:

$$\text{Re} = \frac{\rho_w v_s d_p}{\mu} = \frac{(999.7\ \text{kg/m}^3)(0.00172\ \text{m/s})(50 \times 10^{-6}\ \text{m})}{1.307 \times 10^{-3}\ \text{kg/m} \cdot \text{s}} = 0.07$$

   Because Re < 2, laminar flow exists and Stokes' law is valid. The settling velocity of a 50-μm sand particle in water is 0.00172 m/s (6.19 m/h).

2. Calculate the settling velocity and Reynolds number for the 190-μm sand particles:
   a. Calculate the settling velocity using Eq. 6-13:

$$v_s = \frac{(9.81\ \text{m/s}^2)(2650 - 999.7\ \text{kg/m}^3)(1.90 \times 10^{-4}\ \text{m})^2}{18(1.307 \times 10^{-3}\ \text{kg/m} \cdot \text{s}}$$

$$= 0.0248\ \text{m/s}$$

   b. Check the Reynolds number using Eq. 6-8:

$$\text{Re} = \frac{(999.7\ \text{kg/m}^3)(0.0248\ \text{m/s})(1.90 \times 10^{-4}\ \text{m})}{1.307 \times 10^{-3}\ \text{kg/m} \cdot \text{s}} = 3.61$$

   Since Re > 2.0, Stokes' law is not valid.

   c. Calculate the settling velocity using Eq. 6-14:

$$v_s = \left[\frac{(9.81\ \text{m/s}^2)(2650 - 999.7\ \text{kg/m}^3)(1.90 \times 10^{-4}\ \text{m})^{1.6}}{13.9(999.7\ \text{kg/m}^3)^{0.4}(1.307 \times 10^{-3}\ \text{kg/m} \cdot \text{s})^{0.6}}\right]^{1/1.4}$$

$$= 0.0207\ \text{m/s}$$

d. Check the Reynolds number using Eq. 6-8:

$$\mathrm{Re} = \frac{(999.7\ \mathrm{kg/m^3})(1.90 \times 10^{-4}\ \mathrm{m})(0.0207\ \mathrm{m/s})}{1.307 \times 10^{-3}\ \mathrm{kg/m \cdot s}} = 3.0$$

Because Re > 2, transition flow exists and Eq. 6-14 is valid. The settling velocity of a 190-μm sand particle in water is 0.0207 m/s (74.5 m/h).

If the particles are hard spheres, the settling velocity as a function of particle size does follow Eq. 6-13 or 6-14. However, flocculated particles have snowflake or fractal morphology and are composed of many flocculated small particles. Consequently, fractal particles do not settle as rapidly as would be estimated using a hard-sphere model. Attempts have been made to quantify the irregularity of fractal particles and their deviation from spherical geometry. A detailed discussion of fractals can be found in Logan (1999).

6-2 Discrete Settling in Ideal Rectangulor Sedimentation Basins

Particle settling is dependent on the nature of the particle and geometry of the sedimentation process. As introduced in Sec. 6-1, there are four main types of particle settling (see Fig. 6-2). The analysis of discrete particle settling in sedimentation basins (Type I settling), based on the principles presented in Sec. 6-1, is introduced in this section.

Camp (1936) developed a rational theory for the removal of discrete particles in an ideal sedimentation basin. Camp divided a settling tank into four zones, as illustrated on Fig. 6-5a. The inlet, sludge, and outlet zones are under the influence of entrance, exit, and wall effects, so water and particle flow is not smooth and ideal settling does not occur. However, there is a large region of the tank where conditions are more ideal and settling can be calculated using the fundamental equations that have been developed. This is the settling zone. In addition, the following assumptions were made by Camp to develop a theoretical basis for the removal of discrete particles: (1) plug flow conditions exist in the settling zone, (2) there is uniform horizontal velocity in the settling zone, (3) there is uniform concentration of all size particles across a vertical plane at the inlet end of the settling zone, (4) particles are removed once they reach the bottom of the settling zone, and (5) particles settle discretely without interference from other particles at any depth.

Particle trajectories have two components in the settling zone: the settling velocity v_s and the fluid velocity v_f, as shown on Fig. 6-5b. For a rectangular sedimentation basin the fluid velocity is constant. The settling velocity for discrete particles is also constant because the particles do not flocculate or

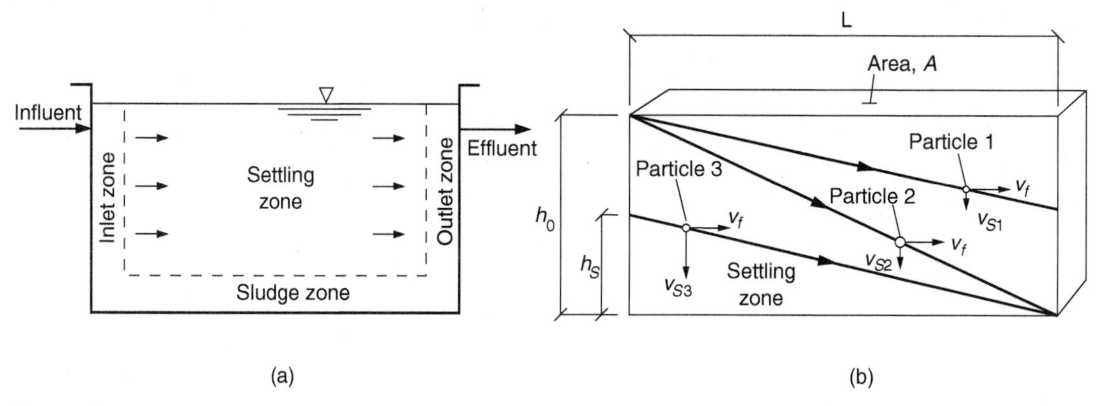

Figure 6-5
Horizontal-flow rectangular sedimentation basin illustrating (a) functional regions within the basin and (b) discrete particle trajectories in the settling zone.

interfere with one another. Since both horizontal and vertical components of the velocity are constant, the particle trajectories are linear. As noted above, every particle that enters the sludge zone is removed. A particle from the inlet zone that enters at the top of the basin and settles to the sludge zone just before the outlet is assigned a settling velocity of v_c, or a critical settling velocity (particle 2 on Fig. 6-5b). The critical particle settling velocity is given by the following equation:

$$v_c = \frac{h_o}{\tau} \tag{6-15}$$

where v_c = particle settling velocity such that particle at surface of inlet is removed in sludge zone just before outlet, m/h
h_0 = depth of sedimentation basin, m
τ = hydraulic detention time of sedimentation basin, h

The critical settling velocity may be defined as the overflow rate using the relationships

$$v_c = \frac{h_0}{\tau} = \frac{h_o Q}{h_o A} = \frac{Q}{A} = v_{\mathrm{OF}} \tag{6-16}$$

where v_{OF} = overflow rate, $m^3/m^2 \cdot h$ (equal to v_c)
A = area of top of basin settling zone (see Fig. 6-5b), m^2
Q = process flow rate, m^3/h

The inlet zone is assumed to be homogenous; therefore particles may enter the settling zone at any height h_s. Any particles in the inlet zone with a settling velocity v_s greater than or equal to the critical settling velocity v_c

will be removed regardless of the starting position because their trajectories will take them into the sludge zone before they exit the basin.

Particles with a settling velocity less than v_c may also be removed, depending on their position at the inlet, as shown on Fig. 6-5b. Particles at the top of the basin will pass through the settling zone and exit in the outlet zone and will not be removed. However, particles starting at position h_s and lower will enter the sludge zone before exiting the basin and will be removed. The fraction of particles that will be removed is given by the expression

$$\text{Fraction of particles removed} = \frac{h_s}{h_0} = \frac{h_s/\tau}{h_0/\tau} = \frac{v_s}{v_c}(v_s < v_c) \qquad (6\text{-}17)$$

where h_s = height of particle from bottom of tank at position entering settling zone, m

v_s = particle settling velocity smaller than v_c, m/h

Other terms are as defined above. Removal of particles as a function of size is demonstrated in Example 6-2.

Example 6-2 Particle removal in sedimentation basin

Calculate the particle removal efficiency in a rectangular sedimentation basin with a depth of 4.5 m, width of 6 m, length of 35 m, and process flow rate of 525 m^3/h. Compute the required sedimentation basin design parameters and plot the influent and effluent particle concentrations as a function of particle size using a histogram. Assume the following influent particle-settling characteristics [adapted from Tchobanoglous et al. (2003)].

Settling Velocity, m/h	Number of Particles, #/mL
0–0.4	511
0.4–0.8	657
0.8–1.2	876
1.2–1.6	1168
1.6–2.0	1460
2.0–2.4	1314
2.4–2.8	657
2.8–3.2	438
3.2–3.6	292
3.6–4.0	292
Total	7665

Solution

1. Compute the sedimentation basin overflow rate and critical settling velocity using Eq. 6-16:

$$v_{OF} = v_c = \frac{Q}{A} = \frac{(525\ \text{m}^3/\text{h})}{(6\ \text{m})(35\ \text{m})} = 2.5\ \text{m}^3/\text{m}^2 \cdot \text{h}$$

2. Compute the percent removal of particles in each size range using a data table:
   a. Compute the average settling velocity for each particle size range; see column 2 in the table below.
   b. Compute the fraction of particles removed using Eq. 6-17. For particles with an average settling velocity of 1.0 m/h, the fraction of particles removed is $(1.0\ \text{m/h})/(2.5\ \text{m}^3/\text{m}^2 \cdot \text{h}) = 0.4$; see column 4. Note that for particle-settling ranges with a fraction removed greater than 1, a value of 1 should be used.
   c. Estimate the number of particles that will be removed and remaining in each size range. The number of particles removed is determined by multiplying the influent particle concentration for a given settling velocity range by the corresponding percent removal, $(876)(0.4) = 350$; see column 5. The number of remaining particles is determined by subtracting the removed particles from the influent particles for each size range, $876 - 350 = 526$, for the range 0.8 to 1.2 m/h; see column 6.
   d. The remaining values are summarized in the following table:

Settling Velocity, m/h (1)	Average Settling Velocity, m/h (2)	Number of Influent Particles, #/mL (3)	Fraction of Particles Removed (4)	Number of Particles Removed, #/mL (5)	Number of Particles in Effluent, #/mL (6)
0–0.4	0.2	511	0.08	41	470
0.4–0.8	0.6	657	0.24	158	499
0.8–1.2	1.0	876	0.40	350	526
1.2–1.6	1.4	1168	0.56	654	514
1.6–2.0	1.8	1460	0.72	1051	409
2.0–2.4	2.2	1314	0.88	1156	158
2.4–2.8	2.6	657	1	657	0
2.8–3.2	3.0	438	1	438	0
3.2–3.6	3.4	292	1	292	0
3.6–4.0	3.8	292	1	292	0
Total		7665		5090	2575

3. Compute the overall particle removal efficiency:

$$\text{Removal efficiency} = \frac{5090}{7665} = 0.664 = 66.4\%$$

4. Plot the influent and effluent particle concentrations for each settling velocity range using a histogram.

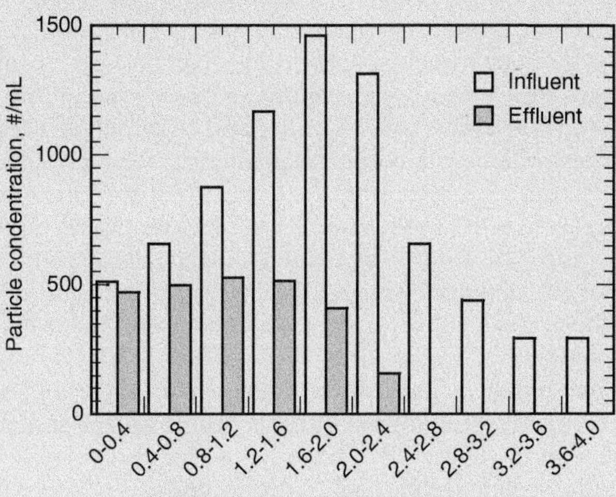

6-3 Principles of Flocculant (Type II) Particle Settling

Type II settling typically occurs in conventional sedimentation basins following coagulation. There are two principal mechanisms of flocculation during sedimentation: (1) differences in the settling velocities of particles whereby faster settling particles overtake those that settle more slowly and coalesce with them and (2) velocity gradients within the liquid that cause particles in a region of a higher velocity to overtake those in adjacent stream paths moving at slower velocities.

Advantages of Flocculant Settling

Flocculation within a sedimentation basin is considered beneficial for two principal reasons. First, the combination of smaller particles to form larger particle aggregates results in faster settling particles because of the increase in size. Second, flocculation tends to have a sweeping effect in which large particles settling at a velocity faster than smaller particles tend to sweep some of the smaller particles from suspension. Consequently, many tiny particles and particles that settle slowly are removed. The net effect of flocculation during settling is a reduction in the size of the sedimentation basin necessary for effective clarification or improved water quality exiting the sedimentation basin.

Analysis of Flocculant Settling

Design equations for Type II suspensions using the flocculation equations have proven to be impractical for sedimentation basin design. Design of sedimentation basins is usually based on overflow rates and detention times that have been reported in design manuals as guidelines or by regulatory agencies. For waters with unusual settling characteristics, a number of investigators have developed design equations based on column experiments. In a technique developed by O'Connor and Eckenfelder (1958), measured solids concentrations taken at regular intervals throughout the depth of a quiescent settling column, slightly deeper than the proposed sedimentation basin, are related to the overall percent removal at a particular basin residence time. The water to be treated is placed in the column and allowed to settle for the detention time of the basin. The effluent concentration is equal to the average concentration in the column. The average concentration can be obtained by draining off the settled solids and then mixing the particles remaining in the column (typically with air) and then sampling the mixed liquid. The concept behind this approach is that the column represents a fluid element that travels as a plug through the basin and has a settling time equal to the basin residence time.

Several fundamentals of sedimentation basin design that are different from design principles arrived at through discrete particle settling have been established. The depth of the basin is important because flocculent particles tend to grow in size during their downward movement through the basin. A greater depth facilitates floc growth and allows for sweep flocculation at high solids concentrations at the bottom of the basin. In general, more flocculant particles are removed in deeper basins.

6-4 Principles of Hindered (Type III) Settling

Type III settling, also known as zone settling, occurs when the settling velocities of particles are affected by the presence of other particles. When particles are dispersed in solution, the movement of the fluid that is displaced by the particle motion has little impact on the drag force. However, when particle concentrations are high enough to restrict the fluid velocity fields around individual particles, a settling particle experiences increased frictional forces. In water treatment, hindered settling typically occurs in the lower regions of the sedimentation basin, where the concentration of suspended particles is highest. When Type III settling occurs, particle aggregates form a blanket of particles with a distinct interface with the clarified liquid in the basin. Zone settling is of primary importance in water treatment in sludge thickening and dewatering operations, as discussed in Chap. 14.

Solids Flux Analysis

The solids flux in a sedimentation basin or solids thickener (see Fig. 6-6) is comprised of the downward movement of particles due to gravity settling and the downward movement of particles due to fluid flow toward the underdrain, as shown in the expression

$$J_T = J_s + J_u \tag{6-18}$$

where J_T = total solids flux toward the bottom of the basin, $kg/m^2 \cdot h$
J_s = solids flux due to particle settling by gravity, $kg/m^2 \cdot h$
J_u = solids flux due to fluid flow from the underflow, $kg/m^2 \cdot h$

To determine the solids flux from gravity settling J_s, the depth of the blanket interface is measured as a function of time in a laboratory column that is initially uniformly mixed with a specified solids concentration C. Figure 6-7a displays a plot of data from a settling column test. The settling velocity is determined from the initial slopes of the concentration curves. The solids flux values due to particle settling is determined by multiplying the concentrations of particles by their respective initial settling velocities as

$$J_s = v_s C \tag{6-19}$$

where v_s = settling velocity for particle concentration C, m/h
C = suspended solids concentration, kg/m^3

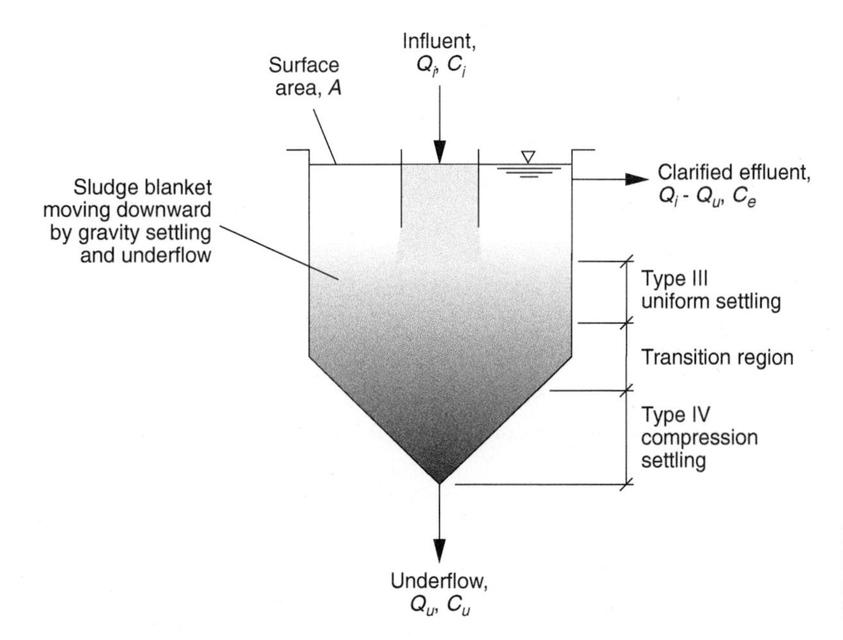

Figure 6-6
Diagram of sludge thickener or sedimentation basin where thickening is taking place.

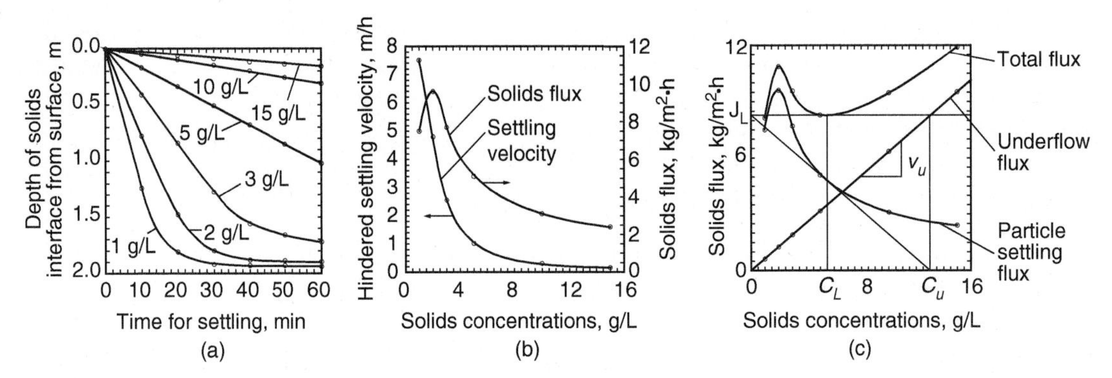

Figure 6-7
Analysis of zone settling data shown in Table 6-1: (a) interfacial settling velocity as function of concentration, (b) solids flux due to settling as function of concentration, and (c) limiting solids flux analysis components for Type III settling. [Adapted from Tchobanoglous et al. (2003).]

The resultant settling velocity and solids flux values are reported in Table 6-1 for the data presented on Fig. 6-7a. The initial settling velocities and the values for solids flux as a function of solids concentration are presented graphically on Fig. 6-7b.

The solids flux due to the fluid flow to the underdrain, J_u, is defined as

$$J_u = \frac{Q_u C}{A} = v_u C \tag{6-20}$$

where Q_u = flow rate leaving the bottom of basin/thickener, m^3/h
A = cross-sectional area of basin, m^2
v_u = bulk downward fluid velocity, m/h

The total flux at a suspended solids concentration C can be written in terms of bulk fluid velocity and sludge blanket settling velocity by substituting Eqs. 6-19 and 6-20 into Eq. 6-18, resulting in the equation

$$J_T = (v_s + v_u)C \tag{6-21}$$

Table 6-1
Settling velocity and solids flux values

Solids Concentration, C, g/L	Initial Settling Velocities, v_i m/min	Initial Settling Velocities, v_i m/h	Solids Flux, J_s, kg/m^2 · h
1	0.125	7.50	7.5
2	0.080	4.80	9.6
3	0.043	2.55	7.7
5	0.017	1.02	5.1
10	0.005	0.31	3.1
15	0.003	0.16	2.4

where terms are as defined previously. The use of the solids flux equations to size solids thickening basins is discussed below.

Limiting Flux Rate

The limiting flux is the point at which the mass flow of solids entering the thickener is equal to the mass flow of solids leaving the thickener. If solids loading exceed the limiting flux rate, solids will accumulate and eventually overflow. If the mass flow of solids entering the thickener is less than the mass flow of solids leaving the thickener and there is continuous constant underflow pumping, all the thickened solids will be removed from the thickener. The solids loading for a basin can be determined from an analysis of the limiting flux rate. To determine the limiting flux rate, an underdrain solids concentration C_u must be selected. On Fig. 6-7c an underdrain concentration of 13 g/L is shown. The underdrain solids concentration is typically determined based on the requirements of downstream residuals processing operations. The limiting solids flux J_L for a given C_u can be determined by drawing a line from the desired underdrain concentration on the x axis and through the tangent to the particle-settling flux curve. The tangent point of the particle-settling flux curve corresponds to the limiting particle concentration C_L, about 5.5 g/L for the case shown on Fig. 6-7c. The intersect of the tangent line with the ordinate axis is the value of the limiting solids flux J_L for the given particle-settling flux curve and selected underdrain concentration C_u. The limiting solids flux shown on Fig. 6-7c is 8.25 kg/m^2 · h. The downward velocity of the bulk fluid may be determined using the relationship

$$v_u = \frac{J_L}{C_u} \tag{6-22}$$

where v_u = downward velocity of bulk fluid, m/h
J_L = limiting solids flux, kg/m^2 · h
C_u = concentration of solids in underflow, kg/m^3

Area Required for Solids Thickening

The flow rate through the underdrain can be estimated using the following mass balance analysis. For the solids thickener shown on Fig. 6-6, a solids mass balance is given by the expression

Suspended solids entering thickener
= suspended solids leaving thickener in effluent
+ settled solids leaving thickener in underflow (6-23)

$$Q_i C_i = (Q_i - Q_u) C_e + Q_u C_u \tag{6-24}$$

where Q_i = influent flow rate to basin/thickener, m^3/h
C_i = influent suspended solids concentration, mg/L
Q_u = flow rate leaving the bottom of basin/thickener, m^3/h

C_u = solids concentration leaving bottom of basin/thickener, mg/L
C_e = effluent solids concentration, mg/L

If it is assumed that $C_e \ll C_u$ and $C_e \ll C_i$, C_e may be considered negligible and the following expression is obtained for the flow rate through the underdrain:

$$Q_u = \frac{Q_i C_i}{C_u} \tag{6-25}$$

Once the flow rate of the underflow is determined, the area required for the basin can be determined using Eq. 6-22, substituting Q_u/A for v_u, and solving for A, as shown below:

$$A = \frac{Q_u C_u}{J_L} = \frac{Q_i C_i}{J_L} \tag{6-26}$$

where A = area required for thickening, m^2

Other terms are as defined previously. Sizing of a thickener is demonstrated in Example 6-3.

Example 6-3 Area required for thickening

Determine the area required for thickening for a basin that receives 600 mg/L of solids and a flow rate of 4000 m^3/h for an underdrain concentration of 15,000 mg/L. Assume the settling velocity of the sludge blanket follows the relationship plotted on Fig. 6-7b. Also determine J_L, C_L, and Q_u.

Solution

1. Determine J_L and C_L. From the data plotted on Fig. 6-7b and an underflow solids concentration of $C_u = 15{,}000$ mg/L, the gravity flux is determined by drawing a line from the x axis at a solids concentration of 15,000 mg/L to the y axis such that it is tangent to the solids flux curve and intersects the y axis. The point at which the line intersects the y axis is the limiting gravity flux and is equal to 7.45 $kg/m^2 \cdot h$. The value for C_L can also be determined by drawing a vertical line from the tangent point to the x axis and is equal to 6500 mg/L.

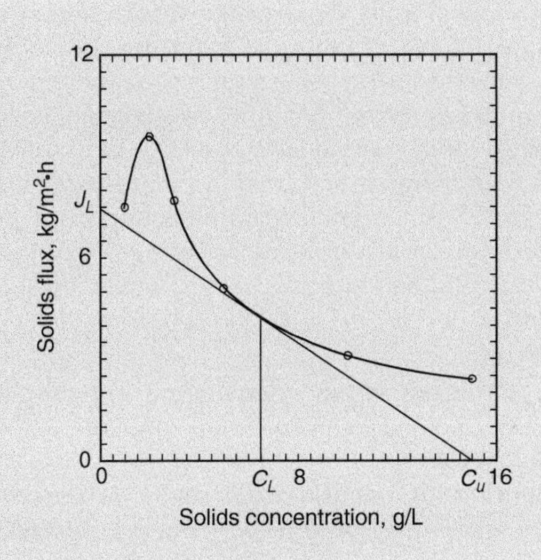

2. Determine Q_u using Eq. 6-25:

$$Q_u = \frac{Q_i C_i}{C_u} = \frac{(4000\ \text{m}^3/\text{h})(600\ \text{mg/L})}{15{,}000\ \text{mg/L}} = 160\ \text{m}^3/\text{h}$$

3. Determine the area for thickening, A, using Eq. 6-26:

$$A = \frac{Q_i C_i}{J_L} = \frac{(4000\ \text{m}^3/\text{h})(600\ \text{g/m}^3)(1\ \text{kg}/10^3\ \text{g})}{7.45\ \text{kg/m}^2 \cdot \text{h}} = 322\ \text{m}^2$$

4. Summary:

$$J_L = 7.45\ \text{kg/m}^2 \cdot \text{h} \qquad C_L = 6500\ \text{mg/L}$$

$$Q_u = 160\ \text{m}^3/\text{h} \qquad A = 322\ \text{m}^2$$

6-5 Conventional Sedimentation Basin Design

Sedimentation basin design is based on applied theoretical principles and practical considerations, including basin location in the overall process treatment train, basin size, and basin geometry. Topics discussed in this section include design considerations for conventional sedimentation processes utilizing rectangular and circular basin configurations. Alternative sedimentation processes for improved performance are described in Sec. 6-6.

Rectangular Sedimentation Basins

Many sedimentation basins are rectangular with horizontal flow, as shown on Figs. 6-1a and 6-8a. A minimum of two basins should be provided so that one may be taken off-line for inspection, repair, and periodic cleaning while the other basin(s) remain in operation. Basins arranged longitudinally side by side, sharing a common wall, have proven to be a cost-effective approach. In addition, a flocculation process may be incorporated into the head end of the sedimentation basin, minimizing piping, improving flow distribution to sedimentation basins, and potentially reducing floc damage during transfer between the flocculation stage and the sedimentation stage.

INLET STRUCTURE

The inlet to a rectangular sedimentation basin should be designed to distribute the flocculated water uniformly over the entire cross section of the basin at low velocity.

When sedimentation basins are fed from a common channel, the basin inlet structure may consist of weirs or submerged ports, with a permeable

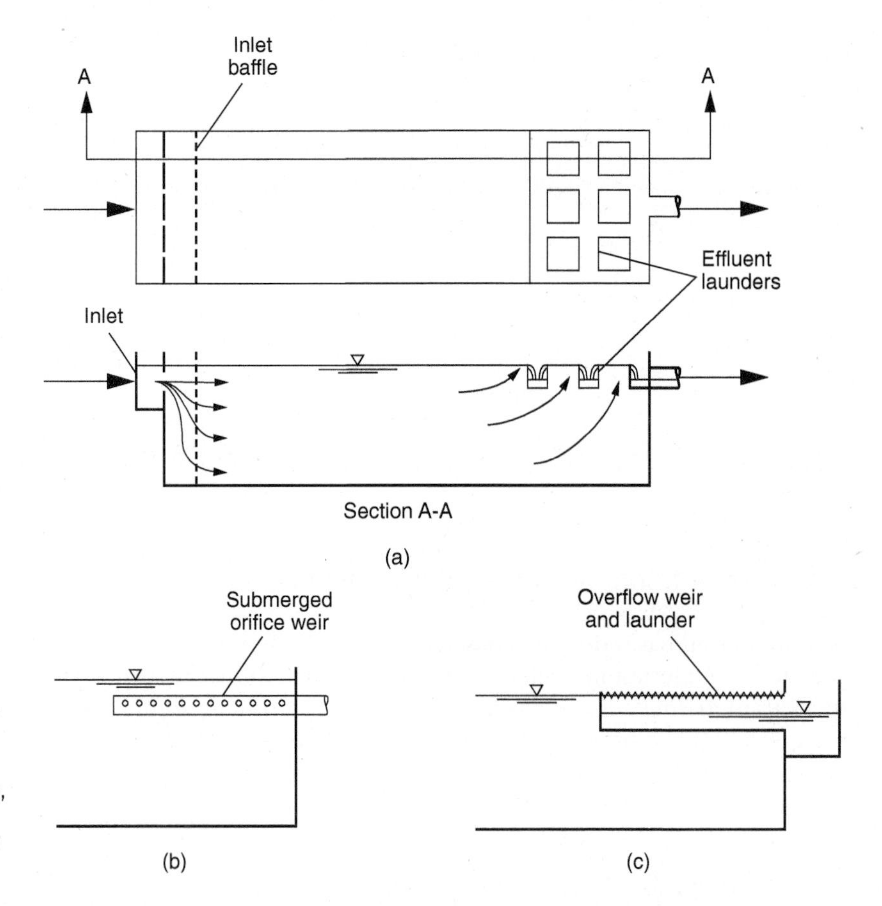

Figure 6-8
Rectangular, horizontal-flow sedimentation basin with various outlet types: (a) inboard effluent launders, (b) submerged orifice withdrawal, and (c) overflow weir and launder.

baffle about 2 m (6.5 ft) downstream in the sedimentation basin. Uniform or equal distribution of flow to each sedimentation basin is also essential.

A well-designed inlet permits water from the flocculation basin to enter directly into the sedimentation basin without channels or pipelines. A diffuser wall is one of the most effective and practical flow distribution methods used at the basin inlet when the flocculation basin is directly attached to the sedimentation basin. A diffuser wall is simply a wall with many small holes strategically placed to uniformly redistribute the flow of water. The openings should be small holes [100 to 200 mm (4 to 8 in.) diameter circular or equivalent] of identical size, evenly distributed on the wall.

SETTLING ZONE

The basic design criteria to be considered for the horizontal-flow settling zone are (1) surface loading rate, (2) effective water depth, (3) detention time, (4) horizontal-flow velocity, and (5) minimum length-to-width ratio. Typical design parameters used for rectangular sedimentation facilities are summarized in Table 6-2 and discussed below.

Surface Loading Rate and Settling Velocity

As discussed in Sec. 6-2, the efficiency of an idealized, horizontal-flow settling tank is solely a function of the settling velocity of discrete particles and of the surface loading rate (the flow rate of the basin divided by the surface area) and is independent of the basin depth and detention time. However, most settling basins treat flocculated suspended matter (not discrete particles) and do not have idealized flow patterns. Furthermore, flocculent particles may increase in size while in the basin, making the depth of the basin an additional design parameter. Factors such as mechanical sludge removal equipment, sun and wind, and flow velocity also affect the minimum basin depth. The settling velocities of selected floc particles are presented in Table 6-3.

Table 6-2
Typical design criteria for horizontal-flow rectangular tanks

Parameter	Units	Value
Minimum number of tanks	Unitless	2
Surface loading rate (overflow rate)	m/h (gpm/ft^2)	1.25–2.5 (0.5–1.0)
Detention time	h	1.5–4
Water depth	m (ft)	3–5 (10–16)
Length-to-width ratio	Dimensionless	minimum 4:1 preferred >5:1
Horizontal mean-flow velocity	m/min (ft/min)	0.3–1.1 (1–3.5)
Reynolds number	Dimensionless	<20,000
Froude number	Dimensionless	$> 10^{-5}$

Source: Adapted from Kawamura (2000).

Table 6-3
Settling velocity of selected floc types

Floc Type	Setting Velocity at 15°C	
	m/h	ft/min
Small fragile alum floc	2–4.5	0.12–0.24
Medium-sized alum floc	3–5	0.18–0.28
Large alum floc	4.0–5.5	0.22–0.30
Heavy lime floc (lime softening)	4.5–6.5	0.25–0.35
Fe floc	2–4	0.12–0.22
PACl floc	2–4	0.12–0.22

Horizontal-Flow Velocity

Settling characteristics, surface loading rate, and detention time are generally the main basis of design. For "high-rate" horizontal-flow basins (detention times less than 2 h) the Reynolds and Froude numbers can be used as a check on turbulence and backmixing. The Reynolds number is determined as

$$\mathrm{Re} = \frac{\rho_w v_f R_h}{\mu} = \frac{v_f R_h}{\nu} \tag{6-27}$$

where
Re = Reynolds number based on hydraulic radius, dimensionless
v_f = average horizontal fluid velocity in tank, m/s
R_h = hydraulic radius, $= A_x/P_w$, m
A_x = cross-sectional area, m^2
P_w = wetted perimeter, m
ρ_w = density of water, $\mathrm{kg/m^3}$
μ = dynamic viscosity, $\mathrm{N \cdot s/m^2}$ or $\mathrm{kg/m \cdot s}$
ν = kinematic viscosity, $\mathrm{m^2/s}$

The Froude number may be determined using the equation

$$Fr = \frac{v_f^2}{gR_h} \tag{6-28}$$

where
Fr = Froude number, dimensionless
g = acceleration due to gravity, 9.81 $\mathrm{m/s^2}$

For high-rate basins, recommended values for settling zone design determined using Eqs. 6-27 and 6-28 are $\mathrm{Re} < 20{,}000$ and $\mathrm{Fr} > 10^{-5}$ (Kawamura, 2000). These dimensionless numbers are useful for design guidelines because a large Reynolds number indicates a high degree of turbulence and a low Froude number implies that the water flow is not dominated by horizontal flow, and backmixing may occur. The criteria for Re and Fe are of less significance and may be exceeded for conservatively designed basins; a basin with an appropriate length-to-width ratio, low overflow rate, and detention time of 3 to 4 h will often achieve satisfactory performance even if the Re and Fr criteria are not met. It is more important to check these criteria for high-rate rectangular basins with detention times of 2 h or less.

Placing longitudinal baffles (in the direction of flow) can help alleviate poor sedimentation basin performance. Adding longitudinal baffles divides the basin into parallel channels that increases the length-to-width ratio, reduces the Reynolds number, and increases the Froude number. To allow for sludge removal equipment, the baffles should be separated by at least 3 m (10 ft) and can be made of wooden planks or concrete. Baffles should never be placed in sedimentation basins where they would cause serpentine flow (180° turns) to occur because the turbulence that is caused by abrupt turns will significantly reduce particle settling.

Length-to-Width Ratio

In general, long, narrow basins are preferred to minimize short circuiting. To promote plug flow in rectangular sedimentation basins, a minimum length-to-width ratio of 4 : 1 should be maintained.

OUTLET STRUCTURE

Water leaving the sedimentation basin should be collected uniformly across the width of the basin. Outlet structures for rectangular tanks are generally composed of launders running parallel to the length of the tank, shown on Fig. 6-8, or a simple weir at the end of the tank. The water level in the sedimentation basin is controlled by the end wall or overflow weirs. V-notch weirs are commonly attached to launders and broad-crested weirs are attached to the end wall. Long weirs have at least three advantages for rectangular sedimentation tanks: (1) a gradual reduction of flow velocity toward the end of the tank, (2) minimization of wave action from wind, and (3) collection of clarified water located in the middle of the tank when a distinct density flow occurs in the basin. A disadvantage of long effluent launders is they are expensive and the support columns for them must be designed so they do not interfere with sludge collection devices. With proper sedimentation basin design, long effluent launders may provide only a marginal improvement in effluent turbidity, and a simple weir at the end of the tank may provide a satisfactory result.

SLUDGE ZONE

Sludge collects in the bottom of the sedimentation basin, and in a rectangular basin, more sludge settles near the inlet than the outlet end of the basin. To facilitate sludge removal, the bottom of the basin is typically sloped toward a sludge hopper.

Manufacturers produce several types of mechanical collectors for rectangular sedimentation basins. The major types of mechanical collectors for

Figure 6-9
Chain-and-flight-type sludge collector.

rectangular basins ranked in order of cost are (1) chain-and-flight (plastic material) collectors (see Fig. 6-9), (2) a traveling bridge with sludge-scraping squeegees and a mechanical cross collector at the influent end of the tank, (3) a traveling bridge with sludge suction headers and pumps, and (4) sludge suction headers supported by floats and pulled by wires.

The standard maximum width of the chain-and-flight sludge collector is 6 m (20 ft), and the operation and maintenance cost usually increases for the chain-and-flight collectors if the length of the basin exceeds 60 m (200 ft). When mechanical scraper units are used, the velocity of the scraper should be kept below 18.0 m/h (60 ft/h) to prevent resuspending the settled sludge. For suction sludge removal units, the velocity can be 60 m/h (200 ft/h) because the principal concern is not the resuspension of settled sludge but the disruption of the settling process.

Traveling bridges can span up to 30 m (100 ft) with widths 12 to 30 m (40 to 100 ft) usually being the most cost effective. Because the width of sedimentation basins is often less than 15 m (50 ft), using one bridge to span two or three tanks can significantly reduce the capital investment for sludge removal equipment. Both the drain and sludge draw-off pipelines should have a minimum diameter of 150 mm (6 in.) to prevent clogging problems. Additionally, traveling bridges are susceptible to high winds, and in cold-weather climates, the pumps and piping need cold-weather protection as they are exposed above the water. Sedimentation basin design is demonstrated in Example 6-4.

Example 6-4 Sedimentation basin design

Design the horizontal-flow rectangular sedimentation basins for a water treatment plant that must treat a flow of 200 ML/d (52.8 mgd). The water is coagulated with alum, and experience with similar plants on the same river suggests that an overflow rate of 1.5 m/h is appropriate. The minimum water temperature is 10°C. Your design should be suitable for chain-and-flight sludge removal equipment. Your design should include the number of basins, length, width, and depth of each basin and the number of baffles within each basin, if any. Verify that your design meets the criteria in Table 6-2 with respect to depth, L:W ratio, detention time, and Reynolds number (if detention time is less than 2 h).

Solution

1. Determine the number of basins. Two basins satisfy the minimum requirement for maintenance purposes. This is a fairly large plant, however, so 4 basins will be selected.
2. Determine the size of each basin.
   a. Determine the basin area using Eq. 6-16. With 4 basins, each basin will treat 50 ML/d.

$$A = \frac{Q}{v_{OF}} = \frac{(50\ \text{ML/d})(1000\ \text{m}^3/\text{ML})}{(1.5\ \text{m/h})(24\ \text{h/d})} = 1390\ \text{m}^2$$

   b. Select the basin width. The basin width is governed by the standard size of sludge removal equipment. The standard maximum width of the chain-and-flight sludge collector is 6 m, so basin widths in increments of 6 m can be considered.
   c. Determine the length using the design guidelines in Table 6-2 for length-to-width ratios. Check L:W ratios for widths of 12 and 18 m:

$$L = \frac{1390\ \text{m}^2}{12\ \text{m}} = 116\ \text{m} \qquad L{:}W = \frac{116}{12} = 9.7:1$$

$$L = \frac{1390\ \text{m}^2}{18\ \text{m}} = 77\ \text{m} \qquad L{:}W = \frac{77}{18} = 4.3:1$$

   Either arrangement of length and width will work, but the 18 m width has a L:W ratio that is only slightly greater than the minimum allowed. Thus the 12 m width and 116 m length is chosen (note that it would be necessary to check with a chain-and-flight equipment manufacturer to verify that they can supply equipment with that length).
   d. Choose the water depth. The hydraulic detention time increases as the water depth increases so choosing a deeper basin will provide a more conservation hydraulic detention time. Choose a depth of 4.5 m and check the hydraulic detention time:

$$\tau = \frac{V}{Q} = \frac{(1390\ \text{m}^2)(4.5\ \text{m})(24\ \text{h/d})}{50{,}000\ \text{m}^3/\text{d}} = 3.0\ \text{h}$$

3. Check the various design parameters listed in Table 6-2. A review of the design criteria developed above indicates that the number of

basins, the overflow rate, detention time, water depth, and L:W ratio are all within the acceptable range. Since the basin design is relatively conservative with respect to overflow rate, L:W ratio, and hydraulic detention time, it is not strictly necessary to check the Reynolds and Froude numbers, but they are checked below using Eqs. 6-27 and 6-28 anyway. From Table C-1 in App. C, $\mu = 1.307\ \text{kg/m} \cdot \text{s}$ and $\rho_w = 999.7\ \text{kg/m}^3$ at 10°C.

$$R_h = \frac{A_x}{P_w} = \frac{(4.5\ \text{m})(18\ \text{m})}{18\ \text{m} + 2(4.5\ \text{m})} = 3.0\ \text{m}$$

$$v_f = \frac{50{,}000\ \text{m}^3/\text{d}}{(18\ \text{m})(4.5\ \text{m})(1440\ \text{min/d})} = 0.43\ \text{m/min} = 0.00715\ \text{m/s}$$

$$\text{Re} = \frac{\rho v_f R_h}{\mu} = \frac{(999.7\ \text{kg/m}^3)(0.00714\ \text{m/s})(3.0\ \text{m})}{(0.00131\ \text{kg/m} \cdot \text{s})} = 16{,}400$$

The Reynolds number of 16,400 is less than the maximum recommended value of 20,000 for a horizontal sedimentation basin.

$$\text{Fr} = \frac{v_f^2}{gR_h} = \frac{(0.00714\ \text{m/s}^2)}{(9.81\ \text{m/s}^2)(3.0\ \text{m})} = 1.73 \times 10^{-6}$$

The Froude number is lower than the recommended value for sedimentation tanks, but because of the conservative design the basin will not be modified to meet this criterion.

Circular Sedimentation Basins and Upflow Clarifiers

Circular sedimentation tanks, also known as upflow clarifiers, provide an opportunity to use relatively trouble-free circular sludge removal mechanisms.

Circular tank diameters are calculated on the basis of overflow rates using approximately the same criteria that are used for rectangular basin design (see Table 6-2). Circular tanks, as shown on Fig. 6-10, may have center feed or peripheral feed. A circular sedimentation basin with center feed and peripheral collection using radial submerged orifice weirs is shown on Fig. 6-11. Inlet weirs provide energy dissipation and direct the flow downward into the depths of the settling tank where particles are removed. Particles settle as the water rises to the outlet structure.

The most significant potential problem of center-feed circular clarifiers is short circuiting of the upward flow of water. Short circuiting occurs when nonuniform flow through the tank causes the up-flow fluid velocity to be greater than the particle-settling velocity. Methods for reducing short circuiting are discussed in detail by Kawamura (2000) and Crittenden et al. (2012).

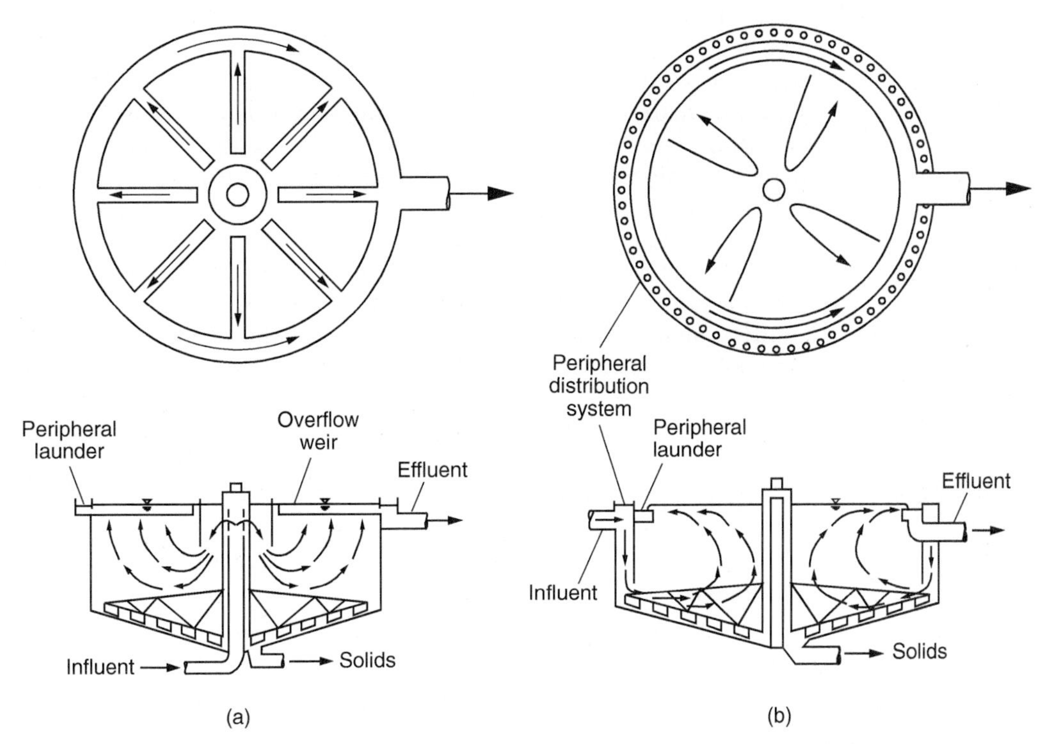

Figure 6-10
Circular sedimentation basins: (a) center feed with radial collection and (b) peripheral feed with peripheral collection.

Figure 6-11
View of circular sedimentation basin with radial collection troughs with submerged orifices.

6-6 Alternative Sedimentation Processes

Large quiescent basins such as those described in Sec. 6-5 require large land areas, which are not always available, and plant upgrades to accommodate increasing water demand may be constrained by the available site area. Increasing the overflow rate in sedimentation basins and achieving the same or better water quality would allow new water treatment plants to fit on smaller sites and existing water treatment plants to expand without having to use additional land area. For example, a high-rate tube settler module, as described below, can be installed under the long launders, significantly increasing the tank loading rate without adding basin volume. Alternative approaches to sedimentation, such as high-rate clarification using parallel-plate or tube settlers, upflow clarifiers, sludge blanket clarifiers, and ballasted sedimentation, are discussed in this section.

Tube and Lamella Plate Clarifiers

Increasing particle size or decreasing the distance a particle must fall prior to removal can accelerate sedimentation of aqueous suspensions. Particle size increase is achieved by coagulation and flocculation prior to sedimentation.

To decrease the distance a particle must fall, the clarification process must be separated from the process of sludge withdrawal and surface current effects. One approach is to provide parallel plates or tubes in the sedimentation basin, permitting solids to reach a surface after a short settling distance. If these settling surfaces (plates or tubes) were oriented in a horizontal direction, they would eventually fill with solids, which would increase the head loss and eventually increase velocities to a point that the suspended materials would be scoured back into suspension. Inclining the surfaces to a degree where the solids can slide from the plate or tube surface results in the settled particles depositing in the sludge zone. Inclined (plate and tube) settlers are illustrated on Figs. 6-1b and 6-12.

The settling characteristics of the suspended particles to be removed and the portion of the total tank surface area that is covered by the settler modules primarily control the surface loading for high-rate settlers. Design criteria for Lamella settlers in rectangular sedimentation basins are provided in Table 6-4. The surface loadings presented in Table 6-4 are based on the portion of the basin area covered by the inclined settlers. In cold regions where alum floc is to be removed, the maximum surface loading should be limited to 6.25 m/h (2.6 gpm/ft^2). Pilot testing may help establish design criteria.

The discussion on the theoretical performance of a sedimentation basin demonstrated that the removal of Type I particles depended on the overflow rate. For a basin depth h_0 and a theoretical detention time τ, particles with a settling velocity v_s would be removed if $v_s \geq h_0/\tau$ (which is equal to the overflow rate). Consequently, if plates or tubes are inserted into a sedimentation basin, then greater particle removal is expected because the

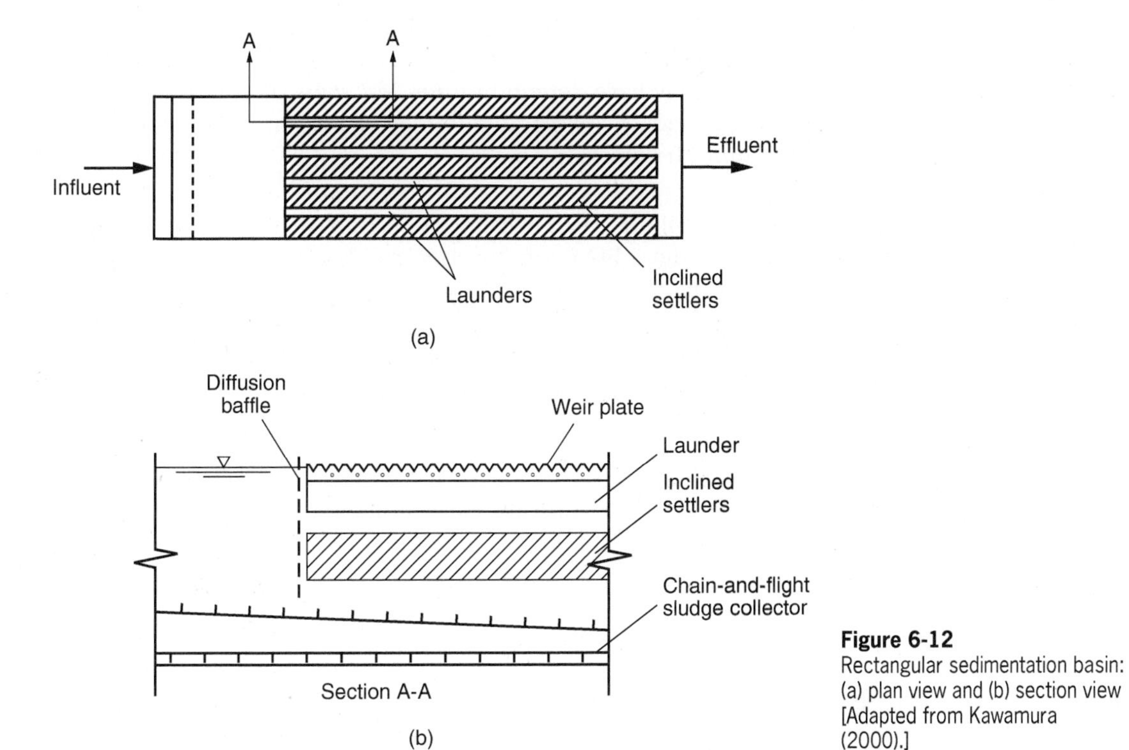

Figure 6-12
Rectangular sedimentation basin: (a) plan view and (b) section view [Adapted from Kawamura (2000).]

Table 6-4
Typical design criteria for horizontal-flow rectangular tanks with tube or plate settlers

Parameter	Units	Value
Minimum number of tanks	Unitless	2
Overflow rate[a]	m/h (gpm/ft^2)	3.8–7.5 (1.5–3.0)
Detention time (with tube settlers)	min	6–10
Detention time (with plate settlers)	min	15–25
Water depth	m (ft)	3–5 (10–16)
Maximum flow velocity in plate or tube settlers	m/min (ft/min)	0.15 (5)
Fraction of basin covered by plate or tube settlers	%	<75
Plate or tube angle	deg	60
Flow direction	—	Normally countercurrent upflow
Reynolds number	Dimensionless	<20,000
Froude number	Dimensionless	$> 10^{-5}$

[a]Based on basin area covered by the settlers.
Source: Adapted from Kawamura (2000).

detention time remains the same, but the distance that particles must settle before they are removed is greatly reduced. Both parallel-plate settlers and tube settlers typically have detention times less than 20 min, but they still have a settling efficiency comparable to that of a rectangular settling tank with a minimum 2 h detention time.

Solids Contact Clarifiers

Solids contact clarifiers can be categorized as reactor clarifiers, sludge blanket reactors, and adsorption reactors. Solids contact units are usually found in industrial and municipal applications, where lime softening or softening clarification is the major treatment process and uniform flow rates and constant water quality prevail. Design criteria and other data for the solids contact units are summarized and compared with conventional rectangular clarifiers in Table 6-5.

REACTOR CLARIFIERS

In a reactor clarifier, the unit operations of rapid mixing, flocculation, and sedimentation are combined in one unit. This combined process has significant advantages, such as reduced cost and more effective use of the sludge blanket. On the other hand, reactor clarifiers reduce, somewhat, the ability of both the designer and the operator to refine the design and operating criteria for each of these operations. Most of these devices are preengineered, packaged proprietary devices that trade reduced flexibility to achieve greater optimization of a particular process option. There are several proprietary reactor clarifier designs. In some circumstances these products are an excellent choice. Common high-rate clarifiers are illustrated on Fig. 6-13 and are described below.

Reactor clarifiers are center-feed clarifiers with a flocculation zone built into the central compartment (see Fig. 6-13a). Generally, these units contain a single motor-driven mixer, with recirculation of the sludge slurry (sometimes optional), followed by a settling zone in a separate outer annular compartment. When slurry recirculation is featured, these are often called solids contact clarifiers and generally include an impeller that provides considerable recirculation. The concentration in the unit is controlled by an adjustable timer on a sludge blow-off line. When using alum, it is common practice to maintain the slurry concentration in the mixing zone at 5 to 20 percent of the sludge volume after 10 min of settling. The slurry concentration is somewhat higher in softening. Reactor clarifiers work well in both clarification and softening, but, in the case of clarification using aluminum or iron salts, sludge recirculation improves performance at the expense of a significantly increased chemical requirement. In the case of lime softening, sludge recirculation improves both performance and chemical consumption.

SLUDGE BLANKET CLARIFIERS

The sludge blanket clarifiers are solids contact clarifiers with a distinct solids layer that is maintained as a suspended filter through which flow

Table 6-5
Typical design criteria for sedimentation processes and their principal applications

Typical Applications	Design Criteria	Advantages	Disadvantages
	Rectangular Basin (Horizontal Flow)		
❑ Many municipal and industrial water works ❑ Particularly suited to large-capacity plants	❑ Surface loading: 1.25–2.5 m/h (0.3–1.0 gpm/ft^2) ❑ Water depth: 3–5 m (10–16 ft) ❑ Detention time: 1.5–4 h ❑ Minimum length-to-width ratio 4 : 1 to 5 : 1 ❑ Weir loading $<$9–13 m^2/h (12–18 gpm/ft)	❑ More tolerance to shock loads ❑ Predictable performance under most conditions ❑ Easy operation and low maintenance costs ❑ Easily adapted for high-rate settler modules	❑ Subject to density flow creation in basin ❑ Requires careful design of inlet and outlet structures ❑ Usually requires separate flocculation facilities
	Upflow (Radial Flow)		
❑ Small to midsize municipal and industrial water treatment plants ❑ Best suited where rate of flow and raw-water quality are constant	❑ Surface loading: 1.25–1.88 m/h (0.5–0.75 gpm/ft^2) ❑ Water depth: 3–5 m (10–16 ft) ❑ Settling time: 1–3 h ❑ Weir loading: 170 $m^3/m \cdot d$ (13,700 gpd/ft)	❑ Economical compact geometry ❑ Easy sludge removal ❑ High clarification efficiency	❑ Problems of flow short circuiting ❑ Less tolerance to shock loads ❑ Need for more careful operation ❑ Limitation on practical size unit ❑ May require separate flocculation facilities

Solids Contact Clarifiers[a]			
❑ Water softening ❑ Flocculation–sedimentation treatment of raw water that has constant quality and rate of flow ❑ Plants treating a raw water with low solids concentration	❑ Flocculation time: ~20 min ❑ Settling time: 1–2 h ❑ Surface loading: 2.1–3.1 m/h (0.85–1.28 gpm/ft^2) ❑ Weir loading: 175–350 $m^3/m \cdot d$ (14,000–28,000 gpd/ft) ❑ Upflow velocity: <10 mm/min (2 in./min) ❑ Higher maintenance costs and need for greater operator skill ❑ Slurry circulation rate: up to 3–5 times raw-water inflow rate	❑ Good softening and turbidity removal ❑ Flocculation and clarification in one unit ❑ Compact and economical design	❑ Sensitive to shock loads and changes in flow rate ❑ Sensitive to temperature change ❑ Two to 3 days required to build up necessary sludge blanket ❑ Plant operation dependent on single mixing motor

[a]Reactor clarifiers and sludge blanket clarifiers are often considered as one category, solids contact clarifiers.

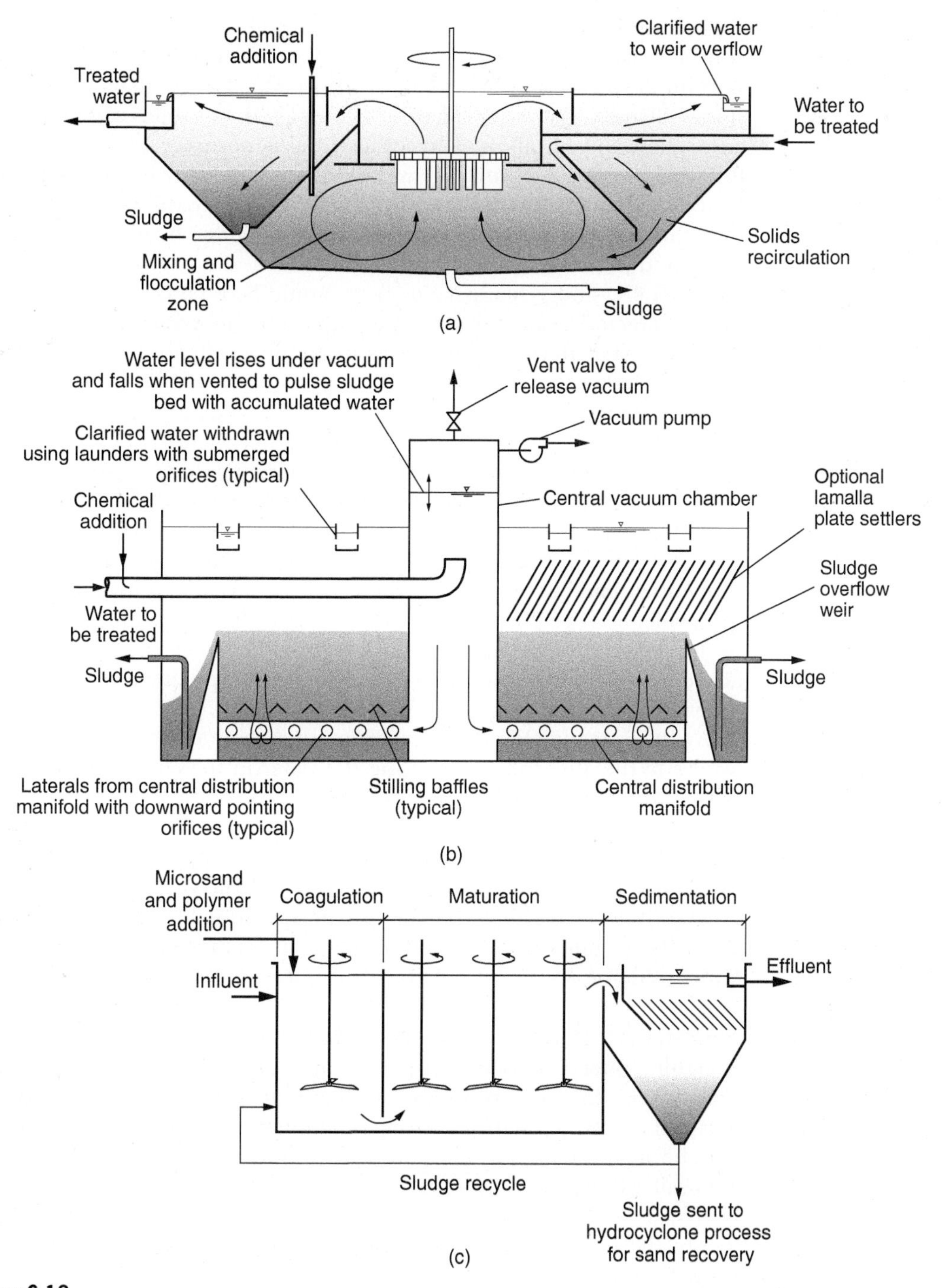

Figure 6-13
Common high rate clarifiers: (a) reactor clarifier (Accelator), (b) sludge blanket clarifier (Pulsator, when optional lamella plates are added, the unit is known as the Superpulsator), and (c) ballasted sedimentation. (Panels (a) and (b) Courtesy of Infilco Degremony, Inc.

passes (see Fig. 6-13b). The sludge blanket unit contains a central mixing zone for partial flocculation and a fluidized sludge blanket in the lower portion of the settling zone. Partially flocculated water flows through the sludge blanket where flocculation is completed and solids are retained by adsorption and filtration. The sludge level is normally 1.5 to 2 m (4.5 to 6 ft) below the water surface, and clarified water is collected in launder troughs along the top of the unit. Sludge blanket clarifiers are made with or without sludge recirculation mechanisms. Sludge blanket clarifiers are compared with other processes in Table 6-5.

Generally, sludge blanket clarifiers should be used only where the raw-water characteristics and flow rates are relatively uniform. Given these parameters, the most effective applications are for lime softening and clarification of low-turbidity water. These units may also be used for clarification of highly turbid water (exceeding 500 NTU) if a sludge-scraping mechanism is provided.

One of the more difficult problems in operating sludge blanket clarifiers is the management of the sludge blanket itself. Some of the more popular designs accomplish this by simply allowing thc sludgc blanket to fall over a submerged weir that is kept a significant distance below the free surface. The Pulsator and its progeny the Superpulsator are widely used examples of this principle. The Pulsator is shown on Fig 6-13b. Operationally, a portion of the flow is brought in the central vacuum chamber and allowed to rise above the operating water level in the clarifier by pulling a vacuum. When the water level in the vacuum chamber is about 0.5 to 1.0 m (1.6 to 3.3 ft) above the operating level in the clarifier, the vacuum is released by opening a valve to the atmosphere, allowing the water in the chamber to flow as a pulse through the influent distribution system located at the bottom of the tank below the sludge blanket. The pulse of water is used to contact the incoming water with the sludge blanket and to suspend and redistribute the sludge blanket. The depth of the sludge blanket is controlled by the overflow weir. The sludge blanket is typically pulsed once every 60 s (40 s to fill the vacuum chamber and 5 to 20 s to drain into the clarifier). The Super pulsator is similar to the pulsator but employs lamella plate settling.

ROUGHING FILTERS AND ADSORPTION CLARIFIERS

Roughing filters are used to create a zone of laminar flow during clarification. Similar in objective to the plate-and-tube settlers, a bed of granular material is used to establish a zone of laminar flow. The media in the bed may be heavy material like gravel or buoyant plastic media. Suspended material will deposit on the media as the water flows through the channels in the media bed. To remove the sludge, the media must be agitated to loosen the particles, which in turn fall to the bottom of the tank or are flushed from the tank with backwash water.

The adsorption clarifier uses buoyant plastic media as a roughing filter. As the coagulated water travels upward through the media, flocculation takes place as the tortuous path of the water causes mixing and collisions between particles. Collisions between particles and the media causes particles to stick to the media, and most of the flocculated solids can be collected in the media. The media is then occasionally washed by introducing air from below, which reduces the bulk specific gravity of the water surrounding the media and allows it to expand and be cleaned.

Ballasted Sedimentation

As the process name suggests, ballasted sedimentation involves the addition of ballast (usually microsand) that increases the settling velocity of the floc particles by increasing their density (providing ballast). Currently, there are a number of proprietary sedimentation processes that employ the ballasted flocculation principle. Two well-known processes are the Actiflo process and the Densideg dense-sludge process. These processes have been used in water treatment for both the production of potable water and the treatment of filter-to-waste washwater.

A schematic of the Actiflo process is shown on Fig. 6-13c. The Actiflo process involves adding an inorganic coagulant (alum or ferric) to the raw water and allowing floc to form in the first stage of flocculation. Subsequently, a high-molecular-weight cationic polymer and microsand particles (20 to 200 μm) are added to the second stage, and the microsand particles flocculate with the preformed floc particles in the second and third stages. After flocculation, the ballasted floc is settled and the sludge containing the microsand is sent through a hydrocyclone (not shown) where the microsand is recovered and reused and the sludge is sent on for further treatment. The microsand is fed at a rate that is approximately 0.15 to 0.4 percent of the influent flow rate and the sludge ultimately contains 10 to 12 percent sand by weight.

The surface loading rate for an Actiflo unit ranges from 35 to 62 m/h (14 to 25 gpm/ft^2) which can be up to 50 times greater than the surface loading rate for a conventional rectangular sedimentation basin. The small size of the Actiflo unit can be attributed to the use of high mixing energy ($\overline{G}$ values ranging from 150 to 400 s^{-1}), shorter detention times for flocculation (between 9 and 10 min), floc settling velocities 20 to 60 times greater than conventional flocculation and sedimentation, and the use of lamella plate settler modules to accelerate particle removal.

The advantages of the high-rate settling processes include (1) a small footprint requirement at water treatment plants with site constraints; (2) turbidity removal down to the 0.5-NTU level, but treating to 2.0 NTU is more common to reduce polymer usage and potential polymer carryover into the filters; (3) a quick process startup, about 15 min; (4) robust process that is not easily upset by changes in raw-water quality; and (5) potential savings in capital costs based on the small footprint. The disadvantages are (1) a heavy dependence on mechanical equipment and a short processing time; (2) the entire process must be shut down when there is a power outage lasting more than 10 min; (3) a higher coagulant dosage is required than for conventional processes with a high proportion of polymers, which may cause problems in downstream processes such as filter blinding and reduced filter run time; (4) potential for sand carryover (e.g., Actiflo process) into downstream processes; and (5) proprietary processes, which may limit competitive bidding.

6-7 Physical Factors Affecting Sedimentation

Accurate prediction of settling tank performance by mathematical and experimental methods is a challenge to even the best design engineers. Model testing using tracers and settling columns are limited by scale-up, which cannot be expressed adequately by principles of similitude, primarily because solid particles are not easily scaled down. In addition, many of the simplifying assumptions of modeling do not hold true in prototype units. Factors such as temperature gradients, wind effects, inlet energy dissipation, outlet currents, and equipment movement affect tank performance but are not easily modeled. Density currents, inlet energy dissipation, outlet currents, and equipment movement are presented and discussed in this section. Most of the information presented below on the physical factors related to sedimentation is directed toward conventional sedimentation basins and less toward innovative designs.

Density Currents

When feed water is entering the sedimentation basin, it can form a surface or a bottom density current, depending on the relative densities of the feed water and water in the basin. Under these flow conditions, actual flow-through velocity will depart from the theoretical, idealized average basin velocity. The theoretical velocity is equal to the total incoming flow divided by the total cross-sectional area in the basin. Short circuiting caused by density currents has been observed in many water treatment plants (Camp, 1946; Harleman, 1961).

Basin inlet and outlet arrangements should be designed to provide some degree of control in minimizing the effects of density currents. At the inlet, the following techniques have been used: (1) feed flow is distributed uniformly through the basin cross section in the plane perpendicular to the flow by employing diffuser walls and (2) devices that will break up the feed stream and dissipate the energy by turbulence.

Improvements can be made in the basin to control the density currents. These improvements include tube settlers, redistribution baffles, or intermediate diffuser walls. Launders extending into sedimentation basins have been used to control the effluent flow distribution, which is more effective for controlling bottom density currents than surface density currents.

TEMPERATURE DIFFERENTIALS

The addition of warm influent water to a sedimentation basin containing cooler water can lead to a short-circuiting phenomenon in which the warm water rises to the surface and reaches the effluent launders in a fraction of the nominal detention time. Conversely, cold water added to a basin containing warm water tends to force the incoming water to dive to the bottom of the basin, flow along the bottom, and rise at the basin outlet. Temperature differences as small as 0.3°C have been observed to cause density gradients. Proper inlet design can minimize temperature effects.

SOLIDS CONCENTRATION EFFECTS

Density current problems similar to those discussed above may also be caused by changes in influent solids concentrations resulting from flash floods or strong winds on lake water surfaces. A rapid increase in turbidity increases the density of the influent and causes it to plunge as it enters the sedimentation basin.

The remediation of problems with varying influent turbidity are similar to those for incoming temperature differences and include diffuser walls in the basin. Additionally, the source of water should be carefully selected and the method of removing water from the source should minimize quality variations. It should be noted that changes in water density resulting from variable dissolved solids (salinity) concentration may also lead to density flow and short circuiting.

Wind Effects

Wind can have a pronounced effect on the performance of large, open gravity settling basins. High wind velocity tends to push the water to the downwind side of a basin and produces a surface current moving in the direction of the wind. An underflow current in the opposite direction is also created, which moves along the bottom of the tank. The resulting circulating current, can lead to short circuiting of the influent to the effluent weir and scouring of settled particles from the sludge zone. For open sedimentation basins with length or diameter greater than 30 m (100 ft), wind effects can be significant and result in reduced effluent quality.

When long and shallow rectangular settling basins are used, orienting the basin with the local prevailing wind direction should be considered. In areas with strong predictable winds, sedimentation basins should be positioned so that the water flow parallels the wind, and wave breakers (launders or baffles) should be placed at approximately 20- to 30-m (65- to 100-ft) intervals. Changes in water surface elevation are minimized when the wind blows across the length of the rectangular settling basin, as opposed to across the width, and the effects of wind currents on sedimentation basin performance are minimized.

Outlet Currents

Outlet currents in a sedimentation basin are often related to design details of effluent weirs and launders. Initially, these weirs were simply flat plates across the end of a rectangular basin. The width of the basin established the length of the weir. When tanks were designed in a long, narrow configuration, the weir length was relatively short and was believed to contribute to formation of outlet currents that, if severe, could sweep settleable particles into the tank effluent. The problem of currents was compounded in early designs because the flat weir plates were sometimes not level. Concern for this led to the development of V-notch weirs, which provide better lateral distribution of outlet flow when leveling is imperfect.

For upflow clarifiers, such as solids contact basins, launders carefully spaced across the surface are considered of vital importance to good performance. The launders, which are often arranged in a radial pattern, serve an important role in directing the vertical flow through the solids contact zone. As solids contact tanks become larger, strategic location of radial weirs becomes more critical.

In general, for most water treatment sedimentation basins, performance is primarily a function of density currents and inlet energy dissipation rather than outlet currents. Careful design of effluent weirs will not solve problems associated with density currents created by other design deficiencies.

Equipment Movement

Another potential effect on sedimentation basin performance is the movement of equipment within the basin. Sludge collection mechanisms, normally consisting of chain-and-flight scrapers, bridge-mounted scrapers, or hydraulic vacuum units, must move through the liquid contents of the tank to remove settled sludge. If equipment movement is excessive, currents may be introduced that upset the sedimentation process. Most equipment moves at a rate of 15 to 30 m/h (50 to 100 ft/h) and has a minimal effect. However, equipment movement in the vicinity of the effluent launders is important because of the potential for disturbed settled solids to be caught in the effluent currents and carried over the effluent weir.

6-8 Energy and Sustainability Considerations

From an operational standpoint, sedimentation is one of the more energy-efficient processes in water treatment because gravity is used to separate the solids from water. For conventional sedimentation basins, energy consumption is due to the head loss through the basin, mechanical solids removal devices in the sludge zone, and pumping of the solids to the solids handling facility. Typical head loss through sedimentation basins is about two-thirds of a meter or less. The energy required to overcome 0.6 m (2 ft) of head loss is 0.0016 kWh/m^3. Additional energy is consumed by mechanical sludge withdrawal systems. Sludge scraper systems typically use small horsepower motors to drive the scrapers at very slow speeds and do not require large amounts of energy. In addition, the sludge must be pumped to the sludge processing facilities for treatment. The engineer should consider minimizing the pumping distance to decrease the energy consumption.

While environmental impacts of construction of conventional water treatment plants is typically small compared to plant operation, in some cases design engineers should consider minimizing the plant footprint. Methods to reduce the footprint of sedimentation basins is the use of common walls between basins, incorporation of plate or tube settlers in the design, and minimizing the use of safety factors that cause overdesigned and redundant basin designs. In particular, high-rate settlers such as plate-and-tube configurations can increase the sedimentation efficiency and reduce the plant footprint. These settlers can also be used to increase the efficiency of existing sedimentation basins, if needed, without adding to the plant footprint. In some cases, however, alternative sedimentation processes that minimize construction impacts have greater energy consumption and operating impacts.

An example of the trade-off between construction impacts and operating impacts is a comparison between the ballasted sedimentation process and conventional horizontal-flow sedimentation. A ballasted sedimentation process uses less land area, requires less construction materials, but has higher operating energy costs than a conventional sedimentation system. An environmental life cycle assessment using local information and data would be necessary to determine which process had overall lower impacts.

6-9 Summary and Study Guide

After studying this chapter, you should be able to:

1. Define the following terms and phrases and describe the significance of each in the context of gravity sedimentation in water treatment:

ballasted sedimentation
compression settling
discrete particle settling
drag coefficient
effluent launders
flocculant settling
hindered settling
inclined plate settlers
inlet zone
limiting flux
outlet zone
overflow rate
overflow weir
reactor clarifier
settling zone
sludge zone
sludge blanket
solids contact clarifier
terminal settling velocity
total flux
tube settlers
upflow clarifier
underflow flux

2. Explain the purpose of sedimentation in water treatment and give a general description of the process of sedimentation.
3. Identify and describe the four types of settling.
4. Identify the key assumptions used in developing Stokes' law.
5. Calculate the terminal settling velocity of a particle given the particle size, density, and water temperature.
6. Calculate the overflow rate of a sedimentation basin.
7. Calculate the particle removed efficiency is a rectangular sedimentation basin.
8. Calculate the area of a thickner.

9. Explain the benefit of a high $L{:}W$ ratio in horizontal-flow sedimentation basin design.
10. Describe various strategies for accelerating the sedimentation process.
11. Describe and explain the various zones associated with a conventional sedimentation basin.
12. Explain the principal causes of flocculation during sedimentation and why flocculation is beneficial in a sedimentation process.
13. Evaluate possible reasons for poor performance in a sedimentation basin.
14. Design a conventional rectangular sedimentation basin.
15. Explain ways to promote energy conservation and sustainability in the design and operation of sedimentation basins.

Homework Problems

6-1 Calculate the terminal settling velocity and the Reynolds number of the particle given (to be selected by instructor).

Parameter	A	B	C	D	E
Particle diameter, μm	50	500	300	150	210
Particle density, kg/m^3	2650	1050	1050	2600	1700
Water temperature, °C	10	15	5	20	15

6-2 Consider the particle shown below with the values in the table (to be selected by instructor). Calculate the overflow rate that corresponds to the setting velocity of the particle on the trajectory shown (report your answers in m/h and gpm/ft^2). If it is desired to achieve complete removal of particles of this size, what adjustment in the length of the basin would be required?

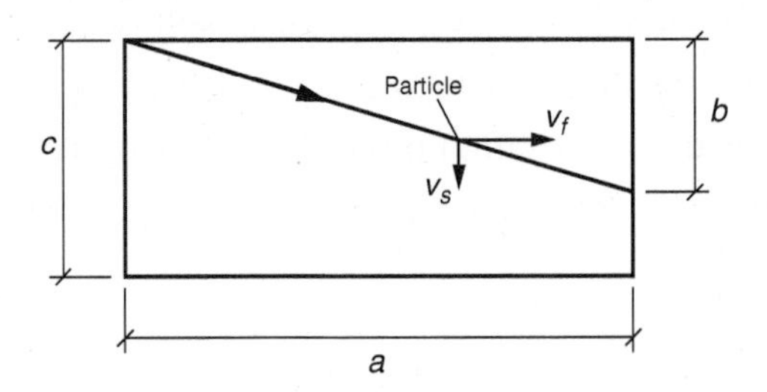

Parameter	A	B	C	D	E
Fluid velocity, cm/s	20	1.4	0.5	1	0.28
Dimension a, m	4	100	72	80	30
Dimension b, m	0.6	3.5	1.7	0.85	3.6
Dimension c, m	0.9	4.2	3.5	1	4.2

6-3 For the rectangular horizontal-flow sedimentation basin and influent particle-settling characteristics given (to be selected by instructor), calculate the particle removal efficiency, and plot the influent and effluent particle concentrations as a function of particle size.

Parameter	A	B	C	D	E
Flow rate, mL/d	7.57	19	19	56.8	56.8
Length, m	30	72	60	100	80
Width, m	5	12	8	18	12

Settling Velocity,	Number of Particles, #/mL				
m/h	A	B	C	D	E
0–0.4	511	511	460	560	255
0.4–0.8	657	657	578	720	314
0.8–1.2	876	876	891	880	454
1.2–1.6	1168	1168	1285	1110	584
1.6–2.0	1460	1460	1748	1320	761
2.0–2.4	1314	1314	1577	1110	639
2.4–2.8	657	657	719	620	321
2.8–3.2	438	438	436	440	219
3.2–3.6	292	292	263	320	141
3.6–4.0	292	292	241	160	116
Total	7665	7665	8198	7240	3804

6-4 For the particle-settling data shown in Example 6-3, plot the removal efficiency as a function of overflow rate for overflow rates ranging from 0.5 to 4 m/h. Determine the overflow rate required to achieve 75 percent removal. If the depth of the basin is 4 m, what is the corresponding detention time.

6-5 Determine the area of a clarifier required for solids thickening for the parameters given below (to be selected by instructor). The settling velocity of the sludge blanket follows the data given in Table 6-1 and plotted on Fig 6-7. Also determine J_L, C_L, and Q_u.

Parameter	A	B	C	D	E
Influent flow rate, m^3/h	3,000	1,500	2,500	3,300	4,500
Influent solids conc., mg/L	500	800	400	500	800
Underflow solids conc., mg/L	10,000	12,000	14,000	14,000	15,000

6-6 A water treatment plant is to be designed to treat water with the maximum daily flow and design temperature shown in the table below. For the given overflow rate, design a horizontal rectangular sedimentation basin. Your design should include the number of basins, length, width, and depth of each basin, and number of baffles within each basin, if any. Your design should be suitable for chain-and-flight sludge removal equipment. Verify that your design meets the criteria in Table 6-2 with respect to depth, $L:W$ ratio, detention time, and Reynolds number (if detention time is less than 2 h).

Parameter	A	B	C	D	E
Influent flow rate, ML/d	15	380	90	220	220
Overflow rate, m/h	1.10	2.15	2.6	1.65	2.0
Water temperature, °C	10	15	20	20	10

References

Brown, P. P., and Lawler, D. F. (2003) "Sphere Drag and Settling Velocity Revisited," *J. Eng. Div. ASCE*, **129**, 222–231.

Camp, T. R. (1936) "A Study of the Rational Design of Settling Tanks," *Sewage Works J.*, **8**, 9, 742–758.

Camp, T. R. (1946) "Sedimentation and the Design of Settling Tanks," Paper no. 2285, *ASCE Trans.*, **3**, 895–936.

Clark, M. M., (1996) *Transport Modeling for Environmental Engineers and Scientists*, Wiley, New York.

Crittenden. J. C., Trussell, R. R., Hand, D. W., Howe, K. J., and Tchobanoglous, G. (2012) *MWH's Water Treatment: Principles and Design*, 3rd ed., Wiley, Hoboken, NJ.

Harleman, D. F. (1961) Stratified Flow, in V. S. (ed.), *Handbook of Fluid Mechanics*, McGraw-Hill, New York.

Kawamura, S. (2000) *Integrated Design and Operation of Water Treatment Facilities*, 2nd ed., Wiley-Interscience, New York.

Logan, B. E. (1999) *Environmental Transport Processes*, Wiley-Interscience, New York.

O'Connor, D. J., and Eckenfelder, W. W., Jr. (1958) Evaluation of Laboratory Settling Data for Process Design, in W. W. Eckenfelder and B. J. McCabe (eds.), *Biologic Treatment of Sewage and Industrial Wastes*, Reinhold, New York.

Tchobanoglous, G., Burton, F. L., and Stensel, H. D. (2003) *Wastewater Engineering: Treatment, and Reuse*, 4th ed., McGraw-Hill, New York.

7 Rapid Granular Filtration

Filtration is widely used for removing particles from water. Filtration can be defined as any process for the removal of solid particles from a suspension (a two-phase system containing particles in a fluid) by passage of the suspension through a porous medium. In granular filtration, the porous medium is a thick bed of granular material such as sand. The most common granular filtration technology in water treatment is *rapid filtration.* The name arises to distinguish it from slow sand filtration, an older filtration technology with a filtration rate 50 to 100 times lower than rapid filtration. Key features of rapid filtration include granular media processed to a more uniform size than typically found in nature, coagulation pretreatment, backwashing to remove accumulated particles, and a reliance on depth filtration as the primary particle removal mechanism. In *depth filtration,* particles accumulate throughout the depth of the filter bed by colliding with and adhering to the media. Captured particles can be many times smaller than the pore spaces in the bed.

Nearly all surface water treatment facilities and some groundwater treatment facilities use filtration. Most surface waters contain algae, sediment, clay, and other organic or inorganic particles. Filtration improves the clarity of water by removing these particles. All surface waters also contain microorganisms that can cause illness, and filtration is nearly always required in conjunction with chemical disinfection to assure that water is free of these pathogens. Groundwater is often free of significant concentrations of microorganisms or particles but may require filtration when other treatment processes (such as oxidation or softening) generate particles that must be removed.

This chapter starts with two sections that present a physical description of a rapid gravity granular filter and a process description of rapid granular filtration. The next sections describe particle capture and hydraulic flow in granular filters, a section on modeling of performance that integrates particle capture and fluid flow, and a section on backwash hydraulics. Finally, energy and sustainability considerations are discussed.

7-1 Physical Description of a Rapid Granular Filter

A typical configuration for a rapid filter is illustrated on Fig. 7-1. The filter bed is contained in a deep structure that is typically constructed of reinforced concrete and open to the atmosphere. Water enters at the top of the filter box from an influent channel. The water then flows down through the granular media, where it is captured by the underdrain system and carried to the filtered water storage tank, known as the clearwell. The media bed is typically 0.6 to 1.8 m (2 to 6 ft) deep. After filtering for a period of time, the filter is backwashed. During *backwashing*, the upward-flowing water fluidizes and expands the filter bed and washes away the collected particles. The components in the filter system are described in more detail below.

Filter Media

The most common granular filter bed in water treatment in North America consists of a layer of anthracite over the top of a layer of sand. *Anthracite* is coal, the type of coal with the fewest impurities and the highest carbon content. Coal is used because it is less dense than sand (the importance of this will become evident later in the chapter) and anthracite specifically is used because it is the hardest of coals and least likely to be worn down by abrasion. In addition to this dual-media configuration, monomedia filters with only anthracite or only sand are sometimes used. In other situations, garnet or ilmenite, which are minerals denser than sand, are incorporated into the filter bed as a third layer below the sand. Granular activated carbon (GAC) is sometimes used as a filter material when adsorption or biodegradation is combined with filtration in a single unit process. Adsorption is addressed in Chap. 10.

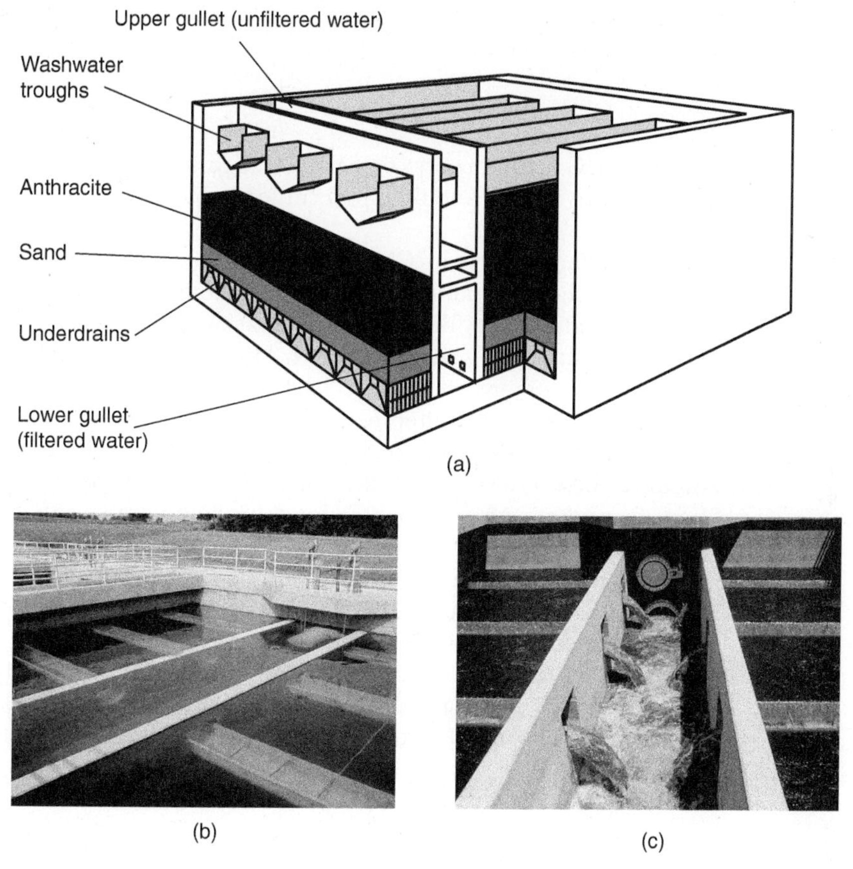

Figure 7-1
Typical dual-media rapid filter. (a) Schematic representation of dual-media filter. (b) View of an operating rapid filter. Washwater troughs are visible below the water surface. Influent water enters through the central channel, flows through the wall openings for the washwater troughs, and then down through the filter media, which is about 2.75 m (9 ft) below the water surface. (c) Rapid filter during the backwash cycle. Washwater flows up through the media, pours over into the troughs, and then runs into the central channel. The influent valve, visible at the far end of the central channel, is closed, and the waste washwater flows out through the open washwater valve.

The size of the media grains is a key parameter in filter performance. In North America, the standard method for characterizing the media size distribution is by effective size and uniformity coefficient. The *effective size* (ES or d_{10}) is the media grain diameter at which 10 percent of the media by weight is smaller, as determined by a sieve analysis. In a sieve analysis, a batch of material is sifted through a stack of calibrated sieves, the weight of material retained on each sieve is measured, and the cumulative weight retained is plotted as a function of sieve size. The *uniformity coefficient* (UC) is the ratio of the 60th percentile media grain diameter to the effective size. A low UC means the media is fairly uniform in size, whereas a high UC means very small and very large grains can be present. The UC is calculated from the equation

$$\text{UC} = \frac{d_{60}}{d_{10}} \tag{7-1}$$

where UC = uniformity coefficient, dimensionless
d_{10}, d_{60} = 10th and 60th percentile media grain diameters, mm

It was demonstrated in Chap. 6 that the settling velocity of particles depends on their size and density. This phenomenon affects the arrangement of media in a filter bed after backwashing because the grains must settle after the bed is fluidized. Fine grains collect at the top of the filter bed, where they cause excessive head loss and reduce overall effectiveness of the filter bed. Large grains settle to the bottom of the bed and are difficult to fluidize during backwash. A low UC can minimize these effects and is a key feature of rapid filters, allowing the filters to operate at a higher hydraulic loading rate, with lower head loss, and for longer time between backwashes. Thus, rapid filter media is processed to remove the largest (by sieving) and smallest (by washing) grain sizes to produce a lower UC than naturally occurring material. An example of sieve analyses of natural sand and filter sand are shown on Fig. 7-2, and the ES and UC of typical filtration materials are provided in Table 7-1 along with typical values of other material properties. Determination of the ES and UC from sieve data is demonstrated in Example 7-1.

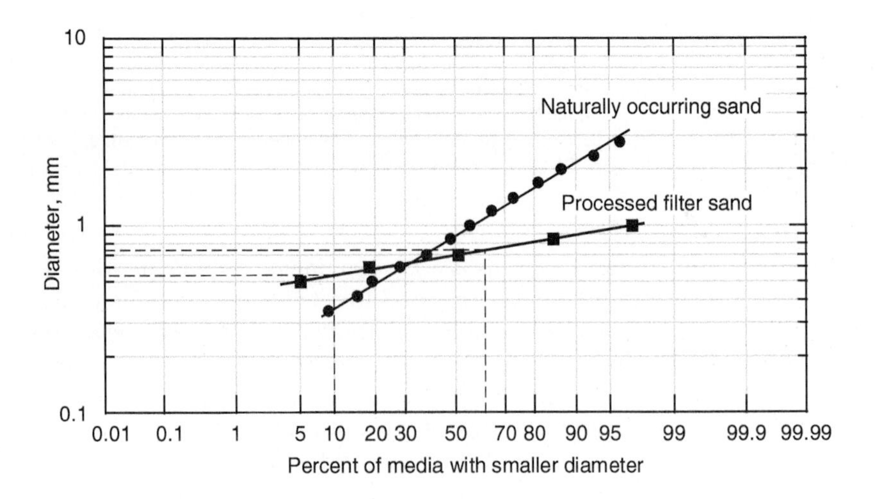

Figure 7-2
Size distribution of typical naturally occurring and processed filter sand.

Table 7-1
Typical properties of filter media used in rapid filters[a]

Property	Unit	Garnet	Ilmenite	Sand	Anthracite	GAC
Effective size, ES	mm	0.2–0.4	0.2–0.4	0.4–0.8	0.8–2.0	0.8–2.0
Uniformity coefficient, UC	UC	1.3–1.7	1.3–1.7	1.3–1.7	1.3–1.7	1.3–2.4
Density, ρ_p	kg/L	3.6–4.2	4.5–5.0	2.65	1.4–1.8	1.3–1.7
Porosity, ε	%	45–58	N/A	40–43	47–52	N/A
Hardness	Moh	6.5–7.5	5–6	7	2–3	Low

[a]N/A = not available.

Example 7-1 Effective size and uniformity coefficient of filter media

Determine the effective size and uniformity coefficient of the processed filter sand shown on Fig. 7-2.

Solution

1. Find the 10th percentile line on the x axis and follow it up to the intersection of the line for the processed filter sand. The corresponding value on the *y* axis is 0.54 mm.
2. The size (*y* axis) corresponding to the 60th percentile (x axis) for the processed filter sand is 0.74 mm.
3. The effective size is $ES = d_{10} = 0.54$ mm. The uniformity coefficient is $UC = d_{60}/d_{10} = 0.74/0.54 = 1.37$.

Comments

Sieve data can be plotted on any type of graph. As long as a smooth curve can be drawn through the data, the d_{10} and d_{60} values can be determined.

Underdrains

Underdrains cover the floor of the filter box, support the filter media, collect and convey filtered water away from the filter system, and distribute backwash water and air. Some types of underdrain systems are shown on Fig. 7-3. The underdrains must capture and distribute water uniformly to avoid spatial variations in filtration rate or backwash rate that would degrade the effectiveness of the filter. Underdrains distribute flow evenly by maintaining low velocity (and therefore low head loss) through the pipes or channels that transport water and higher velocity (and therefore

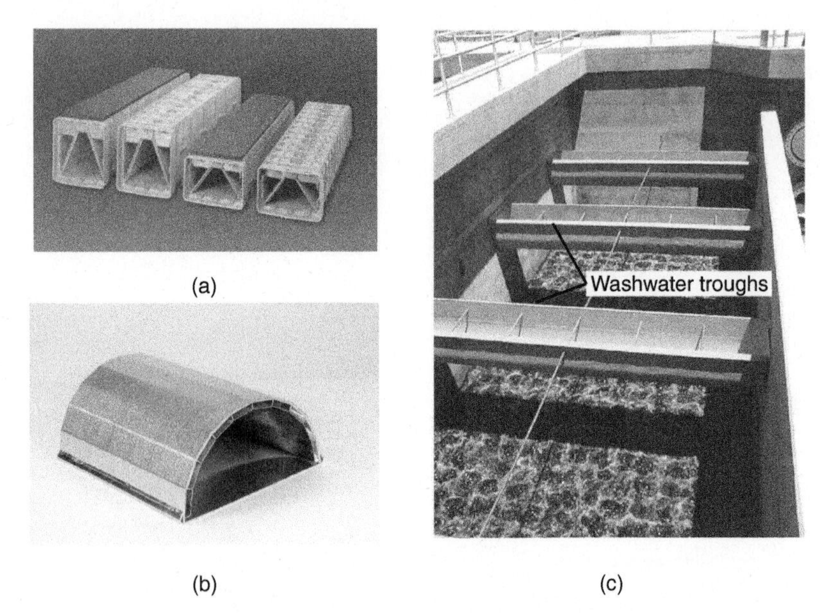

Figure 7-3
Components of rapid filter systems (a, b) underdrains and (c) wash-water troughs. The filter in panel (c) has been drained for the air scour portion of the backwash cycle. Note that air is distributed evenly over the entire filter box by the underdrains.

higher head loss) through small orifices distributed evenly across the floor of the filter. Flow tends to distribute evenly to maintain constant pressure throughout the system. Modern underdrains typically have porous plates or fine mesh screens that retain the filter media directly. Older underdrain systems used layers of different-sized gravel to support the media. The presence of the gravel increased the overall height of the filter box by about 0.5 m (20 in.); the added cost of this height is one reason gravel is less commonly used in modern design. Uniform backwash flow distribution, durability, and cost are the three most important factors in selecting filter underdrains.

Surface Wash

The *surface wash* system is designed to agitate the bed vigorously during backwashing to break deposited solids loose from the media grains. Once the solids are separated from the media grains, the upflowing backwash water can flush the solids from the filter. Surface wash systems typically have water nozzles on a rotating header or on a fixed pipe grid located just above the surface of the bed. As the media fluidizes, it rises above the level of the nozzles, so the surface wash system is able to provide vigorous agitation of the fluidized media. Surface wash systems are effective for cleaning traditional filters with depths of 0.6 to 0.9 m (2 to 3 ft) but are less effective for cleaning deep-bed filters. For deep-bed filters, air scour (discussed in Sec. 7-2) is often used instead of or in addition to surface wash systems.

Wash Troughs

Wash troughs provide a channel to collect the waste wash water that is generated when the filter is backwashed. An important objective of wash troughs is that they should collect the dirty water and particles being washed from the filter without allowing any of the filter media to be washed away. Media washout is prevented by locating the troughs sufficiently above the top of the bed to allow the media to be fluidized without reaching the lip of the troughs. Troughs can be constructed of concrete, stainless steel, or fiberglass, and the most common modern design is a fiberglass trough with a deep U-shaped cross section. Typical wash troughs are shown on Fig. 7-3c. After wash water is collected by the wash troughs, it is discharged to the gullet.

Gullet

The *gullet* is an open channel with appropriate pipe penetrations and valves to manage the flow to and from the filter. Some filters are constructed with an upper and lower gullet. The upper gullet is the channel where influent water enters the filter and waste wash water is collected and carried away, and the lower gullet is where filtered water is collected and backwash water is introduced to the underdrains. Figure 7-1 shows a filter design with filter cells to either side of central upper and lower gullets.

Valves and Piping

Each filter has several pipe connections, each of which needs a valve. During normal filtration, filter influent and effluent lines are open. During backwash, those valves are closed and backwash supply and waste wash water valves are opened. After backwash is complete, the filter influent is opened again but the effluent is directed to the *filter-to-waste* line if the effluent quality is not good enough. After a short time, the filter-to-waste valve can be closed and the filtered water directed to the effluent line again.

Flow Control

Flow control is an important part of any filter system. Flow control systems must accomplish three objectives: (1) distribute flow among individual filters, (2) control the filtration rate of individual filters, and (3) accommodate increasing head loss. Several options for flow control are available. To distribute the flow to individual filters and control the filtration rate, filters can use modulating control valves, influent flow-splitting weirs, or declining-rate filtration (no active flow control or distribution). The total *available head* in a gravity rapid filter system is head available for driving water through the filter and is fixed by the water elevation in the upstream and downstream structures (i.e., sedimentation basins and clearwells). The head loss through the filter bed increases as the filter collects solids, so provisions must be made to accommodate the variation in head loss. Three basic strategies are used: (1) maintain constant head above the filter (e.g., constant water level) and vary the head in the filter effluent by modulating a control valve, (2) maintaining constant head in the filter effluent and vary the head upstream of the filter (allowing the water level to rise), and (3) maintaining nearly constant head loss and allowing the filtration rate to decline as solids accumulate in the bed (declining-rate filtration).

Standard texts and design manuals provide additional details on flow control strategies (Tobiason et al., 2011; Crittenden et al., 2012; Kawamura, 2000). No method of flow control is clearly superior to the others. Selection is typically based on designer and owner preferences. Cost, complexity, and reliability are important issues. Whichever method is used, proper design is important because poor control can cause rapid changes in flow through the filter, which causes detachment of deposited particles and degrades the filter effluent quality.

7-2 Process Description of Rapid Filtration

The rapid-filtration cycle consists of two stages: (1) a filtration stage, during which particles accumulate, and (2) a backwash stage, during which the accumulated material is flushed from the system. During the filtration stage, water flows downward through the filter bed and particles collect within the bed. The filtration stage typically lasts from 1 to 4 days and the backwash typically lasts 10 to 15 min.

The efficiency of particle capture, as reflected by effluent turbidity and head loss, varies during the filtration stage (also called a filter run), as illustrated on Fig. 7-4. Filter effluent turbidity during the filter run follows a characteristic pattern with three distinct segments. During the first segment (immediately after backwash), the filter effluent turbidity rises to a peak and then falls. This segment is called filter ripening (or *maturation*). *Ripening* is the process of media conditioning and occurs as clean media captures particles and becomes more efficient at collecting additional particles. As much as 90 percent of the particles that pass through a well-operating filter

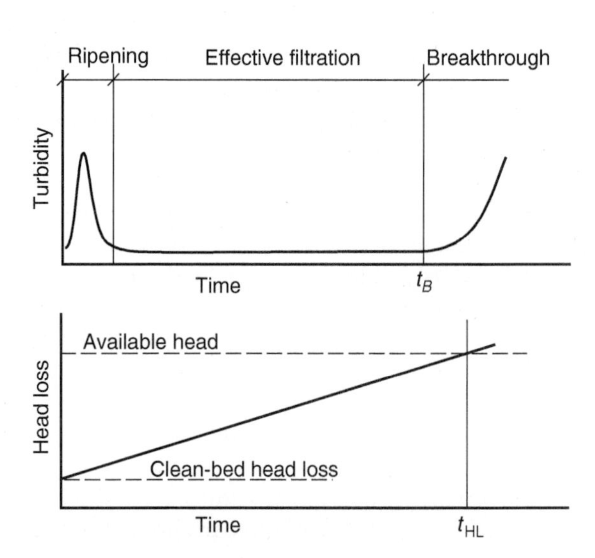

Figure 7-4
Operation of a rapid filter: (a) effluent turbidity versus time and (b) head-loss development versus time.

do so during the initial stage of filtration (Amirtharajah, 1988). Ripening periods are typically between 15 min and 2 h. The magnitude and duration of the ripening peak can be substantially reduced by proper backwashing procedures, such as minimizing the duration of the backwash stage or using filter aid polymers in the backwash water. The water produced during ripening, if of unacceptable quality, can be discharged to the filter-to-waste line, where it is wasted or recycled to the head of the plant.

The particles captured during ripening improve the overall efficiency of the filter by providing a better collector surface than uncoated media grains. After ripening, effluent turbidity typically stays below 0.1 NTU. Even though effluent turbidity is essentially constant after ripening, head loss through the filter continuously increases because of the collection of particles in the filter bed. After the period of effective filtration, the filter can experience *breakthrough*. During breakthrough, the filter contains so many particles than it can no longer filter effectively and the effluent turbidity increases.

Several events can trigger the end of the filter run and lead to backwash. First, if the filter reaches breakthrough, it must be backwashed to prevent high-turbidity water from entering the distribution system. Second, the head loss can increase beyond the available head through the process. Rapid filters typically operate by gravity and are designed with 2 to 3 m (6.6 to 10 ft) of available head. When head loss reaches this available head (also called the *limiting head*), the filter must be backwashed even if it has not reached breakthrough. Some filters do not reach breakthrough or the limiting head within several days. In these cases, utilities backwash filters after a set period to maintain a convenient schedule for plant operators, even though the filter has additional usable capacity.

On Fig. 7-4, the filter reaches breakthrough before reaching the available head, but these events can occur in either order depending on the filter design and raw-water quality. A filter design is optimized when both events occur simultaneously.

During the backwash stage, water flows in the direction opposite to removing the particles that have collected in the filter bed. Effective removal of collected solids is a key component of rapid filtration systems, so while the backwashing stage is short compared to the filtration stage, it is an important part of the filtration cycle. Backwash typically consumes 2 to 5 percent of the filtered water.

Most filters also contain supplemental systems to assist the backwashing process. Supplemental scouring causes vigorous agitation of the bed and causes collisions and abrasion between media grains that break deposited solids loose from the media grains. Once the solids are separated from the media grains, the upflowing wash water can flush the solids from the filter. One option is the surface wash system, discussed earlier. Another option is *air scour*, in which pressurized air is introduced underneath the media with the backwash water. Air and water are introduced simultaneously at

Table 7-2
Typical design criteria for backwash systems

Criteria	Backwash Water	Rotating-Arm Surface Wash	Air Scour
Flow rate	30–60 m/h (12–24 gpm/ft^2)	1.2–1.8 m/h (0.5–0.7 gpm/ft^2)	36–72 m^3/m^2·h (2–4 scfm/ft^2)
Head or pressure	8–10 m (26–33 ft)	5–7 bar (73–100 psi)	0.3–0.5 bar (4.3–7.3 psi)
Duration	10–15 min	4–8 min	4–8 min

the bottom of the filter bed for a portion of the backwash cycle followed by a water-only wash for the remainder of the cycle. The most effective air scouring occurs when the water is flowing between 25 and 50 percent of the minimum fluidization velocity (Amirtharajah, 1993). At this water flow rate, the air forms cavities within the media that subsequently collapse (a phenomenon that has been called *collapse pulsing*), causing a substantial amount of agitation of the bed. For deep filter beds, both air and surface wash are often provided. Typical rates, pressures, and durations for backwash, air scour, and surface wash are shown in Table 7-2.

Filtration Rate

The *filtration rate* is the flow rate through the filter divided by the area of the surface of the filter bed. The filtration rate has units of volumetric flux (reported as m/h in SI units and gpm/ft^2 in U.S. customary units) and is sometimes referred to as the superficial velocity because it is the velocity the water would have in an empty filter box. Filtration rates of 5 to 15 m/h (2 to 6 gpm/ft^2) are typical, although some high-rate filters have been designed with rates as high as 33 m/h (13.5 gpm/ft^2).

Pretreatment Requirements

Coagulation pretreatment is required ahead of rapid filtration. If particles are not properly destabilized, the natural negative surface charge on the particles and filter media grains cause repulsive electrostatic forces that prevent contact between particles and media. The origin of surface charge on particles in nature and the proper use of coagulants for destabilizing particles were discussed in detail in Chap. 5. Properly designed and operated rapid filters can fail quickly if the coagulant feed breaks down or the raw-water quality changes and the coagulant dose is not adjusted accordingly.

Rapid filtration is classified by the level of pretreatment, as presented on Fig. 7-5. The most common configuration in the United States is conventional filtration. The most important factors that determine the required level of pretreatment are the raw-water quality and the preference and resources of the operating utility.

Conventional filtration.
Most common filtration system. Used with any surface water, even those with very high or variable turbidity. Responds well to rapid changes in source water quality.

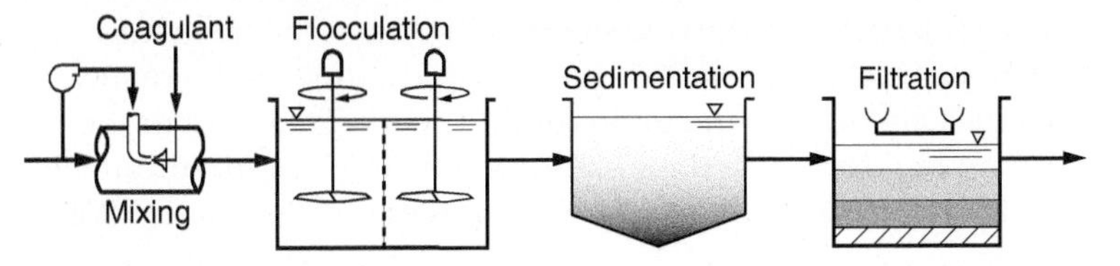

Direct filtration.
Good for surface waters without high or variable turbidity. Typical source waters are lakes and reservoirs, but usually not rivers. Raw-water turbidity < 15 NTU.

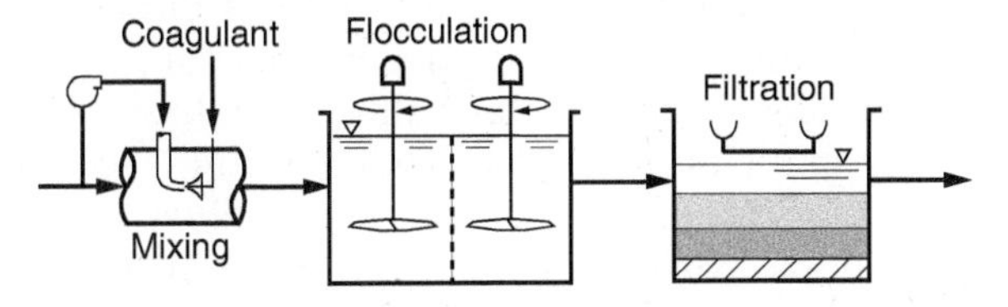

In-line filtration (also called contact filtration).
Requires high-quality surface water with very little variation and no clay or sediment particles. Raw-water turbidity < 10 NTU.

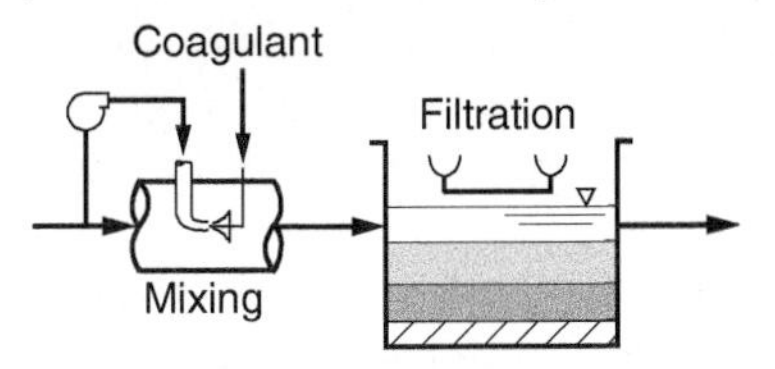

Two-stage filtration.
Preengineered systems used in small treatment plants (also called package plants). Raw-water turbidity < 100 NTU.

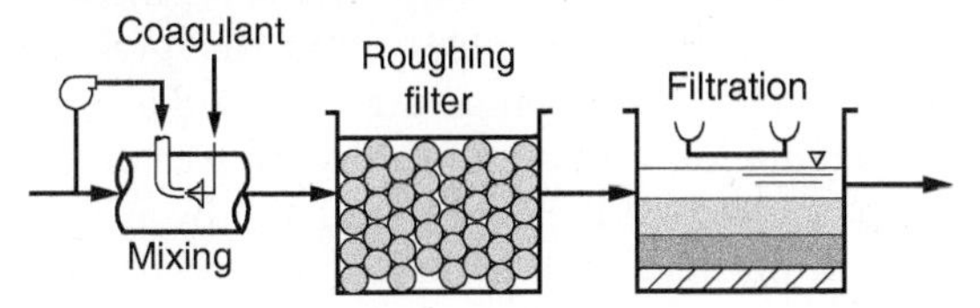

Figure 7-5
Classification of rapid filtration by pretreatment level.

7-3 Particle Capture in Granular Filtration

Filters can remove particles from water by several mechanisms. When particles are larger than the void spaces in the filter, they are removed by *straining*. When particles are smaller than the voids, they can be removed only if they contact and stick to the grains of the media. Transport to the media surface occurs by diffusion, sedimentation, and interception, and attachment occurs by attractive close-range molecular forces such as van der Waals forces.

Straining

Figure 7-6 demonstrates how particles are strained in a granular bed. For uniformly sized spherical media, a close-packed arrangement will cause straining when the ratio of particle diameter to grain diameter is greater than 0.15; smaller particles can pass through the media. The effective size of the smallest media in rapid filters is typically around 0.5 mm; thus, straining is effective only for particles larger than about 75 μm. The vast majority of particles in the influent to rapid filters are smaller. For example, viruses can be more than 1000 times smaller than particles that would be strained in a conventional filter and clearly would not be removed without transport and attachment mechanisms.

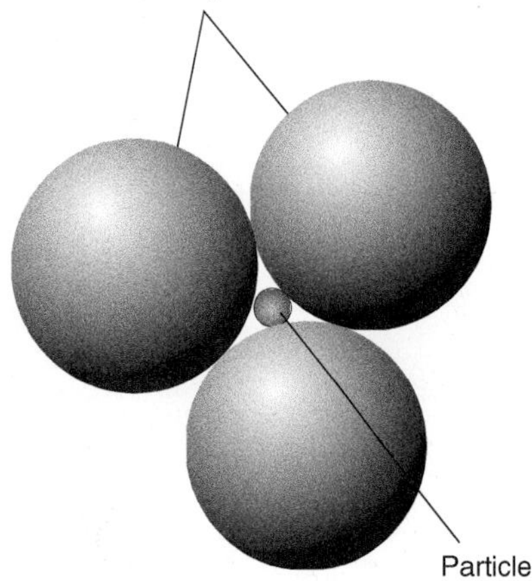

Figure 7-6
Capture of spherical particle by spherical media grains. If the ratio of particle diameter to media diameter is greater than 0.15, the particle will be strained by the media. If it is smaller, straining is not possible and particle capture must occur by other means. For typical rapid filtration, straining is limited to particles about 75 μm and larger.

Straining causes a cake to form at the surface of the filter bed, which can improve particle removal but also increases head loss through the filter. Rapid filters quickly build head loss to unacceptable levels if a significant cake layer forms. In addition, filtration at the surface leaves the bulk of the rapid filter bed unused. Consequently, rapid filters are designed to minimize straining and encourage depth filtration.

The reliance on depth filtration is the key to why coagulation pretreatment is essential in rapid filtration. If particles are stable (see Chap. 5), the repulsive electrostatic forces between the particles and media grains will prevent the particles from contacting the media. Destabilization by coagulation eliminates the repulsive forces and allows the particles to adhere to the media by attractive van der Waals forces (similar to particle agglomeration in flocculation). Without coagulation, particles can pass right through the filter.

Depth Filtration

In depth filtration, particles are removed continuously throughout the filter through a process of transport and attachment to the filter grains. Particle removal within a filter is dependent on the concentration of particles, similar to a first-order rate equation (Iwasaki, 1937), as described by

$$\frac{\partial C}{\partial z} = -\lambda C \tag{7-2}$$

where λ = filtration coefficient, m^{-1}
C = mass or number concentration of particles, mg/L or L^{-1}
z = depth in filter, m

Unfortunately, the filtration coefficient is not a constant, readily available number. Filtration is a complex process, and the filtration coefficient depends on properties of the filter bed (grain shape and size distribution, porosity, depth), influent suspension (turbidity, particle concentration, particle size distribution, particle and water density, water viscosity, temperature, level of pretreatment), and operating conditions (filtration rate).

Fundamental filtration models have been developed to examine the relative importance of mechanisms that cause particles to contact media grains. They describe how particles are removed during depth filtration and the importance of various design and operating parameters under time-invariant conditions. With these models, engineers have gained an understanding of how parameters such as media size and depth, bed porosity, filtration rate, and temperature affect filter performance. For these reasons, fundamental filtration models are valuable to a student acquiring a conceptual understanding of the filtration process.

Although they assist with conceptual understanding, fundamental filtration models are not very effective at quantitatively predicting the effluent turbidity in actual full-scale filters for the following reasons: (1) the models are based on an idealized system in which spherical particles collide with spherical filter grains; (2) the hydrodynamic variability and effect on streamlines introduced by the use of angular media are not addressed; (3) the models predict a single value for the filtration coefficient, which does not change as a function of either time or depth, whereas in real filters the filtration coefficient changes with both time and depth as solids collect on the media; and (4) the models assume no change in grain dimensions or bed porosity as particles accumulate. For these reasons, fundamental depth filtration models are often called clean-bed filtration models, and experimental validation generally focuses on the initial performance of laboratory filters (with spherical particles and media grains).

Formulation of a Filtration Model

The basic model for water filtration was originally developed by Yao et al. (1971). Yao et al.'s theory is based on the accumulation of particles on a single filter grain (termed a "collector"), which is then incorporated into a mass balance on a differential slice through a filter. The accumulation on a single collector is defined as the rate at which particles enter the region of influence of the collector multiplied by a transport efficiency factor and an attachment efficiency factor. The transport efficiency η and the attachment efficiency α are ratios describing the fraction of particles contacting and adhering to the media grain, respectively, as described by the equations

$$\eta = \frac{\text{particles contacting collector}}{\text{particles approaching collector}} \tag{7-3}$$

$$\alpha = \frac{\text{particles adhering to collector}}{\text{particles contacting collector}} \tag{7-4}$$

where η = transport efficiency, dimensionless
α = attachment efficiency, dimensionless

The mass flow of particles approaching the collector is determined by taking a mass balance in a differential unit of depth in the filter and integrating over the total depth. The total accumulation of particles within the control volume is the product of the number of collectors and the accumulation on a single isolated collector. A thorough development of the mass balance equations can be found in the reference book by Crittenden et al. (2012) and the resulting expression is

$$\frac{dC}{dz} = \frac{-3(1-\varepsilon)\eta\alpha C}{2d_C} \tag{7-5}$$

where ε = filter bed porosity, dimensionless
d_C = collector diameter, m

Comparing Eq. 7-5 to Eq. 7-2 will reveal that

$$\lambda = \frac{3(1-\varepsilon)\eta\alpha}{2d_C} \tag{7-6}$$

If the parameters in Eq. 7-6 (ε, η, α, and d_C) are constant with respect to depth in the filter, Eq. 7-5 can be integrated to yield the expression

$$C = C_0 \exp\left[\frac{-3(1-\varepsilon)\eta\alpha L}{2d_C}\right] \tag{7-7}$$

where C_0 = particle concentration in filter influent, mg/L
L = depth of filter, m

The next step in the development of a fundamental filtration model is to evaluate the mechanisms that influence the transport of particles to the media surface. Yao et al. (1971) identified these mechanisms as diffusion, sedimentation, and interception. Additional researchers have expanded and refined the model by using trajectory analysis, developing more sophisticated representations of the region around a media grain, accounting for reduced collisions due to viscous resistance of the water between the particle and collector, and accounting for the attraction between the collectors and particles caused by van der Waals forces. Tufenkji and Elimelech (2004) developed the most current model in use today, which is known as the TE model. Equations for each transport mechanism in depth filtration from the TE model are presented in the next section.

Transport Mechanisms

The mechanisms for transporting particles to media grains are shown on Fig. 7-7. Water approaching a spherical collector in a uniform-flow field under laminar flow conditions follows streamlines to either side of the collector. Some particles will contact the collector because they follow a fluid streamline that passes close to the grain, while others must deviate from their fluid streamline to reach the collector surface. Details for each transport mechanism are as follows.

DIFFUSION

Particles move by Brownian motion and will deviate from the fluid streamlines due to diffusion. For laminar flow, spherical particles, and spherical collectors, particle transport by diffusion is given by the following expression (Tufenkji and Elimelech, 2004):

$$\eta_D = 2.4 A_S^{1/3} N_R^{-0.081} N_V^{0.052} \mathrm{Pe}^{-0.715} \tag{7-8}$$

where η_D = transport efficiency due to diffusion, dimensionless
A_S = porosity parameter that accounts for the effect of adjacent media grains, dimensionless
N_R = relative size number, dimensionless
N_V = van der Waals number, dimensionless
Pe = Peclet number, dimensionless

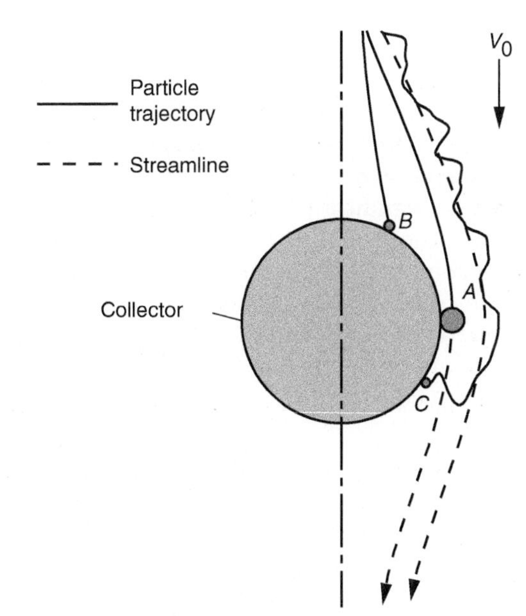

Figure 7-7
Particle transport mechanisms in fundamental filtration theory: (a) interception, particle *A* follows streamline but collides with the collector because of the proximity between the streamline and the collector; (b) sedimentation, particle *B* deviates from the streamline and collides with the collector because of gravitational forces; and (c) diffusion, particle *C* collides with collector due to random Brownian motion.

The terms in Eq. 7-8 are defined as follows:

$$A_S = \frac{2(1-\gamma^5)}{2 - 3\gamma + 3\gamma^5 - 2\gamma^6} \tag{7-9}$$

$$N_R = \frac{d_P}{d_C} \tag{7-10}$$

$$N_V = \frac{\text{Ha}}{k_B T} \tag{7-11}$$

$$\text{Pe} = \frac{3\pi\mu d_P d_C v_F}{k_B T} \tag{7-12}$$

where $\gamma = (1-\varepsilon)^{1/3}$, dimensionless
ε = filter bed porosity, dimensionless
d_P, d_C = particle and collector diameters, respectively, m
Ha = Hamaker constant, J
k_B = Boltzmann constant, 1.381×10^{-23} J/K
T = absolute temperature, K (273 + °C)
μ = liquid viscosity, kg/m·s
v_F = filtration rate (superficial velocity), m/s

The Peclet number is a dimensionless parameter describing the relative significance of advection and dispersion in mass transport. For physically similar systems, a lower value of the Peclet number implies greater significance of diffusion. The formulation of the Peclet number in Eq. 7-12 uses the Stokes–Einstein equation (Clark, 2009) to relate the diffusion coefficient to the diameter of a spherical particle. Transport by diffusion is then further influenced by hydrodynamic interactions caused by adjacent media grains and attractive molecular forces called van der Walls forces. The Hamaker constant is a parameter used in describing van der Waals forces. The theory necessary to calculate a value for the Hamaker constant is beyond the scope of this text, but the value ranges from 10^{-19} to 10^{-20} J.

In rapid filtration, diffusion is most significant for particles less than about 1 μm in diameter. For 0.1-μm particles passing through a filter with 0.5-mm sand under typical filtration conditions, η_D is about 10^{-3}. In other words, only about 1 in 1000 possible collisions with a single collector due to diffusion will actually occur. However, a particle will pass thousands of collectors during its passage through a filter bed, increasing the chance of being removed somewhere in the filter bed.

Sedimentation

Particles with a density significantly greater than water tend to deviate from fluid streamlines due to gravitational forces. The collector efficiency due to gravity is shown in the expression (Tufenkji and Elimelech, 2004):

$$\eta_G = 0.22 N_R^{-0.24} N_V^{0.053} N_G^{1.11} \tag{7-13}$$

where η_G = transport efficiency due to gravity, dimensionless
N_G = gravity number, dimensionless

The gravity number is defined as

$$N_G = \frac{v_S}{v_F} = \frac{g(\rho_P - \rho_W)d_P^2}{18\mu\ v_F} \tag{7-14}$$

where v_S = Stokes settling velocity, m/s
g = gravitational constant, m/s^2
ρ_P, ρ_W = particle and water density, respectively, kg/m^3

The basic prediction of Eq. 7-14 is that particle collection by sedimentation increases as the ratio of the Stokes setting velocity to filtration rate increases; that is, more particles will contact media grains if the Stokes velocity is bigger or the filtration rate is smaller. As with diffusion, this basic effect is then influenced by the hydrodynamics of adjacent media grains and attractive van der Waals forces.

Interception

Particles remaining centered on fluid streamlines that pass the collector surface by a distance of half the particle diameter or less will be intercepted. Particle transport by interception is given by the expression (Tufenkji and Elimelech, 2004)

$$\eta_I = 0.55 A_S\ N_A^{1/8}\ N_R^{1.675} \tag{7-15}$$

where η_I = transport efficiency due to interception, dimensionless
N_A = attraction number accounting for attraction between the collector and particle as the particle gets very close, dimensionless

The attraction number is given as follows:

$$N_A = \frac{N_V}{N_R\,\text{Pe}} = \frac{\text{Ha}}{3\pi\mu\ d_P^2 v_F} \tag{7-16}$$

The most significant term in Eq. 7-15 is N_R. The basic prediction of Eq. 7-15 is that interception increases as the ratio of particle size to collector size increases; that is, more particles will intercept media grains if the particles are bigger or the media grains are smaller.

TOTAL TRANSPORT EFFICIENCY

The relative importance of these various mechanisms for transporting the particle to the surface depends on the physical properties of the filtration system. The model is based on an assumption that the transport mechanisms are additive:

$$\eta = \eta_D + \eta_G + \eta_I \tag{7-17}$$

where η = total transport efficiency, dimensionless

The importance of each mechanism can be evaluated as a function of system properties. The effect of particle diameter on the importance of each mechanism is shown on Fig. 7-8. Small particles are efficiently removed by diffusion, whereas larger particles are removed mainly by sedimentation and interception. This model predicts that the lowest removal efficiency occurs for particles of about 1 to 2 μm in size, which has been verified experimentally (Yao et al., 1971).

Attachment Efficiency

As particles approach the surface of the media, short-range surface forces begin to influence particle dynamics. The attachment efficiency varies from a value of zero (no particles adhere) to a value of 1.0 (every collision between a particle and collector results in attachment). The attachment efficiency is affected by London–van der Waals forces, surface chemical interactions, electrostatic forces, hydration, hydrophobic interactions, or steric interactions (Tobiason and O'Melia, 1988). In water treatment, the focus is to modify the system so that attachment is as favorable as possible, that is, an attachment efficiency value very nearly 1.0. The most important factor in achieving high attachment efficiency is eliminating the repulsive electrostatic forces, that is, proper destabilization of particles by coagulation. The need for high attachment efficiency is exactly why coagulation is a critical part of rapid filtration. Particle stability and destabilization by coagulation was discussed in Chap. 5. An analysis of the impact of lower values for attachment efficiency is described in Tobiason et al. (2011).

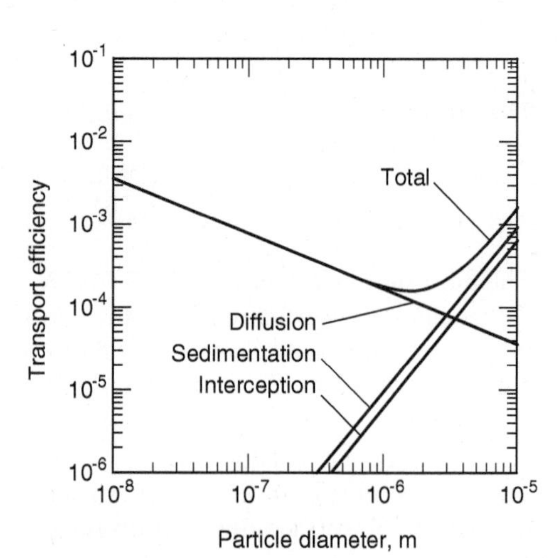

Figure 7-8
Importance of each transport mechanism on particles of different size as predicted by the TE model ($\rho_P = 1050$ kg/m^3, $d_C = 0.5$ mm, $v_F = 10$ m/h, $L = 1$ m, $\varepsilon = 0.42$, Ha $= 1 \times 10^{-20}$ J, $T = 15°$C, $\alpha = 1.0$).

The fundamental filtration model can be used to examine the effect of important variables on filter performance, as demonstrated in Example 7-2.

Example 7-2 Application of the TE fundamental filtration model to evaluate the effect of media diameter on filter removal efficiency

Use the TE model to examine the effect of media diameter (ranging from 0.4 to 2 mm in diameter) on the removal of 0.1-μm particles in a filter bed of monodisperse media under the following conditions: porosity $\varepsilon = 0.50$, attachment efficiency $\alpha = 1.0$, temperature $T = 20^\circ\text{C}$ (293.15 K), particle density $\rho_P = 1050\ \text{kg/m}^3$, filtration ratc $v_F = 15$ m/h, bed depth $L = 1.0$ m, Hamaker constant $\text{Ha} = 10^{-20}$ J (note: $1\ \text{J} = 1\ \text{kg·m}^2/\text{s}^2$), and Boltzmann constant $k_B = 1.381 \times 10^{-23}$ J/K.

Solution

1. Calculate γ and A_S using Eq. 7-9:

$$\gamma = (1 - \varepsilon)^{1/3} = (1 - 0.50)^{1/3} = 0.7937$$

$$A_S = \frac{2(1 - \gamma^5)}{2 - 3\gamma + 3\gamma^5 - 2\gamma^6}$$

$$= \frac{2[1 - (0.7937)^5]}{2 - 3(0.7937) + 3(0.7937)^5 - 2(0.7937)^6} = 21.46$$

2. Calculate N_R for a media diameter of 0.4 mm using Eq. 7-10:

$$N_R = \frac{d_P}{d_C} = \frac{1 \times 10^{-7}}{4 \times 10^{-4}} = 2.5 \times 10^{-4}$$

3. Calculate N_V using Eq. 7-11:

$$N_V = \frac{\text{Ha}}{k_B T} = \frac{1 \times 10^{-20}\ \text{kg·m}^2/\text{s}^2}{(1.381 \times 10^{-23}\ \text{kg·m}^2/\text{s}^2)(293.15\ \text{K})} = 2.47$$

4. Calculate Pe for a media diameter of 0.4 mm using Eq. 7-12. From Table C-1 in App. C, the viscosity of water at 20°C is $\mu = 1.0 \times 10^{-3}$ kg/m·s:

$$\text{Pe} = \frac{3\pi\mu d_P d_C v_F}{k_B T}$$

$$= \frac{3\pi(1 \times 10^{-3}\ \text{kg/m·s})(1 \times 10^{-7}\ \text{m})(4 \times 10^{-4}\ \text{m})(15\ \text{m/h})}{(1.381 \times 10^{-23}\ \text{kg·m}^2/\text{s}^2\text{K})(293.15\ \text{K})(3600\ \text{s/h})}$$

$$= 3.89 \times 10^5$$

5. Calculate η_D using Eq. 7-8:

$$\eta_D = (2.4)(21.46)(2.50 \times 10^{-4})(2.47)^{0.052}(3.89 \times 10^5)$$
$$= 1.38 \times 10^{-3}$$

6. Calculate N_G using Eq. 7-14. From Table C-1 in App. C, the density of water at 20°C is $\rho_W = 998\ \text{kg/m}^3$:

$$N_G = \frac{g(\rho_P - \rho_W)d_P^2}{18\mu v_F}$$
$$= \frac{(1050 - 998\ \text{kg/m}^3)(9.81\ \text{m/s}^2)(1 \times 10^{-7}\ \text{m})^2(3600\ \text{s/h})}{18(1 \times 10^{-3}\ \text{kg/m·s})(15\ \text{m/h})}$$
$$= 6.76 \times 10^{-8}$$

7. Calculate η_G using Eq. 7-13:

$$\eta_G = (0.22)(2.5 \times 10^{-4})^{-0.24}(2.47)^{0.053}(6.76 \times 10^{-8})^{1.11}$$
$$= 1.86 \times 10^{-8}$$

8. Calculate N_A using Eq. 7-16:

$$N_A = \frac{\text{Ha}}{3\pi\mu d_P^2 v_F} = \frac{(1 \times 10^{-20}\ \text{kg·m}^2/\text{s}^2)(3600\ \text{s/h})}{3\pi(1 \times 10^{-3}\ \text{kg/m·s})(1 \times 10^{-7}\ \text{m})^2(15\ \text{m/h})}$$
$$= 2.54 \times 10^{-2}$$

9. Calculate η_I using Eq. 7-15:

$$\eta_I = (0.55)(21.46)(2.54 \times 10^{-2})^{1/8}(2.5 \times 10^{-4})^{1.675}$$
$$= 6.91 \times 10^{-6}$$

10. Calculate η using Eq. 7-17:

$$\eta = 6.91 \times 10^{-6} + 1.86 \times 10^{-8} + 1.38 \times 10^{-3} = 1.39 \times 10^{-3}$$

11. Calculate C/C_0 using Eq. 7-7 and the log removal value (LRV) using Eq. 3-2.

$$\frac{C}{C_0} = \exp\left[\frac{-3(1-0.50)(1.39\times10^{-3})(1.0)(1\text{ m})}{2(4\times10^{-4}\text{ m})}\right] = 0.074$$

$$\text{LRV} = \log(C_i) - \log(C_e) = \log(1) - \log(0.074) = 1.13$$

12. Set up a computation table to determine particle removal for other diameters. Repeat steps 1 through 11 for additional media sizes between 0.4 and 2.0 mm. These calculations are best done with a spreadsheet. The results are as follows:

Media Diameter (mm)	C/C_0	Log Removal
0.4	0.074	1.13
0.6	0.262	0.58
0.8	0.434	0.36
1.0	0.560	0.25
1.2	0.650	0.19
1.4	0.716	0.15
1.6	0.764	0.12
1.8	0.801	0.10
2.0	0.830	0.08

Comment

The initial removal of small particles is highly sensitive to media size. While these particles are removed relatively efficiently by 0.4-mm-diameter media, removal drops dramatically as the media size increases.

7-4 Head Loss through a Clean Filter Bed

Figure 7-4 showed that the length of a filter run can be limited by the buildup of head loss. Head loss has two components to it—the *clean-bed head loss* through the media and the additional head loss that builds up as the bed collects solids. The clean-bed head loss is not insignificant; in fact, it sometimes can have a greater influence on the design of the filter bed than the rate that head accumulates during filtration. This section presents equations for calculating the clean-bed head loss, while the buildup of head loss during filtration is addressed in Sec. 7-5 in the context of optimizing the length of the filter run.

The classic equation for the relationship between head loss and flow through porous media is Darcy's law, which is normally written

$$v_F = k_p \frac{h_L}{L} \tag{7-18}$$

where v_F = filtration rate (superficial velocity), m/s
k_p = hydraulic permeability, m/s
h_L = head loss across media bed, m
L = depth of granular media, m

Darcy's law has two shortcomings that limit its usefulness for filter design. First, the hydraulic permeability is an empirically derived number that is measured to relate flow to head loss in an existing porous medium. For filter design, it is necessary to have a predictive equation that relates head loss to properties of the system: media grain diameter, bed depth, porosity, flow rate, and temperature.

Second, Darcy's law is only valid for laminar flow. Flow in granular media does not experience a rapid transition from laminar to turbulent as in pipes but can be divided into four flow regimes (Trussell and Chang, 1999). Laminar or creeping flow occurs at Reynolds numbers less than about 1. The next regime, called Forchheimer flow, occurs at Reynolds numbers between about 1 and 100. Typical rapid filters have Reynolds numbers ranging from 0.5 to 5. High-rate rapid filters have been designed with filtration rates as high as 33 m/h (13.5 gpm/ft^2), resulting in a Reynolds number of about 18. Backwashing of rapid filters occurs between Reynolds numbers of 3 and 25, completely in the Forchheimer flow regime. The remaining flow regimes, transition flow and turbulent flow, are not relevant in granular filtration.

Unlike laminar flow, Forchheimer flow is influenced by both viscous and inertial forces. In purely viscous flow, momentum is transferred between streamlines solely via molecular interactions. In twisting, irregular voids of a granular media bed, however, the fluid must accelerate and decelerate as void spaces turn, contract, and expand. The complex fluid motion through passageways of varying dimensions complicates the momentum transfer between streamlines, leading to an additional component of head loss due to inertial forces. While Darcy's law predicts a linear relationship between flow and head, Forchheimer flow has two components, with head loss due to viscous forces proportional to v and head loss due to inertial forces proportional to v^2:

$$\frac{h_L}{L} = \kappa_1 v_F + \kappa_2 v_F^2 \tag{7-19}$$

where κ_1 = Forchheimer coefficient for viscous losses, s/m
κ_2 = Forchheimer coefficient for inertial losses, s^2/m^2

Researchers have used experimental data and analogies between granular media and flow through pipes to develop equations relating head loss to properties of the filter bed when filters are operating in the Forchheimer flow regime. A widely used and reliable equation was developed in 1952 and is known as the Ergun equation:

$$h_L = \kappa_V \frac{(1-\varepsilon)^2}{\varepsilon^3} \frac{\mu L v_F}{\rho_W g d^2} + \kappa_I \frac{1-\varepsilon}{\varepsilon^3} \frac{L v_F^2}{gd} \tag{7-20}$$

where κ_V, κ_I = Ergun coefficients for viscous and inertial losses, unitless
ε = filter bed porosity, dimensionless
μ = viscosity of water, kg/m·s
ρ_W = density of water, kg/m^3
g = gravitational constant, m/s^2
d = diameter of media, m

Ergun (1952) correlated data from 640 experiments covering a range of Reynolds numbers between about 1 and 2000 to develop this equation. In conformance to Forchheimer equation, the first term in Eq. 7-20 represents viscous energy losses and the second term represents inertial losses. The dependence on μ, L, v, ρ_W, g, and d in the first term is consistent with equations for laminar flow, while the dependence on these six variables in the second term is consistent with equations for turbulent flow. Although some filters may operate in the Darcy flow regime, the Ergun equation can be used to determine the clean-bed head loss over the full range of values of interest in rapid filtration.

An important issue is the selection of values for κ_V and κ_I in the Ergun equation. A recent study examined head loss through granular media (Trussell and Chang, 1999) and proposed suitable values for sand and anthracite when the diameter is the effective size of the media (e.g., d = ES). These values are shown in Table 7-3. In the absence of pilot data or other site-specific information, the midpoint values in Table 7-3 are recommended for use. Calculation of clean-bed head loss is demonstrated in Example 7-3.

Table 7-3
Coefficients and porosity for use with the Ergun equation, Eq. 7-20[a]

Medium	κ_V	κ_I	ε, %
Sand	110–115	2.0–2.5	40–43
Anthracite	210–245	3.5–5.3	47–52

[a]When effective size as determined by sieve analysis is used for the diameter.

Example 7-3 Clean-bed head loss through rapid filter

Calculate the clean-bed head loss through a deep-bed anthracite filter with 1.8 m of ES = 0.95 mm media at a filtration rate of 15 m/h and a temperature of 15°C.

Solution

The head loss through anthracite is calculated first using Eq. 7-20.

1. No pilot or site-specific information is given, so midpoint values are selected from Table 7-3; $\kappa_V = 228$, $\kappa_I = 4.4$, and $\varepsilon = 0.50$. Values of ρ_W and μ are available in Table C-1 in App. C ($\rho_W = 999\ \text{kg/m}^3$ and $\mu = 1.14 \times 10^{-3}\ \text{kg/m·s}$).
2. Calculate the first term in Eq. 7-20:

$$\frac{(228)(1-0.50)^2(1.14\times10^{-3}\ \text{kg/m·s})(1.8\ \text{m})(15\ \text{m/h})}{(0.50)^3(999\ \text{kg/m}^3)(9.81\ \text{m/s}^2)(0.95\ \text{mm})^2(10^{-3}\ \text{m/mm})^2(3600\ \text{s/h})}$$

$$= 0.44\ \text{m}$$

3. Calculate the second term in Eq. 7-20:

$$\frac{(4.4)(1-0.50)(1.8\ \text{m})(15\ \text{m/h})^2}{(0.50)^3(9.81\ \text{m/s}^2)(0.95\ \text{mm})(10^{-3}\ \text{m/mm})(3600\ \text{s/h})^2} = 0.06\ \text{m}$$

4. Add the two terms together:

$$h_L = 0.44\text{m} + 0.06\text{m} = 0.50\ \text{m} \quad (1.6\ \text{ft})$$

Comments

A relatively small contribution to head loss comes from the inertial term. The inertial term becomes more important for the larger media and higher velocities used in high-rate rapid filters. If the filter is designed with 2.5 m (8.2 ft) of available head, the clean-bed head loss consumes about 20 percent of the available head. Note that if multiple layers of media are present, the head loss through each layer is additive.

7-5 Modeling of Performance and Optimization

As a result of many years of successful operation of rapid filters, it is frequently possible to design an effective rapid-filtration system using experience and the principles presented in this chapter. In some situations, however, pilot testing is used to examine the impact of various design alternatives, optimize performance, verify acceptable performance, or satisfy regulatory agencies. The most common variables considered in filtration pilot studies are media size, media depth, and filtration rate. In addition, mono- and dual-media filters are often compared in pilot studies. Parameters of interest include (1) the duration of ripening and water quality during ripening, (2) the water quality during the effective filtration cycle, (3) the time to breakthrough, and (4) the time to limiting head.

Often pilot filters are run under a limited set of conditions, and it is desirable to predict the performance under other conditions. Phenomenological models have been developed to evaluate pilot data and can be used to predict filter performance for conditions that were not specifically addressed within a pilot study. The specific deposit, which is the mass of accumulated particles per filter bed volume, is used as a master variable. Pilot data analysis is straightforward if a number of simplifying assumptions are used: (1) the specific deposit is averaged over the entire filter bed, (2) solids accumulate at a steady rate over the entire filter run (the reduction in solids capture during ripening is considered negligible), and (3) head loss increases at a constant rate. With these assumptions, the specific deposit can be determined by performing a mass balance over the entire bed:

$$\sigma_t V = C_0 Qt - C_E Qt \tag{7-21}$$

where σ_t = specific deposit at time t, mg/L
V = bed volume, m^3
C_0, C_E = influent and effluent concentrations, mg/L
Q = flow rate, m^3/s
t = time, s

Dividing by the filter bed area and rearranging yields an expression for the specific deposit as a function of time:

$$\sigma_t = \frac{v_F(C_0 - C_E)t}{L} \tag{7-22}$$

where L = filter bed depth, m

The specific deposit increases at a steady rate as solids accumulate in the filter bed. Pilot filters can be operated until breakthrough occurs, and the value of the specific deposit at breakthrough can be related to the time to breakthrough by the expression

$$\sigma_B = \frac{v_F(C_0 - C_E)t_B}{L} \tag{7-23}$$

where σ_B = specific deposit at breakthrough, mg/L
t_B = time to breakthrough, h

Equation 7-23 can be rearranged and expressed as a function of t_B:

$$t_B = \frac{\sigma_B L}{v_F(C_0 - C_E)} \tag{7-24}$$

The value of the specific deposit at breakthrough can be recorded for a number of filter runs in which process parameters such as filtration rate, media diameter, or bed depth are varied. A regression analysis of the data can determine the dependence of the specific deposit at breakthrough, σ_B, on the process parameters (Kavanaugh et al., 1977).

Similarly, the rate of head loss buildup depends on the rate of solids deposition in a filter. If head loss increases at a constant rate, then the head loss at any time during the filtration run can be determined using the expression (Ives, 1967)

$$h_{L,t} = h_{L,0} + k_{\text{HL}}\sigma_t \tag{7-25}$$

where $h_{L,0}, h_{L,t}$ = initial head loss and filter head loss at time t, m
k_{HL} = head loss increase rate constant, L · m/mg

Rearranging yields an expression for the rate constant:

$$k_{\text{HL}} = \frac{h_{L,t} - h_{L,0}}{\sigma_t} \tag{7-26}$$

As before, a regression analysis can be used to determine the dependence of the head-loss rate constant, k_{HL}, on the process parameters. Once the rate constant for head-loss buildup is determined, it can be used to determine the run time before reaching the limiting head by incorporating Eq. 7-22 and rearranging Eq. 7-26:

$$t_{\text{HL}} = \frac{(H_T - h_{L,0})L}{k_{\text{HL}}\, v_F(C_0 - C_E)} \tag{7-27}$$

where t_{HL} = time to limiting head, h
H_T = total available head, m

Once the specific deposit at breakthrough and the rate of head-loss buildup are determined, these equations can be used to determine the duration of filter runs and whether filter runs are limited by breakthrough or limiting head. Use of these equations to analyze pilot data is demonstrated in Example 7-4.

Example 7-4 Determination of optimum media size from pilot data

Four pilot filters with different effective sizes of anthracite (UC < 1.4, ρ_P = 1700 kg/m^3) were operated over multiple runs. The results are summarized in the table below. The media depth in each filter was 1.8 m, the filtration rate was 15 m/h, and the temperature was relatively constant at 20°C. Based on turbidity, you can assume the solids concentration was constant at 2.2 mg/L in the influent and negligible in the effluent.

Media ES (mm)	No. of Runs	Ave. Eff. Turbidity (NTU)	Ave. Time to Breakthrough (h)	Ave. Initial Head (m)	Ave. Final Head (m)
0.73	7	0.08	55	0.77	3.35
0.88	6	0.07	49	0.56	2.38
1.02	9	0.08	41	0.44	1.87
1.23	8	0.13	38	0.29	1.44

Determine: (a) the relationship between specific deposit at breakthrough (σ_B) and the media ES, (b) the relationship between the head-loss rate constant (k_{HL}) and the media ES, and (c) the required available head and optimal media size if the full-scale system is to have a design run length of at least 48 h.

Solution

1. Any type of equation that relates media ES to σ_B and k_{HL} and results in a linear graph can be used. Often the relationships between the media ES and σ_B or k_{HL} can be described by a power function, and that type of equation is used in this example. Thus:

$$\sigma_B = b_1 \, (d)^{m_1} \quad \text{and} \quad k_{\text{HL}} = b_2 \, (d)^{m_2}$$

To find the value of the coefficients b and m, take the log of each equation and plot $\log(\sigma_B)$ and $\log(k_{\text{HL}})$ as a function of $\log(d)$. The slope of the straight line is m and the intercept is $\log(b)$:

$$\log(\sigma_B) = \log(b_1) + m_1 \log(d)$$

and

$$\log(k_{\text{HL}}) = \log(b_2) + m_2 \log(d)$$

2. Calculate the necessary values for the first effective size.
   a. Calculate $\log(d)$.

$$\log(d) = \log(0.73 \text{ mm}) = -0.137$$

   b. Calculate $\log(\sigma_B)$ after calculating σ_B using Eq. 7-23.

$$\sigma_B = \frac{v(C_0 - C_E)\, t_B}{L} = \frac{(15 \text{ m/h})\,(2.2 - 0 \text{ mg/L})\,(55 \text{ h})}{1.8 \text{ m}}$$

$$= 1008 \text{ mg/L}$$

$$\log(\sigma_B) = \log(1008) = 3.00$$

c. Calculate $\log(k_{HL})$ after calculating k_{HL} using Eq. 7-26.

$$k_{HL} = \frac{(3.35 - 0.77\text{ m})(1.8\text{ m})}{15\text{ m/h}\,(2.2 - 0\text{ mg/L})(55\text{ h})} = 0.00256\text{ L}\cdot\text{m/mg}$$

$$\log(k_{HL}) = \log(0.00256) = -2.59$$

3. Repeat step 2 for the remaining effective sizes. The results are summarized in the following table.

ES	log(*d*)	σ_B	log(σ_B)	k_{HL}	log(k_{HL})
0.73	−0.137	1008	3.00	0.00256	−2.59
0.88	−0.056	898	2.95	0.00203	−2.69
1.02	0.0086	752	2.88	0.00190	−2.72
1.23	0.090	697	2.84	0.00165	−2.78

4. Plot log(σ_B) against log(*d*).

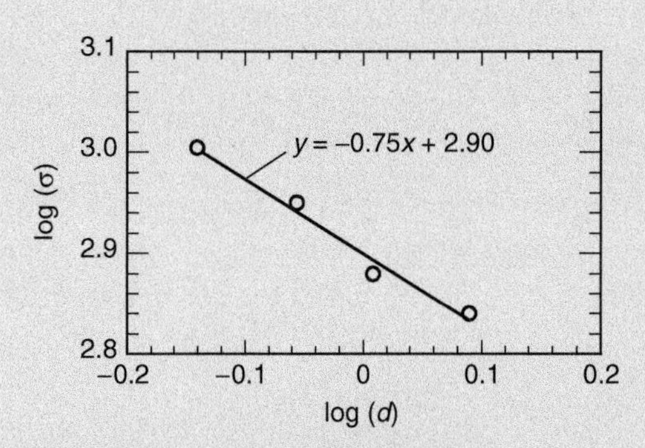

5. Perform a linear regression of the data in the graph in step 4 using the Excel® trendline function. Determine the slope and intercept of the regression line. From the graph in step 4, $m_1 = -0.75$ and $\log(b_1) = 2.90$. Therefore, $b_1 = 794$. The relationship between σ_B and d is

$$\sigma_B = 794\,(d)^{-0.75} \tag{1}$$

when the units of σ_B are mg/L and the units of d are mm.

6. Plot $\log(k_{HL})$ against $\log(d)$.

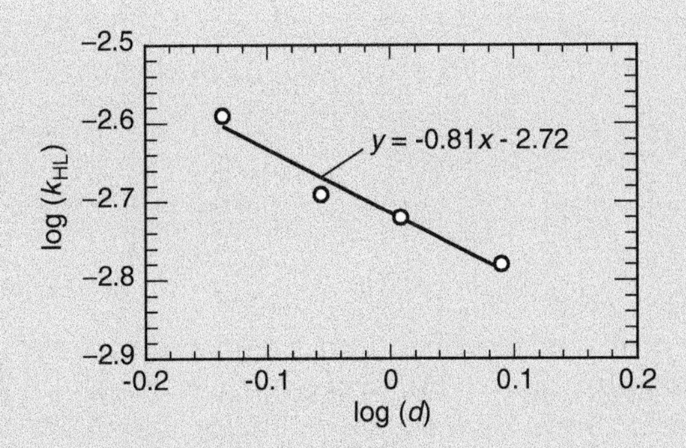

7. Perform a linear regression of the data in the graph in step 6 using the Excel® trendline function. Determine the slope and intercept of the regression line. From the graph in step 6, $m_2 = -0.81$ and $\log(b_2) = -2.72$. Therefore, $b_2 = 0.00191$. Thus, the relationship between k_{HL} and d is

$$k_{HL} = 0.00191\,(d)^{-0.81} \tag{2}$$

when the units of k_{HL} are L · m/mg and the units of d are mm.

8. Calculate the required size to reach 48 h before breakthrough by substituting Eq. 1 above into Eq. 7-24 and solving for the media size.

$$t_B = \frac{794\,(d)^{-0.75}L}{v_F(C_0 - C_E)}$$

$$(d)^{-0.75} = \frac{t_B v_F(C_0 - C_E)}{794L} = \frac{(48\text{ h})(15\text{ m/h})(2.2 - 0\text{ mg/L})}{794(1.8\text{ m})} = 1.108$$

$$d = (1.108)^{\frac{1}{-0.75}} = 0.87\text{ mm}$$

9. Calculate the required head to reach 48 h before reaching the limiting head by substituting Eq. 2 above into Eq. 7-27 and solving for the available head.

$$t_{HL} = \frac{(H_T - h_{L,0})L}{0.00181\,(d)^{-0.81}\,v_F(C_0 - C_E)}$$

$$H_T = \frac{t_{HL}0.00181\,(d)^{-0.81}\,v_F(C_0 - C_E)}{L} + h_{L,0}$$

$$= \frac{(48\text{ h})(0.00181)(0.87\text{ mm})^{-0.81}(15\text{ m/h})(2.2 - 0\text{ mg/L})}{1.8\text{ m}}$$

$$+ 0.53\text{ m} = 2.3\text{ m}$$

where the initial head loss was calculated with the Ergun equation (see Eq. 7-20 and Example 7-3). Thus, a run time of 48 h can be achieved with media that has an effective size of 0.87 mm and a total available head of 2.3 m.

OPTIMIZATION

The length of a filter run is optimized when the time to reach limiting head is equal to the time to breakthrough. Optimization of the filter design presented in Example 7-4 for two conditions of available head with respect to media size is shown on Fig. 7-9. Increasing the media depth will tend to increase the time to reach breakthrough (t_B) but decrease the time to reach the limiting head (t_{HL}). For 2.5 m (8.2 ft) of available head, the optimum design is achieved at a media size of 1.0 mm. An increase in available head to 3.0 m (9.8 ft) would have no effect on the run length if the media stayed the same size, but would increase the run length by about 5 h if the media effective size were decreased to 0.90 mm.

The effect of significant design parameters on t_B and t_{HL} is summarized in Table 7-4. The effects can be predicted from the theory presented earlier in the chapter and have generally been observed in actual filter operation. Some of the parameters, such as media size, media depth, and flow rate, are selected by the design engineer. Other variables, such as influent solids concentration, depend on raw-water quality and pretreatment. Variables such as floc strength and deposit density are difficult to control, but the use of polymers can be employed to improve floc strength (see Chap. 5).

UNIT FILTER RUN VOLUME

An important concept in assessing the length of a filter run is the unit filter run volume (UFRV) (Trussell et al., 1980). The UFRV is used to assess the

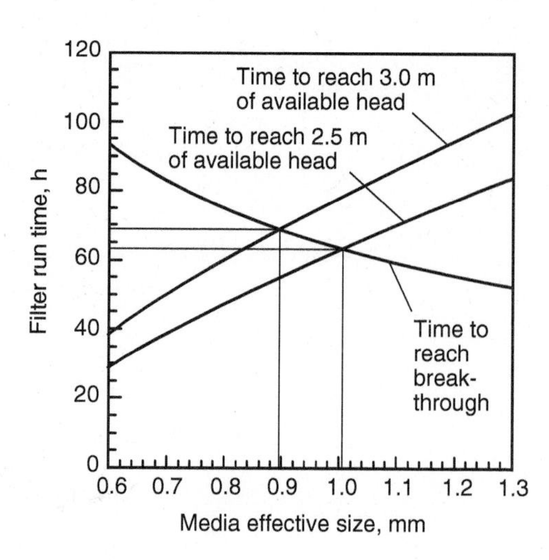

Figure 7-9
Optimization of media size with respect to time to breakthrough and time to limiting head.

Table 7-4
Effect of design parameters on time to breakthrough and limiting head loss

	Effect of Parameter Increase on	
Parameter	**Time to Breakthrough, t_B**	**Time to Limiting Head Loss, t_{HL}**
Effective size	Decrease	Increase
Media depth	Increase	Decrease
Filtration rate	Decrease	Decrease
Influent particle concentration	Decrease	Decrease
Floc strength	Increase	Decrease
Deposit density	Decrease	Decrease
Porosity	Decrease	Increase

recovery achieved by the filter. Recovery is the ratio between the net and total quantity of water filtered. Portions of the filtered water are used for backwashing and discharged as filter-to-waste, so the net water production is lower than the total volume of water processed through the filter. The UFRV is the volume of water that passes through the filter during a run and the unit backwash volume (UBWV) is the volume required to backwash the filter, defined as

$$\text{UFRV} = \frac{V_F}{a} = v_F t_F \tag{7-28}$$

$$\text{UBWV} = \frac{V_{\text{BW}}}{a} = v_{\text{BW}} t_{\text{BW}} \tag{7-29}$$

$$\text{UFWV} = \frac{V_{\text{FTW}}}{a} = v_F t_{\text{FTW}} \tag{7-30}$$

where UFRV = unit filter run volume, m^3/m^2
UBWV = unit backwash volume, m^3/m^2
UFWV = unit filter-to-waste volume, m^3/m^2
V_F, V_{BW}, V_{FTW} = volumes of water filtered during one filter run, used for backwash, and discharged as filter-to-waste, respectively, m^3
v_F, v_{BW} = filtration rate and backwash rate, respectively, m/h
t_F, t_{BW}, t_{FTW} = durations of filter run, backwash cycle, and filter-to-waste period, h
a = filter cross-sectional area, m^2

The ratio of net to total water filtered is the recovery:

$$r = \frac{V_F - V_{\text{BW}} - V_{\text{FTW}}}{V_F} = \frac{\text{UFRV} - \text{UBWV} - \text{UFWV}}{\text{UFRV}} \tag{7-31}$$

where r = recovery, expressed as a fraction

Filters should be designed for a recovery of at least 95 percent. Typical wash-water quantities are about 8 m^3/m^2 (200 gal/ft^2). Thus, to achieve a recovery greater than 95 percent, a UFRV of at least 200 m^3/m^2 (5000 gal/ft^2) is required.

7-6 Backwash Hydraulics

At the end of a filter run, rapid filters are backwashed. The backwash flow rate must be great enough to flush captured material from the bed, but not so high that the media is flushed out of the filter box. To prevent loss of media, it is important to determine the bed expansion that occurs as the filter media is fluidized.

The forces on an individual particle (either a particle from the influent or a media grain) in upward-flowing water are the same as were discussed for settling particles in Chap. 6. The particle will settle (or fail to fluidize) when downward forces predominate, be washed away when upward forces predominate, and remain suspended (fluidized) when the forces are balanced. The downward force is equal to the buoyant weight of the media ($F_G - F_B$) and the upward force is the drag force (F_D), which is caused by head loss from the backwash flow. The sum of forces on a particle is given by the expression

$$\sum F = F_G - F_B - F_D \tag{7-32}$$

where F_G, F_B, F_D = gravitational, buoyant, and drag forces on a particle, N

A filter bed is fluidized when a state of equilibrium is established between gravitational and drag forces (i.e., when $\Sigma F = 0$). During backwash, the velocity in a filter bed is higher than for an isolated particle due to the space taken up by the media, causing higher drag forces that lift the media. As the media rises, increasing porosity reduces the velocity until the drag force is balanced by the net gravitational force. The relationship between bed expansion and porosity is described in the following equation and on Fig. 7-10:

$$\frac{L_E}{L_F} = \frac{1 - \varepsilon_F}{1 - \varepsilon_E} \tag{7-33}$$

where L_F, L_E = depth of fixed (at rest) and expanded bed, respectively, m
$\varepsilon_F, \varepsilon_E$ = porosity of fixed and expanded bed, respectively, dimensionless

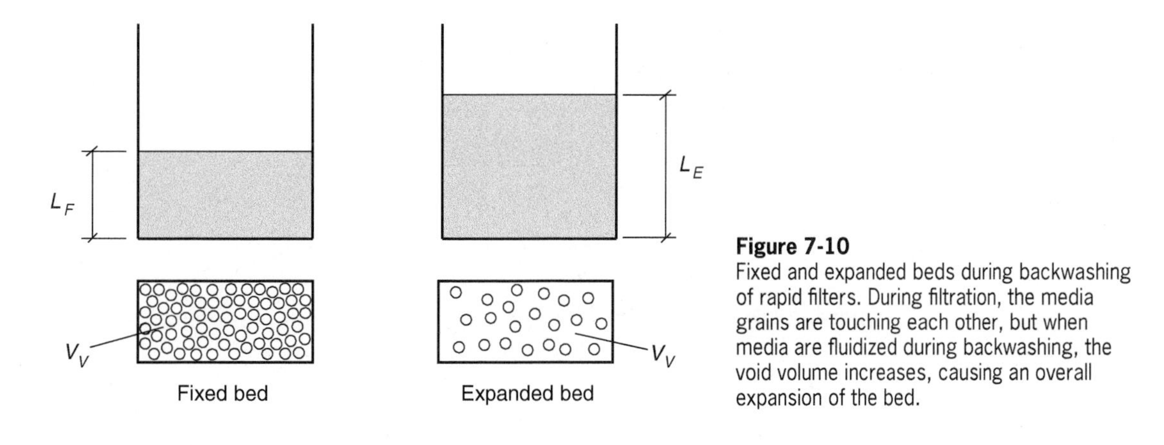

Figure 7-10
Fixed and expanded beds during backwashing of rapid filters. During filtration, the media grains are touching each other, but when media are fluidized during backwashing, the void volume increases, causing an overall expansion of the bed.

The net gravitational force is the fluidized weight of the entire bed as shown in the expression

$$F_G - F_B = mg = (\rho_P - \rho_W)(1 - \varepsilon) aLg \tag{7-34}$$

where F_G = weight of entire filter bed, N
m = mass, mg
g = gravitational constant, m/s^2
ρ_P, ρ_W = density of particles and water, respectively, kg/m^3
ε = filter bed porosity, dimensionless
a = cross-sectional area of filter bed, m^2

The weight of the bed must be divided by the filter area to convert the weight of the bed to units of pressure (i.e., convert N to N/m^2) and divided by $\rho_W g$ to convert units of pressure (N/m^2) to units of head (m) as follows:

$$h_L = \frac{F_G - F_B}{a\rho_W g} = \frac{(\rho_P - \rho_W)(1 - \varepsilon)L}{\rho_W} \tag{7-35}$$

The drag force is the head loss calculated from the Ergun equation. Equating Eqs. 7-20 and 7-35 yields the expression

$$\kappa_V \frac{(1-\varepsilon)^2}{\varepsilon^3} \frac{\mu L v}{\rho_W g d^2} + \kappa_I \frac{1-\varepsilon}{\varepsilon^3} \frac{L v^2}{g d} = \frac{(\rho_P - \rho_W)(1-\varepsilon)L}{\rho_W} \tag{7-36}$$

where v = backwash rate (superficial velocity), m/s
μ = viscosity of water, kg/m·s
d = media grain diameter, m

An analytical solution for Eq. 7-36 in terms of v allows the backwash rate to be calculated directly for any set of filter conditions (Akgiray and Saatçi,

2001). Equation 7-36 can be rearranged as

$$\kappa_I \text{Re}^2 + \kappa_V(1-\varepsilon)\text{Re} - \beta = 0 \tag{7-37}$$

$$\beta = \frac{g\rho_W(\rho_P - \rho_W)d^3\varepsilon^3}{\mu^2} \tag{7-38}$$

$$\text{Re} = \frac{\rho_W v d}{\mu} \tag{7-39}$$

where Re = Reynolds number for flow around a sphere, dimensionless
β = backwash calculation factor, dimensionless

Equation 7-37 is a quadratic equation in terms of Re. One root is necessarily negative because both κ_I and κ_V are positive. The remaining meaningful solution of the quadratic equation is

$$\text{Re} = \frac{-\kappa_V(1-\varepsilon)}{2\kappa_I} + \frac{1}{2\kappa_I}\sqrt{\kappa_V^2(1-\varepsilon)^2 + 4\kappa_I\beta} \tag{7-40}$$

Once the Reynolds number is solved from Eq. 7-40, the backwash flow rate that will maintain the bed in an expanded state corresponding to a specific porosity value can be determined from Eq. 7-39. The minimum fluidization backwash rate can be calculated by determining the velocity that produces head loss equal to the buoyant weight of the media at the fixed-bed porosity. The values of κ_I and κ_V from Table 7-3 are suitable for backwash expansion calculations as well as clean-bed head loss. Calculation of the backwash rate to achieve a certain level of bed expansion is illustrated in Example 7-5.

Example 7-5 Backwash rate for bed expansion

Find the backwash flow rate that will expand an anthracite bed by 30 percent given the following information: $L_F = 2$ m, $d = 1.3$ mm, $\rho_P = 1700$ kg/m^3, $\varepsilon = 0.52$, and $T = 15°$C.

Solution

1. Calculate L_E that corresponds to a 30 percent expansion:

$$L_E = L_F + 0.3L_F = 2\text{ m} + 0.3(2\text{ m}) = 2.6\text{ m}$$

2. Calculate ε_E using Eq. 7-33:

$$\varepsilon_E = 1 - \left[\frac{L_F}{L_E}(1-\varepsilon_F)\right] = 1 - \left[\left(\frac{2\text{ m}}{2.6\text{ m}}\right)(1-0.52)\right] = 0.63$$

3. Calculate β using Eq. 7-38. From Table C-1 in App. C, $\rho_W = 999\ \text{kg/m}^3$ and $\mu = 1.139 \times 10^{-3}$ kg/m·s:

$$\beta = \frac{g\rho_W(\rho_P - \rho_W)d^3\varepsilon^3}{\mu^2}$$

$$= \frac{(9.81\ \text{m/s}^2)(999\ \text{kg/m}^3)(1700 - 999\ \text{kg/m}^3)(0.0013\ \text{m})^3(0.63)^3}{(1.139 \times 10^{-3}\ \text{kg/m·s})^2}$$

$$= 2910$$

4. Calculate Re using Eq. 7-40. Because no pilot or site-specific data are given, use values of κ_V and κ_I from midpoint values from Table 7-3 (e.g., $\kappa_V = 228$ and $\kappa_I = 4.4$):

$$\text{Re} = \frac{-\kappa_V(1-\varepsilon)}{2\kappa_I} + \frac{1}{2\kappa_I}\sqrt{\kappa_V^2(1-\varepsilon)^2 + 4\kappa_I\beta}$$

$$= \frac{-228(1-0.63)}{2(4.4)} + \frac{1}{2(4.4)}\sqrt{(228)^2(1-0.63)^2 + 4(4.4)(2910)}$$

$$= 17.9$$

5. Calculate v using Eq. 7-39:

$$v = \frac{\mu \text{Re}}{\rho_W d} = \frac{(1.139 \times 10^{-3}\ \text{kg/m·s})(17.9)(3600\ \text{s/h})}{(999\ \text{kg/m}^3)(0.0013\ \text{m})}$$

$$= 56.5\ \text{m/h} \quad (23.1\ \text{gpm/ft}^2)$$

Alternatively, it is frequently necessary to determine the bed expansion that occurs for a specific backwash rate. Equation 7-36 is a cubic equation in porosity. Akgiray and Saatçi (2001) showed that two roots of the cubic equation are complex numbers, leaving only one meaningful solution as follows:

$$\varepsilon = \sqrt[3]{X + (X^2 + Y^3)^{1/2}} + \sqrt[3]{X - (X^2 + Y^3)^{1/2}} \tag{7-41}$$

where X, Y = backwash calculation factors, dimensionless

The factors X and Y are defined as

$$X = \frac{\mu v}{2g(\rho_P - \rho_W)d^2}\left(\kappa_V + \frac{\kappa_I \rho_W v d}{\mu}\right) \tag{7-42}$$

$$Y = \frac{\kappa_V \mu v}{3g(\rho_P - \rho_W)d^2} \tag{7-43}$$

The targeted expansion rate is about 25 percent for anthracite and about 37 percent for sand (Kawamura, 2000). The procedure for calculating the expansion of media during backwashing is demonstrated in Example 7-6.

Example 7-6 Filter bed expansion during backwash

Find the expanded bed depth of a sand filter at a backwash rate of 40 m/h given the following information: $L = 0.9$ m, $d = 0.5$ mm, $\rho_P = 2650$ kg/m^3, and $T = 15°$C.

Solution

1. Calculate X using Eq. 7-42. From Table C-1 in App. C, $\rho_W = 999$ kg/m^3 and $\mu = 1.139 \times 10^{-3}$ kg/m·s. Because no pilot or site-specific data are given, use values of κ_V and κ_I from midpoint values in Table 7-3 (e.g., $\kappa_V = 112$ and $\kappa_I = 2.25$):

$$X = \frac{\mu v}{2g(\rho_P - \rho_W)d^2}\left(\kappa_V + \frac{\kappa_I \rho_W v d}{\mu}\right)$$

$$= \frac{(1.14 \times 10^{-3}\ \text{kg/m·s})[(40\ \text{m/h})/(3600\text{s/h})]}{2(9.81\ \text{m/s}^2)(2650 - 999\ \text{kg/m}^3)[0.5\ \text{mm}/(10^3\ \text{mm/m})]^2}$$

$$\times \left[112 + \frac{(2.25)(999\ \text{kg/m}^3)[(40\ \text{m/h})/(3600\ \text{s/h})][0.5\ \text{mm}/(10^3\ \text{mm/m})]}{1.14 \times 10^{-3}\ \text{kg/m} \cdot \text{s}}\right]$$

$$= 0.1921$$

2. Calculate Y using Eq. 7-43:

$$Y = \frac{k_V \mu v}{3g(\rho_P - \rho_W)d^2}$$

$$= \frac{(112)(1.14 \times 10^{-3}\ \text{kg/m·s})(40\ \text{m/h})(10^3\ \text{mm/m})^2}{3(9.81\ \text{m/s}^2)(2650 - 999\ \text{kg/m}^3)(0.5\ \text{mm})^2(3600\ \text{s/h})}$$

$$= 0.1167$$

3. Calculate porosity using Eq. 7-41:

$$\varepsilon_E = \sqrt[3]{X + (X^2 + Y^3)^{1/2}} + \sqrt[3]{X - (X^2 + Y^3)^{1/2}}$$

$$= \sqrt[3]{0.1921 + [(0.1921)^2 + (0.1167)^3]^{1/2}}$$

$$+ \sqrt[3]{0.1921 - [(0.1921)^2 + (0.1167)^3]^{1/2}} = 0.57$$

4. Calculate the expanded bed depth using Eq. 7-33. Because no site-specific porosity value is given, the fixed-bed porosity is taken from Table 7-3 and is assumed to be $\varepsilon_F = 0.42$.

$$L_E = L_F \frac{1 - \varepsilon_F}{1 - \varepsilon_E} = 0.9\text{ m}\left(\frac{1 - 0.42}{1 - 0.57}\right) = 1.21\text{ m}$$

5. Calculate the percent expansion of the bed:

$$\left(\frac{L_E}{L_F} - 1\right) \times 100 = \left(\frac{1.21\text{ m}}{0.9\text{ m}} - 1\right) \times 100 = 34\%$$

Comment

The bed expansion under the example conditions is 34 percent, which is about equal to the desired expansion rate of 37 percent for sand.

Backwash hydraulics depends on the viscosity of water, which varies with temperature. To achieve the same expansion, it is necessary to use a higher backwash rate in summer, when the water is warmer, than the backwash rate used when the water is cold.

The backwash flow rate affects the size of particles that can be removed during backwashing. Particles that have a settling velocity less than the upward velocity of water will be washed away with the backwash water, whereas particles with a greater settling velocity will remain in the filter bed. The settling velocity of particles was introduced in Sec. 6-1, and the same principles apply for backwashing. To determine the size of particles removed during backwashing at a particular backwash rate, the backwash velocity would be set equal to the settling velocity in either Eq. 6-13 or 6-14 and calculating the particle size from that equation. The use of Eq. 6-13 or 6-14 depends on the flow regime, and it is necessary to check the Reynolds number after the particle size has been calculated to determine whether the right equation was used.

Several aspects of rapid filter design and operation result directly from requirements for effective backwashing. These include selection of a low uniformity coefficient to minimize stratification, skimming to remove fines, and selecting media for dual- and multimedia filters, as discussed in the following sections.

Stratification

Stratification is an important side effect of backwashing of rapid media filters. The settling velocity of individual grains of filter media depends on density and diameter, with more dense and larger grains requiring a greater fluidization velocity. When a graded media filter bed (of constant grain density) is backwashed at a uniform rate, the smallest particles fluidize most and rise to the top of the filter bed, while the largest particles collect near the bottom of the bed.

Stratification has several adverse effects on filter performance. First, the accumulation of small grains near the top of the bed causes excessive head loss in the first few centimeters of bed depth. Second, the ability of a filter to remove particles is also a function of grain size, so small grains at the top of a bed cause all particles to be filtered in the first few centimeters of bed depth, which means the entire bed depth is not being used effectively.

Stratification is minimized by proper selection of filter media; a low value of the uniformity coefficient is recommended specifically to minimize stratification of the filter bed during backwashing. A uniformity coefficient less than 1.4 is recommended for all rapid filter media.

The issue of stratification also explains why dual-media filters are desirable. A properly designed dual-media filter reverses the effect of natural stratification by placing larger diameter media above smaller diameter media. The larger diameter media can capture most of the particles, which allows the smaller media to capture the remaining particles without developing excessive head loss. The result is a more effective use of the filter bed. Larger media can be positioned above smaller media if the larger media is less dense. The reason anthracite is used as a filtration media is because it is less dense than sand.

Multimedia Filters

Backwash hydraulics have important implications for the selection of media in dual- and trimedia filters. The media in multimedia filters must be matched so that all media fluidize at the same backwash rate. Otherwise, one media may be washed out of the filter during attempts to fluidize the other media, or, alternatively, one media may fail to fluidize. Fluidization of media can be balanced by selecting a ratio of grain sizes that is matched to the ratio of grain densities so that both media have the same fluidization velocity. Equating an equivalent fluidization velocity for two media in Eq. 6-14 and solving for the ratio of particle sizes yields the expression

$$\frac{d_1}{d_2} = \left(\frac{\rho_2 - \rho_W}{\rho_1 - \rho_W}\right)^{0.625} \tag{7-44}$$

where d_1, d_2 = effective size of each filter media, m
ρ_1, ρ_2 = density of each filter media, kg/m^3

Removal of Fines

Stratification is particularly problematic if the media has excessive fines (particles considerably smaller than the effective size), even if a low uniformity coefficient has been specified. Fines are normally removed by backwashing and skimming immediately after the installation of new media. After media is installed, backwashing at a low rate will bring fines to the top of the bed, where they are then skimmed with a flat-bladed shovel after the filter is drained. It is usually necessary to repeat the backwashing and skimming several times to remove the fines. Dual-media filters should be backwashed and skimmed after each layer of media is installed.

7-7 Energy and Sustainability Considerations

As with other treatment processes, environmental life-cycle assessment of conventional water treatment facilities have consistently found that environmental impacts of the construction phase are relatively minor (typically less than 10 to 20 percent) compared to those from the operating phase. Environmental impacts due to construction can be reduced by minimizing the size of the facility and the number of filters. The single most useful way to decrease the size of the system is to increase the filtration rate. An increase of filtration rate from 10 to 15 m/h (4 to 6 gpm/ft^2) decreases the total area of the filter bed by 33 percent and the environmental impacts by a commensurate amount.

The environmental impact during operation is almost entirely caused by energy consumption during filtration and backwash. Although there is no electrical energy input to the filters directly during filtration, the hydraulic profile of the plant must accommodate the total head loss through the filter system (i.e., the difference in water surface elevations at the upstream and downstream structures). Including the head loss through the pipes and valves as well as the media, the total head through a filter system is typically 3 to 4 m. Using calculations similar to Example 3-3, this head corresponds to electrical energy consumption of 0.010 to 0.014 kWh/m^3, assuming pump efficiency of 80 percent. Environmental impacts might be reduced by reducing the total head through the filter system, but the benefits gained by a small reduction in energy consumption might be outweighed by a significant reduction in the filter run length.

Energy consumption during backwash includes the electricity to run the backwash pumps, air scour blowers, and surface wash pumps. Energy can be calculated from typical flow rates, pressures, and durations shown in Table 7-2, and the specific energy depends on the volume of water processed during the filter run. Assuming a filtration rate of 10 m/h (4 gpm/ft) and run length of 24 h, total energy consumption by backwash pumps, air scour blowers, and surface wash pumps would be 0.0007 to 0.0031 kWh/m^3. Note that the energy consumed during backwash is only 10 to 20 percent of the energy consumed during the filtration stage. Optimization of pump efficiency and backwash process criteria may be effective ways of reducing the environmental impact of backwashing. In addition, increasing the filtration rate or filter run length (i.e., increasing the volume of water filtered between backwashes) is an effective way to reduce the impact of backwashing. Increasing the filtration rate to 15 m/h (6 gpm/ft) and run length to 72 h reduces the total energy consumption by backwash pumps, air scour blowers, and surface wash pumps to 0.0002 to 0.0007 kWh/m^3, or just 2 to 5 percent of the energy consumed during filtration.

Compared to other water treatment processes, these impacts are small. In a municipal water system with raw-water pumping and distribution pumping, the energy consumption by filtration may be less than 1 percent of the overall energy consumption at the plant.

A final consideration in the environmental impact of filtration is the waste stream. The waste wash water is a product of the entire conventional filtration process and is discussed further in Chap. 14.

7-8 Summary and Study Guide

After studying this chapter, you should be able to:

1. Define the following terms and phrases and describe the significance of each in the context of filtration and water treatment:

air scour	conventional filtration	ripening
anthracite	depth filtration	maturation
available head	direct filtration	straining
backwashing	dual-media filter	stratification
breakthrough	effective size	surface wash
clean-bed head loss	filter-to-waste	limiting head
collapse pulsing	filtration rate	underdrain
contact filtration	gullet	uniformity coefficient
	rapid filtration	wash trough

2. Explain the purpose of filtration in water treatment and give a general description of the process of rapid granular filtration.
3. Describe the purpose of various components of rapid filters, including media, underdrains, surface wash, wash troughs, gullet, and flow control systems.
4. Explain why coagulation is an integral part of rapid granular filtration; that is, explain how coagulation prevents particles from passing through a rapid granular filter.
5. Draw a graph of turbidity versus time during a filter cycle, identifying the stages of ripening, effective filtration, and breakthrough. Explain why ripening and breakthrough occur. Give three reasons for terminating a filter run.
6. Draw process flow diagrams for conventional, direct, contact, and two-stage filtration and explain the conditions for which each should be used.
7. Identify the common materials used for granular filtration media.
8. Calculate the effective size and uniformity coefficient for filter media.

9. Explain why a low uniformity coefficient is important.
10. Explain the advantage of having more than one layer of media in a filter (i.e., dual-media and multimedia filters) and how the media are arranged within the filter with respect to effective size and material density.
11. Calculate the size of the second media in a dual media filter that would be matched to the first media.
12. Explain the difference between straining and depth filtration.
13. Describe the mechanisms for particle transport to the media grain surface in depth filtration and identify which mechanisms have the greatest impact on particle removal as a function of particle size.
14. Calculate the clean-bed head loss through a filter bed.
15. Explain the purpose of backwashing. Explain the limitations on the backwash flow rate (what happens if too high or too low).
16. Calculate the expansion of a filter bed at a specified backwash rate and the backwash rate necessary to achieve a specified amount of expansion.
17. Describe the cause of stratification of a rapid filter bed, the consequences of this stratification, and methods to minimize the consequences.
18. Calculate the optimal duration of a filter run (time to breakthrough or limiting head), given pilot data.
19. Evaluate the effect of the following parameters on particle capture and head-loss development in a filter: media effective size, uniformity coefficient, media depth, media porosity, filtration rate, temperature, and influent particle concentration.

Homework Problems

7-1 Samples of filter media were sifted through a stack of sieves and the weight retained on each sieve is recorded below. For a given sample (to be selected by instructor), determine the effective size and uniformity coefficient for the media.

Sieve Designation	Sieve Opening, mm	Weight of Retained Media, g				
		A	B	C	D	E
8	2.36	0				4
10	2.00	35	0			11
12	1.70	178	11		0	60
14	1.40	216	315		4	227

Sieve Designation	Sieve Opening, mm	Weight of Retained Media, g				
		A	B	C	D	E
16	1.18	242	242		16	343
18	1.00	51	116	0	33	216
20	0.85	12	55	23	75	40
25	0.71	5	26	217	285	16
30	0.60	3	14	325	270	3
35	0.50	0	2	151	121	1
40	0.425		0	71	21	0
45	0.355			49	8	
50	0.300			4	3	
Pan	—			20	4	

7-2 A filter is designed with the following specifications. The anthracite and sand have densities of 1700 and 2650 kg/m^3, respectively, and the design temperature is 10°C. For a given sample (to be selected by instructor), calculate the clean-bed head loss.

Item	A	B	C	D	E
Bed type	Monomedia	Monomedia	Dual media	Dual media	Dual media
Filtration rate (m/h)	8	15	15	10	10
Anthracite specifications:					
Effective size (mm)		1.0	1.0	1.2	1.6
Depth (m)		1.8	1.5	1.4	1.2
Sand specifications:					
Effective size (mm)	0.55		0.5	0.55	0.55
Depth (m)	0.75		0.3	0.4	0.7

7-3 A filter contains 0.55-mm sand that has a density of 2650 kg/m^3. Calculate the effective size of 1550 kg/m^3 anthracite that would be matched to this sand.

7-4 For the media specification given in Problem 7-2 (C, D, or E, to be selected by instructor), determine if the two media layers are matched to each other.

7-5 Using the TE filtration model, examine the effect of filtration rate on filter performance for particles with diameters of 0.1, 1.0, and 10 μm. Assume a monodisperse media of 0.5 mm diameter, porosity 0.42, particle density 1020 kg/m^3, filter depth 1 m, temperature 20°C, Hamaker constant Ha $= 10^{-20}$ J, and attachment efficiency 1.0. Plot the results as C/C_0 as a function of filtration rate over a range from 1 to 25 m/h. Comment on the effect of filtration rate and particle size on filter performance.

7-6 The results of pilot filter experiments are summarized in the tables below. The parameter of interest (shown in the second column of each data set) was the media size is experiments A and B and the media depth in experiments C and D. For a given set of experiments (to be selected by instructor), determine: (a) the relationship between specific deposit at breakthrough (σ_B) and the parameter of interest (b) the relationship between the head-loss rate constant (k_{HL}) and the parameter of interest and (c) the optimal value of the parameter of interest and the corresponding run length. For all problems, assume $C_0 = 2.0$ mg/L and $C_E = 0$ mg/L.

a. Design conditions: $v_F = 15$ m/h, media = anthracite, depth = 1.75 m, max available head = 2.8 m.

A Filter	Media Size, m	Time to Breakthrough, h	Initial Head Loss, m	Head Loss When Breakthrough Occurred, m
1	0.8	112	0.65	4.6
2	1.0	85	0.39	2.9
3	1.1	72	0.33	2.4
4	1.2	71	0.30	2.0
5	1.4	58	0.24	1.5

b. Design conditions: $v_F = 15$ m/h, media = GAC, depth = 2.0 m, max available head = 3.0 m.

B Filter	Media Size, m	Time to Breakthrough, h	Initial Head Loss, m	Head Loss When Breakthrough Occurred, m
1	0.83	54	0.65	6.1
2	1.05	43	0.40	4.3
3	1.25	38	0.33	2.9
4	1.54	32	0.22	2.0

c. Design conditions: $v_F = 33.8$ m/h, media = anthracite, ES = 1.55 mm, max available head = 3 m (adapted from pilot results for the Los Angeles Department of Water and Power Aqueduct Filtration Plant).

C Filter	Media Depth, m	Time to Breakthrough, h	Initial Head Loss, m	Head Loss When Breakthrough Occurred, m
1	0.6	4.0	0.16	1.0
2	1.0	6.7	0.30	1.7
3	1.8	11.9	0.50	3.2
4	2.0	13.4	0.58	3.6
5	2.2	14.5	0.65	4.1

d. Design conditions: $v_F = 25$ m/h, media = anthracite, ES = 1.50 mm, max available head = 3 m (adapted from pilot results for Portland, Oregon's Bull Run water supply).

D Filter	Media Depth, m	Time to Breakthrough, h	Initial Head Loss, m	Head Loss When Breakthrough Occurred, m
1	2.0	41	0.43	1.8
2	2.3	49	0.51	2.0
3	2.5	55	0.51	2.5
4	3.0	65	0.63	2.9

7-7 A monomedia anthracite filter has an effective size of 1.0 mm and media density of 1650 kg/m^3.

a. Calculate backwash rate to get a 25 percent expansion at the design summer temperature of 25°C.

b. Calculate the expansion that occurs at the backwash rate determined in part (a) at the minimum winter temperature of 5°C.

c. Discuss the implications of these results on backwash operations for plants that experience a large seasonal variation in water temperature.

7-8 Calculate the largest sand particle (specific gravity = 2.65) that would be removed from a filter during backwashing if the backwash rate is 45 m/h and the water temperature is 15°C.

References

Akgiray, Ö., and Saatçi, A. M. (2001) "A New Look at Filter Backwash Hydraulics," *Water Sci. Technol. Water Supply*, **1**, 2, 65–72.

Amirtharajah, A. (1988) "Some Theoretical and Conceptual Views of Filtration," *J. AWWA*, **80**, 12, 36–46.

Amirtharajah, A. (1993) "Optimum Backwashing of Filters with Air Scour: A Review," *Water Sci. Technol.*, **27**, 10, 195–211.

Clark, M. M. (2009) *Transport Modeling for Environmental Engineers and Scientists,* 2nd ed., Wiley-Interscience, Hoboken, NJ.

Crittenden. J. C., Trussell, R. R., Hand, D. W., Howe, K. J., and Tchobanoglous, G. (2012) *MWH's Water Treatment: Principles and Design,* 3rd ed., Wiley, Hoboken, NJ.

Ergun, S. (1952) "Fluid Flow through Packed Columns," *Chem. Eng. Prog.*, **48**, 2, 89–94.

Ives, K. J. (1967) "Deep Filters," *Filtration and Separation,* **4**, 3/4, 125–135.

Iwasaki, T. (1937) "Some Notes on Sand Filtration," *J. AWWA*, **29**, 10, 1592–1602.

Kavanaugh, M., Evgster, J., Weber, A., and Boller, M. (1977) "Contact Filtration for Phosphorus Removal," *J. WPCF*, **49**, 9, 2157–2171.

Kawamura, S. (2000) *Integrated Design and Operation of Water Treatment Facilities*, Wiley, New York.

Tobiason, J. E., and O'Melia, C. R. (1988) "Physicochemical Aspects of Particle Removal in Depth Filtration," *J. AWWA*, **80**, 12, 54–64.

Tobiason, J. E., Cleasby, J. L.,. Logsdon, G. S., and O'Melia C. R. (2011) Granular Media Filtration, Chap. 10, in J. K. Edzwald (ed.), *Water Quality and Treatment, A Handbook on Drinking Water*, 6th ed., American Water Works Association, Denver, CO.

Trussell, R. R., and Chang, M. (1999) "Review of Flow through Porous Media as Applied to Head Loss in Water Filters," *J. Environ. Eng.*, **125**, 11, 998–1006.

Trussell, R. R., Trussell, A., Lang, J. S., and Tate, C. (1980) "Recent Developments in Filtration System Design," *J. AWWA*, **73**, 12, 705–710.

Tufenkji, N., and Elimelech, M. (2004) "Correlation Equation for Predicting Single-Collector Efficiency in Physicochemical Filtration in Saturated Porous Media," *Environ. Sci. Technol.*, **38** (2), 529–536.

Yao, K.-M., Habibian, M. T., and O'Melia, C. R. (1971) "Water and Waste Water Filtration: Concepts and Applications," *Environ. Sci. Technol.*, **5**, 11, 1105–1112.

8 Membrane Filtration

Membrane filtration is one of the two membrane-based physicochemical processes commonly used in water treatment. The objective of membrane filtration is the same as rapid granular filtration—the removal of microorganisms and other particles from water—but physicochemically the processes are very different. Instead of a thick bed of granular material, the filter media in membrane filtration is a thin synthetic material, typically less than 1 mm thick. The material contains tiny pores through which water can pass. During filtration, water passes through the pores, but particles are physically strained at the surface of the material because they are too large. The difference in filtration mechanism—straining versus depth filtration—is a key difference between membrane and rapid granular filtration that provides advantages that will be discussed later in this chapter. Thus, although membrane filtration is much newer, it has rapidly become a mainstream technology that competes effectively with rapid granular filtration when new water treatment plants are designed.

Membrane filtration design concepts are changing rapidly, so the application of this technology presents unique challenges for the design engineer. Design based on previous projects or ''tried-and-true'' rules of thumb may fail to capitalize on recent technological advancements. On the other hand, undue reliance on manufacturers' claims about unproven technologies may lead to failure. A critical role for the design engineer is to stay abreast of technical advancements and provide appropriate guidance to facility owners. An understanding of the fundamentals of membrane materials, modules, fouling, and performance is necessary to evaluate new technologies with the objective of allowing the design to capitalize on valuable technological advancements while avoiding unproven alternatives that have an unreasonable chance of failure.

The other common membrane process used in water treatment is reverse osmosis. The next section provides a detailed comparison of the similarities and differences between membrane filtration and reverse osmosis. The remainder of this chapter focuses solely on membrane filtration, starting with a comparison to rapid granular filtration, a description of membrane filtration equipment, and a description of the operation of membrane filters. After that, the chapter addresses particle capture, hydraulics of flow, membrane fouling, sizing of membrane systems, and, finally, energy and sustainability considerations.

8-1 Classification of Membrane Processes

As noted in the introduction, membranes are used in two distinct physicochemical processes in water treatment: (1) membrane filtration and (2) reverse osmosis. A general schematic of a membrane process is shown on Fig. 8-1. Water that passes through the membrane is called permeate. *Membrane filtration* is a pressure- or vacuum-driven membrane separation process in which particles are removed from a suspension (a two-phase system consisting of particles in a fluid) by straining as the fluid passes through a porous material. *Reverse osmosis* is a pressure-driven membrane separation process in which dissolved constituents are separated from a solution (a single-phase system consisting of solutes in a solvent) by preferential diffusion as the solvent and solute molecules pass through a permeable material (which may or may not be porous). In reverse osmosis, the constituents targeted for removal are truly dissolved solutes (ions and molecules such as sodium, chloride, calcium, magnesium, dissolved NOM, and synthetic organic chemicals). Reverse osmosis membranes are used to produce potable water from ocean or brackish water, to soften hard waters (remove calcium and magnesium ions), reduce the concentration of NOM to control disinfection by-product (DBP) formation, and to remove

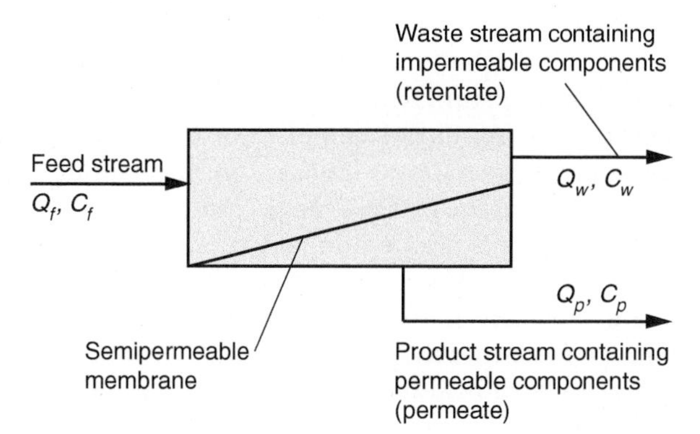

Figure 8-1
Schematic of separation process through semipermeable membrane.

specific dissolved contaminants (e.g., pesticides, pharmaceuticals, arsenic, nitrate, radionuclides).

Within each physicochemical process, manufacturers sell multiple membrane products. Membrane filtration systems are marketed as containing either *microfiltration* (MF) or *ultrafiltration* (UF) membranes, with UF membranes having a smaller pore size than MF membranes. Reverse osmosis membranes are marketed under a number of names (including seawater RO, brackish water RO, low-pressure RO, etc.) and a particular class of low-pressure RO products is sold as *nanofiltration* (NF) membranes. The hierarchy of membrane products used in water treatment is shown on Fig. 8-2. The distinction between membrane products is somewhat arbitrary, but they are loosely identified by the types of materials rejected, operating pressures, and nominal pore dimensions (which are identified

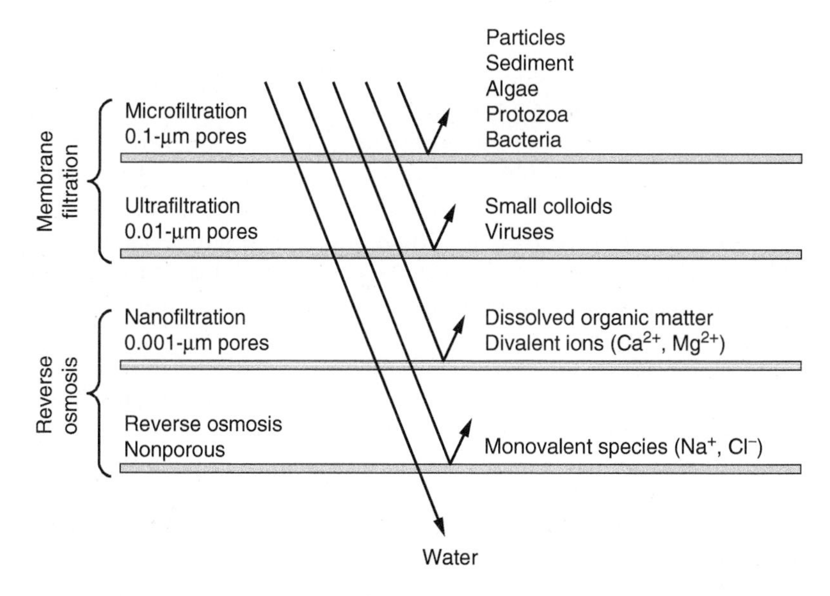

Figure 8-2
Hierarchy of pressure-driven membrane processes.

on an order-of-magnitude basis on Fig. 8-2). A ''loose'' NF membrane marketed by one manufacturer might be substantially similar to a ''tight'' UF membrane marketed by another manufacturer.

The differences between membrane filtration and reverse osmosis are substantial. Because the predominant removal mechanism in membrane filtration is straining, or size exclusion, the process can theoretically achieve perfect exclusion of particles regardless of operational parameters such as influent concentration and pressure. Mass transfer in reverse osmosis, however, involves a diffusive mechanism so that separation efficiency is dependent on influent solute concentration, pressure, and water flux rate. Differences between membrane filtration and reverse osmosis are evident in the materials used for the membranes, the configuration of the membrane elements, the equipment used, the flow regimes, and the operating modes and procedures. Comparisons between membrane filtration and reverse osmosis are detailed in Table 8-1. It should be noted that membranes are used for many purposes in a wide variety of fields and industries, and the distinction between membrane products as used in water treatment may not be appropriate in other industries. For instance, products marketed as UF membranes are used in food-processing and pharmaceutical industries for purifying, concentrating, and fractionating concentrated solutions of macromolecules such as proteins and polysaccharides; UF membrane use in those industries involves phenomena (such as concentration polarization) reserved for reverse osmosis in this book.

8-2 Comparison to Rapid Granular Filtration

A comparison between rapid granular filtration and membrane filtration, including typical permeate flux, operating pressure, and duration of filter and backwash cycles, is presented in Table 8-2. As shown in the table, the flux through a membrane filter is typically about two orders of magnitude lower than the flux through a rapid granular filter; consequently, a membrane filtration plant needs 100 times the filter area of a rapid granular filtration plant to produce the same quantity of water. Membrane filtration plants, however, are frequently smaller than granular filtration plants. This apparent contradiction is possible because of the packing density; thus, 1 m^2 of floor space at a membrane plant may contain much more than 100 m^2 of membrane area.

Membrane filtration has several advantages over granular filtration. Rapid granular filters rely on depth filtration, whereas membrane filters rely on straining. As a result, membrane filtration plants do not require coagulation, flocculation, and sedimentation facilities for effective particle removal. These differences can reduce requirements for chemical storage and handling and residual-handling facilities and allow membrane plants to be more compact and automated. The more compact installation can

Table 8-1
Comparison between membrane filtration and reverse osmosis

Process Characteristic	Membrane Filtration	Reverse Osmosis
Objectives	Particle removal, microorganism removal	Seawater desalination, brackish water desalination, softening, NOM removal for DBP control, specific contaminant removal
Targeted contaminants	Particles	Dissolved solutes
Membranes types	Microfiltration, ultrafiltration	Nanofiltration, reverse osmosis
Typical source water	Fresh surface water (TDS <1000 mg/L)	Ocean or seawater, brackish groundwater (TDS = 1000–20,000 mg/L), colored groundwater (TOC >10 mg/L)
Membrane structure	Homogeneous or asymmetric	Asymmetric or thin-film composite
Most common membrane configuration	Hollow fiber	Spiral wound
Dominant exclusion mechanism	Straining	Differences in solubility or diffusivity
Removal efficiency of targeted contaminants	Frequently 99.9999% or greater	Typically 50–99%, depending on objectives
Most common flow pattern	Dead end	Cross flow
Operation includes backwash cycle	Yes	No
Influenced by osmotic pressure	No	Yes
Influenced by concentration polarization	No	Yes
Noteworthy regulatory issues	Challenge testing and integrity monitoring	Concentrate management
Typical transmembrane pressure[a]	0.2–1 bar (3–15 psi)	5–85 bars (73–1200 psi)
Typical permeate flux[b]	30–170 $L/m^2 \cdot h$ (18–100 $gal/ft^2 \cdot d$)	1–50 $L/m^2 \cdot h$ (0.6–30 $gal/ft^2 \cdot d$)
Typical recovery[c]	>95%	50% (for seawater) to 90% (for colored groundwater)
Competing processes	Granular filtration	Carbon adsorption, ion exchange, precipitative softening, distillation

[a]Transmembrane pressure is the difference between the feed and permeate pressures.
[b]Flow through membrane systems is reported as volumetric flux, or flow per unit area of membrane surface. See Sec. 4-5 for discussion of flux.
[c]See Eq. 8-16.

result in considerable cost savings in densely populated areas or other areas where land costs are high.

The most significant advantage is that the filtered water turbidity from membrane filters is independent of the concentration of particulate matter in the feed. Rapid granular filtration is sensitive to fluctuations in raw-water quality and the experience of the plant operators. Changes in

Table 8-2
Comparison between membrane filtration and rapid granular filtration

Criteria	Membrane Filtration	Rapid Granular Filtration
Filtration rate (permeate flux)	0.03–0.17 m/h[a] (0.01 – 0.07 gpm/ft^2)	5–15 m/h[a] (2 – 6 gpm/ft^2)
Operating pressure	0.2–1 bar (7–34 ft)	0.18–0.3 bar (6–10 ft)
Filtration cycle duration	30–90 min	1–4 d
Backwash cycle duration	1–3 min	10–15 min
Ripening period	None	15–120 min
Recovery	>95 %	>95 %
Filtration mechanism	Straining	Depth filtration

[a]Conventional units for membrane permeate flux are L/m^2·h and gal/ft^2·d. The conversions to the units shown in this table are 10^3 L/m^2·h = 1 m/h and 1440 gal/ft^2·d = 1 gpm/ft^2.

raw-water chemistry without changes in pretreatment (i.e., adjustment of the coagulant dose) can cause the rapid granular filtration process to fail. Membrane filtration is more robust from a finished-water quality perspective.

8-3 Principal Features of Membrane Filtration Equipment

Membrane filters are typically configured as hollow fibers that look like little straws. The fibers are bundled into modules that may contain thousands of individual fibers. Numerous modules (anywhere from 2 to 100) are then assembled with pumps, piping, valves, and other ancillary equipment into treatment units. This section describes basic features of membrane filtration equipment. Important features of the membranes themselves include the geometry, flow orientation, materials, and internal structure. Detailed guidance and design manuals for membrane filtration systems have recently been published by the EPA and AWWA (U.S. EPA, 2005; AWWA, 2005b, 2010) that provide more detail about membrane filter equipment and operation.

Membrane Geometry

At the level of the actual filtration barrier, membrane filters used in water treatment are typically fabricated in one of two basic geometries: hollow fiber or tubular. *Hollow fibers* look like flexible little straws, as shown on Fig. 8-3. The fibers have an outside diameter ranging from about 0.65 to 2 mm (0.026 to 0.08 in.) and a wall thickness (i.e., membrane thickness) ranging from about 0.1 to 0.6 mm (0.004 to 0.02 in.). Unfiltered water can be either inside or outside the fiber, and filtration occurs as water passes through the wall of the fibers to the other side. Water that passes through

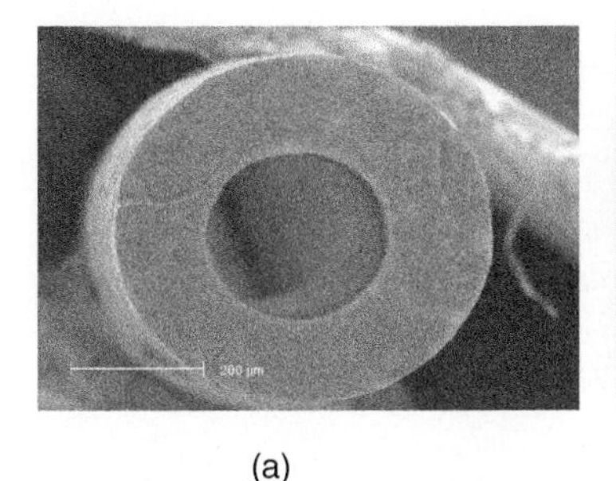

(a)

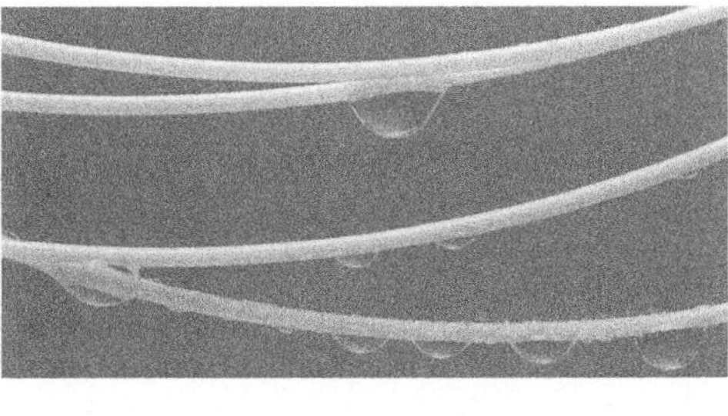

(b)

Figure 8-3
(a) Scanning electron microscope image of end view of a hollow-fiber membrane. (Courtesy of US Filter Memcor Products.) (b) Water permeating hollow-fiber membranes. (Courtesy of Suez Environnement.)

the membrane is called filtrate or *permeate* and water that stays on the feed side is called *retentate*. The fibers range from 1 to 2 m (3.3 to 6.6 ft) long, and thousands of fibers will be packed together to construct a membrane module. The goals of this geometry are to create a thin material that is structurally strong and to pack a large amount of surface area into a small volume. The ratio of surface area to volume is known as the *packing density* and can range from 750 to 1700 m^2/m^3 (230 to 520 ft^2/ft^3) for hollow-fiber modules.

Tubular membranes are rigid monolithic structures with one or more channels through the structure, as shown on Fig 8-4. With tubular membranes, the unfiltered water is always inside the channels and the water is filtered as the water passes to the outside of the monolith. As with hollow fibers, a high packing density is desirable and a monolith with many parallel channels is able to achieve a higher value. The module packing density for tubular membranes can be as high as 400 to 800 m^2/m^3 (120 to 240 ft^2/ft^3).

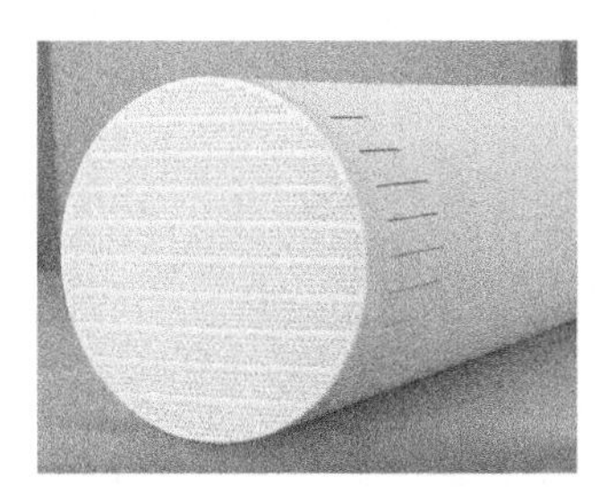

Figure 8-4
End view of a ceramic tubular membrane. (Courtesy of NKG.)

As of 2011, all commercial membrane filtration systems used for drinking water treatment in the United States used hollow-fiber membranes. Tubular membranes constructed of ceramic material are used for some membrane systems in Japan and may eventually penetrate other markets.

Filtration Direction through Hollow Fibers

Hollow-fiber membranes can be designed to filter through the fiber wall from outside to inside or in the opposite direction (inside out). Tubular membranes always operate inside out. By keeping the unfiltered water on the outside of the fibers, outside-in operation is less sensitive to large solids in the feed water, whereas large particles might clog the *lumen* (the inner bore of the fiber) of inside-out membranes. However, inside-out operation allows the flexibility to operate in cross-flow mode, which may allow for maintaining higher flux when filtering high-turbidity feed water.

A key advantage of outside-in operation is that a membrane system can produce more filtrate than inside-out operation with the same number of fibers and operating at the same flux. The difference in flow that can be achieved is demonstrated in Example 8-1.

Example 8-1 Comparison of outside-in and inside-out filtration

A membrane module contains 5760 fibers. The fibers are 1.87 m long with an outside diameter of 1.3 mm and inside diameter of 0.7 mm. Calculate the water production from one module if the volumetric flux is 75 L/m^2·h and the flow direction is (1) outside in and (2) inside out. Compare the two answers.

Solution

1. Compute the product water flow for outside-in flow.
   a. Determine the outside surface area per fiber:

$$A(\text{per fiber}) = \pi dL = \pi\left(1.3\ \text{mm}\right)\left(1.87\ \text{m}\right)\left(10^{-3}\ \text{m/mm}\right)$$
$$= 7.64 \times 10^{-3}\ \text{m}^2/\text{fiber}$$

   b. Compute the product water flow:

$$Q = JA = \left(75\ \text{L/m}^2{\cdot}\text{h}\right)\left(7.64 \times 10^{-3}\ \text{m}^2/\text{fiber}\right)\left(5760\ \text{fibers}\right)$$
$$= 3300\ \text{L/h}$$

2. Compute the product water flow for inside-out flow.
   a. Determine the inside surface area per fiber:

$$A(\text{per fiber}) = \pi dL = \pi\left(0.7\ \text{mm}\right)\left(1.87\ \text{m}\right)\left(10^{-3}\ \text{m/mm}\right)$$
$$= 4.11 \times 10^{-3}\ \text{m}^2/\text{fiber}$$

   b. Compute the product water flow:

$$Q = JA = \left(75\ \text{L/m}^2{\cdot}\text{h}\right)\left(4.11 \times 10^{-3}\ \text{m}^2/\text{fiber}\right)\left(5760\ \text{fibers}\right)$$
$$= 1780\ \text{L/h}$$

3. Compare the outside-in and inside-out flow configurations:

$$\text{Ratio} = (3300/1780) \times 100\% = 186\%$$

Comment

Operating at the same flux, the outside-in system produces 86 percent more water than the inside-out system. Based on the results presented in this example, membrane systems cannot be compared or specified on the basis of flux if the flow configuration is different (the total flow per module and cost per module would be more important design parameters than flux).

Material Properties

Membrane performance is affected strongly by the physical and chemical properties of the material. The ideal membrane material is one that can produce a high flux without clogging or fouling and is physically durable, chemically stable, nonbiodegradable, chemically resistant, and inexpensive. Important characteristics of membrane materials, methods of determination, and effects on membrane performance are described in Table 8-3.

One of the most important characteristics in Table 8-3 is hydrophobicity. Hydrophilic materials, which like contact with water, tend to have low fouling tendencies, whereas hydrophobic materials may foul extensively. Hydrophobicity is quantified by contact angle measurements in which a droplet of water or bubble of air is placed against a membrane surface, and the angle between the surface and water or air is measured. Hydrophobic surfaces have a high contact angle (the water beads like on a freshly waxed car), whereas hydrophilic surfaces have a low contact angle (the water droplets spread out).

Hydrophobicity is affected strongly by the chemical composition of the polymer comprising the material. Polymers that have ionized functional groups, polar groups (water is very polar), or oxygen-containing and hydroxyl groups (for hydrogen bonding) tend to be very hydrophilic.

Material Chemistry

Lacking the existence of a perfect material, a variety of materials has been used. The two most common materials in early commercial membrane filtration systems were cellulose acetate (CA) and polypropylene (PP), but their use has been declining. Celluose acetate membranes have been known to compact over time and have lower resistance to harsh cleaning chemicals and high temperatures. Polypropylene does not have good resistance to chlorine, which is often used as a disinfectant in water treatment. The most common synthetic organic polymers currently used in water treatment membranes are polyvinylidene fluoride (PVDF), polysulfone (PS), and polyethersulfone (PES). These materials have very good resistance to harsh cleaning chemicals, chlorine, and moderately high temperatures, and tolerate a wide pH range for cleaning solutions. Some membrane manufacturers consider the composition of their membranes to be proprietary and do not release information on their material chemistry.

Ceramic membrane may also be gaining in popularity. Ceramic membranes are configured as tubular membranes. The material is hydrophilic, rough, and can withstand high operating pressure and temperature. It has excellent chemical and pH tolerance. Aggressive cleaning and disinfecting is possible.

Internal Membrane Structure

Membrane filters are not constructed of woven or fibrous materials. They are cast as a continuous polymeric structure with tortuous interconnecting voids, as shown in the scanning electron microscope (SEM) images on

Table 8-3
Important properties of membrane materials

Property	Method of Determination	Impact on Membrane Performance
Retention rating[a] (pore size or molecular weight cut-off)	Bubble point, challenge tests	Controls the size of material retained by the membrane, making it one of the most significant parameters in membrane filtration. Also affects head loss.
Hydrophobicity	Contact angle	Reflects the interfacial tension between water and the membrane material. Hydrophobic materials "dislike" water; thus, constituents from the water accumulate at the liquid–solid interface to minimize the interfacial tension between the water and membrane. In general, hydrophobic materials will be more susceptible to fouling than hydrophilic materials.
Surface or pore charge	Streaming potential	Reflects the electrostatic charge at the membrane surface. Repulsive forces between negatively charged species in solution and negatively charged membrane surfaces can reduce fouling by minimizing contact between the membranes and fouling species. In UF, electrostatic repulsion can reduce the passage of like-charged solutes. Membranes fabricated of uncharged polymers typically acquire some negative charge while in operation.
Surface roughness	Atomic force microscopy	Affects membrane fouling; some studies have shown rough materials will foul more than smooth materials.
Porosity (surface and bulk)	Thickness/weight measurements	Affects the head loss through the membrane; higher porosity results in lower head loss.
Thickness	Thickness gauge, electron microscopy	Affects the head loss through the membrane; thinner membranes have lower head loss.
Surface chemistry	ATR/FTIR, SIMS, XPS[b]	Affects fouling and cleaning by influencing chemical interactions between the membrane surfaces and constituents in the feed water.
Chemical and thermal stability	Exposure to chemicals and temperature extremes	Affects the longevity of the membrane; greater chemical and temperature tolerance allows more aggressive cleaning regimes with less degradation of the material.
Biological stability	Exposure to organisms	Affects the longevity of the membrane; low biological stability can result in the colonization and physical degradation of the membrane material by microorganisms.

Table 8-3
(Continued)

Property	Method of Determination	Impact on Membrane Performance
Chlorine/oxidant tolerance	Exposure to chlorine/oxidants	Affects the ability to disinfect the membrane equipment. Routine disinfection prevents microbial growth on membrane surfaces and prevents biological degradation of membrane materials (increasing the longevity of the membrane).
Mechanical durability	Mechanical tests	Affects the ability of the material to withstand surges due to operation of valves and pumps.
Internal physical structure, tortuosity	Electron microscopy	Affects the hydrodynamics of flow and particle capture. There are no standard procedures for quantifying the tortuosity or internal structure of membranes.
Cost	Material cost	Affects the cost of the membrane system.

[a]See Sec. 8-5.
[b]Abbreviations: ATR/FTIR = attenuated total reflectance Fourier transform infrared spectrometry, SIMS = secondary ion mass spectrometry, and XPS = X-ray photoelectron spectrometry.

Fig. 8-5. Most MF membranes have a *homogenous structure*, which means that the structure, porosity, and transport properties are relatively constant throughout their depth. In contrast, UF membranes have an *asymmetric structure* (also called anisotropic or "skinned"), which means that the morphology varies significantly across the depth of the membrane. A homogeneous membrane was shown on Fig. 8-6 and the structure of an asymmetric membrane, consisting of an active layer and a support layer, is shown on Fig. 8-6. The active and support layers have separate functions.

Filtration occurs at the active layer in asymmetric membranes, which is a thin skin with low porosity and very small void spaces. The low porosity and small pores generate significant resistance to flow, which must be minimized by making the active layer as thin as possible. The active layer is so thin that it has no mechanical durability. Thus, the remainder of the membrane is a highly porous layer that provides support but produces very little hydraulic resistance. Filtration through an asymmetric membrane is not the same in both directions. Filtration in the "wrong" direction would cause the voids in the support layer to become clogged and may cause the active layer to separate from the rest of the membrane. To prevent clogging, some commercial asymmetric membranes have active layers on both surfaces of the membrane with a support layer sandwiched between the two active layers.

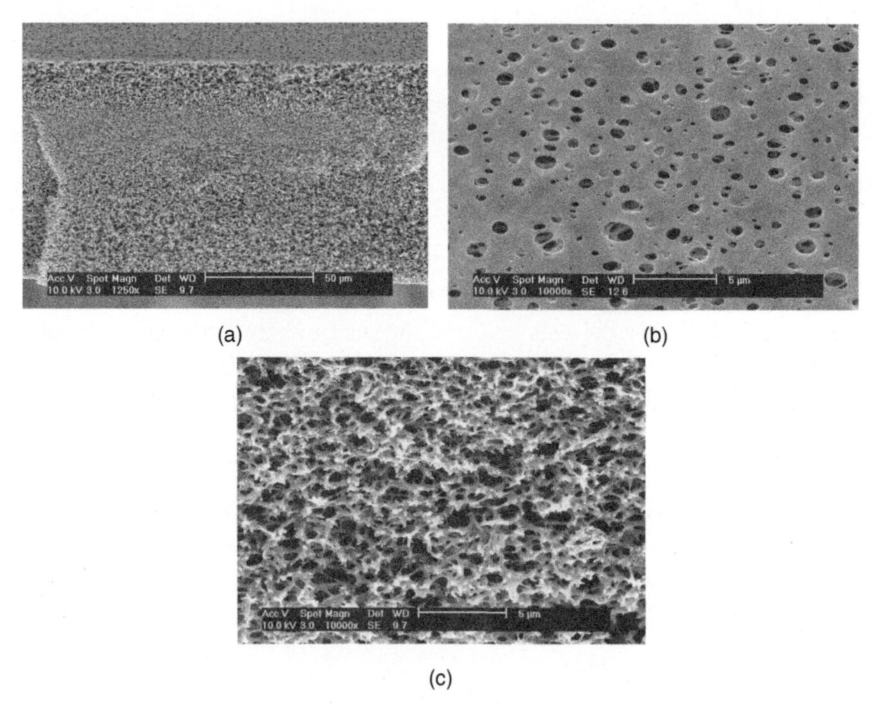

Figure 8-5
Scanning electron microscope images of a 0.2-μm polyethersulfone microfiltration membrane: (a) cross section of the entire membrane; (b) high magnification of the membrane surface, and (c) high magnification of the membrane internal structure.

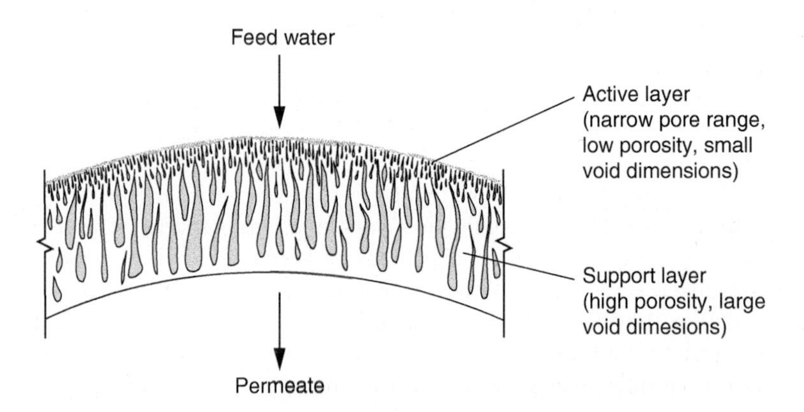

Figure 8-6
Structure of an asymmetric UF membrane.

Module Configuration

To create a filtration system, the individual membrane fibers are packed together and assembled into modules. Currently, membrane filtration systems are available as modular systems from several manufacturers. Current vendors of membrane filtration equipment are listed in AWWA (2005b). Other suppliers are expected to enter the market as the technology evolves, including suppliers experienced in the reverse osmosis (see Chap. 9) market. Membrane modules are available in two basic configurations: pressure-vessel systems or submerged systems.

PRESSURE-VESSEL CONFIGURATION

Pressure-vessel modules are generally 100 to 300 mm (4 to 12 in.) in diameter, 0.9 to 5.5 m (3 to 18 ft) long, and arranged in skids (also known as racks, banks, or units in some design manuals and regulations). Typical pressure-vessel membrane elements are shown on Fig. 8-7. A single module typically contains between 40 and 80 m^2 (430 and 860 ft^2) of filter area. Skids contain between 2 and 100 modules, depending on capacity requirements. Skids and modules in a full-scale production membrane filtration system are shown on Fig. 8-8. The skid is the basic production unit, and all modules within one skid are operated in parallel simultaneously and can be isolated as a group for testing, cleaning, and repair. Each module must be piped individually for feed and permeate water, so large skids involve a substantial number of piping connections. Feed pumps typically deliver water to a common manifold that supplies each skid. The feed pump increases the feed water pressure, while the permeate stays at near-atmospheric pressure. Pressure-vessel systems typically operate at *transmembrane pressure* (pressure drop between the feed and permeate) between about 0.4 and 1 bar

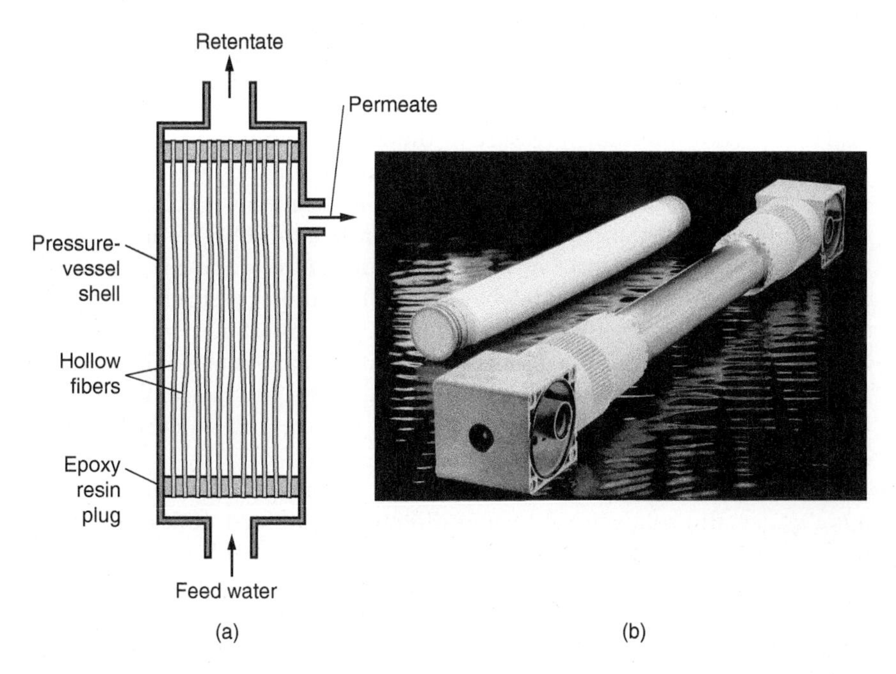

Figure 8-7
Pressure-vessel configuration for membrane filtration: (a) schematic of a single cross-flow membrane module and (b) photograph. (Courtesy of US Filter Memcor Products.)

Figure 8-8
Full-scale membrane filtration facility using the pressure-vessel configuration.

(6 and 15 psi). Pressure vessels can be configured with either outside-in or inside-out membranes.

SUBMERGED CONFIGURATION

Submerged systems (also called immersed membranes) are modules of membranes suspended in basins containing feed water, as shown on Fig. 8-9. The basins are open to the atmosphere, so pressure on the influent side is limited to the static pressure of the water column. Transmembrane pressure is developed by a pump that develops suction on the permeate side of the membranes; thus submerged systems are sometimes called suction- or vacuum-based systems. Net positive suction head (NPSH) limitations on the permeate pump restrict submerged membranes to a maximum transmembrane pressure of about 0.5 bar (7.4 psi), and they typically operate at a transmembrane pressure of 0.2 to 0.4 bar (3 to 6 psi). Submerged systems are configured with multiple basins so that individual basins can be isolated for cleaning or maintenance without shutting down the entire plant. Each basin typically has its own permeate pump. Submerged systems use only outside-in membranes.

Because clean water is extracted from the feed basin through the membranes and solids are returned directly to the feed tank during the backwash cycle, the solids concentration in the feed tank can be significantly higher than in the raw water. A high solids concentration can be advantageous when using treatment additives (i.e., coagulants or PAC) to remove dissolved contaminants but can have an adverse impact on the solids loading on the membrane during filtration. Two basic strategies are used to

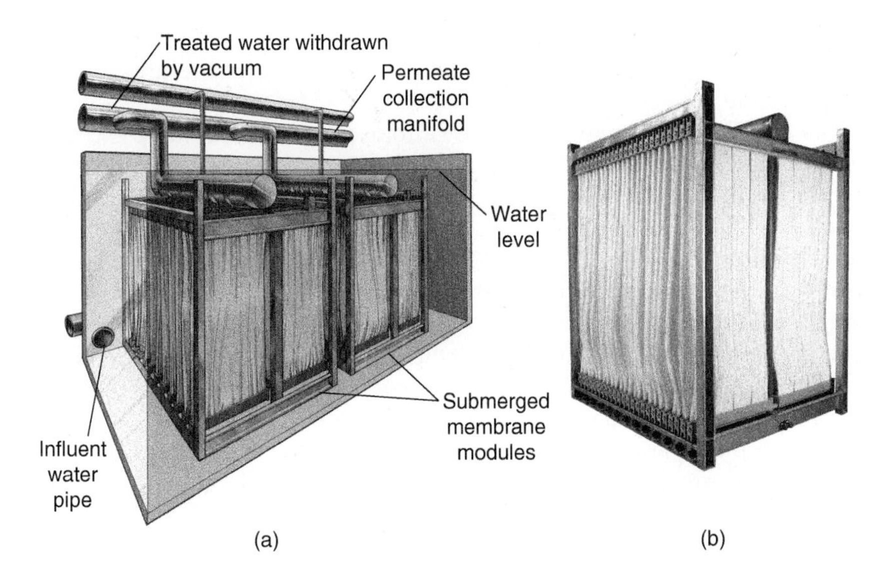

Figure 8-9
Submerged configurations for membrane filtration: (a) schematic of a submerged membrane module and (b) photo of a single module removed from feed tank. (© 2011 General Electric Company. All rights reserved. Reprinted with permission.)

maintain the proper solids concentration in the feed tank, as shown on Fig. 8-10: (1) the feed-and-bleed strategy and (2) the semibatch strategy. In the feed-and-bleed strategy, a small waste stream is continuously drawn from the feed tank. The average solids concentration in the tank will be a function of the size of the waste stream:

$$C_w = \left(\frac{Q_f}{Q_w}\right) C_f \tag{8-1}$$

where C_f, C_w = solids concentration in feed and waste streams, respectively, mg/L
Q_f, Q_w = feed and waste flow rates, respectively, m^3/d or ML/d

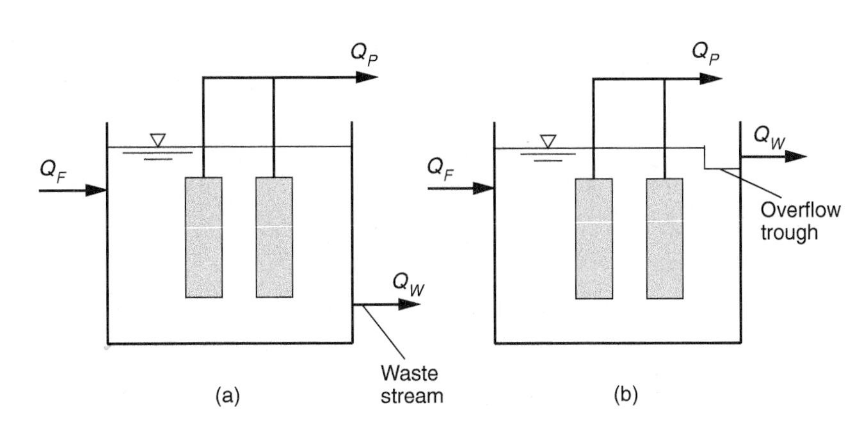

Figure 8-10
Feed-and-bleed and semibatch modes of operation. In feed and bleed, Q_p and Q_w are both continuous, the sum of the two flows equals Q_f. In semibatch, Q_p is continuous and equal to Q_f; Q_w only flows when solids are being wasted.

Some design guides, such as the *Membrane Filtration Guidance Manual* (U.S. EPA, 2005) refer to the ratio C_w/C_f, and therefore the ratio Q_f/Q_w, as the volume concentration factor (VCF).

The semibatch strategy operates without a continuous waste stream, and the feed and permeate flows are at the same rate. As a result, solids accumulate in the feed tank during the filtration cycle. During the backwash cycle, the volume of water in the tank increases due to addition of the backwash flow (raw water continues to flow to the tank during the backwash cycle), and the excess water (and solids) exits the basin through an overflow trough or port.

In currently available equipment, submerged systems tend to accommodate larger modules than pressure-vessel systems. Furthermore, submerged systems have substantially fewer valves and piping connections. As larger membrane plants are designed and built, membrane manufacturers have tried to improve the economy of scale by developing larger modules to reduce the number of individual modules and piping connections necessary in large facilities, and these trends are expected to continue to lead to the development of larger modules.

8-4 Process Description of Membrane Filtration

Membrane filters operate over a cycle consisting of two stages, just like granular filters: (1) a filtration stage, during which particles accumulate, and (2) a backwash stage, during which the accumulated material is flushed from the system. As solids accumulate against the filter medium, the transmembrane pressure to maintain constant flux increases. When a preset time interval or maximum pressure is reached, the system is backwashed.

Although the backwash removes accumulated solids, a gradual loss of performance is observed over a longer period, as shown on Fig. 8-11. The

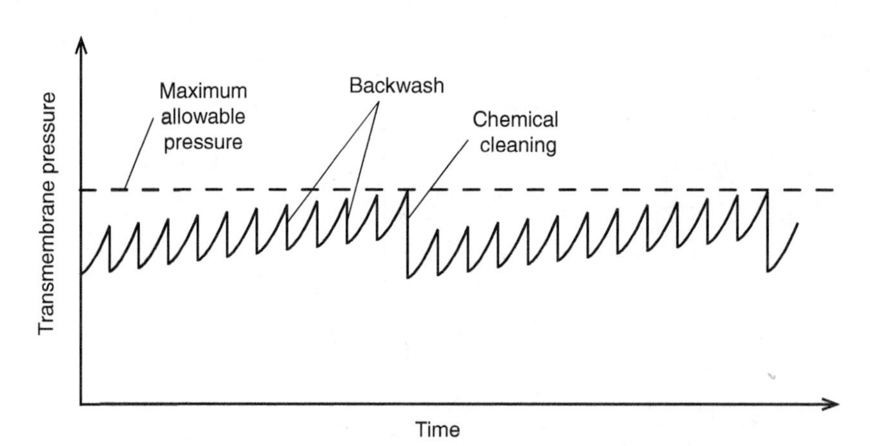

Figure 8-11
Transmembrane pressure development during membrane filtration.

Table 8-4
Typical operating characteristics of membrane filtration facilities

Parameter	Units	Range of Typical Values
Permeate flux		
Pressurized systems	L/m^2·h	30–170
	gal/ft^2·d	18–100
Submerged systems	L/m^2·h	25–75
	gal/ft^2·d	15–45
Normal transmembrane pressure		
Pressurized systems	bar	0.4–1
	psi	6–15
Submerged systems	bar	0.2–0.4
	psi	3–6
Maximum transmembrane pressure		
Pressurized systems	bar	2
	psi	30
Submerged systems	bar	0.5
	psi	7.4
Recovery	%	>95
Filter run duration	min	30–90
Backwash duration	min	1–3
Time between chemical cleaning	d	5–180
Duration of chemical cleaning	h	1–6
Membrane life	yr	5–10

loss of performance, or fouling, is due to slow adsorption or clogging of material that cannot be removed during backwash. Fouling affects the cost effectiveness of membrane filtration and will be discussed in detail later in this chapter. Operational strategies to minimize fouling include pretreatment, chemically enhanced backwash (CEB), chemical wash (CW) operations, and clean-in-place (CIP) operations. In addition, the filtration process includes an integrity testing procedure to validate the reliability of the filtration barrier. These aspects of membrane filtration operation are discussed in the next sections.

Typical operating criteria for membrane filtration facilities are given in Table 8-4.

Cross-Flow and Dead End Flow Regimes

Permeate flux and fouling are affected by the flow regime of the feed water near the membrane surface. Two filtration strategies, cross-flow filtration and dead-end filtration, have been developed to influence this flow regime.

CROSS-FLOW FILTRATION

Cross-Flow filtration is a filtration mode in which the feed water flows continuously through the lumen of inside-out membrane fibers or channels of tubular membranes, parallel to the membrane surface, with a retentate stream that is recycled to the feed water. The cross-flow velocity, typically 0.5 to 1 m/s (1.6 to 3.3 ft/s), is four to five orders of magnitude greater than the superficial velocity of water toward the membrane surface. The cross-flow velocity creates a shear force that reduces the development of a surface cake. Because many solids are carried away with the retentate instead of accumulating on the membrane surface, the system can be operated at a higher flux or with longer intervals between backwashes. Cross-flow filtration requires a substantial recirculation of retentate—the permeate flow is typically only 15 to 20 percent of the feed flow.

DEAD-END FILTRATION

Dead-end filtration is a filtration mode in which all feed water passes through the membrane and there is no recirculated retentate stream. The membrane operates without a defined continuous cross-flow velocity and all solids accumulate on the membrane during the filtration cycle. The greater solids accumulation during the filter run may result in lower average flux values than those achieved with cross-flow filtration.

The dead-end flow regime is most common in membrane filtration for water treatment, in contrast to many industrial applications of MF and UF. Many industrial feed streams have high solids concentrations (e.g., the solids concentration in many food-processing operations can be 1 to 30 percent), and cross-flow operation is critical for achieving reasonable flux and filter run length. Surface waters are fairly dilute (many membrane plants operate with feed water turbidity of 100 NTU or less, which corresponds to a solids concentration of about 0.01 percent) so the advantages of cross-flow filtration are less significant. The piping and pumping costs of recirculating a large fraction of the feed water become prohibitive as the facility size gets larger, and water treatment facilities are built with considerably higher capacity than most industrial applications. Some cross-flow systems are designed to operate in a dead-end mode by closing a valve in the retentate line when raw-water quality conditions permit (turbidity is low) and switch to a cross-flow mode only when necessary to maintain flux.

Pretreatment

When the treatment goals for the facility are only particle and microorganism removal, the pretreatment requirements for membrane filtration are minimal. Pretreatment is necessary to protect the filter fibers from damage or clogging of the lumen (in the case of inside-out membranes). Microscreening or prefiltration to remove coarse sediment larger in diameter than 0.1 to 0.5 mm (0.004 to 0.02 in.), depending on the manufacturer, is required. Prefiltration is accomplished with self-cleaning screens, cartridge filters, or bag filters.

Because the primary removal mechanism is straining, chemical conditioning to destabilize particles is not required. The lack of a requirement for particle destabilization can be an advantage over granular filtration because the elimination of coagulation and flocculation facilities reduces chemical handling and storage facilities and residual management requirements.

When other treatment goals are present, such as the removal of dissolved contaminants, the pretreatment for membrane filters can be similar to the pretreatment for rapid granular filters. Coagulation, flocculation, and sedimentation can be used for high-turbidity water or for DBP precursor removal, PAC pretreatment can be used taste and odor or SOC removal, oxidants can be used for iron and manganese removal, and lime softening can be used for hardness removal. When pretreatment is used, design engineers must consider the impact on membrane fouling and potential for damage to the membrane along with treatment goals. A substantial amount of technical literature about pretreatment for membrane filtration is available, including critical review articles (Farahbakhsh et al., 2004; Huang et al., 2009a) and design manuals (AWWA, 2005b).

Backwash

Backwashing occurs automatically at timed intervals ranging from 30 to 90 min. The increase in transmembrane pressure during the filtration cycle is typically 0.01 to 0.07 bar (0.2 to 1 psi). Most systems will initiate the backwash cycle early if the increase in transmembrane pressure during the filter run exceeds a preset limit. The backwash cycle lasts 1 to 3 min, and the sequence is run entirely by the control system. All modules in a skid are backwashed simultaneously. Backwashing of MF membranes involves forcing either air or permeate water through the fiber wall in the reverse direction at a pressure equal to or higher than the normal filtration pressure. Ultrafiltration membranes are backwashed with permeate water because the air pressure required to force water from the small pores in UF membranes can be excessive. In some pressure-vessel systems, the backwash flow is supplemented by a high-velocity flush in the feed channels to assist with removing the surface cake, and the wastewater is piped to a wash-water handling facility. The backwash water in submerged systems flows directly into the feed tank.

Chemically Enhanced Backwash

Many membrane systems periodically add chemicals to backwash water to improve the backwash process, a sequence called chemically enhanced backwash. CEB chemicals can include hypochlorite or other cleaning chemicals. The CEB is a strategy to reduce the rate of membrane fouling and decrease the required frequency for more extensive cleaning procedures. CEB is typically included on a subset of backwashes (e.g., backwashes occur every 45 min and one backwash per day will be performed as a CEB). Some systems alternate between multiple CEB strategies, such as alternating a citric acid CEB with a hypochlorite CEB.

Chemical Wash Cycle

An alternative to the CEB is a chemical wash (CW) cycle, sometimes known as maintenance wash. The CW is a short cleaning cycle in which cleaning chemicals are introduced into the feed side of the membranes, allowed to soak for 15 to 30 min, and recirculated for an additional 15 to 30 min without forcing water to pass through the membrane wall. The total duration of CW cycles is less than 60 min. The frequency of maintenance CW cycles is similar to CEB cycles; most membrane systems are designed for one or the other, but not both.

Clean-in-Place Cycle

Even with backwashing, chemically enhanced backwashing, and chemical wash cycles, membrane filters gradually lose filtration capacity due to clogging or adsorption of material. When the transmembrane pressure increases to a preset maximum limit or when a preset time interval has elapsed, the membranes are chemically cleaned. The membranes in both pressure-vessel and submerged systems are typically cleaned without removing the membranes from the modules, so the process is typically called the clean-in-place (CIP) cycle. CIP frequency typically ranges from a couple weeks to several months depending on the membrane system characteristics and source water quality. The CIP procedure typically takes several hours and involves circulating cleaning solutions that have been heated to 30 to 40°C. Cleaning solutions are proprietary mixtures provided by membrane manufacturers but are often high-pH solutions containing detergents or surfactants, which are effective for removing organic foulants. Low-pH solutions such as citric acid can be used for removing inorganic foulants.

Integrity Testing and Monitoring

Membrane integrity monitoring involves procedures to verify that membrane filters are meeting treatment objectives. Integrity monitoring is important because of the physical characteristics of the filtration barrier. In a granular filtration plant, water is cleaned gradually as it flows through a series of processes ending with a thick bed of filter media; clean water and dirty water are separated in both time and space. In a membrane filtration plant, water is cleaned nearly instantaneously as it flows through a thin membrane; clean water and dirty water are separated by a distance less than 1 mm and time less than 1 s. In addition, broken fibers or leaking O-ring connectors may compromise the filtration system.

Integrity monitoring for membrane filtration has both direct and indirect components. Pressure-based direct integrity tests involve pressurizing one side of the membrane with air and monitoring the change in air pressure, flow of air, or volume of displaced water. The equipment, instrumentation, and procedures for conducting direct integrity tests are built into the skid and implemented automatically. In a membrane with no breaches, air will diffuse through the water in the membrane pores, and pressure will decay slowly. Air can flow more rapidly through holes or broken fibers. Acceptable rates of pressure decay vary with the system being monitored according to calculations in the *Membrane Filtration Guidance Manual* (U.S. EPA, 2005). Decay rates of 0.007 to 0.03 bar/min (0.1 to 0.5 psi/min) are typical limits

(U.S. EPA, 2001). Direct integrity testing is required once per day unless the state approves less frequent testing (U.S. EPA, 2005).

Indirect integrity monitoring is the continuous (at least every 15 min) monitoring of a water quality parameter that is indicative of particle removal, such as turbidity or particle counts. Indirect integrity monitoring is not as sensitive as direct integrity testing, but it has the advantages that it can be applied continuously and uses commercially available equipment that can be used with any membrane system (whereas most direct integrity testing equipment is proprietary). Therefore, it is complementary to direct integrity testing in an overall integrity verification program.

Posttreatment

The membrane filtration process has no inherent posttreatment requirements. Fluoridation or pH adjustment may be added after membrane filtration to fulfill other treatment objectives. Although membrane filtration is capable of completely removing microorganisms, disinfection is normally practiced after filtration as part of the multibarrier concept and to provide a disinfectant residual in the distribution system. Most state regulatory agencies have specific regulations for chemical disinfection following filtration.

Residual Handling

Residual handling from membrane filters is similar in many respects to residual handling from granular filters. However, the reduced or eliminated use of coagulants reduces the generation of sludge and simplifies sludge disposal in some cases. Some utilities discharge the waste wash water to the wastewater collection system and allow the sludge to be handled at the wastewater treatment plant rather than have separate sludge-handling facilities at the water treatment plant. Waste wash water can be clarified and returned to the plant influent or the source water, depending on regulatory constraints. The sludge can be thickened and dewatered similar to sludge from granular filters, and when coagulants are not used, the sludge is generally easier to thicken and dewater. Residual management is discussed further in Chap. 14.

8-5 Particle Capture in Membrane Filtration

For regulatory purposes in the United States, membrane filtration is defined as "a pressure or vacuum driven separation process in which particulate matter larger than 1 μm is rejected by an engineered barrier, primarily through a size exclusion mechanism, and which has a measurable removal efficiency of a target organism that can be verified through the application of a direct integrity test" (U.S. EPA, 2006, P. 702). The principles by which membranes are rated, particles are captured, and performance is demonstrated is discussed in this section.

Retention Rating

One of the most significant parameters in membrane filtration is the size of material retained. Microfiltration and UF membranes are currently rated with different systems, making them difficult to compare. The retention rating for MF membranes is called the *pore size* or nominal pore diameter. The retention rating for MF membranes used in water treatment is typically between 0.1 and 1 μm. As was shown on Fig. 8-5, however, the "pores" in MF membranes are tortuous voids with a wide size distribution, not cylindrical holes of a particular diameter. Thus, the nominal pore diameter reflects the size of material that will be retained by the membrane, not actual dimensions of pores in the membrane.

Membrane manufacturers use two approaches for defining the retention rating of UF membranes. Some manufacturers use a pore size rating similar to MF membranes, with pore sizes of 0.01 to 0.04 μm being common. For others, the retention rating for UF membranes is based on the molecular weight of material retained by the membrane and is called the *molecular weight cutoff (MWCO)* or nominal molecular weight limit (NMWL). This classification system arose because the first applications of UF membranes were for fractionating macromolecules, where molecular weight is more important than size. Membrane filtration for water treatment is principally concerned with retaining materials of a particular size, so a size-based classification would be more appropriate. Unfortunately, the diameter of solids retained by a UF membrane is only loosely related to the MWCO value and depends on various physical and chemical properties (shape, electrostatic charge, etc.) of the solid. The MWCO for UF membranes range from about 1000 daltons (Da) to about 500,000 Da. These MWCO values correspond to an ability to retain particles ranging from about 0.001 to 0.03 μm in diameter (Cheryan, 1998).

It should be noted that design manuals and regulations define MF and UF membranes as having particular pore size ranges similar to the discussion above, but there are no rigorous standard specifications that classify a particular product as one or the other.

Rejection and Log Removal

The fraction of material removed (see Eq. 3-1) from the permeate stream is called rejection:

$$R = 1 - \frac{C_p}{C_f} \tag{8-2}$$

where R = rejection, dimensionless
C_p, C_f = permeate and feed water concentrations, mol/L or mg/L

Rejection can be calculated for bulk measures of particulate matter (e.g., turbidity, particle counts) or individual components of interest (e.g., *Cryptosporidium oocysts*). In membrane filtration, the concentration of some components in the permeate can be several orders of magnitude lower than in the feed. Many significant figures must be retained to quantify rejection if Eq. 8-2 is used. In these cases, the log removal value defined in Eq. 3-2 is used:

$$\text{LRV} = \log(C_f) - \log(C_p) = \log\left(\frac{C_f}{C_p}\right) \tag{8-3}$$

where LRV = log removal value, dimensionless

A comparison of the calculation of rejection and LRV is demonstrated in Example 3-1.

Filtration Mechanisms

The primary mechanism for removing particles from solution in membrane filtration is straining, but removal is also affected by adsorption and cake formation. These removal mechanisms are depicted on Fig. 8-12.

STRAINING

Straining (also called sieving, steric exclusion, or size exclusion) is the dominant filtration mechanism in membrane filtration. Nominally, particles much larger than the retention rating of the membrane collect at the surface while water and much smaller particles pass through. When particles are near the pore size rating of the membrane, however, a fraction of the particles will be captured, resulting in partial removal. Partial capture

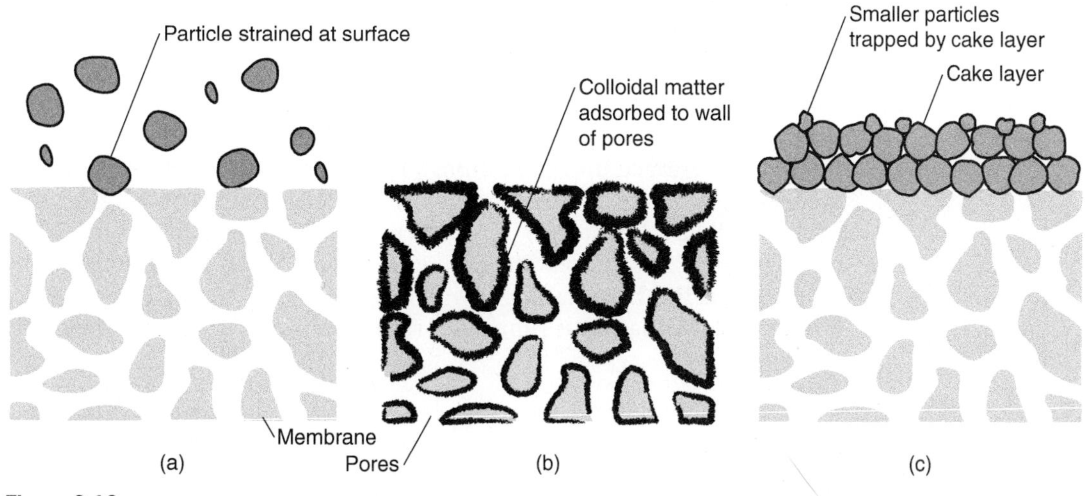

Figure 8-12
Mechanisms for rejection in membrane filtration. (a) Straining occurs when particles are physically retained because they are larger than the pores. (b) Adsorption occurs when material small enough to enter pores adsorbs to the walls of the pores. (c) Cake filtration occurs when particles that are small enough to pass through the membrane are retained by a cake of larger material that collects at the membrane surface.

is caused by the variability of pore size dimensions, nonspherical shape of the particles, and other interactions such as electrostatic repulsion.

As is evident from Fig 8-5, the tortuous interconnecting voids in membrane filters have a distribution of sizes, including some larger than the retention rating. Thus, particles smaller than the retention rating may be trapped in smaller passageways and larger particles may pass through the membrane in other areas.

Particles in natural systems can have shape characteristics significantly different from the materials used to determine the retention rating. Rod-shaped bacteria and linear macromolecules may be very long in one dimension and considerably smaller in others and may not be adequately described by an average diameter. Thus, particles that appear to be slightly larger than the retention rating may pass through the membrane.

Typically, both particles and membrane surfaces are negatively charged. Electrostatic interactions may prevent the particles from entering the pores even if the physical size would permit passage.

ADSORPTION

Natural organic matter adsorbs to membrane surfaces. Thus, these soluble materials may be rejected even though their physical dimensions are orders of magnitude smaller than the membrane retention rating. Adsorption may be an important rejection mechanism during the early stages of filtration with a clean membrane. The adsorption capacity is quickly exhausted, however, and adsorption is not an effective mechanism in the long-term operation of membrane filters. However, adsorbed material may reduce the size of pores at the membrane surface and improve the ability of the membrane to retain smaller material by straining.

CAKE FORMATION

During filtration, a clean membrane will quickly accumulate a cake of solids at the surface due to straining. This surface cake acts as a filtration medium, providing another mechanism for rejection. The surface cake is often called a ''dynamic'' filter since its filtering capability varies with time, growing in thickness during filtration but being partially or wholly removed during backwashing. While this cake can improve membrane filtration performance, it cannot be relied upon since it is removed with every backwash.

Removal of Microorganisms

The principal microorganisms of concern in water treatment are (1) *Giardia lamblia, Cryptosporidium parvum,* and other protozoa, (2) bacteria, and (3) viruses. *Giardia lamblia* cysts are 11 to 15 μm in diameter and *C. parvum* oocysts are 3 to 5 μm in diameter. Thus, both are significantly larger than the pore size ratings of MF and UF membranes and should be completely rejected, as long as there are no integrity problems.

Bacteria range in size from 0.1 to 100 μm. This size is considerably larger than the retention rating for UF membranes and so complete rejection is expected. Most species of bacteria should be completely rejected by MF membranes as well, although a few species of bacteria are near the pore size ratings of MF membranes and less than complete rejection may be possible.

The smallest viruses have a diameter of about 0.025 μm. At this size, viruses are considerably smaller than the retention rating of MF membranes and are similar to that of UF membranes. Many studies have demonstrated that MF membranes are not an effective barrier for viruses, although some virus removal can occur due to adsorption, cake filtration, or capture in the smaller pore spaces of an MF membrane. Despite these possible removal mechanisms, regulatory agencies generally will not allow any credit for virus removal by MF membranes. UF membranes with low MWCO ratings may be able to achieve complete rejection of viruses, but UF membranes with higher MWCO ratings might not. It was noted earlier that the pore size of UF membranes may range from 0.001 to 0.04 μm. Thus, specifying that a treatment system should contain UF membranes will not guarantee that the system can remove viruses; the characteristics of the specific UF membrane product must be considered.

To validate the ability of MF and UF membranes to remove specific microorganisms, challenge testing is performed. Challenge testing is a process conducted by or for membrane equipment manufacturers to verify that a membrane product can remove specific organisms. The test involves spiking the membrane feed water with a high concentration of the actual microorganisms or a suitable surrogate (with similar physicochemical properties) and then measuring the concentration of the microorganisms in the filter effluent to determine the actual log removal value that can be achieved. Specific requirements for challenge testing are included in the Long Term 2 Enhanced Surface Water Treatment Rule (LT2ESWTR) (U.S. EPA, 2005).

8-6 Hydraulics of Flow through Membrane Filters

The relationship for the flow of water through porous media under laminar flow conditions is known as Darcy's law:

$$v = k_P \frac{h_L}{L} \tag{8-4}$$

where v = superficial fluid velocity, m/s
k_P = hydraulic permeability coefficient, m/s
h_L = head loss across porous media, m
L = thickness of porous media, m

The hydraulic permeability coefficient in Darcy's law is an empirical parameter that is used to describe the proportionality between head loss and fluid velocity and is dependent on media characteristics such as porosity and specific surface area. Although flow through membranes follows this linear proportionality between head loss and velocity, the standard equation for membrane flow is written in a substantially different form. Flow is expressed in terms of volumetric flux J rather than superficial velocity, the driving force is expressed as transmembrane pressure ΔP rather than head loss (which are related by $\Delta P = \rho_w g h_L$), and media characteristics are expressed as a resistance coefficient (the inverse of a permeability coefficient). In addition, the membrane flow equation includes the fluid viscosity explicitly (Darcy's law buries it in the permeability coefficient) because viscosity has a significant impact on flux and is easy to determine (via temperature). Finally, the membrane flux equation incorporates the membrane thickness into the resistance coefficient. The equation for membrane flux is

$$J = \frac{Q}{A} = \frac{\Delta P}{\mu \kappa_m} \tag{8-5}$$

where J = volumetric water flux through membrane, $L/m^2 \cdot h$ or m/s
Q = flow rate, L/h
A = membrane area, m^2
ΔP = differential pressure across membrane, bar
μ = dynamic viscosity of water, kg/m·s
κ_m = membrane resistance coefficient, m^{-1}

The membrane resistance coefficient can be calculated from laboratory experiments so that flux through a new membrane can be determined for other pressure or temperature conditions.

The linear relationship between flux and pressure in Eq. 8-5 suggests that the flux can be maximized by operating at the highest possible transmembrane pressure. While that may be true for deionized water, high-pressure operation is not recommended for filtration of natural waters. Fouling can be exacerbated by high-pressure operation, so a balance must be struck between flux and fouling. Studies have found that fouling can increase rapidly when transmembrane pressure is above 1 bar.

Ideally, it would be desirable to calculate flux from measurable parameters that describe MF and UF membranes, such as porosity, nominal pore diameter, specific surface area, and membrane thickness, as is done for clean-bed head loss in granular filtration. These parameters, however, are difficult to measure and the amorphous internal structure of MF and UF membranes (refer to Fig. 8-5) cannot be described mathematically with any great accuracy. In addition, it will be shown later in this chapter that the volumetric flux through a full-scale membrane filter is influenced more by fouling than by the intrinsic clean-membrane resistance. As a result, currently no reliable models allow flux to be predicted from fundamental properties of commercial membranes. Calculation of the membrane resistance coefficient from experimental data is demonstrated in Example 8-2.

Example 8-2 Calculation of membrane resistance coefficient

An MF membrane is tested in a laboratory by filtering clean, deionized water, and the flux is found to be 180 $L/m^2 \cdot h$ at 20°C and 0.9 bar. Calculate the membrane resistance coefficient.

Solution

Rearrange Eq. 8-5 to solve for the membrane resistance coefficient. The dynamic viscosity of water at 20°C, from App. C, is 1.00×10^{-3} kg/m·s. Also recall that 1 bar = 100 kPa = 10^5 N/m^2 = 10^5 $kg/s^2 \cdot m$:

$$\kappa_m = \frac{\Delta P}{\mu J} = \frac{\left(0.9 \times 10^5 \text{ kg/s}^2\cdot\text{m}\right)\left(3600 \text{ s/h}\right)\left(10^3 \text{ L/m}^3\right)}{(1.00 \times 10^{-3} \text{ kg/m}\cdot\text{s})\,(180 \text{ L/m}^2\cdot\text{h})} = 1.79 \times 10^{12} \text{ m}^{-1}$$

Temperature and Pressure Dependence

During operation, changes in permeate flux due to fouling are monitored to determine when cleaning is necessary. Because flux is dependent on pressure and water viscosity, determining the extent of fouling is confounded by simultaneous changes in pressure and temperature (which changes viscosity). In temperate climates, water temperatures can vary by more than 20°C, leading to a 70 percent increase in flux in the summer compared to the winter. Temperature variations are usually accommodated by calculating the equivalent flux at a standard temperature:

$$J_s = J_m\left(\frac{\mu_m}{\mu_s}\right) \tag{8-6}$$

where J_m, J_s = flux at measured and standard (typically 20°C) temperatures, $L/m^2 \cdot h$

μ_m, μ_s = dynamic viscosity of water at measured and standard temperatures, kg/m·s

The dynamic viscosity can be obtained from tabular data or calculated from one of a variety of expressions that relate the viscosity of water to temperature. A relationship often used in membrane operations is (ASTM, 2001)

$$J_s = J_m (1.03)^{T_s - T_m} \tag{8-7}$$

where T_m, T_s = measured and standard temperatures, °C

When using a standard temperature of 20°C, Eq. 8-7 is accurate to within 5 percent over a temperature range of 1 to 28°C, which covers most natural waters. More accurate correlations between viscosity and temperature are available in reference books or on the Internet. Some manufacturers provide their own temperature correction formulas that account for changes in material properties as well as water viscosity.

Flux is normalized for pressure by calculating specific flux, which is the flux at a standard temperature divided by the transmembrane pressure:

$$J_{\text{sp}} = \frac{J_s}{\Delta P} \tag{8-8}$$

where J_{sp} = specific flux at standard temperature, $\text{L/m}^2\cdot\text{h}\cdot\text{bar}$

The specific flux is called membrane permeability when clean water is being filtered through a new, unused membrane in laboratory experiments. Specific flux and membrane permeability are typically reported in units of $\text{L/m}^2\cdot\text{h}\cdot\text{bar}$ or $\text{gal/ft}^2\cdot\text{d}\cdot\text{atm}$.

When flux has been normalized to account for temperature and pressure variations, the effect of fouling can be determined, as illustrated in Example 8-3.

Example 8-3 Calculation of specific flux

A membrane plant has a measured flux in March of 80 $\text{L/m}^2\cdot\text{h}$ at 0.67 bar and 7°C. Four months later, in July, the measured flux is 85 $\text{L/m}^2\cdot\text{h}$ at 0.52 bar and 19°C. Has a change in specific flux occurred? What is the change in percent? Has fouling occurred?

Solution

1. Calculate the specific flux in March.
   a. Calculate the flux in March at a standard temperature of 20°C using Eq. 8-7:

$$J_s = J_m \left(1.03\right)^{T_s - T_m} = \left(80\ \text{L/m}^2\cdot\text{h}\right)\left(1.03\right)^{20^\circ\text{C} - 7^\circ\text{C}} = 117\ \text{L/m}^2\cdot\text{h}$$

b. Calculate the specific flux in March using Eq. 8-8:

$$J_{sp} = \frac{J_s}{\Delta P} = \frac{117 \text{ L/m}^2\text{·h}}{0.67 \text{ bar}} = 175 \text{ L/m}^2\text{·h·bar}$$

2. Calculate the specific flux in July.
   a. Calculate the flux in July at a standard temperature of 20°C using Eq. 8-7:

$$J_s = J_m \left(1.03\right)^{T_s - T_m} = \left(85 \text{ L/m}^2\text{·h}\right)\left(1.03\right)^{20°\text{C}-19°\text{C}} = 87.6 \text{ L/m}^2\text{·h}$$

   b. Calculate the specific flux in July using Eq. 8-8:

$$J_{sp} = \frac{J_s}{\Delta P} = \frac{87.6 \text{ L/m}^2\text{·h}}{0.52 \text{ bar}} = 168 \text{ L/m}^2\text{·h·bar}$$

3. Calculate the percent loss of performance due to fouling:

$$\frac{175 \text{ L/m}^2\text{·h·bar} - 168\text{L/m}^2\text{·h·bar}}{175 \text{ L/m}^2\text{·h·bar}} \times 100 = 4\% \text{ loss of flux due to fouling}$$

Comment

The specific flux at 20°C has declined from 175 to 168 L/m^2·h·bar. Thus, although the plant is operating at a higher flux with a lower pressure in July than it was in March, there has been a 4 percent loss of performance due to fouling.

8-7 Membrane Fouling

The pressure required to maintain flow through a membrane increases as materials collect on and within the membrane. When the resistance through the membrane exceeds the pressure capabilities of the feed pumps, water will no longer flow through the membranes at the required rate. This loss or performance, or *membrane fouling* (defined as a decline in specific flux from the initial conditions), is one of the most significant issues affecting the design and operation of membrane filtration facilities (AWWA, 2005a). Although backwashing and cleaning can restore performance, having to clean too frequently is not cost effective and may eventually degrade the membranes. Fouling is characterized by the mechanism (pore blockage, pore constriction, and cake formation), by whether it can be removed (i.e., reversible or irreversible), and by the material causing it (particles, biofouling, and natural organic matter). Additional details of membrane fouling are presented in Crittenden et al. (2012).

Mechanisms of Fouling

Membrane fouling is traditionally visualized as occurring through three mechanisms—pore blocking, pore constriction, and cake formation. These mechanisms are analogous to the particle retention mechanisms of straining, adsorption, and cake formation, and Fig. 8-12 can be viewed from the perspective of both particle retention mechanisms and fouling mechanisms.

Pore blocking occurs when the entrance to a pore is completely sealed by a particle. As was shown in Fig. 8-5c, commercial membrane filters for water treatment have an interior that is a matrix of tortuous voids. Hydraulic resistance to flow occurs throughout the thickness of the membrane. Sealing of a pore would prevent flow through that portion of the surface, but the flow would simply redistribute in the interior of the membrane. As a result, pore blocking probably has minimal significance in the fouling of commercial membranes for water treatment.

Pore constriction is the reduction of the void volume within a membrane due to adsorption of materials within the pores. Several essential elements must take place for pore constriction to occur. First, the materials must be smaller than the pore size of the membrane so they can penetrate into the membrane matrix instead of being sieved at the surface. Second, they must be transported to the pore walls by either diffusion or hydrodynamic conditions. Third, materials must have an affinity for attaching to the pore walls, without which they would pass through the membrane. Research has demonstrated that hydrophobic membranes foul more than hydrophilic ones, and hydrophobic materials in the feed water can cause greater fouling. Concepts of particle stability presented in Chap. 5 are also relevant here. Finally, the attached material must be sufficiently large to constrict the pore dimensions. Research has shown that high-MW and colloidal organics cause more fouling than low-MW dissolved materials. Low-MW dissolved materials would not have as much of an impact on pore dimensions as colloidal materials.

Particles that are too large to enter the pores collect on the membrane surface in a porous mat called a filter cake. The cake layer generates hydraulic resistance to flow as the thickness builds up. The cake layer can prevent particles smaller than the retention rating from reaching the membrane, improving filtration effectiveness and possibly minimizing fouling from pore constriction.

Reversibility of Fouling

Fouling can be characterized as irreversible or reversible. The specific flux declines during each filter run (normally recorded as an increase in transmembrane pressure) but a significant portion can be recovered during backwashing. This loss of flux that can be recovered during backwashing is called hydraulically reversible fouling. Fouling due to cake formation is largely reversible during backwash. The longer term, slower decline in specific flux over multiple filter runs is due to the slow adsorption and clogging of materials within the membrane matrix (pore constriction), which can be dissolved and removed during chemical cleaning. The loss of flux that can be recovered during cleaning is called chemically reversible fouling. Depending on the source water quality and the type of membrane used, some material can permanently adhere to the membrane and cannot be removed regardless of how aggressive the cleaning is. This permanent flux loss is called irreversible fouling.

Fouling by Natural Organic Matter

Membrane fouling can also be classified by the type of constituent that causes fouling. Three common materials that can foul membranes include particles, biofilms, and natural organic matter. Fouling by particles can be managed by proper backwashing and biofouling can be managed with proper disinfection. The most problematic and least controllable membrane fouling is due to the adsorption of natural organic matter (NOM). Fouling by NOM (or the dissolved fraction, DOM) has been confirmed with laboratory experiments. The relationship between DOM adsorption and flux has not been successfully described mathematically, and there are currently no models that can predict the specific loss of flux due to DOM fouling as a function of water quality measurements. Fouling depends on characteristics of the DOM, the membrane material, and the solution properties, although the size and stability of the DOM appear to be the most important factors.

Research suggests that only a fraction of DOM causes the majority of fouling in membrane filtration and that the high-MW and colloidal fractions are the necessary components because they have the necessary dimensions to constrict membrane pores (Howe, 2001). Chemical properties and particle stability are also important (Huang et. al, 2008) because fouling will not occur unless the colloids have an affinity for attachment to the membrane pore walls.

Resistance-in-Series Model

As noted in the previous sections, several factors may contribute to reduction to flow. The resistance-in-series model applies a resistance value to each component of membrane fouling, assuming that each contributes to hydraulic resistance and that they act independently from one another. The typical form of the resistance-in-series model is

$$J = \frac{\Delta P}{\mu \left(\kappa_m + \kappa_{fc1} + \kappa_{fc2}\right)} \tag{8-9}$$

where J = volumetric water flux through membrane, $L/m^2 \cdot h$ or m/s

ΔP = differential pressure across membrane, bar

μ = dynamic viscosity of water, kg/m·s
κ_m = membrane resistance coefficient, m^{-1}
κ_{fc1} = resistance coefficient for fouling component 1, m^{-1}
κ_{fc2} = resistance coefficient for fouling component 2, m^{-1}

The resistance-in-series equation can be applied to any number of individual resistances, which may be due to irreversible and reversible components, specific fouling materials, or fouling mechanisms. Individual resistance coefficients can be calculated by selecting operating conditions in which individual forms of fouling can be isolated. Alternatively, the resistance-in-series model can be used to develop a membrane fouling index, as shown in the next section.

Membrane Fouling Index

In the absence of fundamental models that predict full-scale performance, it is useful to have empirical models that can compare fouling under different conditions, such as with different source waters, different membrane products, or at different scale. A fouling index can be derived using the resistance-in-series model with two resistance terms: one for clean membrane resistance and another for fouling resistance (Nguyen et al., 2011):

$$J = \frac{\Delta P}{\mu\left(\kappa_m + \kappa_f\right)} \tag{8-10}$$

where κ_f = resistance due to all forms of fouling, m^{-1}

If the fouling resistance is directly proportional to the mass of foulants that have been transported to the membrane surface with the feed water, the fouling resistance can be related to the amount of water filtered per unit of membrane area, that is,

$$\kappa_f = kV_{sp} \tag{8-11}$$

where k = resistance proportionality constant, m^{-2}
V_{sp} = specific throughput, volume of water filtered per membrane area, m^3/m^2

By dividing Eq. 8-10 by ΔP and converting to a standard temperature using Eq. 8-7, the performance can be written in terms of specific flux:

$$J_{sp} = \frac{J_s}{\Delta P} = \frac{1}{\mu\left(\kappa_m + kV_{sp}\right)} \tag{8-12}$$

where J_{sp} = specific flux at standard temperature, $L/m^2{\cdot}h{\cdot}bar$

For a new membrane, $V_{sp} = 0$ so $\kappa_f = 0$, so

$$J_{sp0} = \frac{1}{\mu \kappa_m} \tag{8-13}$$

where J_{sp0} = specific flux of an unused membrane, $L/m^2 \cdot h \cdot bar$

Membrane filtration performance is typically evaluated by comparing the flux over time to the initial flux through the membrane when it was new. Clean-membrane permeability can vary from one membrane sample to another due to slight variations in membrane pore dimensions, thickness, or porosity because of manufacturing variability. Normalizing against new membrane performance eliminates membrane sample variability when comparing experiments. Dividing by clean-membrane specific flux yields

$$J'_{sp} = \frac{J_{sp}}{J_{sp0}} = \frac{1/\left[\mu\left(\kappa_m + kV_{sp}\right)\right]}{1/\left(\mu\kappa_m\right)} = \frac{\kappa_m}{\kappa_m + kV_{sp}} \tag{8-14}$$

where J'_{sp} = normalized specific flux, dimensionless

A fouling index can be defined as the slope of the line when the inverse of J'_{sp} is plotted as a function of the specific throughput:

$$\frac{1}{J'_{sp}} = 1 + (\text{MFI})\, V_{sp} \tag{8-15}$$

where MFI $= k/\kappa_m$ = membrane fouling index, m^{-1}

The MFI is an empirical fouling index that can be used to compare the rate of fouling between experiments, or between bench- and pilot-scale results. The MFI has been used to compare fouling between different membrane products and source waters, and studies have shown reasonably good agreement between MFI values using bench-scale and pilot-scale data with the same membrane and source water (Huang et al., 2009b).

The MFI can be calculated using either a linear regression of flux data or the slope of the line between two points, depending on the data available. Calculation of the MFI is demonstrated in Example 8-4.

Example 8-4 Calculation of the membrane fouling index

A laboratory membrane experiment using a backwashable single-fiber membrane module was carried out. The membrane had a total area of 23.0 cm^2 and the initial permeability of the new membrane was 225.0 $L/m^2 \cdot h \cdot bar$. The test was run at a constant pressure of 1.023 bars and temperature of 22°C. The membrane was backwashed every 30 min. Time and volume filtered were recorded at 2-min intervals and the data from a filter run is shown in the first two columns of Table 1 below. Calculate the fouling index during this filter run.

Solution

1. Divide the volume filtered by the membrane area to determine the specific throughput. Results are in the third column in Table 1. For the second row,

$$V_{sp} = \frac{(743.92 \text{ mL})\left(10^4 \text{ cm}^2/\text{m}^2\right)}{(23.0 \text{ cm}^2)\,(10^3 \text{ mL/L})} = 323.4 \text{ L/m}^2$$

2. Calculate the volume filtered in each time increment by subtracting the previous volume. Results are in the fourth column in Table 1. For the second row:

$$\Delta V = 743.92 \text{ mL} - 732.63 \text{ mL} = 11.29 \text{ mL}$$

3. Divide the volume filtered in each increment by membrane area and time to determine flux. Then correct for temperature and pressure using Eqs. 8-7 and 8-8 to determine specific flux. Results are in the fifth column in Table 1. For the second row,

$$J_m = \frac{(11.29 \text{ mL})\left(10^4 \text{ cm}^2/\text{m}^2\right)(60 \text{ min/h})}{(23.0 \text{ cm}^2)\,(2 \text{ min})\,(10^3 \text{ mL/L})} = 147.3 \text{ L/m}^2\cdot\text{h}$$

$$J_{sp} = \frac{J_m\,(1.03)^{T_s - T_m}}{\Delta P} = \frac{147.3 \text{ L/m}^2\cdot\text{h}\,(1.03)^{20-22}}{1.023 \text{ bars}} = 135.7 \text{ L/m}^2\cdot\text{h}\cdot\text{bar}$$

4. Divide the specific flux (J_{sp}) by the initial specific flux (J_{sp0}). Results are in the sixth column in Table 1. For the second row:

$$J'_{sp} = \frac{135.7}{225.0} = 0.60$$

5. Invert the normalized flux from column 6. Results are in the seventh column in Table 1.

Table 1

(1) Filtration Time, min	(2) Volume Filtered mL	(3) Specific Throughput, L/m^2	(4) Delta volume, mL	(5) Specific Flux, L/m^2·h	(6) Normalized Specific Flux, J'_{sp}	(7) Inverse Normalized Specific Flux, $1/J'_{sp}$
0	732.63					
2	743.92	323.4	11.29	135.7	0.60	1.66
4	754.79	328.2	10.87	130.6	0.58	1.72
6	765.26	332.7	10.47	125.8	0.56	1.79
8	775.40	337.1	10.14	121.9	0.54	1.85
10	785.17	341.4	9.77	118.4	0.53	1.90
12	794.63	345.5	9.46	113.7	0.51	1.98
14	803.79	349.5	9.16	110.1	0.49	2.04
16	812.70	353.3	8.91	107.1	0.48	2.10
18	821.34	357.1	8.64	103.8	0.46	2.17
20	829.73	360.8	8.39	100.8	0.45	2.23
22	837.88	364.3	8.15	97.9	0.44	2.30
24	845.85	367.8	7.97	95.8	0.43	2.35
26	853.62	371.1	7.77	93.4	0.42	2.41
28	861.22	374.4	7.60	91.3	0.41	2.46

6. Plot the inverse of the normalized specific flux ($1/J'_{sp}$) as a function of the specific throughput (V_{sp}), as shown in the following figure:

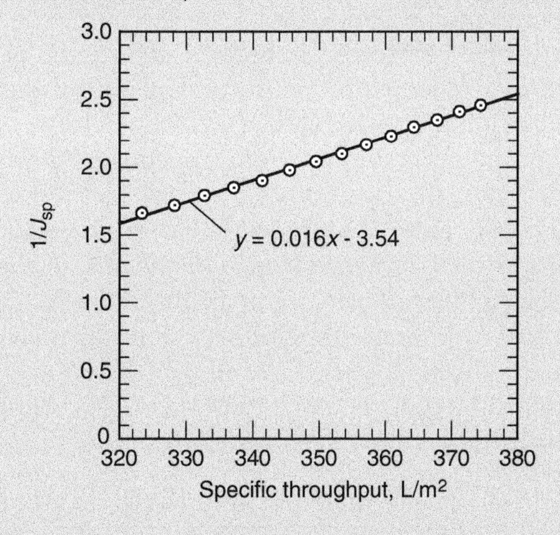

The slope of the line in the membrane fouling index for the filter run is 0.016 L/m^2 = 16 m^{-1}. Note that the intercept of the graph is not 1.0 as is suggested by Eq. 8-15. This result is because previous backwashes removed foulants and reset membrane performance to a higher flux, whereas the specific volume progresses continuously. For an initial filter run (i.e., before any backwashes or cleanings), the intercept is very close to 1.0.

8-8 Sizing of Membrane Skids

Plant capacity is governed by the anticipated water demand at the end of the design life. Summer and winter demand must be considered separately because of the effect of temperature on permeate flux. In most locales, summer water demand is higher than winter demand, which fortunately corresponds to the seasonal variation in water temperatures. For each season, required plant size should be determined for the peak-day demand and minimum water temperature, which are worst-case conditions.

Recovery is the ratio of net water production to gross water production over a filter run:

$$r = \frac{Q_p}{Q_f} = \frac{V_f - V_{\text{bw}}}{V_f} \tag{8-16}$$

where r = recovery
Q_p, Q_f = permeate and feed flow rates, ML/d
V_f = volume of water fed to membrane over filter run, m^3
V_{bw} = volume of water used during backwash, m^3

Recovery in membrane filtration is typically 95 to 98 percent, which is comparable to rapid granular filters. If waste wash water is recovered, processed, and recycled to the feed stream, even higher recovery (greater than 99 percent) can be achieved.

As demonstrated previously, long-term membrane performance is controlled not by intrinsic membrane properties but by the fouled state of the membrane after it has been in contact with natural water. Thus, pilot testing is often part of the process evaluation procedure. Pilot testing can be used to demonstrate the effectiveness of innovative technologies or to provide a basis for comparing alternative systems. Pilot testing should incorporate all pretreatment processes that are being considered for the full-scale facility.

The data generated during pilot testing can be used to design the full-scale facility. Membrane systems are routinely taken off line for backwashing, integrity testing, and cleaning, which reduces the time available for permeate production. The percent of time that permeate is produced, or online production factor, is expressed as

$$\eta = \frac{1440 \text{ min} - t_{\text{bw}} - t_{\text{dit}} - t_{\text{cip}}}{1440 \text{ min}} \tag{8-17}$$

where η = online production factor, dimensionless
$t_{\text{bw}}, t_{\text{dit}}, t_{\text{cip}}$ = time per day for backwashing, direct integrity testing, and cleaning (prorated per day), min

Other factors that may significantly reduce the time available for water production can be incorporated into Eq. 8-17 (AWWA, 2005b). The water produced during each pilot filter run can be determined from the flux, pilot membrane area, and run duration:

$$V_f = JAt_f \tag{8-18}$$

where J = permeate flux, L/m^2·h
A = membrane area, m^2
t_f = duration of filter run (excluding backwash, testing, and cleaning time), min

The water consumed during backwashing should be recorded during the pilot testing. With that information and the volume of water filtered from Eq. 8-18, the recovery and the required feed flow rate can be calculated with Eq. 8-16. The amount of time that the system is not producing permeate and the quantity of water that must be used for backwashing both increase the required membrane area for the full-scale membrane plant:

$$A = \frac{Q_f}{J\eta} = \frac{Q_p}{J\eta r} \tag{8-19}$$

Once the total membrane area for the full-scale plant is determined, the number of skids and modules per skid can be determined by relating the total required membrane area to the capabilities of the system. An example of the sizing of a full-scale membrane system from pilot data is demonstrated in Example 8-5.

Example 8-5 Determining system size from pilot data

A treatment plant is to be designed to produce 75.7 ML/d (20 mgd) of treated water at 20°C. Pilot testing demonstrates that it can operate effectively at a flux of 65 L/m^2·h at 20°C with a 2-min backwash cycle every 45 min and cleaning once per month. The membrane modules have 50 m^2 of membrane area. The pilot unit contained 3 membrane modules and the full-scale skids can contain up to 100 modules. Backwashes for the pilot unit consumed 300 L of treated water. Cleaning takes 4 h. Regulations require direct integrity testing, which takes 10 min, once per day.

Determine the following: (a) the online production factor, (b) system recovery, (c) feed flow rate, (d) total membrane area, (e) number of skids, and (f) number of modules per skid.

Solution

1. Determine the fraction of time the system is producing permeate using Eq. 8-17:

$$t_{bw} = (2 \text{ min})\left(\frac{1440 \text{ min/d}}{45 \text{ min}}\right) = 64 \text{ min/d}$$

$$t_{dit} = 10 \text{ min/d}$$

$$t_{cip} = \frac{(4 \text{ h})(60 \text{ min/h})}{30 \text{ d}} = 8 \text{ min/d}$$

$$\eta = \frac{1440 - t_{bw} - t_{dit} - t_{cip}}{1440} = \frac{1440 - 64 - 10 - 8 \text{ min/d}}{1440 \text{ min/d}} = 0.943$$

2. Determine the system recovery. The system recovery is the same for one element as for all elements and can be calculated using Eq. 8-16. For one element that filters for 43 min per cycle (2 min out of every cycle is backwash), the volume from Eq. 8-18 is

$$V_f = JAt_f = \frac{\left(65 \text{ L/m}^2\cdot\text{h}\right)\left(50.0 \text{ m}^2\right)\left(43 \text{ min}\right)}{60 \text{ min/h}} = 2330 \text{ L}$$

$$V_{bw} = \frac{300 \text{ L}}{3 \text{ modules}} = 100 \text{ L}$$

$$r = \frac{V_f - V_{bw}}{V_f} = \frac{2330 \text{ L} - 100 \text{ L}}{2330 \text{ L}} = 0.957$$

3. Calculate required feed flow by solving Eq. 8-16 for Q_f: (Note $75.7 \text{ ML/d} = 75,700 \text{ m}^3\text{/d}$:

$$Q_f = \frac{Q_p}{r} = \frac{75,700 \text{ m}^3\text{/d}}{0.957} = 79,100 \text{ m}^3\text{/d}$$

4. Calculate the total membrane area required using Eq. 8-19:

$$A = \frac{Q_f}{J\eta} = \frac{\left(79,100 \text{ m}^3\text{/d}\right)\left(10^3 \text{ L/m}^3\right)}{\left(65 \text{ L/m}^2\cdot\text{h}\right)\left(24 \text{ h/d}\right)\left(0.943\right)} = 53,800 \text{ m}^2$$

5. Calculate the total number of modules required:

$$N_{MOD} = \frac{\text{area required}}{\text{surface area per module}} = \frac{53,800 \text{ m}^2}{50 \text{ m}^2} = 1076$$

6. Determine the number of skids and modules/skid. Since the skids can accommodate up to 100 modules, at least 11 skids will be required. Dividing the required modules evenly among skids is preferred. In addition, leaving space in the skids is recommended as an inexpensive way to provide flexibility to reduce flux or increase capacity by adding additional modules in the future. Twelve skids are chosen in this example:

$$N_{\text{Racks}} = 12$$

$$N_{\text{MOD/Rack}} = \frac{1{,}076}{12} = 90$$

The system will have 12 skids that each have 90 modules.

8-9 Energy and Sustainability Considerations

Life-cycle assessments (LCAs) have demonstrated that the environmental impacts of membrane filtration, like many other water treatment processes, are dominated by energy consumption during the operational phase of life. Thus, design decisions that affect energy consumption (largely through operating pressure) will tend to affect sustainability considerations.

An LCA of microfiltration was conducted by Tangsubkul et al. (2006). The study considered both the construction and operating phases and considered 7 environmental indicators using the GaBi software. While that study focused on the filtration of secondary effluent from a wastewater treatment plant, the trends should be applicable to water treatment as well. The study found that operation at low flux was more environmentally favorable. Low flux operation requires more membrane area to produce the same flow, so the main disadvantage is the large increase in equipment fabrication and plant construction impacts and costs. However, the increase in impacts during the construction phase was more than offset by reductions during the operation phase. Low flux operation allows the system to operate at lower pressure, reducing electrical energy consumption. Low flux will also decrease the frequency of backwash and cleaning, reducing environmental impacts associated with chemical production, transportation, and waste disposal. For five indicators (global warming, human toxicity, freshwater aquatic toxicity, marine aquatic toxicity, and terrestrial toxicity potentials), the lowest evaluated flux, 10 $L/m^2{\cdot}h$ (6 $gal/ft^2{\cdot}d$), was the most environmentally favorable operating condition. For photo-oxidant formation potential and eutrophication potential, intermediate fluxes of 30 to 60 $L/m^2{\cdot}h$ (18 – 36 $gal/ft^2{\cdot}d$) were better.

While in practice submerged systems are often designed for lower flux and pressure operation, in reality either pressure or submerged systems can be designed that way. Thus, neither system has inherent advantages from an environmental impact perspective.

At any flux, energy consumption can be affected by system design. The pressure required to maintain constant flow through the membranes increases as the membrane fouls. Increasing pressure requirements can be accommodated several ways. The pump system can be designed to operate continuously at the maximum pressure and the excess pressure can be dissipated through an adjustable valve, or the pump can be equipped with a variable frequency drive (VFD). Operating the pump at maximum pressure wastes energy, and VFDs are the preferred method of flow and pressure control from an energy efficiency perspective. Many membrane manufacturers design their systems with VFDs.

Another design factor that will affect energy consumption is whether the system operates in a cross-flow or dead end mode. Cross-flow filtration requires much larger feed pumps than dead end because a substantial portion of the feed flow is recycled. The specific energy consumption associated with cross-flow pumping can be triple that of dead-end operation (Glucina et al., 1998).

One design decision that probably does not have much effect on sustainability is the specification of MF or UF filters. In full-scale operation, MF and UF systems tend to be designed for similar fluxes and operate at similar pressures, suggesting that the environmental impacts are also probably similar.

Sustainability ought to be considered when comparing rapid granular filtration and membrane filtration as alternate filtration strategies. Considering only direct electrical consumption of the membrane feed pumps, a membrane system that averaged 0.6 bar (9 psi) feed pressure at 95 percent recovery and 80 percent pump efficiency would have specific energy consumption of 0.022 kWh/m^3, compared to 0.01 to 0.014 kWh/m^3 for granular filtration (see Sec. 7-7), suggesting an environmental advantage for rapid granular filtration. The situation is typically more complex because the selection of the filtration technology may influence the selection of other processes within the plant. For instance, in some cases the use of membrane filtration may eliminate the need for coagulation, reducing the environmental impacts associated with chemical production, transportation, and sludge disposal. Granular filtration plants with significant protozoa (*Giardia* and *Cryptosporidium*) removal requirements may need ozonation or UV disinfection facilities that would not be needed with membrane filtration. On the other hand, membrane filtration systems will use cleaning chemicals that are not used in rapid granular filters.

A detailed comparative LCA of conventional granular filtration and membrane filtration considering construction, operation, and decommissioning stages found a mixed situation with respect to the preferred technology (Friedrich, 2002). Comparing a conventional process of coagulation, flocculation, sedimentation, granular filtration, ozonation, and disinfection to a membrane process of prefiltration, membrane filtration, and disinfection, Friedrich found that the conventional process had greater material consumption over all processes and life stages (2.65 kg/m^3 versus 2.53 kg/m^3) but the membrane process had greater energy consumption (0.74 kWh/m^3 versus 0.60 kWh/m^3). In addition, considering eight different environmental indicators from global warming potential to human toxicity potential, the assessment found that the membrane process was more favorable for five indicators and the conventional process was more favorable for the other three. These results indicate that neither filtration technology had a distinct and significant environmental advantage over the other and either might be preferred from a sustainability perspective depending on site-specific design aspects and local environmental concerns.

8-10 Summary and Study Guide

After studying this chapter, you should be able to:

1. Define the following terms and phrases and describe the significance of each in the context of filtration and water treatment:

asymmetric structure	packing density
fouling	permeate
hollow-fiber membrane	pore size
homogeneous structure	retentate
log removal value (LRV)	reverse osmosis
lumen	straining
membrane filtration	transmembrane pressure
microfiltration	tubular membrane
molecular weight cutoff (MWCO)	ultrafiltration
nanofiltration	

2. Explain the purpose of filtration in water treatment and give a general description of the process of membrane filtration.
3. Describe the differences between membrane filtration and reverse osmosis.
4. Compare membrane filtration to rapid granular filtration, describing advantages and disadvantages, similarities and differences, differences in removal mechanisms, and the main features of each.
5. Explain why rapid granular filters must have coagulation pretreatment to be effective but membrane filters do not.
6. Describe the differences between microfiltration and ultrafiltration membranes.

7. Describe the primary features of membrane filtration equipment and operating procedures, including pressure vessel and submerged modules, inside-out and outside-in flow configurations, dead-end and cross-flow filtration, and semibatch and feed-and-bleed operating procedures.
8. Describe the primary function of each of the following aspects of membrane filtration operation: pretreatment, backwashing, chemically enhanced backwash, chemical wash cycle, clean-in-place cycle, and posttreatment. Give a general description of each process.
9. Explain why integrity monitoring is important and how it is done.
10. Calculate rejection and log removal value achieved by a membrane filter.
11. Calculate changes in membrane performance caused by changes in temperature and pressure.
12. Calculate membrane and fouling resistance coefficients.
13. Describe the types of materials that can contribute to membrane fouling.
14. Calculate the membrane fouling index if given data on flow through a membrane over time.
15. Calculate design criteria for a membrane filtration facility, including the membrane surface area required, number of skids and modules, and system recovery.
16. Describe factors that could improve the environmental performance of a membrane filtration system, factors that could degrade performance, and design decisions that have minimal or no effect.

Homework Problems

8-1 An inside-out hollow-fiber membrane system is operated with a cross-flow configuration. Each module contains 10,200 fibers that have an inside diameter of 0.9 mm and a length of 1.75 m. Calculate the following for one module:

a. Feed flow necessary to achieve a cross-flow velocity of 1 m/s at the entrance to the module.

b. Permeate flow rate if the system maintains an average permeate flux of 80 $L/m^2 \cdot h$.

c. Cross-flow velocity at the exit to the module.

d. Ratio of the cross-flow velocity at the entrance of the module to the flow velocity toward the membrane surface. Given the magnitude of this ratio, what effect would you expect cross-flow velocity to have on fouling in cross-flow versus dead-end filtration?

e. Ratio of permeate flow rate to feed flow rate (known as the single-pass recovery). What impact does this ratio have on operational costs in cross-flow versus dead-end filtration?

8-2 Hollow-fiber membranes with a membrane area of 23.3 cm^2 were tested in a laboratory and found to have the clean water flow shown in the table below, at the given temperature and pressure.

	A	B	C	D	E
Flow (mL/min)	4.47	4.22	2.87	6.05	1.22
Temperature (°C)	16	22	23	25	22
Pressure (bar)	0.67	0.80	0.71	1.25	0.21

For the data set selected by your instructor,

a. Calculate the specific flux at 20°C.

b. Calculate the membrane resistance coefficient.

c. Does membrane resistance coefficient depend on the pressure and temperature used for the tests? Why or why not?

8-3 Feed water pressure and temperature and permeate flux at a membrane filtration plant are reported on two dates below. For the plant selected by your instructor, calculate the specific flux on each date, and indicate whether fouling has occurred between the first and second dates.

	A	B	C	D	E
Day 1					
Flux (L/m^2·h)	72	26	31	86	112
Temperature (°C)	21	17	17	22	19
Pressure (bar)	0.62	0.24	0.24	0.72	0.66
Day 2					
Flux (L/m^2·h)	56	26	27	90	120
Temperature (°C)	4	15	10	25	11
Pressure (bar)	0.80	0.29	0.26	0.77	1.05

8-4 A new membrane plant is being designed. Pilot testing indicates that the membrane will be able to operate at a specific flux of 120 L/m^2·h·bar at 20°C. Water demand projections predict a summer peak-day demand of 90 ML/d and a winter peak-day demand of 60 ML/d. Historical records indicate that the source water has a minimum temperature of 3°C in winter and 18°C in summer.

a. Which season will govern the size of the plant?

b. What is the required membrane area, assuming the plant will operate at 0.8 bar, the online production factor is 95 percent, and the recovery is 97 percent?

8-5 Calculate the membrane fouling index for the following data, for the data set specified by your instructor.

a. Experimental flat-sheet laboratory filter, membrane area = 30 cm^2, initial flux = 3560 $L/m^2 \cdot h \cdot bar$, test pressure = 0.69 bar, test temperature = 23.9°C.

Time, min	Permeate Volume, mL	Time, min	Permeate Volume, mL	Time, min	Permeate Volume, mL
0	0	4	345.0	8	552.1
1	108.8	5	404.2	9	594.1
2	199.8	6	458.3	10	634.1
3	277.4	7	506.8	11	670.8

b. Full scale plant operating at constant permeate flow of 15 ML/d, temperature = 20°C, 5800 m^2 of membrane area, pressure each day as shown below. Use day 0 as the initial flux.

Time, Day	Transmemb. Pressure, Bar	Time, Day	Transmemb. Pressure, Bar	Time, Day	Transmemb. Pressure, Bar
0	0.704				
2	0.712	12	0.747	22	0.786
4	0.721	14	0.754	24	0.794
6	0.726	16	0.765	26	0.801
8	0.735	18	0.770	28	0.812
10	0.740	20	0.777	30	0.812

c. Data from a 30-min filter run in the middle of a day of laboratory testing of coagulated feed water, membrane area = 23 cm^2, initial flux = 238 $L/m^2 \cdot h \cdot bar$, test pressure = 2.07 bar, test temperature = 21.5°C.

Time, min	Permeate Volume, mL	Time, min	Permeate Volume, mL	Time, min	Permeate Volume, mL
0	2276.64	10	2354.92	20	2430.04
2	2292.62	12	2370.17	22	2444.76
4	2308.41	14	2385.31	24	2459.35
6	2324.05	16	2400.33	26	2473.88
8	2339.53	18	2415.24	28	2488.26

8-6 A membrane filtration plant is to be designed using results from a pilot study. Treatment plant requirements and pilot results are given in the table below. For the selected system (to be specified by the instructor), determine (a) the online production factor, (b) system recovery, (c) feed flow rate, (d) total membrane area, (e) number of skids, and (f) number of modules per skid. The pilot system contained two membrane elements that had 45 m^2 of membrane area each. In the full-scale plant, integrity testing will be required by regulations once per day and will take 15 min. Chemical cleaning (CIP) will take 4 h.

	A	B	C	D	E
Design capacity (ML/d)	56	115	38	76	227
Memb. area in full-scale modules (m^2)	45	55	45	45	80
Max modules in skid	80	90	80	80	100
Pilot results					
Flux ($L/m^2 \cdot h$)	80	125	40	80	110
Backwash frequency (minutes)	30	25	25	22	30
Backwash duration (minutes)	1.5	0.5	1	2	1
Backwash volume (L)	270	100	200	240	240
Cleaning frequency (day)	45	30	60	30	30

References

ASTM (2001) D5090-90 Standard Practice for Standardizing Ultrafiltration Permeate Flow Performance Data, in *Annual Book of Standards*, Vol. 11.01, American Society for Testing and Materials, Philadelphia, PA.

AWWA (2005a) "Committee Report: Recent Advances and Research Needs in Membrane Fouling," *J. AWWA*, **97**, 8, 79–89.

AWWA (2005b) *Microfiltration and Ultrafiltration Membranes for Drinking Water: Manual of Water Supply Practices M53*, AWWA, Denver, CO.

AWWA (2010) *ANSI/AWWA B110-09 Standard for Membrane Systems*, American Water Works Association, Denver, CO.

Cheryan, M. (1998) *Ultrafiltration and Microfiltration Handbook*, Technomic, Lancaster, PA.

Crittenden. J. C., Trussell, R. R., Hand, D. W., Howe, K. J., and Tchobanoglous, G. (2012) *MWH's Water Treatment: Principles and Design,* 3rd ed., Wiley, Hoboken, NJ.

Farahbakhsh, K., Svrcek, C., Guest, R. K., and Smith, D. W. (2004) "A Review of the Impact of Chemical Pretreatment on Low-Pressure Water Treatment Membranes," *J. Env. Eng. Sci.*, **3**, 4, 237–253.

Friedrich, E. (2002) "Life-Cycle Assessment as an Environmental Management Tool in the Production of Potable Water" *Wat. Sci. and Technol.* **46**, 9, 29–36.

Glucina, K., Lané, J.-M., and Durand-Bourlier, L. (1998) ''Assessment of Filtration Mode for the Ultrafiltration Membrane Process,'' *Desalination*, **118**, 1/3, 205–211.

Howe, K. J. (2001) Effect of Coagulation Pretreatment on Membrane Filtration Performance, Ph.D. Thesis, University of Illinois at Urbana-Champaign, Urbana, IL.

Huang, H., Spinette, R., and O'Melia, C. R. (2008) ''Direct-Flow Microfiltration of Aquasols I. Impacts of Particle Stabilities and Size,'' *J. Memb. Sci.*, **314**, 1–2, 90–100.

Huang, H., Schwab, K., and Jacangelo, J. G. (2009a) ''Pretreatment for Low Pressure Membranes in Water Treatment: A Review,'' *Environ. Sci. Technol.*, **43**, 9, 3011–3019.

Huang, H., Young, T. A., and Jacangelo, J. G. (2009b) ''Novel Approach for the Analysis of Bench-Scale, Low Pressure Membrane Fouling in Water Treatment,'' *J. Memb. Sci.*, **334**, 1–2, 1–8.

Nguyen, A. H., Toblason, J. E., and Howe, K. J. (2011) ''Fouling Indices for Low Pressure Hollow Fiber Membrane Performance Assessment.'' *Water Res.*, **45,** 8, 2627–2637.

Tangsubkul, N., Parameshwaran, K., Lundie, S., Fane, A. G., and Waite, T. D. (2006) ''Environmental Life Cycle Assessment of the Microfiltration Process.'' *J. Memb. Sci.*, **294**, 1–2, 214–226.

U.S. EPA (2001) *Low-Pressure Membrane Filtration for Pathogen Removal: Application, Implementation, and Regulatory Issues*, U.S. Environmental Protection Agency, Cincinnati, OH.

U.S. EPA (2005) *Membrane Filtration Guidance Manual*, EPA 815-R-06-009, U.S. Environmental Protection Agency, Cincinnati, OH.

U.S. EPA (2006) National Primary Drinking Water Regulations: Long Term 2 Enhanced Surface Water Treatment; Final Rule, *Federal Register*, **71**, 3, 654–786.

9 Reverse Osmosis

Reverse osmosis (RO) is a membrane treatment process used to separate dissolved solutes from water. The membrane is a *semipermeable* material that is permeable to some components in the feed stream and impermeable to other components. The feed water to an RO system is pressurized and some water, called *permeate*, passes through the membrane, as shown schematically on Fig. 9-1. As water passes through the membrane, solutes are rejected and the feed stream becomes more concentrated. The permeate is relatively free of targeted dissolved solutes and exits at nearly atmospheric pressure, while the remaining water, called *concentrate*, exits at the far end of the pressure vessel at nearly the feed pressure. Reverse osmosis is a continuous separation process; that is, there is no periodic backwash cycle.

Membrane processes were introduced in Chap. 8, where it was noted that the membranes used in municipal water treatment include microfiltration

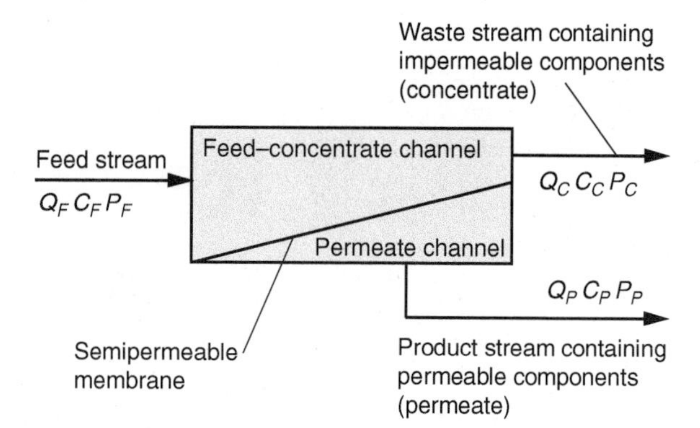

Figure 9-1
Schematic of separation process through reverse osmosis membrane.

(MF), ultrafiltration (UF), nanofiltration (NF), and reverse osmosis (RO) membranes. From a physicochemical perspective, these four types of membranes are used in two distinct physicochemical processes in water treatment: (1) membrane filtration and (2) reverse osmosis. Chapter 8 included details on the classification of membrane processes (Sec. 8-1), including the hierarchy of membranes used in water treatment (Fig. 8-2) and a table of significant differences between membrane filtration and reverse osmosis (Table 8-1).

Reverse osmosis is used for desalinating seawater and brackish groundwater, softening, removing natural organic matter for disinfection by-product control, advanced treatment for potable water reuse, and removing specific contaminants. Uses for RO in water treatment as well as alternative processes are listed in Table 9-1.

Nanofiltration membranes were developed in the 1980s for selective removal of divalent ions and had a separation cutoff size of about 1 nm. Membrane manufacturers have since engineered many RO products with different formulations, permeation capabilities, and rejection characteristics. Some are similar to the original NF membranes and have a variety of names, including softening membranes, brackish water RO membranes, and low-pressure RO membranes. Manufacturers will continue to develop new RO membranes to achieve specific goals, and NF membranes are just one in a succession of many innovative developments in the field of RO. All discussion in this chapter applies equally to NF and RO membranes unless stated otherwise.

Growth in the world population, the urbanization of coastal and arid areas, the scarcity of freshwater supplies, the increasing contamination of freshwater supplies, greater reliance on oceans and poorer quality supplies (brackish groundwater, treated wastewater), and improvements in membrane technology have spurred rapid growth in the number of reverse osmosis installations. By the end of 2008, the total installed capacity of desalination plants was $42 \times 10^6\ m^3/d$ (11 billion gallons per day) worldwide. Over 1100 RO plants are operating in the United States with a total

Table 9-1
Reverse osmosis objectives and alternative processes

Process Objective	RO Process Name	Alternative Processes
Ocean or seawater desalination	High-pressure RO, seawater RO	Multistage flash distillation (MSF), multieffect distillation (MED), vapor compression distillation (VCD)
Brackish water desalination	RO, low-pressure RO, NF	Multistage flash distillation,[a] multieffect distillation,[a] vapor compression,[a] electrodialysis, electrodialysis reversal
Softening	Membrane softening, NF	Lime softening, ion exchange
NOM removal for DBP control	NF	Enhanced coagulation/softening, GAC
Specific contaminant removal[b]	RO	Ion exchange, activated alumina, coagulation, lime softening, electrodialysis, electrodialysis reversal
Potable water reuse	RO	Advanced oxidation
High-purity process water	RO	Ion exchange, distillation

[a]MSF, MED, and VCD are rarely competitive economically for brackish water desalination.
[b]Applicability of alternative processes depends on the specific contaminants to be removed and their concentration.

capacity of around 5.7×10^6 m^3/d (1,500 mgd) (NRC, 2008), which represents about 3 percent of water withdrawn by public water systems. Reverse osmosis plants have been built in every state in the United States. The installation of desalination facilities is expected to double between 2005 and 2015 (Veerapaneni et al., 2010).

Principal features of RO systems—fundamentals such as osmotic pressure, mass transfer, temperature and pressure dependence, and concentration polarization; operating characteristics such as scaling and fouling; procedures for element selection and membrane array design; and energy and sustainability considerations—are presented in this chapter.

9-1 Principal Features of a Reverse Osmosis Facility

A typical RO facility is shown on Fig. 9-2. The facility consists of pretreatment systems, feed pumps, pressure vessels containing membrane elements, and post treatment systems. These components are described in the follow sections.

Membrane Elements

The smallest unit of production capacity in a membrane plant is called a *membrane element*. Reverse osmosis membrane elements are fabricated in either a spiral-wound configuration or a hollow-fine-fiber configuration. Hollow-fine-fiber membranes are similar to the hollow fibers used in

Figure 9-2
Typical reverse osmosis facility.

membrane filtration but are much thinner, and are not commonly used in membrane plants in the United States.

The basic construction of a spiral-wound membrane element and photographs of typical elements are shown on Fig. 9-3. An envelope is formed by sealing two sheets of flat-sheet membrane material along three sides, with the feed water side facing out. A permeate carrier spacer inside the envelope prevents the inside surfaces from touching each other and provides a flow path for the permeate. The open ends of several envelopes are attached to a perforated central tube known as a permeate collection tube. Feed-side mesh spacers are placed between the envelopes to provide a flow path and create turbulence in the feed water. By rolling the membrane envelopes around the permeate collection tube, the feed-side spacer forms a spirally shaped feed channel. This channel, exposed to element feed water at one end and concentrate at the other end, is known as the feed–concentrate channel. Membrane feed water passes through this channel and is exposed to the membrane surface. Spiral-wound elements are typically 1 to 1.5 m (40 to 60 in.) long and 0.1 to 0.46 m (4 to 18 in.) in diameter, although 0.2-m (8-in.) diameter elements are most common. Four to seven elements are arranged in series in a pressure vessel, with the permeate collection tubes of the elements coupled together.

During operation, pressurized feed water enters one side of the pressure vessel and encounters the first membrane element. As the water flows tangentially across the membrane surface, a portion of the water passes through the membrane surface and into the membrane envelope and flows spirally toward the permeate collection tube. The remaining feed water, now concentrated, flows to the next element in series, and the process is repeated until the concentrate exits the pressure vessel.

Individual spiral-wound membrane elements have a permeate recovery (ratio of permeate to feed water flow) of 5 to 15 percent per element. Head

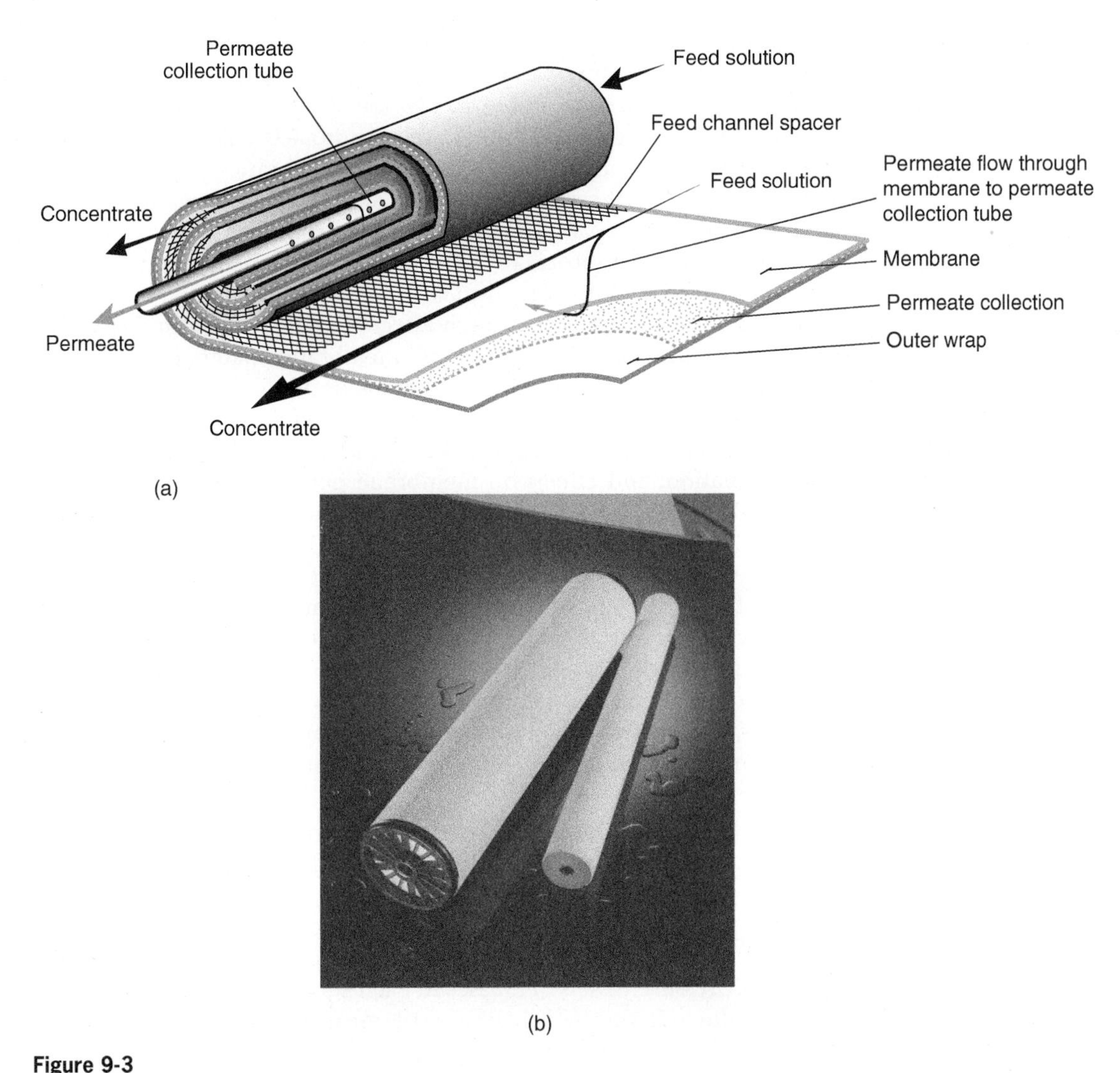

Figure 9-3
(a) Construction of spiral-wound membrane element. (b) Photograph of spiral-wound membrane elements. (Courtesy GE Infrastructure Water Technologies.)

loss develops as feed water flows through the feed channels and spacers, which reduces the driving force for flow through the membrane surface. This feed-side head loss across a membrane element is typically less than 0.5 bar (7 psi) per element.

Membrane Structure and Chemistry

Reverse osmosis membranes are comprised of several layers, with a thin *active layer* that achieves the separation between solutes and water and thicker, more porous layers that provide structural integrity. The active membrane layer must be extremely thin (about 0.1 to 2 μm in RO membranes) to have an effective flux of water, and would not be able to stand the feed pressure without the strength of the support layer. Multilayer membranes are fabricated in two ways. *Asymmetric* membranes are formed from a single material that develops into active and support layers during the casting process. *Thin-film composite* membranes are composed of two or more materials cast on top of one another. Advantage of thin-film membranes are that separation and structural properties can be optimized independently using appropriate materials for each function and the active layer can be deposited in a very thin layer.

Membrane performance is strongly affected by the physical and chemical properties of the material. The ideal membrane material is one that can produce a high flux without clogging or fouling and is physically durable, chemically stable, nonbiodegradable, chemically resistant, and inexpensive. Important characteristics of membrane materials, methods of determination, and effects on membrane performance were discussed in Chap. 8 and shown in Table 8-3. The materials most widely used in RO are cellulose acetate (CA) and polyamide (PA) derivatives. Cellulose acetate membranes are typically asymmetric. Polyamide membranes are typically of thin-film construction. Polyamide membranes are chemically and physically more stable than CA membranes. Under similar pressure and temperature conditions, they typically produce higher water flux and higher salt rejection than CA membranes. However, PA membranes are more hydrophobic and susceptible to fouling than CA membranes and are not tolerant of free chlorine in any concentration.

Membrane Skids, Stages, and Arrays

The membrane elements are enclosed in pressure vessels. A group of pressure vessels operated in parallel is called a *stage*. The concentrate from one stage can be fed to a subsequent stage to increase water recovery (i.e., a two-stage system) or the permeate from one stage can be fed to a second stage to increase solute removal (a two-pass system). In multistaged systems, the number of pressure vessels decreases in each succeeding stage to maintain sufficient velocity in the feed channel as permeate is extracted from the feed water stream. A unit of production capacity, which may contain one or more stages, is called an *array*. Schematics of various arrays are shown on Fig. 9-4.

The recovery from an array ranges from about 50 percent for seawater RO systems to about 90 percent for low-pressure RO systems. Several factors limit recovery, most notably osmotic pressure, concentration polarization, and the solubility of sparingly soluble salts.

A schematic of an entire RO system is shown on Fig. 9-5. Components such as pretreatment, posttreatment, energy recovery, and concentrate management are discussed in the next sections.

Pretreatment

Feed water pretreatment is required in virtually all RO systems. When sparingly soluble salts are present, one purpose of pretreatment is to prevent scaling. Solutes are concentrated as water is removed from the feed

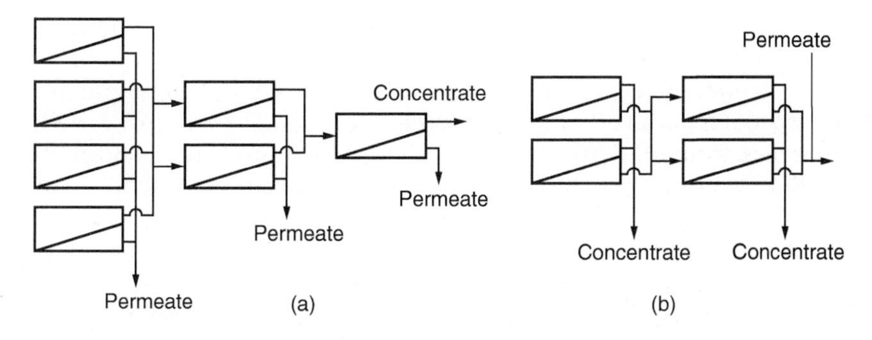

Figure 9-4
Array configurations of reverse osmosis facilities: (a) 4 × 2 × 1 three-stage array and (b) two-pass system.

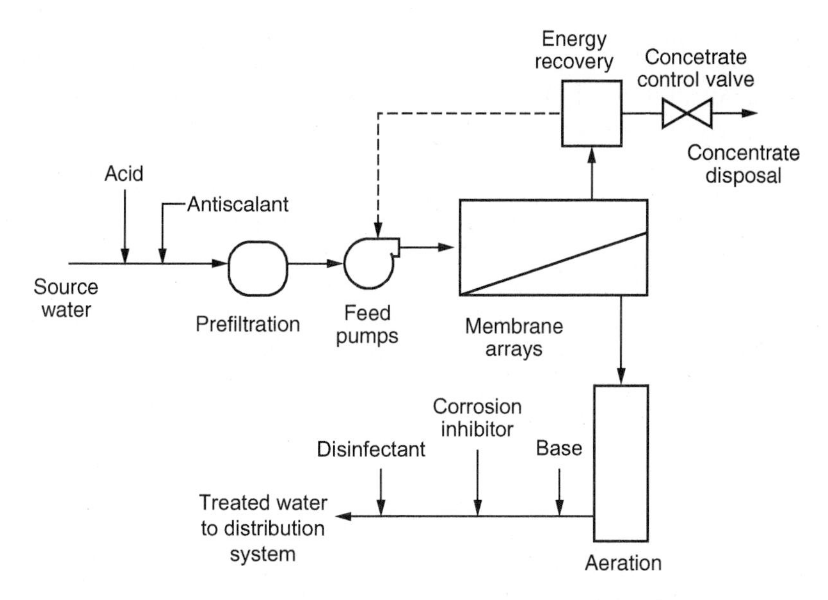

Figure 9-5
Schematic of typical reverse osmosis facility.

stream, and the resulting concentration can be higher than the solubility product of various salts. Without pretreatment, these salts can precipitate onto the membrane surface and irreversibly damage the membrane. Scale control consists of pH adjustment and/or antiscalant addition. Adjusting the pH changes the solubility of precipitates, and antiscalants interfere with crystal formation or slow the rate of precipitate formation.

A second important pretreatment process is filtration to remove particles. Without a backwash cycle, particles can clog the feed channels or accumulate on the membrane surface. As a minimum, cartridge filtration with a 5-μm strainer opening is used, although granular filtration or membrane filtration pretreatment is often necessary for surface water sources. Disinfection is another typical pretreatment step and is used to prevent biofouling. Some membrane materials are incompatible with disinfectants, so the disinfectant can only be applied in specific situations and must be matched to the specific membrane type.

After pretreatment, the feed water is pressurized with feed pumps. The feed water pressure ranges from 5 to 10 bar (73 to 145 psi) for NF membranes, from 10 to 30 bar (145 to 430 psi) for low-pressure and brackish water RO, and from 55 to 85 bar (800 to 1200 psi) for seawater RO.

Posttreatment

Permeate typically requires posttreatment, which consists of removal of dissolved gases and alkalinity and pH adjustment. Membranes do not efficiently remove small, uncharged molecules, in particular dissolved gases. If hydrogen sulfide is present in the source groundwater, it must be stripped prior to distribution to consumers. Permeate is typically low in hardness and alkalinity and frequently has been adjusted to an acidic pH value to control scaling. Consequently, the permeate is corrosive to downstream equipment and piping. Alkalinity and pH adjustments are accomplished with various bases, and corrosion inhibitors are used to control corrosion. The stripping of carbon dioxide raises pH and reduces the amount of base needed to increase the stability (reduce the corrosivity) of the water. Another strategy for producing a stable finished water is to blend the permeate with a bypass stream of raw water that meets all other water treatment requirements (such as filtration if a surface water source is used).

For potable water applications, chlorine is commonly used for disinfection. The RO process is effective at removing DBP precursors; thus, free chlorination can typically be practiced without forming significant quantities of DBPs. Blending with raw water for stability, however, may increase DBP formation when using free chlorine. Disinfection is discussed in Chap. 13.

Concentrate Stream Energy Recovery

The concentrate stream is under high pressure when it exits the final membrane element. This pressure is dissipated through the concentrate control valve, which can be a significant waste of energy. Seawater RO systems utilize energy recovery equipment on the concentrate line, and some brackish water RO systems are starting to use energy recovery as well. More than 90 percent of the energy expended to pressurize the concentrate stream can be recovered. Depending on the price for electricity, capital costs of energy recovery equipment may be recouped within 3 to 5 years. Several types of devices are available, including reverse-running turbines, pelton wheels, pressure exchangers, and electric motor drives.

Concentrate Management

Concentrate management can be a significant issue in the design of RO facilities and the concentrate may require treatment before disposal. Methods for concentrate disposal include ocean, brackish river, or estuary discharge; discharge to a municipal sewer; and deep-well injection. Other concentrate disposal options, including evaporation ponds, infiltration basins, and irrigation, are used by a small number of facilities.

An active area of research and interest in the industry is improved methods of concentrate management that can increase the recovery of water and decrease the quantity of residuals that must be disposed of. One strategy is to provide an intermediate treatment process between two stages of RO membranes. Since calcium is often the limiting cation, lime softening can be an effective intermediate strategy. Softening can also be effective at removing other scale-causing constituents. Brine concentrators and crystallizers are additional technologies to reduce the volume of concentrate, and can lead to zero liquid discharge (ZLD), in which the only residuals from the facility are solids, which are then easier to dispose of. While brine concentrators and crystallizers are used in some industrial processes such as the power generation industry, they are expensive, energy intensive, and have not yet been proven to be cost effective in the municipal water treatment industry.

The following sections of this chapter present important fundamentals of reverse osmosis, including the origin of osmotic pressure and the nature of mass transfer through RO membranes.

9-2 Osmotic Pressure and Reverse Osmosis

An understanding of osmotic pressure is essential to an understanding of reverse osmosis. *Osmosis* is the flow of solvent through a semipermeable membrane, from a dilute solution into a concentrated one. Osmosis reduces the flux through a RO membrane by inducing a driving force for flow in the opposite direction. The physicochemical foundation for osmosis is rooted in the thermodynamics of diffusion.

Diffusion and Osmosis

Consider a vessel with a removable partition that is filled with two solutions to exactly the same level, as shown on Fig. 9-6a. The left side is filled with a concentrated salt solution, the right with pure water, and the partition is gently removed without disturbing the solutions. Initially, the contents are in a nonequilibrium state and the salt will eventually diffuse through the water until the concentration is the same throughout the vessel. With salt ions diffusing from left to right across the plane originally occupied by the partition, conservation of mass requires a flux of water molecules in the opposite direction. Without a flux of water molecules from right to left, mass accumulates on the right side of the container, which is unthinkable with a continuous water surface. Equilibrium requires mass transfer in both directions.

On Fig. 9-6b, the top of the vessel has been closed and fitted with manometer tubes and the removable partition has been replaced with a semipermeable membrane. The semipermeable membrane allows the flow of water but prevents the flow of salt. Filling the chambers with salt solution

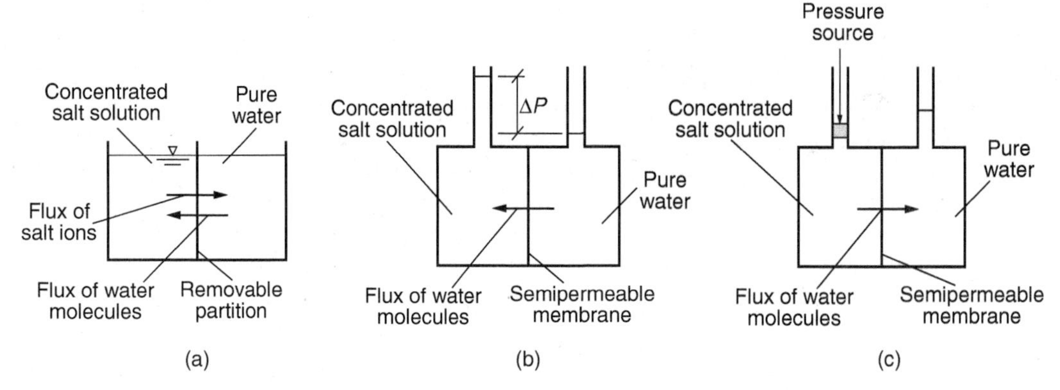

Figure 9-6
Diffusion sketch for reverse osmosis: (a) diffusion, (b) osmosis, and (c) reverse osmosis.

and pure water again creates a thermodynamically unstable system, which must be equilibrated by diffusion. Because the membrane prevents the flux of salt, however, mass accumulates in the left chamber, causing the water level in the left manometer to rise and in the right manometer to drop. This flow of water from the pure side to the salt solution is osmosis. Water flux occurs despite the difference in hydrostatic pressure that develops due to the difference in manometer levels.

Osmotic Pressure

The driving force for diffusion is typically described as a concentration gradient, although a more rigorous thermodynamic explanation is a gradient in Gibbs energy (G). Equilibrium is defined thermodynamically when $\Delta G = 0$. Water stops flowing from right to left when the vessel reaches thermodynamic equilibrium but both pressure and concentration are unequal between the chambers. Although Gibbs energy is constant throughout the second vessel at equilibrium, the Gibbs energy includes components to account for both the pressure and concentration differences, such that the concentration gradient in one direction is exactly balanced by the pressure gradient in the opposite direction. The concept of Gibbs energy (G) and its relationship to osmotic pressure are described in detail in the companion reference book to this textbook (Crittenden et al., 2012).

The pressure required to balance the difference in concentration of a solute is called the *osmotic pressure* and is given the symbol π. When the vessel in the second experiment reaches equilibrium, the difference in hydrostatic pressure between the manometers is equal and opposite to the difference in osmotic pressure between the two chambers. An equation for osmotic pressure can be derived thermodynamically using assumptions of incompressible and ideal solution behavior. In dilute solution, osmotic pressure can be approximated by the van't Hoff equation:

$$\pi = \frac{n_S}{V} RT \qquad \text{or} \qquad \pi = CRT \tag{9-1}$$

where π = osmotic pressure, bar
n_S = total amount of all solutes in solution, mol
V = volume of solution, L
R = universal gas constant, 0.083145 L·bar/mol·K
T = absolute temperature, K (273 + °C)
C = concentration of all solutes, mol/L

The van't Hoff equation is identical in form to the ideal gas law ($PV = nRT$) because it has the same thermodynamic foundation. Equation 9-1 was derived assuming infinitely dilute solutions, which is often not the case in RO systems. To account for the assumption of diluteness, the nonideal behavior of concentrated solutions, and the compressibility of liquid at high pressure, a nonideality coefficient (osmotic coefficient ϕ) must be incorporated into the equation:

$$\pi = \phi CRT \tag{9-2}$$

where ϕ = osmotic coefficient, unitless

Thermodynamically, osmotic pressure is strictly a function of the concentration, or mole fraction, of water in the system. Solutes reduce the mole fraction of water, and the effect of multiple solutes is additive because they cumulatively reduce the mole fraction of water. Solutes that dissociate also have an additive effect on the mole fraction of water (e.g., addition of 1 mol of NaCl produces 2 mol of ions in solution, doubling the osmotic pressure compared to a solute that does not dissociate). If multiple solutes are added on an equal-mass basis, the solute with the lowest molecular weight produces the greatest osmotic pressure. The use of Eq. 9-2 is demonstrated in Example 9-1.

Example 9-1 Osmotic pressure calculations

Calculate the osmotic pressure of 1000-mg/L solutions of the following solutes at a temperature of 20°C assuming an osmotic coefficient of 0.95: (1) NaCl, (2) $SrSO_4$, and (3) glucose ($C_6H_2O_6$).

Solution

1. Determine the osmotic pressure for NaCl, first by calculating the molar concentration of ions and then using Eq. 9-2. A periodic table of the elements is available in App. D.

$$C = \frac{(2\text{ mol ion/mol NaCl})(1000\text{ mg/L})}{(10^3\text{ mg/g})(58.4\text{ g/mol})} = 0.0342\text{ mol/L}$$

$$\pi = \phi CRT = (0.95)(0.0342\text{ mol/L})(0.083145\text{ L}\cdot\text{bar/K}\cdot\text{mol})(293\text{ K})$$

$$= 0.79\text{ bar}$$

2. Determine the osmotic pressure for $SrSO_4$:

$$C = \frac{(2\ \text{mol ion/mol } SrSO_4)(1000\ \text{mg/L})}{(10^3\ \text{mg/g})(183.6\ \text{g/mol})} = 0.0109\ \text{mol/L}$$

$$\pi = (0.95)(0.0109\ \text{mol/L})(0.083145\ \text{L} \cdot \text{bar/K} \cdot \text{mol})(293\ \text{K})$$

$$= 0.25\ \text{bar}$$

3. Determine the osmotic pressure for glucose (no dissociation):

$$C = \frac{1000\ \text{mg/L}}{(10^3\ \text{mg/g})(180\ \text{g/mol})} = 0.0056\ \text{mol/L}$$

$$\pi = (0.95)(0.00556\ \text{mol/L})(0.083145\ \text{L} \cdot \text{bar/K} \cdot \text{mol})(293\ \text{K})$$

$$= 0.13\ \text{bar}$$

Comment

Each solution contains the same mass of solute. Both NaCl and $SrSO_4$ dissociate into two ions, so the molar ion concentration is double the molar concentration of added salt. The NaCl has a higher osmotic pressure because it has a lower molecular weight. Even though $SrSO_4$ and glucose have nearly the same molecular weight, the osmotic pressure of $SrSO_4$ is nearly double that of glucose because it dissociates.

The osmotic pressure of a solution of sodium chloride, calculated with Eq. 9-2 and $\phi = 1$, is shown on Fig. 9-7a along with experimentally determined values. Over the range of salt concentrations of interest in seawater desalination, the osmotic coefficient for sodium chloride ranges from 0.93 to 1.03 and is shown as a function of solution concentration on Fig. 9-7b. Osmotic coefficients for other electrolytes are available in Robinson and Stokes (1959).

Reported values for the osmotic pressure of seawater (Sourirajan, 1970) are about 10 percent below measured values for sodium chloride, as shown on Fig. 9-7a, due to the presence of compounds with a higher molecular weight than sodium chloride. The osmotic pressure for seawater can be calculated with Eq. 9-2 and an equivalent concentration of sodium chloride by using the osmotic coefficient for seawater shown on Fig. 9-7b.

Reverse Osmosis

Two opposing forces contribute to the rate of water flow through the semipermeable membrane on Fig. 9-6b: (1) the concentration gradient and (2) the pressure gradient. These opposing forces are exploited in RO. Consider a new experiment using the apparatus on Fig. 9-6, modified so that it is possible to exert an external force on the left side, as shown on

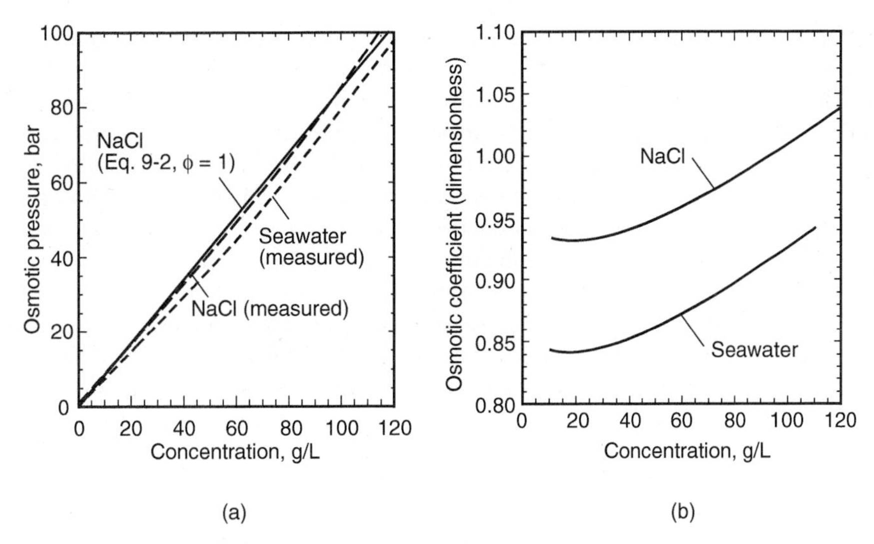

Figure 9-7
(a) Osmotic pressure of aqueous solutions of sodium chloride. (b) Osmotic coefficients for sodium chloride and seawater (calculation of osmotic pressure for seawater with the van't Hoff equation is based on a concentration of NaCl equal to the TDS of the seawater).

Fig. 9-6c. Applying a force equivalent to the osmotic pressure places the system in thermodynamic equilibrium, and no water flows. Applying a force in excess of the osmotic pressure places the system in nonequilibrium, with a pressure gradient exceeding the concentration gradient. Water would flow from left to right, that is, from the concentrated solution to the dilute solution. The process of causing water to flow from a concentrated solution to a dilute solution across a semipermeable membrane by the application of an external pressure in excess of the osmotic pressure is called reverse osmosis.

9-3 Mass Transfer of Water and Solutes through RO Membranes

The active layer of an RO membrane must selectively allow water to pass through the material while rejecting dissolved solutes that may be similar in size to water molecules. Separation cannot occur if water flows through pores in the membrane and small ions are carried convectively with the water. Thus, physical sieving is not a primary separation mechanism. The physicochemical processes that control separation in reverse osmosis are rooted in the principles of mass transfer that were presented in Chap. 4. This section describes a conceptual model of mass transfer through an RO membrane, mechanisms of solute rejection, equations for water and solute flux through the membrane, and equations for pressure and temperature corrections.

Mass Transfer through Dense Materials—The Solution–Diffusion Model

A number of models have been developed to describe mass transfer though RO membranes. The most common model envisions an RO membrane as a *dense* material (meaning a material that is not porous; there are no "holes" in it) that is nonetheless permeable. Water and solutes dissolve into the solid membrane material, diffuse through the solid, and reliquefy on the permeate side of the membrane. Dissolution of water and solutes into a solid material occurs if the solid is loose enough to allow individual water and solute molecules to travel along the interstices between polymer molecules of the membrane. Liquids behave as liquids because of attractive interactions with surrounding liquid molecules. Thus, even if water molecules travel along a defined path (which hypothetically could be called a pore), they are surrounded by polymer molecules and not other water molecules and therefore are dissolved in the solid, not present as a liquid phase. Diffusion occurs by movement of the water and solute molecules in the direction of the pressure and concentration (i.e., Gibbs energy) gradients. Separation occurs when the flux of the water is different from the flux of the solutes. Since the physicochemical phenomena that control separation are dissolution and diffusion in the solid phase, this model is known as the solution–diffusion model (Lonsdale et al., 1965).

Flux through the membrane is determined by both solubility and diffusivity. A review of the basic equation for mass transfer (Eq. 4-114 in Chap. 4) reveals that flux is the product of a mass transfer coefficient and a driving force (usually a concentration gradient). The mass transfer coefficient is a measure of the rate of diffusion through the membrane. The magnitude of the driving force depends on solubility of the material in the solid phase; that is, a material with very low solubility in the membrane will have a low value for the driving force. The solution–diffusion model predicts that separation occurs because the solubility, diffusivity, or both of the solutes are much lower than those of water, resulting in a lower solute concentration in the permeate than in the feed.

Mechanisms of Solute Rejection

With the solution–diffusion model as a framework, the physicochemical properties that allow RO membranes to separate solutes from water can be considered. The basic mechanisms of rejection are differences in the solubility and diffusivity in the membrane, along with electrostatic repulsion at the membrane surface. Solubility and diffusivity are affected by polarity, charge, and size.

Reverse osmosis membranes are often negatively charged in operation because of the presence of ionized functional groups, such as carboxylates, in the membrane material. Negatively charged ions may be rejected at the membrane surface due to electrostatic repulsion, and positively charged ions may be rejected to maintain electroneutrality in the feed and permeate solutions. The presence of polar functional groups in the membrane increases the solubility of polar compounds such as water over nonpolar compounds, providing a mechanism for greater flux of water through the membrane. Large molecules would be expected to be less soluble in the membrane material and also have lower diffusivity through it. If large enough, they may not be able to dissolve into the membrane material at all.

Experimental observations are consistent with these rejection mechanisms. Small polar molecules such as water generally have the highest flux. Dissolved gases such as H_2S and CO_2, which are small, uncharged, and polar, also permeate RO membranes well. Monovalent ions such as Na^+ and Cl^- permeate better than divalent ions (Ca^{2+}, Mg^{2+}) because the divalent ions have greater electrostatic repulsion. Reverse osmosis membranes are capable of rejecting up to 99 percent of monovalent ions. Nanofiltration membranes reject between 80 and 99 percent of divalent ions with considerably poorer rejection of monovalent ions.

Acids and bases (HCl, NaOH) permeate better than their salts (Na^+, Cl^-) because of decreased electrostatic repulsion. Boron (present in water as boric acid, H_3BO_3) and silica (present in water as silicic acid, H_4SiO_4) are weak acids that have poor rejection in their neutral forms. High rejection can be achieved by raising the pH above their pK_a values (9.3 for H_3BO_3 and 9.8 for H_4SiO_4).

Within a homologous series, mass transfer decreases with increasing molecular weight. High-molecular-weight organic materials do not permeate RO membranes at all.

Quantifying Solute Rejection

The rejection capabilities of RO and NF membranes are designated with either a percent salt rejection or a *molecular weight cutoff* (MWCO) value. Salt rejection is typically used for RO membranes:

$$\text{Rej} = 1 - \frac{C_P}{C_F} \tag{9-3}$$

where Rej = rejection, dimensionless (expressed as a fraction)
C_P = concentration in permeate, mg/L or mol/L
C_F = concentration in feed water, mg/L or mol/L

Rejection can be calculated for bulk parameters such as TDS or conductivity. For membrane rating, however, rejection of specific salts is specified. Sodium chloride rejection is normally specified for high-pressure RO membranes, whereas $MgSO_4$ rejection is often specified for NF or low-pressure RO membranes.

Nanofiltration membranes can also be characterized by MWCO. The MWCO of NF membranes is typically determined by measuring the passage of solutes of various size. The MWCO of NF membranes is typically 1000 Daltons (Da), also known as atomic mass units (amu), or less.

Equations for Water and Solute Flux

As noted earlier, the driving force for mass transfer is a difference in Gibbs energy, which includes terms for both pressure and concentration. The van't Hoff equation (Eq. 9-1) describes a relationship between pressure and concentration for the purpose of calculating osmotic pressure. Thus, for a particular constituent, the driving force can be described as either a difference in pressure or a difference in concentration. For mass transfer of water through RO membranes, the driving force is the net pressure difference:

$$\Delta P_{\text{NET}} = \Delta P - \Delta \pi = (P_F - P_P) - (\pi_F - \pi_P) \tag{9-4}$$

where ΔP_{NET} = net transmembrane pressure, bar
P_F, P_P = feed and permeate pressure, respectively, bar
π_F, π_P = feed and permeate osmotic pressure, respectively, bar

The water flux through RO membranes is then given by the expression

$$J_W = k_W(\Delta P - \Delta \pi) \tag{9-5}$$

where J_W = volumetric flux of water, $\text{L/m}^2 \cdot \text{h}$ ($\text{gal/ft}^2 \cdot \text{d}$)
k_W = mass transfer coefficient for water flux, $\text{L/m}^2 \cdot \text{h} \cdot \text{bar}$ ($\text{gal/ft}^2 \cdot \text{d} \cdot \text{atm}$)

The driving force for solute flux is the difference in concentration, and the flux of solutes through RO membranes is expressed as

$$J_S = k_S(\Delta C) \tag{9-6}$$

where J_S = mass flux of solute, $\text{mg/m}^2 \cdot \text{h}$
k_S = mass transfer coefficient for solute flux, $\text{L/m}^2 \cdot \text{h}$ or m/h
ΔC = difference in concentration across membrane, mg/L

The solute concentration in the permeate is the ratio of the fluxes of solutes and water, as shown by

$$C_P = \frac{J_S}{J_W} \tag{9-7}$$

Thus, the lower the flux of solutes or the higher the flux of water, the better removal of solutes is achieved and the permeate will have a lower solute concentration.

The ratio of permeate flow to feed water flow, or recovery, is calculated as

$$r = \frac{Q_P}{Q_F} \tag{9-8}$$

where Q = flow, m^3/s
r = recovery, dimensionless

Using flow and mass balance principles, the solute concentration in the concentrate stream can be calculated from the recovery and solute rejection. The pertinent flow and mass balances using flow and concentration terminology as shown on Fig. 9-1 are

$$\text{Flow balance:} \quad Q_F = Q_P + Q_C \tag{9-9}$$

$$\text{Mass balance:} \quad C_F Q_F = C_P Q_P + C_C Q_C \tag{9-10}$$

where C = concentration, mg/L or mol/L

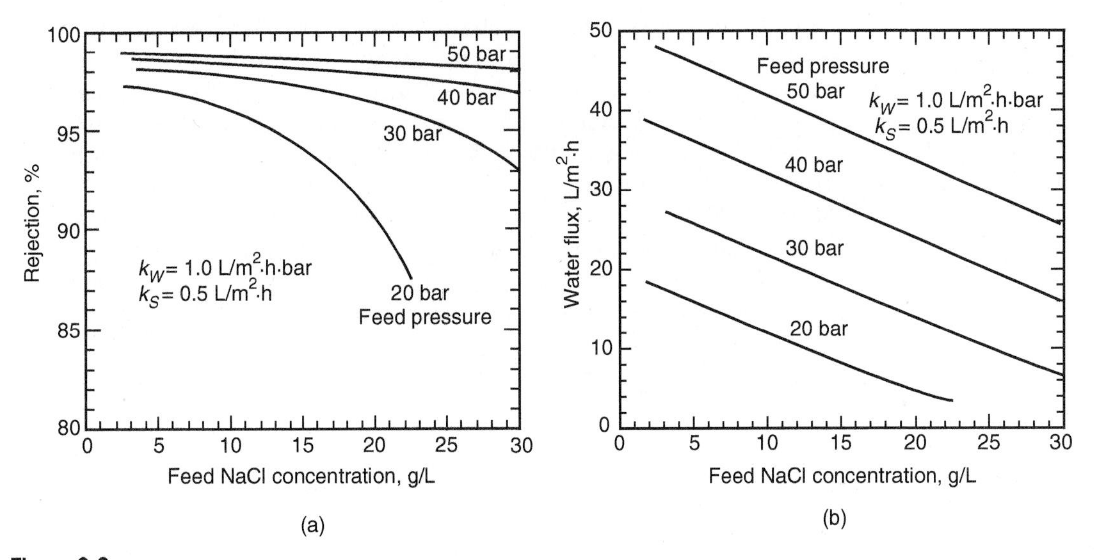

Figure 9-8
Effect of feed water concentration and pressure on (a) percent solute rejection and (b) water flux. The effect of concentration polarization, which is discussed in Sec. 9-5, is not considered in this figure.

Combining the mass and flow balances with Eq. 9-3 (rejection) and Eq. 9-8 (recovery) yields the following expression for the solute concentration in the concentrate stream:

$$C_C = C_F\left[\frac{1-(1-\text{Rej})r}{1-r}\right] \tag{9-11}$$

Rejection is frequently close to 100 percent, in which case Eq. 9-11 can be simplified as follows:

$$C_C = C_F\left(\frac{1}{1-r}\right) \tag{9-12}$$

In some textbooks, the ratio of concentrations in the concentrate and feed is known as the concentration factor (CF).

As shown in Eqs. 9-5 and 9-6, water flux depends on the pressure difference and solute flux depends on the concentration difference. As feed water solute concentration increases at constant pressure, the water flux decreases (because of higher $\Delta\pi$) and the solute flux increases (because of higher ΔC), which reduces rejection and causes a deterioration of permeate quality. As the feed water pressure increases, water flux increases but the solute flux is essentially constant. Therefore, as the water flux increases, the permeate solute concentration decreases, and the rejection increases. These relationships are illustrated on Fig. 9-8.

9-4 Performance Dependence on Temperature and Pressure

Membrane performance declines (water flux decreases, solute flux increases) due to fouling and membrane aging. However, fluxes of water and solute also vary because of changes in feed water temperature, pressure, velocity, and concentration. To evaluate the true decline in system performance due to fouling and aging, permeate flow rate and salt passage must be compared to baseline (standard) conditions. Reverse osmosis design manuals present equations for normalizing RO membrane performance in slightly different ways; the equations presented here are adapted from ASTM (2001b) and AWWA (2007). These procedures incorporate the use of temperature and pressure correction factors, evaluated at standard (subscript S) and measured (subscript M) conditions. The equations for standard permeate flow rate and salt passage are

$$Q_{P,S} = Q_{P,M}\,(\text{TCF})\,\frac{\text{NDP}_S}{\text{NDP}_M} \tag{9-13}$$

$$\text{SP}_S = \text{SP}_M\left(\frac{\text{NDP}_M}{\text{NDP}_S}\right)\left(\frac{C_{\text{FC},S}}{C_{\text{FC},M}}\right)\left(\frac{C_{F,M}}{C_{F,S}}\right) \tag{9-14}$$

where Q_P = permeate flow rate, m^3/h
TCF = temperature correction factor (defined below), dimensionless
NDP = net driving pressure (defined below), bar
SP = salt passage
C_F = feed concentration, mg/L
C_{FC} = average feed–concentrate concentration (defined below), mg/L

Salt passage is defined as the ratio of permeate concentration to feed concentration:

$$\text{SP} = \frac{C_P}{C_F} = 1 - \text{Rej} \tag{9-15}$$

Temperature affects the fluid viscosity and the membrane material. The relationship between membrane material, temperature, and flux is specific to individual products, so TCF values should normally be obtained from membrane manufacturers. If manufacturer TCF values are unavailable, the following relationship is often used:

$$\text{TCF} = (1.03)^{T_S - T_M} \tag{9-16}$$

where T = temperature, °C

The standard temperature is typically taken to be 25 °C for RO operation.

The net driving pressure accounts for changes in feed and permeate pressures, feed channel head loss, and osmotic pressure. In spiral-wound elements, the applied pressure decreases and osmotic pressure increases continuously along the length of the feed–concentrate channel as permeate flows through the membrane. Thus, the net driving pressure must take average conditions into account, as shown in the equation

$$\text{NDP} = \Delta P - \Delta\pi = \left(P_{\text{FC,ave}} - P_P\right) - \left(\pi_{\text{FC,ave}} - \pi_P\right) \tag{9-17}$$

where $P_{FC,ave}$ = average pressure in the feed–concentrate channel, bar
$= \frac{1}{2}(P_F + P_C)$
P_P, P_F, P_C = permeate, feed, and concentrate pressures, respectively, bar
$\pi_{FC,ave}$ = average feed–concentrate osmotic pressure (defined below), bar
π_P = permeate osmotic pressure, bar

Feed, concentrate, and permeate pressures are easily measured using system instrumentation. Osmotic pressure must be calculated from solute concentration using Eq. 9-2. Although osmotic pressure increases continuously along the length of a spiral-wound element, solute concentration normally can only be measured in the feed and concentrate streams. Manufacturers use various procedures for determining the average concentration in the feed–concentrate channel and should be contacted for the correct procedures for specific products. The two most common approaches for determining the average concentration in the feed channel are (1) an arithmetic average (Eq. 9-18) and (2) the log mean average (Eq. 9-19):

$$C_{FC,ave} = \frac{1}{2}(C_F + C_C) \quad (9\text{-}18)$$

$$C_{FC,ave} = \frac{C_F}{r} \ln\left(\frac{1}{1-r}\right) \quad (9\text{-}19)$$

where C_C = concentrate concentration, mg/L

Because head loss is a function of feed flow and osmotic pressure is a function of solute concentration, the system design must establish standard conditions for these parameters in addition to applied pressure.

In multistage systems, it is necessary to standardize the water flux and recovery for each stage independently. The procedures for standardizing RO performance data are shown in Example 9-2.

Example 9-2 Standardization of RO operating data

An RO system uses a shallow brackish groundwater that averages around 4500 mg/L TDS composed primarily of sodium chloride. Permeate flow is maintained constant, but temperature, pressure, and feed concentration change over time as shown in the table below. The operators need to determine whether fouling has occurred between January and May.

Parameter	Unit	January 1	May 31
Permeate flow	m^3/d	7500	7500
Feed pressure	bar	34.5	32.1
Concentrate pressure	bar	31.4	29.1
Permeate pressure	bar	0.25	0.25
Feed TDS concentration	mg/L	4612	4735
Permeate TDS concentration	mg/L	212	230
Recovery	%	69	72
Water temperature	°C	11	18

The pressure vessels contain seven membrane elements. The manufacturer has stated that performance data for this membrane element were developed using the following standard conditions:

Parameter	Unit	Standard
Temperature	°C	25
Feed pressure	bar	30
Permeate pressure	bar	0
Head loss per element	bar	0.4
Feed TDS concentration	mg/L	2000
Permeate TDS concentration	mg/L	100
Recovery	%	80

Determine the change in system performance (permeate flow and salt passage) that has occurred between January 1 and May 31. Assume $\phi = 1.0$.

Solution

1. Calculate the TCF for the January operating condition:

$$\text{TCF}_{\text{Jan}} = (1.03)^{T_S - T_M} = (1.03)^{25-11} = 1.512$$

2. Calculate the NDP for the January operating condition.
   a. Calculate the average molar solute concentration in the feed–concentrate channel using Eq. 9-19. A periodic table of the elements is available in App. D for calculating molar concentration.

$$C_{\text{CF,Jan}} = \frac{C_F}{r} \ln\left(\frac{1}{1-r}\right) = \frac{4612 \text{ mg/L}}{0.69} \ln\left(\frac{1}{1-0.69}\right)$$

$$= 7828 \text{ mg/L}$$

$$C_{\text{CF,Jan}} = \frac{(7828 \text{ mg/L})\,(2 \text{ mol ions/mol NaCl})}{(10^3 \text{ mg/g})(58.4 \text{ g/mol})} = 0.268 \text{ mol/L}$$

b. Calculate the osmotic pressure in the feed–concentrate channel using Eq. 9-2:

$$\pi_{CF,Jan} = \phi CRT$$

$$= (0.268 \text{ mol/L})(0.083145 \text{ L} \cdot \text{bar/K} \cdot \text{mol})(284 \text{ K})$$

$$= 6.33 \text{ bar}$$

c. Calculate the molar concentration and osmotic pressure in the permeate:

$$C_{P,Jan} = \frac{(212 \text{ mg/L})\,(2 \text{ mol ions/mol NaCl})}{(10^3 \text{ mg/g})(58.4 \text{ g/mol})} = 0.0073 \text{ mol/L}$$

$$\pi_{P,Jan} = (0.0073 \text{ mol/L})(0.083145 \text{ L} \cdot \text{bar/K} \cdot \text{mol})(284 \text{ K})$$

$$= 0.17 \text{ bar}$$

d. Calculate the NDP for the January operating condition using Eq. 9-17:

$$P_{FC,ave} = \tfrac{1}{2}\,(P_F + P_C) = \tfrac{1}{2}\,(34.5 + 31.4) = 32.95 \text{ bar}$$

$$\text{NDP} = (32.95 \text{ bar} - 0.25 \text{ bar}) - (6.33 \text{ bar} - 0.17 \text{ bar})$$

$$= 26.5 \text{ bar}$$

3. Repeat the calculations in steps 1 and 2 for the standard condition and the May operating condition. The concentrate pressure is not given for the standard operating condition but can be calculated from the given head loss information:

$$h_L = 0.4 \text{ bar/element (7 element)} = 2.8 \text{ bar}$$

$$P_C = 30 \text{ bars} - 2.8 \text{ bars} = 27.2 \text{ bar}$$

The remaining calculations are summarized in the table below:

Parameter	Unit	Standard Conditions	January 4 Conditions	May 23 Conditions
TCF		1.0	1.51	1.23
$C_{CF,ave}$	mg/L	4024	7828	8372
π_{CF}	bar	3.36	6.33	6.94
π_P	bar	0.08	0.17	0.19
$P_{CF,ave}$	bar	28.6	32.95	30.6
NDP	bar	25.3	26.5	23.6

4. Calculate the standard permeate flow for each date using Eq. 9-13:

$$Q_{W,S(Jan)} = 7500 \text{ m}^3/\text{d}\,(1.51)\left(\frac{25.3 \text{ bar}}{26.5 \text{ bar}}\right) = 10{,}800 \text{ m}^3/\text{d}$$

$$Q_{W,S(May)} = 7500 \text{ m}^3/\text{d}\,(1.23)\left(\frac{25.3 \text{ bar}}{23.6 \text{ bar}}\right) = 9{,}900 \text{ m}^3/\text{d}$$

5. Calculate the actual salt passage for each date using Eq. 9-15:

$$SP_{M,Jan} = \frac{212 \text{ mg/L}}{4612 \text{ mg/L}} = 0.046$$

$$SP_{M,May} = \frac{230 \text{ mg/L}}{4735 \text{ mg/L}} = 0.049$$

6. Calculate the standard salt passage for each date using Eq. 9-14:

$$SP_{S(Jan)} = (0.046)\left(\frac{26.5 \text{ bars}}{25.3 \text{ bars}}\right)\left(\frac{4612 \text{ mg/L}}{2000\text{mg/L}}\right)\left(\frac{4024 \text{ mg/L}}{7828 \text{ mg/L}}\right)$$

$$= 0.057$$

$$SP_{S(May)} = (0.049)\left(\frac{23.6 \text{ bars}}{25.3 \text{ bars}}\right)\left(\frac{4735 \text{ mg/L}}{2000 \text{ mg/L}}\right)\left(\frac{4024 \text{ mg/L}}{8372 \text{ mg/L}}\right)$$

$$= 0.052$$

Comment

Even though the membrane system is producing the same permeate flow with less pressure in May than in January, there has been an 8 percent loss of system performance because the standard permeate flow has declined from 10,800 to 9900 m^3/d. The standard salt passage also decreased between January and May, even though a higher permeate concentration was observed.

9-5 Concentration Polarization

Concentration polarization (CP) is the accumulation of solutes in a boundary layer near the membrane surface and has adverse effects on membrane performance. The flux of water through the membrane brings feed water (containing water and solute) to the membrane surface, and as clean water flows through the membrane, the solutes stay behind and form a boundary layer of higher concentration near the membrane surface. Thus, the concentration in the feed solution becomes polarized, with the concentration at the membrane surface higher than the concentration in the bulk feed water in the feed channel.

The increase of concentration near the membrane surface has several negative impacts on RO performance:

1. Water flux is lower because the osmotic pressure difference through the membrane is higher.
2. Rejection is lower because the flux of solutes through the membrane is higher (because of the increase in the concentration difference) and the water flux of water is lower.
3. Solubility limits of solutes may be exceeded, leading to precipitation and scaling.

Equations for concentration polarization can be derived using the same principles of mass transfer from Chap. 4 that were used to describe water and solute mass transfer through the membrane. In the membrane schematic shown on Fig. 9-9, feed water is traveling vertically on the left side of the membrane and water is passing through the membrane to the right. Water and solutes also move toward the membrane surface. As water passes through the membrane, the solute concentration at the membrane surface increases and creates a boundary layer. The concentration gradient in the boundary layer leads to diffusion of solutes back toward the bulk feed water. During continuous operation, a steady-state condition is reached in which the solute concentration at the membrane surface is constant with respect to time because the convective flow of solutes toward the membrane is

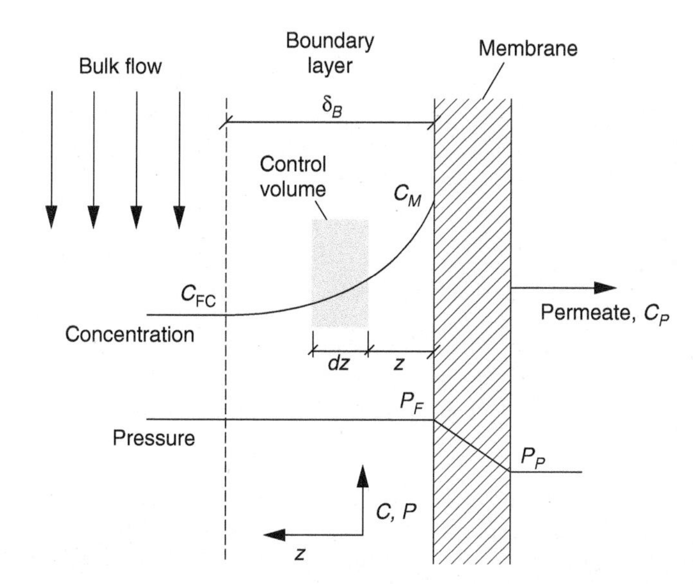

Figure 9-9
Schematic of concentration polarization.

balanced by the diffusive flow of solutes away from the surface. The solute flux toward the membrane surface due to the convective flow of water is described by the expression

$$J_S = J_W C \qquad (9\text{-}20)$$

A mass balance can be developed at the membrane surface as follows:

$$[\text{accum}] = [\text{mass in}] - [\text{mass out}] \qquad (9\text{-}21)$$

With no accumulation of mass at steady state, the solute flux toward the membrane surface must be balanced by fluxes of solute flowing away from the membrane (due to diffusion) and through the membrane (into the permeate) as follows:

$$\frac{dM}{dt} = 0 = J_W Ca - D_L \frac{dC}{dz} a - J_W C_P a \qquad (9\text{-}22)$$

where M = mass of solute, g
t = time, s
D_L = diffusion coefficient for solute in water, m^2/s
z = distance perpendicular to membrane surface, m
a = surface area of membrane, m^2

Equation 9-22 applies not only at the membrane surface but also at any plane in the boundary layer because the net solute flux must be constant throughout the boundary layer to prevent the accumulation of solute anywhere within that layer (the last term in Eq. 9-22 represents the solute that must pass through the boundary layer and the membrane to end up in the permeate). Rearranging and integrating Eq. 9-22 across the thickness of the boundary layer with the boundary conditions $C(0) = C_M$ and $C(\delta_B) = C_{FC}$, where C_{FC} is the concentration in the feed–concentrate channel and C_M is the concentration at the membrane surface, are done in the following equations:

$$D_L \int_{C_M}^{C_{FC}} \frac{dC}{C - C_P} = -J_W \int_0^{\delta_B} dz \qquad (9\text{-}23)$$

Integrating yields

$$\ln\left(\frac{C_M - C_P}{C_{FC} - C_P}\right) = \frac{J_W \delta_B}{D_L} \qquad (9\text{-}24)$$

$$\frac{C_M - C_P}{C_{FC} - C_P} = e^{J_W \delta_B / D_L} = e^{J_W / k_{CP}} \qquad (9\text{-}25)$$

where $k_{CP} = D_L/\delta_B$ = concentration polarization mass transfer coefficient, m/s

The concentration polarization mass transfer coefficient describes the diffusion of solutes away from the membrane surface. Concentration polarization is expressed as the ratio of the membrane and feed–concentrate channel solute concentrations as follows:

$$\beta = \frac{C_M}{C_{FC}} \tag{9-26}$$

where β = concentration polarization factor, dimensionless

Combining Eq. 9-26 with Eqs. 9-3 and 9-25 results in the following expression:

$$\beta = (1 - \text{Rej}) + \text{Rej}\left(e^{J_W/k_{CP}}\right) \tag{9-27}$$

If rejection is high (greater than 99 percent), Eq. 9-27 can be reasonably simplified as follows:

$$\beta = e^{J_W/k_{CP}} \tag{9-28}$$

To predict the extent of concentration polarization, the value of the concentration polarization mass transfer coefficient is needed. As demonstrated in Sec. 4-16, mass transfer coefficients are often calculated using a correlation between Sherwood (Sh), Reynolds (Re), and Schmidt (Sc) numbers. Correlations for mass transfer coefficients depend on physical characteristics of the system and the flow conditions (e.g., laminar or turbulent). To promote turbulent conditions and minimize concentration polarization in RO membrane elements, spiral-wound elements contain mesh feed channel spacers and maintain a high velocity flow parallel to the membrane surface. The feed channel spacer complicates the flow patterns and promotes turbulence. The superficial velocity (assuming an empty channel) in a spiral-wound element typically ranges from 0.02 to 0.2 m/s (0.066 to 0.66 ft/s), but the actual velocity is higher because of the space taken up by the spacer.

Because the mesh spacer affects mass transfer in the feed channel and many feed spacer configurations have been developed, many correlations have been developed to calculate the mass transfer coefficient. Mariñas and Urama (1996) developed a correlation using the channel height and the superficial velocity, which eliminates the task of determining the parameters of the spacer. Their correlation is

$$k_{CP} = \lambda \frac{D_L}{d_H} (\text{Re})^{0.50} (\text{Sc})^{1/3} \tag{9-29}$$

$$\text{Re} = \frac{\rho v d_H}{\mu} \tag{9-30}$$

$$\text{Sc} = \frac{\mu}{\rho D_L} \tag{9-31}$$

where Re = Reynolds number, dimensionless
Sc = Schmidt number, dimensionless
v = velocity in feed channel, m/s
ρ = feed water density, kg/m^3
μ = feed water dynamic viscosity, kg/m·s

d_H = hydraulic diameter, m
λ = empirical parameter ranging from 0.40 to 0.54 for different elements, dimensionless

The hydraulic diameter is defined as

$$d_H = \frac{4(\text{volume of flow channel})}{\text{wetted surface}} \tag{9-32}$$

For hollow-fiber membranes (circular cross section), the hydraulic diameter is equal to the inside fiber diameter. Spiral-wound membranes can be approximated by flow through a slit, where the width is much larger than the feed channel height ($w \gg h$). In an empty channel (i.e., the spacer is neglected), the hydraulic diameter is twice the feed channel height, as shown in the equation

$$d_H = \frac{4wh}{2w + 2h} \approx 2h \tag{9-33}$$

where h = feed channel height, m
w = channel width, m

The feed channel height in typical spiral-wound elements ranges from about 0.4 to 1.2 mm (0.016 to 0.047 in.) and is governed by the thickness of the spacer.

Concentration polarization varies along the length of a membrane element; the parameters that change most significantly are the velocity in the feed channel (v) and the permeate flux (J_W). As might be expected, concentration polarization increases as the permeate flux increases and as the velocity in the feed channel decreases. The maximum concentration polarization allowed for membrane elements is specified by manufacturers; $\beta = 1.2$ is a typical value. Calculation of the concentration polarization factor is illustrated in Example 9-3.

Example 9-3 Concentration polarization

For a spiral-wound element, calculate the concentration polarization factor and the concentration of sodium at the membrane surface given the following information: water temperature 20°C, feed channel velocity 0.15 m/s, feed channel height 0.86 mm, permeate flux 25 L/m^2 · h, sodium concentration 6000 mg/L, and diffusivity of sodium in water 1.58×10^{-9} m^2/s. Use the correlation in Eq. 9-29 and a value of 0.47 for the coefficient. Assume that the rejection is high enough that the impact of sodium flux through the membrane can be ignored.

Solution

1. Calculate the Reynolds and Schmidt numbers using Eqs. 9-30 and 9-31. Because the feed channel height is 0.86 mm, the hydraulic diameter is 1.72 mm. Water density and viscosity at 20°C can be found in App. C, $\rho = 998 \text{ kg/m}^3$, and $\mu = 1.0 \times 10^{-3} \text{ kg/m}\cdot\text{s}$:

$$\text{Re} = \frac{\rho v d_H}{\mu} = \frac{(998 \text{ kg/m}^3)(0.15 \text{ m/s})(1.72 \text{ mm})}{(1.0 \times 10^{-3} \text{ kg/m}\cdot\text{s})(10^3 \text{ mm/m})} = 257$$

$$\text{Sc} = \frac{\mu}{\rho D_L} = \frac{1.0 \times 10^{-3} \text{ kg/m}\cdot\text{s}}{(998 \text{ kg/m}^3)(1.58 \times 10^{-9} \text{ m}^2/\text{s})} = 634$$

2. Calculate k_{CP} using Eq. 9-29:

$$k_{CP} = \frac{\lambda D_L (\text{Re})^{0.5} (\text{Sc})^{1/3}}{d_H}$$

$$= \frac{(0.47)(1.58 \times 10^{-9} \text{ m}^2/\text{s})(257)^{0.5}(634)^{1/3}}{(1.72 \text{ mm})(10^{-3} \text{ m/mm})}$$

$$= 5.95 \times 10^{-5} \text{ m/s}$$

3. Because the rejection is high, β can be calculated using Eq. 9-28 (otherwise, Eq. 9-27 must be used):

$$\beta = \exp\left(\frac{J_W}{k_{CP}}\right) = \exp\left[\frac{(25 \text{ L/m}^2\cdot\text{h})(10^{-3} \text{ m}^3/\text{L})}{(5.95 \times 10^{-5} \text{ m/s})(3600 \text{ s/h})}\right] = 1.12$$

4. Calculate the sodium concentration at the membrane surface using Eq. 9-26:

$$C_M = (1.12)(6000 \text{ mg/L}) = 6720 \text{ mg/L}$$

9-6 Fouling and Scaling

Nanofiltration and RO membranes are susceptible to fouling via a variety of mechanisms. The primary sources of fouling and scaling are particulate matter, precipitation of inorganic salts (known as scaling), oxidation of soluble metals, and biological matter.

Fouling by Particulate Matter

Particulate fouling is a concern in RO because the operational cycle does not include a backwashing step to remove accumulated solids. Both inorganic and organic materials, including microbial constituents and biological debris, can cause particulate fouling. Particles can clog the feed channels and piping. Although the mesh spacers in spiral-wound elements are designed to minimize plugging, an excessive load of particles may cause plugging anyway. Particulate matter forming a cake on the membrane surface also adds resistance to flow and affects system performance.

The tendency for fouling by particles is assessed with an empirical test known as the silt density index (SDI). The SDI (ASTM, 2001a) is a timed filtration test using three time intervals through a 0.45-μm membrane filter at a constant applied pressure of 2.07 bars (30 psi). The first interval is the duration necessary to collect 500 mL of permeate. Filtration continues through the second interval without recording volume until 15 min (total) have elapsed. At the end of 15 min, the third interval is started, during which an additional 500-mL aliquot of water is filtered through the now dirty membrane, and the time is recorded. The SDI is calculated from these time intervals:

$$\mathrm{SDI} = \frac{100(1 - t_I/t_F)}{t_T} \tag{9-34}$$

where SDI = silt density index, min^{-1}
t_I = time to collect first 500-mL sample, min
t_F = time to collect final 500-mL sample, min
t_T = duration of first two test intervals (15 min)

The SDI has been criticized as being too simplistic to accurately predict RO membrane fouling. Other tests, such as the modified fouling index (MFI) and its variants, have attempted to improve on the SDI but still do not accurately predict particulate fouling. These tests might best be considered as screening tools that provide rough guidelines for acceptable feed water quality. A high value is a good indicator of fouling problems in RO systems, but a low value does not necessarily mean the source water has a low fouling tendency. RO manufacturers typically specify a maximum SDI value of 4 to 5 min^{-1}.

Virtually all RO systems require pretreatment to minimize particulate fouling. Prefiltration through 5-μm cartridge filters is considered to be the minimum pretreatment for protection of the membrane elements. Source waters with excessive potential for particulate fouling require advanced pretreatment to lower the particulate concentration to an acceptable level. Coagulation and filtration (using sand, carbon, or other filter media) or membrane filtration are sometimes used for pretreatment. Pilot tests are often necessary to determine the appropriate level of pretreatment. Fouling by residual particulate matter also affects the cleaning frequency.

Scaling from Precipitation of Inorganic Salts

Inorganic scaling occurs when salts in solution are concentrated beyond their solubility limits and form precipitates. Common sparingly soluble salts are listed in Table 9-2. If the ions that comprise these salts are concentrated past the solubility product, precipitation occurs. Precipitation reactions

Table 9-2
Typical limiting salts and their solubility products

Salt	Equation	Solubility Product[a] (pK_{sp} at 25°C)
Calcium carbonate (calcite)	$CaCO_3(s) \rightleftarrows Ca^{2+} + CO_3^{2-}$	8.48
Calcium fluoride	$CaF_2(s) \rightleftarrows Ca^{2+} + 2F^-$	10.5
Calcium orthophosphate	$CaHPO_4(s) \rightleftarrows Ca^{2+} + HPO_4^{2-}$	6.90
Calcium sulfate (gypsum)	$CaSO_4{\cdot}2H_2O(s) \rightleftarrows Ca^{2+} + SO_4^{2-} + 2H_2O$	4.61
Strontium sulfate (celestite)	$SrSO_4(s) \rightleftarrows Sr^{2+} + SO_4^{2-}$	6.62
Barium sulfate (barite)	$BaSO_4(s) \rightleftarrows Ba^{2+} + SO_4^{2-}$	9.98
Silica, amorphous	$SiO_2(s) + 2H_2O \rightleftarrows Si(OH)_4(aq)$	2.71

[a]Data from Gustafsson (2011).

were introduced in Sec. 4-4. The precipitation reaction for a typical salt is as follows:

$$CaSO_4(s) \rightleftarrows Ca^{2+} + SO_4{}^{2-} \tag{9-35}$$

The solubility product is written as

$$K_{SP} = \{Ca^{2+}\}\{SO_4{}^{2-}\} = \gamma_{Ca}[Ca^{2+}]\gamma_{SO_4}[SO_4{}^{2-}] \tag{9-36}$$

where K_{SP} = solubility product
$\{Ca^{2+}\}, \{SO_4{}^{2-}\}$ = activity of calcium and sulfate
$\gamma_{Ca}, \gamma_{SO_4}$ = activity coefficients for calcium and sulfate
$[Ca^{2+}], [SO_4{}^{2-}]$ = concentration of calcium and sulfate, mol/L

As discussed in Sec. 4-2, it is common in water treatment applications involving freshwater sources to assume that activity is equal to concentration and skip the calculation of activity coefficients. The ionic strength of brackish and saline waters, however, is too high to make this simplification, and activity coefficients must be used for all water chemistry calculations in RO systems.

Equation 9-12 demonstrates how the concentration of ions in the concentrate depends on recovery; the higher the recovery, the more concentrated constituents become. Thus, placing a limit on the recovery is often necessary to prevent precipitation. The highest recovery possible before any salts precipitate is the *allowable recovery*, and the salt that precipitates at this condition is the *limiting salt*. The most common scales encountered in water treatment applications are calcium carbonate ($CaCO_3$), calcium sulfate ($CaSO_4$), and silica (SiO_2).

The allowable recovery that can be achieved without pretreatment in RO is determined by performing solubility calculations for each of the possible limiting salts. The highest solute concentrations occur in the final membrane element immediately prior to the water exiting the system as the concentrate stream, so concentrate stream concentrations are used to evaluate solubility limits. In addition, the concentration in the concentrate steam must be adjusted for the level of concentration polarization that is occurring. Incorporating the concentration polarization factor defined in Eq. 9-26 with the expression for the solute concentration in the concentrate stream defined by Eq. 9-11 yields

$$C_M = \beta C_F \left[\frac{1 - (1 - \text{Rej})r}{1 - r} \right] \tag{9-37}$$

Allowable recovery is determined by substituting the activities at the membrane into a solubility product calculation and solving for the recovery, as demonstrated in Example 9-4.

Example 9-4 Allowable recovery from limiting salt calculations

Determine the limiting salt and allowable recovery for a brackish water RO system containing the following solutes: calcium 74 mg/L, barium 0.008 mg/L, and sulfate 68 mg/L. Assume 100 percent rejection of all solutes and a polarization factor of 1.15. While activity coefficients cannot be ignored in actual applications, they are ignored in this example (i.e., activity = concentration) so that the use of solubility to determine the allowable recovery can be demonstrated.

Solution

1. Calculate the molar concentration for each component. A periodic table of the elements is available in App. D.

$$[Ca^{2+}] = \frac{74 \text{ mg/L}}{(40 \text{ g/mol})(10^3 \text{ mg/g})} = 1.85 \times 10^{-3} \text{ mol/L}$$

$$[Ba^{2+}] = \frac{0.008 \text{ mg/L}}{(137.3 \text{ g/mol})(10^3 \text{ mg/g})} = 5.83 \times 10^{-8} \text{ mol/L}$$

$$[SO_4^{2-}] = \frac{68 \text{ mg/L}}{(96 \text{ g/mol})(10^3 \text{ mg/g})} = 7.08 \times 10^{-4} \text{ mol/L}$$

2. Simplify the expression for concentration at the membrane. Let $y = 1 - r$. Because Rej = 1, Eq. 9-37 becomes

$$C_M = \frac{\beta C_F}{y}$$

3. Substitute the concentrations at the membrane surface into the equation for solubility products and calculate recovery. Solubility product constants are available in Table 9-2.

a. For calcium sulfate,

$$K_{sp} = 10^{-4.61} = [Ca^{2+}]_M[SO_4^{2-}]_M = \left(\frac{\beta[Ca^{2+}]_F}{y}\right)\left(\frac{\beta[SO_4^{2-}]_F}{y}\right)$$

$$= \frac{\beta^2}{y^2}[Ca^{2+}]_F[SO_4^{2-}]_F$$

$$y = \left(\frac{\beta^2}{K_{sp}}[Ca^{2+}]_F[SO_4^{2-}]_F\right)^{1/2}$$

$$= \left[\frac{(1.15)^2}{10^{-4.61}}(1.85 \times 10^{-3}\ \text{mol/L})(7.08 \times 10^{-4}\ \text{mol/L})\right]^{1/2}$$

$$= 0.27$$

$$r = 1 - y = 1 - 0.27 = 0.73$$

b. For barium sulfate,

$$y = \left[\frac{(1.15)^2}{10^{-9.98}}(5.83 \times 10^{-8}\ \text{mol/L})(7.08 \times 10^{-4}\ \text{mol/L})\right]^{1/2}$$

$$= 0.72$$

$$r = 1 - y = 1 - 0.72 = 0.28$$

Comments

The allowable recovery before barium sulfate precipitates is 28 percent, compared to 73 percent before calcium sulfate precipitates. Therefore, barium sulfate is the limiting salt and the allowable recovery is 28 percent.

Ignoring activity coefficients in Example 9-4 permitted the allowable recovery to be calculated directly. Had activity coefficients been included, it would have been necessary to use an iterative or simultaneous solution procedure because the ionic strength depends on recovery, so the activity coefficients cannot be calculated until the recovery is known. Additional factors that complicate limiting salt calculations are pH and the formation of ion complexes. The concentrations of two important anions, carbonate and phosphate, depend on pH. Ion complexes increase solubility by decreasing the concentration of the free ions used in solubility product calculations (e.g., calcium and sulfate form a neutral $CaSO_4{}^0$ species that increases the solubility of $CaSO_4(s)$).

Temperature, supersaturation, the use of antiscalants, and the necessity to compare multiple treatment scenarios further complicate these calculations. Thus, the computational requirements of limiting salt calculations can be daunting and membrane manufacturers provide computer programs to perform these calculations. These programs account for the concentration polarization factor and rejection capabilities of specific products, temperature and pH effects, and the degree of supersaturation that can be accommodated with various pretreatment strategies.

In addition to limiting recovery, common pretreatment in virtually all RO systems to prevent scaling includes acid and antiscalant addition. Calcium carbonate precipitation can be prevented by adjusting the pH of the feed stream with acid to convert carbonate to bicarbonate and carbon dioxide. The pH of most RO feed waters is adjusted to a pH value of 5.5 to 6.0. At this pH, most carbonate is in the form of carbon dioxide and passes through the membrane. Antiscalants allow supersaturation without precipitation by preventing crystal formation and growth. The degree of supersaturation allowed because of antiscalant addition depends on properties of the antiscalant, which are often proprietary, and characteristics of specific equipment configurations. It is appropriate to rely on the recommendations of equipment and antiscalant manufacturers when determining appropriate antiscalant selection and doses necessary for a specific feed water analysis and design recovery. Silica scaling can be particularly challenging and if high concentrations are present, high-pH softening may be necessary to remove silica from the feed water to prevent precipitation on the membrane.

Fouling from Oxidation and Precipitation of Soluble Metals

Groundwater used as the source water for RO systems is often anaerobic. Iron and manganese, soluble compounds in their reduced states, can oxidize, precipitate, and foul membranes if any oxidants, including oxygen, enter the feed water system. Fouling may be avoided by preventing oxidation or removing the iron or manganese after oxidation. If iron concentrations are low, precautions to prevent air from entering the feed system may be sufficient because antiscalants often include additives to minimize fouling by low concentrations of iron. Pretreatment to remove iron might include oxidation with oxygen or chlorine followed by adequate mixing and hydraulic detention time and granular media or membrane filtration or greensand filtration in which oxidation and filtration take place simultaneously.

Biological Fouling

Biological fouling refers to the attachment or growth of microorganisms or extracellular soluble material on the membrane surface or in the membrane element feed channels. Biological fouling can lead to lower flux, reduced solute rejection, increased head loss through the membrane modules, contamination of the permeate, degradation of the membrane material, and reduced membrane life (Ridgway and Flemming, 1996). The primary source of microbial contamination is the feed water. Biological fouling is a significant problem in many RO systems.

Biological fouling is prevented by maintaining proper operating conditions, applying biocides, and flushing membrane elements properly when not in use. Many RO feed waters (groundwater in many cases) have low microbial populations. When operated properly, the shear in the feed channels helps to keep bacteria from accumulating or growing to unacceptable levels. When membrane trains are out of service, however, bacteria can quickly multiply. To avoid this problem, membranes should be flushed with permeate periodically or filled with an approved biocide if out of service for any significant period. An excellent review of the issues involved in biological fouling of membranes is provided in Ridgway and Flemming (1996).

9-7 Element Selection and Membrane Array Design

The basis for design of an RO system typically includes characteristics of the feed water (solute concentrations, turbidity, SDI values) from laboratory or historical data, required treated-water quality (established by the client or regulatory limits), and required treated-water capacity (established by demand requirements). Design is typically done with the assistance of manufacturer's design programs and pilot testing.

Manufacturer Design Programs

Membrane array design involves determination of the quantity and quality of water produced by each membrane element in an array. This involves calculation of the flow, velocity, applied pressure, osmotic pressure, water flux, and solute flux in each element, which leads to the determination of the number of stages, number of passes, number of elements in each pressure vessel, and number of vessels in each stage. Membrane array design is a complex and iterative process using a large number of interrelated design parameters. Several important design parameters such as mass transfer coefficients are specific to individual products and available only from membrane manufacturers. Thus, design is an iterative process and typically takes place with the cooperation of several membrane system manufacturers. Because of the complexity of the calculations and dependence on manufacturer information, array design is often done with design software provided by membrane manufacturers. This software is based on the principles presented in this chapter and incorporates issues such as osmotic pressure, limiting salt solubility, mass transfer rates, concentration polarization, and permeate water quality. Other process parameters, such as permeate backpressure and interstage booster pumps, can be incorporated into the design. As such, manufacturers' software is reliable for predicting effluent water quality from a specific membrane system design and a given set of operating conditions. An example of the output from a vendor-supplied RO design program is shown in Table 9-3.

Table 9-3
Example output from vendor-supplied RO design program[a]

Hydranautics Membrane System Design Software, v. 8.00 © 2002 3/11/03
RO program licensed to: K Howe
Calculation created by: K Howe
Project name: Example

HP pump flow:	4666.7 gpm	Permeate flow:	3500.0 gpm
Recommended pump press:	204.4 psi	Raw-water flow:	4666.7 gpm
Feed pressure:	175.4 psi	Booster pump pressure:	10.0 psi
Feed water temperature:	15.0 C(59F)	Permeate recovery ratio:	75.0%
Raw-water pH:	8.00	Element age:	5.0 years
Acid dosage, ppm (100%):	131.1 H_2SO_4	Flux decline % per year:	7.0
Acidified feed CO_2:	127.3	Salt passage increase, %/yr:	10.0
Average flux rate:	15.8 gfd	Feed type:	Well water

Stage	Perm. Flow, gpm	Flow/Vessel Feed, gpm	Flow/Vessel Conc, gpm	Flux, gfd	Beta	Concentration and Throt. Pressures psi	Concentration and Throt. Pressures psi	Element Type	Element No.	Array
1-1	2623.6	53.0	23.2	17.9	1.16	149.5	0.0	ESPA3	528	88 × 6
1-2	876.4	45.4	25.9	11.7	1.08	133.1	0.0	ESPA3	270	45 × 6

Ion	Raw water mg/L	Raw water $CaCO_3$	Feed water mg/L	Feed water $CaCO_3$	Permeate mg/L	Permeate $CaCO_3$	Concentrate mg/L	Concentrate $CaCO_3$
Ca	8.0	20.0	8.0	20.0	0.27	0.7	31.2	77.7
Mg	2.0	8.2	2.0	8.2	0.07	0.3	7.8	32.1
Na	734.3	1596.3	734.3	1596.3	115.11	250.2	2591.9	5634.5
K	8.0	10.3	8.0	10.3	1.52	2.0	27.4	35.2
NH_4	0.0	0.0	0.0	0.0	0.00	0.0	0.0	0.0
Ba	0.004	0.0	0.004	0.0	0.000	0.0	0.016	0.0
Sr	2.000	2.3	2.000	2.3	0.069	0.1	7.794	8.9
CO_3	3.0	5.0	0.2	0.4	0.00	0.0	0.8	1.4
HCO_3	631.0	517.2	473.5	388.1	174.26	142.8	1371.3	1124.0
SO_4	79.0	82.3	207.5	216.1	7.41	7.7	807.7	841.3
Cl	730.0	1029.6	730.0	1029.6	72.28	101.9	2703.2	3812.6
F	1.1	2.9	1.1	2.9	0.28	0.7	3.6	9.4
NO_3	0.0	0.0	0.0	0.0	0.00	0.0	0.0	0.0
SiO_2	24.0			24.0	5.83		78.5	
TDS	2222.4		2190.6		377.1		7631.2	
pH	8.0		6.8		6.4		7.3	

	Raw Water	Feed Water	Concentrate
$CaSO_4/K_{sp} \times 100$:	0%	0%	2%
$SrSO_4/K_{sp} \times 100$:	2%	5%	29%
$BaSO_{-4}/K_{sp} \times 100$:	7%	17%	97%
SiO_2 saturation:	20%	20%	65%
Langelier saturation index (LSI)	−0.14	−1.47	0.04
Stiff–Davis saturation index	−0.20	−1.53	−0.24
Ionic strength	0.03	0.04	0.13
Osmotic pressure	22.2 psi	21.3 psi	74.2 psi

Table 9-3 *(Continued)*

Stage	Element No.	Feed Pressure, psi	Pressure Drop, psi	Permeate Flow, gpm	Permeate Flux, gfd	Beta	Permeate TDS	Concentrate Osmotic Pressure	Concentrate Saturation Level, %				
									$CaSO_4$	$SrSO_4$	$BaSO_4$	SiO_2	LSI
1-1	1	175.4	6.5	5.7	20.5	1.11	116.6	23.8	1	6	20	22	−0.9
1-1	2	168.9	5.5	5.4	19.4	1.12	126.5	26.7	1	7	23	25	−0.7
1-1	3	163.4	4.6	5.1	18.3	1.12	137.8	30.2	1	8	27	28	−0.6
1-1	4	158.8	3.8	4.8	17.2	1.13	151.0	34.4	2	9	32	32	−0.4
1-1	5	155.0	3.1	4.5	16.1	1.15	166.2	39.6	2	11	38	36	−0.3
1-1	6	151.8	2.5	4.1	14.9	1.16	203.0	45.9	2	14	47	42	−0.1
1-2	1	156.3	5.4	4.1	14.6	1.09	225.4	49.8	3	16	52	45	0.0
1-2	2	150.9	4.7	3.7	13.4	1.09	251.4	54.0	3	18	59	49	0.1
1-2	3	146.3	4.1	3.4	12.2	1.09	279.6	58.5	3	20	66	53	0.1
1-2	4	142.1	3.6	3.1	11.1	1.09	309.1	63.2	4	22	74	56	0.2
1-2	5	138.5	3.2	2.8	10.0	1.09	341.4	68.2	4	25	84	60	0.3
1-2	6	135.4	2.8	2.5	8.9	1.08	374.9	73.3	5	28	94	64	0.3

[a]These calculations are based on nominal element performance when operated on a feed water of acceptable quality. No guarantee of system performance is expressed or implied unless provided in writing by Hydranautics.
[b]The manufacturer's output expresses each concentration in units of mg/L as $CaCO_3$ in addition to mg/L.

Pilot Testing

An important aspect of long-term RO operation is loss of performance due to compaction, fouling, or degradation of the membrane. Unfortunately, fouling cannot be quantitatively predicted from water quality measurements, and parameters such as SDI provide only a general indication of the severity of fouling. Therefore, it is necessary to perform pilot testing for nearly all RO installations. Pilot testing is guided by membrane system selection and operating conditions developed during array design and serves to verify the array design criteria and identify pretreatment requirements to prevent excessive fouling. Reverse osmosis pilot plant systems are typically available from membrane manufacturers or consulting engineering firms. A typical commercially available skid-mount pilot system is shown on Fig. 9-10.

9-8 Energy and Sustainability Considerations

Reverse osmosis has the most significant impact on sustainability and energy consumption of any process in this book. Reverse osmosis uses the most energy and often has the lowest water recovery and highest waste production of any common water treatment process.

The feed pumps consume the most energy in an RO plant. The feed pressure is dictated by the osmotic pressure at the concentrate end of the pressure vessels with enough additional pressure to overcome head loss and provide a driving force for mass transfer through the membranes. For a brackish water system with 2000 mg/L of TDS, the feed water

Figure 9-10
Typical reverse osmosis pilot plant.

osmotic pressure is about 1.7 bar (24 psi). With 80 percent recovery and a concentration polarization factor (β) of 1.1, the osmotic pressure at the membrane surface at the concentrate end of the pressure vessel is 9.4 bar (135 psi). With additional pressure to overcome head loss and to provide a driving force, a feed pressure of at least 10 bar (145 psi) would be needed. An RO system treating seawater with a TDS of 35,000 mg/L at 50 percent recovery and $\beta = 1.1$ has to overcome an osmotic pressure of about 56 bar (810 psi). Some seawater RO systems operate at feed pressures as high as 85 bar (1,230 psi). The feed pumps on such a system would consume 5.6 kWh/m^3, or 30 to 40 times the energy consumption of a conventional surface water filtration plant.

It is important to realize that osmotic pressure is a fundamental thermodynamic limitation that cannot be overcome by advances in membrane technology. Better membranes can improve separation, increase the rate of mass transfer, reduce the size of plants, and improve cost effectiveness, but cannot change the basic thermodynamics of osmotic pressure.

Nevertheless, RO systems can be designed to reduce energy consumption. Two or more stages with booster pumps between stages allow for lower pressure feed pumps, so water in the first stage is produced at lower pressure, and the pressure is increased in each subsequent stage as the osmotic pressure increases. The result is lower overall energy consumption. Energy recovery devices are also an important part of design. Ultimately, however, the best way to reduce energy consumption from an RO system is to select the water supply with the lowest possible TDS (and, therefore, the lowest osmotic pressure) if multiple water supplies are available.

A second significant environmental concern for RO facilities is the low product water recovery compared to other water treatment processes. Most common water treatment processes achieve 95 to 99 percent water recovery. Reverse osmosis systems rarely achieve above 85 percent recovery and seawater systems typically achieve only 50 percent recovery. The remaining 15 to 50 percent of the water is a waste product containing all of the salts of the feed water. For inland systems, the low recovery has two negative consequences. First, an inability to recover a high fraction of the feed water is simply a poor use of scarce natural resources. Second, the unrecovered water becomes the concentrate stream that must be disposed of. The high salinity of the concentrate stream greatly limits the disposal options because of the potential for contaminating the scarce freshwater resources. Thus, increasing recovery of product water and decreasing the volume of concentrate is an area of active research.

Increasing recovery from inland brackish water RO facilities involves preventing the precipitation of sparingly soluble salts. As noted earlier, scale inhibitors are used to prevent precipitation and increase recovery up to a point. However, scale inhibitors are limited in their effectiveness, and more aggressive strategies typically must be employed to achieve recovery of greater than 90 percent.

One strategy is to provide an intermediate treatment process between two stages of RO membranes. Since calcium is often the limiting cation, lime softening can be an effective intermediate strategy. Softening can also be effective at removing other scale-causing constituents, including barium, strontium, and silica. However, the high alkalinity and hardness present after a first stage of RO can lead to high doses of lime or NaOH; doses in excess of 1000 mg/L have been reported in experimental studies. Similarly high doses of acid can be necessary to reduce the pH after softening. The high doses also lead to a large amount of lime sludge, another waste stream. Thus, the cost and waste production of interstage treatment must be balanced against the reduced waste production resulting from higher water recovery.

Brine concentrators and crystallizers are additional technologies to reduce the volume of concentrate and can lead to zero liquid discharge (ZLD), in which the only residuals from the facility are solids (Mickley, 2006). While brine concentrators and crystallizers are used in some industrial processes such as the power generation industry, they are expensive, energy intensive, and have not yet been used in municipal water treatment industry.

9-9 Summary and Study Guide

After studying this chapter, you should be able to:

1. Define the following terms and phrases and describe the significance of each in the context of reverse osmosis and water treatment:

allowable recovery	membrane element	semipermeable membrane
array	nanofiltration	silt density index
asymmetric membrane	osmosis	spiral-wound element
concentrate	osmotic pressure	stage
concentration polarization	permeate	thin-film composite
dense membrane	reverse osmosis	
limiting salt	scaling	

2. Explain key similarities and differences between membrane filtration and reverse osmosis.
3. Identify applications for reverse osmosis in drinking water treatment.
4. Describe the principle features of an RO membrane facility.
5. Describe the processes of osmosis and reverse osmosis.
6. Calculate osmotic pressure and explain the effect that osmotic pressure has on reverse osmosis.
7. Describe the theory for how water and solutes permeate a dense membrane.
8. Explain general trends in rejection by RO membranes, what kinds of constituents are rejected well and what kinds are rejected poorly, and how these trends can be explained by the physicochemical properties of the membranes and constituents.
9. Calculate the flux of water and solutes through a membrane.
10. Calculate membrane performance (permeate flow rate and salt passage) under standard conditions and determine whether any changes in performance have been observed.
11. Describe the concept of concentration polarization, calculate the concentration polarization factor, and describe the effects that concentration polarization has on reverse osmosis.
12. Calculate the silt density index (SDI).
13. Given ion concentrations in feed water, determine the limiting salt and calculate the maximum recovery that can be achieved before scaling occurs.
14. Design an RO system if given water flow rate, raw-water quality, and required effluent concentration.
15. Explain why reverse osmosis has the highest energy consumption of any common water treatment process and what can be done to reduce energy consumption.

Homework Problems

9-1 The following solutions are representative of common applications of reverse osmosis. Calculate the osmotic pressure of each at 20°C. Discuss the importance of osmotic pressure and how it affects the applied pressure for these applications.

a. NaCl = 35,000 mg/L (representative of seawater RO)

b. NaCl = 8000 mg/L (representative of brackish water RO)

c. Hardness = 400 mg/L as $CaCO_3$ (representative of softening NF)

d. Dissolved organic carbon (DOC) = 25 mg/L (representative of using NF to control DBP formation by removing DBP precursors). Assume an average MW of 1000 g/mol.

9-2 Operating data for a low-pressure RO system on two different days are shown in the table below:

	Unit	Day 1	Day 2
Water temperature	°C	13	22
Water flux	$L/m^2 \cdot h$	17.5	18.8
Feed pressure	bar	41.9	38.7
Concentrate pressure	bar	39.0	35.8
Permeate pressure	bar	0.25	0.25
Feed TDS concentration	mg/L	10,500	10,200
Permeate TDS concentration	mg/L	120	120
Recovery	%	66	68

Performance data for this membrane element were developed using the following standard conditions:

	Unit	Standard
Temperature	°C	20
Feed pressure	bar	40
Permeate pressure	bar	0
Head loss per element	bar	0.4
Number of elements	no.	7
Feed TDS concentration	mg/L	10,000
Permeate TDS concentration	mg/L	100
Recovery	%	70

Determine the difference in system performance (water flux and rejection) between the two days using the temperature correction formula in this text and an arithmetic average for the solute concentration in the feed–concentrate channel. Assume the salts in the feed water are sodium chloride for the purpose of calculating osmotic pressures.

9-3 Examine the importance of the diffusion coefficient on concentration polarization by graphing β as a function of the diffusion coefficient for diffusion coefficient values between 10^{-10} m^2/s (typical of NOM with a diameter of 5 nm) and 1.58×10^{-9} m^2/s (sodium chloride). Use feed channel velocity = 0.12 m/s, feed channel height = 0.90 mm, permeate flux = 22 $L/m^2 \cdot h$, and temperature = 20°C. Discuss the implications that the trend shown in this graph has on the accumulation of material at the membrane surface.

9-4 An SDI test was performed to evaluate the fouling tendency of potential RO source water. The time to collect 500 mL of water was measured as 24 s. Filtration continued for a total of 15 min, and then a second 500 mL was collected. The time necessary to collect the second 500-mL sample was 32 s. Calculate the SDI.

9-5 An RO facility is being designed to treat groundwater containing the ions given below. Calculate the allowable recovery before scaling occurs and identify the limiting salt. Assume 100 percent rejection, a concentration polarization factor of 1.08, and $T = 25°C$, and ignore the impact of ionic strength. The water contains calcium = 105 mg/L, strontium = 2.5 mg/L, barium = 0.0018 mg/L, sulfate = 128 mg/L, fluoride = 1.3 mg/L, and silica = 9.1 mg/L as Si.

9-6 Calculate and plot water flux and salt rejection as a function of recovery, for recovery ranging from 50 to 85 percent, given C_F = 10,000 mg/L NaCl, ΔP = 50 bar, $k_W = 2.2$ $L/m^2 \cdot h \cdot bar$, k_S = 0.75 $L/m^2 \cdot h$, $\phi = 1$, and T = 20° C. Comment on the effect of recovery on RO performance.

9-7 A new brackish water RO system is being proposed. The water quality is as shown in the table below. Using RO manufacturer design software (provided by the instructor or obtained from a membrane manufacturer website), develop the process design criteria for the plant. The required water demand is 38,000 m^3/d and the finished-water TDS should be 500 mg/L or lower.

Constituent	Concentration, mg/L	Constituent	Concentration, mg/L
Ammonia	1.3	Bicarbonate	680
Barium	0.04	Chloride	890
Calcium	20	Fluoride	0.7
Iron	0.5	Orthophosphate	0.7
Magnesium	2.5	Sulfate	105
Manganese	0.02	Silica	21.5
Potassium	17	Nitrate	1.2
Sodium	875	Hydrogen sulfide	0.3
Strontium	2.17		
pH	7.8	Turbidity	0.3 NTU
SDI	<1 min^{-1}	Temperature	15°C

9-8 A new seawater RO system is being proposed. The water quality is as shown in the table below. Using RO manufacturer design software (provided by the instructor or obtained from a membrane manufacturer website), develop the process design criteria for the plant. The required water demand is 4000 m^3/d and the finished-water TDS should be 500 mg/L or lower.

Constituent	Concentration, mg/L	Constituent	Concentration, mg/L
Aluminum	0.15	Strontium	6.6
Ammonia	0.092	Bromide	51
Barium	0.00	Bicarbonate	112
Boron	4.3	Chloride	18,900
Calcium	439	Fluoride	0.61
Iron	0.1	Phosphate	0.12
Magnesium	1,240	Sulfate	2380
Potassium	425	Silica	0.86
Sodium	10,100	Hydrogen sulfide	0.0
Strontium	6.6		
pH	8.0	Turbidity	3.3 NTU
SDI	$<1\ min^{-1}$	UV_{254}	0.03/cm
Temperature	15°C		

9-9 A new membrane softening system is being proposed. The water quality is as shown in the table below. Using RO manufacturer design software (provided by the instructor or obtained from a membrane manufacturer website), develop the process design criteria for the plant. The required water demand is 14,200 m^3/d and the finished-water hardness should be between 50 and 75 mg/L as $CaCO_3$.

Constituent	Concentration, mg/L	Constituent	Concentration, mg/L
Ammonia	1.5	Bicarbonate	135.1
Barium	0.0	Bromide	0.0
Calcium	100	Carbonate	0.11
Magnesium	10	Chloride	98.8
Manganese	0.002	Fluoride	0.5
Sodium	60	Phosphate	0.5
Strontium	1.0	Sulfate	167.6
		Silica	15.0
pH	7.0	Temperature	20°C
SDI	$<1\ min^{-1}$		

References

ASTM (2001a) *D4189-95 Standard Test Method for Silt Density Index (SDI) of Water*, American Society for Testing and Materials, Philadelphia, PA.

ASTM (2001b) D4516-00 Standard Practice for Standardizing Reverse Osmosis Performance Data, in *Annual Book of Standards*, Vol. 11.02, American Society for Testing and Materials, Philadelphia, PA.

AWWA (2007) *Reverse Osmosis and Nanofiltration*, AWWA Manual M46, American Water Works Association, Denver, CO.

Crittenden. J. C., Trussell, R. R., Hand, D. W., Howe, K. J., and Tchobanoglous, G. (2012) *MWH's Water Treatment: Principles and Design*, 3rd ed. Wiley, Hoboken, NJ.

Gustafsson, J. P. (2011) *Visual MINTEQ*, Version 3.0, KTH Royal Institute of Technology, Stockholm, Sweden.

Lonsdale, H. K., Merten, U., and Riley, R. (1965) "Transport Properties of Cellulose Acetate Osmotic Membranes," *J. Appl. Polym. Sci.*, **9**, 1341–1362.

Mariñas, B. J., and Urama, R. I. (1996) "Modeling Concentration-Polarization in Reverse Osmosis Spiral-Wound Elements," *J. Environ. Engr – ASCE*, **122**, 4, 292–298.

Mickley, M. (2006) *Membrane Concentrate Disposal: Practices and Regulation*, U.S. Bureau of Reclamation, Denver, CO.

NRC (National Research Council) (2008) *Desalination: A National Perspective*, National Academies Press, Washington, DC.

Ridgway, H. F., and Flemming, H.-C. (1996) Membrane Biofouling, Chap. 9 in J. Mallevialle, P. E. Odendaal, and M. R. Wiesner (eds.), *Water Treatment Membrane Processes*, McGraw-Hill, New York.

Robinson, R. A., and Stokes, R. H. (1959) *Electrolyte Solutions; the Measurement and Interpretation of Conductance, Chemical Potential, and Diffusion in Solutions of Simple Electrolytes*, Butterworths, London.

Sourirajan, S. (1970) *Reverse Osmosis*, Academic, New York.

Veerapaneni, V., Klayman, B., Wang, S., and Ozekin, K. (2010) Desalination Facility Design and Operation for Maximum Energy Efficiency—WRF Project 4038, Presented at the AWWA Annual Conference and Exposition, Chicago, IL.

10 Adsorption and Ion Exchange

Adsorption and ion exchange (IX) are treatment processes in which solutes (dissolved constituents) are removed from water by transferring them to the surface of a solid. Adsorption processes are commonly used in municipal drinking water treatment to remove synthetic organic chemicals (SOCs), taste- and odor-causing organics, color-forming organics, and disinfection by-product (DBP) precursors. Natural organic matter (NOM) and some inorganic constituents such as perchlorate, arsenic, and some heavy metals can be removed by either adsorption or ion exchange. Other inorganic constituents, such as hardness (calcium and magnesium), nitrate, iron, and manganese are effectively removed by ion exchange but not by adsorption.

The most common adsorbent material in drinking water treatment is activated carbon, which can be used in either a granular or powdered form.

Granular ferric hydroxide, activated alumina, and zeolite are other available adsorbent materials. Ion exchange media is typically synthetic resins that have been engineered specifically for that purpose.

The vast majority of IX installations in the United States are small, point-of-use devices at individual households that are used as water softeners. Full-scale systems are in use for industrial applications, such as the demineralization of water for prevention of scale formation in power plant boilers, removal of calcium and magnesium in car-washing facilities, and production of ultrapure water for making pharmaceuticals and semiconductor materials. The application of IX to municipal water treatment has been limited.

It is appropriate to discuss adsorption and ion exchange together because they have some similarities, such as design configurations that include fixed-bed contactors and suspended-media reactors. However, they also have important differences. In ion exchange, ions participate in a two-way transfer between the water and resin; the ions transferred to the resin must be replaced by an equal amount (in equivalents) of ions transferred from the resin to the water so that electroneutrality is maintained in both phases. Adsorption has no such requirement; the target contaminants are transferred to the adsorbent with no accompanying flow of matter from the adsorbent to the water. This difference in behavior leads to important differences in the equilibrium and kinetics of the processes, even though process design tends to be similar. Thus, the first three sections of this chapter present the description, equilibrium, and kinetics of the adsorption process, and the next three sections do the same for the ion exchange process. Section 10-7 presents the design of fixed-bed contactors for both processes, and Sec. 10-8 describes the design of suspended-media reactors. Sustainability and energy considerations are presented in Sec. 10-9.

10-1 Introduction to the Adsorption Process

The constituent that undergoes adsorption onto a surface is called the *adsorbate*, and the solid onto which the constituent is adsorbed is called the *adsorbent*. During the adsorption process, dissolved species diffuse into the porous solid adsorbent granule and are then adsorbed onto the extensive inner surface of the adsorbent. A key feature of adsorbents is a high degree of porosity within the adsorbent granules, which translates into a vast amount of interior surface area onto which adsorption can occur.

Pore Size and Surface Area

Because adsorption takes place on the surface, a large amount of surface area is an essential characteristic of an effective adsorbent. The large surface area is accomplished by using materials that have a vast number of tiny pores in the interior of a granular material. The porosity (ratio of pore volume to total volume) is often near 50 percent. With this porosity, adsorbents can have a pore volume of 0.1 to 0.8 mL/g and an internal surface area ranging from 400 to 1500 m^2/g. As a result, the adsorption capacity can be as high as 0.2 g of adsorbate per gram of adsorbent, depending on the adsorbate concentration and type. Generally, there is an inverse relationship between the pore size and surface area: the smaller the pores for a given pore volume, the greater the surface area that is available for adsorption. In addition, the size of the adsorbate that can enter a pore is limited by the size of the pore. The relationship between pore size and volume is shown in the following example.

Example 10-1 Calculation of the internal surface area of a porous adsorbent

Assume a granule of adsorbent material has cylindrical pores with diameters of either 1 or 5 nm, a porosity of 50 percent, and a particle density of 1 g/cm^3. Determine the internal surface area of the adsorbent.

Solution

1. Develop a relationship for the ratio of surface area to pore volume for the adsorbent.
   a. The volume of cylindrical pores, V_{ad} (m^3/g), can be computed based on the number of pores n (amount/g), the pore radius R (m), and the pore length L (m):

$$V_{ad} = n\pi R^2 L$$

   b. The surface area of the pores, A_{ad} (m^2/g), is also determined assuming a cylindrical pore shape:

$$A_{ad} = 2n\pi RL$$

   c. The surface area–pore volume ratio for the adsorbent, A_{ad}/V_{ad}, can be written by combining the expressions developed in steps 1a and 1b:

$$\frac{A_{ad}}{V_{ad}} = \frac{2}{R}$$

2. Determine the surface area for adsorbents with pore sizes of 1 and 5 nm.
   a. Compute the adsorbent volume using the porosity and adsorbent density provided in the problem statement. By definition, porosity = pore volume/total volume, so 1 g of adsorbent with a porosity of 0.5 would have a total volume of 1 cm^3 and a pore volume of 0.5 cm^3. Therefore $V_{ad} = 0.5\ cm^3/g = 5 \times 10^{-7}\ m^3/g$.

b. For a pore diameter $d_p = 1.0\ \text{nm} = 10^{-9}\ \text{m}$, $R = 5 \times 10^{-10}\ \text{m}$, then

$$A_{ad} = V_{ad}\frac{2}{R} = (5 \times 10^{-7}\ \text{m}^3/\text{g})\frac{2}{5 \times 10^{-10}\ \text{m}} = 2000\ \text{m}^2/\text{g}$$

c. For a pore diameter $d_p = 5.0\ \text{nm} = 5 \times 10^{-9}\ \text{m}$, $R = 2.5 \times 10^{-9}\ \text{m}$, then

$$A_{ad} = V_{ad}\frac{2}{R} = (5 \times 10^{-7}\ \text{m}^3/\text{g})\frac{2}{2.5 \times 10^{-9}\ \text{m}} = 400\ \text{m}^2/\text{g}$$

The porosity of adsorbents generally does not exceed 50 percent because of the manufacturing process and the skeletal strength of the adsorbent. If the porosity is higher, adsorbents become brittle and break apart when transported into and out of adsorption vessels, which can result in significant adsorbent losses and expense.

For the purpose of classifying pore sizes (diameter d_p), the International Union of Pure and Applied Chemistry (IUPAC) uses the following convention:

Micropores: $d_p < 20$ nm.

Mesopores: 20 nm $< d_p <$ 500 nm.

Macropores: 500 nm $< d_p$.

Adsorption Media

Commercially available adsorbents that merit consideration in water treatment include activated carbon, synthetic polymeric adsorbents, activated alumina, zeolites, and granular ferric hydroxide. Most activated carbons have a wide range of pore sizes and can accommodate large organic molecules such as NOM and SOCs such as pesticides, solvents, and fuels. Synthetic polymeric adsorbents usually have only micropores, which prevents them from adsorbing NOM. Actvated alumina and zeolites (aluminosilicates with varying Al-to-Si ratios) tend to have very small pores, which means they cannot remove NOM and larger synthetic organic compounds. Granular ferric hydroxide and iron-impregnated granular activated carbon have been developed to remove arsenic. Properties of several commercially available adsorbents are reported in Table 10-1.

Activated carbon is the most commonly used adsorbent because it is less expensive than the alternatives and it is effective for adsorption of a wide range of contaminants. Activated carbon is manufactured from natural, carbonaceous materials such as coal, peat, and coconuts by several inexpensive processes (e.g., high temperatures ~800°C and steam). Consequently, most of the discussion in this chapter centers on the use of activated carbon; where appropriate, alternative adsorbents are discussed.

Table 10-1
Properties of several commercially available adsorbents

Adsorbent	Manufacturer	Type	Surface Area, m^2/g (BET)[a]	Packed-Bed Density, g/cm	Pore Volume, cm^3/g
Filtrasorb 300 (8 × 30)	Calgon	GAC	950–1050	0.48	0.851
Filtrasorb 400	Calgon	GAC	1075	0.4	1.071
CC-602	US Filter/Wastates	Coconut-shell-based GAC	1150–1250	0.47–0.52	0.564
Aqua Nuchar	MWV	PAC	1400–1800	0.21–0.37	1.3–1.5
Dowex Optipore L493	Dow	Polymeric	>1100	0.62	1.16
Lewatit VP OC 1066	Bayer	Synthetic polymer	700	0.5	0.65–0.8

Sources: Adapted from Sontheimer et al. (1988), Crittenden (1976), Lee et al. (1981), Munakata et al. (2003), and Sigma Aldrich Online Catalog (2004).
[a]BET is the Brunauer, Emmett, and Teller method for measuring surface area based on gas (usually nitrogen) adsorption.

The base materials used in making adsorbents can influence the distribution of the pores. The distribution of pore sizes for several different raw materials that are used to make activated carbon are shown on Fig. 10-1.

Adsorption Contactors

Adsorption takes place in either fixed-bed or suspended-media contactors. Fixed-bed contactors consist of a bed of granular media typically 1 to 3 m deep, either in a pressure vessel or an open basin. Water passes through the bed and adsorption occurs as the water contacts the media. Water flow is typically downward through the media. In some cases, the adsorption media also functions as granular filtration media (see Chap. 7), in which case the media captures particles as well as solutes. When used as a filter, the media must be

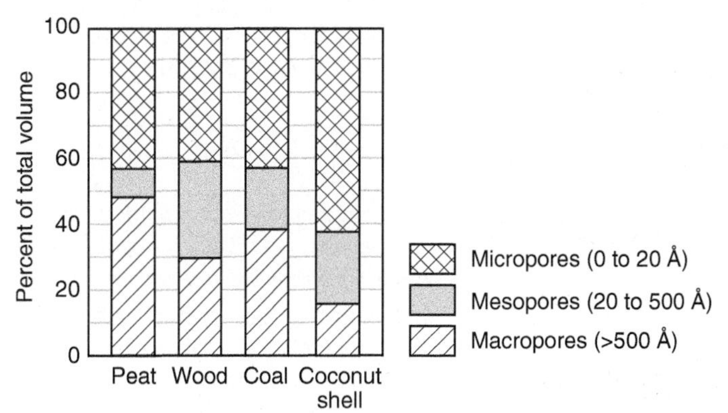

Figure 10-1
Pore size distributions for activated carbons with different starting materials.

backwashed to remove the particles when the head loss becomes excessive. However, fixed-bed adsorption works most effectively when the media is not disrupted by backwashing, and the media can be loaded with contaminants progressively from the top to the bottom without the media being mixed. When the adsorption capacity has been used up, the media must be replaced or regenerated. Fixed-bed contactors are typically designed to be operated for months or years before the adsorption capacity is exhausted. Theory and design of fixed-bed adsorbers is presented in Sec. 10-7.

In suspended-media contactors, the adsorbent is mixed directly into the process water and allowed to travel with the process stream as the water makes its way through the treatment plant. Adsorption takes place as the adsorbent travels with the water. After a sufficient period of time has been allowed for adsorption to take place, the adsorbent is separated from the water, typically by sedimentation and filtration. Theory and design of suspended-media adsorbers is presented in Sec. 10-8.

In the case of activated carbon, the size of the media is different when used in fixed-bed versus suspended-media contactors. The media used in fixed-bed contactors typically has a mean particle diameter of 0.5 to 3 mm and is known as granular activated carbon (GAC). The media used in suspended-media contactors has a mean particle diameter of 20 to 50 μm and is known as powdered activated carbon (PAC). The principal uses, advantages, and disadvantages of using activated carbon in fixed-bed versus suspended-media contactors is summarized in Table 10-2.

Table 10-2
Principal uses, advantages, and disadvantages of using activated carbon in fixed-bed versus suspended-media contactors

Parameter	Fixed-Bed Contactor (GAC)	Suspended-Media Contactor (PAC)
Principal uses	❑ Control of toxic organic compounds in groundwater ❑ Barrier to occasional spikes of toxic organics in surface waters and control of taste and odor compounds ❑ Control of disinfection by-product precursors (NOM)	❑ Seasonal control of taste and odor compounds and strongly adsorbed pesticides and herbicides at low concentration (<10 μg/L)
Advantages	❑ Lower carbon usage rate per volume of water treated compared to PAC	❑ Easily added to existing water intakes or coagulation facilities for occasional control of organics
Disadvantages	❑ Need contactors and yard piping to distribute flow and replace exhausted carbon ❑ Previously adsorbed compounds can desorb and in some cases appear in the effluent at concentrations higher than present in influent	❑ Impractical to recover from sludge from coagulation facilities for reuse or recovery of carbon ❑ Higher carbon usage rate per volume of water treated compared to GAC

A common use of activated carbon is the removal of taste and odor compounds. Taste and odor outbreaks are seasonal, and according to a recent survey in North America, outbreaks usually occur between June and October (Graham et al., 2000). The two principal odor-causing compounds that are not removed by chlorine are geosmin and 2-methyl-isoborneol (MIB). Cyanobacteria are thought to produce and release these compounds into the water. Reported odor threshold concentrations for geosmin and MIB are 4 and 9 ng/L (McGuire et al., 1981). Accordingly, the treatment objective for these compounds must be below these threshold concentrations.

Powdered activated carbon is added to water as a suspension using adsorbent doses in the range of 5 to 25 mg/L. According to a 1994 survey of U.S. water utilities, 90 percent of the plants surveyed used a dose between 0.5 and 18 mg/L and the average dose was 5.1 mg/L (Graham et al., 2000).

Adsorption Mechanisms

Dissolved species are concentrated on the solid surface by chemical reaction (chemisorption) or physical attraction (physical adsorption) to the surface. Key elements of adsorption mechanisms are listed in Table 10-3.

Physical adsorption is a nonspecific reversible reaction; that is, the adsorbate desorbs in response to a decrease in solution concentration or displacement by a more strongly adsorbed species. Physical adsorption is exothermic with a heat of adsorption that is typically 4 to 40 kJ/mol (about 2 times greater than the heat of vaporization or dissolution for gases and liquids, respectively). Chemisorption is more specific because a chemical bond between adsorbent and adsorbate occurs. The heat of adsorption for chemisorption is typically above 200 kJ/mol. Chemisorption is usually not reversible, and desorption, if it occurs, is accompanied by a chemical change in the adsorbate.

In aqueous solution, adsorption involves three interactions: (1) adsorbate–water interactions, (2) adsorbate–surface interactions, and (3) water–surface interactions. The extent of adsorption is determined by the strength of adsorbate–surface interactions as compared to the adsorbate–water and water–surface interactions. Adsorbate–surface

Table 10-3
Comparison of adsorption mechanisms between physical adsorption and chemisorption

Parameter	Physical Adsorption	Chemisorption
Occurrence	Most common mechanism	Rare for constituents in water treatment
Process speed	Rapid, limited by mass transfer	Variable, depends on the reaction rate with the surface
Type of bonding	Nonspecific binding mechanisms such as van der Waals forces, vapor condensation	Specific exchange of electrons, chemical bond at surface
Type of reaction	Reversible, exothermic	Typically nonreversible, exothermic
Heat of adsorption	4–40 kJ/mol	>200 kJ/mol

interactions are determined by surface chemistry, and adsorbate–water are related to the solubility of the adsorbate. Water–surface interactions are determined by the surface chemistry, for example, the graphitic surface of activated carbon is hydrophobic, and oxygen-containing functional groups that are sometime present on the activated carbon surface are hydrophilic. For chemisorption, the primary factor controlling the extent of reaction is the type of reaction that occurs on the surface.

CHEMICAL ADSORPTION AND ION EXCHANGE

Chemical adsorption, or *chemisorption*, occurs when the adsorbate reacts with the surface to form a covalent bond or an ionic bond. In chemisorption, the attraction between adsorbent and adsorbate approaches that of a covalent bond with shorter bond length and higher bond energy. The bond may also be specific to particular sites or functional groups on the surface of the adsorbent.

For ion exchange, charged surface groups contain an ion of the same charge that exchanges for the ion to be removed and these ions are attracted to the counter charge on the surface according to Coulomb's law.

Adsorbates bound by ion exchange or chemisorption to a surface generally cannot accumulate at more than one molecular layer because of the specificity of the bond that is formed between the adsorbate and surface or because electroneutrality must be maintained.

PHYSICAL ADSORPTION

Adsorbates undergo *physical adsorption* if the forces of attraction include only physical forces that exclude covalent bonding with the surface and coulombic attraction of unlike charges. In some cases, the difference between physical adsorption and chemisorption may not be that distinct. Physical adsorption is less specific for which compounds sorb to surface sites, has weaker forces and energies of bonding, operates over longer distances (multiple layers), and is more reversible.

In water treatment, adsorption often involves adsorbing organic adsorbates from water (polar solvent) onto a nonpolar adsorbent (activated carbon). Because activated carbon has crosslinked graphitic crystallite planes that form micropores, the major attractive force between organics and the adsorbent is van der Waals forces that exist between organic compounds and the graphitic carbon basal planes.

In general, attraction between an adsorbate and water (a polar solvent) is weaker for adsorbates that are less polar or have lower solubility. The attraction between an adsorbate and activated carbon surface increases with increasing polarizability and size, which are directly related to van der Waals forces. More nonpolar and larger compounds tend to adsorb more strongly to nonpolar adsorbents such as activated carbon. This form of adsorption is also known as hydrophobic bonding (Nemethy and Scheraga, 1962); hydrophobic ("disliking water") compounds will adsorb more strongly to hydrophobic surfaces.

ADSORBABILITY OF VARIOUS CLASSES OF COMPOUNDS

Applying what is known about the adsorption of organics to determine their adsorbability requires consideration of the summation of the interactions and forces described above. Although these interactions and forces are not readily measurable, in a general sense they can be related to some properties of the adsorbate and solvent. For example, solubility is a direct indication of adsorption strength or magnitude of the adsorption force. The lower the solubility of an adsorbate in the solvent, the higher the adsorption strength. Adsorption strength is inversely proportional to solubility. Unfortunately, all other factors are different for different classes of organics (e.g., aliphatic, aromatic, or polar compounds); consequently, solubility is not the only indicator of adsorbability.

10-2 Adsorption Equilibrium

Adsorption is an equilibrium reaction; at equilibrium, the adsorbate will be distributed between the aqueous and solid phases according to a relationship known as an adsorption *isotherm*. Several isotherm relationships are commonly used in adsorption. The Langmuir isotherm can be derived from equilibrium relationships that were introduced in Sec. 4-2 using relatively straightforward assumptions about the nature of the surface of the adsorbent. When the surface of the adsorbent does not conform to these assumptions, such as for activated carbon, the Freundlich isotherm can provide a better fit to experimental data. These isotherms and an extension of the Freundlich isotherm for multicomponent adsorption are introduced in this section.

Langmuir Isotherm for a Single Solute

The Langmuir adsorption isotherm describes the equilibrium between surface and solution as a reversible chemical reaction. The adsorbent surface has individual fixed sites where molecules of adsorbate may be chemically bound. The following reaction describes the relationship between the adsorbate concentration in solution and bound to surface sites:

$$S_V + A \rightleftarrows S \cdot A \tag{10-1}$$

where S_V = vacant surface sites
A = adsorbate species A in solution
$S \cdot A$ = adsorbate species bound to surface sites

The derivation of the Langmuir isotherm is based on the assumptions that adsorption to every adsorption site has the same free-energy change, and each site is capable of binding only one molecule of adsorbate; that is, the model allows accumulation only up to a monolayer. Accordingly, the equilibrium constant (see Sec. 4-2) may be written as

$$K_{\text{ad}} = \frac{[S \cdot A]}{[S_V]\,[A]} \tag{10-2}$$

where K_{ad} = adsorption equilibrium constant, L/mol
$[S \cdot A]$ = concentration of adsorbate on surface sites, mol/m^2
$[S_V]$ = vacant surface sites, mol/m^2
$[A]$ = concentration of adsorbate A in solution, mol/L

The sum of the vacant and filled sites equals the total number of sites:

$$[S_T] = [S_V] + [S \cdot A] = \frac{[S \cdot A]}{K_{ad}\,[A]} + [S \cdot A] = [S \cdot A]\left(1 + \frac{1}{K_{ad}\,[A]}\right) \quad (10\text{-}3)$$

where $[S_T]$ = total number of sites available, mol/m^2

Rearranging and solving for $S \cdot A$ yields

$$[S \cdot A] = \frac{[S_T]}{1 + (1/K_{ad}\,[A])} = \frac{[S_T]\,K_{ad}\,[A]}{1 + K_{ad}\,[A]} \quad (10\text{-}4)$$

Concentrations on the solid phase are more easily expressed as mass loading per mass of adsorbent than as the concentration of occupied sites in mol/m^2. Multiplying both sides of Eq. 10-4 by the molecular weight of the adsorbate (g/mol) and surface area per gram of adsorbent (m^2/g), and converting from grams to milligrams, results in an expression in terms of mass loading as follows:

$$q_A = [S \cdot A]\,(A_{ad})\,(\mathrm{MW}) = \frac{[S_T]\,(A_{ad})\,(\mathrm{MW})\,K_{ad}\,[A]}{1 + K_{ad}\,[A]} = \frac{Q_M K_{ad}\,[A]}{1 + K_{ad}\,[A]} \quad (10\text{-}5)$$

$$q_A = \frac{Q_M\,b_A\,C_A}{1 + b_A\,C_A} \quad (10\text{-}6)$$

where
q_A = concentration of adsorbate A on adsorbent, mg adsorbate/g adsorbent
A_{ad} = surface area per gram of adsorbent, m^2/g
MW = molecular weight of the adsorbate, g/mol
$C_A = [A]\mathrm{MW}$ = concentration of adsorbate A in solution, mg/L
$Q_M = [S_T]A_{ad}\mathrm{MW}$ = adsorption capacity, concentration of the adsorbate on the solid when all sites are filled, mg/g
$b_A = K_{ad}/\mathrm{MW}$ = Langmuir adsorption constant, L/mg

Equation 10-6 describes the equilibrium concentration of adsorbate on an adsorbent as a function of the concentration in the solution and two coefficients, Q_M and b_A, which can be determined experimentally or obtained from reference books. The coefficients can be determined by conducting a series of experiments in which the adsorbent is placed in jars with the adsorbate at different concentrations and allowed to reach equilibrium; each jar will have different final concentrations of adsorbate on the solid and in solution. Procedures to conduct these experiments are described in Crittenden et al. (2012). Rearranging Eq. 10-6 to a linear form yields

$$\frac{C_A}{q_A} = \frac{1}{b_A Q_M} + \frac{C_A}{Q_M} \tag{10-7}$$

A plot of C_A/q_A versus C_A using Eq. 10-7 results in a straight line with a slope of $1/Q_M$ and intercept $1/b_A Q_M$.

Freundlich Isotherm for a Single Solute

While the Langmuir isotherm has a straightforward derivation, the requirements for constant site energy and monolayer coverage implicit in the derivation are not satisfied for many adsorbents. Consequently, the Freundlich adsorption isotherm, originally proposed as an empirical equation, is used to describe the equilibrium for adsorbents having adsorption sites with differing site energies, such as activated carbon:

$$q_A = K_A C_A^{1/n} \tag{10-8}$$

where K_A = Freundlich adsorption capacity parameter, $(\text{mg/g})(\text{L/mg})^{1/n}$
$1/n$ = Freundlich adsorption intensity parameter, unitless

The linear form of Eq. 10-8 is

$$\log(q_A) = \log(K_A) + (1/n)\log(C_A) \tag{10-9}$$

A log–log plot of q_A versus C_A using the form shown in Eq. 10-9 will result in a straight line, as shown on Fig. 10-2 for tetrachloroethene (TCE) and 1,1,1-trichloroethane.

The Freundlich isotherm provides a better fit to isotherm data than the Langmuir isotherm for activated carbon because many layers of adsorbate can adsorb to the surface and there is distribution of sites with different

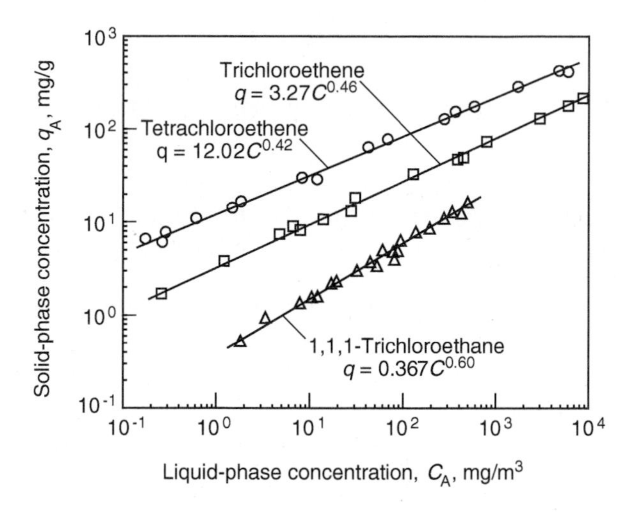

Figure 10-2
Single-solute isotherms for tetrachloroethene, trichloroethene, and 1,1,1-trichloroethane over a wide concentration range. [Adapted from Zimmer et al. (1988).]

Table 10-4
Aqueous-phase Freundlich isotherm parameters K and $1/n$ for selected organic adsorbates[a]

Compound	K^b	$1/n$	pH	T_{min} (°C)	Name of Carbon[c]	References
Atrazine	182	0.18	7.1	20	F 100	3
Benzoic acid	0.7	1.8	7	20	F 300	1
Chlorodibromomethane	45	0.517	6	11	F 400	4
Chloroform	15	0.47	7.1	20	F 100	3
Cyclohexanone	6.2	0.75	7.3	20	F 300	1
Cytosine	1.1	1.6	7	20	F 300	1
1,2-Dichlorobenzene	242.2	0.4	7.1	20	F 100	3
1,3-Dichlorobenzene	458.8	0.63	7.9	24	F 400	2
1,2-*trans*-Dichloroethene	3.1	0.51	6.7	20	F 300	1
2,4-Dichlorophenol	141	0.29	9	20	F 300	1
Ethylbenzene	53	0.79	7.4	20	F 300	1
Methyl ethyl ketone	19.4	0.295	8	24	F 400	2
N-Dimethylnitrosamine	0	0	7.5	20	F 300	1
Pentachlorophenol	150	0.42	7	20	F 300	1
Tetrachloroethene	218.2	0.42	7.1	20	F 100	5
Trichloroethene	55.9	0.48			F 400	2
1,1,1-Trichloroethane	23.2	0.6	7.1	20	F 100	5

1. Dobbs and Cohen (1980); 2. Speth and Miltner (1998); 3. Haist-Gulde (1991); 4. Crittenden et al. (1985); 5. Zimmer et al. (1988)
[a]Additional Freundlich isotherm parameters are available in electronic resource E5 at the website listed in App. E.
[b]Units of K are $(mg/g)(L/mg)^{1/n}$. This means that C_A is in units of mg/L and q_A is in units of mg/g.
[c]Calgon Carbon Corporation.

adsorption energies. Examples of Freundlich isotherm parameters are shown in Table 10-4, and the procedure for determining Freundlich isotherm parameters is demonstrated in Example 10-2.

Example 10-2 Determination of Freundlich isotherm parameters from experimental data

A bench study was conducted to determine Freundlich isotherm parameters. Six jars were each filled with 250 mL of a solution containing 4.23 mg/L of trichlorethene (TCE). Different amounts of F-400 activated carbon were added to each jar as shown in the table below. The jars were sealed and agitated for 31 days at 13°C to allow the system to reach equilibrium, and then the final concentration of solute in each jar was measured. From the equilibrium concentrations given below, calculate the Freundlich isotherm parameters.

Jar	1	2	3	4	5	6
GAC dose, mg	2.93	13.9	22.9	66.1	93.0	112
Equilibrium TCE concentration, mg/L	3.1	0.93	0.43	0.044	0.020	0.015

Solution

The Freundlich isotherm parameters are determined by calculating the concentration of TCE on the GAC at equilibrium, calculating the logarithms, and plotting the concentration on the adsorbent against the concentration in the liquid.

1. The concentration on the GAC is calculated using a mass balance analysis as developed in Eq. 4-158 in Chap 4. For the first jar

$$q_e = \frac{V}{M}(C_0 - C_e) = \frac{(250\ \text{mL})(10^3\ \text{mg/g})}{(2.86\ \text{mg})(10^3\ \text{mL/L})}(4.23 - 3.10\ \text{mg/L})$$
$$= 98.8\ \text{mg/g}$$

2. The logarithms of the liquid and GAC concentrations are calculated as

$$\log(C_e) = \log(3.10) = 0.491$$
$$\log(q_e) = \log(98.8) = 1.99$$

3. The concentrations and log values for the remaining jars are shown in the table below

Sample	C_e, mg/L	q_e, mg/g	log C_e	log q_e
1	3.10	96.4	0.49	1.99
2	0.93	59.4	−0.031	1.77
3	0.43	41.5	−0.366	1.62
4	0.044	16.6	−1.357	1.22
5	0.020	11.3	−1.699	1.05
6	0.015	9.4	−1.824	0.97

4. Log(q_e) is plotted against log(C_e) as shown in the following figure:

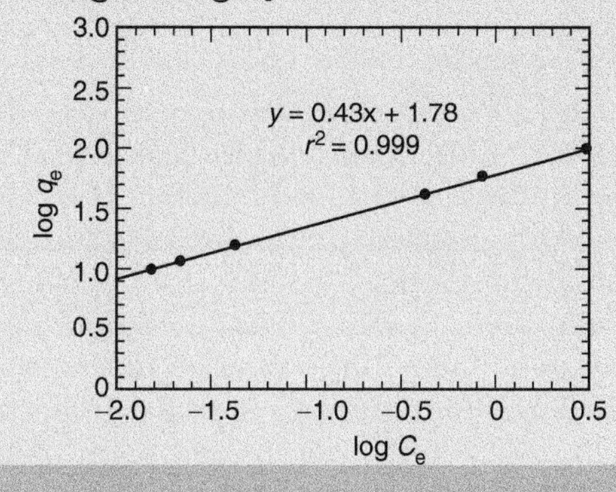

5. The isotherm parameters are obtained from results of the linear regression.

$$1/n = \text{slope} = 0.43$$

$$\log(K) = \text{intercept} = 1.78$$

$$K = 60.7\ \text{mg/g(L/mg)}^{0.43}$$

Multicomponent Adsorption

In water treatment, the ideal case of one adsorbate being removed onto an adsorbent is seldom encountered, and the objective in most real systems is to remove several adsorbates. Since the capacity of the adsorbent is limited, the adsorbates must compete for the available space. This competition complicates both the equilibrium of the adsorbate between the aqueous and solid phases and the ability of the engineer to apply the theory to practice. The Freundlich isotherm for a single solute can be extended to the presence of multiple adsorbates using the ideal adsorbed solution theory (IAST). The derivation of the IAST is beyond the scope of this book and details are presented in Crittenden et al. (2012). A convenient form of the IAST assuming the $1/n$ values for all components are identical is

$$q_i = K_i^n C_i \left(\sum_{j=1}^{N} K_j^n C_j \right)^{1/n-1} \tag{10-10}$$

where q_i = concentration of adsorbate i on adsorbent, mg/g
K_i = Freundlich adsorption capacity parameter for adsorbate i, $(\text{mg/g})(\text{L/mg})^{1/n}$
C_i = concentration of adsorbate i in solution, mg/L
n = inverse of Freundlich adsorption intensity parameter $1/n$, unitless

10-3 Adsorption Kinetics

The kinetics of the adsorption process is controlled by the rate of mass transfer of the solute to the adsorbent surface. Principles of mass transfer were introduced in Secs. 4-13 to 4-17 of this book. When a particle of adsorbent material is immersed in a flowing fluid such as water, a boundary layer forms around the particle. To adsorb onto a surface in the porous interior of the adsorbent, a solute from the liquid must diffuse through the boundary layer and then diffuse through the interior of the particle, as shown on Fig. 10-3. Diffusion through the exterior boundary layer is called film diffusion. Once inside the porous adsorbent particle, the adsorbate can diffuse through the liquid in the pore spaces, which is called pore

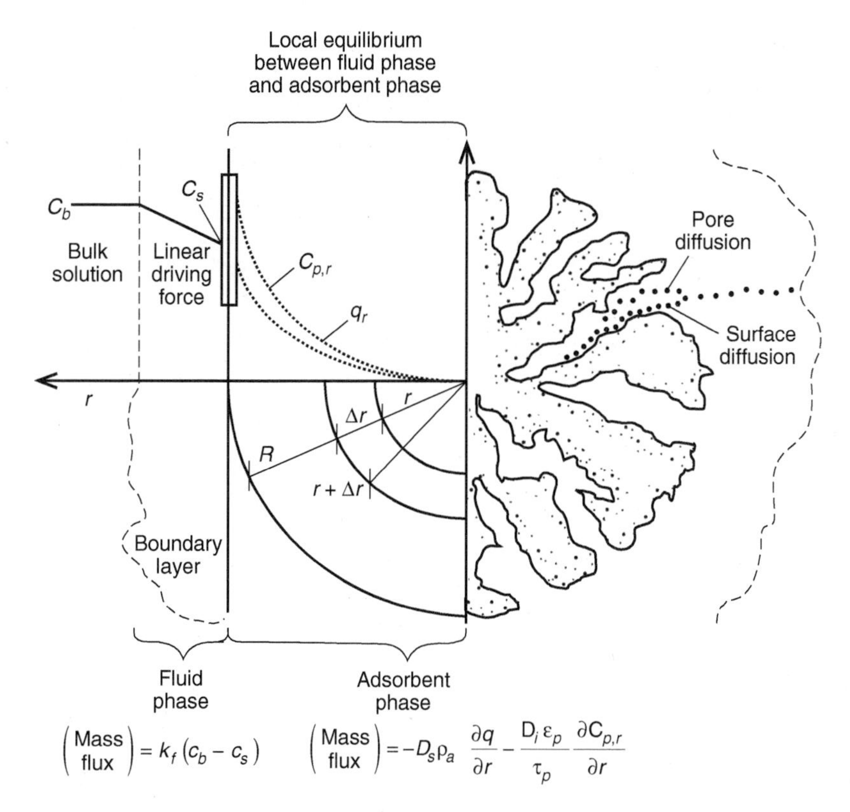

$$\left(\begin{matrix}\text{Mass}\\ \text{flux}\end{matrix}\right) = k_f (C_b - C_s) \qquad \left(\begin{matrix}\text{Mass}\\ \text{flux}\end{matrix}\right) = -D_s \rho_a \frac{\partial q}{\partial r} - \frac{D_l \varepsilon_p}{\tau_p} \frac{\partial C_{p,r}}{\partial r}$$

Figure 10-3
Mechanisms involved in adsorption kinetics.

diffusion, or adhere to the surface and then travel along the surface, which is known as surface diffusion.

The rate of mass transfer is relatively slow. Batch experiments to determine adsorption isotherm parameters can take 2 to 4 weeks to reach equilibrium. In full-scale suspended-media contactors, the media is rarely in contact with the water for sufficient time for equilibrium to be achieved; thus, the amount of adsorption that occurs is dictated by the kinetics rather than the equilibrium of the process. The amount adsorbed will be less than the equilibrium amount, so the kinetics must be considered during design.

In fixed-bed contactors, the media is in contact with the water for many weeks so that the solute on the influent end of the contactor can reach equilibrium with the influent water, but a concentration profile known as the *mass transfer zone* (MTZ) develops in the bed. The MTZ is the length of bed needed for the adsorbate to be transferred from the fluid into the adsorbent. All adsorbate gets removed from the water in the MTZ, so the media in the bed downstream of the MTZ remains unexposed to the adsorbate. As additional adsorption occurs, the MTZ moves through the bed and eventually unadsorbed solutes begin to appear in the column effluent, as shown on Fig. 10-4. When the solute concentration exceeds

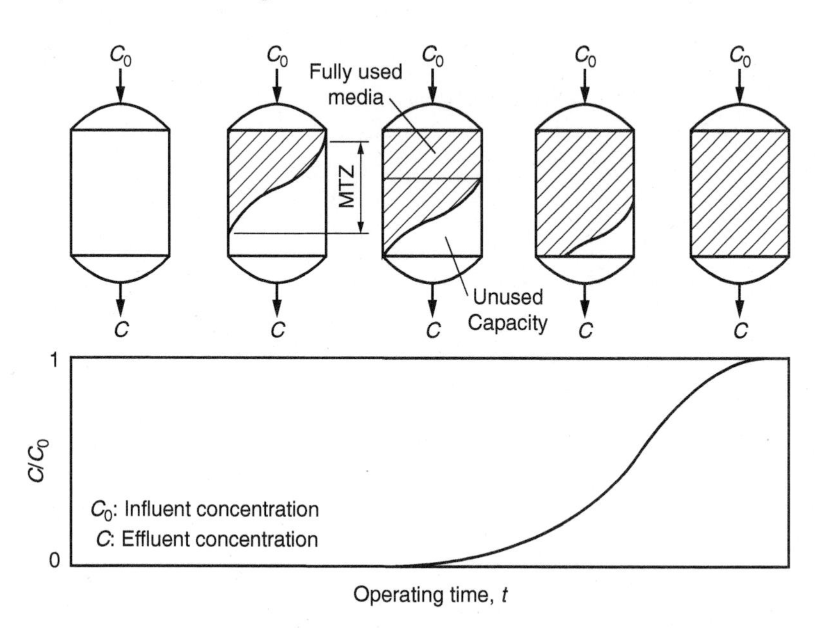

Figure 10-4
Concentration profiles and breakthrough curves for fixed-bed adsorption columns.

the treatment objective (called breakthrough), the media must be replaced or regenerated. The MTZ represents a region of media that has not been fully utilized for adsorption, so if the MTZ is wide, a significant portion of the media bed can go unused. To minimize the unused portion of the bed, fixed-bed adsorbers typically have an empty bed contact time (defined in Sec. 10-7) of 15 to 30 min. The time when the effluent concentration essentially equals the influent is called the point of exhaustion because the bed is no longer able to remove the solutes. Thus, the kinetics of the adsorption process affect the design of fixed-bed contactors as well as suspended-media contactors.

The kinetics of the adsorption process can be modeled by performing a mass balance analysis on the particles of adsorptive media using diffusion of the mass transport mechanism. Diffusion through the boundary layer and the interior of the particles occurs in series. Boundary layer mass transport models were introduced in Sec. 4-16. The intraparticle flux is the sum of the pore and surface diffusion, as shown in the following expression:

$$J = -D_s \rho_a \frac{\partial q}{\partial r} - D_p \frac{\partial C_p}{\partial r} \tag{10-11}$$

where J = mass flux of adsorbate to the adsorbent surface, $\text{mg/m}^2 \cdot \text{s}$
D_s = surface diffusion coefficient, m^2/s
$D_p = D_l \varepsilon_p / \tau_p$ = pore diffusion coefficient, m^2/s
D_l = liquid-phase diffusion coefficient, m^2/s
ρ_a = adsorbent particle density (mass of the carbon divided by the total volume of the particle including pore volume), kg/m^3
q = adsorbent-phase concentration, mg/g

r = radial coordinate, m
ε_p = porosity of the particle, dimensionless
τ_p = tortuosity of the path adsorbate must take as compared to the radius, dimensionless
C_p = liquid-phase concentration of the adsorbate in the adsorbent pores, mg/L

Development and solution of the partial differential equations that result from the mass balance analysis are beyond the scope of this book and are described in Crittenden et al. (2012). The various models that have been developed incorporate several dimensionless parameter groups that characterize important aspects of the mass transfer process. The dimensionless parameter groups are summarized in Table 10-5.

While modeling is a key method for designing adsorption processes, the mathematical models are not completely accurate because organic matter present in water has an impact on the intraparticle diffusion and adsorption capacity that is not completely understood. Thus, alternative design methods are often employed. Pilot testing using the actual source water and adsorbent can be the most reliable way of obtaining design information, but it can be expensive and time consuming. For fixed-bed contactors, an alternative testing strategy using miniature columns has been developed. This method, called rapid small-scale column testing (RSSCT) uses the dimensionless parameters in Table 10-5 to equate the performance of a small column to a full-scale column. Replacing a pilot study with an RSSCT significantly reduces the time and cost of a full-scale design.

Table 10-5
Dimensionless groups that characterize adsorption models

Dimensionless Group	Equation	Definition
Solute distribution parameter, D_g	$\frac{\rho_a q_e(1-\varepsilon)}{\varepsilon C_0}$	$\left.\frac{\text{Mass of solute in solid phase}}{\text{Mass of solute in liquid phase}}\right\|_{\text{equilibrium}}$
Peclet number, Pe	$\frac{Lv}{E}$	$\frac{\text{Solute transfer rate by advection}}{\text{Solute transfer rate by axial dispersion}}$
Stanton number, St	$\frac{k_f\tau(1-\varepsilon)}{\varepsilon R}$	$\frac{\text{Solute liquid-phase mass transfer rate}}{\text{Solute transfer rate by advection}}$
Biot number, Bi	$\frac{k_f R(1-\varepsilon)}{D_s D_g \varepsilon}$	$\frac{\text{Solute liquid-phase mass transfer rate}}{\text{Solute intraparticle mass transfer rate}}$
Surface diffusion modulus, Ed_s	$\frac{D_s D_g \tau}{R^2}$	$\frac{\text{Solute transfer rate by intraparticle surface diffusion}}{\text{Solute transfer rate by advection}}$
Pore diffusion modulus, Ed_p	$\frac{D_p \tau (1-\varepsilon) \varepsilon_p}{R^2 \varepsilon}$	$\frac{\text{Solute transfer rate by intraparticle pore diffusion}}{\text{Solute transfer rate by advection}}$

Table 10-6
Methods for estimating full-scale adsorption performance

Method	Reliability	Advantages	Disadvantages
Pilot studies	Excellent	1. Can predict full-scale GAC performance very accurately.	1. Can take a very long time to obtain results. 2. Expensive and must be conducted onsite.
RSSCTs (for fixed-bed contactors)	Good if scaling factor is known	1. Can predict full-scale GAC performance accurately. 2. Small volume of water is required for the test, which can be transported to a central laboratory for evaluation. 3. Extensive isotherm or kinetic studies are not required. 4. Can be conducted in the fraction of the time and cost that is required to conduct pilot studies.	1. Cannot predict GAC performance for different concentrations. 2. Biological degradation that may prolong GAC bed life is not considered. 3. The impact of NOM on micropollutant removal is less than is observed in full-scale plants.
Models	Good if calibrated; fair if not calibrated	1. Once calibrated, models can be used to predict impact of EBCT and changes in influent concentration. 2. Can predict breakthrough of SOCs with 20–50% error.	1. Cannot predict TOC breakthrough and must be used in conjunction with pilot or RSSCT data. 2. Accurate prediction of SOC removal requires calibration with pilot or RSSCT data.

The advantages and disadvantages of the various methods for predicting adsorption performance are described in Table 10-6.

10-4 Introduction to the Ion Exchange Process

Ion exchange involves the exchange of an ion in the aqueous phase for an ion in the solid phase. In drinking water treatment applications, the ion exchange media is typically a synthetic polymeric resin. This section introduces the types of resins available and the principles of ion exchange selectivity.

Ion Exchange Resin Structure

Polymeric ion exchange resins are composed of a three-dimensional, crosslinked polymer matrix that contains covalently bonded functional groups with fixed ionic charges. Vinyl polymers (typically, polystyrene and polyacrylic) are used for the resin matrix backbone. Divinylbenzene (DVB) is used to crosslink the polymer backbone. There are two major types of resins: macroreticular and gel resins. Macrorecticular resins are solid beads and retain their size when they are dried out because they have a great deal of crosslinking. Gel-type resins contain lots of water and resemble fish scales when they are dried out.

On a microscopic level, ion exchange resins resemble a plate of spaghetti where the spaghetti represents the polymer chains to which cationic or anionic functional groups are attached. As shown on Fig. 10-5, the charged functional groups fixed to the polymer chains have counterions associated with them (shown as A^+ in Fig 10-5a), which are mobile and free to move in the pores of the polymer matrix. Cation A^+ is called the *presaturant* ion. During exchange, some of the A^+ ions will move from the resin into the solution to be replaced by B^+ ions. Cation B^+ is called the exchanging ion.

Classification of Resins by Functional Group

Based on the functional groups bonded to the resin backbone, the four general types of exchange resins are (1) strong-acid cation (SAC), (2) weak-acid cation (WAC), (3) strong-base anion (SBA), and (4) weak-base anion (WBA). The distinctions are based on the pK values of the functional groups as summarized in Table 10-7.

The general exchange and regeneration reactions for these functional groups can be written as

$$n[\mathrm{R}^{\pm}]\mathrm{A}^{\pm} + \mathrm{B}^{n\pm} \rightleftarrows [n\mathrm{R}^{\pm}]\mathrm{B}^{n\pm} + n\mathrm{A}^{\pm} \quad \text{(exchange reaction)} \tag{10-12}$$

$$[n\mathrm{R}^{\pm}]\mathrm{B}^{n\pm} + n\mathrm{A}^{\pm} \rightleftarrows n[\mathrm{R}^{\pm}]\mathrm{A}^{\pm} + \mathrm{B}^{n\pm} \quad \text{(regeneration reaction)} \tag{10-13}$$

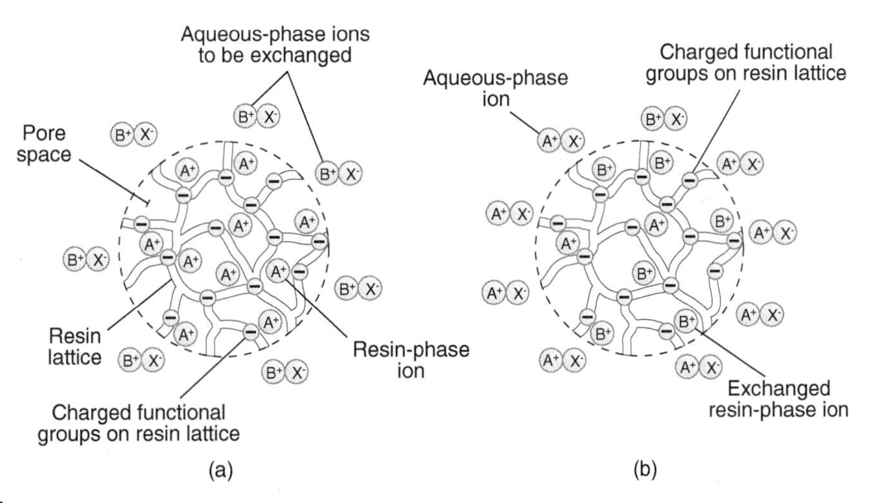

Figure 10-5
Schematic framework of a cation exchange resin: (a) resin with A^+ presaturant ions initially immersed in an aqueous solution containing B^+ cations and X^- anions and (b) cation exchange resin in equilibrium with the aqueous solution of B^+ cations and X^- anions.

Table 10-7
Characteristics of ion exchange resins used in water treatment

Resin Type	Acronym	Fixed Charged Functional Group (R)	Presaturant Ion (A)	pK	Exchange Capacity, meq/mL	Constituents Removed
Strong-acid cation	SAC	Sulfonate, SO_3^-	H^+ or Na^+	<0	1.7–2.1	H^+ form: any cation; Na^+ form: divalent cations
Weak-acid cation	WAC	Carboxylate, COO^-	H^+	4–5	4–4.5	Divalent cations first, then monovalent cations until alkalinity is consumed
Strong-base anion (type 1)	SBA-1	Quaternary amine, $(CH_3)_3\ N^+$	OH^- or Cl^-	>13	1–1.4	OH^- form: any anion; Cl^- form: sulfate, nitrate, perchlorate, etc.
Stong-base anion (type 2)	SBA-2	Quaternary amine, $(CH_3)_2(CH_3CH_2OH)N^+$	OH^- or Cl^-	>13	2–2.5	OH^- form: any anion; Cl^- form: sulfate, nitrate, perchlorate, etc.
Weak-base anion	WBA	Tertiary amine, $H(CH_3)_2N^+$	OH^-	5.7–7.3	2–3	Divalent anions first, then monovalent anions until strong acid is consumed

where $R^{\pm}$ = fixed charged functional group (see Table 10-7)
$A^{\pm}$ = presaturant ion (see Table 10-7)
$B^{n\pm}$ = counterion in solution being exchanged
n = charge on the counterion

Equations 10-12 and 10-13 are general equations for cation exchange resins. The reactions for anion exchange resins are essentially identical except the charge on the fixed functional groups and exchanging ions are reversed. The specific resin types are discussed in more detail in the following sections.

STRONG-ACID CATION EXCHANGE RESIN

In SAC exchange resins, a charged sulfonate group typically acts as the exchange site. The term "strong" in SAC has nothing to do with the physical strength of the resin, but it originates from the ease with which the functional group will lose a proton. For strong acids such as sulfuric acid, functional groups will readily dissociate at any pH. In other words, the resin's low $pK_a (< 0)$ implies SAC resins will readily give up a proton over a wide pH range (1 to 14). For the reaction shown in Eq. 10-12, based on the pK_a of

SAC resins and the large hydrated radius of hydrogen, SAC resins have little affinity for the hydrogen ion and will readily exchange it for another cation. Because the hydrated radius of the H^+ ion in a SAC resin is much larger than other cations, the resin will typically shrink upon exchange ($\approx$7 percent for a gel-type resin, 3 to 5 percent for macroreticular-type resin). The sodium form of a SAC will also behave in a similar manner, although the shrinkage will be less than observed for the hydrogen form.

WEAK-ACID CATION EXCHANGE RESIN

In WAC exchange resins, the functional group on the resin is usually a carboxylate, and the exchange reaction can be written with R = COO in Eqs. 10-12 and 10-13. Weak-acid cation resins have pK_a values in the range of 4 to 5 and thus will not readily give up a proton unless the pH is greater than 6. At a pH less than 6, WAC resins have a great affinity for hydrogen and will not exchange it for another cation; therefore the apparent cation exchange capacity of a WAC resin is a function of pH and the effective operating range for exchange is pH > 7. As the pH increases, the apparent capacity increases to a maximum total capacity between pH values of 10 and 11.

Weak-acid resins usually require alkaline species in the water to react with the more tightly bound hydrogen ions. Because WAC resins exhibit a higher affinity for H^+ than SAC resins do, they exhibit higher regeneration efficiencies. WAC resins do not require as high a concentration of regenerant as that required for regenerating SAC resins to the hydrogen form. The carboxylic functional groups will utilize up to 90 percent of the acid (HCl or H_2SO_4) regenerant, even with low acid concentrations. By comparison, SAC resin regeneration requires a large excess of regenerant solution to provide the driving force for exchange to take place.

Weak-acid resins have been used in water treatment to remove cations in high alkalinity water (e.g., high $CO_3{}^{2-}$, OH^-, and $HCO_3{}^-$ concentrations) with low dissolved carbon dioxide and sodium.

STRONG-BASE ANION EXCHANGE RESIN

Strong-base anion exchange resins typically have a quaternary amine group as the fixed positive charge. Strong-base anion resins have pK_b values of 0 to 1, implying that they will readily give up a hydroxide ion if the pH value is less than 13. The operational pH of SBA resins (pH < 13) makes the apparent anionic exchange capacity independent of pH. Strong-base anion resins in the hydroxide form will shrink upon exchange due to other anions typically having hydrated radii smaller than hydroxide. Type 1 has a slightly greater chemical stability, while type 2 has a slightly greater regeneration efficiency and capacity. SBA resins are less stable than SAC resins and are characterized by the fishy odor of amines even at room temperature.

Strong-base resins traditionally have been used for many years to demineralize water. However, more recently SBA resins are increasingly being used to treat waters contaminated with nitrate, arsenic, and perchlorate ions and are usually operated in the chloride cycle, where the resin is regenerated with NaCl.

WEAK-BASE ANION EXCHANGE RESIN

In WBA exchange resins, the exchange site is a tertiary amine group, which does not have a permanent fixed positive charge. Weak-base anion exchange resins are available in either chloride or free-base forms. The free-base designation indicates that the tertiary amine group is not ionized but has a water molecule (HOH) associated with it. The tertiary amine groups will adsorb ions without the exchange of an ion (Helfferich, 1995).

The weak-base designation is derived from the WBA resin's pK_b values of 5.7 to 7.3. Weak-base anion resins will not readily give up hydroxide ion unless the pOH is greater than the pK_b of the resin (pH values less than 8.3 to 6.7 at 25°C); hence, the effective operating range is $pH < 6$.

Ion Exchange Contactors

Ion exchange processes can be operated using either fixed-bed or suspended contactors, similar to the types of contactors used for adsorption. A key difference between ion exchange and adsorption is that the capacity of ion exchange media is used much more quickly. Ion exchange columns used for applications like softening often operate only for a few days or less before reaching capacity. Fortunately, ion exchange resin is easily regenerated on site, whereas adsorption media typically has to be replaced or taken offsite for regeneration.

After reaching the exchange capacity, ion exchange columns are regenerated by contacting the resin with a concentrated brine solution containing the presaturant ion. If the feed water contains some particulate matter, the resin column may also filter the solids from the water and consequently will need to be backwashed to remove solids prior to regenerating the media. Following regeneration, the media is rinsed to remove the excess brine solution from the bed pore volume prior to being placed back in service. Typical operating parameters for fixed-bed SAC and SBA ion exchange columns, along with properties of those resins, are provided in Table 10-8.

Exchange Capacity

The maximum amount of ions that can be exchanged before the resin must be regenerated is known as the exchange capacity. In most ion exchange literature, the capacity is expressed in terms of a wet-volume capacity. The wet-volume capacity depends upon the moisture content of the resin, which is dependent upon the functional form of the resin and will vary for a given type of resin. The wet-volume capacity is commonly expressed in milliequivalents per milliliter of resin bed (meq/mL), although it may also be expressed in terms of kilograins as $CaCO_3$ per cubic foot (kgr/ft^3) of resin. The conversion is 21.8 meq/mL = 1 kgr/ft^3. As shown in Table 10-8, typical SAC exchange capacities are 1.8 to 2.0 meq/mL in the sodium form, and SBA exchange capacities are 1.0 to 1.3 meq/mL in the chloride form.

Table 10-8
Properties of fixed bed ion exchange columns using styrene–divinylbenzyl, gel-type strong-acid cation and strong-base anion resins

Parameter	Unit	Strong-Acid Cation Resin	Type I, Strong-Base Anion Resin
Screen size, U.S. mesh	—	16 × 50	16 × 50
Shipping weight	kg/m^3	850	700
	(lb/ft^3)	(53)	(44)
Moisture content	%	45–48	43–49
pH range	—	0–14	0–14
Maximum operating temperature	°C	140	OH^- form 60, Cl^- form 100
Turbidity tolerance	NTU	5	5
Iron tolerance	mg/L as Fe	5	0.1
Chlorine tolerance	mg/L Cl_2	1.0	0.1
Backwash rate	m/h	12–20	4.9–7.4
	$(gal/min \cdot ft^2)$	(5–8)	(2–3)
Backwash period	min	5–15	5–20
Expansion volume	%	50	50–75
Regenerant and concentration[a]	%	NaCl, 3.0–14	NaCl, 1.5–14
Regenerant dose	kg $NaCl/m^3$ resin	80–320	80–320
	(lb/ft^3)	(5–20)	(5–20)
Regenerant rate	BV/min	0.067	0.067
	$(gal/min\ ft^3)$	(0.5)	(0.5)
Rinse volume	BV	2–5	2–10
	(gal/ft^3)	(15–35)	(15–75)
Exchange capacity	meq/mL as $CaCO_3$,	1.8–2.0	1–1.3
	$(kgr/ft^3$ as $CaCO_3)$[b]	(39–41)	(22–28)
Operating capacity[c]	meq/mL as $CaCO_3$,	0.9–1.4	0.4–0.8
	$(kgr/ft^3$ as $CaCO_3)$[b]	(20–30)	(12–16)
Service flow rate	BV/h	8–40	8–40
	$(gal/min \cdot ft^3)$	(1–5)	(1–5)

Source: Adapted from Clifford et al. (2011).
[a]Other regenerants such as H_2SO_4, HCl, and $CaCl_2$ can also be used for SAC resins while NaOH, KOH, and $CaCl_2$ can be used for SBA regeneration.
[b]Kilograins $CaCO_3/ft^3$ are the units commonly reported in resin manufacturer literature. To convert kgr $CaCO_3/ft^3$ to meq/mL, multiply by 0.0458.
[c]Operating capacity is based on Amberlite IR-120 SAC resin. Operating capacities depend on method of regeneration and amount of regenerant applied. Manufacturers should provide regeneration data in conjunction with operating capacities for their resins.

Weak-acid cation exchange capacities are about 4.0 meq/mL in the H^+ form and WBA exchange capacities are around 1.0 to 1.8 meq/mL in the free-base form, although WAC and WBA resin capacities are variable due to their partially ionized conditions and because exchange capacity is also a function of pH.

Selectivity

Ion exchange resins have a greater affinity or preference for certain ions. This preference is called *selectivity*. The direction, forward or reverse, of the ion exchange reactions shown in Eqs. 10-12 and 10-13 depends upon the resin selectivity for a particular ion system. For example, if a dilute aqueous solution containing NO_3^- and Cl^- ions are being treated with a type I SBA resin in the OH^- form, both NO_3^- and Cl^- ions will be exchanged over the presaturant ion OH^- because they are preferred by the resin. In this case the reaction proceeds in the forward direction. Type I SBA resins also have a higher selectivity for NO_3^- ions over Cl^- ions so NO_3^- will occupy more exchange sites in a dilute solution.

Resin selectivity depends upon the physical and chemical characteristics of the exchanging ion and resins. Physical properties of the resins that influence selectivity include pore size distribution and the type of functional groups on the polymer chains. Chemical properties of the ions that impact selectivity are the magnitude of the valence and the atomic number of the ion. The following discussion provides insight into these properties.

For dilute aqueous-phase concentrations at temperatures encountered in water treatment, ion exchange resins prefer the counterion of higher valence, as shown below:

$$\text{Cations:} \quad Th^{4+} > Al^{3} > Ca^{2+} > Na^{+}$$

$$\text{Anions:} \quad PO_4^{3-} > SO_4^{2-} > Cl^{-}$$

In the preference shown above, it is assumed that the spacing of the functional groups allow for the exchange of multivalent ions. In other words, there has to be the correct number of cationic functional groups in close proximity to neutralize the charge of the anion or vica versa.

There are some exceptions to the above general rule. For example, divalent CrO_4^{2-} has a lower preference than monovalent I^- and NO_3^- ions, as shown in the following series:

$$SO_4^{2-} > I^{-} > NO_3^{-} > CrO_4^{2-} > Br^{-}$$

Resin selectivity can also be influenced by the degree of swelling or pressure within the resin bead. In an aqueous solution, both resin-phase ions and ions in aqueous solution have water molecules that surround them. The group of water molecules surrounding each ion is called the radius of hydration and is different for different ions. Typically, the radius of hydration becomes larger as the size of the ion decreases, as shown in Table 10-9. When these ions diffuse in solution, the water molecules associated with them move as well. The crosslinking bonds that hold the resin matrix together oppose the osmotic forces exerted by these exchanged ions. These opposing forces cause swelling of the resin. In a dilute aqueous phase containing ion exchange resins, the ions with a smaller hydrated radius are preferred because they reduce the swelling pressure of the resin and are more tightly

Table 10-9
Comparison of ionic, hydrated radii, molecular weight, and atomic number for a number of cations

Ion	Ionic Radii,[a] Å	Hydrated Radii,[b] Å	Molecular Weight	Atomic Number
Li^+	0.60	10.0	6.94	3
Na^+	0.95	7.9	22.99	11
K^+	1.33	5.3	39.10	19
Rb^+	1.48	5.09	85.47	37
Cs^+	1.69	5.05	132.91	55
Mg^{2+}	0.65	10.8	24.30	12
Ca^{2+}	0.99	9.6	40.08	20
Sr^{2+}	1.13	9.6	87.62	38
Ba^{2+}	1.35	8.8	137.33	56

[a]From Mortimer (1975).
[b]From Kunin and Myers (1950).

bound to the resin. For some alkali metals the order of preference for exchange is inversely related to their hydrated radius:

$$Cs^+ > Rb^+ > K^+ > Na^+ > Li^+$$

The selectively is also in reverse order of atomic number. Similarly, for alkaline earth metals the preference for exchange is

$$Ba^{2+} > Sr^{2+} > Ca^{2+} > Mg^{2+} > Be^{2+}$$

For a given series, anion exchange follows the same selectivity relationship with respect to ionic and hydrated radii as cations:

$$ClO_4^- > I^- > NO_3^- > Br^- > Cl^- > HCO_3^- > OH^-$$

Consequently, for a given series of ions, the resin selectivity for ions increases with increasing atomic number, increasing ionic radius, and decreasing hydrated radius.

With the exception of specialty resins, WAC resins with carboxylic functional groups behave similar in preference to SAC resins with the exception that hydrogen is the most preferred ion. In a similar manner, the preference of anions for WBA resins is the same as for SBA resins with the exception that the hydroxide ion is the most preferred ion.

The above general rules for order of selectivity apply to ions in waters that have TDS values less than approximately 1000 mg/L. The preference for divalent ions over monovalent ions diminishes as the ionic strength of a solution increases. For example, in a sulfonic cation exchange resin operating on the sodium cycle, calcium ion is preferred over sodium in dilute concentrations; hence calcium will replace sodium on the resin

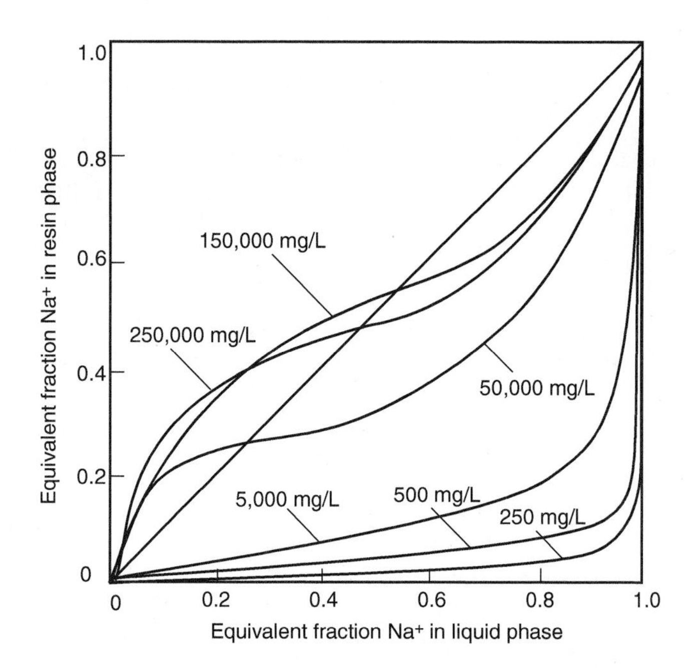

Figure 10-6
The Na^+–Ca^{2+} equilibria for sulfonic acid cation exchange resin. (Courtesy of Rohm and Haas.)

structure. However, at high salt concentrations (≈100,000 mg/L TDS), the preference reverses and this enhances regeneration efficiency, as shown in Fig. 10-6.

The concentrations in Fig. 10-6 are based on equivalent fractions. The equivalent fraction in the aqueous phase is calculated from the following:

$$X_i = \frac{C_i}{C_T} \quad X_j = \frac{C_j}{C_T} \tag{10-14}$$

where C_T = total aqueous ion concentration, eq/L
C_i, C_j = aqueous-phase concentration of counterion and presaturant ion, eq/L

The equivalent fraction in the resin phase is expressed as

$$Y_i = \frac{q_i}{q_T} \quad Y_j = \frac{q_j}{q_T} \tag{10-15}$$

where q_T = total exchange capacity of resin, eq/L

Equilibrium isotherms for Na^+–Ca^{2+} exchange are shown on Fig. 10-6. As the TDS concentration increases, a higher concentration of sodium (or equivalent fraction of Na) can be found in the resin phase. This is because as the salt concentration increases, the sodium concentration increases, and the activity coefficient for calcium decreases such that sodium is preferred over calcium.

10-5 Ion Exchange Equilibrium

Because electroneutrality must be maintained, ion exchange equilibrium relies on equivalence instead of concentration of the ions, thus, equilibrium is expressed in terms of equivalent fractions. The binary *separation factor* α_j^i is a measure of the preference for one ion over another during ion exchange and can be expressed as

$$\alpha_j^i = \frac{Y_i X_j}{X_i Y_j} \tag{10-16}$$

where α_j^i = separation factor of ion i with respect to ion j, unitless
X_i, X_j = equivalent fraction of counterion and presaturant ion in aqueous phase
Y_i, Y_j = equivalent fraction of counterion and presaturant ion in resin

Substituting Eqs. 10-14 and 10-15 into Eq. 10-16 yields

$$\alpha_j^i = \frac{q_i C_j}{C_i q_j} \tag{10-17}$$

where α_j^i = separation factor of ion i with respect to ion j, unitless
concentrations are in eq/L

For process design calculations, binary separation factors are primarily used in ion exchange calculations because they are experimentally determined and account for the solution concentration and the total ion exchange capacity.

It is important to note that the separation factor may not be a constant but rather is influenced by various factors: exchangeable ions (size and charge), properties of the resins, including particle size, degree of crosslinking, capacity, and type of functional groups occupying the exchange sites; water matrix, which includes total concentration, type, and quantity of organic compounds present in solution; reaction period; and temperature. Both binary component systems and isotherms are discussed in the following sections.

Binary Ion Exchange

A binary system involves the exchange of a presaturant ion with only one other component ion in solution. For the binary system, the total aqueous-phase equivalent concentration can be expressed as

$$C_T = C_i + C_j \tag{10-18}$$

where C_T = total aqueous ion concentration, eq/L
C_i, C_j = aqueous concentration of counterion and presaturant ion, eq/L

Total resin-phase equivalent concentration can be expressed as

$$q_T = q_i + q_j \tag{10-19}$$

where q_T = total resin-phase ion concentration, eq/L
q_i, q_j = concentration of counterion and presaturant ion in resin, eq/L

Substitution of Eqs. 10-18 and 10-19 into Eq. 10-17 and rearranging algebraically yields the following expression for calculating the resin-phase concentration of the counterion of interest:

$$q_i = \frac{q_T \alpha_j^i C_i}{C_j + \alpha_j^i C_i} \tag{10-20}$$

For a given counterion concentration, Eq. 10-20 can be used to estimate the resin-phase concentration provided the binary separation factor and the total resin capacity are known. An inspection of Eq. 10-6 reveals that Eq. 10-20 is essentially equivalent to the Langmuir isotherm. This similarity is because the conditions of ion exchange equilibrium are the same as the assumptions used in developing the Langmuir isotherm; that is, exchange is equivalent to monolayer coverage and all exchange sites have the same energy.

Separation factors for commercially available SAC and SBA resins are given in Table 10-10. Based on the definition of Eq. 10-16, a separation factor greater than 1 means that ion *i* is preferred over ion *j*. For example, if $\alpha_{Cl^-}^{NO_3^-} = 2.3$, if the aqueous-phase concentrations expressed in equivalents are equal, NO_3^- is preferred over chloride by 2.3 to 1.0 in the resin. The magnitude of the separation factors is different for WAC and WBA resins from those shown in Table 10-10 for SAC and SBA resins. When separation factors for a given resin are unknown, they may be determined experimentally using binary isotherms. The procedures for performing ion exchange isotherms are essentially identical to adsorption isotherms.

Multicomponent Ion Exchange

The conventional application of ion exchange involves treatment of water containing multiple cations and anions (e.g., Na^+, Ca^{2+}, Mg^{2+}, Cl^-, HCO_3^-, SO_4^{2-}). Some waters may also contain ions of more significant health threat, such as Ba^{2+}, Ra^{2+}, Pb^{2+}, Cu^{2+}, NO_3^-, $HAsO_4^-$, F^-, and ClO_4^-. Consequently, a multicomponent expression is needed to describe the competitive interactions between the ions for the fixed resin sites at equilibrium. In a multicomponent system, the total capacity of the resin and the total concentration of exchanging ions in solution can be expressed as

$$q_T = q_i + q_j + \cdots + q_n \tag{10-21}$$

$$C_T = C_i + C_j + \cdots + C_n \tag{10-22}$$

Table 10-10
Separation factors for commercially available cation and anion exchange resins[a]

Strong-Acid Cation Resin		Strong-Base Anion Resin[b]	
Cation	$\alpha^i_{Na^+}$	Anion	$\alpha^i_{Cl^-}$
Ra^{2+}	13.0	$UO_2(CO_3)_3^{4-}$	3200
Ba^{2+}	5.8	ClO_4^-[c]	150
Pb^{2+}	5.0	CrO_4^{2-}	100
Sr^{2+}	4.8	SO_4^{2-}	9.1
Cu^{2+}	2.6	$HAsO_4^{2-}$	4.5
Ca^{2+}	1.9	NO_3^-	3.2
Zn^{2+}	1.8	Br^-	2.3
Fe^{2+}	1.7	SeO_3^{2-}	1.3
Mg^{2+}	1.67	NO_2^-	1.1
K^+	1.67	Cl^-	1.0
Mn^{2+}	1.6	BrO_3^-	0.9
NH_4^+	1.3	HCO_3^-	0.27
Na^+	1.0	CH_3COO^-	0.14
H^+	0.67	F^-	0.07

Source: Adapted from Clifford et al. (2011).
[a]Values are approximate separation factors for solutions with TDS = 250–500 mg/L.
[b]SBA resin has $-N(CH_3)_3$ functional groups (i.e., a type 1 resin).
[c]ClO_4^-/Cl^- separation factor is for polystyrene SBA resins; on polyacrylic SBA resins, the ClO_4^-/Cl^- separation factor is approximately 5.0.

where q_T = total resin-phase ion concentration, eq/L resin
q_i, q_j, q_n = resin-phase concentrations of counterions i to n (presaturant ion is j), eq/L
C_T = total aqueous-phase ion concentration, eq/L
C_i, C_j, C_n = aqueous concentrations of counterions i to n, (presaturant ion is j), eq/L

Using the same substitutions and algebraic manipulations used to develop Eq. 10-20, the following expression for q_i in terms of n exchanging ions can be developed:

$$q_i = \frac{q_T C_i}{\sum_{k=1}^{n} \alpha_i^k C_k} \tag{10-23}$$

where C_k = aqueous-phase concentration for ion k (presaturant ion when $k = j$), eq/L resin
α_i^k = separation factor for counterion i with respect to ion k

Note that α_i^k assumes the separation factors are known with respect to the ion concentrations being sought on the resin phase for ion i. Since separation factors are reported in terms of the presaturant ion, Eq. 10-23 would be easier to use if the separation factors were with respect to the presaturant instead of each resin-phase ion. Two identities of index notation are useful in manipulating the separation factors:

$$\alpha_i^j = \frac{1}{\alpha_j^i} \tag{10-24}$$

$$\alpha_i^k = \alpha_i^j \alpha_j^k \tag{10-25}$$

If the subscript j is set equal to p where p is equal to the presaturant ion, the following expression for the separation factor in Eq. 10-23 can be obtained:

$$\alpha_i^k = \alpha_i^p \alpha_p^k = \frac{\alpha_p^k}{\alpha_p^i} \tag{10-26}$$

Substitution of Eq. 10-26 into Eq. 10-23 yields the following expression:

$$q_i = \frac{q_T C_i}{\sum_{k=1}^{N} \left(\frac{\alpha_p^k}{\alpha_p^i} C_k \right)} = \frac{q_T C_i}{\frac{1}{\alpha_p^i} \sum_{k=1}^{N} \left(\alpha_p^k C_k \right)} = \frac{q_T \alpha_p^i C_i}{\sum_{k=1}^{N} \left(\alpha_p^k C_k \right)} \tag{10-27}$$

If all the aqueous-phase ion concentrations and the total resin capacity are known, the resin-phase concentrations can be calculated using the separation factors referenced to the presaturant ion as reported in Table 10-10. The use of Eq. 10-27 to calculate resin-phase concentrations at equilibrium is demonstrated in Example 10-3.

Example 10-3 Determination of resin-phase concentrations in multicomponent ion exchange equilibrium

Consider the removal of nitrate from well water using an SBA exchange resin in the chloride form. The major ions contained in the well water are given below. Assuming nitrate is removed completely from solution, calculate the maximum volume of water that can be treated per liter of resin assuming equilibrium conditions. Assume total resin capacity of the SBA is 1.4 eq/L.

Cation		meq/L	Anion		meq/L
Ca^{2+}		0.9	Cl^-		1.0
Mg^{2+}		0.8	SO_4^{2-}		1.5
Na^+		2.6	NO_3^-		1.8
	Total	4.3		Total	4.3

Solution

1. Applying Eq. 10-27 with the use of the separation factors provided in Table 10-10, the summation term in the denominator can be calculated:

$$\sum_{k=1}^{N}\left(\alpha_p^k C_k\right) = (1.0)\,(1\ \text{meq/L}) + (9.1)\,(1.5\ \text{meq/L}) + (3.2)\,(1.8\ \text{meq}\,/\,\text{L}) = 20.41\ \text{meq}\,/\,\text{L}$$

2. Calculate q_i for each ion:

$$q_{\text{Cl}} = \frac{(1.4\ \text{eq}\,/\,\text{L})\,(1.0)\,(1\text{meq}\,/\,\text{L})}{20.41\ \text{meq}\,/\,\text{L}} = 0.069\ \text{eq}\,/\,\text{L}$$

$$q_{\text{SO}_4{}^{2-}} = \frac{(1.4\ \text{eq/L})\,(9.1)\,(1.5\ \text{meq/L})}{20.41\ \text{meq/L}} = 0.936\ \text{eq/L}$$

$$q_{\text{NO}_3{}^{-}} = \frac{(1.4\ \text{eq/L})\,(3.2)\,(1.8\ \text{meq/L})}{20.41\ \text{meq/L}} = 0.395\ \text{eq/L}$$

Check: $0.069 + 0.936 + 0.395 = 1.4$ eq/L total capacity
Note that because the sulfate concentration is higher than nitrate and sulfate is preferred over nitrate ($9.1 \gg 3.2$), the equilibrium capacity of nitrate is low. In other words, nitrate will occupy only about 28 percent (0.395/1.4) of the exchange sites on the resin.

3. Calculate the maximum quantity of water that can be treated per cycle before nitrate breakthrough occurs.

$$V_{\text{max}} = \frac{(0.395\ \text{eq/L resin})(10^3\ \text{meq/eq})}{1.8\ \text{meq/L water}} = 219\ \text{L water/L resin}$$

Comment

When sulfate is present, the capacity of the resin to remove nitrate is reduced significantly. Competitive exchange with sulfate is the primary reason that arsenic removal with ion exchange is rarely economical.

10-6 Ion Exchange Kinetics

The kinetics of ion exchange are similar to those of adsorption as discussed in Sec. 10-3, where mass transfer occurs by boundary layer diffusion followed by intraparticle diffusion. An additional factor in ion exchange is the importance of electroneutrality. When ions diffuse at different rates, charge separation can arise, inducing an electric field that causes ionic migration to satisfy electroneutrality within the resin particle. For example, as cation B diffuses into the resin particle, it is transferring charge to the resin and this charge must be offset by an equivalent charge by another ion (e.g., presaturant ion) or ions diffusing out of the resin particle into solution to satisfy the local electrical balance. If the transfer of charge in opposite directions is not exactly balanced, a net transfer of electric charge would result and violate the requirement of electroneutrality. A small deviation from electroneutrality generates an electric field that produces an additional force that causes all the charged ions in the electric field to move in response to the electrical gradient. The electrical field increases the flux of the slow diffusing ions and decreases the flux of the faster ones, equalizing the net fluxes and so preventing any further buildup of the net charge.

It was noted in Sec. 10-3 that batch adsorption isotherms can take 2 to 4 weeks to reach equilibrium. Ion exchange reaches equilibrium quickly; batch isotherms reach equilibrium in a few minutes. These numbers imply that the rate of mass transfer in ion exchange is several orders of magnitude faster than in adsorption. However, an inspection of Table 4-3 indicates that the diffusion coefficients of ions and neutral molecules are of the same order of magnitude. The rapidity of mass transfer in ion exchange underscores the equal importance of the mass transfer coefficient and concentration gradient in mass transfer processes (see Eq. 4-114). The presaturant ion is present in the resin at mol/L concentrations, whereas constituents being removed by adsorption are present in water at mmol/L or even μmol/L concentrations. The high concentration of presaturant ion in the resin induces a large concentration gradient out of the resin, stimulating a high flux of presaturant ions. Consequently, a large flux of ions into the resin is induced to maintain electroneutrality, with the net result that equilibrium is achieved in a short period of time.

The speed of ion exchange has implications for design. Fixed-bed adsorption columns are typically designed with 15 to 30 min of empty bed contact time, whereas ion exchange columns have a narrow mass transfer zone and are often designed with 2 to 4 min of contact time. Furthermore, equilibrium models are sufficient for predicting IX performance, as opposed to the necessity of using kinetic-based models for adsorption. Moreover, unlike adsorption applications which may operate for months to years, IX columns typically only operate for days before regeneration is needed. Consequently, pilot studies can be conducted in a reasonable period of time to assess field performance of IX processes.

10-7 Fixed-Bed Contactors

Fixed-bed adsorption and ion exchange contactors consist of a bed of media typically 1 to 3 m deep, through which water passes to provide contact between the water and media. Transfer of solutes from the water occurs progressively from the top to the bottom of the column. When the media capacity has been used up, the media is regenerated (in the case of ion exchange) or replaced (in the case of adsorption). The theory and design of fixed beds is presented in this section.

Fixed-Bed Contactor Process Parameters

Process parameters that describe fixed-bed contactor operation include the contact time between the water and media, the loading rate, and the volume of water that can be treated. The contact time between the water being treated and the adsorbent or ion exchange media is characterized by the time empty-bed contact time (EBCT):

$$\text{EBCT} = \frac{V_b}{Q} \tag{10-28}$$

where EBCT = empty-bed contact time, h
V_b = volume occupied by the media bed, m^3
Q = flow rate to the contactor, m^3/h

The EBCT varies from 5 to 60 min for adsorption processes. For removal of SOCs from water by GAC, an EBCT in the range of 5 to 30 min is common. The EBCT for IX contactors is 1.5 to 7.5 min because the mass transfer zone is much shorter for IX than adsorption, as discussed in Sec. 10-6.

The quantity of water treated in a fixed-bed column is often expressed as a ratio to the media volume with units of *bed volumes* (BV):

$$V^* = \frac{V_w}{V_b} = \frac{Qt}{V_b} = \frac{t}{\text{EBCT}} \tag{10-29}$$

where V^* = specific volume of water treated, m^3/m^3 or BV
V_w = volume of water treated, m^3
t = time of operation, h

Ion exchange columns normally operate for hundreds of bed volumes before regeneration, and adsorption columns typically operate thousands or tens of thousands of bed volumes before the media reaches breakthrough. In adsorption, it is also common to express the quantity of water treated in terms of the mass of adsorbent instead of volume:

$$V_{sp} = \frac{V_w}{M} = \frac{Qt}{\rho_b V_b} = \frac{t}{\text{EBCT}\rho_b} \tag{10-30}$$

where V_{sp} = specific throughput, m^3/kg
M = mass of media, kg
$\rho_b = M/V_b$ = media bed density, kg/m^3

The performance of GAC adsorption columns is often quantified as the inverse of the specific throughput, which is amount of carbon used to treat a volume of water and is known as the carbon usage rate (CUR):

$$\text{CUR} = \frac{M}{V_w} = \frac{1}{V_{sp}} \tag{10-31}$$

where CUR = carbon usage rate, kg/m^3

The loading rate through the column is normalized on a surface area or volumetric basis. The surface area loading rate is known as the superficial velocity and is important in determining the head loss through the column:

$$v = \frac{Q}{A_b} \tag{10-32}$$

where v = superficial velocity, m/h
A_b = cross-sectional area of bed perpendicular to flow, m^2

The superficial velocity typically ranges from 5 to 15 m/h (2 to 6 gpm/ft^2) for adsorption columns and from 8 to 80 m/h (3.2 to 32 gpm/ft^2) for ion exchange columns. Noting that $V_b = L \times A_b$ and $Q = v \times A_b$, the EBCT can be related to the superficial velocity as follows:

$$\text{EBCT} = \frac{V_b}{Q} = \frac{A_b L}{A_b v} = \frac{L}{v} \tag{10-33}$$

where L = depth of the column, m

The volumetric loading rate is known as the service loading rate and is defined as the flow rate divided by the volume of the media:

$$\text{SFR} = \frac{Q}{V_b} \tag{10-34}$$

where SFR = service flow rate, $\text{m}^3/\text{m}^3 \cdot \text{h}$ or BV/h

An inspection of Eq. 10-28 reveals that the SFR is the inverse of the EBCT. The SFR is typically 1 to 12 BV/h (0.12 to 1.5 $\text{gal/min} \cdot \text{ft}^3$) for adsorption columns and 8 to 40 BV/h (1 to 5 $\text{gal/min} \cdot \text{ft}^3$) for ion exchange columns. The use of the design equations presented in this chapter are demonstrated in Example 10-4.

Example 10-4 Process design parameters for fixed-bed columns

An adsorption column has a diameter of 3.0 m, a media depth of 2.5 m, and treats a flow of 2.54 ML/d. Calculate the (a) empty-bed contact time, (b) superficial velocity, and (c) service flow rate.

Solution

1. Calculate the cross-sectional area and volume of the media bed:

$$A_b = \frac{\pi}{4}(3.0\text{ m})^2 = 7.07\text{ m}^2$$

$$V_b = A_b L = (7.07\text{ m}^2)(2.5\text{ m}) = 17.7\text{ m}^3$$

2. Calculate the empty-bed contact time using Eq. 10-28:

$$\text{EBCT} = \frac{V_b}{Q} = \frac{(17.7\text{ m}^3)(1440\text{ min/d})}{(2.54\text{ ML/d})(10^3\text{ m}^3/\text{ML})} = 10\text{ min}$$

3. Calculate the superficial velocity using Eq. 10-32:

$$v = \frac{Q}{A_b} = \frac{(2.54\text{ ML/d})(10^3\text{ m}^3/\text{ML})}{(7.07\text{ m}^2)(24\text{ h/d})} = 15\text{ m/h}$$

4. Calculate the service flow rate using Eq. 10-34:

$$\text{SFR} = \frac{Q}{V_b} = \frac{1}{\text{EBCT}} = \frac{60\text{ min/h}}{10\text{ min}} = 6\text{ BV/h}$$

Particle and Bed Porosity

The porosity of a fixed-bed contactor is complicated by the fact that the porosity of the media grains needs to be taken into account when the bed porosity is calculated. The particle and bed porosities are defined as

$$\varepsilon_p = 1 - \frac{\rho_p}{\rho_s} \tag{10-35}$$

$$\varepsilon_b = 1 - \frac{\rho_b}{\rho_p} \tag{10-36}$$

where ε_p = particle porosity, unitless
ε_b = bed porosity, unitless
ρ_p = particle density, kg/L
ρ_s = solid material density, kg/L
ρ_b = bed density, kg/L

Using activated carbon as an example, the solid material density of graphite is about 2.0 to 2.2 kg/L. Activated carbon grains can have a porosity of 0.2 to 0.7; a value of 0.5 results in a particle density of about 1.1 kg/L. If the bed porosity is also around 0.5, the bed density would be about 0.55 kg/L. Bed densities of 0.35 to 0.55 kg/L are common for GAC.

Theoretical Capacity of Fixed-Bed Columns

The maximum specific throughput of a fixed column can be calculated from a mass balance analysis if the MTZ is assumed to be so short that the concentration in the column appears to be a step function. In this case, the media will be completely saturated at the point when the solute reaches the end of the column. In effect, all the solute fed is transferred to the media in the column and the media is in equilibrium with the influent concentration. Relating the total quantity of solute fed to the column to the ultimate capacity of the media in the column, an expression for the maximum specific throughput can be derived as follows:

$$QC_{\text{inf}}\, t_{\text{bk}} = M\, q_e|_{C_{\text{inf}}} \tag{10-37}$$

where Q = flow rate to the contactor, m^3/h
C_{inf} = influent aqueous-phase concentration of the solute, mg/L
t_{bk} = time to breakthrough, h
M = mass of media, kg
$q_e|_{C_{\text{inf}}}$ = solid-phase concentration of the solute in equilibrium with the influent concentration, mg adsorbate/g adsorbent

The maximum specific throughput and minimum carbon usage rate are then given by the expressions

$$V_{sp,\text{max}} = \frac{Qt_{\text{bk}}}{M} = \frac{q_e|_{C_{\text{inf}}}}{C_{\text{inf}}} \tag{10-38}$$

$$\text{CUR}_{\text{min}} = \frac{1}{V_{sp,\text{max}}} = \frac{C_{\text{inf}}}{q_e|_{C_{\text{inf}}}} \tag{10-39}$$

where $V_{sp,\text{max}}$ = maximum specific throughput, m^3/kg
CUR_{min} = minimum carbon usage rate, kg/m^3

Calculation of the volume of water treated and bed life of a fixed-bed column using the maximum specific throughput is demonstrated in Example 10-5.

Example 10-5 Maximum capacity of a fixed-bed adsorption column

The adsorption column in Example 10-4 is used to remove TCE from groundwater. The influent concentration is 1 mg/L and maximum effluent concentration is 0.005 mg/L. The column contains Calgon Filtrasorb 400 (12 × 40 mesh) that has a bed density of 450 g/L. Calculate the (a) maximum specific throughput, (b) minimum carbon usage rate, (c) volume of water treated, and (d) bed life.

Solution

1. Maximum specific throughput can be calculated with Eq. 10-38. The solid-phase concentration in equilibrium with the influent TCE concentration can be calculated using the Freundlich isotherm (Eq. 10-8), using the Freundlich parameters for TCE in Table 10-4:

$$q_{e|C_{\text{inf}}} = K(C_{\text{inf}})^{1/n} = [55.9\ (\text{mg/g})(\text{L/mg})^{0.48}](1.0\text{mg/L})^{0.48} = 55.9\ \text{mg/g}$$

$$V_{sp,\text{max}} = \frac{q_{e|C_{\text{inf}}}}{C_{\text{inf}}} = \frac{55.9\ \text{mg/g}}{1\ \text{mg/L}} = 55.9\ \text{L/g} \quad \text{(L of water treated per g of carbon)}$$

2. The carbon usage rate is the inverse of the specific throughput. Thus, the minimum CUR is the inverse of the maximum V_{sp}:

$$\text{CUR}_{\text{min}} = \frac{1}{V_{sp,\text{max}}} = \frac{1}{55.9\ \text{L/g}} = 0.018\ \text{g/L} \quad \text{(g of carbon used per L of water treated)}$$

3. Calculate the volume of water treated using Eqs. 10-29 and 10-30 (the volume of the bed was determined in Example 10-4):

$$V_W = V_{sp,\text{max}}(M) = V_{sp,\text{max}}(\rho_b V_b) = (55.9\ \text{L/g})(450\ \text{g/L})(17.7\ \text{m}^3) = 4.45 \times 10^5\ \text{m}^3$$

$$V^* = \frac{V_W}{V_b} = \frac{4.45 \times 10^5\ \text{m}^3}{17.7\ \text{m}^3} = 25{,}100\ \text{BV}$$

4. The bed life can be calculated from Eq. 10-29.

$$t = V^*(\text{EBCT}) = \frac{(25{,}100\ \text{BV})(10\ \text{min})}{1440\ \text{min/d}} = 175\ \text{d}$$

Comment

The volume of water treated and bed life assume no appreciable MTZ and no competitive adsorption from other constituents such as NOM. Thus, this example represents the maximum possible bed life and volume of water treated; actual performance may be significantly less.

The MTZ occupies a portion of the column and reduces the time until some of the influent solute begins showing up in the column effluent. For an MTZ of constant size and shape, the fraction of utilized capacity increases for a GAC column as the length of column is increased, as shown on Fig. 10-7. The maximum specific throughput is zero up to a minimum EBCT because the column must be longer than the MTZ or the effluent concentration immediately exceeds the treatment objective. From a cost perspective, it is important to realize that as the specific throughput increases (by increasing the EBCT) the operation and maintenance costs decrease, but it comes at the expense of increasing capital cost because the contactor size needed is larger.

Process Design for Fixed-Bed Contactors

The primary criteria for the design of a fixed-bed adsorption or ion exchange system is the capacity of the system, the influent water quality, and the required effluent water quality. Information needed to complete

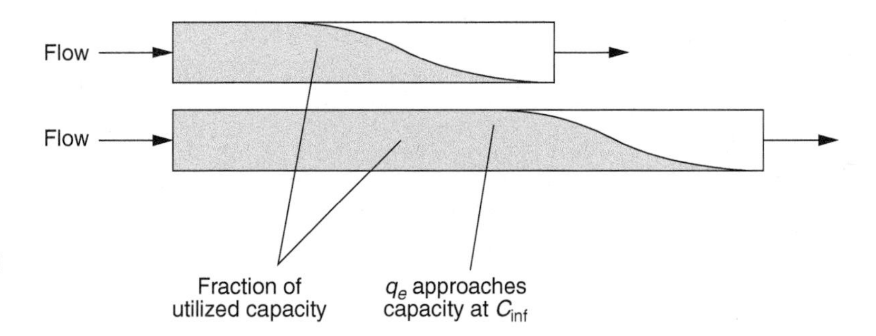

Figure 10-7
Utilized capacity for two fixed-bed columns with constant MTZ lengths.

the design can be obtained from preliminary process analysis, bench-scale studies, and pilot studies, as described in the following sections.

PRELIMINARY PROCESS ANALYSIS

The first step in design of a fixed-bed adsorption or ion exchange system is to define the problem. In addition to determining the concentrations of the solutes to be removed, other water quality parameters such as temperature, pH, and turbidity are needed. Depending upon the specific conditions, the most likely location to apply treatment should be determined so that possible design constraints such as process size, geography, and utility services (sewers, brine waste lines) can be considered in the initial phases of the design. Possible design constraints such as the availability of chemicals, space requirements, regulatory permitting requirements and/or guidelines, and cost limitations should also be considered.

Preliminary studies start with selection of possible media for bench-scale testing. Preliminary calculations and a literature review combined with media manufacturer's performance specifications can be used to assess and choose promising media for bench-scale testing. Properties of several commercial adsorbents are presented in Table 10-1 and ion exchanges resins are presented in Table 10-7. Equilibrium calculations presented in this chapter can be used to assess process capabilities and limitations of each of the media selected. Equilibrium or mass-transfer-based modeling equations and software can also be used to assess preliminary process performance. Adsorption modeling is presented in greater detail in Crittenden et al. (2012).

BENCH-SCALE STUDIES

Bench-scale studies are used to identify media and operating parameters that provide the best possible performance and cost effectiveness over the design life period. For adsorption, the main criteria determined in bench-scale studies is the adsorption capacity of the media, which is determined with adsorption isotherms.

For ion exchange, in addition to exchange characteristics, operating parameters that can be determined in bench studies may include

Figure 10-8
Pilot-scale ion exchange column used to verify bench-scale column tests and obtain data on fouling.

(a) saturation and elution curves to assess ion exchange performance, (b) hydraulic considerations (flow rate, head loss, backwashing rate), and (c) regeneration requirements (i.e., salt requirements, backwash cycle time, rinse requirements, column requirements).

Other variables such as adsorption bed life and IX resin stability under cyclic operation must be monitored over long periods of time and will require pilot-scale testing.

PILOT-SCALE TESTING

The purpose of pilot testing is to assess actual performance of the fixed column prior to the design and construction of full-scale facilities. An example of pilot-scale columns is shown on Fig 10-8. As noted in Sec. 10-3, the kinetics of the adsorption process control the extent of the mass transfer zone, which occupies a portion of the fixed bed. The capacity of the bed will be influenced by the MTZ. Pilot testing can provide a quantitative assessment of the MTZ and bed life provided the pilot testing is designed and operated properly. The MTZ in the pilot column will be similar to the full-scale column if the media diameter and superficial velocity are similar to the full-scale design (i.e., the Reynolds number is maintained between pilot- and full-scale systems). In addition, the amount of water that can be treated and bed life will be similar if the EBCT is the same (see Eq. 10-29). The use of pilot data to predict full-scale performance is demonstrated in Example 10-6.

Example 10-6 Analysis of pilot plant adsorption data

A GAC pilot plant study was performed on a groundwater containing *cis*-1,2-dichloroethene (DCE). The impact of EBCT on GAC performance was evaluated by conducting column experiments for EBCTs of 3, 5, 10, 21, and 32 min. The DCE effluent concentration for each EBCT was plotted in terms of specific throughput using Eq. 10-30 and is displayed below. Using the column data, plot the specific throughput for a treatment objective of 5 μg/L as a function of EBCT and determine a reasonable EBCT for DCE in this groundwater.

Solution

1. Construct a plot of the specific throughput for a treatment objective of 5 μg/L as a function of EBCT.
   a. On the *y* axis, locate the 5-μg/L treatment objective and draw a line parallel to the x axis so it intersects the effluent curves.
   b. Where the 5-μg/L line intersects each effluent curve, draw a line down to the x axis to obtain the specific throughput for each EBCT as shown. For EBCTs of 3, 5, 10, 21, and 32, the specific throughputs are 6.5, 16.0, 22.0, 27.5, and 29.0 m^3 water treated per gram of GAC, respectively.
   c. Plot the specific throughput as a function of EBCT.

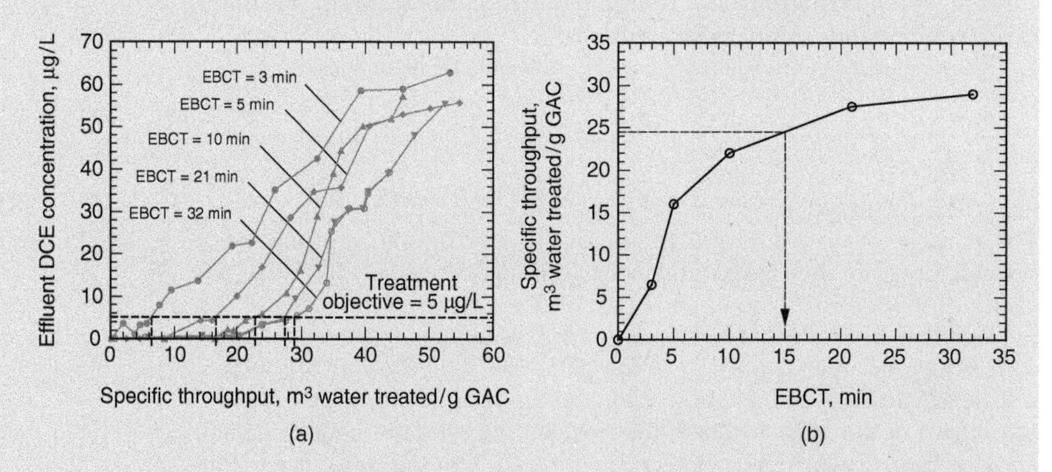

(a) (b)

2. From the plot constructed in step 1c, it is clear that the specific throughput reaches a point of diminishing returns at about EBCT = 15 min.

Comment

The pilot data presented in this example took one year to collect. A pilot test of this duration can be very costly. Accordingly, rapid small-scale column tests may be useful for determining the carbon usage rate.

Pilot testing also allows fouling or changes in media performance to be assessed. The pilot testing can also provide insight into (a) scaleup considerations; (b) column design details, including volume of resin, surface area of columns, number of columns, sidewall height, pressure drop, and inlet and outlet arrangements; (c) overall cycle time; and, in the case of ion exchange (d) regeneration requirements, including volume, salt quantity and concentration, rinse water, and regeneration cycle time.

The disadvantage of pilot testing that operates at the same superficial velocity and EBCT as the full-scale facilities is that the time to breakthrough will also be the same as the full-scale plant. For ion exchange, which can reach breakthrough in several days, the duration is not excessive and many operating cycles can be tested in a reasonable period of time to assess the possibility for fouling or changes in media performance. For adsorption, however, the necessity to operate a pilot plant for more than a year may be prohibitive. Rapid small-scale column tests (RSSCTs) offer an alternative that allows data similar to pilot testing to be collected in a shorter period of time.

RAPID SMALL-SCALE COLUMN TESTS FOR ADSORPTION

Small, scaled-down fixed-bed contactors that use the actual raw water can be used to predict the performance of full-scale contactors if the transport processes scale according to the dimensionless groups that appear in the fixed-bed adsorption models described in Sec. 10-3. Three primary advantages of using RSSCTs to predict performance are (1) the RSSCT may be conducted in a fraction of the time required to conduct pilot studies; (2) unlike predictive mathematical models, extensive isotherm or kinetic studies are not required; and (3) an RSSCT can be conducted with a small volume of water, which can be transported to a central laboratory for evaluation. Consequently, replacing a pilot study with an RSSCT significantly reduces the time and cost of a full-scale design. However, the results from an RSSCT are site specific and only valid for the raw-water conditions that are tested.

RSSCTs rely on scaling equations to relate the results of the RSSCT to the performance of a full-scale model. The RSSCT columns use smaller media than the full-scale columns, and then scale EBCT, run length, and superficial velocity according to the difference in media size. The scaling equations are developed from the mass-transfer-based mathematical models of adsorption processes. Details of the models and the derivation of the

RSSCT scaling equations are presented in Crittenden et al. (2012). The form of the models vary depending on whether diffusion coefficients are assumed to be the same in the RSSCT and full-size columns. The final set of design equations for a constant-diffusivity RSSCT design is given as

$$\frac{\text{EBCT}_{\text{SC}}}{\text{EBCT}_{\text{LC}}} = \frac{t_{\text{SC}}}{t_{\text{LC}}} = \frac{d_{\text{SC}}^2}{d_{\text{LC}}^2} \tag{10-40}$$

$$\frac{v_{\text{SC}}}{v_{\text{LC}}} = \frac{d_{\text{LC}}}{d_{\text{SC}}} \tag{10-41}$$

where $\text{EBCT}_{\text{SC}}, \text{EBCT}_{\text{LC}}$ = empty bed contact time in small-scale and large-scale column, respectively, min

$t_{\text{SC}}, t_{\text{LC}}$ = operating time in small-scale and large-scale column, respectively, h

$d_{\text{SC}}, d_{\text{LC}}$ = media particle diameter in small-scale and large-scale column, respectively, mm

$v_{\text{SC}}, v_{\text{LC}}$ = superficial velocity in small-scale and large-scale column, respectively, m/h

The use of the RSSCT scaling equations is demonstrated in Example 10-7.

Example 10-7 Development of the design and operating parameters of an RSSCT

Calculate the design and operating parameters of an RSSCT that has a particle diameter of 0.21 mm compared to a full-scale unit that has a particle diameter of 1.0 mm. The RSSCT is to be designed using constant-diffusivity RSSCT design. The RSSCT column diameter is 1.10 cm. Typical operating conditions for pilot-scale columns are given in the following table:

Design Parameters	Unit	Pilot Scale
Particle diameter	mm	1.0 (12 × 40)
Bulk density	g/mL	0.49 (F-400)
EBCT	min	10.0
Loading rate	m/h	5.0
Flow rate	mL/min	170.1
Column diameter	cm	5.1
Column length	cm	83.3
Mass of adsorbent	g	833.8
Time of operation	d	100.0
Water volume required	L	24,501

Solution

1. Calculate the $EBCT_{sc}$ using Eq. 10-40:

$$EBCT_{SC} = EBCT_{LC}\frac{d_{SC}^2}{d_{LC}^2} = 10\left(\frac{0.21}{1.0}\right)^2 = 0.44 \text{ min}$$

2. Calculate the hydraulic loading rate using Eq. 10-41:

$$v_{SC} = v_{LC}\frac{d_{LC}}{d_{SC}} = 5.0\left(\frac{1.0}{0.21}\right) = 23.8 \text{ m/h}$$

3. Calculate the run time using Eq. 10-40:

$$t_{SC} = t_{LC}\frac{d_{SC}^2}{d_{LC}^2} = 100\left(\frac{0.21}{1.0}\right)^2 = 4.4 \text{ d}$$

4. Calculate the bed length, flow rate, and mass of carbon using the RSSCT column diameter, superficial velocity, and EBCT:

$$L_{SC} = v_{SC}\, EBCT_{SC} = \frac{(23.8 \text{ m/h})(100 \text{ cm/m})(0.44 \text{ min})}{60 \text{ min/h}} = 17.4 \text{ cm}$$

$$Q_{SC} = v_{SC}A_{SC} = v_{SC}\left(\frac{\pi D_{SC}^2}{4}\right) = \frac{(23.8 \text{ m/h})(\pi)(1.10 \text{ cm})^2(100 \text{ cm/m})}{(4)(60 \text{ min/h})}$$

$$= 37.7 \text{ cm}^3/\text{min} = 37.7 \text{ mL/min}$$

$$M_{SC} = Q_{SC}\, EBCT_{SC}\rho_{SC} = (37.7 \text{ mL/min})(0.44 \text{ min})(0.49 \text{ g/mL}) = 8.1 \text{ g}$$

5. Calculate the volume of water required to run the RSSCT

$$V_W = Q_{SC}t_{SC} = \frac{(37.7 \text{ mL/min})(4.4 \text{ d})(1440 \text{ min/d})}{10^3 \text{ mL/L}} = 239 \text{ L}$$

The design parameters for the RSSCT are:

$D = 1.1$ cm	EBCT $= 0.44$ min	$Q = 37.7$ mL/min
$L = 17.4$ cm	$t = 4.4$ d	$M = 8.1$ g
$d = 0.21$ mm	$v = 23.4$ m/h	$V = 239$ L

Comment

The quantity of water required to simulate 100 days of pilot column operation is 239 L, which can be transported to an off-site laboratory to conduct the test. The RSSCT will be complete in 4.4 days.

SMALL-DIAMETER COLUMNS FOR ION EXCHANGE

Small-diameter IX columns can also be used to develop meaningful process data for ion exchange if operated properly. Because the main issues of concern are mass transfer and operating exchange capacity, small (1.0- to 5.0-cm-inside-diameter) columns using the same media can be scaled directly to full-scale design if the superficial velocity and EBCT are the same. An examination of Eqs. 10-28 and 10-32 indicate that the depth of the small-scale column should be the same as the depth intended for full-scale operation. However, if full-scale depth is not possible to match in the preliminary studies, a minimum bed depth of 0.6 to 0.9 m (2 to 3 ft) should be adequate to properly design a laboratory or pilot IX column. While the depth should be similar to full-scale design, the cross-sectional area can be small provided the ratio of column diameter to particle diameter is larger than 25 to minimize the error due to channeling of the water down the walls of the column. Column studies are used primarily to evaluate and compare resin performance in terms of capacity and ease of regeneration. For example, an automated small-column system used to perform laboratory studies for the removal of perchlorate from a groundwater is shown on Fig. 10-9.

For most commercially available resins pressure drop curves versus flow rate and temperature and bed expansion (for backwash) versus flow rate and temperature can be obtained from the manufacturer. For example,

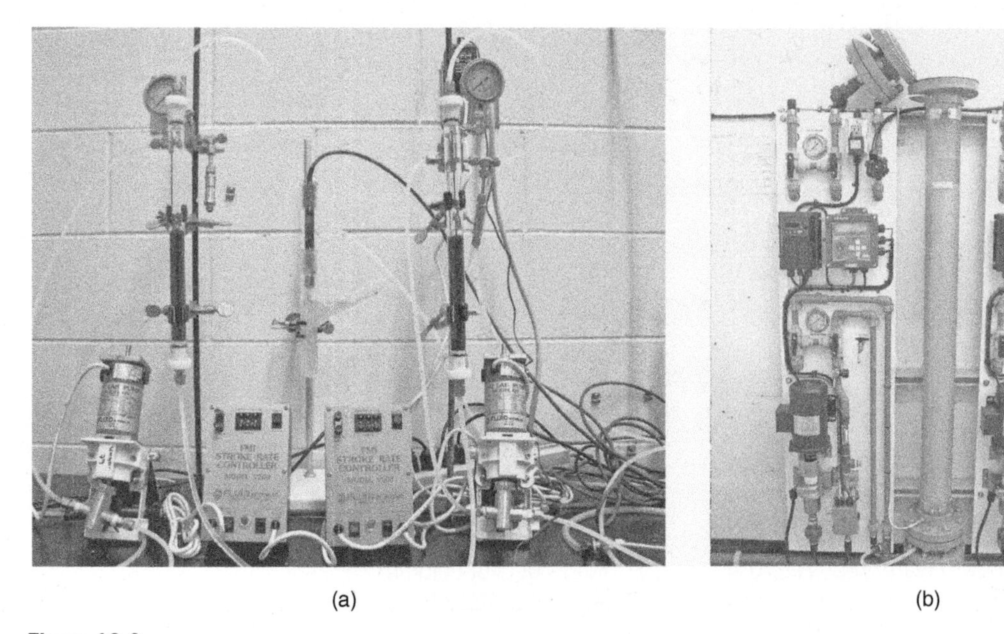

(a) (b)

Figure 10-9
Ion exchange system used to perform preliminary experiments: (a) small-scale laboratory columns and (b) larger laboratory-type ion exchange column.

performance curves for pressure drop and bed expansion as a function of flow rate are given on Fig. 10-10.

The two main types of data collected from small-scale column testing are saturation loading curves and elution curves. Data developed from these curves form the basis for pilot plant studies and/or for the development of full-scale designs.

The *saturation loading curve* is obtained by passing the process stream or a simulated stream of the same chemical composition through a fully regenerated column of resin. During the runs, samples of the effluent are collected and analyzed until the effluent concentration of the contaminant of interest equals the influent concentration. The effluent concentration is plotted as a function of the number of bed volumes of process stream treated to develop a saturation loading curve.

Generalized saturation loading curves for water containing three ions (A, B, and C) that were treated through an exchange column are presented on Fig. 10-11. As shown on Fig. 10-11, each anion has an effluent profile with the less preferred ions (i.e., A and B) appearing first in the effluent followed by the preferred anion (i.e., C). The chromatographic effect, known as *chromatographic peaking* shown on Fig. 10-11 depends upon the equilibrium and mass transfer conditions within the column. Percentage concentrations greater than 100 are possible because of the competitive effects among the

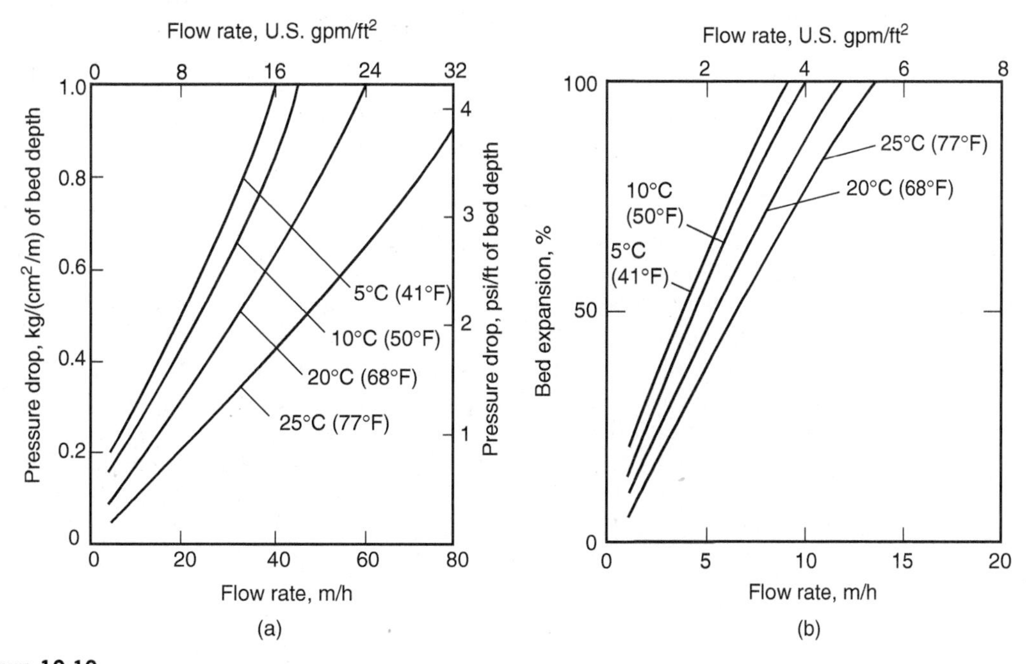

Figure 10-10
(a) Pressure drop and (b) resin bed expansion curves at various water temperature as function of flow rate for strong-base type I acrylic anion exchange resin (A-850, Purolite).

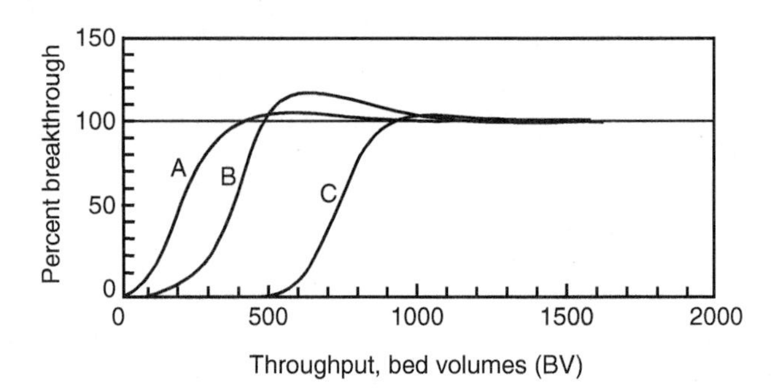

Figure 10-11
Generalized saturation loading curves for compounds A, B, and C.

competing ions, which force previously exchanged ions off the resin. For example, the highest observed effluent concentration for ion B is about 120 percent, or 1.2 times its average influent concentration. In the previous sections, both binary and multicomponent equilibria were discussed and mathematical descriptions were developed. The chromatographic effect within a column can be described when these equilibrium descriptions are incorporated into mass balance expressions. Saturation loading curves provide the performance data necessary to size the columns and determine the operational aspects of the column design.

To determine the optimum SFR, the rate must be varied during the saturation loading tests over a range of choices to see if any noticeable maximum in breakthrough capacity is achieved. Typically, the volumetric flow rate is the criterion used because it is directly related to the film mass transfer rate. The main goal in determining the optimum SFR is to reduce the capital cost of equipment. The optimum SFR will minimize the impact of the film mass transfer resistance and consequently shorten the length of the MTZ. The higher the acceptable flow rate, the smaller the contactor can be for a given treatment flow because the MTZ length can be contained in a smaller column.

After completing each saturation loading curve, the resin must be eluted with an excess of regenerant to fully convert it back to its presaturant form. An curve is obtained, similar to a breakthrough curve, by collecting sample volumes of regenerant after it has passed through the bed and determining the concentrations of the ions of interest in each sample volume. The bed volumes of regenerant used can be converted in terms of a salt loading rate by multiplying it by the salt concentration used and dividing by the volume of the resin bed. These data can be used to choose a regeneration level that will be optimum with respect to operating capacity (resin conversion) and regenerant efficiency.

Generalized regeneration curves for ions A, B, and C for the regeneration of a resin are presented on Fig. 10-12. Notice that with a salt loading of about 240 kg/m^3 all of ion A elutes rapidly and is replaced by chloride ions if the resin is an SBA form and sodium if the resin is an SAC form.

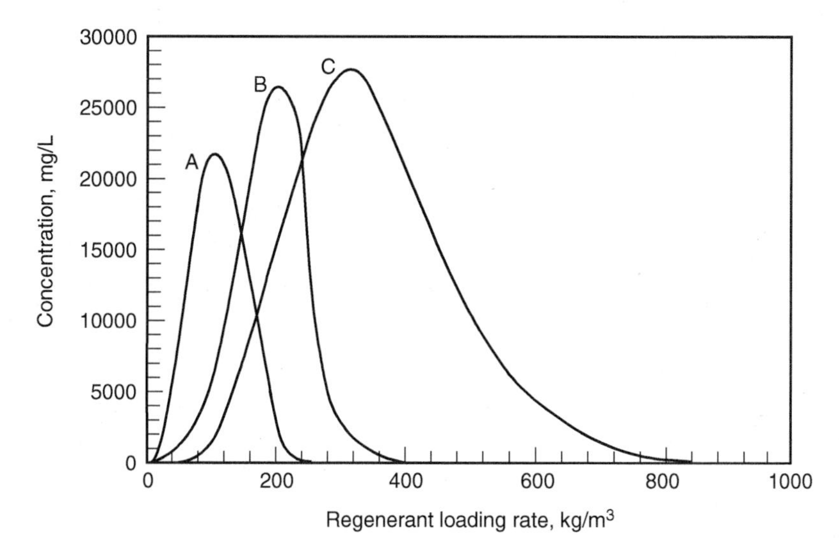

Figure 10-12
Generalized regeneration curves for regeneration of a resin loaded with compounds A, B, and C.

Ion B requires a little longer to be removed and requires about 350 kg/m^3. Ion C requires about 850 kg/m^3 to ensure that a significant fraction is removed. From equilibrium theory it is known that divalent ions (i.e., ion A on Fig. 10-12) will not be preferred in concentrated solutions and hence are easily replaced by sodium or chloride ions.

Backwashing of Fixed-Bed Adsorption Contactors

To obtain the best performance, adsorption contactors should be operated in the postfiltration mode or receive low-turbidity water because backwashing will greatly reduce their performance. The MTZ will be disrupted due to backwashing, which in turn causes premature breakthrough of contaminants. Backwashing is usually not needed for treatment of groundwater from deep wells as long as there is no potential for precipitation of calcium carbonate or metals. When treating turbid surface waters, turbidity must be removed prior to the fixed-bed adsorption process, otherwise backwashing will be required. Based on operating experience it has been found that backwashing does not appear to affect removal of NOM because high degrees of removal cannot be achieved with reasonable EBCTs.

Parallel and Series Column Operation

The performance of fixed-bed columns can be influenced by operating multiple columns in either a parallel or series configuration.

BEDS IN SERIES

The operation of two beds in series is illustrated on Fig. 10-13. During cycle 1, the MTZ forms in bed I and moves into bed II. Once the effluent concentration from bed I equals the influent concentration, cycle 2 begins. During the first phase of cycle 2, bed I is taken offline and the media is regenerated (for IX) or replaced (for adsorption) and bed II is switched

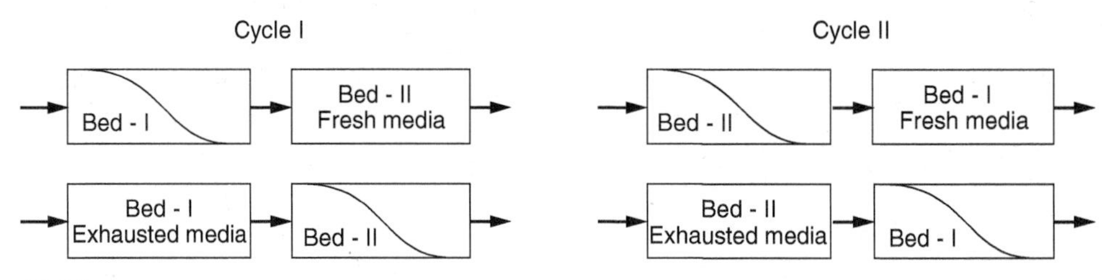

Figure 10-13
Operation of two beds in series.

to the influent. The operation continues until the MTZ moves from bed II into bed I and the effluent from bed II equals the influent concentration. At this point, cycle 3 begins and bed I receives the influent, and bed II is regenerated or replaced with fresh adsorbent and put into operation just as shown in cycle I. If the length of beds I and II are greater than the length of the MTZ, then the media will be saturated fully and the maximum specific throughput can be calculated using Eq. 10-38. The largest specific throughput is obtained for EBCTs around 10 to 20 min for the removal of SOCs onto GAC with stringent treatment objectives. Two beds that are operated in series may increase the specific throughput by 20 to 50 percent.

BEDS IN PARALLEL

Beds in parallel can be used to increase the flow capacity of an adsorption or ion exchange system. Parallel-bed operation can also increase the specific throughput for adsorption systems that do not require a stringent treatment objective ($C_{\text{eff}}/C_{\text{inf}} > \sim 0.3$), such as the removal of dissolved organic carbon (DOC). Typically, 30 to 70 percent of the DOC can be removed using GAC using reasonable specific throughputs. Adsorption beds operated in parallel can significantly increase specific throughput and reduce the amount of GAC that is required.

The blending of effluent from three beds operating in parallel after startup and after several cycles is shown on Fig. 10-14. At startup, all three beds have similar profiles; once the treatment objective is exceeded, the first bed is regenerated or replaced with fresh adsorbent. After replacement, the treatment objective can be met with blended effluent from the beds. Operation continues until the treatment objective cannot be met and then the second bed is replaced. At this point, there are three beds that have different degrees of saturation, and the treatment objective is still being met because effluent from nearly exhausted beds is blended with effluent from fresh beds. After the treatment objective is exceeded, the third bed is regenerated or replaced, and the cycle begins again by replacing the first column, which will be the column that has been online for the longest period of time.

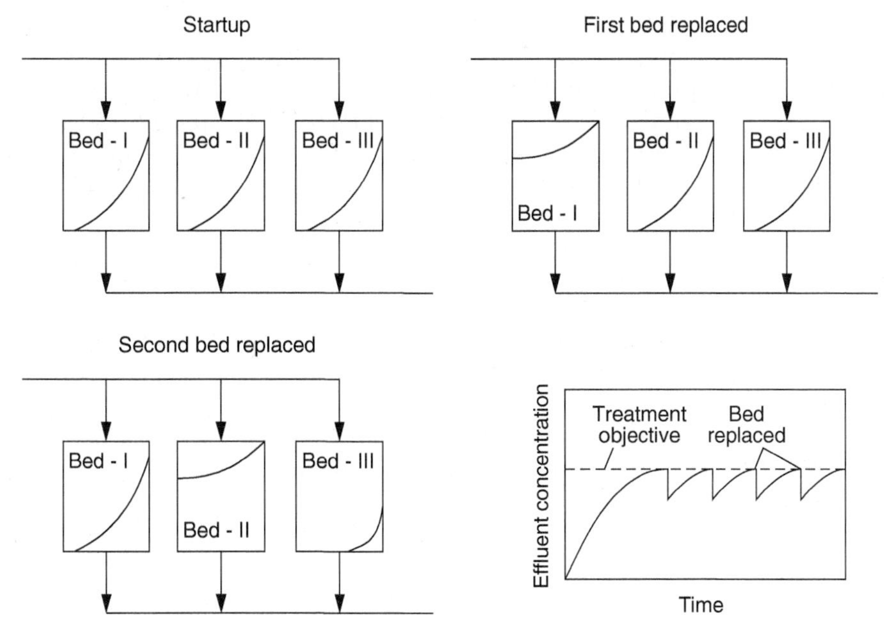

Figure 10-14
Operation of three beds in parallel.

Regeneration of Ion Exchange Columns

The regeneration steps of an ion exchange resin are important to the overall efficiency of the process. There are two methods for regenerating an ion exchange resin: (1) co-current, where the regenerant is passed through the resin in the same flow direction as the solution being treated, and (2) countercurrent, where the regenerant is passed through the resin in the opposite direction as the solution being treated. The selection of the best regeneration method depends on the tolerance for unwanted ions in the effluent and the location within the bed of the target exchanged ion.

CO-CURRENT REGENERATION

In co-current regeneration, the direction of the service and regeneration flows are usually both downward. The concurrent regeneration method can be effective for minimizing the concentration of unwanted ions in the effluent (referred to as leakage) if the ions have intermediate separation factors and accumulate toward the effluent end of the bed. The location of ions within the bed depends on the ions in the water matrix and their separation factors for a given resin. For example, for many SBA resins, sulfate has a higher affinity than either nitrate or arsenate. Consequently, the sulfate will push most of the exchanged arsenate and nitrate toward the effluent end of the column. Regenerating in the countercurrent direction will flush these ions back through the column and some of the arsenic and nitrate will stay in the column unless large amounts of regenerant solution are used. During the next operating cycle, leakage of the arsenic or nitrate left in the column during the previous regeneration will occur. If regenerated in a co-current direction, the arsenic and nitrate will be pushed completely from the bed. While some leakage of sulfate may occur during the next operating cycle, sulfate is not the target ion and leakage is not a concern.

COUNTERCURRENT REGENERATION

In most cases, countercurrent regeneration will result in lower leakage levels and higher chemical efficiencies than co-current regeneration. In situations where (1) high-purity water is necessary, (2) chemical consumption must be reduced to a minimum, or (3) the least waste volume is produced, the countercurrent method of regeneration is used. Countercurrent regeneration can be operated with either the service flow or the regeneration flow in the upward direction.

With flow in the upward direction, it is important to prevent the resin from fluidizing. Any resin movement during the upflow cycle will destroy the ionic interface (exchange front) that ensures good exchange. A number of methods have been devised to prevent resin particle movement during upflow operation. These methods include operating with a completely full column or filling the column's headspace with compressible inert granules to prevent the upward movement of the resin media. A small reservoir is used periodically to withdraw the inert granules to backwash the resin.

Example Development of Full-Scale Design Criteria

An example of a study for the design of an ion exchange facility to remove perchlorate from a groundwater can be used to demonstrate the design process. In this example, design criteria for a full-scale ion exchange treatment plant were developed based on the results of the bench-scale and pilot plant study. The pilot was operated for 31 cycles and perchlorate breakthrough in the pilot plant study consistently occurred at 560 BV for each cycle, at which time the resin was regenerated. The plant is sized for a maximum finished-water capacity of 0.160 m^3/s (2500 gpm). The plant is designed with one extra column that is in the regeneration mode or on standby while the others are in the operational mode. Results from the pilot study demonstrated that an SFR of 28 BV/h (3.5 gpm/ft^3) was appropriate for the full-scale design.

ION EXCHANGE COLUMN DESIGN

Design of the ion exchange columns involves the determination of the volume of resin, the surface area of resin required, the number of columns, the sidewall height, and the pressure drop.

The number of columns can be found by first calculating the total volume of resin required for the specified SFR of 28 BV/h (3.5 gpm/ft^3) using Eq. 10-34:

$$V_{b,\text{total}} = \frac{Q}{\text{SFR}} = \frac{0.160\ \text{m}^3/\text{s}}{(28\ \text{BV/h})(1\ \text{h}/3600\text{s})} = 20.6\ \text{m}^3\,(727\ \text{ft}^3)$$

As discussed above, the EBCT of the pilot plant should be about the same as the EBCT used in the full-scale design. Because a resin depth of 0.863 m (2.83 ft) was used in the pilot plant study, a similar full-scale design with a depth of 1.0 m (3 ft) will be used. Consequently, the total ion exchange surface area required is determined to be

$$A_{b,\text{total}} = \frac{V_{b,\text{total}}}{L} = \frac{20.6\ \text{m}^3}{1.0\ \text{m}} = 20.6\ \text{m}^2\,(223\ \text{ft}^2)$$

Ion exchange columns come in standard sizes from the manufacturer. Typically, they may have column diameters of 1.0 m (3.3 ft), 2.0 m (6.6 ft), 3.0 m (9.8 ft), 4.0 m (13.1 ft), and 5.0 m (16.4 ft). If a 3-m column diameter is chosen for the design, the column would provide 7.1 m^2 (76.4 ft^2) of cross-sectional area and the volume occupied by the resin would be 7.1 m^3. If the total column area is divided by the area of one column, the number of columns required can be calculated as

$$\text{Number of columns} = \frac{A_{b,\text{total}}}{A_b} \frac{20.6\ \text{m}^2}{7.1\ \text{m}^2} = 2.9 \approx 3$$

With one column in the regeneration or standby mode a total of four 3.0-m diameter columns are required.

PRESSURE DROP

Before continuing the design calculations, the column pressure drop needs to be checked. The maximum pressure drop for the ion exchange resin bed should not exceed 172 kPa (25 psi). Manufacturers provide pressure drop curves for commercially available resins such as shown on Fig. 10-10a. The superficial velocity for this system is 28 m/h, the initial pressure drop through the resin is 0.62 kg/cm^2/m of bed depth, as shown on Fig. 10-10a. For 1.0 m of resin depth, the clean-water pressure drop is 0.62 kg/cm^2, or 60.8 kPa (8.8 lb/in^2.). In this case, the clean-water pressure drop column design is well below the maximum allowable pressure drop (60.8 kPa $\ll$ 172 kPa). If these curves are not available, the column head loss can be calculated. Typically, the pressure drop can be determined in the pilot plant studies if the loading rate and EBCT used in the pilot columns are the same as those in the full-scale design.

OVERALL CYCLE TIME

The time for each column loading cycle can be calculated using Eqs. 10-29 and 10-34:

$$t = V^* (\text{EBCT}) = \frac{V^*}{\text{SFR}} = \frac{560 \text{ BV}}{28 \text{ BV/h}} = 20 \text{ h}$$

If the columns are staggered or started at different times, then each column will be regenerated every 20 h, and the blended effluent will not exceed 4 μg/L perchlorate concentration, based on the pilot study results.

REGENERATION REQUIREMENTS

Based on the results of the pilot plant studies, it was found that the perchlorate-loaded columns could be regenerated fully using 480 kg NaCl/m^3 (30 lb NaCl/ft^3) of resin with a salt strength of 10 percent (specific gravity 1.07). The full-scale design will use the same regeneration requirements. The salt solution can be calculated from the specific gravity of the salt and the salt strength as

$$10\% \text{ salt solution} = (0.1 \text{ kg NaCl/kg soln})\ (1070 \text{ kg soln/m}^3 \text{ soln})$$

$$= 107 \text{ kg NaCl/m}^3 \text{ soln}$$

The regeneration volume can be calculated by dividing the salt requirements per volume of resin by the salt solution concentration:

$$\text{Regeneration volume} = \frac{480 \text{ kg NaCl/m}^3 \text{ resin}}{107 \text{ kg NaCl/m}^3 \text{ soln}} = 4.5 \text{ m}^3\text{soln/m}^3 \text{ resin}$$

$$= 4.5 \text{ BV}$$

The total quantity of salt required on an annual basis can be calculated by multiplying the number of regenerations in a year by the quantity of salt required per regeneration. The number of regenerations can be calculated by dividing the number of hours in a year by the loading cycle time per column:

$$\text{Number of regenerations per column per year} = \frac{(365 \text{ d/yr})\,(24 \text{ h/d})}{20 \text{ h/regen}}$$

$$= 438/\text{yr}$$

The quantity of salt per regeneration per column is calculated as

$$\text{Salt quantity per column regeneration}$$

$$= (7.1 \text{ m}^3 \text{ resin/regen})\,(480 \text{ kg NaCl/m}^3 \text{ resin})$$

$$= 3408 \text{ kg NaCl (7531 lb)}$$

The annual salt consumption requirement per column is given as

$$\text{Annual salt quantity required per column}$$

$$= (438 \text{ regen/yr})\,(3408 \text{ kg NaCl/regen})$$

$$= (1.5 \times 10^6 \text{ kg NaCl/yr})\,(3.3 \times 10^6 \text{ lb/yr})$$

The volume of spent regeneration solution per column regeneration is given as

$$\begin{aligned}&\text{Spent regeneration solution per column}\\&\quad= (7.1\ \text{m}^3\ \text{resin/BV})(4.5\ \text{BV})\\&\quad= 32\ \text{m}^3/\text{column}\ (1130\ \text{gal/column})\end{aligned}$$

The total annual volume of spent regeneration solution per column is calculated as

$$\begin{aligned}&\text{Total annual spent regeneration solution per column}\\&\quad= (32\ \text{m}^3/\text{column})\ (438\ \text{columns/yr})\\&\quad= 14{,}016\ \text{m}^3/\text{yr}\ (3.7\ \text{Mgal/yr})\end{aligned}$$

The total annual quantity of salt required and regeneration solution generated for the whole plant will be three times the above quantities because within every 20-h period each of the three columns in service will be regenerated. The total plant quantity values are shown in Table 10-11.

RINSE WATER REQUIREMENT

The quantity of rinse water can be determined based on the rinse quantity used in the pilot plant study. In the pilot plant study, 2 to 6 BV were used to reduce the conductivity of the rinse water below 700 μS/cm. To be conservative, 6 BV will be used for the full-scale design. The quantity of rinse volume per regeneration is calculated as

$$\text{Rinse volume per column} = (7.1\ \text{m}^3\text{resin/BV})(6\ \text{BV})(43\ \text{m}^3/\text{column})$$

The total annual rinse volume is given as

$$\begin{aligned}&\text{Annual rinse volume per column}\\&\quad= (43\ \text{m}^3/\text{column})(438\ \text{columns/yr})\\&\quad= 18{,}834\ \text{m}^3/\text{yr}\ (5.0\ \text{Mgal/yr})\end{aligned}$$

REGENERATION CYCLE TIME

The cycle time for the salt regeneration is calculated by multiplying the EBCT by the number of bed volumes of regeneration solution per column.

$$\text{EBCT} = \frac{1}{\text{SFR}} = \frac{60\ \text{min/h}}{28\ \text{BV/h}} = 2.14\ \text{min}$$

$$\begin{aligned}\text{Regeneration time per column} &= \text{EBCT}\left(\frac{\text{BV}}{\text{regen}}\right)\\&= (2.14\ \text{min/BV})(4.5\ \text{BV}) = 9.6\ \text{min}\end{aligned}$$

Table 10-11
Summary of design criteria for perchlorate removal case study

Parameter	SI Units	Value	U.S. Customary Units	Value
Design product water capacity	m^3/s	0.160	gpm	2,536
Minimum water temperature	°C	15	° F	59
Resin type	—	SBA, polyacrylic, type I	—	SBA, polyacrylic, type I
Effective resin size	mm	0.6	in.	0.024
SFR	BV/h	28	gpm/ft^3	3.6
EBCT	min	2.14	min	2.14
Resin depth	m	1.0	ft	3.0
Total minimum sidewall depth	m	3.15	ft	10.3
Required resin volume	m^3	20.6	ft^3	728
Column diameter	m	3.0	ft	10
Number of columns	—	4	—	4
BVs to perchlorate breakthrough (single column)	BV	560	BV	560
Salt loading rate (NaCl)	kg/m^3	480	lb/ft^3	30
Salt strength	%	10	%	10
Rinse volume	BV	6	BV	6
Clean-water head-loss rate	kPa/m	60.8	psi/ft	2.7
Clean-water head loss	kPa	60.8	psi	8.8
Regeneration volume per column	BV	4.5	BV	4.5
Number of regenerations for each column per year	—	438	—	438
Spent regeneration solution volume per column	m^3	32	Gal	8,454
Annual regeneration solution volume per column	m^3/yr	14,016	Mgal/yr	3.7
Salt quantity required per column	kg	3,408	lb	7,513
Annual salt quantity required per column	kg/yr	1.5×10^6	lb/yr	3.3×10^6
Rinse volume required per column	m^3	43	gal	11,360
Annual rinse volume per column	m^3/yr	18,834	Mgal/yr	5.0
Total annual salt requirements	kg/yr	4.50×10^6	lb/yr	9.9×10^6
Total annual regeneration solution volume	m^3/yr	42,048	Mgal/yr	11.1
Total annual rinse requirements	m^3/yr	56,502	Mgal/yr	15.0
Total regeneration cycle time	min	32.4	min	32.4

The cycle time for the rinse step is calculated as

$$\text{Rinse time per column} = \text{EBCT}\left(\frac{\text{BV}}{\text{regen}}\right) = (2.14\ \text{min/BV})(6\ \text{BV})$$

$$= 12.8\ \text{min}$$

Typical backwash times range from 5 to 20 min, so choosing a backwash time of 10 min, the total time a column will be out of service for the regeneration cycle can be estimated to be

Total regeneration cycle time per column

= regeneration time per column + rinse time per column

+ backwash time per column

= 9.6 min + 12.8 min + 10 min

= 32.4 min

The design criteria for the full scale plant are summarized in Table 10-11.

10-8 Suspended-Media Reactors

Suspended-media reactors consist of a basin, channel, or pipeline where contact between the adsorption or ion exchange media and the water can take place. The media (adsorbent or IX resin) is mixed directly into the process water and allowed to travel with the process stream as the water makes its way through the treatment facility. Transfer of solutes (adsorbates or ions) takes place as the media travels with the water. After a period of time transpires to allow the solutes to transfer to the media, the media is separated from the water, typically by sedimentation and/or filtration. Basic features of suspended media adsorption and ion exchange processes are introduced in this section.

Theoretical Suspended-Media Dose Requirements

The dose of media needed to achieve a desired effluent concentration of the solutes of interest can be determined using a mass balance analysis. The mass of solute entering the reactor with the water is partitioned between the aqueous and solid phases as shown on Fig. 10-15 as follows:

$$QC_{\text{inf}} = QC_{\text{eff}} + q_{\text{eff}}\dot{M} \qquad (10\text{-}42)$$

where Q = water flow rate, L/d

$C_{\text{inf}}, C_{\text{eff}}$ = influent and effluent concentrations of solute in aqueous phase, mg/L or meq/L

q_{eff} = effluent concentration of solute in solid phase, mg/g or meq/L

$\dot{M}$ = media dosing rate (mass or volume of solid added per unit time), g/d or L/d

Note that conventional units for adsorption and ion exchange media are different, the solid-phase concentration is typically determined in units of meq/L for IX and mg/g for adsorption so the media dosing rate would be measured in appropriate units of volume or mass.

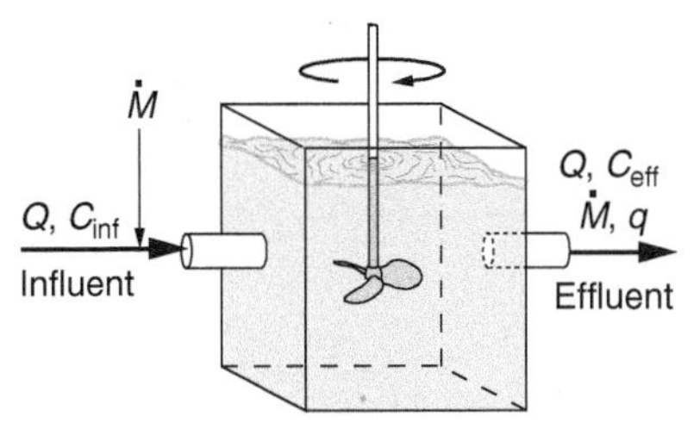

Figure 10-15
Sketch of CMFR suspended-media reactor.

If equilibrium is achieved between the solid- and aqueous-phase concentrations before the media is separated from the water, the following expression is obtained:

$$QC_{\text{inf}} = QC_{\text{eff}} + q_{e|C_{\text{eff}}}\dot{M} \tag{10-43}$$

where $q_{e|C_{\text{eff}}}$ = concentration of solute in solid phase in equilibrium with C_{eff}, mg/g or meq/L

Rearranging Eq. 10-43 results in an expression for the required dose of suspended adsorbent or IX media:

$$D = \frac{\dot{M}}{Q} = \frac{C_{\text{inf}} - C_{\text{eff}}}{q_{e|C_{\text{eff}}}} \tag{10-44}$$

where D = suspended media dose, g/L or L/L

It is important to note that the dose calculated from Eq. 10-44 is based on achieving equilibrium. In conventional water treatment plants, the time available for contact between the media and water is typically between 30 min and 2 h. As noted earlier in the chapter, ion exchange is sufficiently rapid for equilibrium to be achieved, but adsorption typically cannot be achieved in the time that is available. As a result, Eq. 10-44 gives the lowest possible dose for adsorbents such as PAC. Competitive adsorption will also increase the required dose.

Comparison of Dose Requirements for Suspended- and Fixed-Media Processes

A comparison of Eqs. 10-37 and 10-43 indicates that the media in fixed-bed contactors reaches equilibrium with the influent water and suspended-media reaches equilibrium with the effluent water. Since the concentration in the influent water is greater than the concentration in the effluent water, the concentration of solute on the solid phase at equilibrium will also be higher. Thus, the amount of media required to achieve a given effluent aqueous concentration would be expected to be less for fixed-bed processes than for suspended-media processes. The difference in media usage can be evaluated by calculating the ratio of the minimum carbon usage rate (Eq. 10-39) to the equilibrium-based suspended-media dose (Eq. 10-44) as follows:

$$\frac{\text{Eq. 10-44}}{\text{Eq. 10-39}} = \frac{D}{\text{CUR}_{\text{min}}} = \frac{(C_{\text{inf}} - C_{\text{eff}})/q_{e|C_{\text{eff}}}}{C_{\text{inf}}/q_{e|C_{\text{inf}}}} = \frac{1 - C_{\text{eff}}/C_{\text{inf}}}{q_{e|C_{\text{eff}}}/q_{e|C_{\text{inf}}}} \tag{10-45}$$

where D = suspended-media dose, g/L
CUR_{min} = minimum fixed-bed carbon usage rate, g/L
$C_{\text{inf}}, C_{\text{eff}}$ = influent and effluent concentrations of solute in aqueous phase, respectively, mg/L
$q_{e|C_{\text{inf}}}, q_{e|C_{\text{eff}}}$ = concentration of solute in solid phase in equilibrium with C_{inf} and C_{eff}, respectively, mg/g

Thus, the difference in media usage between fixed-bed and suspended-media processes depends on both the removal efficiency (C_{eff}/C_{inf}) and the solid-phase equilibrium with the influent and effluent aqueous-phase concentrations. In the case of adsorption, if adsorption equilibrium can be described by the Freundlich isotherm (Eq. 10-8), then the ratio of the PAC dose to the GAC dose can be evaluated by substituting Eq. 10-8 into Eq. 10-45; the resulting solid-phase equilibrium depends only on $1/n$:

$$q_e = KC_e^{1/n} \qquad \text{(Eq. 10-8)}$$

$$\frac{D_{PAC}}{D_{GAC}} = \frac{1 - (C_{eff}/C_{inf})}{(C_{eff}/C_{inf})^{1/n}} \qquad (10\text{-}46)$$

where D_{PAC}, D_{GAC} = dose of PAC and GAC, respectively, mg/L

The ratio of the PAC to GAC dramatically increases for a higher percentage removal, but the increase decreases as the value of $1/n$ decreases, as shown in Fig. 10-16, because the PAC is in equilibrium with the effluent concentration and the GAC is in equilibrium with the influent concentration. The difference in capacity becomes smaller as $1/n$ becomes smaller. As a point of reference, most $1/n$ values are around 0.5 to 0.7 for the compounds that are considered for removal using adsorption. Thus, if removals of less than 95 percent are required and the problem is seasonal, PAC may be the most economical solution. It should be noted that the curves apply to organic-free water and that the presence of NOM in natural waters can reduce the adsorption capacity of GAC and PAC significantly.

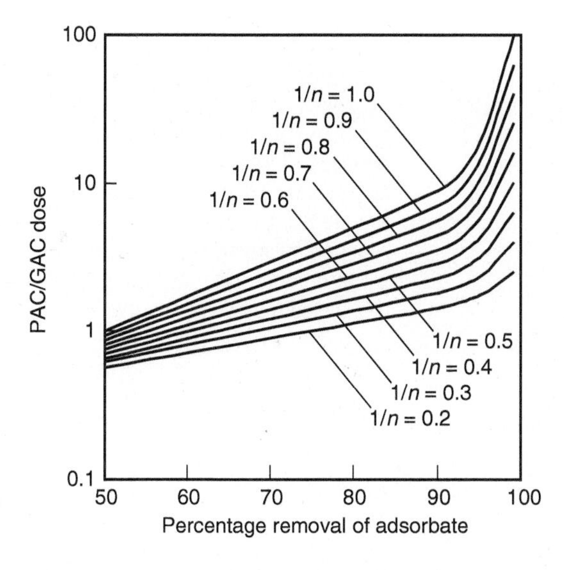

Figure 10-16
Comparison of adsorption capacity for PAC and GAC.

Factors That Influence Suspended-Media Performance

The dose of suspended media required to achieve a given effluent concentrate of solute depends on the type of media, location of media addition, contact time, and presence of competing compounds and oxidants.

Suspended media can be added (1) at the raw-water intake, (2) in the rapid-mix facilities for coagulation, and (3) in a dedicated completely mixed flow reactor (specially designed for suspended media). The advantages and disadvantages of the common points of PAC addition are summarized in Table 10-12. With respect to the point of addition of PAC in a water plant, jar test studies optimizing PAC performance for taste and odor removal of MIB and geosmin show that PAC should be added before coagulation (termed precoagulation time).

The addition of other chemicals such as oxidants and coagulants can interfere with the adsorption or exchange process; consequently, it is generally recommended that suspended media not be added to the process train at the same location as other chemicals.

Equilibrium can often be achieved for ion exchange resins but typically cannot be achieved for adsorption in the time available in conventional water treatment facilities. For instance, the impact of MIB removal as a

Table 10-12
Advantages and disadvantages of different points of addition of suspended media

Point of Addition	Advantages	Disadvantages
Raw-water intake	Long contact time, good mixing	Interferes with preoxidation process (Cl_2 or $KMnO_4$). Some substances may be adsorbed that would otherwise probably be removed by coagulation, thus increasing carbon usage rate (this still needs to be demonstrated).
Rapid mix	Good mixing during rapid mix and flocculation, reasonable contact time	Interferes with preoxidation process (Cl_2 or $KMnO_4$). Possible reduction in rate of adsorption because of interference by coagulants; contact time may be too short for equilibrium to be reached for some contaminants; some competition may occur from molecules that would otherwise be removed by coagulation.
Completely mixed flow reactor	Excellent mixing for design contact time, no interference by coagulants, additional contact time possible during flocculation and sedimentation	A new basin and mixer may have to be installed; some competition may occur from the molecules that may otherwise be removed by coagulation.

Source: Adapted from Graham et al. (2000).

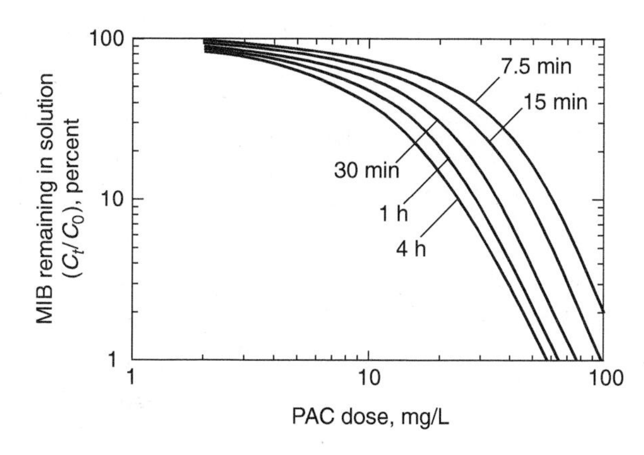

Figure 10-17
MIB remaining in solution as function of PAC dose and contact time.

function of PAC dose for various contact times is plotted on Fig. 10-17. As the contact time increases for a given removal efficiency, the PAC dose decreases. For example, given an MIB removal efficiency of 90 percent (or 10 percent remaining), the PAC dose for 7.5 min contact time is about 65 mg/L as compared to only about 25 mg/L for a contact time of 4 h.

Bench Testing for Determining Suspended-Media Doses

Because adsorption media generally will not achieve equilibrium and competitive adsorption from NOM and other constituents occurs, the appropriate dose for media generally is greater than predicted by Eq. 10-44. Thus, other methods of determining the correct dose are necessary. Models of the adsorption process by PAC have been developed and are presented in Crittenden et al., (2012). However, the models are complex and sometimes do not properly account for competitive adsorption by NOM and other factors. Bench-scale tests can also be performed to determine the correct media dose using standard jar testing procedures such as those described for coagulation in Chap. 5. Since the contact time between the media and water in a treatment facility will be between 30 min and 2 h, bench tests only need to be of that duration. Raw water from the source is placed in jars, and the mixing velocities, timing of chemical addition, retention times, and doses are selected to mimic conditions in the full-scale facility. Measurement of effluent concentrations after the contact time are used to develop dose–response curves. An example of dose–response curves obtained for five different types of PAC removing 40 ng/L of geosmin and MIB is presented on Fig. 10-18. The dose–response curves are given in terms of contaminant percent removal as a function of PAC dose. If the treatment objective is 80 percent removal of geosmin (8 ng/L), PAC type B provides the best result with a 34-mg/L PAC dose.

Magnetic Ion Exchange (MIEX) Resin

The Orica Limited Company of Australia developed the MIEX process for removal of dissolved organic carbon (DOC) from drinking water supplies. The process consists of an SBA ion exchange resin, usually in the chloride

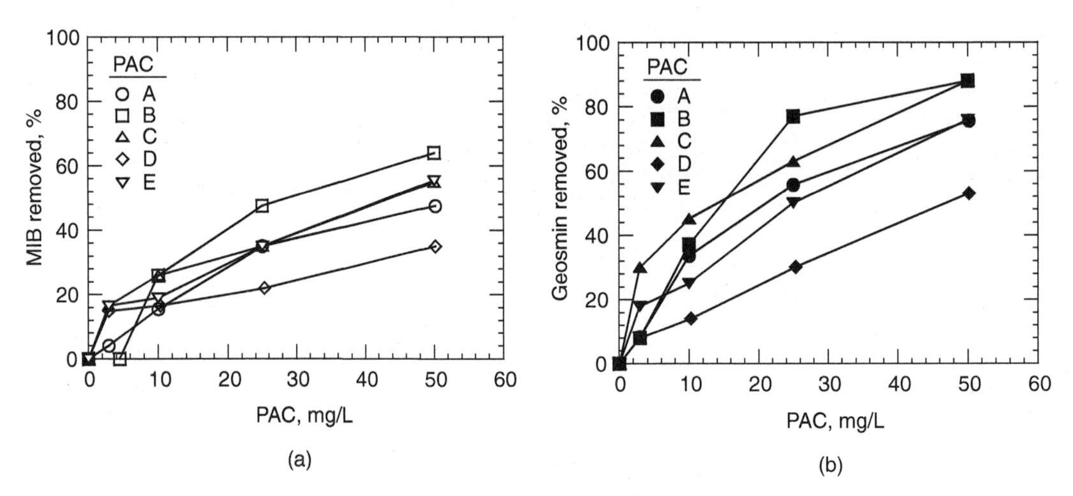

Figure 10-18
Percent removal of MIB and geosmin using Manatee Lake water and testing protocol and 40 ng/L initial contaminant concentrations. Letters A through E correspond to different types of PAC. [Adapted from Graham et al. (2000).]

form, with a magnetic component built into it. The IX resin beads, which are smaller than the conventional resin beads (i.e., diameter $\approx$ 180 μm), are contacted with the water in a completely mixed flow reactor. A typical process flow diagram employing the MIEX resin is shown in Fig. 10-19. The negatively charged DOC molecules exchange with presaturant chloride ion on the resin and are removed from the water. The resin and water are then separated in an upflow clarifier as the resin beads will agglomerate due to their magnetic properties and rapidly settle out of the water. The settling rate can be as high as 15 m/h. The treated water goes on to further treatment. The settled resins are recovered and recycled to the front of the process. A portion of the recovered resin beads (5 to 10 percent) is removed for regeneration. The resin is regenerated with about 10 percent by weight NaCl for 30 min. The regenerated resin beads are stored and reintroduced into the process as needed. An important advantage of the MIEX resin, compared to other ion exchange resins, is its apparent abrasion-resistant properties.

Because the DOC removal remains constant in the reactor, the DOC leakage is controlled at a predetermined level. Also, because the resin has a high selectivity for DOC, the only inorganic anion that is exchanged is SO_4^{2-}.

Based on preliminary test results, it appears that the removal of DOC on the resin is a surface phenomenon. While other ion exchange resins may be suitable, the time it takes for the DOC to diffuse into the resin may limit its applicability. The performance of MIEX depends on the resin dose, the concentration and nature of the DOC, and the contact time. Reported DOC removal values have been as high as 80 percent, but site-specific

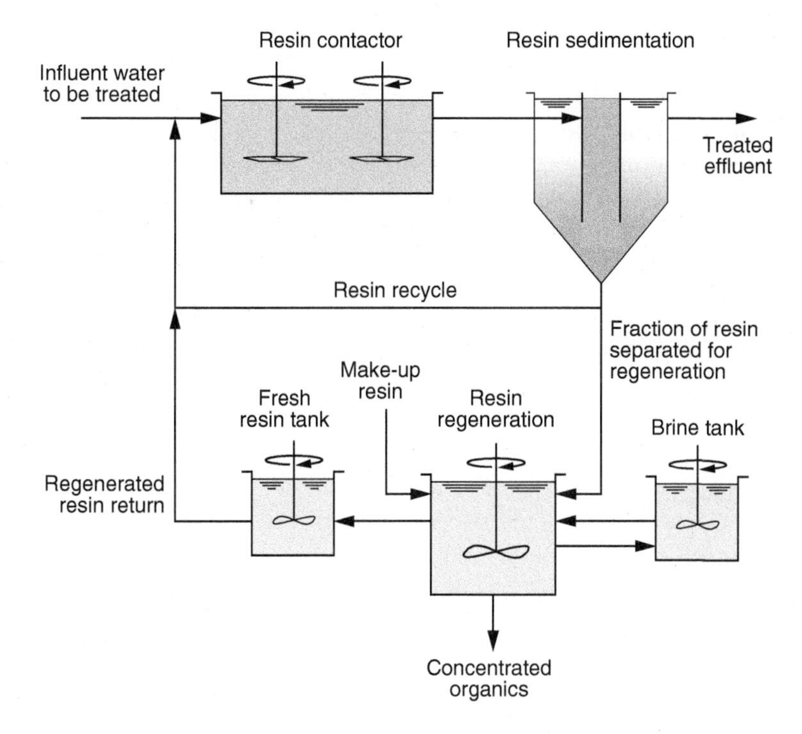

Figure 10-19
Schematic process flow diagram for use of MIEX ion exchange resin for pretreatment of surface water to reduce concentration of natural organic matter (NOM) before addition of coagulating chemical.

testing is required. A pilot study for the City of West Palm Beach, Florida, achieved 67 percent TOC removal with MIEX, compared to 57 percent TOC removal with enhanced coagulation (MWH, 2010). Use of MIEX also reduced coagulant use and sludge production by about 80 percent compared to enhanced coagulation alone.

MIEX is a relatively new technology; as of the end of 2010, about 15 MIEX systems had been installed at treatment plants greater than 3785 m^3/d (1 mgd) in North America.

10-9 Energy and Sustainability Considerations

Adsorption and ion exchange are relatively low pressure processes. The maximum pressure drop through fixed adsorption or ion exchange beds is typically 1.7 bar (25 psi). Using calculations similar to Example 3-2, this head corresponds to electrical energy consumption of 0.06 kWh/m^3, assuming pump efficiency of 80 percent. However, a significant amount of energy is required to produce and reactivate GAC. For example, about 2 to 5 kg of carbon dioxide is released per kilogram of GAC that is reactivated.

For ion exchange, the disposal of the regeneration brine has a significant environmental impact. In some cases, brine can be disposed of in the ocean.

In some places such as California, brine lines transport waste brines to the ocean. If the concentration is high, diffusers are required to reduce the concentration of the brine to avoid impacts on marine organisms. If marine disposal is not available, then various methods (e.g., evaporation ponds, falling film evaporators, etc.) must be explored to concentrate the brine into salt. When IX is used for nitrate removal, the brine can be regenerated using biological treatment such as denitrification. Other strategies that may be used for brine control is to use gaseous carbon dioxide to produce bicarbonate, which is used as a presaturant ion. Bicarbonate can be later removed from the brine by off-gassing. Also clean seawater can be used to regenerate IX, but it is not as effective as a concentrated brine solution which is made from salt.

10-10 Summary and Study Guide

After studying this chapter, you should be able to:

1. Define the following terms and phrases and describe the significance of each in context of adsorption and ion exchange in water treatment:

adsorbate	exchanging ion	presaturant
adsorbent	Freundlich isotherm	resin
bed volume	ion	RSSCT
chemisorption	isotherm	saturation loading curve
chromatographic peaking	Langmuir isotherm	selectivity
elution curve	mass transfer zone	separation factor
empty-bed contact time	physical adsorption	

2. List and describe the applications for adsorption and ion exchange processes in water treatment and the types of contactors and reactors used for each application.
3. Compare similarities and differences between adsorption and ion exchange, addressing issues such as mechanism for removing constituents from water, types of contactors, rate of mass transfer, time to reach equilibrium, typical operation time before reaching exhaustion, method for restoring the capacity, and waste stream generated.
4. Explain why surface area is such an important parameter for an adsorbent.
5. Explain the differences in the assumptions that were used in the development of the Langmuir and Freundlich isotherms. Which isotherm has wider applicability?

6. Evaluate experimental data to determine which isotherm best describes the equilibrium distribution between solid and liquid phases.
7. If given isotherm data, calculate the equilibrium concentration of solutes (adsorbates or ions) in the solid and liquid phases.
8. Describe the major types of ion exchange resins, and determine the appropriate type of resin to use if given raw water quality data and treatment goals.
9. Predict the order of preference for ions partitioning into the resin (the selectivity) if given physical and chemical characteristics of various ions and resins.
10. Calculate the distribution of adsorbates or ions on an adsorbent or ion exchange resin in a multicomponent system.
11. Describe how mass transfer controls the rate of adsorption and ion exchange, and explain why ion exchange is so much faster than adsorption.
12. Explain the cause of chromatographic peaking in an ion exchange process and how to design an ion exchange system to prevent it.
13. Calculate the basic parameters needed to design a fixed bed adsorption or ion exchange system, including empty bed contact time, superficial velocity, media characteristics, bed depth, and specific throughput.
14. Design an ion exchange column (resin capacity, resin bed dimensions, regeneration cycle time, and regenerant requirements, including salt used, brine production rate, and volume of brine storage tank), if given raw water quality data and water demand requirements.
15. Calculate design parameters for a full-scale GAC column using data from a rapid small scale column test (RSSCT).
16. Calculate the ratio of GAC usage rate to PAC usage rate to remove a specific contaminant.

Homework Problems

10-1 A bench study is conducted to determine Freundlich isotherm parameters. Six jars are each filled with 500 mL of a solution containing a contaminant, and then different amounts of adsorbent are added to each jar. The jars are sealed and agitated for 2 weeks at 20°C to allow the system to reach equilibrium, and then the final concentration of solute in each jar is measured. For the problem listed below (to be selected by instructor), calculate the Freundlich isotherm parameters.

a. Adsorbent: F-300 GAC. Solute: ethylbenzene. Initial concentration: 1.5 mg/L.

Jar	1	2	3	4	5	6
Adsorbent dose, mg	5.0	7.4	11.2	16.7	21.4	27.2
Final aqueous solute concentration, mg/L	0.97	0.81	0.68	0.49	0.42	0.31

b. Adsorbent: F-100 GAC. Solute: chloroform. Initial concentration: 3.6 mg/L.

Jar	1	2	3	4	5	6
Adsorbent dose, mg	30	55	75	100	160	200
Final aqueous solute concentration, mg/L	2.2	1.6	1.2	0.9	0.4	0.3

c. Adsorbent: F-300 GAC. Solute: benzene. Initial concentration: 12.2 mg/L.

Jar	1	2	3	4	5	6
Adsorbent dose, mg	20	35	50	75	88	100
Final aqueous solute concentration, mg/L	10.6	9.53	8.71	7.25	6.62	6.1

d. Adsorbent: powdered activated carbon. Solute: 1,4-dimethylbenzene. Initial concentration: 0.965 mg/L.

Jar	1	2	3	4	5	6
Adsorbent dose, mg	0.25	0.81	1.61	3.15	4.05	4.78
Final aqueous solute concentration, mg/L	0.87	0.66	0.45	0.22	0.15	0.11

e. Adsorbent: powdered activated carbon. Solute: 2,4,6-trichlorophenol. Initial concentration: 5.51 mg/L.

Jar	1	2	3	4	5	6
Adsorbent dose, mg	8.07	10.41	12.01	14.13	16.42	18.12
Final aqueous solute concentration, mg/L	1.51	0.96	0.73	0.50	0.32	0.25

10-2 Isotherm experiments were conducted in bottles with two different initial concentrations to measure the adsorption isotherm of MIB on PAC in a natural water and the following data were obtained (Gillogly et al., 1998). Plot the percentage of MIB remaining in the solution as a function of PAC dose, and determine the PAC dose corresponding to 90 percent removal of MIB in a batch reactor for an initial concentration of 200 ng/L. Calculate the Freundlich isotherm parameters for MIB on this PAC.

C_0, ng/L	PAC dose, mg/L	C_e, ng/L
150	2.2	137.7
	4.1	122.7
	9.9	81.6
	32.4	16.2
	45.7	5.85
1245	2.1	1088.13
	4	949.94
	14.6	329.68
	40.2	51.04
	60.3	14.94

10-3 A contaminated groundwater contains 100 μg/L each of chloroform, trichloroethene, and tetrachlorethene. Calculate the equilibrium concentration (in mg/g) of each compound on activated carbon if the Freundlich isotherm K value for each compound is as given in Table 10-4 and the $1/n$ value is assumed to be 0.45 for the following conditions (a) assuming each is the only contaminant present in the groundwater (single-solute adsorption) and (b) that all three are present in the groundwater simultaneously (multicomponent adsorption). Calculate the solid-phase concentration of each under multicomponent adsorption conditions as a percentage of its solid-phase concentration under single-solute adsorption conditions. How is the solid-phase concentration of each affected by the presence of other compounds? What in the impact of the value of K with regard to the effect of competition from other solutes?

10-4 For the ion concentrations in the problem below (to be selected by instructor), calculate the concentration of each ion on the resin at equilibrium using the separation factors in Table 10-10. Prepare pie charts showing the distribution of ions in the water phase and in the resin phase on an equivalence basis.

	A	B	C	D	E
Total exchange capacity of resin, eq/L	2.0	1.8	1.9	4.2	2.1
Sodium (Na^+), mg/L	119	100	68	216	85
Potassium (K^+), mg/L	4.1	3.1	—	3.5	2.7
Magnesium (Mg^{2+}), mg/L	8.5	24	3.2	5.8	3.2
Calcium (Ca^{2+}), mg/L	35	84	54	119	20
Barium (Ba^{2+}), mg/L	11.3	—	0.7	2.5	—
Radium (Ra^{2+}), mg/L	—	—	—	—	1.6

10-5 For the ion concentrations in the problem below (to be selected by instructor), calculate the concentration of each ion on the resin at equilibrium using the separation factors in Table 10-10. Prepare pie charts showing the distribution of ions in the water phase and in the resin phase on an equivalence basis.

	A	B	C	D	E
Total exchange capacity of resin, eq/L	1.4	1.4	2.2	2.3	1.4
Chloride (Cl^-), mg/L	112	151	32	185	115
Bicarbonate (HCO_3^-), mg/L	151	160	195	425	27
Sulfate (SO_4^{2-}), mg/L	82	145	54	112	61
Nitrate (NO_3^-), mg/L	2.15	—	—	—	1.1
Arsenic ($HAsO_4^{2-}$), mg/L	—	—	0.085	—	0.061
Perchorate (ClO_4^-), mg/L	—	—	—	0.17	0.17

10-6 Design a fixed-bed adsorption system to treat the contaminant below (to be selected by instructor) based on the information given below and the Freundlich isotherm parameters in Table 10-4. Assume the adsorption system removes the contaminant to below the detection limit and the bed density is 450 kg/m^3. Your design should include the (a) concentration of the contaminant on the carbon at equilibrium, (b) maximum specific throughput, (c) operating time to reach exhaustion of the media (bed life), (d) volume of water treated in bed volumes, (e) volume of water treated in cubic meters, (f) volume of media, (g) cross-sectional area of bed, and (h) depth of bed.

Contaminant	Concentration, μg/L	Plant capacity, ML/d	EBCT, min	Superficial velocity, m/h
A Ethylbenzene	85	15	10	15
B Chloroform	120	3.8	15	12
C Trichlorethene	650	20	12	8
D Tetrachloroethene	650	25	20	5
E Atrazine	56	10	25	5

10-7 RSSCT columns were used to determine the bed life of a fixed-bed adsorption system for the removal of methyl-*tert*-butyl ether (MTBE) from a raw water source. The RSSCT column had a media particle diameter of 0.19 mm, superficial velocity of 45.0 m/h, and EBCT of 27 s. Under those conditions, breakthrough of MBTE occurred in 12.28 d. If the full-scale adsorber is designed with a media particle diameter of 1.10 mm, calculate the appropriate EBCT and superficial velocity of the full scale column, and the predicted operating time before breakthrough of MBTE occurs.

10-8 For the water quality in Problem 10-4, design an ion exchange system for the removal of calcium (waters A and B), barium (waters C and D), or radium (water E), as selected by your instructor. The flowrate to be treated is 5.45 ML/d. The system should be sized so that the minimum time between regenerations is 72 h. The column diameter should be 3 m and there should be at least 2 columns. Pilot testing indicates optimal regeneration efficiency corresponded to using a 10 percent NaCl solution (specific gravity = 1.07) at a salt usage rate of 310 kg NaCl per m^3 of resin at a flow rate of 10 m/h in a countercurrent mode. The slow rinse after regeneration should run for 2 bed volumes at the regeneration flow rate and the fast rinse for 3 bed volumes at the service flow rate. Summarize your design in a table that includes: (a) plant capacity, (b) water treated per cycle, (c) total resin volume, (d) service (volumetric) flow rate, (e) empty bed contact time, (f) number, diameter, and depth of columns, (g) surface area loading rate, (h) regeneration volume and time, (i) slow rinse volume and time, (j) fast rinse volume and time, (k) total waste volume produced per month, and (l) net water production rate, assuming treated water is used for regeneration and rinsing.

10-9 Calculate the dose of activated carbon to reduce an influent concentration of 300 μg/L of chloroform to 100 μg/L (treatment objective) using powdered (PAC) and granular activated carbon (GAC). Assume for the GAC and PAC processes that the carbons are saturated at the influent concentration and treatment objective, respectively, and that the Freundlich isotherm parameters in Table 10-4 apply to both carbons.

References

Clifford, D. A., Sorg, T. J., and Ghurye, G. L. (2011) Ion Exchange and Adsorption of Inorganic Contaminants, Chap. , in J. E. Edzwald (ed.), *Water Quality and Treatment: A Handbook on Drinking Water*, 6th ed., American Water Works Association, McGraw-Hill, New York.

Crittenden, J. C. (1976) Mathematic Modeling of Fixed Bed Adsorber Dynamics–Single Component and Multicomponent, Dissertation, University of Michigan, Ann Arbor, MI.

Crittenden, J. C., Luft, P. J., Hand, D. W., Oravitz, J., Loper, S., and Ari, M. (1985) "Prediction of Multicomponent Adsorption Equilibria Using Ideal Adsorbed Solution Theory," *Environ. Sci. Technol.*, **19**, 11, 1037–1043.

Crittenden, J. C., Trussell, R. R., Hand, D. W., Howe, K. J., and Tchobanoglous, G. (2012) *MWH's Water Treatment: Principles and Design*, 3rd ed., Wiley, Hoboken, NJ.

Dobbs, R.A., and Cohen, J. M. (1980) "Carbon Adsorption Isotherms for Toxic Organics," U.S. Environmental Protectioin Agency, EPA-600/8-80-023.

Gillogly, T. E. T., Snoeyink, V. L., Elarde, J. R., Wilson, C. M., and Royal, E. P. (1998) "Kinetic and Equilibrium Studies of ^{14}C-MIB Adsorption on PAC in Natural Water," *J. AWWA*, **90**, 98–108.

Graham, M., Najm, I., Simpson, M., Macleod, B., Summers, S., and Cummings, L. (2000) *Optimization of Powdered Activated Carbon Application for Geosmin and MIB Removal*, American Water Works Association Research Foundation, Denver, CO.

Haist-Gulde, B. (1991) Zur Adsorption von Spurenverunreinigungen aus Oberflächenwässern, Ph.D. Dissertation, University of Karlsrush, Germany.

Helfferich, F. (1995) *Ion Exchange*, Dover, New York.

Kunin, R., and Myers, R. J. (1950) *Ion Exchange Resins*, Wiley, New York.

Lee, M. C., Snoeyink, V. L., and Crittenden, J. C. (1981) "Activated Carbon Adsorption of Humic Substances," *J. AWWA*, **73**, 8, 440–446.

McGuire, M. J., Krasner, S. W., Hwang, C. J., and Lzaguirre, G. (1981) "Closed-Loop Stripping Analysis as a Tool for Solving Taste and Odor Problems," *J. AWWA*, **73**, 10, 530–537.

Mortimer, C. E. (1975) *Chemistry: A Conceptual Approach*, 4th ed., D. Van Norstrand Co., New York.

Munakata, K., Kanjo, S., Yamatsuki, S., Koga, A., and Lanovski, D. (2003) "Adsorption of Noble Gases on Silver-Mordenite," *J. Nucl. Sci. Tech.*, **40**, 9, 695–697.

MWH (2010) Work Authorization No. 3, Task 4, Pilot Plant Report, Phase One Operations, Final Report submitted to City of West Palm Beach.

Nemethy, G., and Scheraga, H. A. (1962) "Structure of Water and Hydrophobic Bonding in Proteins. I. A Model for the Thermodynamic Properties of Liquid Water," *J. Chem. Phys.*, **36**, 3382–3401.

Sigma Aldrich Online Catalog (2004, June) Available at: http://www.sigmaaldrich .com/Brands /Supelco_Home/Datanodes.html?cat_path = 982049,1005395, 1005413&supelco_ name = Liquid%20Chromatography&id = 1005413.

Sontheimer, H., Crittenden, J. C., and Summers, R. S. (1988) *Activated Carbon for Water Treatment*, 2nd ed., DVGW-Forschungsstelle, University of Karlsruhe, Karlsruhe, Germany. Distributed in the U.S. by the American Water Works Association.

Speth T. F., and Miltner, R. J., (1998) "Technical Note: Adsorption Capacity of GAC for Synthetic Organics," *J. AWWA*, **90**, 4, 171–174.

Zimmer, G., Crittenden, J. C., and Sontheimer, H. (1988) Design Considerations for Fixed-Beds Adsorbers That Remove Synthetic Organic Chemicals in the Presence of Natural Organic Matter, paper presented at the American Water Works Association Annual Conference, Orlando, FL.

11 Air Stripping and Aeration

Air stripping and aeration are two water treatment unit processes that utilize the principles of mass transfer to move volatile substances between liquid and gaseous phases. These treatment processes bring air and water into intimate contact to transfer volatile substances from the water into the air (e.g., hydrogen sulfide, carbon dioxide, volatile organic compounds) or from the air into the water (e.g., carbon dioxide, oxygen). The mass transfer process involving the removal of volatile substances from water into the air is known as *desorption. Air stripping* is one of the most common desorption processes used in water treatment. The addition of gases from air into water is the mass transfer process known as *absorption. Aeration* involving the addition of oxygen to water is a commonly used absorption process.

An understanding of the principles of the underlying mass transfer processes, including how to calculate diffusion coefficients and the basis for mass transfer correlations (presented in Chap. 4), is necessary to design air strippers and aerators effectively. In this chapter, the focus is on the application of the aforementioned mass transfer principles to water treatment unit processes. Specific topics considered in this chapter include (1) an introduction to air stripping and aeration including the various types of systems, (2) gas–liquid equilibrium (Henry's law), (3) the fundamentals of packed-tower air stripping, and (4) analysis and design for packed tower air stripping.

11-1 Types of Air Stripping and Aeration Contactors

Water treatment objectives that can be achieved through gas–liquid mass transfer are summarized in Table 11-1. In both air stripping and aeration, air–water contactors are used to increase the contact between the gas and liquid phases. By increasing the air–water interface, the desorption or absorption mass transfer process is accelerated above the rate that would occur naturally, meaning volatile substances move more rapidly from the water into the air, or soluble gases move more rapidly from the air into the water.

Several methods have been developed to bring about effective air–water contact. Gas transfer devices can be broadly divided into two categories: (1) gas-phase contactors, which disperse droplets of water into a continuous gas phase, and (2) flooded contactors, which disperse bubbles of air into a continuous liquid phase. Several types of typical gas-phase contactors are shown in Fig. 11-1 and several flooded contactors are shown in Fig. 11-2.

Table 11-1
Applications of air–water mass transfer in water treatment

Examples	Water Treatment Objectives
	Adsorption
O_2	Oxidation of Fe^{2+}, Mn^{2+}, S^{2-}; lake destratification
O_3	Disinfection, color removal, oxidation of selected organic compounds
Cl_2	Disinfection; oxidation of Fe^{2+}, Mn^{2+}, H_2S
ClO_2	Disinfection
CO_2	pH control
SO_2	Dechlorination
NH_3	Chloramine formation for disinfection
	Desorption
CO_2	Corrosion control
O_2	Corrosion control
H_2S	Odor control
NH_3	Nutrient removal
Volatile organics (e.g., $CHCl_3$)	Taste and odor control, removal of potential carcinogens

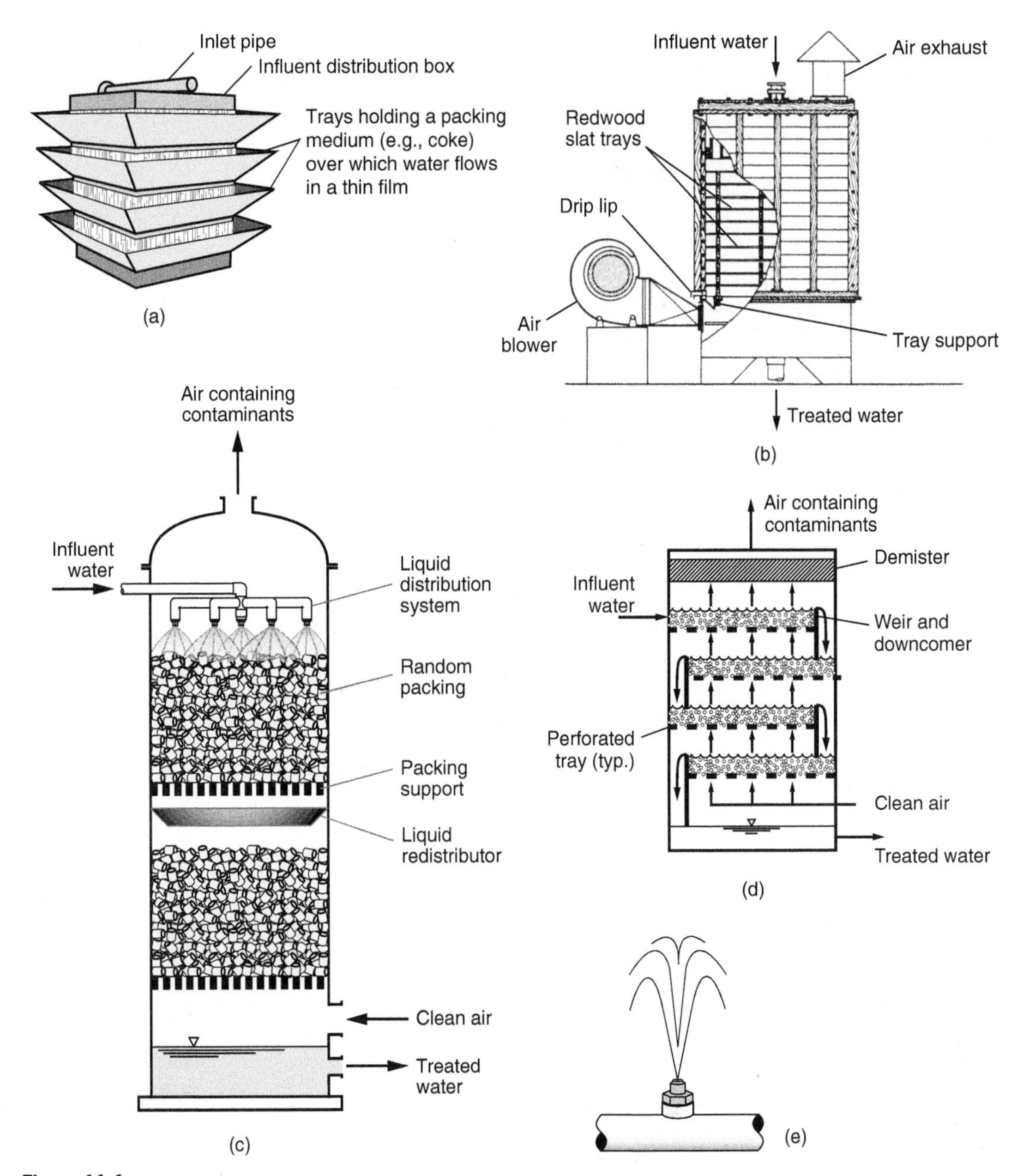

Figure 11-1
Typical gas-phase contactors: (a) multiple tray aerator, (b) cascade aerator, (c) countercurrent packed tower, (d) low-profile or sieve tray aerator, and (e) spray aerator.

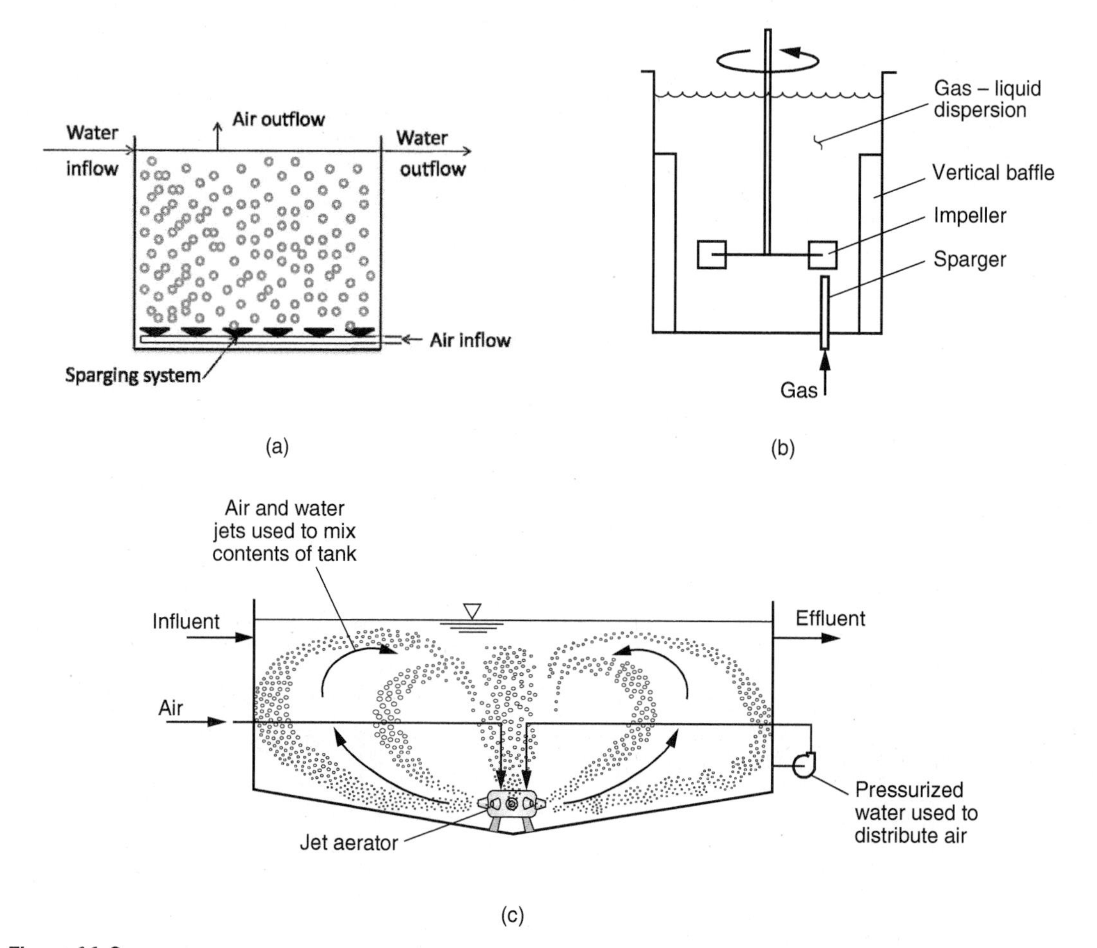

Figure 11-2
Typical flooded contactors: (a) fine-bubble diffuser, (b) mechanical aspirator, and (c) dispersed air contactor.

Key features of these contactors are summarized in Table 11-2. Gas-phase contactors such as packed towers or slat countercurrent flow towers are typically used to remove (or strip) gases or volatile chemicals from water. Flooded contactors are typically used to add gases (e.g., O_2, CO_2, O_3) into water.

Despite the name, aerators can be used to accomplish air–water contact in both air-stripping and aeration processes. In general, aerators are a relatively simple method for increasing the air–water ratio by (1) spraying water into the air or (2) introducing air into the water through surface turbines, or submerged nozzles and diffusors (bubble columns). Thus, aerators allow both of the mass transfer procesess, desorption and absorption, to occur in a relatively cost-effective manner. However, because backmixing can occur in aeration systems, a high degree of removal may be difficult to achieve.

Table 11-2
Characteristics of some gas–liquid contacting systems

Type of Contacting Device	Process Description	Method of Gas Introduction	Typical Applications
Multiple tray aerator (Fig. 11-1a)	Water to be treated trickles by gravity through trays containing media [layers 0.1–0.15 m (4–6 in.) deep] to produce thin-film flow. Typical media used include coarse stone or coke [50–150 mm (2–6 in.) in diameter] or wood slats.	Natural or forced-draft aeration	H_2S, CO_2 removal, taste and odor control
Cascade aerator (Fig. 11-1b)	Water to be treated flows over the side of sequential pans, creating a waterfall effect to promote droplet-type aeration.	Aeration primarily by natural convection	CO_2 removal, taste and odor control, aesthetic value, oxygenation
Countercurrent packed tower (Fig. 11-1c)	Water to be treated is sprayed onto high-surface-area packing to produce a thin-film flow.	Forced-draft aeration	H_2S, CO_2, and VOC removal; taste and odor control
Low-profile (sieve tray) aerator (Fig. 11-1d)	Water flows from entry at the top of the tower horizontally across series of perforated trays. Large air flow rates are used, causing frothing upon air–water contact, which provides large surface area for mass transfer. Units are typically less than 3 m (10 ft) high.	Air introduced under pressure at bottom of tower	VOC removal
Spray aerator (Fig. 11-1e)	Water to be treated is sprayed through nozzles to form disperse droplets; typically a fountain configuration. Nozzle diameters usually range from 2.5 to 4 cm (1 to 1.6 in.) to minimize clogging.	Natural aeration through convection	H_2S, CO_2, and marginal VOC removal; taste and odor control, oxygenation
Fine bubble diffuser (Fig. 11-2a)	Fine bubbles are supplied through porous diffusers submerged in the water to be treated; tank depth is typically restricted to 4.5 m (15 ft).	Compressed air or ozone	Fe and Mn removal, CO_2 removal, taste and odor control, oxygenation, ozonation
Mechanical aspirator (Fig. 11-2b)	A hollow-blade impeller rotates at a speed sufficient to aspirate and discharge a gas stream into the water.	Compressed air or ozone	Ozonation, CO_2 addition
Dispersed air (Fig. 11-2c)	Compressed air is supplied through a stationary sparger orifice-type dispersion apparatus located directly below a submerged high-speed turbine.	Compressed air or ozone	Ozonation, especially when high concentrations of Fe and Mn are present due to clogging of porous diffusers

Two major types of air–water contactors are used for air stripping: (1) towers and (2) aerators. Two principal factors that control the selection of the type of air–water contactor for stripping are (1) the desired degree of removal of the compound and (2) the Henry's constant of the compound. Towers are used when either a high degree of removal is desired or the compound has a high affinity for water (is not very volatile so it has a low Henry's constant), as shown on Fig. 11-3. Aerators are used when either the desired degree of removal is not very high or the gas has a low affinity for water (high volatility or low solubility). When removals less than 90 percent are required, both spray and diffused aeration systems, including mechanical aeration, may be economically attractive.

Aeration is used to increase the oxygen content in the water by adding air into water through (1) diffusors in a pipe, channel, or process basin; (2) cascading water over stacked trays; or (3) surface turbines and wheels that mix air into water at the top of basins. Oxygenation can also be accomplished using pure oxygen.

The process description including typical applications of the various types of both air stripping and aeration systems are summarized in Table 11-2 and are covered in more detail in the companion reference book for this text (Crittenden et al., 2012). For most of these processes, design equations are developed by incorporating equilibrium and mass transfer principles into mass balance expressions to describe the process performance. Equilibrium and mass transfer principles applied to air-stripping and aeration processes are presented below. The fundamentals and practical application of countercurrent packed-tower air stripping is also presented.

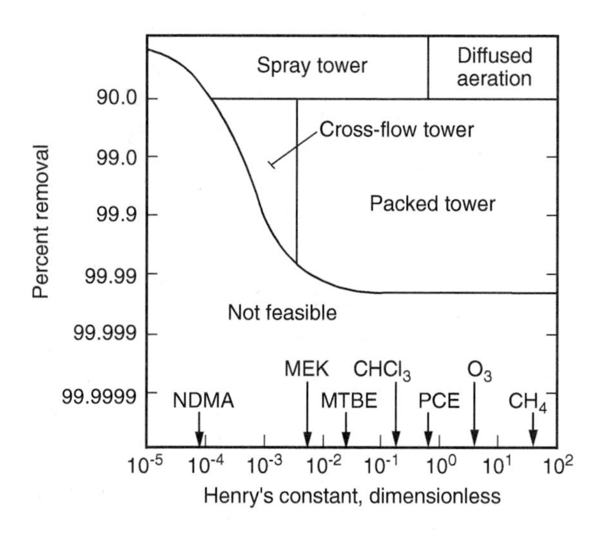

Figure 11-3
Schematic diagram for selection of feasible aeration process for control of volatile compounds. [adapted from Kavanaugh and Trussell, (1981).]

11-2 Gas–Liquid Equilibrium

When gas-free water is exposed to air, compounds such as oxygen and nitrogen will diffuse from the air into the water until the concentration of these gases in the water reaches equilibrium with the gases in the air. Conversely, if water in deep wells is brought to the ground surface, dissolved gases such as methane or carbon dioxide will be released to the air because their concentrations in groundwater typically exceed equilibrium conditions with air. The eruption of a carbonated beverage after it is opened is a more familiar example of carbon dioxide release after a pressure change. In each case, the driving force for mass transfer is the difference between the existing and equilibrium concentrations in the two phases, as discussed in Sec. 4-16.

Vapor Pressure and Raoult's Law

Consider water poured into a closed container that contains some headspace as shown on Fig. 11-4a. Some water molecules will have enough energy to overcome the attractive forces among the liquid water molecules and escape into the headspace above the liquid water, which is called evaporation. At the same time, some water molecules that have escaped into the gas-phase above the liquid water may lose energy and move back into the liquid water, which is called condensation. When the rates of evaporation and condensation are equal, the system is at equilibrium. The partial pressure exerted by the water vapor above the liquid water in the container at equilibrium is called the vapor pressure. Vapor pressure is dependent on temperature and increases with increasing temperature. For example, the vapor pressure of water is 1.23 kPa at 10°C and 3.17 kPa at 25°C. Other volatile liquids (e.g., acetone, benzene) behave the same way and also have a vapor pressure.

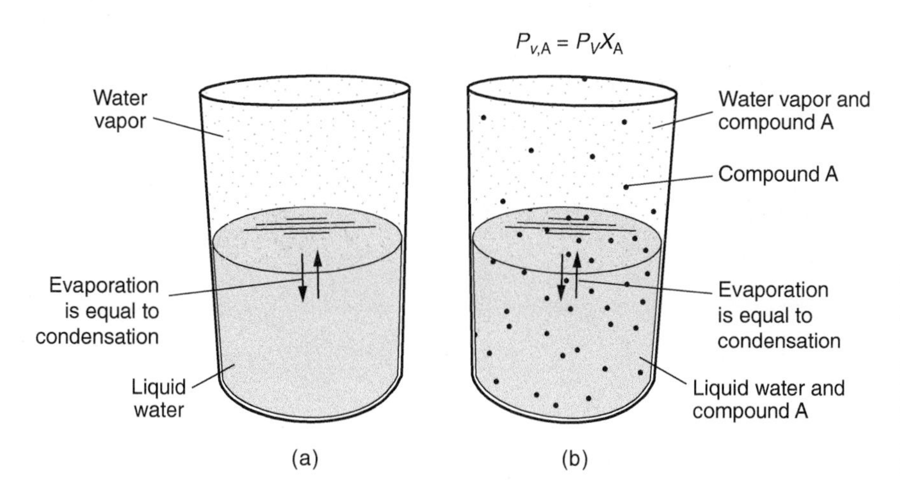

Figure 11-4
Schematic diagram for solution equilibrium description of vapor pressure with (a) vapor pressure of water, and (b) partial pressure of compound A in the presence of water.

If a volatile compound (A) is placed in the same closed container with the water and forms a solution as shown on Fig. 11-4b, the volatile compound would also come to equilibrium between the liquid and gas phases and exert a partial pressure above the liquid. If the solution is assumed to behave ideally in which the molecular forces between the solute (A) and the solvent (water) are idential to the solvent–solvent forces, and the solute (or solvent) molecule behaves identically regardless of whether it is surrounded by solute or solvent molecules, then the partial pressure of the solute would be a function of its vapor pressure and the mole fraction of the solute. The partial pressure of solute A can be calculated from the following expression known as Raoults's law:

$$P_{\mathrm{A}} = P_{V,\mathrm{A}} X_{\mathrm{A}} \tag{11-1}$$

where P_{A} = partial pressure of solute A, bar
$P_{V,\mathrm{A}}$ = vapor pressure of pure liquid A, bar
X_{A} = mole fraction of solute A in water, dimensionless

The mole fraction of A was introduced in Eq. 4-2 and is defined as

$$X_{\mathrm{A}} = \frac{n_{\mathrm{A}}}{\sum_{i=1}^{N} n_i} = \frac{n_{\mathrm{A}}}{n_{\mathrm{A}} + n_{\mathrm{H_2O}}} \tag{11-2}$$

where n = amount of A (solute) and water (solvent), mol
N = number of components in system

The relationship between partial pressure and mole fraction for solute A is illustrated on Fig. 11-5; ideal solutions follow Raoult's law and the slope is $P_{V,\mathrm{A}}$. For nonideal solutions the molecular forces between the solute and solvent are not identical to the solvent–solvent forces because the molecular forces between water molecules are very strong, so the solute–solvent attractions are generally smaller than the solvent–solvent attractions. Since there are smaller attractive forces holding the solute in solution, it is pushed out of solution and into the gas phase. Consequently,

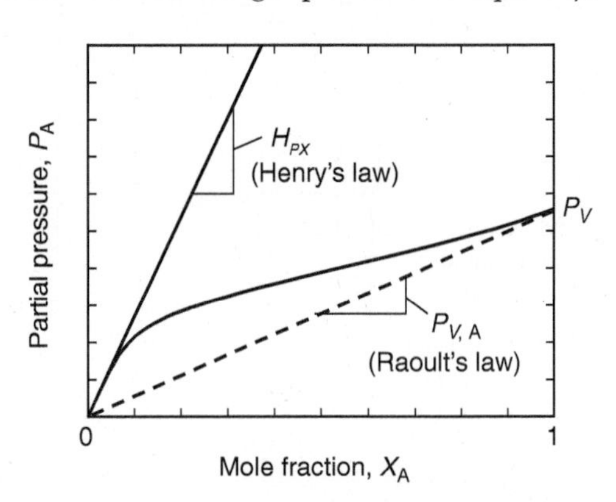

Figure 11-5
Relationship between partial pressure of a volatile compound and the mole fraction of the volatile compound in solution.

as shown, the partial pressure of the solute is higher than predicted by Raoult's law (a positive deviation from Raoult's law).

Henry's Law

For dilute solutions often found in environmental applications, the molecular interactions don't change significantly as additional solute is added, so partial pressure is proportional to mole fraction as shown on Fig. 11-5; this relationship is known as Henry's law. The equilibrium partitioning of a chemical solute between a liquid and gas phase is governed by Henry's law when the solute is very dilute in the mixture. Henry's law in equation form is

$$P_A = H_{PX}\,X_A \tag{11-3}$$

where H_{PX} = is Henry's law constant for solute A in solvent (water) when the liquid concentration is a mole fraction and the gas concentration is a partial pressure, bar

Henry's law is valid and constant up to mole fractions of about 0.01 and has been shown to be valid for concentrations up to 0.1 mol/L (Rogers, 1994). Solvent–solvent forces are unaffected by small amounts of solute and the solvent follows Raoult's law for dilute solutions. Henry's law constants are valid for binary systems (e.g., component A in water). For systems where there are several solutes in a solvent (water) and the solution is still considered dilute, Henry's law will be valid for each solute (i.e., because each solute is dilute, interactions between them are generally negligible). The presence of air does not affect the Henry's law constant for volatile organic chemicals (VOCs) or gases because the constituents of interest have low concentrations in air.

Other Units for Henry's Law

The units of Henry's law constant, H_{PX}, in Eq. 11-3 are in bar because the units for the gas-phase pressure and the liquid-phase concentration are given as bar and mole fraction, respectively. Henry's law constants can also be expressed in terms of concentration or partial pressure of A for the gas phase and mole fraction or concentration for the liquid or water phase. The gas-phase concentration expressed as either partial pressure (bar) or concentration in mol/L is related through the ideal gas law as shown below:

$$P_A V = n_A RT \quad \text{or} \quad Y_A = \frac{n_A}{V} = \frac{P_A}{RT} \tag{11-4}$$

where R = universal gas constant, 0.083145 L·bar/mol·K
T = temperature, K
$Y_A = n_A/V$ = gas-phase concentration, mol/L
V = volume of gas, L

The liquid-phase concentration can be expressed as either mole fraction (mol/mol) or concentration (mol/L) as

$$X_A = \frac{n_A}{n_A + n_W} \approx \frac{n_A}{n_W} = \frac{C_A}{C_W} \tag{11-5}$$

Table 11-3
Unit conversions for Henry's law constants

Form of Henry's Law[a]	Units for Henry's Constant	Conversion to H_{YC}
$P_A = H_{PX} X_A$	bar	$H_{YC} = \dfrac{H_{PX}}{RT(55.56\ \text{mol/L})}$
$P_A = H_{PC} C_A$	bar·L/mol	$H_{YC} = \dfrac{H_{PC}}{RT}$
$Y_A = H_{YC} C_A$	L_{H_2O}/L_{air}[b]	—

where P_A = partial pressure of solute A, bar
X_A = liquid-phase mole fraction of solute A, dimensionless
C_A = liquid-phase concentratio of solute A, mol/L
Y_A = gas-phase concentration of solute A, mol/L

[a]Subscripts on H correspond to units as follows: P = partial pressure, X = mole fraction, Y = gas phase concentration, and C = liquid phase concentration.
[b]Because the units of H_{YC} are volume in both the numerator and denominator, H_{YC} is often known as the dimensionless form of the Henry's constant.

where n_W = amount of water in solution, mol

$$C_W = \frac{\text{density of water}}{\text{molecular weight of water}} = \frac{1000\ \text{g/L}}{18\ \text{g/mol}} = 55.56\ \text{mol/L}$$

$C_A = X_A C_W$ = concentration of solute A, mol/L

Applying these relationships results in three common forms of expressing Henry's law that are summarized in Table 11-3. A particularly useful set of units is when the solute is expressed as concentration (either mass or molar) in both the gas and liquid phases. These units are known as a "dimensionless" form of Henry's law and are widely used in environmental engineering. Use of the relationships displayed in Table 11-3 is illustrated in the Example 11-1.

Example 11-1 Converting the units of Henry's law constants

Calculate the dimensionless Henry's law constant, H_{YC}, for a compound that has a H_{PX} value of 250 bar. Also calculate the Henry's law constant in bar·L/mol for a compound that has a dimensionless Henry's law constant of 0.0545. The temperature is 25°C.

Solution

1. Calculate the dimensionless Henry's law constant using the relationship shown in Table 11-3 for converting H_{PX} to H_{YC}. Note 25°C = 298 K.

$$H_{YC} = \frac{H_{PX}}{RT(55.56 \text{ mol/L})} = \frac{250 \text{ bar}}{(0.083145 \text{ L·bar/mol·K})(298 \text{ K})(55.56 \text{ mol/L})}$$

$$= 0.181$$

2. Determine Henry's law constant in bar·L/mol by rearranging the expression for converting H_{PC} to H_{YC} and solving for H_{PC} for an H_{YC} of 0.0545:

$$H_{PC} = H_{YC}RT = (0.0545)(0.083145 \text{ L·bar/mol·K})(298 \text{ K}) = 1.35 \text{ bar·L/mol}$$

Sources of Henry's Law Constants

Experimental methods have been developed to determine Henry's law constant for volatile compounds (Gossett, 1987; Ashworth et al., 1988; Robbins et al., 1993; Dewulf et al., 1995; Heron et al., 1998; Ayuttya et al., 2001). Table 11-4 displays some experimentally determined values of Henry's law constants for some VOCs and gases encountered in water supplies. Henry's constants for a large number of volatile chemicals can be readily found in a number of Internet databases, including sites maintained by NIST (2011) and SRC (2011). Additionally, methods have been developed to estimate

Table 11-4
Dimensionless Henry's Law constants for selected volatile organic chemicals[a]

	Henry's Law Constants, *H*				
Component	**10°C**	**15°C**	**20°C**	**25°C**	**30°C**
Benzene	0.142	0.164	0.188	0.216	0.290
Carbon tetrachloride	0.637	0.808	0.96	1.210	1.520
Chloroform	0.0741	0.0968	0.1380	0.1720	0.2230
Cis-1,2-Dichloroethylene	0.116	0.138	0.150	0.186	0.231
Dibromochloromethane	0.0164	0.0190	0.0428	0.0483	0.0611
1,2-Dichlorobenzene	0.0702	0.0605	0.0699	0.0642	0.0953
1,3-Dichlorobenzene	0.0952	0.0978	0.1220	0.1170	0.1700
1,2-Dichloropropane	0.0525	0.0533	0.0790	0.1460	0.1150
Ethylbenzene	0.140	0.191	0.250	0.322	0.422
Methyl ethyl ketone	0.01210	0.01650	0.00790	0.00532	0.00443
Methyl *t*-butyl ether*	0.0117	0.0177	0.0224	0.0292	0.0387
m-Xylene	0.177	0.210	0.249	0.304	0.357
n-Hexane	10.3	17.5	36.7	31.4	62.7
o-Xylene	0.123	0.153	0.197	0.199	0.252
1,1,2,2-Tetrachloroethane	0.01420	0.00846	0.03040	0.01020	0.02820
Tetrachloroethylene (PCE)	0.364	0.467	0.587	0.699	0.985
Toluene	0.164	0.210	0.231	0.263	0.325
Trichloroethylene (TCE)	0.237	0.282	0.350	0.417	0.515

[a]Adapted from Ashworth et al. (1988).

Henry's law constants when experimental values are not available; details on these methods may be found in Crittenden et al. (2012).

Factors Influencing Henry's Constant

Temperature, ionic strength, surfactants, and solution pH (for ionizable species such as NH_3 and CO_2) can influence the equilibrium partitioning between air and water. The impact of total system pressure on H_{YC} is negligible because other components in air have limited solubility in water. For water supplies that contain multiple VOCs in low concentrations (<10 mg/L), their Henry's constant values are not impacted by the other VOCs present in the water.

EFFECT OF TEMPERATURE

As shown in Table 11-4, for typical water temperatures encountered in drinking water treatment, Henry's constant values increase with temperature. As the temperature increases the volatility of the compound increases significantly as compared to the compound solubility, thereby increasing Henry's constant. Henry's constant is an equilibrium constant for a reaction between the gaseous and dissolved forms of a volatile species; thus, the temperature dependence is governed by the van't Hoff equation for temperature dependence of chemical reactions (Eq. 4-24), as was presented in Chap. 4. Thus, Henry's constant can be calculated at different temperatures if the enthalpy of dissolution is known.

IONIC STRENGTH

Natural waters used for drinking water may contain concentrations of dissolved solids (50 to 600 mg/L TDS) and natural organic matter (0.5 to 15 mg/L as DOC). The value of Henry's constants is not impacted by the range of these dissolved constituents in natural waters (Nicholson et al., 1984). Gases or synthetic organic chemicals (SOCs) have a higher apparent Henry's law constant ($H_{YC,\text{app}}$) when the dissolved solids are high, such as the retentate from a reverse osmosis process or seawater. In cases where the dissolved solids are high enough to impact Henry's law constant, methods are available to determine $H_{YC,\text{app}}$ (Crittenden et al., 2012; Gossett, 1987; Schwarzenbach et al., 1993).

EFFECT OF SURFACTANTS

Surfactants can impact the volatility of compounds. In most natural waters, surfactant concentrations are relatively low; consequently, surfactants do not affect the design of most aeration devices. However, when surfactants are present in relatively high concentrations, the volatility of other compounds may decrease by several mechanisms. The dominant mechanism is the collection of surfactant molecules at the air–water interface, decreasing the mole fraction of the volatile compound at the interfacial area, thereby decreasing the apparent Henry's law constant. For example, the solubility of oxygen in water can decrease by 30 to 50 percent due to the presence of surfactants.

IMPACT OF PH

The pH does not affect Henry's constant directly, but it does affect the distribution of species between ionized and un-ionized forms, which influences the overall gas–liquid distribution of the compound because only the un-ionized species are volatile. Acid–base chemistry is presented in Sec. 4-4.

11-3 Fundamentals of Packed Tower Air Stripping

Packed towers are either cylindrical columns or rectangular towers containing packing that disrupts the flow of liquid, thus producing and renewing the air–water interface, as shown in Fig. 11-1c and described in Table 11-2. Packing material is available in a wide variety of sizes and shapes depending on the manufacturer. Operationally, water is pumped to the top of the tower and into a liquid distributor where it is dispersed as uniformly as possible across the packing surface, and then it flows by gravity through the packing material and is collected at the bottom of the tower. A blower is used to introduce fresh air into the bottom of the tower and the air flows countercurrent to the water up through the void spaces between the wetted packing material. Packed towers have high liquid interfacial areas and void volumes greater than 90 percent, which minimizes air pressure drop through the tower.

The random packing material is important to the efficient transfer of volatile contaminants from the water to the air because it provides a large air–water interfacial area. Various types of packing shapes, sizes and their physical properties are available commercially, as shown on Fig. 11-6. The packing can be structured packing or individual pieces that are randomly placed in the tower.

Pilot plant and full-scale studies are the most conservative approach to the design of packed towers. However, in many cases they can be time consuming and expensive. In most cases, mathematical models are used to design countercurrent packed-tower air-strippings systems as well as describe their performance. The following sections provide development and application of the design equations used in the design of these systems.

Mass Balance Analysis for a Countercurrent Packed Tower

The design equations for countercurrent packed towers are developed from two mass balances. The first mass balance describes the relationship between the bulk water-phase concentration and the bulk air-phase concentration at any point in the tower. A schematic of a countercurrent packed tower is

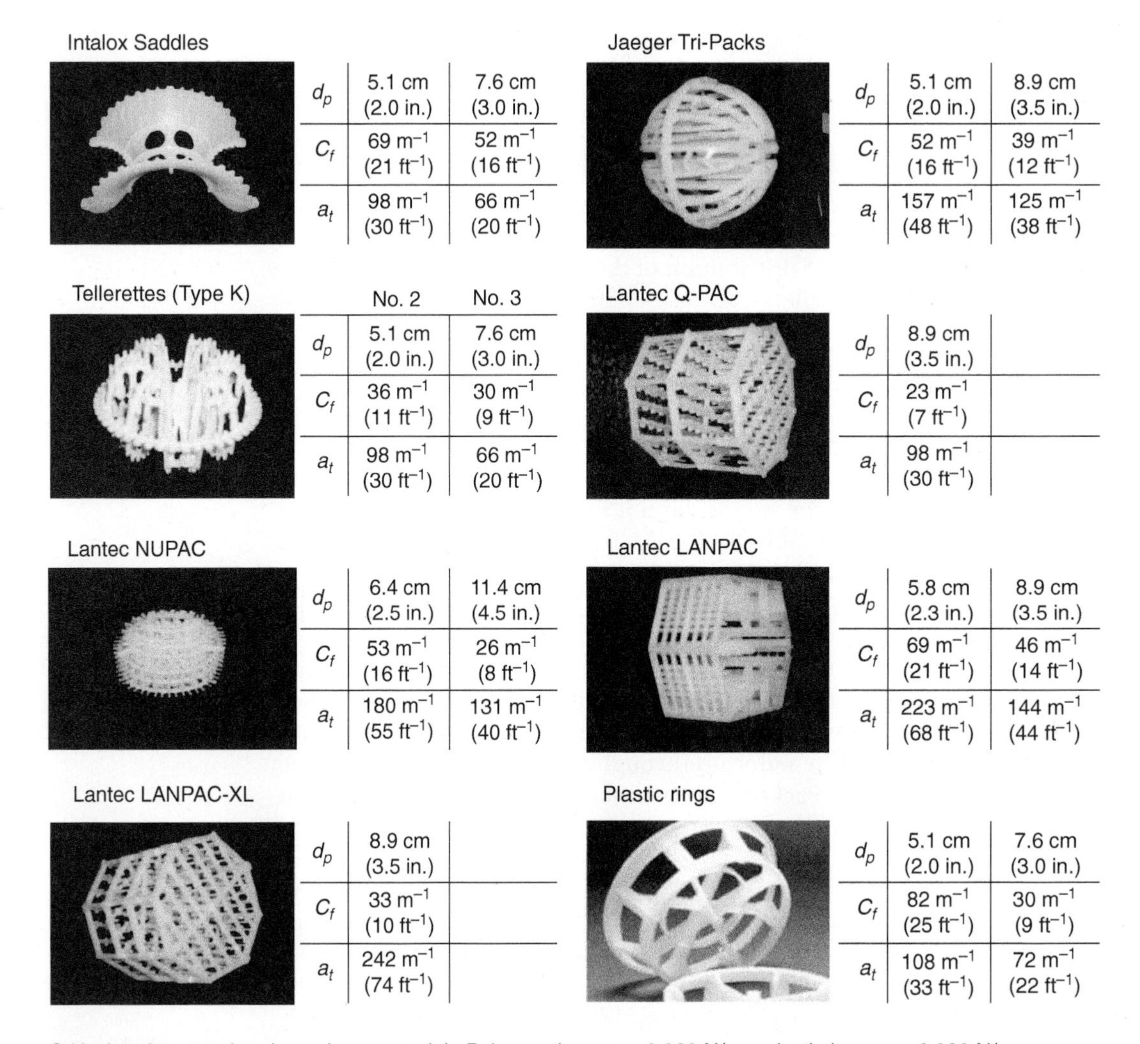

Intalox Saddles

d_p	5.1 cm (2.0 in.)	7.6 cm (3.0 in.)
C_f	69 m^{-1} (21 ft^{-1})	52 m^{-1} (16 ft^{-1})
a_t	98 m^{-1} (30 ft^{-1})	66 m^{-1} (20 ft^{-1})

Jaeger Tri-Packs

d_p	5.1 cm (2.0 in.)	8.9 cm (3.5 in.)
C_f	52 m^{-1} (16 ft^{-1})	39 m^{-1} (12 ft^{-1})
a_t	157 m^{-1} (48 ft^{-1})	125 m^{-1} (38 ft^{-1})

Tellerettes (Type K)

	No. 2	No. 3
d_p	5.1 cm (2.0 in.)	7.6 cm (3.0 in.)
C_f	36 m^{-1} (11 ft^{-1})	30 m^{-1} (9 ft^{-1})
a_t	98 m^{-1} (30 ft^{-1})	66 m^{-1} (20 ft^{-1})

Lantec Q-PAC

d_p	8.9 cm (3.5 in.)	
C_f	23 m^{-1} (7 ft^{-1})	
a_t	98 m^{-1} (30 ft^{-1})	

Lantec NUPAC

d_p	6.4 cm (2.5 in.)	11.4 cm (4.5 in.)
C_f	53 m^{-1} (16 ft^{-1})	26 m^{-1} (8 ft^{-1})
a_t	180 m^{-1} (55 ft^{-1})	131 m^{-1} (40 ft^{-1})

Lantec LANPAC

d_p	5.8 cm (2.3 in.)	8.9 cm (3.5 in.)
C_f	69 m^{-1} (21 ft^{-1})	46 m^{-1} (14 ft^{-1})
a_t	223 m^{-1} (68 ft^{-1})	144 m^{-1} (44 ft^{-1})

Lantec LANPAC-XL

d_p	8.9 cm (3.5 in.)	
C_f	33 m^{-1} (10 ft^{-1})	
a_t	242 m^{-1} (74 ft^{-1})	

Plastic rings

d_p	5.1 cm (2.0 in.)	7.6 cm (3.0 in.)
C_f	82 m^{-1} (25 ft^{-1})	30 m^{-1} (9 ft^{-1})
a_t	108 m^{-1} (33 ft^{-1})	72 m^{-1} (22 ft^{-1})

Critical surface tension depends on material. Polypropylene σ_c = 0.029 N/m, polyethylene σ_c = 0.033 N/m.

Figure 11-6
Typical examples of plastic packing materials used in air-stripping towers and their physical characteristics.

shown on Fig. 11-7. A contaminant can exist in the gas and liquid phases but does not accumulate in the tower (i.e., the system is at steady state). With the assumption that no reactions occur in the tower and that mass can only be transferred between gas and liquid phases, the mass balance concept presented in Sec. 4-5 can be simplified to

$$[\sout{\text{accum}}] = [\text{mass in}] - [\text{mass out}] + [\sout{\text{rxn}}] \tag{11-6}$$

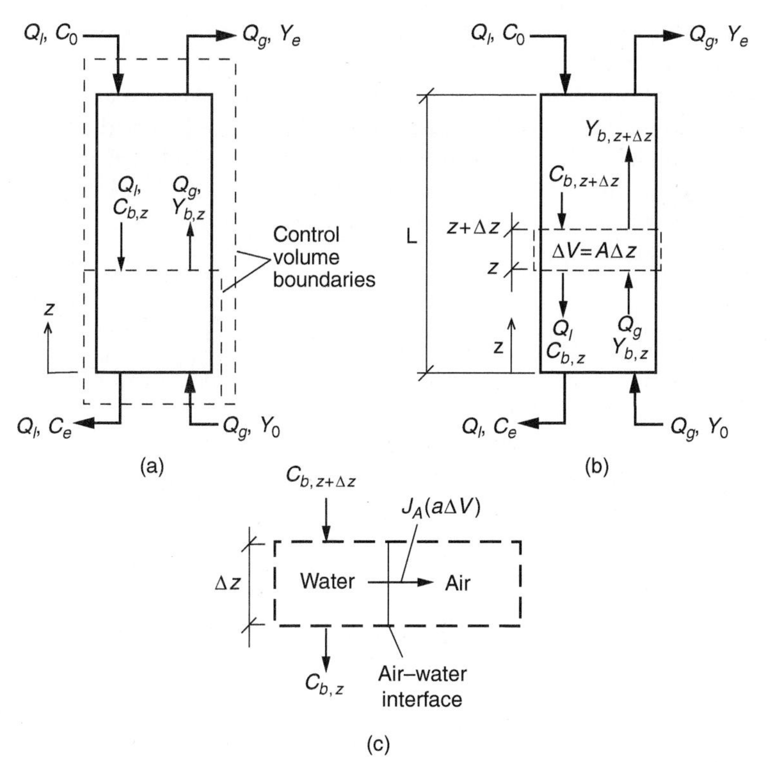

Figure 11-7
Packed-tower design equation definition drawing: (a) definition drawing for mass balances on a packed tower and (b) schematic of differential element used in liquid-side mass balance.

Writing the mass balance around the volatile constituent in the lower half of the tower using the symbols on Fig. 11-7 yields

$$Q_l C_{b,z} + Q_g Y_0 = Q_l C_e + Q_g Y_{b,z} \tag{11-7}$$

where Q_l, Q_g = liquid and gas flow rates, m^3/s
$C_{b,z}$ = bulk liquid-phase concentration at axial position z along tower, mg/L
C_e = effluent liquid-phase concentration, mg/L
Y_0 = gas-phase concentration entering tower, mg/L
$Y_{b,z}$ = bulk gas-phase concentration at axial position z along tower, mg/L

Rearranging algebraically, Eq. 11-7 can be written

$$Y_{b,z} = \left(\frac{Q_l}{Q_g}\right)\left(C_{b,z} - C_e\right) + Y_0 \tag{11-8}$$

Equation 11-8 relates the concentration of the volatile constituent in the gas phase to the corresponding concentration in the liquid phase at every position in the tower. Since the gas and liquid flow rates are constant through the tower (the volatile constituent is so dilute that its transfer has no impact on the flow rates), Eq. 11-8 is a straight line with a slope of Q_l/Q_g. This line is known as the *operating line* equation for countercurrent packed-tower aeration. The *operating diagram* for packed-tower aeration is presented on Fig. 11-8, which is known as a McCabe–Thiele diagram (McCabe and Thiele, 1925). Operating diagrams were introduced in Sec. 4-17. The operating line is labeled 1 on Fig. 11-8, and the equilibrium line is labeled 2. The *equilibrium line* is described by a straight line known as Henry's law (see Sec. 11-2):

$$Y = H_{YC} C \tag{11-9}$$

where Y = gas-phase concentration, mg/L
H_{YC} = Henry's law constant when concentration in gas and liquid phases are both mg/L, dimensionless
C = liquid-phase concentration in equilibrium with gas-phase concentration Y, mg/L

The operating and equilibrium lines are an important concept in separation processes, such as air stripping, because they can be used to determine the minimum amount of extracting phase (e.g., air in packed-tower aeration), in terms of mass or volume required to remove a component (e.g., from water in packed-tower aeration) to a desired removal efficiency.

Stripping Factor

A parameter commonly used in the evaluation of packed towers is the stripping factor (S), where S is defined as the ratio of the slope of the equilibrium line to the operating line slope. As shown in Fig. 11-8, the equilibrium line divided by the operating line yields the following expression for the S:

$$S = \frac{\text{slope of equilibrium line}}{\text{slope of operating line}} = \frac{H_{YC}}{Q_l/Q_g} = \left(\frac{Q_g}{Q_l}\right) H_{YC} \tag{11-10}$$

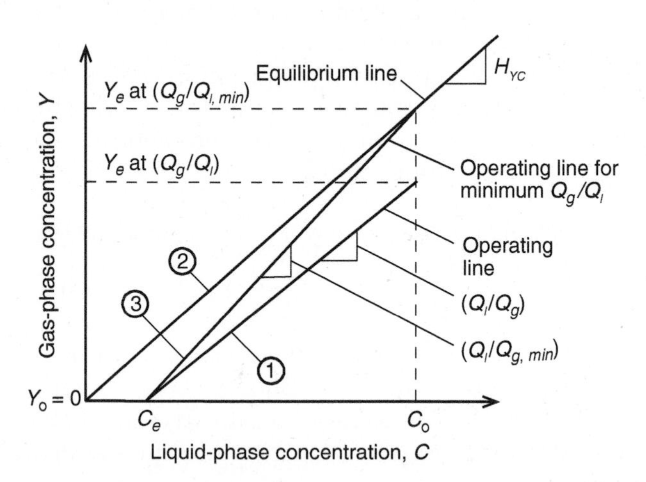

Figure 11-8
Operating diagram for a countercurrent packed tower.

where S = stripping factor, dimensionless
Q_g/Q_l = operating air-to-water ratio of tower
Q_g = air flow rate, m^3/s
Q_l = water flow rate, m^3/s

When $S = 1$, the slopes of the equilibrium and operating lines are parallel, and removal of the volatile constituent to any treatment objective is possible. However, a low effluent concentration requires the equilibrium and operating lines to be very close to each other, resulting in a low driving force for mass transfer and thus a very tall stripping tower (theoretically, infinite if the required effluent concentration is zero). If $S < 1$, the slope of the operating line is greater than the slope of the equilibrium line, the desired removal will be equilibrium limited and the treatment objective cannot be obtained if a very low effluent concentration is needed. When $S > 1$, the slope of the operating line is less than the slope of the equilibrium line. In this case, the equilibrium and operating lines diverge, resulting in a favorable driving force that leads to effective mass transfer, and the treatment objective can be met using stripping. Since the slope of the operating line is the ratio of the liquid and gas flow rates, the operating diagram and the stripping factor demonstrate the importance of the gas flow rate as a key operating parameter for countercurrent packed-tower aeration.

Minimum Air-to-Water Ratio

A special case of the operating line shown in Fig. 11-8 is line 3. This line intersects the equilibrium line where the influent concentration, C_0, is in equilibrium with the exiting gas-phase concentration (i.e., $Y_e = H_{YC}C_0$). The slope of this line represents the inverse of the minimum air-to-water ratio that can meet the treatment objective if the packed-tower length is infinite. If it is assumed the influent gas-phase concentration, Y_0, is equal to zero, and the influent liquid-phase concentration is in equilibrium with the exiting air, Eq. 11-8 can be rearranged to yield the following expression for the minimum air-to-water ratio:

$$\left(\frac{Q_g}{Q_l}\right)_{\min} = \frac{C_0 - C_e}{H_{YC}C_0} \qquad (11\text{-}11)$$

where $(Q_g/Q_l)_{\min}$ = minimum air-to-water ratio, dimensionless
C_0 = influent liquid-phase concentration, mg/L
C_e = treatment objective, mg/L

The minimum air-to-water ratio $(Q_g/Q_l)_{\min}$ represents the theoretical minimum air-to-water ratio that can be applied for a packed tower to meet its treatment objective C_e. If the air-to-water ratio applied is less than the minimum air-to-water ratio, it will not be possible to design a packed tower capable of meeting the treatment objective because equilibrium will be established in the tower before the treatment objective is reached.

With respect to the selection of the optimum air-to-water ratio, it has been demonstrated that minimum tower volume and power requirements are achieved using approximately 3.5 times the minimum air-to-water ratio for contaminants with Henry's law constants greater than 0.05 for high percentage removals, corresponding to a stripping factor of 3.5 (Hand et al., 1986).

The stripping factor can be related to the minimum air-to-water ratio when the treatment efficiency is very high, and Eq. 11-11 can be approximated as

$$\left(\frac{Q_g}{Q_l}\right)_{\min} = \frac{C_0 - C_e}{H_{YC} C_0} \approx \frac{1}{H_{YC}} \quad (C_e \ll C_0) \tag{11-12}$$

Substitution of Eq. 11-12 into Eq. 11-10 yields a relationship for stripping factor in terms of minimum air-to-water ratio.

$$S = \frac{Q_g/Q_l}{(Q_g/Q_l)_{\min}} \tag{11-13}$$

When $C_e \ll C_0$, the stripping factor is approximately equal to the ratio of the actual air flow rate to the minimum air flow rate for treating a given flow of water. Use of the stripping factor and minimum air-to-water ratio is demonstrated in Example 11-2.

Example 11-2 Calculating minimum and operating air-to-water ratios, required air flow rate, and stripping factors

Calculate the minimum air-to-water ratio and operating air-to-water ratio for 1,2-dichloropropane (DCP) and tetrachloroethylene (PCE) with 90 percent removal at 10°C for a countercurrent packed tower, assuming the optimal operating ratio is 3.5 times the minimum ratio for each contaminant. If a tower must treat a liquid flow rate of 8.64 ML/d that contains both compounds, determine the required air flow rate for the tower and the stripping factor for each contaminant.

Solution

1. Calculate the minimum air-to-water ratio for each compound using Eq. 11-11. H_{PC} for each compound are available in Table 11-4.
   a. DCP:

$$\left(\frac{Q_g}{Q_l}\right)_{\min,\text{DCP}} = \frac{C_0 - C_e}{H_{YC,\text{DCP}} C_0} = \frac{C_0 - 0.1C_0}{0.0525C_0} = 17.14$$

   b. PCE:

$$\left(\frac{Q_g}{Q_l}\right)_{\min,\text{PCE}} = \frac{C_0 - C_e}{H_{YC,\text{PCE}} C_0} = \frac{C_0 - 0.1C_0}{0.364C_0} = 2.47$$

2. To calculate the operating air-to-water ratio that minimizes tower volume and power consumption, multiply the minimum air-to-water ratio by 3.5.
   a. DCP:

$$\left(\frac{Q_g}{Q_l}\right)_{\text{DCP}} = (17.14)\,(3.5) = 60$$

   b. PCE:

$$\left(\frac{Q_g}{Q_l}\right)_{\text{PCE}} = (2.47)\,(3.5) = 8.65$$

3. Since DCP has the greatest operating air-to-water ratio, the removal of DCP will control the air flow rate. Thus

$$Q_l = 8.64\ \text{ML/d} = \frac{8.64 \times 10^6\ \text{L/d}}{(10^3\ \text{L/m}^3)(86{,}400\ \text{s/d})} = 0.10\ \text{m}^3/\text{s}$$

$$Q_g = Q_l\left(\frac{Q_g}{Q_l}\right)_{\text{DCP}} = (0.10\ \text{m}^3/\text{s})\,(60) = 6.0\ \text{m}^3/\text{s}$$

4. The stripping factor for each compound is calculated using Eq. 11-10:

$$S_{\text{DCP}} = \frac{Q_g H_{YC}}{Q_l} = \frac{(6.0\ \text{m}^3/\text{s})\,(0.0525)}{0.1\ \text{m}^3/\text{s}} = 3.15$$

$$S_{\text{PCE}} = \frac{(6.0\ \text{m}^3/\text{s})\,(0.364)}{0.1\ \text{m}^3/\text{s}} = 21.8$$

Comment

The compound with the lower Henry's law constant (DCP) requires a higher air-to-water ratio to achieve the desired removal. This is expected because a smaller Henry's constant indicates lower volatility; that is, a greater preference of the compound for the water phase and a lower tendency for stripping from the water phase to the air phase.

Design Equation for Determining Packed-Tower Height

Predicting the required height of a packed tower to meet a given air-stripping treatment objective is one of the goals of packed-tower design. The design equation for tower height can be derived using these assumptions: (1) steady-state conditions prevail in the tower, (2) air flow rate and water flow rate are constant through the column, (3) no chemical reactions occur, and (4) plug flow conditions prevail for both the air and water.

LIQUID-PHASE MASS BALANCE AROUND A DIFFERENTIAL ELEMENT

A liquid-phase mass balance around the differential element surrounded by a dashed box on Fig. 11-7a serves as the basis for the design equation. A schematic of the differential element applicable to the case of a liquid-side mass balance is presented on Fig. 11-7b. As with the previous mass balance, this system is at steady state because no volatile constituent accumulates in the control volume. The volatile contaminant can enter the liquid phase in the differential element with the water entering the element, and can leave either by exiting with the water exiting the element or by being transferred to the air within the element. In words, this mass balance can be expressed as

$$\begin{pmatrix}\text{mass entering}\\ \text{with liquid}\end{pmatrix} - \begin{pmatrix}\text{mass exiting}\\ \text{with liquid}\end{pmatrix} - \begin{pmatrix}\text{mass transferred}\\ \text{to air phase}\end{pmatrix} = 0 \quad (11\text{-}14)$$

Equation 11-14 can be written symbolically as

$$Q_l C_{b,z+\Delta z} - Q_l C_{b,z} - J_A(a\,\Delta V) = 0 \quad (11\text{-}15)$$

where Q = water flow rate, m^3/s
C_b = bulk liquid-phase concentration, mg/L
z = axial position along tower, m
Δz = height of differential element, m
J_A = flux across air–water interface, $mg/m^2{\cdot}s$
a = area available for mass transfer divided by vessel volume, m^2/m^3
ΔV = volume of differential element, m^3

As shown in Eq. 4-137 in Sec. 4-16, the term J_A in Eq. 11-15 is obtained from the two-film theory:

$$J_A = K_L\left(C_{b,z} - C^*_{s,z}\right) \quad (11\text{-}16)$$

where K_L = overall liquid-phase mass transfer coefficient, m/s
$C^*_{s,z}$ = liquid-phase concentration in equilibrium with the bulk gas-phase concentration, mg/L

Inserting Eq. 11-16 and $\Delta V = A\,\Delta z$ into Eq. 11-15 yields

$$Q_l C_{b,z+\Delta z} - Q_l C_{b,z} - K_L\left(C_{b,z} - C^*_{s,z}\right)(aA\,\Delta z) = 0 \quad (11\text{-}17)$$

where A = cross-sectional area of packed tower, m^2

Rearranging Eq. 11-17 and dividing by $A\,\Delta z$ yields the equation

$$\frac{Q_l}{AK_L a}\left(\frac{C_{b,z+\Delta z} - C_{b,z}}{\Delta z}\right) = C_{b,z} - C^*_{s,z} \quad (11\text{-}18)$$

where $K_L a$ = overall liquid-side mass transfer rate constant, s^{-1}

Taking the limit as Δz approaches zero results in

$$\frac{Q_l}{AK_L a} \lim_{\Delta z \to 0} \left(\frac{C_{b,z+\Delta z} - C_{b,z}}{\Delta z} \right) = \frac{Q_l}{AK_L a} \frac{dC_b}{dz} = C_{b,z} - C^*_{s,z} \quad (11\text{-}19)$$

Separating variables in Eq. 11-19 and setting up both sides of the equation for integration from the bottom of the tower to the top results in

$$\frac{Q_l}{AK_L a} \int_{C_e}^{C_0} \frac{dC_b}{C_b - C^*_s} = \int_0^L dz = L \quad (11\text{-}20)$$

where L = height of packed tower, m
C_0 = influent liquid-phase concentration, mg/L
C_e = treatment objective, mg/L

The left side of Eq. 11-20 contains two components, which are known in gas transfer literature as the height of a transfer unit (HTU) and the number of transfer units (NTU), so that Eq. 11-20 can be expressed as

$$L = (\text{HTU})\,(\text{NTU}) \quad (11\text{-}21)$$

where

$$\text{HTU} = \frac{Q_l}{AK_L a} \quad (11\text{-}22)$$

$$\text{NTU} = \int_{C_e}^{C_0} \frac{dC_b}{C_b - C^*_s} \quad (11\text{-}23)$$

The HTU is the ratio of the superficial velocity or liquid loading rate (Q_l/A) to the overall liquid-side mass transfer rate constant ($K_L a$). For packed towers, the HTU is a measure of the stripping effectiveness of particular packings for a given stripping process. Packing that is typically smaller in size has higher specific surface area, causing more efficient transfer of solute from one phase to another, there by increasing $K_L a$ and decreasing the HTU. The HTU and tower length will decrease as the superficial velocity decreases or the rate of mass transfer increases.

RELATING CONCENTRATION AT AIR–WATER INTERFACE TO CONCENTRATION IN BULK LIQUID

Equation 11-23 cannot be integrated directly because C^*_s varies over the depth of the tower. Before integration, it is necessary to express C^*_s in terms of C_b. In the development of the two-film theory in Sec. 4-16, it was noted that C^*_s is the liquid-phase concentration that is in equilibrium with the bulk gas phase. Thus, Henry's law defines the value of C^*_sas a function of gas-phase concentration at any position within the tower:

$$C^*_s = \frac{Y_{b,z}}{H_{YC}} \quad (11\text{-}24)$$

where $Y_{b,z}$ = bulk gas-phase concentration at position z, mg/L

The operating line, Eq. 11-8, describes the relationship between the concentrations in the gas and liquid phases. Substituting Eq. 11-24 into Eq. 11-8 yields the desired relationship between C_s^* and C_b:

$$C_s^* = \frac{Y_b}{H_{YC}} = \frac{(Q_l/Q_g)(C_b - C_e) + Y_0}{H_{YC}} \tag{11-25}$$

where Q_g = air flow rate, m^3/s

In most stripping operations, the concentration of the target contaminant in the influent gas to the tower, Y_0, is zero. The parameter group $Q_l/Q_g H_{YC}$ is equal to the inverse of the stripping factor, so Eq. 11-25 can be simplified to

$$C_s^* = \frac{C_b - C_e}{S} \tag{11-26}$$

Substituting Eq. 11-26 into Eq. 11-23 and setting $Y_0 = 0$, followed by some algebraic rearranging, yields

$$\text{NTU} = \int_{C_e}^{C_0} \frac{dC_b}{(S - 1/S)\, C_b + \dfrac{C_e}{S}} \tag{11-27}$$

Integrating and rearranging yields

$$\text{NTU} = \left(\frac{S}{S-1}\right) \ln\left[\frac{1 + (C_0/C_e)(S-1)}{S}\right] \tag{11-28}$$

Equation 11-28 describes an important result that relates the number of transfer units to the stripping factor and the required removal efficiency (C_0/C_e). The NTU can be thought of as a measure of the difficulty of stripping a solute from the liquid to the gas phase. The more difficult it is to strip the solute, the more NTUs are needed to achieve a given removal efficiency. The relationship between NTU, C_0/C_e, and S is demonstrated on Fig. 11-9, which is a plot of numerous solutions of Eq. 11-28. For a

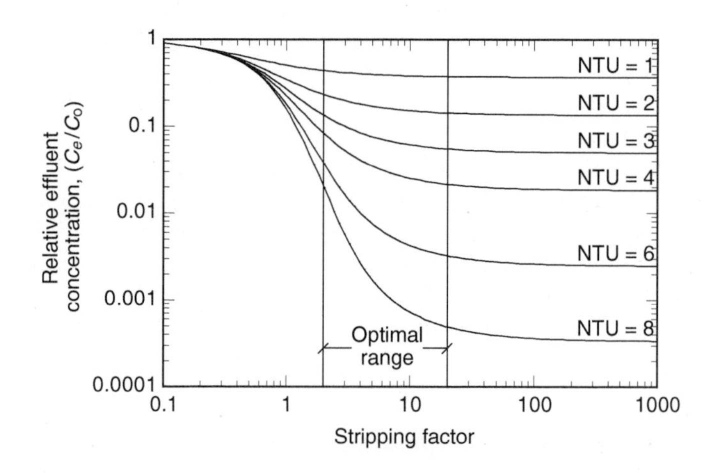

Figure 11-9
Dependence of relative effluent concentration on NTU and stripping factor.

given value of S, the removal efficiency increases with increasing NTU. In addition, for a given removal efficiency, increasing S (by increasing the air-to-water ratio) will decrease the NTU required. As shown on Fig. 11-9, the optimal range for the stripping factor might be considered between about 1 and 20 because high removal efficiency is not possible at S less than 1 and no additional improvement in removal occurs at values of S greater than about 20 for a given value of NTU. The best efficiency point for minimum power requirements and tower volume tend to occur at an air-to-water ratio of 3.5 times the minimum air-to-water ratio required for stripping, which would correspond to a low value of the stripping factor (Hand et al., 1986).

DETERMINING TOWER HEIGHT

Having determined both HTU and NTU, the depth of packing in a courntercurrent packed tower can be determined by substituting Eqs. 11-22 and 11-28 into Eq. 11-21 to yield the following equation:

$$L = \left(\frac{Q_l}{AK_La}\right)\left(\frac{S}{S-1}\right)\ln\left[\frac{1+(C_0/C_e)(S-1)}{S}\right] \tag{11-29}$$

where L = packed tower height, m
A = cross-sectional area of packed tower, m^2
K_La = overall liquid-side mass transfer rate constant, 1/s
S = stripping factor, dimensionless
C_0 = influent liquid-phase concentration, mg/L
C_e = treatment objective, mg/L

DETERMINING EFFLUENT CONCENTRATION

For an existing packed tower the following variables are typically known: (1) tower height, (2) tower diameter, (3) type of packing, (4) water flow rate, (5) air flow rate, (6) pressure, (7) temperature, (8) influent concentration, and (9) mass transfer coefficient. Knowing these variables, it is possible to determine effluent concentration for the tower. The effluent concentration can be found by rearranging Eq. 11-29 and solving for effluent concentration C_e:

$$C_e = \frac{C_0(S-1)}{S\exp\left[\frac{LAK_La}{Q_l}\frac{(S-1)}{S}\right]-1} \tag{11-30}$$

where C_e = effluent liquid-phase concentration, mg/L

11-4 Design and Analysis of Packed-Tower Air Stripping

The two main design activities for packed-tower air stripping are (1) designing new towers (design analysis) and (2) modifications to existing towers (rating analysis). The procedure for determining the tower packing depth is decribed in the following section, and the procedure for the rating analysis is described after that.

To determine the packed-tower depth as described by Eq. 11-29, the following properties are needed: (1) the air flow rate, (2) gas pressure drop, (3) cross-sectional area of the tower, and (4) overall liquid-side mass transfer rate coefficient. The procedure for determining the air flow rate was demonstrated earlier in Example 11-2. Determination of the remaining properties required to calculate packed-tower depth is discussed below.

Gas Pressure Drop

The gas pressure drop in packed columns is an important design and operational parameter because the electrical costs of the blower account for a significant fraction of the operational costs. Consequently, it is important to operate at a low gas pressure drop to minimize the power consumption and blower costs. A common method of estimating the gas pressure drop through random packing in towers is the use of the generalized Eckert pressure drop correlation (see Fig. 11-10). The Eckert correlation relates the gas pressure drop to the capacity parameter on the ordinate (y axis) as a function of the flow parameter on the abscissa (x axis). For high gas loading rates, entrainment of the liquid by the rising gas can occur, characterized by a sudden rapid increase in the gas pressure drop, and eventually the column will become a flooded contactor because of the back pressure caused by the rising gas. However, as discussed above, most all air-stripping applications operate at low gas pressure drops to minimize enery costs associated with the blower operation and flooding is rarely a problem.

The Eckert correlation shown in Fig. 11-10 was developed based on data for packings such as small intalox saddles, rashig, and pall rings.

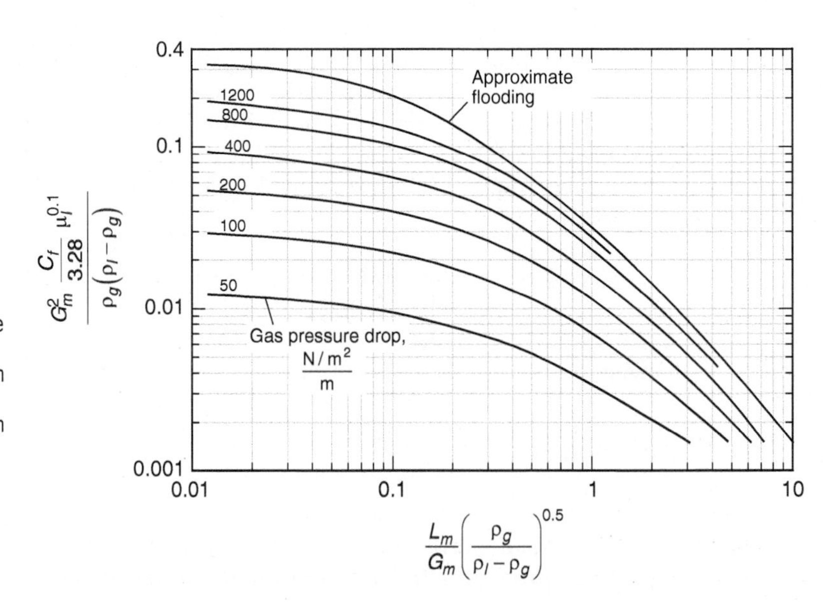

Figure 11-10
Generalized Eckert gas pressure drop and liquid and gas loading correlation in SI units for random packed tower. The coefficient 3.28 is a conversion factor when the packing factor in SI units (m^{-1}) is used because the Eckert diagram was originally developed in English units (Adapted from Eckert, 1961; Treybal, 1980).

Incorporated in the capacity parameter on the ordinate scale is an empirical parameter characteristic of the shape, size, and material property of the packing type and is called the packing factor (C_f). Packing factor C_f has units of inverse length and is used to relate the packing type to the relative gas pressure drop through the packing in the tower. Figure 11-6 displays C_f values for several commonly used plastic packing types. Since C_f is incorporated in the numerator of the capacity parameter on the ordinate scale, packing materials with a higher C_f value will have a higher gas pressure drop than packing materials with a lower C_f value. In general, the gas pressure drop will increase with increasing packing factor.

Tower Cross-Sectional Area

The cross-sectional area of a packed tower can be estimated from the generalized Eckert pressure drop curves shown on Fig. 11-10 (see above discussion of gas pressure drop). The gas loading rate, liquid loading rate, and tower area may be determined from Fig. 11-10 using the following procedure:

1. Specify the following design parameters:
   a. Packing factor for the media (see Fig. 11-6)
   b. Air-to-water ratio [determined by the stripping factor for the least volatile contaminant (i.e., the contaminant with the lowest Henry's constant)]
   c. Gas pressure drop (typically 50 to 100 $N/m^2/m$)

2. Determine the value on the x axis on the Eckert curve shown on Fig. 11-10:

$$x = \left(\frac{1}{G_m/L_m}\right)\left(\frac{\rho_g}{\rho_l - \rho_g}\right)^{0.5} \tag{11-31}$$

where x = value on x axis on Eckert curve
G_m = air mass loading rate, $kg/m^2{\cdot}s$
L_m = water mass loading rate, $kg/m^2{\cdot}s$
ρ_g = air density, kg/m^3
ρ_l = water density, kg/m^3

The value of G_m/L_m can be determined knowing the air-to-water ratio, water density, and air density:

$$\frac{G_m}{L_m} = \left(\frac{Q_g}{Q_l}\right)\left(\frac{\rho_g}{\rho_l}\right) \tag{11-32}$$

3. Graphically determine the numerical value y on the y axis on the Eckert curve shown on Fig. 11-10 knowing the gas pressure drop and x.

4. Determine the gas loading rate based on the following relationship for the y-axis value on the Eckert curve shown on Fig. 11-10:

$$y = \frac{G_m^2 (C_f/3.28)\, \mu_l^{0.1}}{\rho_g(\rho_l - \rho_g)} \tag{11-33}$$

where y = numerical value on y axis of Eckert curve determined in step 3
C_f = packing factor, m^{-1}
μ_l = dynamic viscosity of water, kg/m·s

Rearrange Eq. 11-33 and solve for G_m:

$$G_m = \left[\frac{y\rho_g(\rho_l - \rho_g)}{(C_f/3.28)\, \mu_l^{0.1}}\right]^{0.5} \tag{11-34}$$

5. Determine the water mass loading rate by rearranging Eq. 11-32, which yields the following relationship:

$$L_m = \frac{G_m}{(Q_g/Q_l)(\rho_g/\rho_l)} \tag{11-35}$$

6. Determine the cross-sectional area of the packed tower based on the following relationship:

$$A = \frac{Q_l \rho_l}{L_m} \tag{11-36}$$

where A = cross-sectional area of packed tower, m^2
Q_l = water flow rate, m^3/s

Correlations describing the Eckert pressure drop curves to predict gas loading rate and tower area were fit by Cummins and Westrick (1983). These pressure drop correlations are useful for performing packed-tower aeration design calculations using spreadsheets or computer programs, but the correlations are beyond the scope of this book (see Crittenden et al., 2012).

Example 11-3 Tower diameter, area, and pressure drop of a packed tower

Determine the cross-sectional area and tower diameter for a packed-tower design based on the removal of 1,2-dichloropropane (DCP) at 10°C for a water flow rate Q_l of 8.64 ML/d. The basis for design is given by the operating air-to-water ratio of 60 (determined in Example 11-2), gas pressure drop $\Delta P/L = 50$ N/m^2·m, and the 8.9-cm (3.5-in.) nominal diameter Jaeger Tri-Packs.

Solution

1. Determine G_m/L_m using Eq. 11-32. From App. B and C, the density of air and water at 10°C is $\rho_g = 1.247\ \text{kg/m}^3$ and $\rho_l = 999.7\ \text{kg/m}^3$:

$$\frac{G_m}{L_m} = \left(\frac{Q_g}{Q_l}\right)\left(\frac{\rho_g}{\rho_l}\right) = 60\left(\frac{1.247}{999.7}\right) = 0.075\ \text{kg air/ kg water}$$

2. Determine the value on the x axis on the Eckert curve shown on Fig. 11-10 and Eq. 11-31:

$$x = \left[\frac{1}{G_m/L_m}\right]\left(\frac{\rho_g}{\rho_l - \rho_g}\right)^{0.5} = \left(\frac{1}{0.075}\right)\left(\frac{1.247}{999.7 - 1.247}\right)^{0.5}$$
$$= 0.47$$

3. Graphically determine the numerical value y on the y axis on the Eckert curve shown on Fig. 11-10 knowing the gas pressure drop and x. At the location on Fig. 11-10 where $x = 0.470$ and $\Delta P/L = 50\ \text{N/m}^2\cdot\text{m}$,

$$y = 0.0051$$

4. Determine the gas loading rate based on the relationship for the y-axis value on the Eckert curve shown on Fig. 11-10. Solve for G_m using Eq. 11-34. From App. C, water viscosity at 10°C is $\mu_l = 1.307 \times 10^{-3}\ \text{kg/m}\cdot\text{s}$. The packing factor is found in Fig. 11-6; C_f for 8.9-cm (3.5-in.) plastic tripacks is $39.0\ \text{m}^{-1}$.

$$G_m = \left[\frac{y\rho_g(\rho_l - \rho_g)}{(C_f/3.28)\,\mu_l^{0.1}}\right]^{0.5} = \left[\frac{0.0051(1.247)(999.7 - 1.247)}{(39.0/3.28)(1.307 \times 10^{-3})^{0.1}}\right]^{0.5}$$
$$= 1.02\ \text{kg/m}^2\cdot\text{s}$$

5. The water mass loading rate is determined using Eq. 11-35:

$$L_m = \frac{G_m}{(Q_g/Q_l)(\rho_g/\rho_l)} = \frac{1.02\ \text{kg/m}^2\cdot\text{s}}{(60)(1.247\ \text{kg/m}^3/999.7\ \text{kg/m}^3)}$$
$$= 13.5\ \text{kg/m}^2\cdot\text{s}$$

6. Determine the cross-sectional area of the packed tower using Eq. 11-36. Note 8.64 ML/d = 0.10 m^3/s from step 3 in Example 11-2.

$$A = \frac{Q_l\rho_l}{L_m} = \frac{(0.10\ \text{m}^3/\text{s})(999.7\ \text{kg/m}^3)}{13.5\ \text{kg/m}^2\cdot\text{s}} = 7.4\ \text{m}^2$$

7. Determine the tower diameter assuming a circular tower area:

$$D = \sqrt{\frac{4A^2}{\pi}} = \sqrt{\frac{4 \times 7.4\ \text{m}^2}{\pi}} = 3.07\ \text{m}$$

Standard tower sizes of 1.22 m (4 ft), 1.83 m (6 ft), 2.44 m (8 ft), 3.048 m (10 ft), 3.66 m (12 ft), and sometimes 4.27 m (14 ft) in diameter are common for most packed-tower equipment manufacturers in the United States. For this case we will use a 3.048-m (10-ft) diameter tower, which yields a tower area of 7.3 m^2. The actual loading rates are $G_m = 1.02\ \text{kg/m}^2\cdot\text{s}$ and $L_m = 13.7\ \text{kg/m}^2\cdot\text{s}$.

If multiple compounds are to be removed, the compound with the lower Henry's law constant in the water to be treated is used as the basis for determining the cross-sectional area of the tower because it will require the highest Q_g/Q_l ratio to have a stripping factor in the optimal range.

Mass Transfer Coefficient

The general equation for calculating the overall liquid-side mass transfer coefficient $K_L a$ in aeration processes was derived in Chap. 4 based on the two-film theory of mass transfer. Equation 4-147 in Chap. 4 is

$$\frac{1}{K_L a} = \frac{1}{k_\ell a} + \frac{1}{H_{YC} k_g a} \tag{11-37}$$

where $K_L a$ = overall liquid-side mass transfer rate constant, s^{-1}
k_ℓ = liquid-phase mass transfer coefficient, m/s
k_g = gas-phase mass transfer coefficient, m/s
a = area available for mass transfer divided by vessel volume, m^2/m^3

The $K_L a$ values for packed towers can be determined by performing pilot plant studies or taken from previously reported field studies. They can also be estimated from the following equations (Onda et al., 1968):

$$k_l = 0.0051 \left(\frac{L_m}{a_w \mu_l}\right)^{2/3} \left(\frac{\mu_l}{\rho_l D_l}\right)^{-0.5} (a_t d_p)^{0.4} \left(\frac{\rho_l}{\mu_l g}\right)^{-1/3} \tag{11-38}$$

$$k_g = 5.23\,(a_t D_g) \left(\frac{G_m}{a_t \mu_g}\right)^{0.7} \left(\frac{\mu_g}{\rho_g D_g}\right)^{1/3} (a_t d_p)^{-2} \tag{11-39}$$

$$a_w = a_t \left\{1 - \exp\left[-1.45 \left(\frac{\sigma_c}{\sigma}\right)^{0.75} \left(\frac{L_m}{a_t \mu_l}\right)^{0.1} \left(\frac{L_m^2 a_t}{\rho_l^2 g}\right)^{-0.05} \left(\frac{L_m^2}{\rho_l a_t \sigma}\right)^{0.2}\right]\right\} \tag{11-40}$$

where
a_t = total specific surface area of the packing material, m^2/m^3
a_w = wetted specific surface area of the packing material, m^2/m^3
d_p = nominal diameter of the packing material, m
D_g, D_l = gas- and liquid-phase diffusion coefficients of the contaminant, m^2/s
g = gravitational constant, m/s^2
G_m, L_m = gas and liquid mass loading rates, $kg/m^2{\cdot}s$
μ_g, μ_l = gas and liquid viscosity, kg/m·s
ρ_g, ρ_l = gas and liquid density, kg/m^3
σ = surface tension of liquid, kg/s^2
σ_c = critical surface tension of the packing material, kg/s^2

It is recommended that a safety factor of 0.70 $(K_L a_{actual}/K_L a_{Onda})$ be applied for packing diameters greater than 2.5 cm (1 in.) as a conservative estimate of packing height required. Note that a in Eq. 11-37 is the wetted specific surface area a_w calculated from Eq. 11-40. The following example illustrates the application of the Onda mass transfer correlation for pack towers.

Example 11-4 Calculating overall liquid-side mass transfer rate constants

Determine the overall liquid-side mass transfer rate constants for DCP and PCE at 10°C in packed-tower aeration for the air and water mass loading rates determined in Examples 11-2 and 11-3 using the Onda correlations and a safety factor of 0.70 for 8.9 cm (3.5 in.) polyethylene Jaeger Tri-Packs. The water flow rate, Q_l is 8.64 ML/d. From Example 11-3, the air loading rate $G_m = 1.02\ kg/m^2{\cdot}s$ and the water loading rate $L_m = 13.5\ kg/m^2{\cdot}s$.

Solution

1. Obtain the necessary values for the physical properties of water and air, packing properties, and Henry's constants.
   a. The physical properties of air and water from Apps. B and C at 10°C are water density $\rho_l = 999.7\ kg/m^3$, dynamic viscosity of water $\mu_l = 1.307 \times 10^{-3}$ kg/m·s, water surface tension σ = 0.0742 N/m, air density $\rho_g = 1.247\ kg/m^3$, and air viscosity $\mu_g = 1.79 \times 10^{-5}$ kg/m·s.
   b. The properties of the packing material from Fig. 11-6 are nominal diameter of packing $d_p = 0.0889$ m, specific surface area of packing $a_t = 125.0\ m^2/m^3$, and critical surface tension of packing $\sigma_c = 0.033$ N/m.

c. The dimensionless Henry's law constants of DCP and PCE at 10°C are obtained in Table 11-4 and found to be $H_{YC,DCP} = 0.0525$ and $H_{YC,PCE} = 0.364$.

2. Calculate the liquid diffusion coefficient for DCP and PCE using the Hayduk–Laudie correlation (Eq. 4-121) as demonstrated in Example 4-13. Using this correlation, the liquid diffusion coefficients are calculated to be

$$D_{l,DCP} = 6.08 \times 10^{-10}\ \text{m}^2/\text{s} \quad \text{and} \quad D_{l,PCE} = 5.86 \times 10^{-10}\ \text{m}^2/\text{s}$$

3. Calculate the gas diffusion coefficient for DCP and PCE using the Wilke–Lee correlation (Eq. 4-123) as demonstrated in Example 4-15. Using this correlation, the gas diffusion coefficients are calculated to be

$$D_{g,DCP} = 7.65 \times 10^{-6}\ \text{m}^2/\text{s} \quad \text{and} \quad D_{g,PCE} = 7.13 \times 10^{-6}\ \text{m}^2/\text{s}$$

4. Calculate the specific surface area available for mass transfer a_w, using Eq. 11-40.

$$a_w = a_t\left\{1 - \exp\left[-1.45\left(\frac{\sigma_c}{\sigma}\right)^{0.75}\left(\frac{L_m}{a_t\mu_l}\right)^{0.1}\left(\frac{L_m^2 a_t}{\rho_l^2 g}\right)^{-0.05}\left(\frac{L_m^2}{\rho_l a_t \sigma}\right)^{0.2}\right]\right\}$$

$$= 125\left\{1 - \exp\left[\begin{array}{l}-1.45\left(\dfrac{0.0330}{0.0742}\right)^{0.75}\left(\dfrac{13.5}{125.0 \times 1.307 \times 10^{-3}}\right)^{0.1}\\ \times\left(\dfrac{13.5^2 \times 125.0}{999.7^2 \times 9.81}\right)^{-0.05}\\ \times\left(\dfrac{13.5^2}{999.7 \times 125.0 \times 0.0742}\right)^{0.2}\end{array}\right]\right\}$$

$$= 67\ \text{m}^2/\text{m}^3$$

5. Calculate the liquid-phase mass transfer coefficient k_l using Eq. 11-38:
   a. DCP:

$$k_l = 0.0051\left(\frac{L_m}{a_w\mu_l}\right)^{2/3}\left(\frac{\mu_l}{\rho_l D_l}\right)^{-0.5}(a_t d_p)^{0.4}\left(\frac{\rho_l}{\mu_l g}\right)^{-1/3}$$

$$= 0.0051\left\{\begin{array}{l}\left[\dfrac{13.5}{67 \times (1.307 \times 10^{-3})}\right]^{2/3}\left[\dfrac{1.307 \times 10^{-3}}{999.7 \times (6.08 \times 10^{-10})}\right]^{-0.5}\\ \times (125.0 \times 0.0889)^{0.4}\left[\dfrac{999.7}{(1.307 \times 10^{-3}) \times 9.81}\right]^{-1/3}\end{array}\right\}$$

$$= 1.95 \times 10^{-4}\ \text{m/s}$$

b. PCE:

$$k_l = 0.0051\left(\frac{L_m}{a_w\mu_l}\right)^{2/3}\left(\frac{\mu_l}{\rho_l D_l}\right)^{-0.5}(a_t d_p)^{0.4}\left(\frac{\rho_l}{\mu_l g}\right)^{-1/3}$$

$$= 0.0051\left\{\begin{array}{l}\left[\dfrac{13.5}{67\times(1.307\times10^{-3})}\right]^{2/3}\left[\dfrac{1.307\times10^{-3}}{999.7\times(5.87\times10^{-10})}\right]^{-0.5}\\ \times(125.0\times0.0889)^{0.4}\left[\dfrac{999.7}{(1.307\times10^{-3})\times9.81}\right]^{-1/3}\end{array}\right\}$$

$$= 1.92\times10^{-4}\ \text{m/s}$$

6. Calculate the gas-phase mass transfer coefficient k_g using Eq. 11-39:
   a. DCP:

$$k_g = 5.23(a_t D_g)\left(\frac{G_m}{a_t\mu_g}\right)^{0.7}\left(\frac{\mu_g}{\rho_g D_g}\right)^{1/3}(a_t d_p)^{-2}$$

$$= 5.23\left\{\begin{array}{l}\left[125.0\times(7.65\times10^{-6})\right]\left[\dfrac{1.02}{125.0\times(1.79\times10^{-5})}\right]^{0.7}\\ \times\left[\dfrac{1.79\times10^{-5}}{1.247\times(7.65\times10^{-6})}\right]^{1/3}(125.0\times0.0889\ \text{m})^{-2}\end{array}\right\}$$

$$= 3.63\times10^{-3}\ \text{m/s}$$

   b. PCE:

$$k_g = 5.23(a_t D_g)\left(\frac{G_m}{a_t\mu_g}\right)^{0.7}\left(\frac{\mu_g}{\rho_g D_g}\right)^{1/3}(a_t d_p)^{-2}$$

$$= 5.23\left\{\begin{array}{l}\left[125.0\times(7.13\times10^{-6})\right]\left[\dfrac{1.02}{125.0\times(1.79\times10^{-5})}\right]^{0.7}\\ \times\left[\dfrac{1.79\times10^{-5}}{1.247\times(7.13\times10^{-6})}\right]^{1/3}(125.0\times0.0889\ \text{m})^{-2}\end{array}\right\}$$

$$= 3.46\times10^{-3}\ \text{m/s}$$

7. Calculate the overall liquid-side mass transfer rate constant $K_L a$ based on a_w, k_l, and k_g from the Onda correlations using Eq. 11-37.
   a. DCP:

$$\frac{1}{K_L a} = \frac{1}{k_l a_w} + \frac{1}{k_g a_w H_{YC}} = \frac{1}{(1.95\times10^{-4})(67)} + \frac{1}{(3.63\times10^{-3})(67)(0.0525)}$$

$$\Rightarrow K_L a_{\text{Onda}} = 6.45\times10^{-3}\text{s}^{-1}$$

b. PCE:

$$\frac{1}{K_L a} = \frac{1}{k_l a_w} + \frac{1}{k_g a_w H_{YC}} = \frac{1}{(1.92 \times 10^{-4})(67)} + \frac{1}{(3.46 \times 10^{-3})(67)(0.364)}$$

$$\Rightarrow K_L a_{\text{Onda}} = 0.011 \text{s}^{-1}$$

8. Calculate design $K_L a$ applying a safety factor (SF) of 0.70 on $K_L a_{\text{Onda}}$.
   a. DCP:

$$K_L a = K_L a_{\text{Onda}} (\text{SF})_{K_L a} = (6.45 \times 10^{-3} \text{ s}^{-1}) \times 0.70$$
$$= 4.52 \times 10^{-3} \text{ s}^{-1}$$

   b. PCE:

$$K_L a = K_L a_{\text{Onda}} \times (\text{SF})_{K_L a} = (0.011 \text{s}^{-1}) \times 0.70$$
$$= 7.7 \times 10^{-3} \text{s}^{-1}$$

Design versus Rating Analysis of a Packed Tower

Once the packing material has been selected and the air flow rate (see Example 11-2), tower diameter (Example 11-3), and mass transfer coefficients (Example 11-4) have been calculated, the height of the tower can be calculated. In a design analysis, it is desired to *size a new packed tower* to meet the treatment objective C_{TO}. The depth of packing is determined by substituting $C_{TO} = C_e$ into Eq. 11-29.

In a rating analysis, the effluent concentrations of various compounds *for an existing tower* can be determined. Modifications are made to existing towers to either treat greater volumes of water or modify constituent removal (e.g., lower levels, different constituents using Eq 11-29). Process efficiency may be improved by increasing the air-to-water ratio, replacing the packing with a more efficient packing type, or increasing the depth of packing. The following variables are known in a rating analysis: (1) tower height, (2) tower diameter, (3) type of packing, (4) water flow rate, (5) air flow rate, (6) pressure, (7) temperature, (8) influent concentration, and (9) mass transfer coefficient. Knowing these variables, it is possible to determine effluent concentration and gas pressure drop for the tower. The effluent concentration is calcuated using Eq. 11-30. The use of Eq. 11-29 to calculate the depth of packing and Eq. 11-30 to determine effluent concentrations is demonstrated in Example 11-5.

Example 11-5 Depth of packing in a packed tower

Determine the depth of packing required to remove DCP and PCE at 10°C for a water flow rate Q_l of 8.64 ML/d if the influent concentrations are $C_{0,DCP} = 40$ μg/L and $C_{0,PCE} = 35$ μg/L and the treatment objective for both is $C_e = 5$ μg/L. Complete the design using 8.9-cm (3.5-in.) Jaeger Tri-Packs, the stripping factors from Example 11-2, the tower diameter from Example 11-3, and the K_La values from Example 11-4.

Solution

The depth of packing required to achieve the treatment goal will be different for each contaminant. The best approach is to calculate the depth of packing for each contaminant, select the largest value as the required packing depth, and then calculate the effluent concentrations for the selected packing depth.

1. Calculate the packing height for DCP using Eq. 11-29. Note 8.64ML/d = 0.10 m^3/s. From previous examples, the stripping factor for DCP is 3.15 (Example 11-2), the tower area is 7.3 m^2 (Example 11-3), and the $K_La_{DCP} = 0.00452$ s^{-1} (Example 11-4).

$$L = \frac{Q_l}{AK_La_{DCP}}\left(\frac{S}{S-1}\right)\ln\left[\frac{1+(C_{0,DCP}/C_{e,DCP})(S-1)}{S}\right]$$

$$= \frac{0.10\ \text{m}^3/\text{s}}{(7.3\ \text{m}^2)(0.00452\ \text{s}^{-1})}\left(\frac{3.15}{3.15-1}\right)\ln\left[\frac{1+(40/5)(3.15-1)}{3.15}\right]$$

$$= 7.8\ \text{m}$$

2. Calculate the packing height for PCE using Eq. 11-29. From previous examples, the stripping factor for PCE is 21.8 (Example 11-2) and the $K_La_{DCP} = 0.0077$ s^{-1} (Example 11-4).

$$L = \frac{0.10\ \text{m}^3/\text{s}}{(7.3\ \text{m}^2)(0.0077\ \text{s}^{-1})}\left(\frac{21.8}{21.8-1}\right)\ln\left[\frac{1+(80/5)(21.8-1)}{21.8}\right]$$

$$= 3.6\ \text{m}$$

3. Compare the depths. Since the depth required to meet the treatment goal for DCP is greater, the required tower packing depth is 7.8 m.

4. Calculate the effluent concentration of PCE for the given tower height using Eq. 11-30.

$$C_e = \frac{C_0\,(S-1)}{S\exp\left[\dfrac{LAK_L a}{Q_l}\dfrac{S-1}{S}\right]-1}$$

$$= \frac{(35\ \mu g/L)(21.8-1)}{(21.8)\exp\left[\dfrac{(7.8\ m)(7.3\ m^2)(0.0077\ s^{-1})}{0.1\ m^3/s}\dfrac{21.8-1}{21.8}\right]-1}$$

$$= 0.51\ \mu g/L$$

Comment

The design based on DCP for this example resulted in both components meeting their treatment objectives. In many cases, the compound with the highest removal requirement will control the height of the tower.

Factors Influencing Packed-Tower Performance

Packed-tower performance may be impacted by environmental conditions such as water temperature and water quality such as dissolved solids.

TEMPERATURE

Temperature influences both the rate of mass transfer and Henry's constant and thus impacts equipment size, as well as the removal efficiency, in an existing packed tower. A packed tower that is designed to meet treatment objectives at one temperature may not be able to achieve the same treatment objectives at a lower temperature, as shown in Table 11-5. For example, if the temperature decreases from 15 to 5°C, the effluent concentration increases threefold.

DISSOLVED SOLIDS

During operation of a packed tower, dissolved inorganic chemicals such as calcium, iron, and manganese may precipitate onto packing media, which

Table 11-5
Effect of temperature on packed-tower operation

Temperature T, °C	$C_{e,T}/C_{e,15°C}$
0	5.2
5	3.3
10	2.0
15	1
20	0.45

can cause a pressure drop increase and a void volume decrease in the tower. The main methods for controlling the negative effects of chemical precipitates are cleaning the precipitate off the packing and controlling precipitate formation.

Precipitate Potential

The potential for fouling of packing material by precipitates is especially great in waters containing appreciable amounts of carbon dioxide. Groundwater often contains 30 to 50 mg/L of carbon dioxide. Carbon dioxide can be removed in an air stripping tower, particularly at high air-to-water ratios, but removal of carbon dioxide tends to raise the pH of the water. As pH increases, bicarbonate is converted to carbonate. In natural waters containing significant quantities of calcium ion, calcium carbonate will precipitate when the carbonate ion concentration is high enough that the solubility product of calcium carbonate is surpassed.

Cleaning

Plastic packing can be removed periodically and put into a tumbler so that the precipitate can be broken off. Acid treatment dramatically deteriorates the plastic packing (making it brittle) over time and is not recommended. In some instances, conditioning chemicals may be necessary to add to the cleaning process because precipitates can form within weeks in hard water.

Controlling Precipitate

Larger packing size, which has smaller specific surface area, may be preferable because there is less surface area upon which precipitate can form as well as larger spaces for airflow. Special structured packings can also minimize fouling. Controlling precipitation with scale inhibitors represents a significant cost in certain situations; therefore, the potential for precipitation must be carefully analyzed.

Additional Gas Transfer Devices

Figures 11-1 and 11-2 and Table 11-2 described a variety of other gas transfer devices. The concepts presented in this chapter, including the significance of equilibrium via Henry's law, air–water ratios, maximizing interfacial area, and mass transfer as a controlling factor for gas transfer, are equally important for other gas transfer devices. Design procedures for other gas transfer devices are available in reference books such as Crittenden et al. (2012) and Hand et al. (2011).

11-5 Energy and Sustainability Considerations

Air stripping with countercurrent packed towers is a cost-effective treatment process for removing VOCs from water if off-gas treatment is not required. The towers usually have a small footprint but require a large vertical dimension; the towers can be as high as 10 to 15 m depending upon the treatment objective and the volatility of the compound. Based on calculations similar to Example 3-3, the energy cost to pump water to the top of a 15-m tower is 0.051 kWh/m^3 if the pump has an efficiency of 80 percent. Energy for the air blower depends on the gas pressure drop and the amount of air required, which can vary significantly depending on the volatility of the contaminants but often can add 50 to 100 percent to the energy consumption for the tower. The gas pressure drop is dependent on the air-to-water ratio and the tower diameter. Typically, most towers are designed for a low gas pressure drop (50 to 100 N/m^2·m) to reduce the energy costs associated with the blower operation. As discussed in the chapter, the optimum tower design for a given gas pressure drop is obtained at an operating air-to-water ratio of about 3.5 times the minimum air-to-water ratio required to meet the specified treatment objective. Energy considerations should be incorporated into the design of a packed-tower system. Hand et al. (1986) provided an in-depth analysis of the design and operational aspects of an optimum tower design.

An important consideration regarding sustainability is that, as a treatment process, air stripping does not destroy contaminants but merely transfers them from one phase to another. In other words, a water pollution problem might be solved by creating an air pollution problem. In designing an air-stripping tower, it is necessary to consider the fate of the contaminants once they enter the air. In some cases, the contaminants may be dilute enough or degrade fast enough in the presence of sunlight that no further treatment is needed. In other cases, it is necessary to treat the effluent gas from an air-stripping tower. Often, however, destruction of volatile organic contaminants is easier in the gas phase using thermal destruct units or adsorption onto activated carbon than it is in the liquid phase, so air stripping can be an effective treatment technology even when off-gas treatment is required.

11-6 Summary and Study Guide

After studying this chapter, you should be able to:

1. Define the following terms and phrases and describe the significance of each in the context of air stripping and aeration used in water treatment:

absorption	Eckert diagram	operating line
aeration	equilibrium line	packing factor
air stripping	Henry's law	Raoult's law
air-to-water ratio	$K_L a$	stripping factor
countercurrent packed tower	Onda correlations	vapor pressure
desorption	operating diagram	

2. Discuss the advantages and disadvantages, and most appropriate uses for, flooded contactors versus gas-phase contactors. Describe at least four types of gas–liquid contacting devices.
3. Explain the equilibrium distribution of a solute between gas and liquid phases. Describe the relationship between vapor pressure, Raoult's law, and Henry's law.
4. Explain from a molecular perspective why, for dilute solutions, Henry's law is valid for the solute and Raoult's law is valid for the solvent.
5. Describe the impact of temperature, ionic strength, surfactants, and pH on Henry's constant.
6. Convert Henry's constant from one set of units to another set of units.
7. Draw an operating diagram for a countercurrent packed tower, labeling the equilibrium line, operating line, and concentration gradient.
8. Calculate the stripping factor and describe how it affects the efficiency of the stripping process.
9. Calculate the overall liquid-phase mass transfer coefficient from the Onda correlations.
10. Design an air-stripping tower (tower diameter and height, packing material, gas flow rate) when given water flow rate, contaminant concentration, and required effluent contaminant concentration.
11. Explain some possible causes of precipitation buildup on tower packing material during tower operation.

Homework Problems

Note: Several of these problems pertain to the design of countercurrent packed towers, which is a computationally intensive process. The spreadsheet identified as Resource E10 at the website listed in App. E can be used to perform the calculations.

11-1 Calculate the dimensionless Henry's law constant for a compound if Henry's law constant in other units is the value given below (to be selected by instructor), at $T = 25°C$.

a. $H_{PX} = 400$ bar.

b. $H_{PX} = 53$ bar.

c. $H_{PC} = 1.1$ bar·L/mol.

d. $H_{PC} = 21.5$ bar·L/mol.

e. $H_{PC} = 3.63$ bar·L/mol.

11-2 Calculate the minimum air-to-water ratio for a countercurrent packed tower for 95 percent removal of chloroform and benzene at 10°C.

11-3 Determine the cross-sectional area and diameter for a packed-tower design based on the conditions given below (to be selected by instructor).

	Contaminant	Temp, °C	Water Flow Rate, ML/d	Stripping Factor	Pressure Drop, N/m² · m	Packing
A	chloroform	20	13	3.5	50	5.1 cm Intalox saddles
B	benzene	10	4.3	3.5	50	5.8 cm LANPAC
C	carbon tetrachloride	20	2.5	4	400	5.1 cm Jaeger Tri-Packs
D	trichloroethylene	15	10	6	100	8.9 cm LANPAC-XL
E	tetrachloroethylene	20	10	3.5	200	8.9 cm Jaeger Tri-Packs

11-4 Determine the overall liquid-side mass transfer rate constant for the compound and tower design given in Problem 11-3 using the Onda correlations and a safety factor of 0.75. Determine the liquid-phase diffusion coefficient D_l using the Hayduk–Laudie correlation (Eq. 4-121) and the gas-phase diffusion coefficient using the Wilke–Lee correlation (Eq. 4-123). The packing material is polyethylene.

11-5 Determine the packed-tower height (packing depth) required to remove the compound given in Problem 11-3. Use the tower area determined in Problem 11-3 and the overall liquid-side mass transfer rate constants determined in Problem 11-4 in the solution of the problem. The influent and effluent concentrations are given below.

	Contaminant	Influent Conc., μg/L	Effluent Conc., μg/L
A	chloroform	100	5
B	benzene	50	5
C	carbon tetrachloride	75	4
D	trichloroethylene	150	5
E	tetrachloroethylene	40	2

11-6 Design a packed-tower aeration system to treat 6.48 ML/d of water at 20°C and remove benzene ($C_0 = 40$ μg/L), chloroform ($C_0 = 60$ μg/L), and carbon tetrachloride ($C_0 = 30$ μg/L) to a treatment objective for each concentration that equals 5 μg/L. Select an appropriate stripping factor, gas pressure drop, and factor of safety for the overall mass transfer rate constant. Use 5.1-cm polyethylene Jaeger Tri-Packs as the packing material. Determine an appropriate gas flow rate, tower diameter, and tower length. Calculate the effluent liquid concentration of each constituent for the completed design condition.

References

Ashworth, R. A., Howe, G. B., Mullins, M. E., and Rogers, T. N. (1988) "Air-Water Partitioning Coefficients of Organics in Dilute Aqueous Solutions," *J. Hazardous Materials*, **18**, 1, 25–36.

Ayuttaya, P. C. N., Rogers, T. N., Mullins, M. E., and Kline, A. A., (2001) "Henry's Law Constants Derived from Equilibrium Static Cell Measurements for Dilute Organic-Water Mixtures," *Fluid Phase Equilibria*, **185**, 359–377.

Crittenden, J. C., Trussell, R. R., Hand, D. W., Howe, K. J, and Tchobanoglous, G. (2012) *MWH's Water Treatment: Principles and Design*, 3rd ed., Wiley, Hoboken, NJ.

Cummins, M. D., and Westrick, J. J. (1983) "Trichlorethylene Removal by Packed Column Air Stripping: Field Verified Design Procedure," in Proceedings, American Society of Civil Engineers Environmental Engineering Conference, Boulder, CO, pp. 442–449.

Dewulf, J., Drijvers, D., and Langenhove, H. V., (1995) "Measurement of Henry's Law Constant as Function of Temperature and Salinity for the Low Temperature Range," *Atmos. Environ.*, **29**, 4, 323–331.

Eckert, J. S. (1961) "Design Techniques for Sizing Packed Towers," *Chem. Eng. Progr.*, **57**, 9, 54–58.

Gossett, J. M. (1987) "Measurement of Henry's Law Constants for C1 and C2 Chlorinated Hydrocarbons," *Environ. Sci. Technol.*, **21**, 2, 202–208.

Hand, D. W., Crittenden, J. C., Gehin, J. L., and Lykins, B. W., Jr. (1986) "Design and Evaluation of an Air Stripping Tower for Removing VOCs from Groundwater," *J. AWWA*, **78**, 9, 87–97.

Hand, D. W., Hokanson, D. R., and Crittenden, J. C. (2011) Gas-Liquid Processes: Principles and Applications, Chapter 6, in J. K. Edzwald, Ed. *Water Quality and Treatment, A Handbook on Drinking Water,* 6th ed., American Water Works Association, Denver, CO.

Heron, G., Christensen, T. H., and Enfield, C. G. (1998) "Henry's Law Constant for Trichloroethylene between 10 and 95 C", *Environ. Sci. Technol.*, **32**, 10, 1433–1437.

Kavanaugh, M. C., and Trussell, R. R. (1981) "Air Stripping as a Treatment Process," in *Proceedings of AWWA Symposium on Organic Contaminants in Groundwater,* St. Louis, MO. American Water Works Association, Denver, CO. Paper S2-6, pp. 83–106.

McCabe, W. L., and Thiele, E. W. (1925) ''Graphical Design of Fractionating Columns,'' *Ind. Eng. Chem.*, **17**, 6, 605–611.

Nicholson, B. C., Maguire, B. P., and Bursell, D. B. (1984) ''Henry's Law for the Trihalomethanes: Effects of Water Composition and Temperature,'' *Environ. Sci. Technol.*, 18, 7, 518–521.

NIST (2011) Available at <http://webbook.nist.gov/chemistry/> accessed on Jan 3, 2011.

Onda, K., Takeuchi, H., and Okumoto, Y. (1968) ''Mass Transfer Coefficients between Gas and Liquid Phases in Packed Columns,'' *J. Chem. Eng. Jpn.*, **1**, 1, 56–62.

Robbins, G. A., Wang, S., and Stuart, J. D. (1993) ''Using Static Headspace Method to Determine Henry's Law Constants,'' *Anal. Chem.*, **65**, 21, 3113–3118.

Rogers, T. N. (1994) Predicting Environmental Physical Properties from Chemical Structure Using a Modified Unifac Model, Ph.D. Dissertation, Michigan Technological University, Houghton, MI.

Schwarzenbach, R. P., Gschwend, P. M., and Imboden, D. M. (1993) *Environmental Organic Chemistry*, Wiley, New York.

SRC (2011) Available at: <http://www.syrres.com/what-we-do/databaseforms.aspx?id=386>; accessed on Jan. 3, 2011.

Treybal, R. E. (1980) *Mass-Transfer Operations*, 3rd ed., McGraw-Hill, New York.

12 Advanced Oxidation

The previous chapters of this book have focused on separation processes that effectively remove contaminants from water. An alternative is to chemically transform compounds such that the compounds themselves have been destroyed, the undesirable properties of the compounds have been eliminated, or the properties have been modified to make the compounds more amenable to separation by one of the other processes. A primary method for chemical transformation is the oxidation–reduction (redox) reaction. As introduced in Chap. 4, an oxidation–reduction reaction involves the transfer of electrons between one reactant and another. In water treatment, undesirable contaminants that can be transformed are nearly always reduced species; therefore, the treatment approach generally involves adding an oxidant to the water, and the unit process is simply known as oxidation.

Oxidation processes commonly employed in water treatment can be separated into two categories: (1) conventional oxidation and (2) advanced oxidation. Conventional oxidation involves the addition of an oxidant to the water that then reacts directly with the target contaminant. Conventional oxidants are selective; specific oxidants must be used to remove specific contaminants. The conventional chemical oxidants used in water treatment include chlorine, chlorine dioxide, ozone, potassium permanganate, and hydrogen peroxide. Several of these oxidants are also disinfectants and are described in detail in Chap. 13.

A common use of conventional oxidation is to transform soluble reduced metal species such as Fe(II) and Mn(II) to their oxidized forms Fe(III) and Mn(IV), which are insoluble. The insoluble species then precipitate and can be removed by sedimentation and filtration. Other common uses of conventional oxidation are taste and odor control, color removal, and hydrogen sulfide removal. Because conventional oxidation relies on specific reactions between oxidants and target compounds, the kinetics of the reaction control the effectiveness of the process, and in some cases the oxidation reaction can be relatively slow.

This chapter, however, is focused on advanced oxidation, which differs from conventional oxidation in several important ways. First, advanced oxidation involves the addition or presence of multiple reagents in the water to form a highly reactive species known as the hydroxyl radical. It is this transitory species, rather than the original reagents, that performs the oxidation process. Second, the hydroxyl radical is so reactive that it is nonselective; virtually any reduced species can be oxidized by it. Of particular interest are synthetic organic chemicals (SOCs), which can include agricultural pesticides and herbicides, fuels, solvents, human and veterinary drugs, and other potential endocrine disruptors. The highest oxidation state for carbon is C(IV), the oxidation state in inorganic carbon compounds such as carbon dioxide (CO_2) and carbonate (CO_3^{2-}). All organic compounds have carbon in a reduced form and can be destroyed by the hydroxyl radical; with sufficiently high doses, nearly any SOC can be converted completely to carbon dioxide, water, and mineral ions (e.g., Cl^-). Third, the reactive nature of the hydroxyl radical results in extremely rapid kinetics such that, with sufficiently high doses, SOCs can be completely oxidized in a short time.

Advanced oxidation processes (AOPs) also have several inherent advantages over other treatment processes, such as reverse osmosis, adsorption onto activated carbon, or air stripping: (1) the contaminants can be destroyed completely, (2) contaminants that are not adsorbable or volatile can be destroyed, and (3) mass transfer processes such as adsorption or stripping only transfer the contaminant to another phase, which becomes a residual and may require additional treatment.

The purpose of this chapter is to introduce the general subject of advanced oxidation. Following a section that introduces basic concepts of AOPs, three subsequent sections focus on specific commercially available advanced oxidation processes; namely, ozonation as an AOP, ozone combined with hydrogen peroxide, and UV light combined with hydrogen peroxide.

12-1 Introduction to Advanced Oxidation

A basic understanding of advanced oxidation includes an introduction to the hydroxyl radical, estimating AOP performance, the factors that affect AOP performance, feasibility of AOPs, and by-products of AOPs. These topics are presented in this section.

The Hydroxyl Radical

In chemical nomenclature, *radicals* are species that have an unpaired electron. The hydroxide ion (OH^-) has a complete outer orbital of eight elections and is stable; the hydroxyl radical ($HO\cdot$) has only seven electrons in the outer orbital and is extremely unstable. The "dot" written as part of the hydroxyl and other radical species designates an unpaired electron. Hydroxyl radicals participate as an oxidant in redox reactions by gaining an electron from another species to fill the outer orbital. Hydroxyl radicals are effective in destroying organic chemicals because they are reactive electrophiles (electron preferring) that react rapidly and nonselectively with nearly all electron-rich organic compounds.

The second-order hydroxyl radical rate constants for most organic pollutants in water are on the order of 10^8 to 10^9 L/mol·s (Buxton and Greenstock, 1988), which is about the magnitude of diffusion-limited acid–base reactions ($\sim 10^9$; Stumm and Morgan, 1996). Acid–base reactions are considered to be the fastest aqueous-phase chemical reactions because they only involve the transfer of a hydrated proton. These rate constants are three to four orders of magnitude greater than the rate constants for conventional oxidants.

Estimating AOP Performance

One of the most important considerations in advanced oxidation is the quantity of oxidants that are required to destroy the organics that are targeted for destruction. The influence of background matter on AOP performance is discussed later, but insight into the type of compounds that may be degraded in a reasonable time can be evaluated by using typical $HO\cdot$ concentrations and reported rate constants.

Full-scale advanced oxidation processes generate $HO\cdot$ concentrations between 10^{-11} and 10^{-9} mol/L. The second-order hydroxyl radical rate constants for several commonly encountered water pollutants are provided in Table 12-1. A more comprehensive list is provided in the electronic Table E-4 available at the website listed in App. E. The reaction mechanism and the rate law for $HO\cdot$ that reacts with an organic compound is given by these expressions:

$$HO\bullet + R \rightarrow \text{products} \tag{12-1}$$

$$r_R = -k_R C_{HO\cdot} C_R \tag{12-2}$$

Table 12-1
Reaction rate constants and half-lives for degradation of selected compounds by hydroxyl radicals [a]

Compound	HO· Rate Constant, L/mol · s	Half-Life, min		
		[HO·] = 10^{-9} M	[HO·] = 10^{-10} M	[HO·] = 10^{-11} M
Acetate ion	7.0×10^7	0.2	2	17
Acetone	1.1×10^8	0.11	1.1	11
Ammonia	9.0×10^7	0.13	1.3	13
Atrazine	2.6×10^9	0.004	0.04	0.44
Benzene	7.8×10^9	0.001	0.01	0.1
Chloroacetic acid	4.3×10^7	0.3	2.7	27
Chloroform	5.0×10^6	2	23	231
Geosmin	$(1.4 \pm 0.3) \times 10^{10}$	0.00083	0.0083	0.083
Hydrogen peroxide	2.7×10^7	0.43	4.3	43
Methyl ethyl ketone	9.0×10^8	0.01	0.1	1
Methyl *tert*-butyl ether	1.6×10^9	0.01	0.1	1
MIB	$(8.2 \pm 0.4) \times 10^9$	0.0014	0.014	0.14
Ozone	1.1×10^8	0.11	1	11
Phenol	6.6×10^9	0.002	0.02	0.2
Tetrachloroethylene	2.6×10^9	0.004	0.04	0.4
1,1,1-Trichloroethane	4.0×10^7	0.3	3	29
1,1,2-Trichloroethane	1.1×10^8	0.11	1	11
Trichloroethylene	4.2×10^9	0.003	0.03	0.3
Trichloromethane	5.0×10^6	2	23	231
Vinyl chloride	1.2×10^{10}	0.001	0.01	0.1

[a] Additional values are available in the electronic Table E-4 at the website listed in App. E.

where r_R = destruction rate of R with HO•, mol/L·s
k_R = second-order rate constant for destruction of R with HO• radicals, L/mol·s
$C_{HO\bullet}$ = concentration of hydroxyl radical, mol/L
C_R = concentration of target organic R, mol/L

The half-life of the target organic compounds may be calculated assuming that the concentration of HO• is constant and equal to a typical value that is encountered in the field. If the HO• concentration is constant during the reaction, the value can be multiplied by the second-order rate constant and the resulting parameter is known as the pseudo-first-order rate constant. The expression for the half-life of an organic compound is obtained by substituting the rate expression into a mass balance on a batch reactor whose contents are mixed completely and solving and rearranging the result, as follows:

$$[\text{accum}] = [\cancel{\text{mass in}}] - [\cancel{\text{mass out}}] + [\text{rxn}]$$

$$\frac{dC_R}{dt} = -k_R C_{HO\cdot} C_R \tag{12-3}$$

$$t_{1/2} = \frac{\ln(2)}{k_R C_{HO\cdot}} \tag{12-4}$$

where $t_{1/2}$ = half-life of organic compound R, s.

The half-lives of selected compounds for HO• concentrations of 10^{-9}, 10^{-10}, and 10^{-11} mol/L are provided in Table 12-1. Based on the reported half-life, it is possible to mineralize many organic compounds completely within a matter of minutes. However, if reactions with background matter reduce the HO• concentration to 10^{-11} mol/L, then AOPs may not be effective. The influence of NOM, carbonate, bicarbonate, and pH on AOPs is considered later in this chapter.

Example 12-1 Half-life and required reaction time for advanced oxidation of MTBE

Methyl *tert*-butyl ether (MTBE) was used as an octane enhancer and has been found in groundwater underneath a gasoline station at a concentration of 100 μg/L. Calculate the (a) half-life, (b) time it would take to lower the concentration of MTBE to 5 μg/L in an ideal batch reactor, and (c) detention time for an ideal plug flow reactor (PFR) to achieve a treatment objective of 5 μg/L. Assume an HO• concentration of 10^{-11} mol/L.

Solution

1. From Table 12-1, the second-order rate constant of HO• for MTBE is 1.6×10^9 L/mol·s. If $C_{HO\cdot}$ is constant, it can be combined with the second-order rate constant to form a pseudo-first-order rate constant; thus $k = k_R C_{HO\cdot} = (1.6 \times 10^9 \text{ L/mol·s})(10^{-11} \text{ mol/L}) = 1.6 \times 10^{-2} \text{ s}^{-1}$. Calculate the half-life of MTBE from Eq. 12-4:

$$t_{1/2} = \frac{\ln(2)}{k} = \frac{0.693}{1.6 \times 10^{-2} \text{ s}^{-1}} = 43.3 \text{ s}$$

2. Calculate the time it would take to achieve a concentration of 5 μg/L in a batch reactor using Eq. 4-67 in Sec. 4-7. Rearranging the equation $C = C_0 e^{-kt}$ and solving for t yields

$$t = \frac{1}{k} \ln \frac{C_0}{C} = \frac{1}{1.6 \times 10^{-2} \text{ s}^{-1}} \ln\left(\frac{100}{5}\right) = 187 \text{ s} \qquad (3.1 \text{ min})$$

3. The residence time, τ, for a PFR would also be 3.1 min because the elapsed time in a completely mixed batch reactor is equivalent to residence time in an ideal PFR (see Sec. 4-9).

Comment

Many AOPs have much shorter residence times than 3 min. Consequently, the hydroxyl radical concentration must be much higher than 10^{-11} M for AOPs to be feasible.

Two common reactions of HO• with organic compounds are addition reactions with double bonds and extraction of hydrogen atoms. Double-bond addition is much more rapid than hydrogen abstraction. For example, trichloroethylene (TCE) reacts much more rapidly than 1,1,2-trichloroethane (TCA) as shown in Table 12-1. Other examples of double-bond addition reactions in Table 12-1 include chloroform, and 1,1,1-trichloroethane. These compounds will require longer reaction times and/or high concentrations of HO•.

Factors Affecting AOP Performance

The performance of advanced oxidation processes can be affected by factors that (1) reduce the generation of hydroxyl radicals or (2) reduce the ability of HO• to react with the target compounds. The UV/H_2O_2 process will generate fewer hydroxyl radicals if the water has lower transitivity for UV light. The effect of UV light transmission is presented in Sec. 13-5. The rate of HO• generation will also be affected by the pH of the water. The ability of HO• to react with the target compounds is affected primarily by competing reactions and pH, which are discussed in this section.

IMPACT OF COMPETING REACTIONS

Because the hydroxyl radical is a nonspecific oxidant, it can react with constituents in addition to the target compounds. The consumption of HO• by nontarget constituents is known as scavenging; the constituents in water that contribute the most to HO• scavenging are carbonate species (HCO_3^- and CO_3^{2-}), natural organic matter, and reduced metal ions (iron and manganese). The extent to which it reacts with the target compounds versus the nontarget constituents depends on the rates of reaction. The ratio between the rate of reaction with the target compound and the rates of all hydroxyl radical reactions in the solution describes the reduction in the rate of target compound oxidation resulting from the presence of other constituents. This ratio, known as the quenching rate, is shown in the expression

$$Q_R = \frac{k_R C_R}{k_R C_R + \sum k_i C_i} \qquad (12\text{-}5)$$

where Q_R = quenching rate, dimensionless
k_R = second-order rate constant for destruction of R with HO•, L/mol·s

k_i = second-order hydroxyl radical rate constants for water matrix i, L/mol·s
C_i = concentration of water matrix i, mol/L

The rate constants between HO• and background matter are summarized in Table 12-2. The rate constants for HCO_3^- and CO_3^{2-} are much lower than for many organic compounds that are shown in Table 12-2. Unfortunately, the concentrations of HCO_3^- and CO_3^{2-} are often three orders of magnitude higher than the organic pollutants targeted for destruction. Depending on the reactivity of the parent compound, the destruction rate of the parent compound can be significantly reduced by background matter. NOM can reduce the destruction rate of the parent compound by a factor of 100 for compounds with a second-order rate constant of 10^9 L/mol·s or 1000 for compounds with a second-order rate constant of 10^8 L/mol·s.

Example 12-2 Impact of NOM on the oxidation of MTBE

Evaluate the impact of NOM on the rate of oxidation of MTBE in a PFR if the NOM concentration is 3 mg/L as C (known as dissolved organic carbon, DOC) and the initial concentration of MTBE is 100 μg/L. Calculate the (1) quenching rate, (2) the HO• concentration required to get the same amount of oxidation as in Example 12-1, when NOM was not present, and (3) the increase in the residence time of the PFR due to the presence of NOM if the HO• concentration stays the same as Example 12-1.

Solution

1. Calculate Q_R for MTBE.
   a. Convert the DOC and MTBE concentrations from mg/L to mol/L:

$$C_{DOC} = \frac{3 \text{ mg/L as C}}{(12 \text{ g/mol})(1000 \text{ mg/ g})} = 2.50 \times 10^{-4} \text{ mol/L as C}$$

$$C_R = \frac{100 \text{ μg/L}}{(88 \text{ g/mol})(10^6 \text{ μg/ g})} = 1.14 \times 10^{-6} \text{ mol/L}$$

   b. Calculate Q_R using Eq. 12-5:

$$Q_R = \frac{(1.6 \times 10^9 \text{ L/mol·s})(1.14 \times 10^{-6} \text{ mol/L})}{(3.0 \times 10^8 \text{ L/mol·s})(2.50 \times 10^{-4} \text{ mol/L}) + (1.6 \times 10^9 \text{ L/mol·s})(1.14 \times 10^{-6} \text{ mol/L})}$$

$$= 0.0237$$

2. From Example 12-1, a PFR with $\tau = 3.12$ min achieved a effluent MTBE concentration of 5 μg/L. Using $k = k_R Q_R C_{HO\cdot}$ as the rate constant in Eq. 4-93, and rearranging to solve for the HO· concentration yields

$$C = C_I e^{-k_R C_{HO\cdot} \tau}$$

$$C_{HO\cdot} = \frac{\ln(C_I/C)}{k_R Q_R \tau} = \frac{\ln(100/5)}{(1.6 \times 10^9 \text{ L/mol·s})(0.0237)(187 \text{ s})}$$

$$= 4.22 \times 10^{-10} \text{ mol/L}$$

3. When $C_{HO\cdot} = 10^{-11}$ mol/L, calculate the time to reduce 100 μg/L of MTBE to 5 μg/L, using $k = k_R Q_R C_{HO\cdot}$ in Eq. 4-93, and rearranging to solve for the HO· concentration:

$$\tau = \frac{\ln(C_I/C)}{k_R Q_R C_{HO}} = \frac{\ln(100/5)}{(1.6 \times 10^9 \text{ L/mol·s})(0.0237)(10^{-11} \text{ mol/L})}$$

$$= 7900 \text{ s} \qquad (132 \text{ min})$$

Comment

The presence of NOM has a significant impact on the rate of MTBE oxidation. A PRF designed to reduce the concentration to 5 μg/L needs a residence time 42 times larger (132 min versus 3.1 min), compared to Example 12-1. Alternatively, an increase of the HO· concentration by 42× can achieve the same amount of oxidation in the same amount of time. Note that in both cases, the increase is equal to $1/Q_R$.

IMPACT OF PH

The performance of AOPs is affected by pH in three ways: (1) pH affects the concentration of HCO_3^- and CO_3^{2-} (HCO_3^- and CO_3^{2-} have pK_a values of 6.3 and 10.3, respectively); (2) the concentration of HO_2^- (H_2O_2 has a

Table 12-2
Rate constant with various background species that affect performance of AOPs

Background species	HO· Rate Constant (L/mol·s)	Reference
Bicarbonate (HCO_3^-)	8.5×10^6	Buxton and Greenstock (1988)
Carbonate (CO_3^{2-})	3.9×10^8	Buxton and Greenstock (1988)
Natural organic matter (mol/L as C)	1.4 to 4.5×10^8	Westerhoff et al. 2007
Iron, Fe(II)	3.2×10^8	Buxton and Greenstock (1988)
Manganese, Mn(II)	3.0×10^7	Buxton and Greenstock (1988)

pK_a of 11.6); and (3) pH affects the charge on the organic compounds if they are weak acids or bases.

For the O_3/UV and H_2O_2/O_3 processes, the rate-limiting step is the reaction between O_3 and HO_2^- to form HO•. Accordingly, low pH (<5.0) greatly reduces the concentration of HO_2^- and hence the rate of production of HO•. High pH (>11) is also thought to catalyze the formation of HO• radicals directly from O_3, but significant rates of organics destruction have not been observed with O_3 at high pH.

The reactivity and light absorption properties of the compound can be affected by its charge. For example, in the H_2O_2/UV process, HO_2^- has about 10 times the UV molar absorptivity at 254 nm (228 L/mol·cm) than does H_2O_2 (19.3 L/mol·cm); consequently, H_2O_2/UV may be more effective at higher pH, especially if the background water matrix absorbs a lot of UV light (this would only be practical if the pH was raised for other purposes and carbonate was removed, as would occur in lime soda ash softening).

By-products of AOPs

Both hydrogen abstraction and double-bond addition produce reactive organic radicals that rapidly undergo subsequent oxidation and most often combine with dissolved oxygen to form peroxy organic radicals (ROO•). These peroxy organic radicals undergo radical chain reactions that produce a variety of oxygenated by-products. The following general pattern of oxidation is observed for AOPs (Bolton and Carter, 1994):

$$\text{Organic pollutant} \rightarrow \text{aldehydes} \rightarrow \text{carboxylic acids} \rightarrow CO_2 \text{ and mineral ions} \tag{12-6}$$

Some of the significant by-products and the highest yields observed are listed in Table 12-3. The most significant observed by-products are the carboxylic acids, due to the fact that the second-order rate constants for these compounds are much lower than those for most other organics. However, if adequate reaction time is provided, all by-products (>99 percent as measured by a TOC mass balance) are destroyed (Stefan and Bolton, 1998, 1999, 2002; Stefan et al., 2000). Other by-products of concern are the halogenated acetic acids, formed from the oxidation of halogenated alkenes such as TCE. The rate constant and half-life for chloroacetic acid is reported in Table 12-1, and longer retention time and/or higher HO• concentrations are needed to destroy this compound. For example, it has been demonstrated that it is possible to completely mineralize TCE in a few minutes of reaction time using an AOP that uses TiO_2, O_3, and UV light, which produces higher HO• concentrations (Zhang et al., 1994).

Another by-product of advanced oxidation processes (and processes that use ozone) is the production of brominated by-products and bromate in waters containing bromide ion.

Table 12-3
By-products observed following advanced oxidation for four selected organic compounds

Target Compound	Observed By-products	Approximate Yield: Mol By-product per Mol Compound, %
Acetone[a]	Acetic, pyruvic, and oxalic acids, pyruvaldehyde	10–30
	Formic and glyoxylic acids, hydroxyacetone, formaldehyde	2–5
Methyl *tert*-butyl ether[b]	Acetone, acetic acid, formaldehyde, *tert*-butyl formate (TBF), pyruvic acid, *tert*-butyl alcohol (TBA), 2-methoxy-2-methyl propionaldehyde (MMP), formic, methyl acetate	10–30
	Hydroxy-*iso*-butyraldehyde, hydroxyacetone, pyruvaldehyde and hydroxy-*iso*-butyric, oxalic acid	2–5
Dioxane[c]	1,2-Ethanediol diformate, formic acid, oxalic acid, glycolic acid, acetic acid formaldehyde, 1,2-ethanediol monoformate	10–30
	Methoxyacetic acid glyoxal	2–5
	Acetaldehyde	<1
Trichloroethene[d]	Formic acid	10–40
	Oxalic acid	2–5
	1,1-Dichloroacetic acid, 1-mono acetic acid	<1

[a]Stefan and Bolton (1999).
[b]Stefan et al. (2000).
[c]Stefan and Bolton (1998).
[d]Stefan and Bolton (2002).

Assessing Feasibility of AOPs

To assess the feasibility of AOPs, the following parameters should be measured: (1) alkalinity, (2) pH, (3) chemical oxygen demand (COD), (4) total organic carbon (TOC), (5) Fe, (6) Mn, and (8) light transmission. Once these parameters are known, this information can be used to interpret and plan treatability studies for AOPs and investigate pretreatment options that may be needed. In addition, these parameters can be used in the simple models presented in Secs.12-2 and 12-3 to assess the feasibility of AOPs.

12-2 Ozonation as an Advanced Oxidation Process

Approximately 10 percent of the water treatment plants in the United States use ozone for disinfection, taste and odor control, and target compound destruction. The production of ozone, ozone contactor design, and disinfection using ozone are presented in Chap. 13. In this section, the focus is on the destruction of organic compounds including MIB, geosmin, and atrazine. Ozone can react with constituents present in natural waters to form hydroxyl radicals; during ozonation, then, organic compounds can be oxidized by ozone directly or by hydroxyl radicals. The rate of HO• formation, however, depends on the concentration of radical-forming compounds in the water, and, if these are low, water utilities may achieve additional target compound destruction by adding hydrogen peroxide after their disinfection requirements (Ct) are met. The addition of hydrogen peroxide, which reacts with ozone to produce hydroxyl radicals, is discussed in Sec. 12-3. Target compound destruction using ozone is discussed in this section.

Hydroxyl Radical Production from O_3 and OH^-

High pH values ($\approx$11) are thought to catalyze the formation of HO• radicals directly from O_3. The complete set of reactions for HO• production from OH^- is shown in reactions 2 to 5 in Table 12-4.

The overall stoichiometry of the reaction is given by the following reaction:

$$3O_3 + OH^- + H^+ \rightarrow 4O_2 + 2HO\cdot \qquad (12\text{-}7)$$

Based on this stoichiometry, the high-pH ozone process requires 1.5 mol of ozone to produce 1 mol of HO•.

The first step in the reaction sequence, reaction 2 in Table 12-4, is only fast at high pH and is the rate-limiting step in the overall reaction. For example, the half-life of ozone for reaction 2 in Table 12-4 is 1650 min at pH = 7, 16.5 min at pH = 9, and 0.165 min at pH = 11. Consequently, the reaction does not proceed rapidly unless the pH is 11 or above. Unfortunately, high pH values are detrimental to the production of HO•, as shown in Eq.12-7 (H^+ is required), and carbonate species quench the hydroxyl radicals that are formed from subsequent reactions. As a direct result of the low relative reaction rates at high pH (pH >9), significant destruction rates of target compounds have not been observed with the high-pH O_3 process.

Hydroxyl Radical Production from O_3 and NOM

When ozone reacts with NOM, it produces low levels of hydroxyl radical via the reaction

$$O_3 + \text{NOM} \rightarrow \text{HO}\cdot + \text{by-products} \qquad (12\text{-}8)$$

As discussed in Sec. 12-1, the hydroxyl radical produced from Eq. 12-8 may also be quenched by the reaction with NOM, as shown in the reaction

$$\text{HO}\cdot + \text{NOM} \xrightarrow{k_{13}} \text{by-products} \qquad (12\text{-}9)$$

where k_{13} = second-order rate constant between hydroxyl radical and NOM (measured in mol/L as C), L/mol·s

The quenching of the hydroxyl radical with NOM is usually more important than quenching by bicarbonate and carbonate or metal species (see Sec. 12-1 for details, including rate constants).

Table 12-4
Important elementary reactions involved in H_2O_2/O_3 and H_2O_2/UV processes at near-neutral pH and acid base dissociation reactions

Reaction Number	Reactions	Rate Constants at 25°C, L/M·s	References
		Reactions Specifically for H_2O_2/O_3	
1	$HO_2^- + O_3 \xrightarrow{k_1} O_3^- \cdot + HO_2 \cdot$	$k_1 = 2.8 \times 10^6$	Staehelin and Hoigné (1982)
2	$OH^- + O_3 \xrightarrow{k_2} HO_2^- + O_2$	$k_2 = 70$	Staehelin and Hoigné (1982)
3	$O_2^- \cdot + O_3 \xrightarrow{k_3} O_3^- \cdot + O_2$	$k_3 = 1.6 \times 10^9$	Buhler et al. (1984)
4	$O_3^- \cdot + H^+ \xrightarrow{k_4} HO_3 \cdot$	$k_4 = 5.2 \times 10^{10}$	Buhler et al. (1984)
5	$HO_3 \cdot \xrightarrow{k_5} HO \cdot + O_2$	$k_5 = 1.1 \times 10^5\ s^{-1}$	Buhler et al. (1984)
6	$O_3 + R \xrightarrow{k_6}$ products	k_6—see note[a]	
		Reactions Specifically for H_2O_2/UV Process	
		$r_{UV,H_2O_2} = r_{HO\cdot}/2 = -\phi_{H_2O_2} P_{u+v} f_{H_2O_2} \left(1 - e^{-A}\right)$	
		$A = 2.303b\left(\varepsilon_{H_2O_2} C_{H_2O_2} + \varepsilon_{R1} C_{R1} + \varepsilon_{NOM} C_{NOM}\right)$	
7	$H_2O_2 + h\nu \rightarrow 2HO\cdot$	$f_{H_2O_2} = 2.303b\left(\varepsilon_{H_2O_2} C_{H_2O_2} + \varepsilon_{HO_2^-} C_{HO_2^-}\right)/A$	
		$\varepsilon_{H_2O_2,254\ nm} = 17.9 \sim 19.6\ M^{-1}\ cm^{-1}$	
		$\phi_{H_2O_2} = \phi_{HO_2^-} = 0.5$	
8	$R + h\nu \rightarrow$ products	$r_{UV,R} = -\phi_R P_{u\text{-}v}(1 - e^{-A})$ $f_R = 2.303b\ \varepsilon_R C_R/A$	

	Reactions Common for Both H_2O_2/O_3 and H_2O_2/UV Process		
9	$HO\cdot + HO_2^- \xrightarrow{k_9} OH^- + HO_2\cdot$	$k_9 = 7.5 \times 10^9$	Christensen et al. (1982)
10	$HO\cdot + H_2O_2 \xrightarrow{k_{10}} H_2O + HO_2\cdot$	$k_{10} = 2.7 \times 10^7$	Buxton and Greenstock (1988)
11	$HO\cdot + HCO_3^- \xrightarrow{k_{11}} CO_3^-\cdot + H_2O$	$k_{11} = 8.5 \times 10^6$	Buxton and Greenstock (1988)
12	$HO\cdot + R \xrightarrow{k_{12}}$ products	k_{12}—see Table 12-1	
13	$HO\cdot + NOM \xrightarrow{k_{13}}$ products	$k_{13} = 1.4 \times 10^8$ to 4.5×10^8 L/mol·s as C	Westerhoff et al. (2007)
	Acid Dissociation Constants		
14	$H_2CO_3^* \rightleftarrows H^+ + HCO_3^-$	$pK_{a1} = 6.3$	Stumm and Morgan (1996)
15	$HCO_3^- \rightleftarrows H^+ + CO_3^{2-}$	$pK_{a2} = 10.3$	Stumm and Morgan (1996)
16	$H_2O \rightleftarrows H^+ + OH^-$	$pK_{a3} = 14$	Stumm and Morgan (1996)
17	$H_2O_2 \rightleftarrows H^+ + HO_2^-$	$pK_{a5} = 11.75$	Behar et al. (1970)
18	$HO_2\cdot \rightleftarrows H^+ + O_2^-\cdot$	$pK_{a6} = 4.8$	Staehelin and Hoigné (1982)

[a]See resource E2 at the website listed in App. E.

For target compound destruction, ozonation can destroy organic compounds by either direct reactions with O_3 or indirect reactions with HO•, as shown in the following:

$$O_3 \rightarrow \begin{cases} \xrightarrow[O_3]{\text{direct pathway}} & O_3 + R \rightarrow \text{product 1} \\ \xrightarrow[NOM]{\text{indirect pathway}} & HO\cdot + R \rightarrow \text{product 2} \end{cases} \tag{12-10}$$

The rate of destruction of a target compound R is given by the equation

$$r_R = -k_{O_3}\,[O_3]\,[R] - k_{HO\bullet}\,[HO]\,[R] \tag{12-11}$$

where

r_R = rate of disappearance of target compound R, mol/L·s

$[O_3]$, $[HO\bullet]$, $[R]$ = concentrations of ozone, hydroxyl radical, and target compound R, respectively, mol/L

$k_{HO\bullet}$ = rate constant between hydroxyl radical and R, L/mol·s

k_{O_3} = rate constant between ozone and R, L/mol·s

The relative importance of the direct reaction with ozone and the indirect reaction with HO• can be assessed using the expression

$$f_{HO\bullet} = \frac{k_{HO\bullet}\,[HO\bullet]}{k_{HO\bullet}\,[HO\bullet] + k_{O_3}\,[O_3]} = \frac{k_{HO\bullet}\,([HO\bullet]/[O_3])}{k_{HO\bullet}\,([HO\bullet]/[O_3]) + k_{O_3}} \tag{12-12}$$

$$= \frac{k_{HO\bullet}\,C_{[HO\bullet]/[O_3]}}{k_{HO\bullet}\,C_{[HO\bullet]/[O_3]} + k_{O_3}} \tag{12-13}$$

where

$f_{HO\bullet}$ = fraction of target compound destruction due to indirect reaction with HO•, dimensionless

$C_{[HO\bullet]/[O_3]}$ = ratio of hydroxyl radical concentration to ozone concentration, dimensionless

It has been reported that the ratio of the concentrations of HO• to ozone $\left(C_{[HO\bullet]/[O_3]}\right)$ was relatively constant during the decomposition of ozone in the presence of NOM, with typical values ranging from 10^{-7} to 10^{-10} (Elovitz and von Gunten, 1999). Second-order rate constants for ozone are useful in assessing possible reactions and reaction kinetics. The rate constants for organics depend on the type of organic being oxidized. The trend in the rate of the indirect pathway is from high for amine-substituted benzenes to low for aliphatics without nucleophilic sites (electron rich or donating sites). In contrast, many of the rate constants for the direct reaction with ozone appear to be low.

When ozone is added to natural waters containing NOM, the indirect pathway can be the most important mechanism to destroy target compounds (Elovitz and von Gunten, 1999). For example, it has been demonstrated that 83 percent of MIB and 90 percent of geosmin were degraded by the hydroxyl radical for an ozonated natural water (Bruce et al., 2002). This finding is important because, if ozone is able to remove taste- and odor-causing organics, it is probably due to HO• production from the reaction of ozone with NOM. The following example demonstrates when the indirect and direct pathways are important for target compound destruction.

Example 12-3 Fraction of target compound destruction

Determine the fraction of the reaction that is carried out by the indirect reaction with HO· for second-order HO· rate constants of 10^7, 10^8, and 10^9 L/mol·s. For the calculation, use $C_{[HO\bullet]/[O_3]}$ values of 10^{-7}, 10^{-8}, 10^{-9}, and 10^{-10} and a rate constant for the direct reaction with ozone of $k_{O_3} = 10$ L/mol•s.

Solution

1. Calculate the fraction of target compound destruction due to the indirect reaction with HO· using Eq. 12-12: For $k_{HO\bullet} = 10^7$ L/mol·s and $C_{[HO\bullet]/[O_3]} = 10^{-7}$,

$$f_{HO\bullet} = \frac{(10^7 \text{ L/mol·s})(10^{-7})}{(10^7 \text{ L/mol·s})(10^{-7}) + 10 \text{ L/mol·s}} = 0.0909$$

2. Repeat the calculation for the other conditions given and plot the results. The fraction of target compound destruction due to indirect reaction with HO· for k_{O_3} of 10 L/mol·s is shown in the figure below.

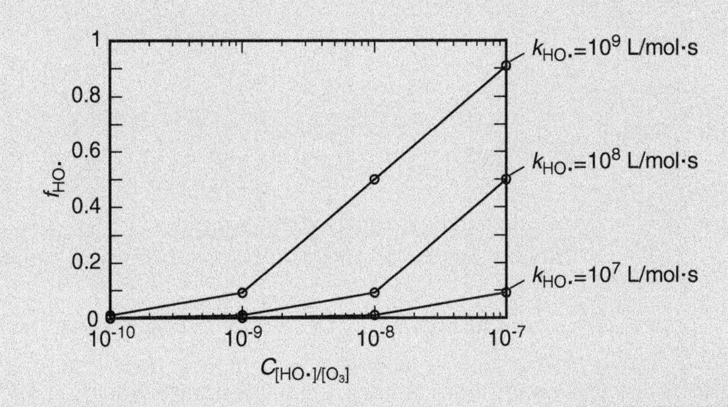

Comment

The indirect pathway becomes more significant relative to the direct pathway as the HO· concentration and HO· rate constant increase. More than half of the degradation occurs by the indirect pathway when $k_{HO\bullet} > 10^9$ L/mol·s and $C_{[HO\bullet]/[O_3]} > 10^{-8}$.

Performance of a Batch Reactor or Plug Flow Reactor

A simple model for the concentration in a batch reactor or the effluent concentration from a PFR may be developed using Eq. 12-12. The degradation of ozone and target compound may be described by the following pseudo-first-order reactions:

$$r_{O_3} = -k\,[O_3]$$

$$r_R = -\left(k_{O_3} + k_{HO\bullet} C_{[HO\bullet]/[O_3]}\right)[O_3]\,[R] \tag{12-14}$$

where r_{O_3} = rate of loss of ozone, mol/L·s
k = decay rate constant for ozone, s^{-1}

For a batch reactor with an elapsed time t or a PFR with detention time τ, a mass balance analysis using Eqs.12-13 and 12-14 may be written and solved as shown below:

$$[\text{accum}] = [\cancel{\text{mass in}}] - [\cancel{\text{mass out}}] + [\text{rxn}]$$

$$\frac{d\,[O_3]}{dt} = -k\,[O_3] \tag{12-15}$$

$$[O_3] = [O_3]_0\,e^{-kt} \tag{12-16}$$

$$\frac{d\,[R]}{dt} = -\left(k_{HO\bullet} C_{[HO\bullet]/[O_3]}\,[R] + k_{O_3}\,[R]\right)[O_3]_0\,e^{-kt} \tag{12-17}$$

$$[R] = [R]_0\,e^{\left[\left([O_3]_0/k\right)\left(k_{HO\bullet} C_{[HO\bullet]/[O_3]} + k_{O_3}\right)\left(e^{-kt} - 1\right)\right]} \tag{12-18}$$

where $[O_3]_0$ = initial concentration of ozone, mol/L
$[R]_0$ = initial concentration of target compound R, mol/L

For a PFR, the time t in Eqs. 12-16 to 12-18 is replaced with the detention time τ.

The following example is presented to illustrate how Eq. 12-18 can be used to predict the destruction of target compounds.

Example 12-4 Time required for destruction of target compound

Calculate the time required for 95 percent destruction of MIB in a batch reactor. Use $C_{[HO\bullet]/[O_3]}$ ranging from 10^{-9} to 10^{-7} and an initial ozone concentration of 3 mg/L. The ozone decay rate constant is 0.1 min^{-1}, and the rate constant for the direct reaction of MIB with ozone is 10 L/mol·s.

Solution

1. Convert initial ozone concentration from mg/L to mol/L:

$$[O_3]_0 = \frac{3 \text{ mg/L}}{(48 \text{ g/mol})(10^3 \text{ mg/ g})} = 6.25 \times 10^{-5} \text{ mol/L}$$

2. Rearrange Eq. 12-18 to solve for t:

$$t = -\frac{1}{k} \ln \left[1 + \frac{k \ln([R]/[R]_0)}{\left(k_{HO\bullet} C_{[HO\bullet]/[O_3]} + k_{O_3}\right)[O_3]_0} \right]$$

3. Calculate the time required for 95 percent destruction of a target compound when $C_{[HO\bullet]/[O_3]} = 10^{-7}$.
   a. Calculate the overall rate constant for MIB destruction. From Table 12-1, the rate constant for MIB decay by HO• is 8.2×10^9 L/mol·s:

$$\begin{aligned} k_{HO\bullet} C_{[HO\bullet]/[O_3]} + k_{O_3} &= \left[\left(8.2 \times 10^9 \text{ L/mol·s}\right)\left(10^{-7}\right) \right. \\ &\quad \left. + 10 \text{ L/mol·s}\right](60 \text{ s/min}) \\ &= 49{,}800 \text{ L/mol·min} \end{aligned}$$

   b. Calculate t using the equation in step 2:

$$t = -\frac{1}{0.1 \text{ min}^{-1}} \ln \left[1 + \frac{\left(0.1 \text{ min}^{-1}\right) \ln(0.05)}{(49{,}800 \text{ L/mol·min})(6.25 \times 10^{-5} \text{ mol/L})} \right]$$

$$= 1.01 \text{ min}$$

4. Repeat step 3 for additional values and tabulate the results for $10^{-9} \leqslant C_{[HO\bullet]/[O_3]} \leqslant 10^{-7}$. The results are shown below:

$C_{[HO\bullet]/[O_3]}$	t, min
1.00×10^{-7}	1.01
5.00×10^{-8}	2.11
1.00×10^{-8}	20.27
8.55×10^{-9}	58.8
8.50×10^{-9}	∞
1.00×10^{-9}	∞

Comment

The destruction of MIB is possible if $C_{[HO\bullet]/[O_3]}$ is greater than 8.55×10^{-9}.

Determination of Destruction of Target Compounds from Bench-Scale Tests

While Eq. 12-18 can be used to describe the destruction rate of target compounds, there is no way to predict the value of $C_{[HO\bullet]/[O_3]}$ or the decay rate constant for ozone, k from water quality measurements. Consequently, batch tests are required to determine $C_{[HO\bullet]/[O_3]}$ and k. The basics of the batch test method is described in detail in the companion reference book for this text (Crittenden et al., 2012).

12-3 Hydrogen Peroxide/Ozone Process

When hydrogen peroxide (H_2O_2) in added in conjunction with ozone, the ozone and hydrogen peroxide react to form hydroxyl radical. The amount of HO• that forms can be significant compared to the amount that forms from the reaction between ozone and NOM. When H_2O_2 is added, the reaction between NOM and ozone to form HO• can be ignored because it is insignificant compared to the reaction between ozone and H_2O_2. However, the quenching of HO• by NOM is significant and must be considered.

Reaction Mechanisms

The elementary reactions that are involved in the production of HO• from H_2O_2/O_3 are listed in Table 12-4. The following discussion of the reaction mechanisms will cover the H_2O_2/O_3 process at neutral pH (values near 7). The H_2O_2/O_3 reaction sequence begins with dissociation of the H_2O_2 to form HO_2^-, as shown in reaction 17 in Table 12-4. The HO_2^- then reacts with O_3 to form the ozonide ion radical, $O_3^-\bullet$, and the superoxide radical, $HO_2\bullet$, as shown in reaction 1.

The rate-limiting step in the formation of HO• is reaction 1 and the rate decreases at low pH. Consequently, low reaction rates have been observed at pH values of 5 or less, and the H_2O_2/O_3 process may not be a viable option for the destruction of organics if the pH is less than 5. The superoxide radical, $HO_2\cdot$, can dissociate according to reaction 18 to form $O_2\cdot$, and the $O_2\cdot$ reacts with additional ozone to form additional ozonide ion radical as shown in reaction 3. The ozonide ion radical then proceeds rapidly through reactions 4 and 5 to form $HO\cdot$. The overall reaction for $HO\cdot$ radical formation is

$$H_2O_2 + 2O_3 \rightarrow 2HO\cdot + 3O_2 \tag{12-19}$$

Proper Ratio of Hydrogen Peroxide to Ozone

According to Eq. 12-19, 0.5 mol of H_2O_2 is needed for every mole of O_3, corresponding to a mass ratio of 0.354 kg H_2O_2 per kilogram of O_3. However, several factors affect the proper ratio of H_2O_2 to O_3. First, the relevant O_3 dose is the transferred dose, not the applied dose, although O_3 mass transfer efficiency is usually greater than 95 percent. Second, O_3 tends to be more reactive with background organic matter and inorganic species than H_2O_2. As a result, some O_3 will degrade immediately and will not be available to react with H_2O_2. The applied O_3 dose will have to be higher than estimated from stoichiometry to achieve the optimum ratio. However, excess O_3 has the potential to waste O_3 and scavenge $HO\cdot$ via the reaction

$$O_3 + HO\cdot \rightarrow HO_2\cdot + O_2 \tag{12-20}$$

The $HO_2\cdot$ radical formed as shown in Eq. 12-20 may produce more $HO\cdot$ via reactions through reactions 3 to 5 in Table 12-4 if adequate ozone remains in solution. Excess H_2O_2 is also detrimental to the H_2O_2/O_3 process because it may scavenge $HO\cdot$ via reactions 9 and 10 in Table 12-4.

Furthermore, the H_2O_2 residual can be more problematic than ozone because hydrogen peroxide is more stable than ozone and may interfere with downstream processes and equipment. Some vendors have attempted to overcome the problem of H_2O_2 quenching of $HO\cdot$ by adding H_2O_2 at multiple points in a single reactor or by using multiple reactors in series.

Modeling the H_2O_2/O_3 Process

The elementary reactions for the O_3/H_2O_2 process are listed in Table 12-4. The elementary reactions include the initiation (reactions 1, 3, 4, and 5), propagation (reactions 9 and 10), and termination reactions of the radical chain reaction. Termination reactions involve recombination of radical species and are not shown because they have a low probability of occurrence (e.g., $HO\cdot + HO\cdot \rightarrow H_2O_2$). The elementary reactions also include the oxidation of the target organic compound (R) and the scavenging of the hydroxyl radical by bicarbonate, carbonate, and NOM, as discussed in Sec. 12-1.

The net rates of formation of various radicals are given by the expressions

$$\begin{aligned} r_{HO\cdot} = {} & k_5[HO_3\cdot] - k_9[HO\cdot][HO_2^-] - k_{10}[HO\cdot][H_2O_2] \\ & - k_{11}[HO\cdot][HCO_3^-] - k_{12}[HO\cdot][R] - k_{13}[HO\cdot][NOM] \end{aligned} \tag{12-21}$$

$$r_{HO_3\cdot} = k_4[O_3^-\cdot][H^+] - k_5[HO_3\cdot] \quad (12\text{-}22)$$

$$r_{O_3^-\cdot} = k_1[O_3][HO_2^-] + k_3[O_2^-\cdot][O_3] - k_4[O_3^-][H^+] \quad (12\text{-}23)$$

$$r_{HO_2\cdot/O_2^-\cdot} = k_1[HO_2^-][O_3] + k_9[HO\cdot][HO_2^-] + k_{10}[HO\cdot][H_2O_2] - k_3[O_3][O_2^-\cdot] \quad (12\text{-}24)$$

Equation 12-21 ignores the quenching of the hydroxyl radical by the carbonate ion, which is valid for pH less than 8.0. Because radical species are so reactive, they participate in decay reactions as soon as they are formed; consequently, their concentrations are small and the net rates of formation and decay are equal to each other. The equivalence between formation and decay rates results in a zero net rate of formation and is known as the pseudo-steady-state approximation. Invoking the pseudo-steady-state approximation for the various radical intermediates, four algebraic equations are obtained. After solving the system of equations and eliminating all radical species other than HO•, Eq. 12-21 can be rearranged to obtain the following expression for HO•:

$$[HO\cdot]_{ss} = \frac{2k_1[HO_2^-][O_3]}{k_{11}[HCO_3^-] + k_{12}[R] + k_{13}[NOM]} \quad (12\text{-}25)$$

where $[HO\cdot]_{ss}$ = pseudo-steady-state concentration of HO•, mol/L

When the H_2O_2/O_3 ratio is close to the stoichiometric optimum, the liquid-phase reactions in reactions 2 through 5 in Table 12-4 occur so fast that the ozone transfer is the limiting factor in the reaction rate. The O_3 concentration can be assumed to be constant and is much lower than the saturation concentration. The saturation concentration is given by:

$$[O_3]_s = \frac{P_{O_3}}{H_{O_3}} \quad (12\text{-}26)$$

where $[O_3]_s$ = saturation concentration of ozone, mol/L
P_{O_3} = partial pressure of ozone in inlet gas, bar
H_{O_3} = Henry's law constant for ozone, bar•L/mol

Combining the equation for the mass transfer of O_3 into water (see Chap. 11) and the decay of O_3 by reactions 1 and 3 from Table 12-4, the resulting rate expression for O_3 is

$$r_{O_3} = (K_La)_{O_3}\left(\frac{P_{O_3}}{H_{O_3}} - [O_3]\right) - k_1[HO_2^-][O_3] - k_3[O_2^-\cdot][O_3] \quad (12\text{-}27)$$

where $(K_La)_{O_3}$ = overall mass transfer coefficient for ozone, s^{-1}

The pseudo-steady-state approximation for the rate of formation and decay of ozone is invoked (e.g., $r_{O_3} = 0$) and Eq. 12-27 may be rearranged to the form

$$(K_L a)_{O_3}\left(\frac{P_{O_3}}{H_{O_3}} - [O_3]\right) = k_1[HO_2^-][O_3] + k_3[O_2^-\cdot][O_3] \qquad (12\text{-}28)$$

The rate of formation of $HO_2\cdot/O_2\cdot$ shown in Eq. 12-24 may be rearranged to the form

$$k_3[O_2^-\cdot][O_3] = k_1[HO_2^-][O_3] + k_9[HO_2^-][HO\cdot] + k_{10}[H_2O_2][HO\cdot] \qquad (12\text{-}29)$$

Substituting Eq. 12-28 into Eq. 12-29 and rearranging yields

$$2k_1[HO_2^-\cdot][O_3] = (K_L a)_{O_3}\left(\frac{P_{O_3}}{H_{O_3}} - [O_3]\right) - k_9[HO_2^-][HO\cdot] - k_{10}[H_2O_2][HO\cdot] \qquad (12\text{-}30)$$

The following expression is obtained after substituting Eq. 12-30 into Eq. 12-25 and rearranging:

$$[HO\cdot]_{ss} = \frac{(K_L a)_{O_3}\left(P_{O_3}/H_{O_3} - [O_3]\right)}{k_9[HO_2^-] + k_{10}[H_2O_2] + k_{11}[HCO_3^-] + k_{12}[R] + k_{13}[NOM]} \qquad (12\text{-}31)$$

The initial pseudo-steady-state concentration of HO• is obtained by neglecting $[O_3]$ as compared to P_{O_3}/H_{O_3} as shown in the expression

$$[HO\cdot]_{ss,0} = \frac{(K_L a)_{O_3}\left(P_{O_3}/H_{O_3}\right)}{k_9[HO_2^-]_0 + k_{10}[H_2O_2]_0 + k_{11}[HCO_3^-]_0 + k_{12}[R]_0 + k_{13}[NOM]_0} \qquad (12\text{-}32)$$

where $[HO\cdot]_{ss,0}$ = initial steady-state concentration of HO•, mol/L
$[HO_2^-]_0$ = initial concentration of anion of hydrogen peroxide, mol/L
$[H_2O_2]_0$ = initial concentration of hydrogen peroxide, mol/L
$[HCO_3^-]_0$ = initial concentration of bicarbonate, mol/L
$[R]_0$ = initial concentration of target organic compound R, mol/L
$[NOM]_0$ = initial concentration of NOM, mol/L

Hydrogen peroxide is a weak acid that dissociates to form HO_2^- as shown in reaction 17 in Table 12-4. As shown in Sec. 4-4, the equilibrium constant for this reaction is

$$K_{H_2O_2} = \frac{[H^+][HO_2^-]}{[H_2O_2]} \qquad (12\text{-}33)$$

Expressing the HO_2^- concentration as a function of the initial H_2O_2 concentration yields

$$[HO_2^-]_0 = \frac{[H_2O_2]_0 K_{H_2O_2}}{[H^+]} \tag{12-34}$$

Expressing the hydrogen ion and equilibrium constant as pH and pK values (see Sec. 4-1, and 4-2) yields

$$[HO_2^-]_0 = \frac{[H_2O_2]_0 10^{-pK_{H_2O_2}}}{10^{-pH}} = [H_2O_2]_0 \left(10^{pH - pK_{H_2O_2}}\right) \tag{12-35}$$

where $pK_{H_2O_2}$ = acid dissociation constant for hydrogen peroxide (pK_{a5} in Table 12-4).

The initial steady-state O_3 concentration can be estimated by substituting Eq. 12-32 into Eq. 12-30 and rearranging:

$$[O_3]_{ss,0} = \frac{K_L a\left(P_{O_3}/H_{O_3}\right) - k_9[HO\bullet]_{ss,0}[HO_2^-]_0 - k_{10}[HO\bullet]_{ss,0}[H_2O_2]_0}{(K_L a)_{O_3} + 2k_1[HO_2^-]_0} \tag{12-36}$$

The rate laws for the decay of the target compound R and H_2O_2 are given by the equations

$$r_R = -k_6[R][O_3] - k_{12}[R][HO\bullet] \tag{12-37}$$

$$r_{H_2O_2} = -k_1[HO_2^-][O_3] - k_9[HO_2^-][HO\bullet] - k_{10}[H_2O_2][HO\bullet] \tag{12-38}$$

where r_R = rate of target compound R destruction, mol/L·s
$r_{H_2O_2}$ = rate of hydrogen peroxide loss, mol/L·s
k_6 = second-order rate constant between target compound R and ozone, L/mol·s

For the situation where the direct ozonation rate of a target compound is much lower than the reaction rate with hydroxyl radicals (the most common situation), that is, $k_6[O_3] \ll k_{12}[HO\bullet]$ (e.g., $[O_3]/[HO\bullet] \approx 10^4 \sim 10^6$), the first term in Eq. (12-37) can be ignored.

The equations developed above include equations for the rate of decay of hydrogen peroxide and target compounds, and steady-state-concentrations of ozone and HO• as a function of measurable water quality parameters.

Simplified Model for H_2O_2/O_3 Process

A simplified model of the H_2O_2/O_3 process can be developed for various cases to provide an estimate of the destruction rates of the parent compound and hydrogen peroxide. The following two cases are considered: (1) H_2O_2 and O_3 are added together and (2) H_2O_2 is added to water containing O_3.

H_2O_2 AND O_3 ARE ADDED SIMULTANEOUSLY

A simplified analysis for the H_2O_2/O_3 process can be obtained by assuming that the hydroxyl radical concentration does not change with time and is equal to the initial steady-state hydroxyl radical concentration. This assumption yields a pseudo-first-order rate law, which results in the prediction of reaction rates that are faster than would be observed. The pseudo-first-order rate law and coefficient are given by the expressions

$$r_R = -k_R [R] \tag{12-39}$$

$$k_R = k_{12} [HO]_{ss,0} \tag{12-40}$$

where k_R = pseudo-first-order rate constant for target compound R, s^{-1}

Other terms are as defined previously.

The residual hydrogen peroxide concentration is important because if hydrogen peroxide enters the distribution system, it will react with chlorine to produce oxygen. Therefore, hydrogen peroxide must be removed. The following pseudo-first-order rate law and rate coefficient can be obtained by assuming that the hydroxyl radical and ozone concentrations do not change with time and are equal to their initial steady-state concentration, which are defined in Eqs. 12-32 and 12-36, respectively. Substituting Eq. 12-35 and rearranging yields

$$r_{H_2O_2} = -k_{H_2O_2} [H_2O_2] \tag{12-41}$$

$$k_{H_2O_2} = k_1 [O_3]_{ss,0} \times 10^{pH-pK_{H_2O_2}} + k_9 [HO\bullet]_{ss,0} \times 10^{pH-pK_{H_2O_2}} + k_{10} [HO\bullet]_{ss,0} \tag{12-42}$$

where $k_{H_2O_2}$ = pseudo-first-order rate constant for hydrogen peroxide, s^{-1}

The above model, termed the *simplified pseudo-steady-state model*, overestimates the destruction rates of the target compound and hydrogen peroxide. Consequently, when these expressions are used to assess the feasibility of destroying organic compounds, they will predict lower effluent concentrations of hydrogen peroxide and target compound than will be observed. Experience has demonstrated that the models predict removal of 12 percent higher than experimental data for most cases.

H_2O_2 ADDED TO WATER CONTAINING O_3

Some utilities add ozone for disinfection; then, when *Ct* disinfection credit is obtained, they add hydrogen peroxide for the destruction of target micropollutants. In this situation, the residual ozone concentration $[O_3]_{res}$ is known, and hydrogen peroxide is added to produce the hydroxyl radical. The rate law for O_3 decay is given by the equation

$$r_{O_3} = -k_1 [HO_2^-][O_3]_{res} - k_3 [O_2^-\bullet][O_3]_{res} \tag{12-43}$$

Substituting Eq. 12-29 into Eq. 12-43 yields

$$r_{O_3} = -\left(2k_1[HO_2^-][O_3]_{res} + k_9[HO\cdot][HO_2^-] + k_{10}[HO\cdot][H_2O_2]\right) \quad (12\text{-}44)$$

Substituting Eq. 12-35 into Eq. 12-25 and using $[O_3]_{res}$ as the ozone concentration, the initial pseudo-steady-state concentration of $HO\cdot$ is given by the equation

$$[HO\cdot]_{ss,0} = \frac{2k_1[H_2O_2]_0 \times 10^{(pH - pK_{H_2O_2})}[O_3]_{res}}{k_{11}[HCO_3^-]_0 + k_{12}[R]_0 + k_{13}[NOM]_0} \quad (12\text{-}45)$$

The rate laws for the target compound R and H_2O_2 decay are given by Eqs. 12-37 and 12-38. In most cases, because $k_6\,[O_3] \ll k_{12}[HO\cdot]$, the first term in Eq. 12-37 can be ignored. A simplified model can be obtained by assuming that the hydroxyl radical does not change within the time and is equal to the initial steady-state hydroxyl radical concentration, which is given by Eq. 12-45. The pseudo-first-order rate law and coefficient are given by the expressions in Eqs. 12-39 and 12-40.

The pseudo-first-order rate law and coefficient for hydrogen peroxide can be obtained by assuming that the hydroxyl radical and ozone concentrations do not change with time and are equal to their initial steady-state concentration. The initial concentration of ozone is equal to $[O_3]_{res}$ in Eq. 12-45.

Because the initial concentrations are used, the above model predicts reaction rates that are faster than would be observed. For example, the simple pseudo-steady-state (Sim-PSS) model can be compared to data provided by Glaze and Kang (1989). The measured data are predicted well by the Sim-PSS model when the H_2O_2/O_3 mass ratio is from 0.3 to about 0.6, which is around the stoichiometric optimum of 0.35. For a mass ratio less than 0.3, the predicted rate constants are higher than the measured values, and when the ratio exceeds 0.6, the predicted values are less than the measured values. The observed variations are due to the complexity of the H_2O_2/O_3 reaction system; in particular, different mechanisms control the overall reaction rate from O_3 control to H_2O_2 control as the H_2O_2/O_3 ratio changes. Consequently, to predict process performance more accurately, a sophisticated model is required. However, the Sim-PSS model is precise enough to examine the feasibility of the process. Moreover, pilot testing is necessary to evaluate the technology in the field once process feasibility has been assessed using the Sim-PSS model.

Disadvantages of H_2O_2/O_3 Process

Several problems are associated with the hydrogen peroxide/ozone process. One problem is the stripping of volatile species into the off-gas from the ozone contactor. Stripping phenomenon is not significant for the more reactive volatile species but can be for species that are less reactive with hydroxyl radical, such as carbon tetrachloride. Another problem with the use of the hydrogen peroxide/ozone process is the production of bromate when the water being treated contains significant amounts of bromide ion. Significant bromate formation (above the U.S. EPA regulated value of 10 μg/L) can occur with ozone addition; raising pH, adding ammonia and chlorine are strategies for reducing bromate formation.

Reactor Sizing for the H_2O_2/O_3 Process

The equations presented previously in this section can be used to estimate the hydraulic retention time of a reactor to achieve a specified level of removal of a target contaminant by the H_2O_2/O_3 process. Using the reactor analysis principles presented in Chap. 4, the reaction kinetics can be applied to any type of ideal or real reactor. The application of these equations to a real reactor that can be described with the tanks-in-series reactor model is demonstrated in Example 12-5.

Example 12-5 Hydrogen peroxide/ozone process

A small city has recently discovered that one of its wells is contaminated with 200 μg/L TCE. To continue using the well as a drinking water source, the TCE effluent concentration must be reduced to less than 5 μg/L. The HCO_3^-, pH, and DOC concentrations are 480 mg/L as CO_3, 7.5, and 0.7 mg/L, respectively. The physicochemical properties of TCE and NOM are as follows:

Compound	MW, g/mol	HO· Rate Constant, $k_{HO\cdot}$, L/mol·s
TCE	131.4	4.20×10^9
NOM[a]	NA	3.90×10^8

[a]For NOM, the unit of $k_{HO\cdot}$ is L/mol·s when the concentration of NOM is measured as mol/L as C.

For simplicity, a proprietary reactor will be used. It has been determined by conducting tracer studies on the reactor that its hydraulic performance can be described using four completely mixed flow reactors in series. Given the following information: (1) the H_2O_2 dosage is 0.8 mg/L, (2) O_3 is generated onsite and the ozone flow rate is 1 mg/L·min, (3) the partial pressure of ozone in the inlet gas is 0.07 bar, (4) the Henry's law constant for O_3 at 23°C is 83.9 L·bar/mol, and (5) the overall mass transfer coefficient for O_3, K_La, was measured to be 7×10^{-4} s^{-1}. Determine the hydraulic retention time (τ) and H_2O_2 residual to obtain 5 μg/L of the effluent TCE concentration.

Solution

1. Calculate the initial steady-state concentration of hydroxyl radical using Eq. 12-32.
   a. Obtain the reaction rate constants and acid dissociation constants from Table 12-4.
   b. Calculate the concentration of each component from reactions 19 through 13 in Table 12-4:

$$[HO_2^-]_0 = [H_2O_2]_0 \times 10^{pH - pK_{H_2O_2}} = \frac{(0.8 \text{ mg/L})\left(10^{7.5-11.75}\right)}{(34 \text{ g/mol})(10^3 \text{ mg/g})}$$

$$= 1.32 \times 10^{-9} \text{ mol/L}$$

$$[HCO_3^-]_0 = \frac{480 \text{ mg/L}}{(60 \text{ g/mol})(10^3 \text{ mg/ g})} = 0.008 \text{ mol/L}$$

$$[R]_0 = [TCE]_0 = \frac{200 \text{ }\mu\text{g/L}}{(131.4 \text{ g/mol})(10^6 \text{ }\mu\text{g/g})}$$

$$= 1.52 \times 10^{-6} \text{ mol/L}$$

$$[NOM]_0 = \frac{0.7 \text{ mg/L as C}}{(12 \text{ g/mol})(10^3 \text{ mg/g})} = 5.83 \times 10^{-5} \text{ mol/L as C}$$

   c. Calculate the product of the rate constant and initial concentration of each component needed in Eq. 12-32:

$$k_9 [HO_2^-]_0 = \left(7.5 \times 10^9 \text{ L/mol·s}\right)\left(1.32 \times 10^{-9} \text{ mol/L}\right) = 9.9 \text{ s}^{-1}$$

$$k_{10} [H_2O_2]_0 = \left(2.7 \times 10^7 \text{ L/mol·s}\right)\left(2.35 \times 10^{-5} \text{ mol/L}\right) = 634.5 \text{ s}^{-1}$$

$$k_{11} [HCO_3^-]_0 = \left(8.5 \times 10^6 \text{ L/ mol·s}\right)(0.008 \text{ mol/L}) = 68000 \text{ s}^{-1}$$

$$k_{12} [R]_0 = k_{12} [TCE]_0 = \left(4.2 \times 10^9 \text{ L/mol·s}\right)$$

$$\times \left(1.52 \times 10^{-6} \text{ mol/L}\right) = 6384 \text{ s}^{-1}$$

$$k_{13} [NOM]_0 = \left(3.9 \times 10^8 \text{ L/mol·s}\right)\left(5.83 \times 10^{-5} \text{ mol/L as C}\right)$$

$$= 22737 \text{ s}^{-1}$$

d. Calculate the initial steady-state concentration of the hydroxyl radical using Eq. 12-32:

$$[\text{HO}\bullet]_{ss,0} = \frac{(7 \times 10^{-4}\ \text{s}^{-1})(0.07\ \text{bar})}{\left(9.9 + 634.5 + 68000 + 6384 + 22737\ \text{s}^{-1}\right) \times (83.9\ \text{L}\cdot\text{bar/mol})}$$

$$= 5.97 \times 10^{-12}\ \text{mol/L}$$

2. Caculate the hydraulic retention time ($\tau = V/Q$, see Eq. 4-73) when TCE effluent concentration is 5 μg/L:
   a. Dertermine K_{TCE} using the pseudo-first-order rate law presented in Eq. 12-40

$$k_R = k_{TCE} = k_{12}[\text{HO}\bullet]_{ss,0}$$

$$= \left(4.2 \times 10^9\ \text{L/mol}\cdot\text{s}\right)\left(5.97 \times 10^{-12}\ \text{mol/L}\right) = 0.025\ \text{s}^{-1}$$

   b. Determine τ when the effluent TCE concentration is 5 μg/L using the TIS model (Eq. 4-113 in Sec. 4-12):

$$[\text{TCE}] = [\text{R}] = \frac{[\text{R}]_0}{(1 + k_{TCE}\tau/n)^n}$$

$$= \frac{(200\ \mu\text{g/L})}{[1 + (0.025\ \text{s}^{-1})(\tau\ \text{min})(60\ \text{s/min})/4]^4}$$

$$= 5.0\ \mu\text{g/L}$$

   τ is 4.04 min.

3. Estimate the initial steady-state concentration of O_3 using Eq. 12-36:
   a. From Table 12-4,

$$k_1 = 2.8 \times 10^6\ \text{L/mol}\cdot\text{s}$$

   b. From steps 2a, 2b, and 2c,

$$k_9\,[\text{HO}_2^-]_0 = 9.9\ \text{s}^{-1}$$

$$k_{10}\,[\text{H}_2\text{O}_2]_0 = 634.5\ \text{s}^{-1}$$

$$[\text{HO}_2^-]_0 = 1.32 \times 10^{-9}\ \text{mol/L}$$

c. Calculate for $K_L a\left(P_{O_3}/H_{O_3}\right)$:

$$K_L a \frac{P_{O_3}}{H_{O_3}} = \left(7 \times 10^{-4}\ \mathrm{s}^{-1}\right)\left(\frac{0.07\ \mathrm{bar}}{83.9\ \mathrm{L \cdot bar/mol}}\right)$$

$$= 5.84 \times 10^{-7}\ \mathrm{mol/L \cdot s}$$

d. Combine results of 4b with result of 2d:

$$k_9 \left[HO_2^-\right]_0 \left[HO\cdot\right]_{ss,0} = \left(9.9\ \mathrm{s}^{-1}\right)\left(5.97 \times 10^{-12}\ \mathrm{mol/L}\right)$$

$$= 5.9 \times 10^{-11}\ \mathrm{mol/L \cdot s}$$

$$k_{10} \left[H_2O_2\right]_0 \left[HO\cdot\right]_{ss,0} = \left(634.5\ \mathrm{s}^{-1}\right)\left(5.97 \times 10^{-12}\ \mathrm{mol/L}\right)$$

$$= 3.79 \times 10^{-9}\ \mathrm{mol/L \cdot s}$$

e. Calculate for $[O_3]_{ss,0}$:

$$[O_3]_{ss,0} = \frac{\left[\left(5.84 \times 10^{-7}\right) - \left(5.9 \times 10^{-11}\right) - \left(3.79 \times 10^{-9}\right)\right]\ \mathrm{mol/L \cdot s}}{\left(7 \times 10^{-4}\ \mathrm{s}^{-1}\right) + \left[2\left(2.8 \times 10^{6}\ \mathrm{L/\ mol \cdot s}\right) \times \left(1.32 \times 10^{-9}\ \mathrm{mol/L}\right)\right]}$$

$$= 7.17 \times 10^{-5}\ \mathrm{mol/L} \quad (3.44\ \mathrm{mg/L})$$

4. Estimate H_2O_2 residual:
   a. Estimate the pseudo-first-order rate constant for hydrogen peroxide using Eq. 12-42:

$$k_{H_2O_2} = k_1 \left[O_3\right]_{ss,0} \times 10^{pH - pK_{H_2O_2}} + k_9 \left[HO\cdot\right]_{ss,0} \times 10^{pH - pK_{H_2O_2}} + k_{10} \left[HO\cdot\right]_{ss,0}$$

   i. Calculate the values of the rate constant times concentration needed in Eq. 12-42:

$$k_1 \left[O_3\right]_{ss,0} \times 10^{pH - pK_{H_2O_2}} = \left(2.8 \times 10^{6}\ \mathrm{L/\ mol \cdot s}\right)(7.17 \times 10^{-5}\ \mathrm{mol/L}) \times 10^{(7.5-11.75)} = 0.0112\ \mathrm{s}^{-1}$$

$$k_9 \left[HO\cdot\right]_{ss,0} \times 10^{pH - pK_{H_2O_2}} = \left(7.5 \times 10^{9}\ \mathrm{L/\ mol \cdot s}\right)(5.97 \times 10^{-12}\ \mathrm{mol/L}) \times 10^{(7.5-11.75)} = 2.52 \times 10^{-6}\ \mathrm{s}^{-1}$$

$$k_{10}[\text{HO}\cdot]_{ss,0} = \left(2.7 \times 10^7 \text{ L/ mol·s}\right)\left(5.97 \times 10^{-12} \text{ mol/L}\right)$$

$$= 1.61 \times 10^{-4} \text{s}^{-1}$$

ii. Calculate $k_{H_2O_2}$:

$$k_{H_2O_2} = \left(0.0112 \text{ s}^{-1}\right) + \left(2.52 \times 10^{-6} \text{ s}^{-1}\right)$$

$$+ \left(1.61 \times 10^{-4} \text{ s}^{-1}\right) = 0.0114 \text{ s}^{-1}$$

b. Estimate the H_2O_2 residual using the τ and TIS model (Eq. 4-113 in Sec. 4-12):

$$[H_2O_2] = \frac{[H_2O_2]_0}{(1 + k_{H_2O_2}\tau/n)^n}$$

$$= \frac{(0.8 \text{ mg/L})}{[1 + (0.0114 \text{ s}^{-1})(4.04 \text{ min})(60 \text{ s/min})/4]^4}$$

$$= 0.10 \text{ mg/L}$$

Comment

The initial ozone concentration is only an approximate estimate because the model assumed that the reactor contents are mixed completely, but the example was for a real reactor that fits the TIS model with $n = 4$. The effluent hydrogen peroxide concentration is only an estimate, and, based on the reactions that were considered, it is the lowest expected effluent concentration. Measurements will have to be taken to ensure that this residual does not interfere with disinfection (e.g., consume chlorine) and is not transmitted to the distribution system.

12-4 Hydrogen Peroxide/UV Light Process

The UV/hydrogen peroxide process includes hydrogen peroxide injection and mixing followed by a reactor that is equipped with UV lights. As shown on Fig. 12-1 a typical UV reactor is a stainless steel column that contains UV lights in a crisscrossing pattern. The details of the reactor are discussed in Sec 13-5.

The UV/H_2O_2 process cannot be used for potable water treatment because it has high effluent H_2O_2 concentrations. High effluent H_2O_2 concentrations are unavoidable because high initial dosages of H_2O_2 are

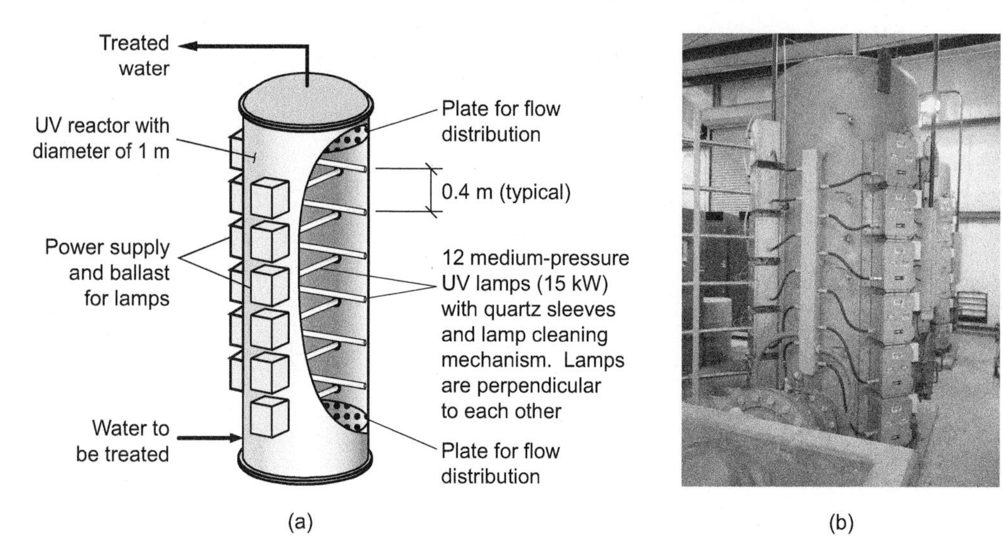

Figure 12-1
UV reactor used for advanced oxidation: (a) schematic and (b) photograph.

required in order to efficiently utilize the UV light and produce hydroxyl radicals. Aside from the health issues associated with high effluent H_2O_2 concentrations in the finished water, the residual H_2O_2 will consume chlorine and interfere with disinfection. This challenge will have to be overcome before the UV/ H_2O_2 process is used in drinking water treatment.

Elementary Reactions for the Hydrogen Peroxide/UV Process

The complex elementary radical reactions that are involved with the H_2O_2/UV process have been discovered. It is now possible to predict the destruction of the target compound using these reactions and gain insight into the factors that impact the H_2O_2/UV process. The mechanisms that may be considered are: (1) photolysis of H_2O_2 with a multichromatic light source, (2) UV absorption by the background components in the water matrix, (3) scavenging of hydroxyl radical by NOM and carbonate species, and (4) direct photolysis of NOM and the target compound. A rigorous AOP model was developed to predict the destruction of target compounds and the effluent H_2O_2 concentration using the complete radical reaction pathway presented by Crittenden et al. (1999).

However, reasonable predictions of target compound destruction and residual H_2O_2 can be obtained by using a simplified pseudo-steady-state model, although some accuracy will be lost (Crittenden et al., 1999). The most important elementary reactions in the H_2O_2/UV process at neutral pH are shown in Table 12-4. The reaction pathway is extremely simplified and ignores radical–radical reactions, the reactions between HO_2^- and CO_3^{2-} and other species (due to the large pK_a values) and unimportant radical species ($CO_3^{-}\bullet$ etc.).

The elementary reactions that are involved in the H_2O_2/UV process that are shown in Table 12-4 include initiation (reaction 7), propagation (reactions 9 and 10), and termination reactions of the radical chain reactions (termination reactions involve recombination of radical species and are not shown because they have a low probability of occurrence, e.g., $HO\bullet + HO\bullet \rightarrow H_2O_2$). The elementary reactions also include the oxidation of the target organic compound (R) and the scavenging of hydroxyl radicals by bicarbonate, carbonate, and NOM. As shown in Table 12-4, the production of hydroxyl radicals is initiated via the reaction

$$H_2O_2 + h\nu \rightarrow 2HO\bullet \tag{12-46}$$

UV LIGHT TRANSMISSION

The ability of H_2O_2 to absorb UV light and produce $HO\bullet$ via the reaction is dependent upon the wavelength and quantum yield and the UV light absorbance of the background components in the water.

AOPs that utilize UV light for the production of $HO\bullet$ must have reasonable light transmission in the ultraviolet region of the spectrum because any light that is not absorbed by the oxidant is wasted, and the generation of UV light represents a significant operational cost. Accordingly, it is important to evaluate the influence of pretreatment effectiveness and cost (e.g., particle removal and the removal of certain UV absorbing species) on UV light transmission. For example, when considering the UV/H_2O_2 process, a preliminary evaluation would include an estimate of the fraction of UV light that would be available to activate the H_2O_2 and the influence that pretreatment would have on the available light transmission. In a groundwater highly contaminated, an absorbance of 0.385 for a 1-cm depth at 254 nm was measured. The light absorption coefficient for H_2O_2 is about 19 $M^{-1}cm^{-1}$ at 254 nm, and the quantity of light and the fraction of light that produces the hydroxyl radical may be estimated from this equation:

$$f_{oxidant} = \frac{\varepsilon_{H_2O_2} C_{H_2O_2} L}{\varepsilon_{H_2O_2} C_{H_2O_2} L + \varepsilon_{bac} C_{bac} L} = \frac{\varepsilon_{H_2O_2} C_{H_2O_2}}{\varepsilon_{H_2O_2} C_{H_2O_2} + \varepsilon_{bac} C_{bac}} \tag{12-47}$$

where $f_{oxidant}$ = fraction of light absorbed by the oxidant, dimensionless
$C_{H_2O_2}$ = concentration of hydrogen peroxide, mol/L
C_{bac} = concentration of background, mol/L
L = reactor depth, cm
$\varepsilon_{H_2O_2}$ = extinction coefficients for hydrogen peroxide, L/(mol·cm)
ε_{bac} = extinction coefficients for background, L/(mol·cm)

For example, the fraction of light absorbed by an H_2O_2 concentration of 80 mg/L is only 10.7 percent, thus 90 percent of the light is wasted. When considering pretreatment options, it is useful to know the light absorption of certain dissolved species in water because this can form the basis for pretreatment.

QUANTUM YIELD

The fraction of adsorbed photos that result in a photolysis reaction is called the quantum yield. The primary quantum yield for H_2O_2 is 0.5 for wavelengths in the UV region (Volman and Chen, 1959), but the primary quantum yield of H_2O_2 depends slightly on temperature. For example, the quantum yield is 0.41 at 5°C. However, temperature is not important because the temperature in a UV reactor generally achieves room temperature due to heat produced during lamp illumination.

SIMPLIFIED PSEUDO-STEADY-STATE MODEL

Based on the reactions that are presented in Table 12-4, the rate expression for HO• is given by the expression

$$r_{HO\bullet} = 2\phi_{H_2O_2} P_{UV} f_{H_2O_2}(1 - e^{-A}) - k_{10}[HO\bullet][H_2O_2] - k_{11}[HO\bullet][HCO_3^-] - k_{12}[HO\bullet][R] - k_{13}[HO\bullet][NOM] \quad (12\text{-}48)$$

where
r_{HO} = rate of hydroxyl radical formation, mol/L·s
$\phi_{H_2O_2}$ = quantum yield of hydrogen peroxide, mol/einstein
P_{UV} = photonic intensity per unit volume, einsteins/cm^3·s
$f_{H_2O_2}$ = fraction of light absorbed by hydrogen peroxide, dimensionless
A = absorbance, dimensionless
k_{10} = second-order rate constant between hydroxyl radical and hydrogen peroxide, L/mol·s ($M^{-1}\ s^{-1}$)
k_{11} = second-order rate constant between hydroxyl radical and carbonate, L/mol·s ($M^{-1}\ s^{-1}$)
k_{12} = second-order rate constant between hydroxyl radical and target organic compound R, L/mol·s ($M^{-1}\ s^{-1}$)
k_{13} = second-order rate constant between hydroxyl radical and NOM, L/mol·s ($M^{-1}\ s^{-1}$)
[HO•] = concentration of hydroxyl radical, mol/L
$[H_2O_2]$ = concentration of hydrogen peroxide, mol/L
$[HCO_3^-]$ = concentration of carbonate, mol/L
[R] = concentration of target compound R, mol/L
[NOM] = concentration of NOM, mol C/L

The photonic intensity per unit volume of reactor, P_{UV}, can be calculated using the the following expression:

$$P_{UV} = \frac{P\eta}{N_{av} V h\nu} \quad (12\text{-}49)$$

where
η = efficiency of the UV lamp, dimensionless
V = volume of reactor solution, L
P = lamp power, W
N_{av} = Avogadro's number, 6.023×10^{23} molecules/mol
h = Planck's constant, 6.62×10^{-34} J·s
ν = frequency of light, s^{-1}

According to the pseudo-steady-state assumption, the change of hydroxyl radical concentration with time is negligible because the rate of reactions involving HO• are very fast and HO• concentration is very small as compared to other compounds. Consequently, the formation rate of the hydroxyl radical can be set equal to zero. By setting the formation rate of HO• equal to zero, the pseudo-steady-state concentration of hydroxyl radical can be determined:

$$[HO\bullet]_{ss} = \frac{2\phi_{H_2O_2} P_{UV} f_{H_2O_2}\,(1 - e^{-A})}{k_{10}[H_2O_2] + k_{11}[HCO_3^-] + k_{12}[R] + k_{13}[NOM]} \tag{12-50}$$

where $[HO\bullet]_{ss}$ = pseudo-steady-state concentration of hydroxyl radical, mol/L

A further simplification of the UV/H_2O_2 process model that can be used to show trends and estimate process feasibility is obtained by assuming that the NOM, R, and H_2O_2 concentrations are constant and equal to their initial concentration, when calculating the pseudo-steady-state concentration of hydroxyl radical. This version of the model is called the Sim-PSS model and the hydroxyl radical concentration becomes

$$[HO\bullet]_{ss,0} = \frac{2\phi_{H_2O_2} P_{UV} f_{H_2O_2}\,(1 - e^{-A})}{k_{10}[H_2O_2]_0 + k_{11}[HCO_3^-]_0 + k_{12}[R]_0 + k_{13}[NOM]_0} \tag{12-51}$$

where $[HO\bullet]_{ss,0}$ = initial pseudo-steady-state concentration of hydroxyl radical, mol/L

$[H_2O_2]_0$ = initial concentration of hydrogen peroxide, mol/L

$[HCO_3^-]_0$ = initial concentration of carbonate, mol/L

$[R]_0$ = initial concentration of target compound R, mol/L

$[NOM]_0$ = initial concentration of NOM, mol/L

Accordingly, the rate law for the disappearance of the target compound and hydrogen peroxide are given by the following expressions:

$$r_R = -k_{12}[R][HO\bullet]_{ss,0} - \phi_R P_{UV} f_R (1 - e^{-A}) \tag{12-52}$$

$$r_{H_2O_2} = -\phi_{H_2O_2} P_{UV} f_{H_2O_2} (1 - e^{-A}) - k_{10}[HO\bullet]_{ss,0}[H_2O_2] \tag{12-53}$$

In many cases, the photolysis rate of the target compound is small, and the second term in Eq. 12-52 can be neglected; and photoreactors are designed so all the light is absorbed. For this situation, Eqs. 12-52 and 12-53 simplify to the following equations.

$$r_R = -k'_{12}[R] \tag{12-54}$$

$$r_{H_2O_2} = -\phi_{H_2O_2} P_{UV} f_{H_2O_2} - k_{10}[HO\bullet]_{ss,0}[H_2O_2] \tag{12-55}$$

where $k'_{12} = k_{12}[HO\bullet]_{ss,0}$ = pseudo-first-order rate constant, s^{-1}

Equation 12-55 may be further simplified by assuming that $f_{H_2O_2}$ does not change with time. Substituing $\phi_{H_2O_2} = 0.5$ and Eq. 12-46 for $f_{H_2O_2}$, the first term in Eq. 12-55 can be written

$$\phi_{H_2O_2} P_{UV} f_{H_2O_2} = \frac{0.5 P_{UV} \varepsilon_{H_2O_2} [H_2O_2]}{\sum \varepsilon_i C_i} \tag{12-56}$$

where $\varepsilon_{H_2O_2}$ = extinction coefficient for hydrogen peroxide, L/(mol·cm)
ε_i = extinction coefficient for component i, L/(mol·cm)
C_i = concentration of component i, mol/L

If the major background chromophores are Fe(II) and NOM and their concentrations are assumed to be constant and equal to their initial concentration, Eq. 12-56 simplifies to the following expression:

$$\phi_{H_2O_2} P_{UV} f_{H_2O_2} = \frac{0.5 P_{UV} \varepsilon_{H_2O_2} [H_2O_2]}{\varepsilon_{H_2O_2} [H_2O_2]_0 + \varepsilon_{NOM} [NOM]_0 + \varepsilon_{Fe(II)} [Fe(II)]_0} \tag{12-57}$$

where $[Fe(II)]_0$ = initial concentration of ferrous ion, mol/L

The key assumption for Eq. 12-57 is that $\varepsilon_{H_2O_2}[H_2O_2]$ is a constant equal to $\varepsilon_{H_2O_2}[H_2O_2]_0$, and this assumption will predict a lower photolysis rate. However, the effluent concentration that is predicted using the Sim-PSS is lower than what will be actually observed because the psuedo-steady-state concentration of the hydroxyl radical is taken to be the initial value in the Sim-PSS model. Accordingly, if the predicted concentration is too high, then the process may be considered infeasible.

The final rate expression for loss of H_2O_2 using the Sim-PSS model is given by these expressions:

$$r_{H_2O_2} = -k'_{10} [H_2O_2] \tag{12-58}$$

$$k'_{10} = \frac{0.5 P_{UV} \varepsilon_{H_2O_2}}{\varepsilon_{H_2O_2} [H_2O_2]_0 + \varepsilon_{NOM} [NOM]_0 + \varepsilon_{Fe(II)} [Fe(II)]_0} + k_{10} [HO\cdot]_{ss,0} \tag{12-59}$$

where k'_{10} = pseudo-first-order rate constant for the destruction of hydrogen peroxide, s^{-1}

Describing Reactor Performance

The steady-state mass balances for a completely mixed flow reactor (CMFR), CMFRs in series, and a plug flow reactor (PFR) for a first-order reaction are provided in Chap. 4. The identical equations may be used for pseudo-first-order reactions. Another model for nonideal mixing, the segregated flow model, is described in Crittenden et al. (2012).

The comparison of pseudo-first-order rate constants of 1,2-dibromo-3-chloropropane (DBCP) degradation from experimental data, the AdOx model, the pseudo-steady-state (PSS) model the Sim-PSS model, and discussions are shown in Crittenden et al. (2012).

Comparison of the Simplified Model to Data and Its Limitations

SELECTION OF HYDROGEN PEROXIDE DOSAGE

An important design issue for the UV/H_2O_2 process is proper selection of the appropriate dose of H_2O_2. The predicted trichloroethene concentration versus time for hydrogen peroxide dosages of 0.1, 0.5, 1.0, and 2.0 mM (3.4, 17.0, 34, 68 mg/L), alkalinity of 100 mg/L as $CaCO_3$, and a pH of 7 using the fully dynamic model (AdOx), and the sim-PSS model is shown on Fig. 12-2. The initial TCE concentration is 100 μg/L. The rate of destruction increases until the hydrogen peroxide concentration increases to 1 mM, and then it decreases slightly because of increased scavenging of the hydroxyl radical by hydrogen peroxide. It appears that the optimum hydrogen peroxide dosage is in the range of 0.5 to 2 mM. The predicted results using the Sim-PSS model were very close to the fully dynamic model, which does not invoke the pseudo-steady state assumption (AdOx; Li et al. 1999); consequently, it could be used to examine the impact of hydrogen peroxide dosage.

ELECTRICAL EFFICIENCY PER ORDER OF TARGET COMPOUND DESTRUCTION

Most photoreactors are designed to absorb all the UV light. For these reactors, the destruction of the target compound will only depend on the total radiant energy that is received by the reactor. EE/O is an effective metric for evaluating the electrical efficiency of the UV/H_2O_2 process. It is defined by this equation:

$$\text{EE/O} = \frac{P}{Q \log\left(C_i/C_f\right)} \tag{12-60}$$

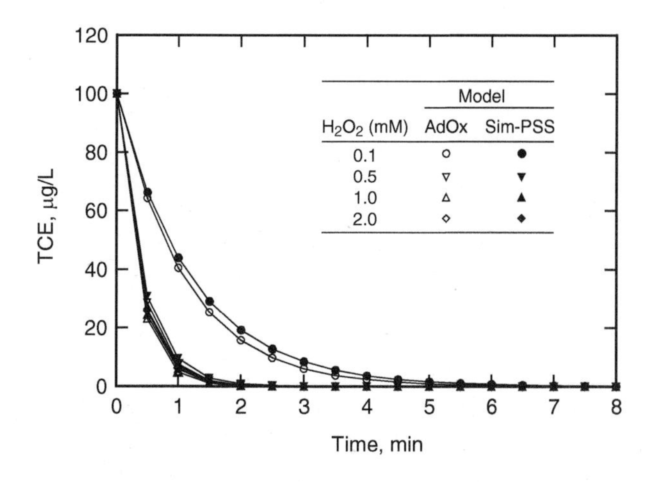

Figure 12-2
Comparison of predicted TCE concentration versus time for hydrogen peroxide dosages of 0.1, 0.5, 1.0, and 2.0 mM, alkalinity of 100 mg/L as $CaCO_3$, a pH of 7 using AdOx, and the simplified psuedo-steady-state model.

where EE/O = electrical energy use per order of target compound destruction and volume of solution treated, kWh/m^3
P = power, kW
Q = flow rate, m^3/h

The EE/O versus hydrogen peroxide concentration is plotted on Fig. 12-3, and the optimum hydrogen peroxide concentration can be determined. Predictions using the Sim-PSS model and AdOx are identical. Accordingly, the optimum hydrogen peroxide dosage is between 0.5 to 2 mM and optimum energy consumption about 0.03 kWh/m^3 of water treated (0.1 kWh/1000 gallons) for an order of magnitude reduction in TCE concentration. This is a low value for energy consumption but the influence of NOM has not been included.

Generally, EE/O values less than 0.265 kWh/m^3 (1.0 kWh/1000 gal) of water treated are considered favorable, but the process has been used in cases where much higher energy consumption is required when there are no other treatment options (Bolton and Carter, 1994). A value of 0.265 kWh/m^3 would correspond to electrical energy costs of $\$0.13/m^3$ of water treated ($0.50/1000 gal) for an order of magnitude reduction in concentration, assuming that electric power costs are 10 per kWh and the lamps have an electrical efficiency of 20 percent. Given the price of hydrogen peroxide versus the cost of electricity, EE/O is the most important design parameter, and the optimum hydrogen peroxide dosage must be selected on the basis of EE/O.

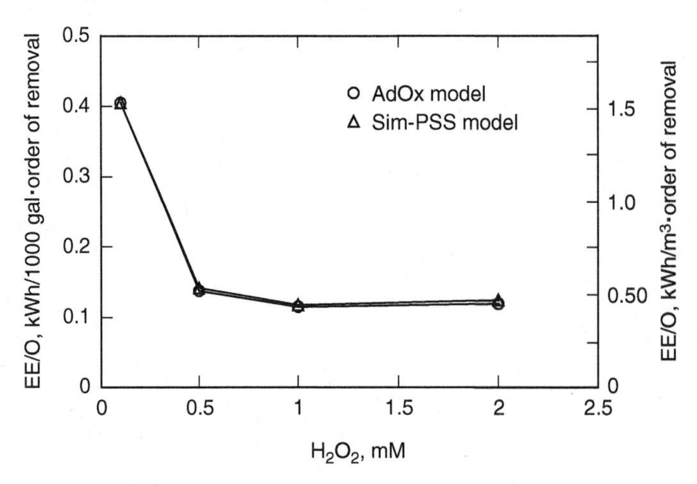

Figure 12-3
Impact of H_2O_2 dosage on EE/O for (TCE) destruction using H_2O_2/UV process (operating conditions: $[TCE]_0 = 100\ \mu g/L$, alkalinity = 100 mg/L $CaCO_3$, [NOM] = 0 mg/L, UV light intensity = 1.04×10^{-6} einstein $L^{-1}\ s^{-1}$ at 254 nm, reactor size = 70 L with 15.8 cm of the effective path length and the total lamp power is 160 W (assuming 20 percent efficiency).

Example 12-6 Lamp power requirement

Calculate the lamp wattage for a flow rate of 0.03 m^3/s (500 gal/min), 1 order of magnitude of destruction, and a EE/O of 0.25 kWh/m^3 (0.95 kWh/1000 gal). The lamp efficiency is 30 percent.

Solution

1. Calculate lamp power output by rearranging Eq. 12-60:

$$P = (EE/O)(Q)\log\left(\frac{C}{C_0}\right)$$

$$= (0.25\ \text{kWh/m}^3)(0.03\ \text{m}^3/\text{s})(3600\ \text{s/h})\log(10) = 27\ \text{kW}$$

2. Calculate the lamp power requirement:

$$\text{Power requirement} = \frac{\text{power output}}{\text{efficiency}} = \frac{27\ \text{kW}}{0.30} = 90\ \text{kW}$$

Comment

High-output low-pressure lamps are more efficient than medium-pressure lamps. High-output lamps require about 400 W, and the medium-pressure lamps can be 15 kW. If 15-kW lamps are used, only 6 such lamps would be required for this example. A reactor that uses 400-W lamps would need about 225 lamps.

Impact of NOM and Compound Type on Target Compound Destruction

Another important factor in the H_2O_2/UV AOP process is the reactivity of the compounds. Compounds with double bonds tend to react more quickly than saturated compounds because saturated compounds must undergo hydrogen abstraction, whereas compounds with double bonds undergo addition reactions. Consequently, more energy and hydrogen peroxide are required to destroy saturated compounds than compounds with double bonds. For instance, the EE/O for TCA, DBCP and TCE are shown on Fig. 12-4. The optimum EE/O for TCE, DBCP, and TCA are on the order of 0.052, 2.4, and 10.2 kW/m^3, respectively. The EE/O for DBCP is lower than TCA because there are more hydrogen atoms on the molecule for attack by hydrogen abstraction. As expected, TCA requires a great deal more radiant energy and hydrogen peroxide than does TCE. Further, the Sim-PSS model can describe most situations at one wavelength and is useful to assess the feasibility of the process. NOM has a significant impact on the UV/H_2O_2 process because it not only scavenges hydroxyl radicals, but also

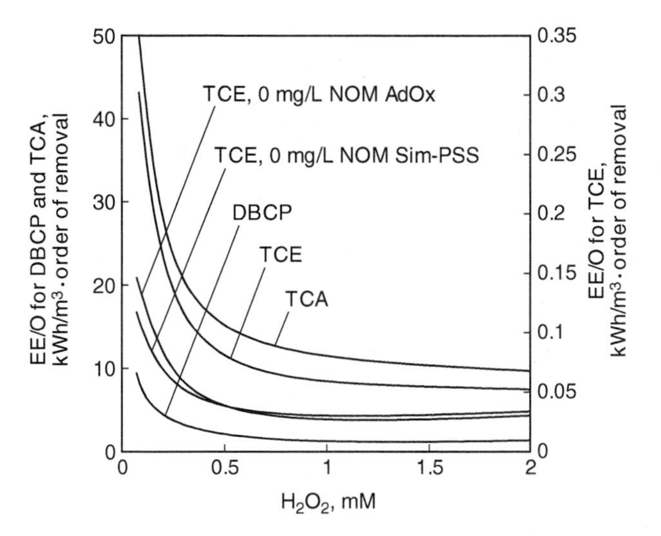

Figure 12-4
Comparison of EE/O values for 1,1,1-trichloroethane (TCA), dibromochloropropane (DBCP), and trichloroethene (TCE) (initial concentrations = 100 μg/L, pH of 7, and alkalinity = 100 mg/L as $CaCO_3$. NOM = 1 mg/L except where noted. Results are both AdOX and Sim-PSS models except where noted).

absorbs UV that may otherwise be absorbed by the hydrogen peroxide to create hydroxyl radicals.

Example 12-7 Using the (Sim-PSS) model to estimate the effluent concentration

The city of Eagle River recently discovered that one of its wells was contaminated with 200 μg/L (1.52 μmol/L) TCE. Calculate the effluent concentration of TCE for an H_2O_2 dosage of 2.5 mM (85 mg/L) and estimate the residual of H_2O_2 in the effluent. The treatment objective for TCE is 5.0 μg/L. During normal pumping of the well field, the flow rate is 0.20 m^3/s (3200 gpm). The pH, HCO_3^-, and DOC concentrations are 6.8, 480 mg/L as CO_3, and 0.7 mg/L, respectively. The following table shows some important physicochemical properties of H_2O_2, TCE, and NOM.

Compound	MW (g/mol)	HO· Rate Constant, k_{OH}, (L/mol·s)	Extinction coefficient, ε, (L/mol·cm)	Quantum Yield, φ (mol/einstein)
Trichloroethylene	131.389	4.20×10^9	Ignored	0
NOM[a]	NA	3.90×10^8	0.0196	0
H_2O_2	34.015	-	19.6	0.5

[a]For NOM, the units for $k_{OH\bullet}$ are L/mol·s based on moles of C in NOM and the units for ε are L/mg·cm based on milligrams of C in NOM.

For simplicity, a proprietary reactor will be used. A dye study on the reactor has shown that four CMFRs in series describes mixing that occurs in the reactor. The reactor size is 1 m in diameter by 3 m in height and the volume is approximately 2300 L with twelve15-kW medium-pressure lamps. To simplify the calculations, it can be assumed that the UV light intensity is monochromatic at 254 nm and that the lamps are 20 percent efficient. Assume that all the UV light is absorbed and $[HO_2^-]$ and $[CO_3^{2-}]$ can be neglected at pH 6.8.

Solution

1. Calculate the hydraulic detention time (τ):

$$\tau = \frac{V}{Q} = \left[\frac{2300\text{ L}}{(0.20\text{ m}^3/\text{s})(1000\text{ L/m}^3)(60\text{ s/min})} \right] = 0.19\text{ min}$$

2. Calculate the fraction of light absorbed by H_2O_2 according to Eq. 12-47: To simplify the calculation, it will be assumed that all the light is absorbed by the water matrix, and the walls of the vessel absorb no light when that is reflected off the walls.

$$f_{H_2O_2} = \frac{\varepsilon_{H_2O_2} C_{H_2O_2}}{\varepsilon_{H_2O_2} \times C_{H_2O_2} + \varepsilon_{NOM} \times C_{NOM}}$$

$$= \frac{(19.6\text{ L/mol·cm})(2.5 \times 10^{-3}\text{ mol/L})}{(19.6\text{ L/mol·cm})(2.5 \times 10^{-3}\text{ mol/L}) + (0.0196\text{ L/mg·cm})\,0.7\text{ mg/L}} = 0.78$$

3. Determine the UV light intensity using Eq. 12-49:
   a. Calculate the frequency of light:

$$\nu = \frac{c}{\lambda} = \frac{(3 \times 10^8\text{ m/s})(10^9\text{ nm/m})}{254\text{ nm}} = 1.18 \times 10^{15}\text{ s}^{-1}$$

   b. Calculate UV intensity: Assume 20 percent efficiency and 12 lamps turned on. The UV intensity can be calculated from Eq.12-49:

$$P_{uv} = \frac{\left[\begin{array}{l} (180\text{ kW})(1000\text{ W/kW})[(1\text{ J/s})/\text{W}] \\ \times (0.2)(1\text{einstein/ mol}) \end{array} \right]}{\left[\begin{array}{l} (6.023 \times 10^{23}\text{ photons/ mol})(2300\text{ L}) \\ \times (6.62 \times 10^{-34}\text{ J·s})(1.18 \times 10^{15}\text{ 1/s}) \end{array} \right]}$$

$$= 3.3 \times 10^{-5}\text{ einsteins/}L\text{·s}$$

4. Calculate the effluent concentration of TCE:
   a. Convert the concentration of each component from mg/L to mol/L:

$$[HCO_3^-]_0 = \frac{480\ mg/L}{(60\ g/mol)\ (1000\ mg/g)} = 0.008\ mol/L$$

$$[NOM]_0 = \frac{0.7\ mg/L}{(12\ g/mol\ of\ carbon)\ (1000\ mg/g)}$$

$$= 5.83 \times 10^{-5}\ mol/l\ of\ carbon$$

   b. Obtain k from Table 12-4 and the problem statement.
   c. Determine values of the product of rate constant and concentration:

$$k_{10}[H_2O_2]_0 = \left(2.7 \times 10^7\ L/mol{\cdot}s\right)\left(2.5 \times 10^{-3}\ mol/L\right)$$

$$= 67500\ s^{-1}$$

$$k_{11}[HCO_3^-]_0 = \left(8.5 \times 10^6\ L/\ mol{\cdot}s\right)(0.008\ mol/L)$$

$$= 68000\ s^{-1}$$

$$k_{12}[R]_0 = k_{12}[TCE]_0 = \left(4.2 \times 10^9\ L/\ mol{\cdot}s\right)$$

$$\times \left(1.52 \times 10^{-6}\ mol/L\right) = 6384\ s^{-1}$$

$$k_{13}[NOM]_0 = \left(3.9 \times 10^8\ L/mol{\cdot}s\right)\left(5.83 \times 10^{-5}\ mol/L\right)$$

$$= 22737\ s^{-1}$$

   d. Calculate $[HO]_{ss,0}$ using Eq. 12-51. Assuming that all the light was absorbed and $[HO_2^-]$ and $[CO_3^{2-}]$ can be neglected at pH 6.8, the psuedo-steady-state concentration of the hydroxyl radical is given by the following expression:

$$[HO\bullet]_{ss,0} = \frac{2\phi_{H_2O_2} P_{UV} f_{H_2O_2}\,(1 - e^{-A})}{k_{10}[H_2O_2]_0 + k_{11}[HCO_3^-] + k_{12}[TCE]_0 + k_{13}[NOM]_0}$$

$$= \frac{2\,(0.5\ mol/einstein)\left(3.3 \times 10^{-5}\ einstein/Ls\right) \times 0.78}{(67{,}500 + 68{,}000 + 6384 + 22737)\ 1/s}$$

$$= 1.58 \times 10^{-10}\ mol/L$$

e. Calculate pseudo-first-order rate constant for TCE:

$$k'_{12} = k_{12}[\text{HO}\cdot]_{ss,0} = \left(4.2 \times 10^9 \text{ L/ mol·s}\right)\left(1.58 \times 10^{-10} \text{ mol/L}\right)$$

$$= 0.66 \text{ s}^{-1}$$

f. Calculate TCE effluent concentration using the tanks-in-series model (see Sec. 4-11):

$$[\text{TCE}] = \frac{[\text{TCE}]_0}{(1 + k'_{12}\tau/n)^n}$$

$$= \frac{200 \ \mu\text{g/L}}{[1 + (0.661/\text{s})(0.19 \text{ min})(60 \text{ s/min})/4]^4}$$

$$= 2.9 \ \mu\text{g/L}$$

5. Calculate the residual hydrogen peroxide concentration:

a. Estimate pseudo-first order rate constant for hydrogen peroxide assuming that NOM is the major background chromophore:

$$k'_{10} = \frac{0.5P_{\text{u-v}}\varepsilon_{\text{H}_2\text{O}_2}}{\varepsilon_{\text{H}_2\text{O}_2}[\text{H}_2\text{O}_2]_0 + \varepsilon_{\text{NOM}}[\text{NOM}]_0} + k_{10}[\text{HO}\cdot]_{ss,0}$$

i. Determine $0.5P_{uv}\varepsilon_{H_2O_2}$, $\varepsilon_{H_2O_2}[H_2O_2]_0$, $\varepsilon_{NOM}[NOM]_0$, and $k_{10}[HO\cdot]_{ss,0}$:

$$0.5P_{\text{UV}}\varepsilon_{\text{H}_2\text{O}_2} = (0.5 \text{ mol/einstein})(19.6 \text{ L/ mol·s})$$

$$\times \left(3.3 \times 10^{-5} \text{ einstein/L·s}\right) = 3.23 \times 10^{-4} \text{ s}^{-2}$$

$$\varepsilon_{\text{H}_2\text{O}_2}[\text{H}_2\text{O}_2]_0 = (19.6 \text{ L/ mol·s})\left(2.5 \times 10^{-3} \text{ mol/L}\right)$$

$$= 0.049 \text{ s}^{-1}$$

$$\varepsilon_{\text{NOM}}[\text{NOM}]_0 = (0.7 \text{ mg/L})(0.0196 \text{ L/ mg·s}) = 0.01372 \text{ s}^{-1}$$

$$k_{10}[\text{HO}\cdot]_{ss,0} = \left(2.7 \times 10^7 \text{ L/mol·s}\right)\left(1.58 \times 10^{-10} \text{ mol/L}\right)$$

$$= 0.004266 \text{ s}^{-1}$$

ii. Determine k'_{10}:

$$k'_{10} = \frac{3.23 \times 10^{-4} \text{ s}^{-2}}{(0.049 + 0.01372) \text{ s}^{-1}} + \left(0.004266 \text{ s}^{-1}\right)$$

$$= 9.42 \times 10^{-3} \text{ s}^{-1}$$

b. Estimate H_2O_2 residual using the tanks-in-series model:

$$[H_2O_2] = \frac{[H_2O_2]_0}{(1 + k'_{10}\tau/n)^n}$$

$$= \frac{2.5 \times 10^{-3}\ \text{mol/L}}{[1 + (9.42 \times 10^{-3}\ 1/\text{s})\ (0.19\ \text{min})\ (60\ \text{s/min})/4]^4}$$

$$= 2.25 \times 10^{-3}\ \text{mol/L}\ (76.5\ \text{mg/L})$$

Comment

While the treatment objective for TCE can be met, the residual hydrogen peroxide concentration is too high to use the process for water treatment. The residual hydrogen peroxide concentration is the lowest possible value because the psuedo-state-state concentration of the hydroxyl radical is taken to be the initial value in the Sim-PSS model. However, this approach is still useful because it can be used to calculate the lowest expected residual hydrogen peroxide concentration, and, if the residual is unacceptable, then the process is not a viable option. The effluent concentration of hydrogen peroxide predicted by the rigorous AOP model is 2.39×10^{-3}mol/L (81.3 mg/L).

12-5 Energy and Sustainability Considerations

Advanced oxidation processes are able to degrade contaminants in a short amount of time; consequently most AOPs use reactors with relatively small hydraulic retention times (several minutes or less). As a result, the construction of the reactor has a relatively small environmental impact compared with the operation of the process. The energy consumption associated with the operation of the process depends on the energy required to produce the oxidants and the dose of the oxidants. Modern ozone generation equipment using liquid oxygen as the feed gas can generate ozone for about 6–10 kWh per kilogram of ozone. Hydrogen peroxide can be produced for about 2 to 4 kWh per kilogram of H_2O_2. When ozone is used as an AOP, the doses range from about 4 to 8 mg/L, which corresponds to specific energy consumption of 0.04 to 0.08 kWh/m^3. The H_2O_2/O_3 process uses similar ozone doses and, based on the optimal mass ratio, hydrogen peroxide doses between 1.4 and 2.8 mg/L, which adds an additional 0.003 to 0.01 kWh/m^3 of energy consumption. Thus, the specific energy consumption of the H_2O_2/O_3 process ranges from 0.043 to 0.09 kWh/m^3.

The relationship between UV dose and energy consumption depends on the design and hydraulic characteristics of the reactor and cannot be predicted easily. Data in the literature of equipment manufacturers indicates that the energy required to produce a UV dose of 40 mJ/cm^2 ranges from 0.003 to 0.025 kWh/m^3. UV doses for advanced oxidation are significantly higher than the doses used for disinfection and can range from 100 to 1000 mJ/cm^2. The specific energy consumption can thus vary significantly from 0.0075 to 0.63 kWh/m^3. The H_2O_2 dose used with the UV/H_2O_2 is often from 3 to 5 mg/L but can be up to 100 mg/L in some applications, such as remediation, and thus adds an additional 0.004 to 0.28 kWh/m^3 to the energy requirements of the UV/H_2O_2 process, resulting in an overall process requirement ranging from about 0.01 to 0.9 kWh/m^3. A comparison of the energy requirements of the ozone versus UV-based advanced oxidation processes indicates that the UV process may consume as much as 10 times as much energy. Thus, the H_2O_2/O_3 process appears to be a more energy-efficient method for producing hydroxyl radicals than the UV/H_2O_2 process.

Advantages of advanced oxidation processes, compared to other processes for removing synthetic organic chemicals, is that they allow full recovery of the water, do not transfer the contaminants to a separate phase, and do not produce a waste stream. Air stripping transfers contaminants to the air, reverse osmosis transfers the contaminants to the concentrate, which must be disposed of, and adsorption with activated carbon transfers the contaminants to the surface of a solid, which may need to be disposed of once it reaches exhaustion. With high enough doses, hydroxyl radicals that are produced in AOPs are capable of mineralizing organic contaminants. Although complete mineralization frequently does not occur, AOPs can be designed with the subsequent biological treatment process that treats by-products from AOPs. The biological process is typically a filter operated with gravity flow that consequently has low energy consumption. As a result, AOPs can have advantages over other processes for removing synthetic organic chemicals.

12-6 Summary and Study Guide

After suding this chapter, you should be able to:

1. Define the following terms and phrases and describe the significance of each in the context of advanced oxidation processes:

advanced oxidation process	quantum yield
EE/O	quenching rate
hydroxyl radical	radical
oxidation–reduction reaction	

2. Explain the key differences between and advantages of advanced oxidation processes over conventional oxidation processes.
3. Identify the advanced oxidation processes that are commercially available for full-scale water treatment plants.

4. Calculate the half-life for oxidation of a compound at a given HO· concentration.
5. Apply the kinetics of AOP processes to various types of reactors to determine the hydraulic residence time needed to achieve a given effluent concentration.
6. Describe the major factors that affect AOP performance.
7. Calculate the quenching rate due to competing reactions, and the increase in hydraulic detention time of a reactor or the increase in HO· concentration required to achieve the level of oxidation that would occur without competing reactions.
8. Describe the alternate pathways that can cause the oxidation of a compound when ozone is used as an AOP and the conditions that result in one of the pathways being the most significant.
9. Design the size of a reactor for destruction of a target compound using an AOP process.
10. Describe the advantages and disadvantages of the advanced oxidation processes presented in this chapter.

Homework Problems

12-1 Calculate the half-life of the oxidation of the given compound by hydroxyl radicals using the rate constants in Table 12-1 (compounds to be selected by instructor).

a. Methyl ethyl ketone; $[HO\bullet] = 10^{-12}$ mol/L

b. Benzene; $[HO\bullet] = 5 \times 10^{-10}$ mol/L

c. Tetrachloroethylene; $[HO\bullet] = 3.5 \times 10^{-10}$ mol/L

d. 1,1,1-Trichloroethane; $[HO\bullet] = 9.2 \times 10^{-10}$ mol/L

e. Atrazine; $[HO\bullet] = 10^{-9}$ mol/L

12-2 For the target compound in Problem 12-1, calculate the quenching rate due to competing reactions listed below (compounds to be selected by instructor).

a. NOM = 3 mg/L as C

b. HCO_3^- = 114 mg/L and CO_3^{2-} = 1.05 mg/L (corresponding to alkalinity 95 mg/L as $CaCO_3$ at pH = 8.3)

c. HCO_3^- = 113 mg/L and CO_3^{2-} = 16.4 mg/L (corresponding to alkalinity = 120 mg/L as $CaCO_3$ at pH = 9.5)

d. Fe(II) = 1.5 mg/L, Mn(II) = 0.78 mg/L

e. Fe(II) = 1.3 mg/L, Mn(II) = 0.90 mg/L, NOM = 2.7 mg/L as C

12-3 Calculate the hydraulic retention time of a plug flow reactor required to achieve 90 percent removal (by oxidation with hydroxyl radicals) of the compound in Problem 12-1 without and with the competing reactions in Problem 12-2. Is the process feasible without competing reactions? Is it feasible with competing reactions?

12-4 Determine the fraction of the reaction that is carried out by the indirect reaction with HO• versus the direct reaction with O_3 for the oxidation of geosmin and MIB. For the calculation, use $C_{[HO\bullet]/[O_3]}$ values of 10^{-7}, 10^{-8}, and 10^{-9} and a rate constant for the direct reaction with ozone of 10 L/mol·s.

12-5 Calculate the time required for 99 percent destruction of MIB using ozone as an AOP in a batch reactor. For the calculation, use $C_{[HO\bullet]/[O_3]}$ ranging from 10^{-9} to 10^{-7} and an initial ozone concentration of 3 mg/L. The rate constant for the direct reaction with ozone is 10 L/mol·s and the ozone decay rate constant is $0.1\ min^{-1}$.

12-6 A municipality recently discovered that one of its wells was contaminated with the compounds listed in the following:

Compound	Influent Concentration, C_0, μg/L	Objective, Treatment C_{TO}, μg/L
Trichloroethylene (TCE)	130	5.0
Tetrachloroethylene (PCE)	75	5.0
Vinyl chloride	15	2.0
Benzene	80	5.0

To continue using the well as a drinking water resource, the compounds shown in the above table need to be removed to meet the treatment objectives shown in the table. During normal pumping operations, the well produces about 2.18 ML/d, and further expansion of the well field may be considered depending on the efficacy of the ozone/hydrogen peroxide process. The pH, and NOM concentrations are 7.5, 400 mg/L as $CaCO_3$, and 1.2 mg/L as C, respectively (at this pH and alkalinity, HCO_3^- = 487 mg/L and CO_3^{2-} = 0.71 mg/L). Important physicochemical properties for the compounds that need to be removed are as follows:

Compound	MW, g/mol	HO• Rate Constant, $k_{HO\bullet}$, L/mol·s
Trichloroethylene	131.4	4.20×10^9
Tetrachloroethylene	165.8	2.60×10^9
Vinyl chloride	62.5	1.20×10^{10}
Benzene[a]	78.1	7.80×10^9
NOM[b]	NA	17,666

[a]Molar extinction coefficient is high but quantum yield is very low; consequently, photolysis can be ignored.
[b]For NOM, the unit of $k_{HO\bullet}$ is L/mg·s.

For simplicity, a proprietary reactor will be used. Based on dye studies, it has been found that the reactor can be modeled as four completely mixed reactors in series. The reactor is 1 m in diameter and 3 m in height, and the volume is approximately 2300 L. For the given conditions, determine the optimum H_2O_2 dosage to achieve the treatment objectives based on the simplified model (Sim-PSS). Consider ozone dosages of 1, 3, and 5 mg/L.

References

Behar, D., Czapski, G., and Duchovny, I. (1970) "Carbonate Radical in Flash Photolysis and Pulse Radiolysis of Aqueous Carbonate Solutions," *J. Phys. Chem.*, **74**, 2206–2210.

Bolton, J. R., and Carter, S. R. (1994) Homogeneous Photodegradation of Pollutants in Contaminated Water: An Introduction, Chap. 33, in G. R. Helz, R. G. Zepp, and D. G. Crosby (eds.), *Aquatic and Surface Photochemistry*, CRC Press, Boca Raton, FL.

Bruce, D., Westerhoff, P., and Brawley-Chesworth, A. (2002) "Removal of 2-Methylisoborneol and Geosmin in Surface Water Treatment Plant in Arizona," *J. Water Supply: Res. Technol.—Aqua*, **51**, 4, 183–197.

Buhler, R. F., Staehelin, J., and Hoigné, J. (1984) "Ozonation Decomposition in Water Studied by Radiolysis," *J. Phys. Chem.*, **8**, 12, 2560–2564.

Buxton, G. V., and Greenstock, C. L. (1988) "Critical Review of Rate Constants for Reactions of Hydrated Electrons, Hydrogen Atoms and Hydroxyl Radicals (•OH/H•) in Aqueous Solution," *J. Phys. Chem. Ref. Data*, **17**, 2, 513–586.

Christensen, H. S., Sehested, K., and Corftizan, H. (1982) "Reaction of Hydroxyl Radicals with Hydrogen Peroxide at Ambient Temperatures," *J. Phys. Chem.*, **86**, 15–88.

Crittenden, J., Hu, S., Hand, D., and Green, S., (1999) "A Kinetic Model for H_2O_2/UV Process in a Completely Mixed Batch Reactor," *Water Res.*, **33**, 10, 2315–2328.

Crittenden. J. C., Trussell, R. R., Hand, D. W., Howe, K. J., and Tchobanoglous, G. (2012) *MWH's Water Treatment: Principles and Design,* 3rd ed., Wiley, Hoboken, NJ.

Elovitz, M. S., and von Gunten, U. (1999) "Hydroxyl Radical/Ozone Ratios during Ozonation Processes," *Ozone: Sci. Eng.*, **21**, 3, 239–260.

Glaze, W. H., and Kang, J.-W. (1989) "Advanced Oxidation Processes. Description of a Kinetic Model for the Oxidation of Hazardous Materials in Aqueous Media with Ozone and Hydrogen Peroxide in a Semibatch Reactor," *Ind. Eng. Chem. Res.*, **28**, 11, 1573–1580.

Li, K., Crittenden, J. C., Hand, D. W., and Hokanson, D. R. (1999) Advanced Oxidation Process Simulation Software (AdOxTM) Version 1.0, Michigan Technological University, Houghton, MI.

Staehelin, J., and Hoigné, J. (1982) "Decomposition of Ozone in Water: Rate of Initiation by Hydroxide Ions and Hydrogen Peroxide," *Environ. Sci. Technol.*, **16**, 10, 676–681.

Stefan, M. I., and Bolton, J. R. (1998) "Mechanism of the Degradation of 1,4-Dioxane in Dilute Aqueous Solution Using the UV/Hydrogen Peroxide Process," *Environ. Sci. Technol.*, **32**, 1588–1595.

Stefan, M. I., and Bolton, J. R. (1999) "Reinvestigation of the Acetone Degradation Mechanism in Dilute Aqueous Solution by the UV/H_2O_2 Process," *Environ. Sci. Technol.*, **33**, 870–873.

Stefan, M. I., and Bolton, J. R. (2002) Personal communication.

Stefan, M. I., Mack, J., and Bolton, J. R. (2000) "Degradation Pathways During the Treatment of Methyl *tert*-Butyl Ether by the UV/H_2O_2 Process," *Environ. Sci. Technol.*, **34**, 650–658.

Stumm, W., and Morgan, J. J. (1996) *Aquatic Chemistry*, 3rd ed., Wiley-Interscience, New York.

Volman, D. H., and Chen, J. C. (1959) The Photochemical Decomposition of Hydrogen Peroxide in Aqueous Solutions of Allyl Alcohol at 2537A. *JACS* **81** (16), 4141–4144.

Westerhoff, P., Mazyk, S. P., Cooper, W. J., and Minakata, D. (2007) "Electron Pulse Radiolysis Determination of Hydroxyl Radical Rate Constants with Suwannee River Fulvic Acid and Other Dissolved Organic Matter Isolates," *Environ. Sci. Technol.* **41**, 4640–4646.

Zhang, Y., Crittenden, J. C., Hand, D. W., and Perram, D. L. (1994) "Fixed-Bed Photocatalysts for Solar Decontamination of Water," *Environ. Sci. Technol.*, **28**, 3, 435–442.

13 Disinfection

Disinfection is an essential element of the overall strategy for providing water that is safe to drink. Providing water free from pathogenic organisms is accomplished using several complementary strategies: (1) selecting a water source that is free from microbiological contamination, such as groundwater, (2) protecting surface water sources to minimize microbiological contamination, (3) treating water to remove microorganisms or eliminate their pathogenicity, and (4) preventing recontamination of water as it is delivered to customers through the distribution system. Disinfection is an element of the last two actions. Treatment can include removing microorganisms, primarily through filtration as discussed in Chaps. 7 and 8, or inactivating them. Inactivation is the process in which microorganisms are transformed so that they are unable to cause disease (inactivation can include eliminating the ability of microorganisms to reproduce in a host organism; while they are not necessarily dead, they still are unable to cause disease). Inactivation is sometimes called primary disinfection and occurs at a treatment facility. Disinfection also includes residual maintenance, which is sometimes called secondary disinfection and occurs in the distribution system.

Most groundwater is free from pathogenic organisms. If wells are constructed properly and are not influenced by surface water, groundwater can be pumped into the distribution system without primary disinfection. Water from private wells for individual households in rural areas is commonly consumed without disinfection. Surface waters, however, virtually always contain pathogenic organisms and must be disinfected. Many different types of microorganisms can be present in surface water, but for purposes of disinfection they can be grouped in broad classes that include viruses, bacteria, and protozoa. Regulations in the United States are based on removal and inactivation of organisms that are considered particularly challenging with the reasoning that if disinfection is able to inactivate the difficult organisms successfully, easier organisms will also be inactivated. The current target organisms in U.S. drinking water regulations are viruses and the protozoa *Giardia lamblia* and *Cryptosporidium parvum.*

The fundamentals of disinfection are introduced in this chapter. The basic features of disinfection systems are introduced in Sec. 13-1. The capabilities, chemistry, production, and use of each of the primary disinfectants used in water treatment are addressed in Secs. 13-2 through 13-5. The concepts of disinfection kinetics are introduced in Sec 13-6 and extended to the discussion of the kinetics of nonideal reactors in Sec. 13-7. The design of disinfection contactors with low dispersion is considered in Sec. 13-8. The chapter ends with a presentation of material on disinfection by-products, Sec. 13-9; residual maintenance, Sec. 13-10; and energy and sustainability considerations, Sec. 13-11.

13-1 Disinfection Agents and Systems

The disinfection process involves the use of a disinfecting agent and some means of contacting the disinfecting agent with the water to be treated. The commonly used disinfecting agents and the design of disinfection systems are introduced and described in this section.

Disinfecting Agents

Five disinfection agents are commonly used in drinking water treatment today: (1) free chlorine, (2) combined chlorine (chlorine combined with ammonia, also known as chloramines), (3) chlorine dioxide, (4) ozone,

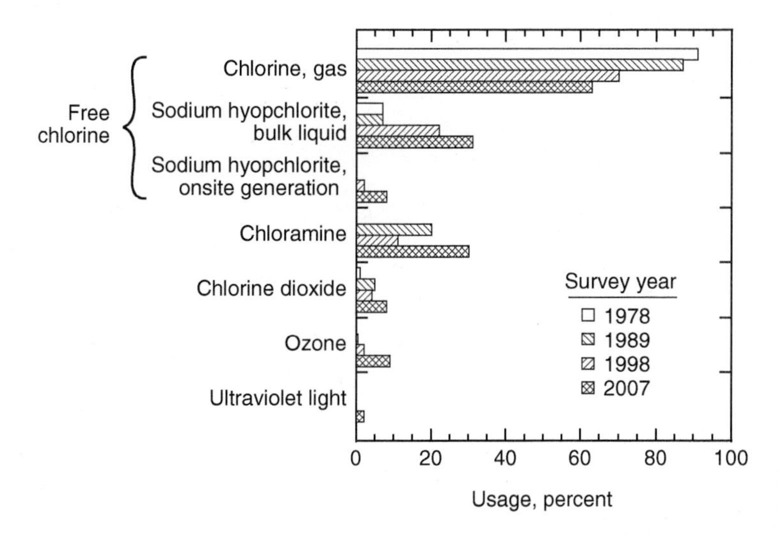

Figure 13-1
Chemicals used for disinfection.

and (5) ultraviolet (uv) light. The first four are chemical oxidants, whereas UV light involves the use of electromagnetic radiation. Of the five, the most common in the United States is the use of free chlorine. As shown on Fig. 13-1, in 1978, 91 percent of utilities used chlorine gas to apply free chlorine to the water and 7 percent used sodium hypochlorite (i.e., bleach). By 2007, however, only 63 percent of utilities were using chlorine gas and nearly 40 percent were using either bulk liquid or onsite generation of sodium hypochlorite. The transition from chlorine gas to hypochlorite is primarily because of safety and security reasons because chlorine gas is highly toxic.

The number of utilities using chloramines (free chorine combined with ammonia) for disinfection has increased to 30 percent by 2007. Its use, however, is often limited to residual maintenance and typically a different disinfectant is used for primary disinfection when chloramine is used.

The use of ozone, the strongest of the four oxidants, use has increased from less than 1 percent of utilities in 1989 to 9 percent in 2007. The increasing use is in part because of its stronger disinfecting properties and in part because it controls taste and odor compounds, specifically geosmin and methyl isoborneol.

Ultraviolet light is not used frequently for disinfecting in drinking water applications, with only 2 percent of utilities reporting to use it in 2007. Its use will continue to increase in the future because of its lack of by-product generation and its effectiveness against protozoa. Information on each of these common disinfectants is summarized in Table 13-1.

Disinfectant System Design

Designing a disinfection system includes three primary activities: (1) selecting a suitable disinfectant and dose, (2) designing a system to inject or introduce the disinfectant into the water, and (3) designing contactors that

Table 13-1
Characteristics of five most common disinfectants

	Disinfectant				
Issue	**Free chlorine**	**Combined chlorine**	**Chlorine dioxide**	**Ozone**	**Ultraviolet light**
Effectiveness for disinfecting:					
Bacteria	Excellent	Good	Excellent	Excellent	Good
Viruses	Excellent	Fair	Excellent	Excellent	Fair
Protozoa	Fair to poor	Poor	Good	Good	Excellent
Endospores	Good to poor	Poor	Fair	Excellent	Fair
Regulatory limit on residuals	4 mg/L	4 mg/L	0.8 mg/L	—	—
Formation of chemical by-products					
Regulated by-products	Forms 4 THMs[a] and 5 HAAs[b]	Traces of THMs and HAAs	Chlorite	Bromate	None
By-products that may be regulated in future	Several	Cyanogen halides, NDMA	Chlorate	Biodegradable organic carbon	None known

Typical application dose, mg/L (kg/ML)	1–6	2–6	0.2–1.5	1–5	20–100 mJ/cm^2
Typical application dose, lb/MG	8–50	17–50	2–13	8–42	—
Chemical source	Delivered: as liquid gas in tank cars, 1 tonne and 68-kg (150-lb) cylinders, or as liquid bleach. Onsite generation from salt and water using electrolysis. Calcium hypochlorite powder is used for very small applications.	Same sources for chlorine. Ammonia is delivered as aqua ammonia solution, liquid gas in cylinders, or solid ammonium sulfate. Chlorine and ammonia are mixed in treatment process.	ClO_2 is manufactured with an onsite generator from chlorine and chlorite. Same sources for chlorine. Chlorite as powder or stabilized liquid solution.	Manufactured onsite by passing pure oxygen or dry air through an electric field. Oxygen is usually delivered as a liquid. Oxygen can also be manufactured onsite.	Uses low-pressure or low-pressure, high-intensity UV (254-nm) or medium-pressure UV (several wavelengths) lamps in the contactor itself.

[a]THMs = trihalomethanes.
[b]HAAs = haloacetic acids.

provide a sufficient amount of time for the disinfectant reactions to take place. In many cases, disinfectant injection and contact take place within a single system.

DISINFECTANT DOSE

Disinfectant doses depend on whether the disinfectant is being used for inactivation, residual maintenance, or both. When chemical disinfectants are added to water, some of the chemical will be consumed during rapid oxidation of reduced compounds in the water; this consumption is known as the initial demand. Once the initial demand has been satisfied, additional chemical addition leads to a residual concentration in the water.

For inactivation, the dose is based (in the United States) on requirements established in a series of regulations starting with the Surface Water Treatment Rule (SWTR) and leading to, most recently, the Long Term 2 Enhanced Surface Water Treatment Rule (LT2ESWTR). These regulations set specific disinfection requirements for specific target organisms based on the source water quality and the treatment process being used. As one example, a conventional filtration treatment plant treating high-quality surface water must achieve 2-log reduction of *C. parvum*, 3-log removal of *G. lamblia*, and 4-log removal of viruses, where reduction includes both removal and inactivation (refer to Sec. 3-2 for definition of log removal and how to calculate it). If the plant meets effluent turbidity requirements, it is awarded credit for physical removal (filtration) of 2 log of *C. parvum*, 2.5 log of *G. lamblia*, and 2 log of viruses, leaving 0.5 log of *Giardia lamblia* and 2 log of viruses that must be eliminated by inactivation. Requirements are different for poorer quality water or different treatment processes and can be found in the regulations on the U.S. EPA website.

Achieving a specific log inactivation of a specific microorganism is accomplished by maintaining a disinfectant residual for a specific amount of time. The residual concentration and the time are of equal importance, so the regulations specify Ct values, the product of concentration and time, where C refers to the disinfectant residual and t refers to the contact time. Tables of Ct values for different log removal values, disinfectants, microorganisms, and water quality conditions are provided in the regulations on the U.S. EPA website. The fundamental basis for basing regulations on the product of concentration and time is made evident when disinfection kinetics is introduced in Sec. 13-6.

For residual maintenance, the necessary dose is based on the requirement to have a specific residual present in the distant edges of the distribution system. Disinfectant concentrations decay over time; thus, the dose for residual maintenance will depend on conditions within the distribution system (temperature, residence time, presence of biofilms, and other water quality parameters) that affect the rate of decay.

DISINFECTANT ADDITION

Equipment for adding disinfectants to the process stream is straightforward. For aqueous solutions of disinfectants, the liquid can be injected directly into the water just upstream of a mixing device, such as a static mixer, or a location with significant turbulence, such as flow over a weir. Ozone is a gas; one common design is to bubble ozone into the water at the bottom of a deep basin. Another design that is becoming more common is to withdraw a portion of the process water into a side stream, which then flows through a venturi. The low pressure in the throat of the venturi is used to aspirate the ozone into the side-stream flow, which is then reinjected into the main process stream.

Ultraviolet light addition and contact is accomplished together in proprietary engineered systems available from UV disinfection manufacturers. The contactors are essentially a short section of pipeline with tubular UV lamps arranged either parallel or perpendicular to the process flow. Water flowing past the lamps is illuminated by the UV light, resulting in rapid disinfection. The hydraulic residence time in a UV reactor is only a few seconds, so the hydraulic characteristics are carefully designed to maximize interaction between the water and light and minimize the opportunity for short circuiting or dispersion.

DISINFECTION CONTACTORS

Throughout much of the twentieth century, chorine was added early in the treatment process, and the chlorine residual was carried throughout the plant to provide sufficient contact time; the design of specialized disinfectant contactors was not of particular concern. In 1979, however, a regulation was passed that placed a maximum contaminant level (MCL) on trihalomethanes, a class of disinfection by-products formed during the reaction of chlorine with natural organic matter (NOM). Because of this regulation, many utilities moved the point of disinfectant addition to after the filters, after as much NOM as possible had been removed from the water.

Moving the chlorine application point necessitated the design of engineered disinfectant contactors because, as was demonstrated in Sec. 4-9, reactor hydraulics influence the extent to which reactions occur. Reactor hydraulics are particularly important for disinfection. As a result, regulations do not use the hydraulic residence time, τ, as the value for t in the Ct product, but instead use t_{10}, the residence time at which 10 percent of a conservative tracer would exit a nonideal reactor (see Sec. 4-11). Maximizing disinfectant contact time for a reactor of a given size (i.e., maximizing t_{10} with respect to τ) requires designing contactors with low dispersion (see Sec. 13-8). Engineered disinfectant contactors are typically of three types: (1) pipelines, (2) serpentine basins, and (3) over–under baffled contactors. Chlorine, combined chlorine, and chlorine dioxide contactors are typically pipelines or serpentine basins; ozone contactors can be any of the three.

13-2 Disinfection with Free and Combined Chlorine

Until approximately World War II, free and combined chlorine (chlorine combined with ammonia, also known as chloramines) were both commonly used and viewed as effective disinfectants. In 1943, the U.S. Public Health Service (PHS) demonstrated that free chlorine exhibits more rapid kinetics in the disinfection of several bacteria (Wattie and Butterfield, 1944). As a result, the use of combined chlorine declined between 1943 and the mid-1970s. In the mid-1970s, it became widely recognized that free chlorine formed chemical by-products and that combined chlorine did so to a much lesser degree. Since that realization, the use of combined chlorine has increased, particularly the addition of ammonia to convert a free-chlorine residual to a combined chlorine residual once primary disinfection has been accomplished. By 2007, about 30 percent of the utilities in the United States were using combined chlorine (see Fig. 13-1).

Chemistry of Free Chlorine

Free-chlorine disinfection can be accomplished using either chlorine gas or sodium hypochlorite. When chlorine gas is injected into water, it dissolves into the water and then rapidly reacts with the water to form hypochlorous acid and hydrochloric acid:

$$Cl_2(g) + H_2O \rightarrow HOCl + HCl \tag{13-1}$$

Similarly, when sodium hypochlorite is added to water, it reacts rapidly to form hypochlorous acid and sodium hydroxide:

$$NaOCl + H_2O \rightarrow HOCl + Na^+ + OH^- \tag{13-2}$$

The species that contributes the greatest disinfecting power is the hypochlorous acid (HOCl), so chlorine gas and sodium hypochlorite have the exact same disinfecting capabilities on a molar basis. Hypochlorous acid is a weak acid (see Sec. 4-4) that dissociates to form hypochlorite ion (OCl^-). The extent of dissociation depends on pH:

$$HOCl \rightleftarrows H^+ + OCl^- \tag{13-3}$$

$$K_a = \frac{[H^+][OCl^-]}{[HOCl]} \tag{13-4}$$

The pK_a for HOCl is 7.6 at 20°C; thus, HOCl is the predominant form below this pH value, and OCl^- is the predominant form above it (see Fig. 13-2). HOCl and OCl^- both have disinfecting capabilities, but HOCl exhibits faster disinfection kinetics and, therefore, is a stronger disinfectant than OCl^-. Consequently, a pH of 7 or less is desirable where disinfection alone is concerned.

The disinfection reactions are oxidation reactions that convert the chlorine to chloride ion while microorganisms are being inactivated:

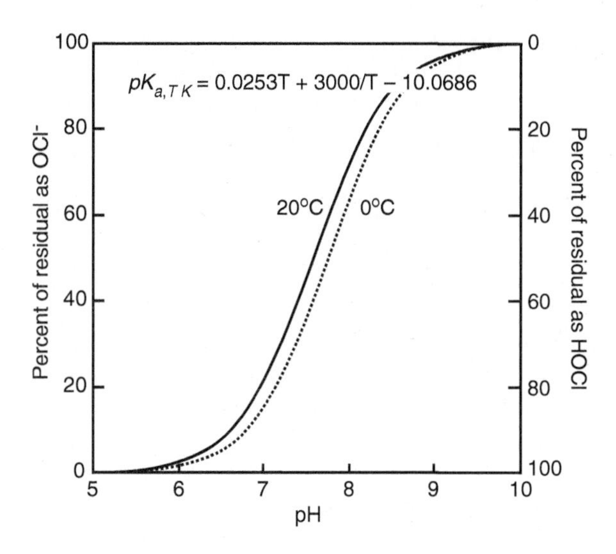

Figure 13-2
Effect of temperature and pH on fraction of free chlorine present as hypochlorous acid. [Adapted from Morris (1966)]

$$HOCl + \begin{pmatrix} \text{pathogenic} \\ \text{microrganisms} \end{pmatrix} + H^+ + 2e^- \rightarrow H_2O + Cl^- + \begin{pmatrix} \text{inactivated} \\ \text{microrganisms} \end{pmatrix} \quad (13\text{-}5)$$

$$OCl^- + \begin{pmatrix} \text{pathogenic} \\ \text{microrganisms} \end{pmatrix} + H^+ + 2e^- \rightarrow OH^- + Cl^- + \begin{pmatrix} \text{inactivated} \\ \text{microrganisms} \end{pmatrix} \quad (13\text{-}6)$$

The other species formed during the addition of chlorine gas and sodium hypochlorite, hydrochloric acid (HCl) and sodium hydroxide (NaOH), are a strong acid and strong base, respectively. They dissociate completely in water; in doing so, hydrochloric acid causes a reduction in alkalinity and pH and sodium hydroxide causes an increase in alkalinity and pH:

$$HCl \rightarrow H^+ + Cl^- \quad (13\text{-}7)$$

$$NaOH \rightarrow Na^+ + OH^- \quad (13\text{-}8)$$

Thus, while chlorine gas and sodium hypochlorite have identical disinfecting capabilities, their addition to water will have opposite effects with respect to pH and alkalinity. Sodium hypochlorite solutions often contain excess sodium hydroxide, which causes an additional increase in alkalinity and pH.

Chemistry of Combined Chlorine

When ammonia is present in water, chlorine reacts to form species that combine chlorine and ammonia, known as chloramines. As chlorine is added, it reacts successively with ammonia to form the three chloramine species.

$$NH_3 + HOCl \rightarrow NH_2Cl + H_2O \quad \text{(monochloramine)} \quad (13\text{-}9)$$

$$NH_2Cl + HOCl \rightarrow NHCl_2 + H_2O \quad \text{(dichloramine)} \quad (13\text{-}10)$$

$$NHCl_2 + HOCl \rightarrow NCl_3 + H_2O \quad \text{(trichloramine)} \quad (13\text{-}11)$$

The sum of these three reaction products is called combined chlorine. The total chlorine residual is the sum of the combined residual and any free-chlorine residual. A summary of these definitions is provided below:

$$\text{Free chlorine} = \text{HOCl} + \text{OCl}^- \tag{13-12}$$

$$\text{Combined chlorine} = \text{NH}_2\text{Cl} + \text{NHCl}_2 + \text{NCl}_3 \tag{13-13}$$

$$\text{Total chlorine} = \text{free chlorine} + \text{combined chlorine} \tag{13-14}$$

All chlorine species are expressed as milligrams per liter as Cl_2 and the ammonia concentration is expressed as mg/L as nitrogen (i.e., mg/L NH_3-N). When small amounts of chlorine are added to water, the reactions are much like the simple model above. However, as the amount of chlorine added increases, the reactions become more complex. These reactions and their behavior are partially illustrated by the three zones on Fig. 13-3. At first, as depicted in zone A, the total chlorine residual increases by approximately the amount of chlorine added until the mole ratio of chlorine to ammonia approaches 1 (a weight ratio of 5.07 as Cl_2 to NH_3-N), assuming no other species that consume chlorine are present.

Beyond a molar ratio of 1, the addition of more chlorine decreases, rather than increases, the total chlorine residual (zone B) because the chlorine is oxidizing some of the chloramine species. Eventually, essentially all of the chloramine species are oxidized. The point at which the oxidation of chloramine species is complete is called the *breakpoint* and is the beginning of zone C. The exact locations of maximum residual and breakpoint (minimum residual) are influenced by the presence of dissolved organic matter, organic nitrogen, and reduced substances [e.g., S^{2-}, Fe(II), Mn(II)]. The presence of any of these will shift all three zones to the right. The shift in the point of maximum residual depends on how easily they are oxidized. The shift in the breakpoint corresponds to their stoichiometric chlorine demand. After the breakpoint is reached, the free-chlorine residual increases in proportion to the amount of additional chlorine added. Prior to concerns about disinfection by-products, breakpoint chlorination was often used as a simple means of ammonia removal.

In zone A, monochloramine forms rapidly and with little interference. Nevertheless, the species present in zone A are influenced by pH. At low pH values, dichloramine can form via the following reactions:

$$\text{NH}_2\text{Cl} + \text{H}^+ \rightleftarrows \text{NH}_3\text{Cl}^+ \tag{13-15}$$

$$\text{NH}_3\text{Cl}^+ + \text{NH}_2\text{Cl} \rightleftarrows \text{NHCl}_2 + \text{NH}_4^+ \tag{13-16}$$

Monochloramine is the only chloramine present in zone A at pH 8 but significant amounts of dichloramine can be present at pH 6 (Palin, 1975). In zone B, which has more chlorine, some dichloramine will be present

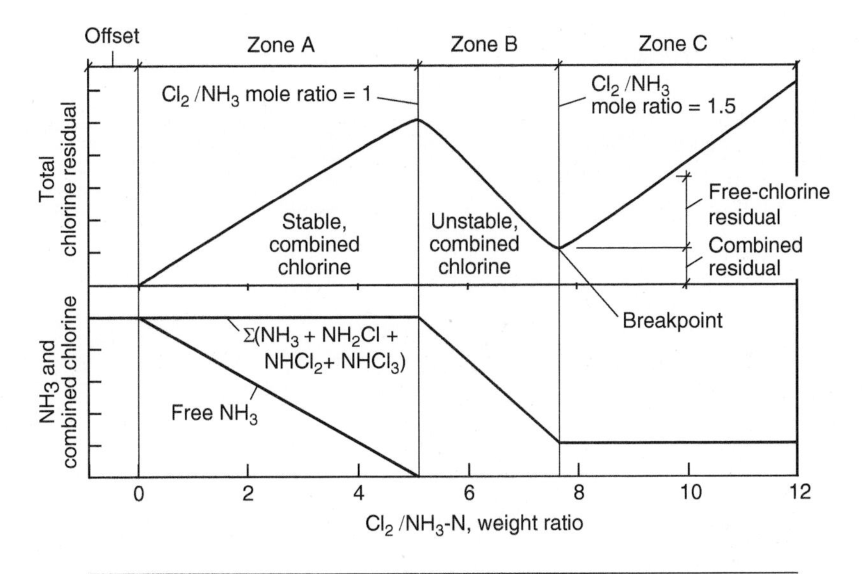

Parameter	Offset	Zone A	Zone B	Zone C
Time to metastable equilibrium	Fraction of a second	Seconds to a few minutes	10 to 60 min	10 to 60 min
Composition of metastable residual.	Reduction of readily oxidizable substances such as Fe(II), Mn(II), and H_2S.	Mostly monochloramine, some dichloramine, and traces of trichloramine at neutral or acid pH or at high Cl_2/NH_3 ratios.	A mixture of monochloramine and dichloramine, some free chlorine, and traces of trichloramine at low pH.	Mostly free chlorine, trichloramine can be significant (aesthetically, but not as fraction of residual) at neutral pH, but especially in acid region.

Figure 13-3
Overview of chlorine breakpoint stoichiometry.

even at pH 8 (Palin, 1975). In zone B, hypochlorous acid can oxidize ammonia to nitrogen gas and nitrate ion, resulting in the complete loss of chlorine residual. Between these, the dominant reaction is the conversion to nitrogen gas:

$$2NH_3 + 3HOCl \rightarrow N_2(g) + 3H_2O + 3HCl \quad \text{(ammonia to nitrogen gas)} \tag{13-17}$$

$$NH_3 + 4HOCl \rightarrow H^+ + NO_3^- + H_2O + 4HCl \quad \text{(ammonia to nitrate ion)} \tag{13-18}$$

Example 13-1 Estimating breakpoint chlorine requirements

Ammonia is added to pure water in the laboratory to reach a concentration of 1 mg /L as N. Estimate the chlorine dose needed to reach breakpoint for the following conditions: (1) all the ammonia is converted to nitrogen gas and (2) all the ammonia is converted to nitrate ion. Which reaction requires less chlorine?

Solution

1. Determine the chlorine dose needed to convert ammonia to nitrogen gas. From Eq. 13-17, 3 mol of HOCl is needed for every 2 mol of NH_3:

$$\text{Weight ratio} = (1.5\ \text{mol/mol})\frac{71\ \text{g Cl}_2}{14\ \text{g N}} = 7.61\ \text{mg Cl}_2/\text{mg N}$$

$$\text{Required dose} = 7.61\ \text{mg Cl}_2/\text{mg N} \times 1\ \text{mg N/L} = 7.61\ \text{mg/L as Cl}_2$$

2. Determine the chlorine dose to convert ammonia to nitrate. From Eq. 13-18, 4 mol of HOCl is needed for each mole of NH_3:

$$\text{Weight ratio} = (4\ \text{mol/mol})\frac{71\ \text{g Cl}_2}{14\ \text{g N}} = 20.2$$

$$\text{Required dose} = 20.2\ \text{mg Cl}_2/\text{mg N} \times 1\text{mg N/L} = 20.2\ \text{mg/L as Cl}_2$$

3. The reaction to nitrogen gas uses less chlorine.

Although breakpoint chlorine can be described with equilibrium reactions, the behavior of the $Cl_2{-}NH_3$ system is actually quite dynamic, and the breakpoint curve shown on Fig. 13-4 should be considered more of a metastable than an equilibrium state. As a result, laboratory studies to construct a breakpoint curve require precise timing to be reproducible, especially for Cl_2/NH_3 mole ratios above 1. Above this ratio, the reactions proceed rapidly until the metastable state is reached. Anywhere along the curve, the rate at which the reaction progresses is strongly influenced by the pH (Fig. 13-4), particularly in the vicinity of the breakpoint itself. Near the breakpoint, the reaction is at its maximum rate at a pH between 7 and 8. The rate decreases rapidly at pH values outside that range.

Sodium Hypochlorite

When chlorine was first used for disinfection, it was often applied in the form of hypochlorite. Sodium hypochlorite (NaOCl), or liquid bleach, came into use near the beginning of the Great Depression in the late 1920s. Later, chlorination using liquid chlorine became predominant because

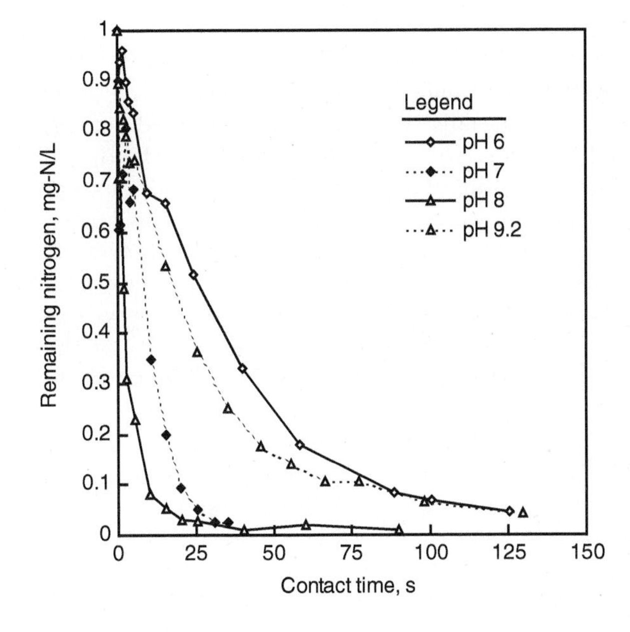

Figure 13-4
Effect of pH on breakpoint chlorination. (Data from Saunier and Selleck (1979), Temp. 15 – 18.5° C, $[NH_3]_O = 1$ mg/L, and $[Cl_2/NH_3]_O \sim 10$.)

of its lower cost, but now hypochlorite is again becoming more common because of the hazardousness of liquid chlorine.

Sodium hypochlorite is the most widely used form of hypochlorite today. It is widely used not only in disinfection of water but also for a myriad of other household and industrial uses.

Whereas chlorine gas is prepared by an electrolytic process that breaks sodium chloride solution into chlorine gas and sodium hydroxide, ironically, sodium hypochlorite is generally prepared by mixing sodium hydroxide and chlorine gas together:

$$Cl_2 + 2NaOH \rightarrow NaOCl + NaCl + H_2O \tag{13-19}$$

On a weight basis, 1.128 kg of NaOH reacts with 1 kg of chlorine to produce 1.05 kg of NaOCl and 0.83 kg of NaCl. The process is complicated by the fact that the reaction generates a significant amount of heat. It is common practice to add an excess of NaOH because hypochlorite is more stable at higher pH values. As a result, the density of one hypochlorite solution is not necessarily the same as another, even if both have the same strength (percent Cl_2). This density difference occurs because the final density depends on the amount of excess NaOH added during manufacture. Liquid bleach usually has a pH between 11 and 13. Hypochlorite can also be manufactured via onsite generation; this process is becoming more common.

13-3 Disinfection with Chlorine Dioxide

When the regulation of the chlorination by-products began, chlorine dioxide and ozone were viable alternative disinfectants. Chlorine dioxide is widely used in continental Europe, particularly Germany, Switzerland, and France, and produces almost no identifiable organic by-products, except low levels of a few aldehydes and ketones. Chlorine dioxide was known to produce two inorganic by-products, chlorite and the chlorate ion. As a result, most applications of chlorine dioxide were on low-TOC (total organic carbon) waters that did not require a high dose to overcome oxidant demand. Late in the 1980s, concern about the toxicity of the chlorite ion and chlorine dioxide itself reached a peak. Also, based on field experience, it was found that the use of chlorine dioxide was sometimes responsible for a very undesirable ''cat urine'' odor. As a precautionary measure, the State of California banned the use of chlorine dioxide for the disinfection of drinking water and several other states followed.

Eventually, when the disinfectant by-product rule was promulgated (U.S. EPA, 1998), an MCL of 0.8 mg/L was set for the chlorite ion and a maximum disinfectant residual limit (MDRL) of 1 mg/L was set for chlorine dioxide. No MCL was placed on the chlorate ion, but utilities were encouraged to be cautious about the production of chlorate and, again as a precautionary measure, the State of California has set an action level of 0.8 mg/L. About 10 percent of water utilities in the United States use chlorine dioxide (see Fig. 13-1); it is also often used in low doses for the oxidation of iron, manganese, and taste and odor rather than for disinfection.

13-4 Disinfection with Ozone

Ozone (O_3) is the strongest of the chemical disinfectants and its use is becoming increasingly common. Ozone is generated at the treatment plant site as a gas and is then injected into water. Once dissolved in water, ozone begins a process of decay that results in the formation of the hydroxyl radical ($HO\cdot$). Ozone reacts in two ways with contaminants and microbes: (1) by direct oxidation and (2) through the action of hydroxyl radicals generated during its decomposition. The consensus is that the action of ozone as a disinfectant is primarily dependent on its direct reactions; hence it is the residual of the ozone itself that is important.

Ozone Demand, Decay, and Disinfection Reaction

The ozone demand is the ozone dose that must be added before any ozone residual is measured in the water. It corresponds to the amount of ozone consumed during rapid reactions with readily degradable compounds. Ozone decay is the rate at which the residual ozone concentration decreases over time when the ozone dose is greater than the ozone demand. The overall rate of ozone decay in water is generally consistent with first-order kinetics.

An introduction to ozone decay based on the models developed by Staehelin and Hoigné (1982) is provided on Fig. 13-5. The cyclic nature of the ozone decay process in pure water is illustrated on Fig. 13-5a. The process must be initiated by a reaction between ozone and the hydroxide ion to form superoxide radicals ($HO_2\cdot$) and peroxide ions (HO_2^-), a slow process. As a result, decay is accelerated at higher pH. Once completed, the decay reactions enter a cyclic process represented in the figure by a circle. The cyclic reactions are promoted by ozone itself. If the concentration of ozone is increased, the cycle is accelerated. The importance of other materials in promoting ozone decay is illustrated on Figs 13-5b and 13-5c.

The disinfection reaction with ozone is an oxidation reaction in which ozone is converted to oxygen and water while microorganisms are being inactivated:

$$O_3 + \begin{pmatrix} \text{live} \\ \text{microorganisms} \end{pmatrix} + 2H^+ + 2e^- \rightarrow O_2 + H_2O + \begin{pmatrix} \text{inactivated} \\ \text{microorganisms} \end{pmatrix} \tag{13-20}$$

Generation of Ozone

At high concentrations (>23 percent) ozone is unstable (explosive) and under ambient conditions it undergoes rapid decay. Therefore, unlike chlorine gas, it cannot be stored inside pressurized vessels and transported to the water treatment plant. It must be generated onsite. Ozone can be generated by photochemical, electrolytic, and radiochemical methods, but the corona discharge method is the most commonly used in water treatment (see Fig. 13-6b). In this method, oxygen is passed through an electric field that is generated by applying a high-voltage potential across two electrodes separated by a dielectric material. The dielectric material prevents arcing and spreads the electric field across the entire surface of the electrode. As the oxygen molecules pass through the electric field, they are broken down to highly reactive oxygen singlets ($O\cdot$), which then react with other oxygen molecules to form ozone. The thickness of the gap through which the oxygen-rich gas stream passes is 1 to 3 mm wide.

Oxygen Source

Ozone can be generated directly from the oxygen in air or from pure oxygen. Pure oxygen is generated onsite from ambient air at larger plants or provided through the use of liquid oxygen (commonly referred to as LOX), which is generated offsite and transported to the plant. The most suitable method for providing oxygen for ozone generation in a particular plant depends on economic factors, the principal ones being the scale of the facility and the availability of industrial sources of liquid oxygen.

USE OF PREPARED, AMBIENT AIR

The most accessible oxygen source is ambient air, which contains about 21 percent oxygen by volume. Ambient air used to be the most common

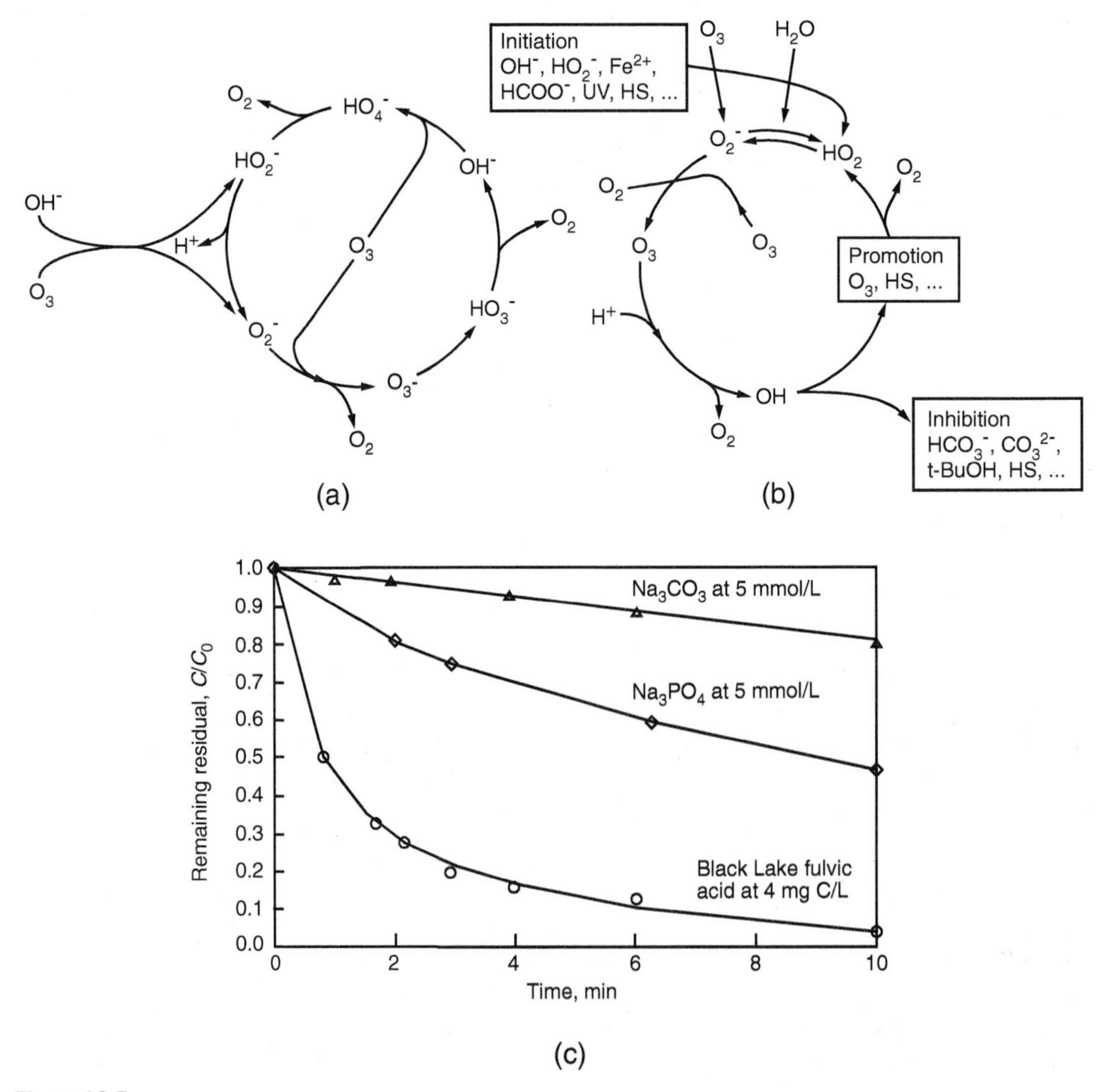

Figure 13-5
Understanding ozone reaction pathways and decay of residual ozone in natural waters: (*a*) the ozone decay wheel—reaction pathways in pure water (adapted from Hoigné and Bader, 1976); (*b*) influence of initiators, promotors, and inhibitors (adapted from Hoigné and Bader, 1976); and (*c*) effect of fulvic acid and carbonate on ozone decay—all tests conducted at 20°C with GAC filtered, deionized tap water adjusted to pH 7, and $C_0 \sim 8$ mg/L (adapted from Reckhow et al., 1986).

source of oxygen for ozone systems, but it has largely been replaced by liquid oxygen except for small, remote systems. Ambient air contains significant levels of particulates and water vapor, which must be removed. Water vapor is detrimental to corona discharge ozone generators for two reasons: (1) the presence of water vapor significantly reduces the ozone generation

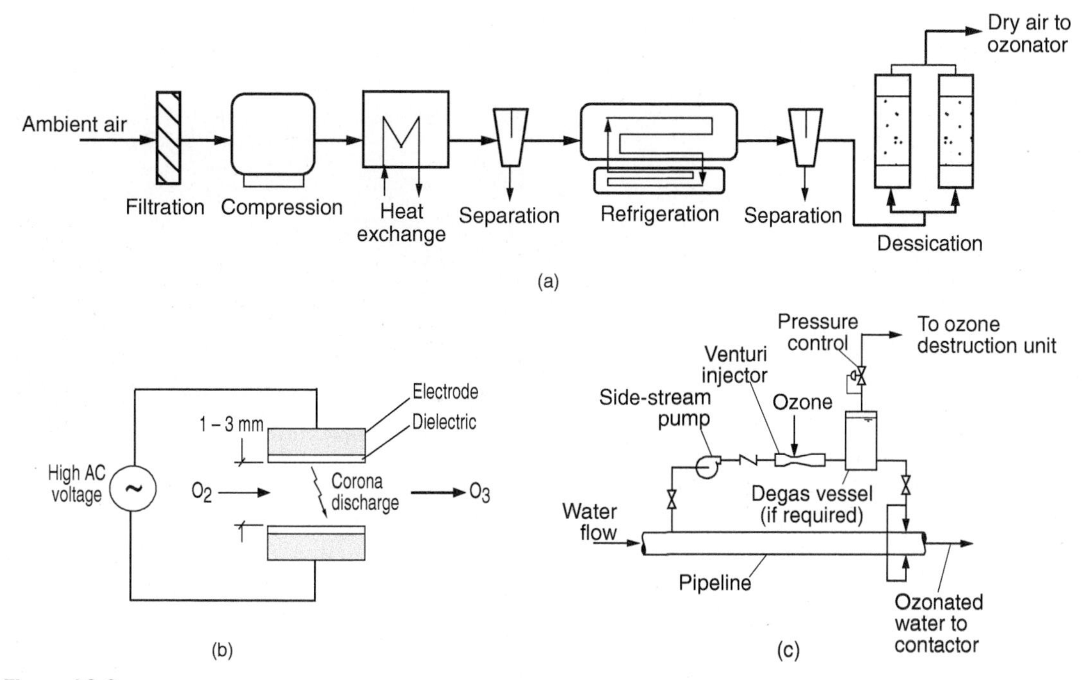

Figure 13-6
Components of an ozone disinfection system: (a) preparation system for ozone generation from ambient air, (b) generation of ozone by corona discharge, and (c) side-stream injection of ozone.

efficiency and (2) trace levels of water can react with the nitrogen present in the air and the generated ozone to form nitric acid.

Drying ambient air level is usually accomplished by a three-step process of compression, refrigeration, and desiccant drying. Compression and refrigeration help because the water vapor capacity of air decreases with increased pressure and decreased temperature, reducing the load on the desiccant system. Desiccant drying, however, is required to achieve the specifications for ozone generation. A schematic of all the components of such a system is shown on Fig. 13-6.

PURE OXYGEN

Liquid oxygen is widely available as a commercial, industrial-grade chemical and is the most common source of oxygen for ozone systems. Water treatment plants can purchase commercially available LOX, store it at the plant, and use it as the oxygen source for ozone generation. Liquid oxygen is delivered in trucks and stored in insulated pressurized tanks. Liquid oxygen is then drawn from the tank and piped to a vaporizer that warms and converts the oxygen to the gaseous form. Commercially available LOX is inherently low in contaminants and water vapor as a result of the manufacturing process. Therefore, minimal additional processing of the

Figure 13-7
Liquid oxygen (LOX) storage container tanks at a large water treatment plant.

oxygen stream is required before it is introduced to the ozone generator. A LOX storage system at a large water treatment plant is shown on Fig. 13-7.

Oxygen can also be generated at the plant site using pressure swing or vacuum swing adsorption processes or cryogenic oxygen generation processes, but these methods generally are not economical compared to delivery of liquid oxygen.

Ozone Injection Systems

The addition of ozonation in a water treatment plant requires two components in the process treatment train: (1) a device for injecting the ozone into the water and (2) a contact chamber in which the disinfection reaction takes place. For several decades, the most common approach to ozonation has been to combine these components by introducing the ozone into the water in large, deep basins using porous diffusers. More recently, the injection and contact systems are designed separately. For injection systems, side-stream injection using venturi injectors with or without side-stream degassing has become more common than fine-bubble diffusers (see Fig. 13-6c). Ozone contactors can be pipeline contactors, serpentine basins, or over–under baffled contactors and are described in Sec. 13-8. Details of the design of side-stream ozone injection systems can be found in Rakness (2005) and are described briefly below.

In side-stream injection (see Fig. 13-6c), a portion of the process flow is withdrawn from the main process line and pumped through a venturi injector. Low pressure in the throat of the injector draws ozone gas in from the ozone generator. After dissolution of the ozone gas, the side stream is injected back into the process stream through nozzles that provide good blending of the ozonated side stream into the main flow. In some systems, the side stream passes through a degassing tower before being injected into the process stream. After the ozone is injected, the process water flows to a pipeline or serpentine basin contactor. Design of contactors is presented in Sec. 13-8.

13-5 Disinfection with Ultraviolet Light

The disinfectants discussed previously in this chapter are oxidizing chemicals. Disinfection can also be accomplished by other means, heat and electromagnetic radiation among them. Heat is used to disinfect, or "pasteurize," beverages and even to disinfect water through boiling. Electromagnetic radiation, specifically radiation and UV radiation, is also used for disinfection: radiation in the case of food products and UV radiation in the case of air, water, and some medical surfaces. Of these, only UV radiation has so far found a place in the disinfection of drinking water. UV disinfection is not common for drinking water disinfection in the United States, as was shown on Fig 13-1. It is used more commonly in other countries, however, and its use is growing in the United States.

The definition of UV light, the sources of UV light, and UV equipment configurations are introduced in the following discussion. This material will serve as an introduction to the analysis of the UV disinfection process and its application to disinfection.

WHAT IS ULTRAVIOLET LIGHT?

Ultraviolet light is the name used to describe electromagnetic radiation having a wavelength between 100 and 400 nm. As illustrated on Fig. 13-8, electromagnetic radiation of slightly shorter wavelength has been named "X-rays" and electromagnetic radiation of slightly longer wavelength, visible to the human eye, is referred to as "visible light." Light in the UV spectrum is often further subdivided into four segments, vacuum UV, short-wave UV (UV-C), middle-wave UV (UV-B), and long-wave UV (UV-A). These classifications can also be described as follows:

1. Both UV-A and UV-B activate the melanocytes in the skin to produce melanin ("a tan").

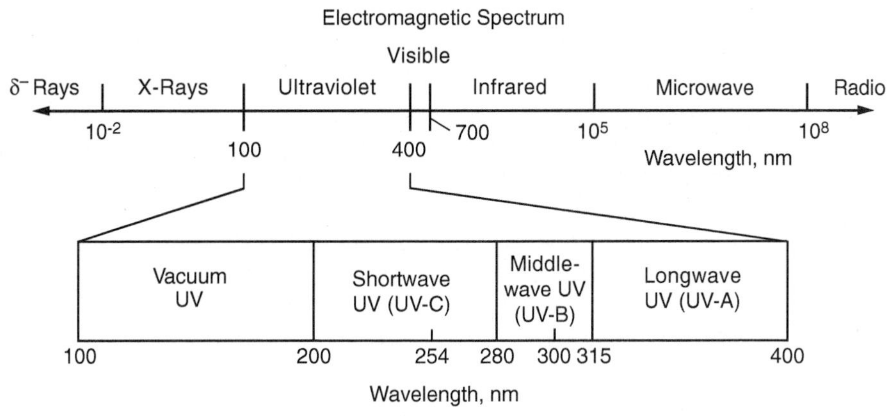

Figure 13-8
Location of the ultraviolet light region within the electromagnetic spectrum.

2. UV-B radiation also causes "sunburn."
3. UV-C radiation is absorbed by the DNA (deoxyribonucleic acid) and is the most likely of the three to cause skin cancer.

If electromagnetic radiation is thought of as photons, then the energy associated with each photon is related to the wavelength of the radiation:

$$E = \frac{hc}{\lambda} \tag{13-21}$$

where E = energy in each photon, J
h = Planck's constant (6.6×10^{-34} J · s)
c = speed of light, m/s
λ = wavelength of radiation, m

As a general rule, the more energy associated with each photon in electromagnetic radiation, the more dangerous it is for living organisms. Thus, visible and infrared light have relatively little affect on organisms, whereas both x-rays and γ-rays can be quite dangerous. Beyond these broad considerations, there are other factors that determine the fraction of the UV spectrum that is effective in disinfection. The portion of the UV spectrum that is more effective in disinfection is called the *germicidal range*. On the lower end, the germicidal range is limited by the absorption of UV radiation by water. As wavelengths decrease, water becomes an increasingly efficient barrier for UV. Vacuum UV, the fraction of UV with a wavelength below 200 nm, cannot penetrate water effectively. As a result, radiation having a wavelength of 200 nm or less is not considered germicidal. It is also well established that UV inactivates microorganisms by transforming their DNA. This transformation cannot happen unless the UV is at a wavelength at which DNA will absorb it, and this absorption does not occur above wavelengths of approximately 300 nm. Therefore, the germicidal range for UV is between approximately 200 and 300 nm (Fig. 13-9a).

SOURCES OF ULTRAVIOLET LIGHT

The UV disinfection units used most commonly in the water industry employ three different types of UV lamps: (1) low-pressure low-intensity lamps; (2) low-pressure high-intensity lamps (also called low-pressure high-output lamps), and (3) medium-pressure high-intensity lamps. The design of these lamps closely approximates that of the common fluorescent lightbulb. Low- and medium-pressure, high-intensity lamps are able to achieve a higher UV output in an equivalent space. Of the three technologies, medium-pressure UV has the greatest output. The spectrum of the UV light output by both types of low-pressure lamps is essentially the same, a very small amount of the light energy emanating at a wavelength of 188 nm and the vast majority of it emanating at a wavelength of 254 nm. The spectrum of the UV light output by medium-pressure lamps includes a number of wavelengths.

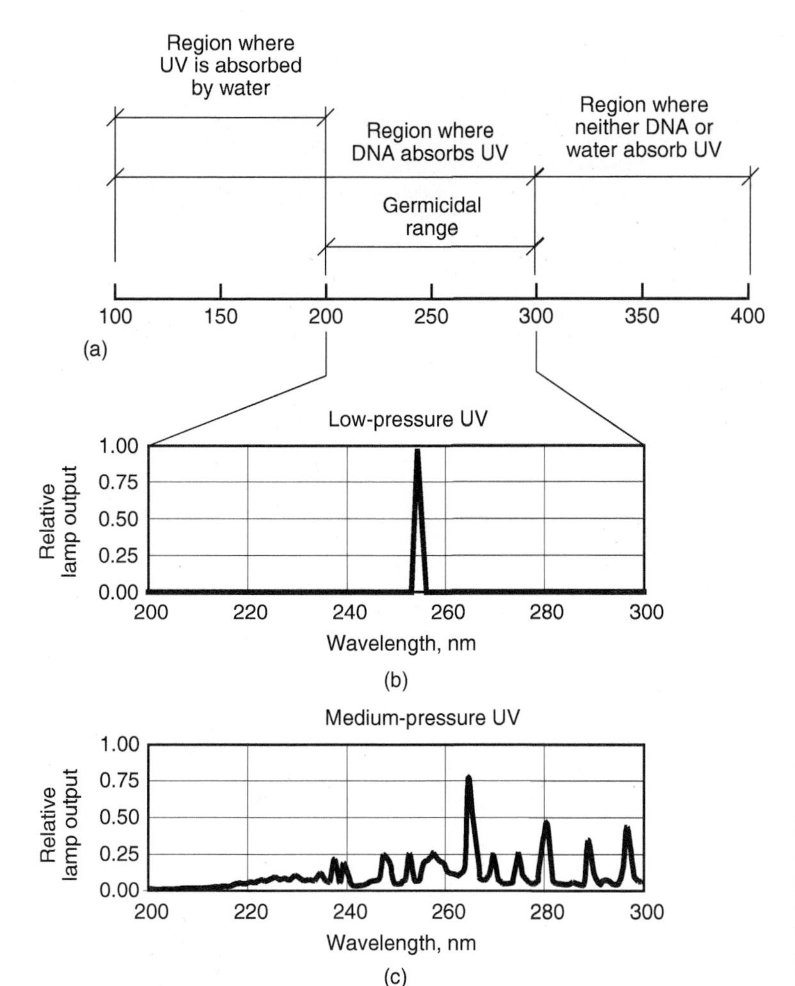

Figure 13-9
Ultraviolet sources and germicidal range: (a) ultraviolet portion of electromagnetic spectrum, (b) output from low-pressure UV lamp, and (c) output from medium-pressure UV lamp.

These spectra are illustrated and compared with the germicidal range on Fig. 13-9b and c. Several important characteristics of each of these UV lamps are compared in Table 13-2.

UV EQUIPMENT CONFIGURATIONS

Before discussing the fundamentals of UV disinfection, it will be useful to consider the types of reactors used for UV disinfection, as many of the factors that affect the effectiveness of UV disinfection are related to the reactor configuration. The components of a UV disinfection system consists of (1) the UV lamps; (2) transparent quartz sleeves that surround the UV lamps, protecting them from the water to be disinfected; (3) the structure that supports the lamps and sleeves and holds them in place; (4) the power supply for the UV lamps and cleaning system; (5) online UV

Table 13-2
Characteristics of three types of UV lamps

		Type of lamp		
Item	**Unit**	**Low-pressure low-intensity**	**Low-pressure high-intensity**	**Medium-pressure**
Power consumption	W	40–100	200–500[a]	1,000–10,000
Lamp current	ma	350–550	Variable	Variable
Lamp voltage	V	220	Variable	Variable
Germicidal output/input	%	30–40	25–35	10–15[b]
Lamp output at 254 nm	W	25–27	60–400	
Lamp operating temperature	°C	35–45	60–100	600–900
Partial pressure of Hg vapor	kPa	0.00093	0.0018–0.10	40–4,000
Lamp length	m	0.75–1.5	Variable	Variable
Lamp diameter	mm	15–20	Variable	Variable
Sleeve life	yr	4–6	4–6	1–3
Ballast life	yr	10–15	10–15	1–3
Estimated lamp life	h	8,000–10,000	8,000–12,000	4,000–8,000
Decrease in lamp output at estimated lamp life	%	20–25	25–30	20–25

[a]Up to 1200 W in very high output lamp.
[b]Output in the most effective germicidal range (~255–265 μm).

dose monitoring sensors and associated equipment, and (6) the cleaning system used to maintain the transparency of the quartz sleeves.

Cleaning systems are necessary for low-pressure high-intensity and medium-pressure UV lamps because they operate at such high temperatures (see Table 13-2) that salts with inverse solubility can precipitate, fouling the outer surface of the quartz sleeve and reducing the net UV output. These UV system components are installed in closed-vessel pressurized systems or as open-channel gravity flow systems, as shown on Fig. 13-10. Closed-vessel systems are used most commonly for the disinfection of drinking water, whereas open-channel systems are more common in wastewater disinfection.

Mechanism of Inactivation

More is known about the specific mechanisms of disinfection by UV than for any other disinfectant used in water treatment. The photons in UV light react directly with the nucleic acids in the target organism, damaging them. The genetic code that guides the development of every living organism is made up of nucleic acids. These nucleic acids are either in the form of DNA or ribonucleic acid (RNA). The DNA serves as the databank of life while the RNA directs the metabolic processes in the cell. Ultraviolet light damages DNA by dimerizing adjacent thymine molecules, inhibiting further transcription of the cell's genetic code. While not usually fatal to the organism, such dimerization will prevent its successful reproduction.

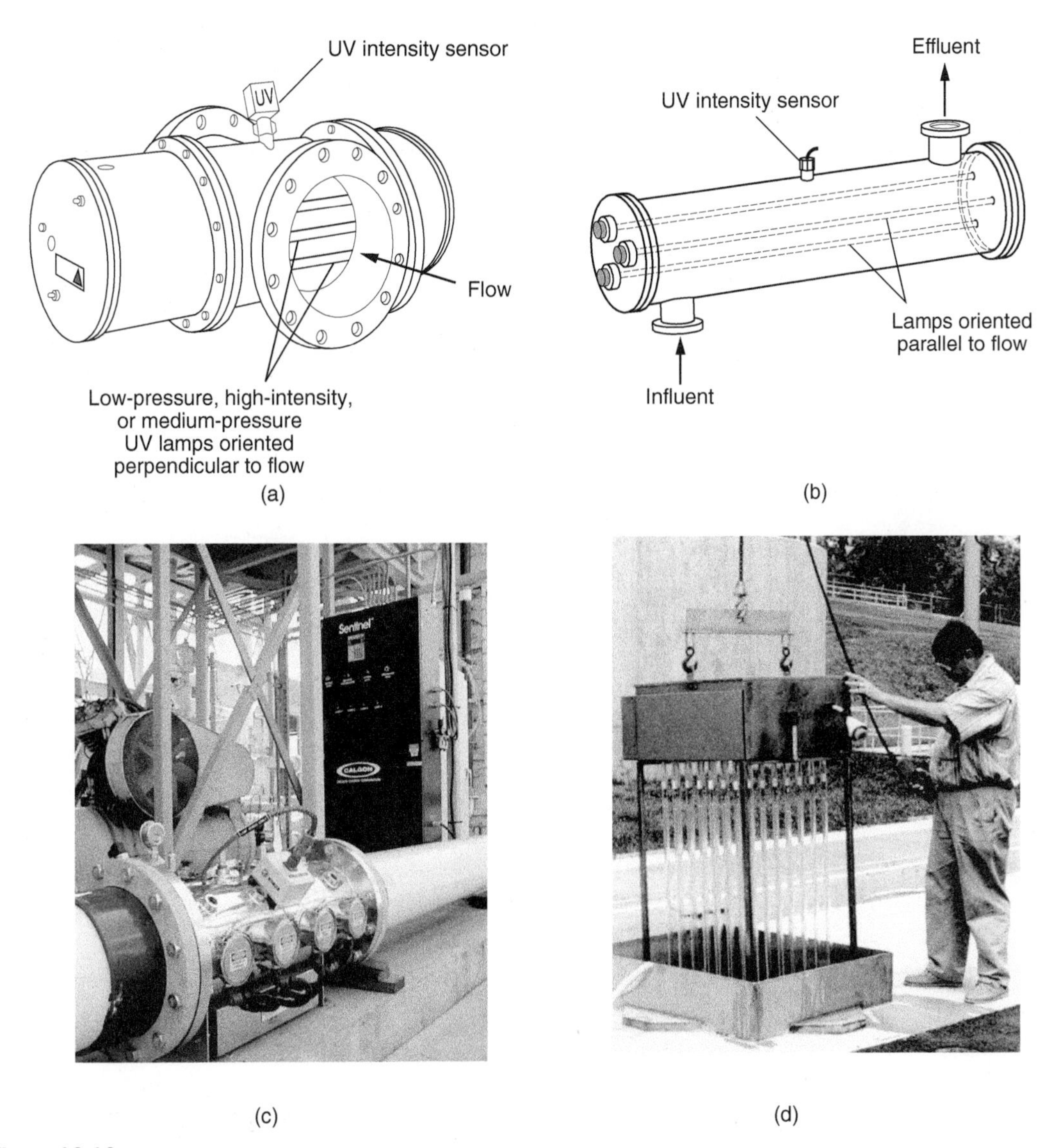

Figure 13-10
Common UV configurations: (a) medium-pressure lamps placed perpendicular to the flow in a closed reactor, (b) low- pressure high-intensity lamps placed parallel to flow, (c) view of medium-pressure closed reactor, and (d) view of vertical low-pressure lamp arrangement in open reactor.

Concept of Action Spectrum

There is no reason to expect that light will have the same disinfecting power at each wavelength. Earlier, the boundaries of the germicidal range of wavelengths were broadly established, the lower boundary (200 nm) being defined by the absorption of light by water and the upper boundary (300 nm) being defined by the lack of absorption of light by DNA. To gain

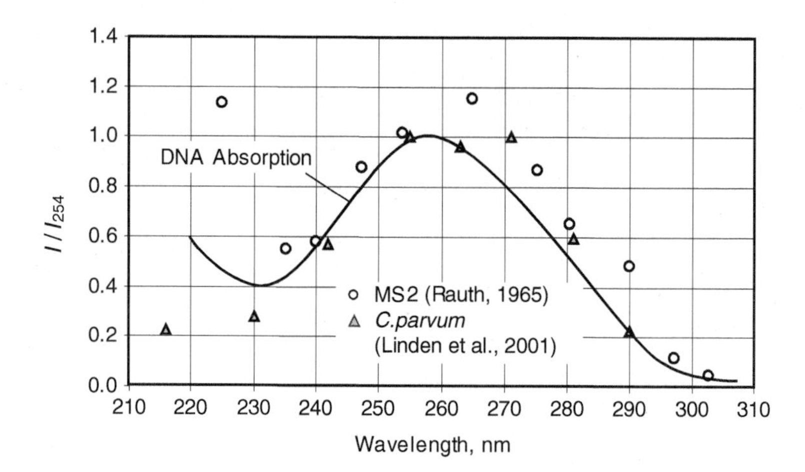

Figure 13-11
Comparing action spectra for *C. parvum* and MS2 coliphage with absorption spectrum for DNA.

an understanding of the possible significance of UV radiation at different wavelengths, action spectra have been developed for UV light of diffenent wavelengths. The action spectrums for *C. parvum* (Linden et al., 2001) and MS2 bacteriophase spores (Rauth, 1965) are compared with the absorption spectrum for DNA on Fig. 13-11. A close correlation between Λ_λ and DNA absorption is observed. The action spectra of a number of organisms have been determined and are similar to the results shown on Fig. 13-11.

UV Dose

The effectiveness of UV disinfection is based on the UV dose to which the microorganisms are exposed. The UV dose, *D*, is defined as

$$D = I_{\text{avg}} \times t \tag{13-22}$$

where D = UV dose, mJ/cm^2 (note mJ/cm^2 = mW · s/cm^2)
I_{avg} = average UV intensity, mW/cm^2
t = exposure time, s

Note that the UV dose term is analogous to the dose term used for chemical disinfectants (i.e., *Ct*). As given by Eq. 13-22, the UV dose can be varied by changing either the average UV intensity or the exposure time.

Influence of Water Quality

The quality of the water being treated can have an important influence on the performance of UV disinfection systems. The impact of dissolved and suspended substances on average UV intensity, and ultimately dose, are discussed below.

DISSOLVED SUBSTANCES

Pure water absorbs light in the lower UV wavelengths. A number of dissolved substances also have important influence on the absorption of UV radiation as it passes through the water on its way to the target organism. Among the more significant are iron, nitrate, and natural organic matter. Chlorine, hydrogen peroxide, and ozone can also have important effects.

The absorptivity of the water is an important aspect of UV reactor design. Waters with higher absorptivity absorb more UV light and need a higher energy input for an equivalent level of disinfection. Absorbance is measured using a spectrophotometer, typically using a fixed sample path length of 1.0 cm. The absorbance of water is typically measured at a wavelength of 254 nm.

In the application of UV radiation for microorganism inactivation, transmittance, which reflects the amount of UV radiation that can pass through a specified length at a particular wavelength, is the water quality parameter used in the design and monitoring of UV systems. The transmittance of a solution is defined as

$$\text{Transmittance, } T\,\% = \left(\frac{I}{I_0}\right) \times 100 \qquad (13\text{-}23)$$

where I = light intensity at distance x from light source, mW/cm^2
I_0 = light intensity at light source, mW/cm^2

The transmittance at a given wavelength can also be derived from absorbance measurements using the following relationship:

$$T = 10^{-A(\lambda)} \qquad (13\text{-}24)$$

where A = absorbance at wavelength λ

Thus, for a perfectly transparent solution $A(\lambda) = 0$, $T = 1$, and for a perfectly opaque solution $A(\lambda) \to \infty$, $T = 0$.

The term *percent transmittance*, commonly used in the literature, is

$$\text{UVT}_{254}, \% = 10^{-A_{254}} \times 100 \qquad (13\text{-}25)$$

where UVT_{254} = transmittance at a wavelength of 254 nm
A_{254} = absorbance at a wavelength of 254 nm

Typical absorbance and transmittance values for various waters are presented in Table 13-3.

Particulate Matter

Particulate matter can also interfere with the transmission of UV light. Two mechanisms of particular importance are shading and encasement, as shown on Fig. 13-12. The effect of shading can be integrated into models for the absorption of light. Beyond that, the number of organisms is dominated

Table 13-3
Typical absorbance and transmittance values for various waters

Type of wastewater	UV254 absorbance, a.u./cm[a]	Transmittance, UVT254, %
Groundwater	0.0706–0.0088	85–98
Surface water, untreated	0.3010–0.0269	50–94
Surface water, after coagulation, flocculation, and sedimentation	0.0969–0.0132	80–97
Surface water, after coagulation, flocculation, sedimentation, and filtration	0.0706–0.0088	85–98
Surface water after microfiltration	0.0706–0.0088	85–98
Surface water after reverse osmosis	0.0458–0.0044	90–99

[a]a.u. = absorbance unit.

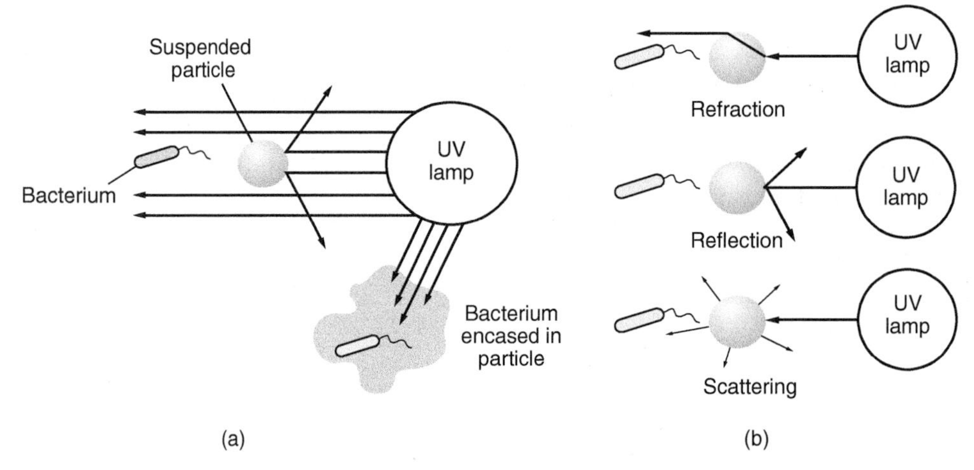

Figure 13-12
Illustration of mechanisms for interference in disinfection by particles: (a) overview of mechanisms for interference and (b) mechanisms of "shading."

by the effect of organisms associated with particles. Particles can "shade" target organisms from UV light via three mechanisms: refraction, reflection, and scattering. Where filtration is used, these effects are not very important, but in the treatment of unfiltered water supplies and unfiltered wastewater effluents, these effects can be quite significant.

Influence of UV Reactor Hydraulics

Ultraviolet disinfection systems, particularly medium-pressure systems, are characterized by overall residence times that are much shorter than other kinds of disinfection systems. In these systems short circuiting and dispersion are important and difficult design issues. Designing these systems to achieve good performance requires a greater appreciation of the factors that influence dispersion and short circuiting than is required for the design of most other disinfection systems. The issues are the same as those discussed in Sec. 13-7 with contactors for disinfection with chlorine, chloramines, chlorine dioxide, and ozone; however, with UV disinfection contactors, the time spent in transition zones becomes much more important.

Use Collimated Beam to Determine UV Dose and to Develop UV Dose–Response Curves

To determine UV dose both the UV intensity and exposure time must be known. In disinfection applications, UV intensity is determined using collimated beam device under controlled conditions. The use of the collimated beam device to determine UV intensity and dose and to develop dose–response curves is described in the following discussion.

DETERMINATION OF UV DOSE

The most common procedure for determining the required UV dose for the inactivation of challenge microorganisms involves the exposure of a well-mixed water sample in a small batch reactor (i.e., a Petri dish) to a collimated beam of UV light of known UV intensity for a specified period of time, as illustrated on Fig. 13-13. Use of a monochromatic low-pressure low-intensity lamp in the collimated beam apparatus allows for accurate characterization of the applied UV intensity. Use of a batch reactor allows for accurate determination of exposure time. The applied UV dose, as defined by Eq. 13-22, can be controlled either by varying the UV intensity or the exposure time. Because the geometry is fixed, the depth-averaged UV intensity within the Petri dish sample (i.e., the batch reactor) can be computed using the following relationship, which also takes into account other operational variables that may affect the UV dose:

$$D_{CB} = E_s t(1 - R)P_f\left[\frac{(1 - 10^{-k_{254}d})}{2.303(k_{254}d)}\right]\left(\frac{L}{L + d}\right) \quad (13\text{-}26)$$

$$D_{CB} = E_s t(1 - R)P_f\left[\frac{(1 - e^{-2.303k_{254}d})}{2.303(k_{254}d)}\right]\left(\frac{L}{L + d}\right) \quad (13\text{-}27)$$

where D_{CB} = average UV dose, mW/cm^2

E_S = incident UV intensity at the center of the surface of the sample, before and after sample exposure, mW/cm^2

t = exposure time, s

R = reflectance at the air-water interface at 254 nm

P_f = Petri dish factor

k_{254} = absorptivity, a.u./cm (base 10)

d = depth of sample, cm

L = distance from lamp centerline to liquid surface, cm

The term $1 - R$ on the right-hand side accounts for the reflectance at the air–water interface. The value of R is typically about 2.5 percent. The term

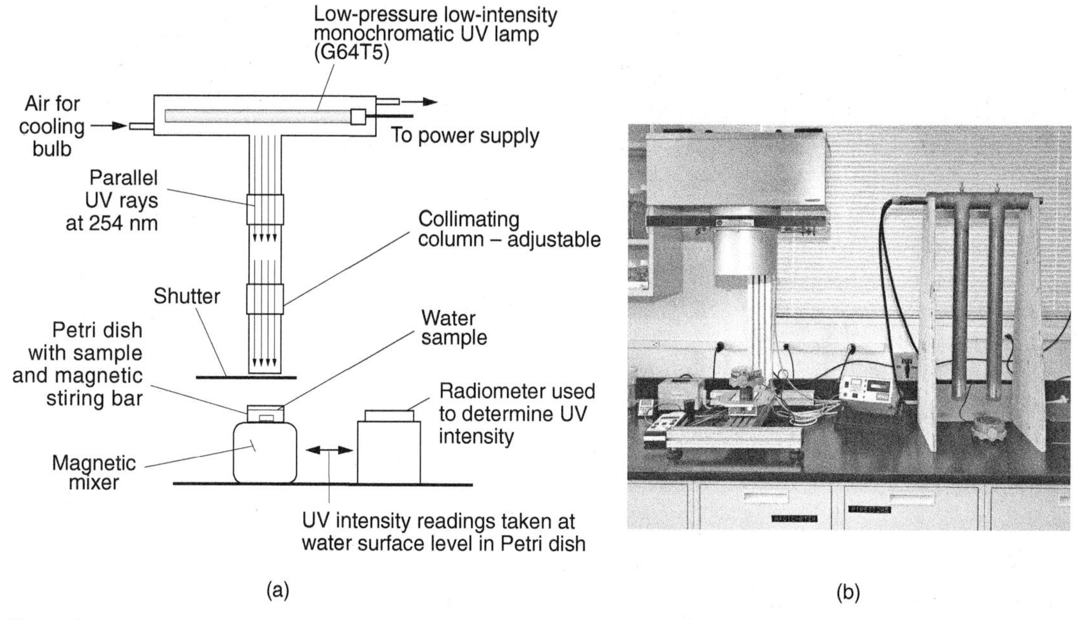

Figure 13-13
Collimated beam devices used to develop dose–response curves for UV disinfection: (a) schematic of the key elements of a collimated beam setup and (b) view of two different types of collimated beam devices. The collimated beam on the left is of European design; the collimated beam on the right is of the type shown in the schematic on the left.

P_f accounts for the fact that the UV intensity may not be uniform over the entire area of the Petri dish. The value of P_f is typically greater than 0.9. The term within the brackets is the depth averaged UV intensity within the Petri dish and is based on the Beer-Lambert law. The final term is a correction factor for the height of the UV light source above the sample. The application of Eq. 13-26 is illustrated in Example 13-2.

Example 13-2 Estimation of UV dose using collimated beam

A collimated beam, with the following characteristics, is to be used for biodosimetry testing. Using these data estimate the average UV dose delivered to the sample:

$E_S = 15\ \text{mW/cm}^2 \quad t = 10\ \text{s} \quad R = 0.025 \quad p_f = 0.94$

$k_{254} = 0.065\ \text{cm}^{-1} \quad d = 1\ \text{cm} \quad L = 40\ \text{cm}$

Solution

1. Using Eq. 13-26 estimate the delivered dose:

$$D = E_s t(1 - R)P_f \left[\frac{(1 - 10^{-k_{254}d})}{2.303(k_{254}d)}\right]\left(\frac{L}{L + d}\right)$$

$$D = (15 \times 10)(1 - 0.025)(0.94)\left[\frac{(1 - 10^{-0.065\times 1})}{2.303(0.065 \times 1)}\right]\left(\frac{40}{40 + 1}\right)$$

$$D = (150)(0.975)(0.94)(0.928)(0.976) = 124.6 \text{ mJ/cm}^2$$

Development of UV Dose–Response Curve

To assess the degree of inactivation that can be achieved at a given UV dose, the concentration of microorganisms is determined before and after exposure in a collimated beam apparatus (see Fig. 13-13). Microorganism inactivation is measured using a most probable number (MPN) procedure for bacteria, a plaque count procedure for viruses, or an animal infectivity procedure for protozoa. The development of a dose–response curve is illustrated in Example 13-3.

Example 13-3 Develop dose–response curve for bacteriophage MS2 using a collimated beam

Bacteriophage MS2 (American Type Culture Collection (ATCC) 15597) is to be used to validate the performance of a full-scale UV reactor. The following collimated beam test results were obtained for MS2 in a phosphate buffer solution with a UVT_{254} in the range from 95 to 99 percent (data courtesy B. Cooper, BioVir Labs). Estimate the UV dose required to achieve 2-log of inactivation.

Dose, mJ/cm^2	Surviving Concentration, phage/mL	Log Survival, log (phage/mL)	Log Inactivation
0	5.00×10^6	6.70	0.00
20	4.00×10^5	5.60	1.10[a]
40	4.30×10^4	4.63	2.07
60	6.31×10^3	3.80	2.9
80	8.70×10^2	2.94	3.76
100	1.20×10^2	2.08	4.62

[a]Sample calculation. Log inactivation = 6.70 − 5.60 = 1.10.

Solution

1. Plot the collimated beam test results. The results are plotted in the figure given below.

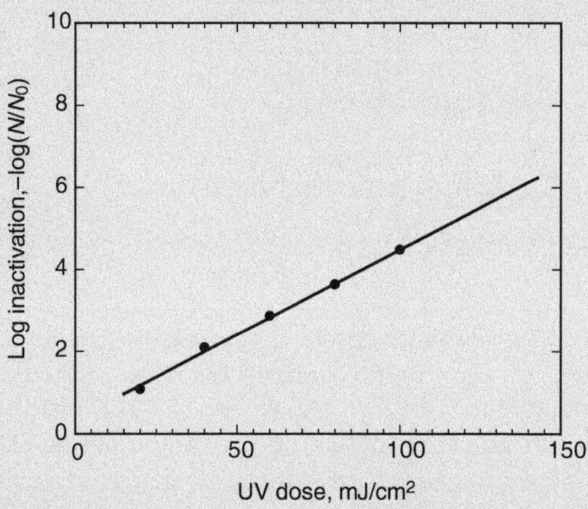

2. Dose–response curve for bacteriophage MS2. The equation of the line, based on a linear fit, is

$$y = 0.269 + 0.04365x$$

which corresponds to

$$-\log(N/N_0) = 0.269 + (0.04365\ \text{cm}^2/\text{mJ})\ (\text{UV dose, mJ/cm}^2)$$

3. UV dose required for two logs of inactivation of MS2. Using the equation from step 2, the required UV dose is

$$\text{UV dose} = \frac{-\log(N/N_0) - 0.269}{0.04365\ \text{cm}^2/\text{mJ}} = \frac{2 - 0.269}{0.04365\ \text{cm}^2/\text{mJ}}$$

$$= 39.6\ \text{mJ/cm}^2$$

Validation Testing of UV Reactors

At the present time there are a number of UV manufacturers that produce UV reactors suitable for the inactivation of microorganisms. Because of the interest in utilizing UV by the water industry to obtain partial inactivation credit for *Cryptosporidium, Giardia,* and viruses (in some cases) and the need to protect public health, the United States and many other countries have established regulations and guidelines for the use of UV radiation for water and wastewater treatment. The general procedure used to validate a UV reactor (i.e., specifically the delivered UV dose) is illustrated on Fig. 13-14. The inactivation values obtained in the field are compared

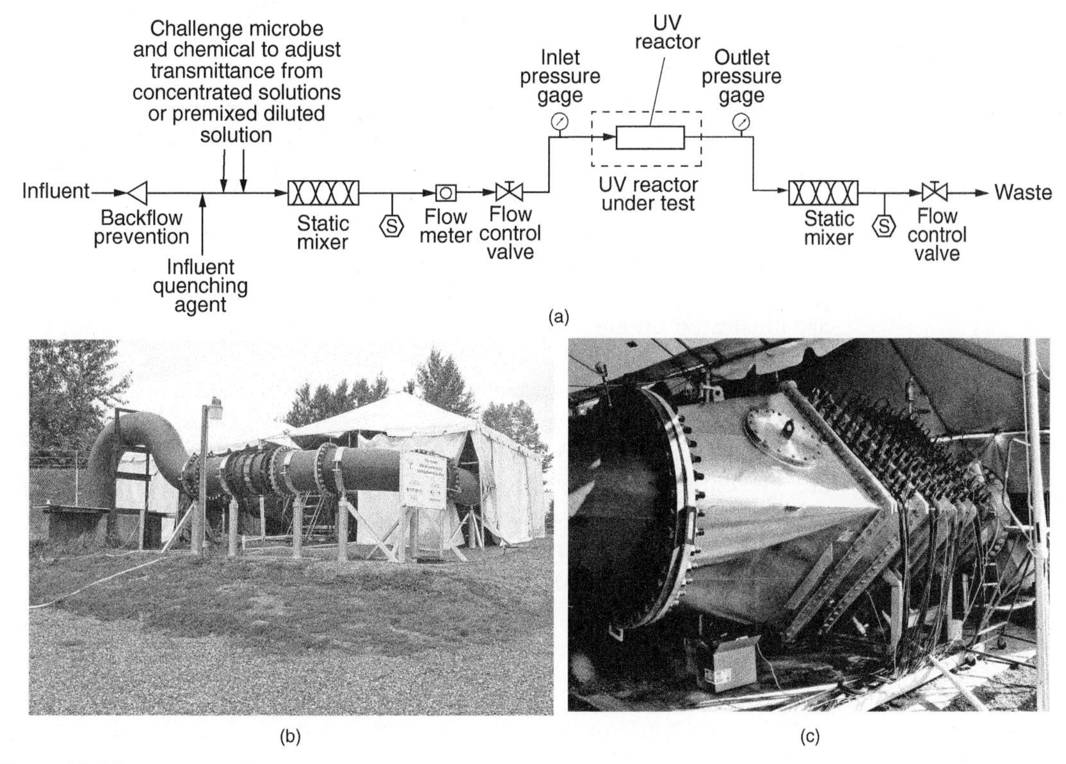

Figure 13-14
Experimental setup for validation of UV reactors under controlled conditions: (a) schematic of setup requirements for testing full-scale UV reactor, (b) view of test facility at Portland, OR, and (3) large UV reactor instrumented for UV dose validation by dosimetry.

to the values obtained with the collimed beam to establish the delivered UV dose.

13-6 Disinfection Kinetics

For chemical disinfectants, the specific mechanisms of microorganism inactivation are not well understood. Inactivation depends on the properties of each microorganism, the disinfectant, and the water. The reaction rates that have been observed can vary by as much as six orders of magnitude from one organism to the next, even for one disinfectant. Even for disinfection reactions where the reaction mechanism is well understood, for example, UV light, reaction rates vary by one and one-half orders of magnitude. In the following discussion, the form of disinfection data resulting from laboratory experiments is examined by considering both classical disinfection kinetics as well as with a more contemporary phenomenological kinetic model.

The *Ct* model, derived from disinfection kinetic models, which is used by regulatory agencies, is also considered.

Observed Disinfection Data

Over a period of many years, a number of different anomalies have been observed in plots of disinfection data obtained for a variety of disinfectants and waters. Substantially different kinetics mechanisms may control the rate of inactivation of different microorganisms with the same disinfectant or the same microorganisms with different disinfectants. The form of the disinfection plots can be generalized into three typical forms: pseudo-first-order, accelerating rate, and decelerating rate. These forms are illustrated on Fig. 13-15. Reasons often cited in the literature for these particular curve shapes and the circumstances (organism, disinfectant, and magnitude of disinfection) under which each type of curve is sometimes found are also given. With respect to the curves shown on Fig. 13-15, there is extensive literature on disinfection modeling. Two of these models are presented in the following discussion. These models have been used to model disinfection data that can be described with a pseudo-first-order reaction and for reactions with accelerating rates on a semilog plot (Figs. 13-15a and 13-15b). Additional details on the decelerating rate model, commonly encountered in the disinfection of wastewater, may be found in Asano et al. (2007) and Crittenden et al. (2012).

Chick-Watson Model

Early in the twentieth century, Dr. Harriet Chick, a research assistant at the Lister Institute of Preventive Medicine in Chelsea, England, proposed that disinfection could be modeled as a first-order reaction with respect to the concentration of the organisms. Chick demonstrated her concept by plotting the concentration of viable organisms versus time on a semilog graph for disinfection data for a broad variety of disinfectants and organisms (Chick, 1908). Chick worked with disinfectants such as phenol, mercuric chloride, and silver nitrate and organisms such as *Salmonella typhi, Salmonella paratyphi, Escherichia coli, Staphylococcus aureus, Yersinia pestis,* and *Bacillus anthracis*. Since then, "Chick's law" has been shown to be broadly applicable to disinfection data. Chick's law takes the form

$$r = -k_c N \tag{13-28}$$

where r = reaction rate for the decrease in viable organisms with time, org/L·min
k_c = Chick's law rate constant, min^{-1}
N = concentration of organisms, org/L

While Chick's law has broad applicability, an important effect not addressed in the model is the concentration of the disinfectant. Frequently, different concentrations of disinfectant will lead to different rates in the decrease in viable organisms, as illustrated on Fig. 13-16. Note that there is a different slope for each concentration of bromine and, using Eq. 13-28, the reaction

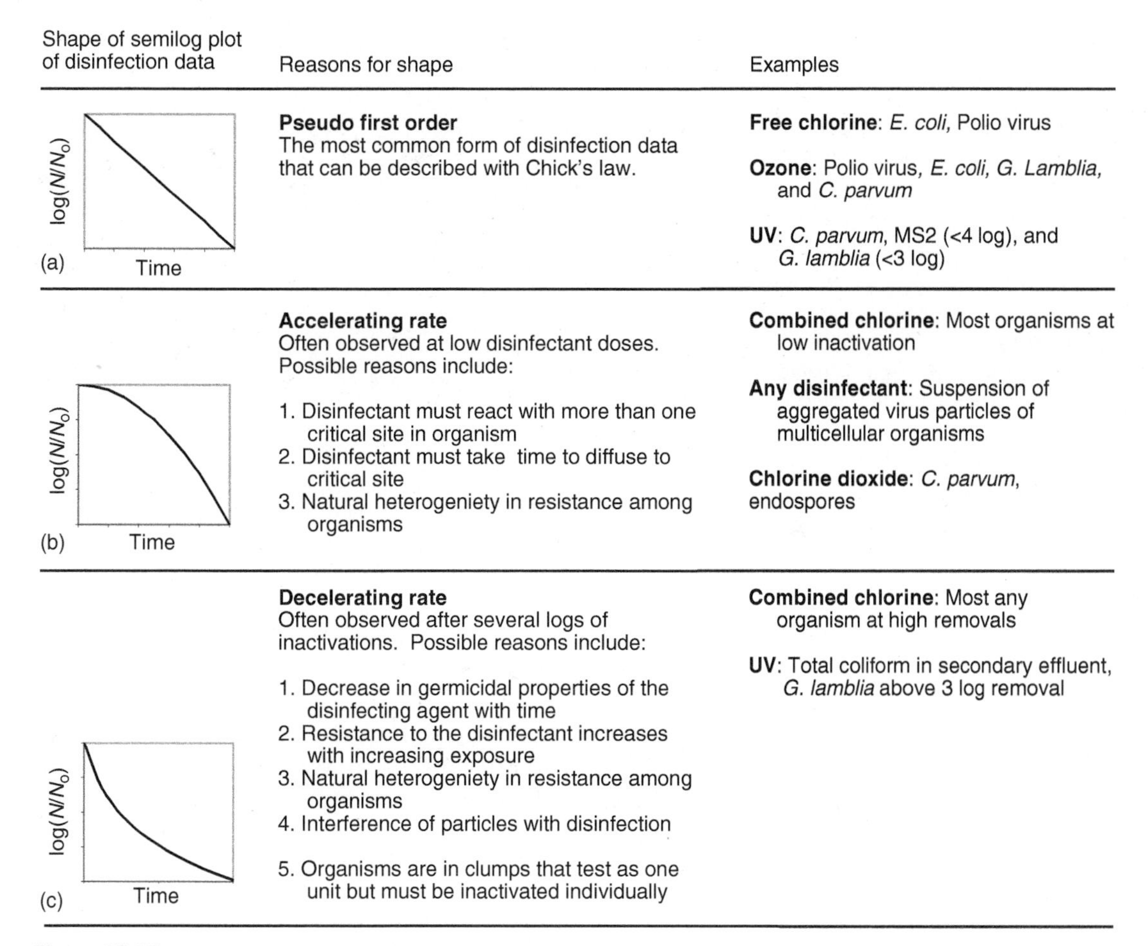

Figure 13-15
Graphical forms of disinfection data.

has a different rate constant for each concentration. Thus, while Chick's first-order concept is consistent with the data, a better means for accounting for disinfectant concentration is necessary.

In the same year that Dr. Chick proposed her model, Herbert Watson proposed that the time needed to reach a specific level of disinfection was related to the disinfectant concentration by the equation (Watson, 1908)

$$C^n t = \text{const} \tag{13-29}$$

where C = concentration of disinfectant, mg/L
n = dilution coefficient, unitless

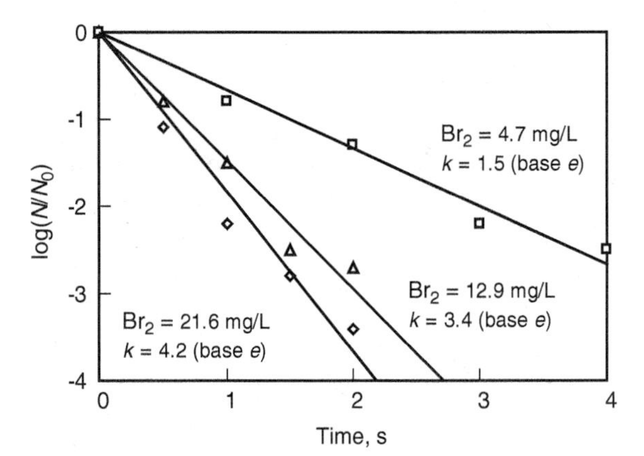

Figure 13-16
Inactivation of poliovirus Type I with three concentrations of bromine in a batch reactor (adapted from Floyd et al., 1978).

t = time required to achieve a constant percentage of inactivation (e.g., 99%)
const = value for given percentage of inactivation, dimensionless

Watson demonstrated the concept by plotting data showing equal inactivation on a plot of $\log(C)$ versus $\log(t)$. The slope of the log–log plot, n, is the coefficient of dilution and reflects the effect of diluting the disinfectant (Morris, 1975). Such plots are still used today, and an example is shown on Fig. 13-17. As a matter of convention, Watson plots are generally constructed with data corresponding to a removal of 99 percent. In such plots, the dilution coefficient is generally found to be approximately 1,

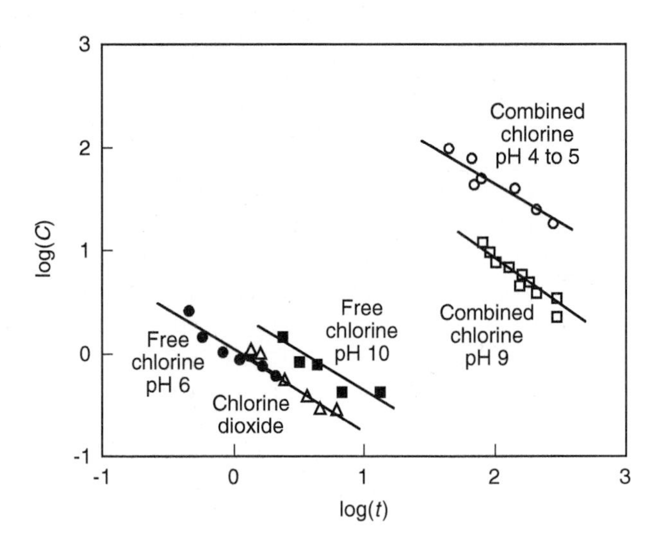

Figure 13-17
Watson plot of requirements for 99 percent inactivation of poliovirus Type I (adapted from Scarpino et al., 1977).

and given the inaccuracies involved in collecting disinfection data, there is little evidence for a dilution coefficient other than unity. When the dilution coefficient is equal to 1, the disinfection concentration and time are of equal importance for inactivating microorganisms.

With the knowledge that disinfection concentration and time are of equal importance, Chick's law and the Watson equation can be combined and are often referred to as the "Chick–Watson model":

$$r = -\Lambda_{CW} CN \tag{13-30}$$

where Λ_{CW} = coefficient of specific lethality (disinfection rate constant), L/mg·min
C = concentration of disinfectant, mg/L
N = concentration of organisms, org/L

Most laboratory disinfection studies are conducted using completely mixed batch reactors (CMBR). Using concepts presented in Chap. 4, a mass balance analysis on a batch reactor can be written and integrated, leading to

$$\ln\left(\frac{N}{N_0}\right) = -\Lambda_{CW}\, Ct \tag{13-31}$$

where N_0 = initial (time = 0) concentration of organisms, org/L
t = time, min

It is important to note that even though laboratory disinfection studies typically use batch reactors, the rate equation (Eq. 13-30) can be applied to other reactors using the concepts presented in Chap. 4.

When Chick did her work, she plotted the organism concentration directly against time on a semilog graph [$\log(N)$ vs. t]. Now that Eq. 13-31 has received broad recognition, it is more common to plot the log or natural log of the survival ratio, where $S = N/N_0$, versus time [$\ln(N/N_0)$or $\log(N/N_0)$ vs. t]. In disinfection studies, however, it is typically difficult to get an accurate measurement of the initial concentration of organisms, N_0, even with several replicates of the tests. As a result, a line fit through the data may not pass through zero [i.e., $\ln(N/N_0)_{t=0} \neq 0$]. Although it is not consistent with the definition of N_0 [at $t = 0$, $\ln(N/N_0) \equiv 0$], it is often best to find the coefficient of specific lethality without forcing the regression line to pass through zero.

Equation 13-31 was derived using calculus so the term on the left is the natural logarithm (i.e., base e). However, disinfection effectiveness is typically expressed using the log removal value (LRV), which uses base 10 logarithms. Thus, it is necessary to convert between natural and base 10 logarithms when evaluating disinfection data. The use of Eq. 13-31 to determine the coefficient of specific lethality for a disinfection reaction is demonstrated in Example 13-4.

Example 13-4 Application of the Chick–Watson model

Plot the data given below according to Eq. 13-31. Determine the coefficient of specific lethality and the coefficient of determination (r^2). The data for the inactivation of poliovirus Type I with bromine (Floyd et al., 1978) are provided in the following table:

C, mg/L	Time, s	$\log(N/N_0)$	C, mg/L	Time, s	$\log(N/N_0)$
21.6	0	0	12.9	1.5	−2.5
21.6	0.5	−1.1	12.9	2	−2.7
21.6	1	−2.2	4.7	1	−0.8
21.6	1.5	−2.8	4.7	2	−1.3
21.6	2	−3.4	4.7	3	−2.2
12.9	0.5	−0.8	4.7	4	−2.5
12.9	1	−1.5			

Solution

1. Determine the values of Ct and $\ln(N/N_0)$ for each organism survival value.
   a. Ct is calculated simply by multiplying C by t.
   b. To convert from base 10 to base e logarithms, recall the logarithmic identity $\log_b(x) = \log_a(x)/\log_a(b)$, thus:

$$\ln(N/N_0) = \frac{\log(N/N_0)}{\log(e)} = 2.303 \log\left(\frac{N}{N_0}\right)$$

   c. The required data table is shown below:

Time, s	C, mg/L	Ct, mg · s/L	$\ln(N/N_0)$	Time, s	C, mg/L	Ct, mg · s/L	$\ln(N/N_0)$
0.5	21.6	10.8	−2.53	1.5	12.9	19.4	−5.76
1	21.6	21.6	−5.07	2	12.9	25.8	−6.22
1.5	21.6	32.4	−6.45	1	4.7	4.7	−1.84
2	21.6	43.2	−7.83	2	4.7	9.4	−2.99
0.5	12.9	6.5	−1.84	3	4.7	14.1	−5.07
1	12.9	12.9	−3.45	4	4.7	18.8	−5.76

2. Prepare a plot of $\ln(N/N_0)$ as a function of Ct and fit a linear trendline through the data. Select trendline options to display the equation and r^2 value.

3. The required plot is shown below.

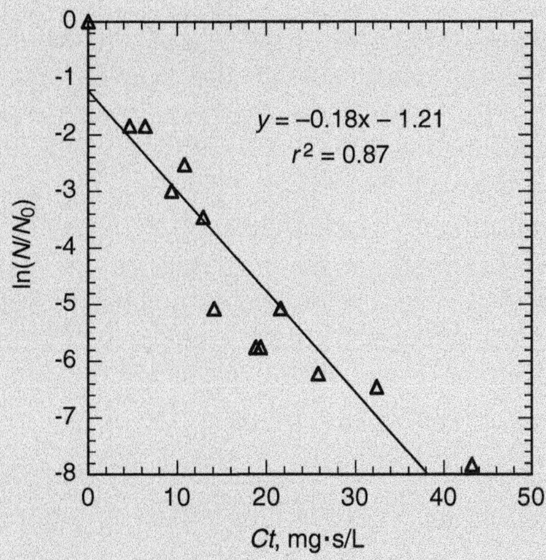

4. The slope of the line in the above plot corresponds to the coefficient of specific lethality, Λ_{CW}. From the plot $\Lambda_{CW} = 0.18$ and $r^2 = 0.87$.

Rennecker–Mariñas Model (Accelerating Rate)

Some organisms do not exhibit significant inactivation until a certain Ct value has been exceeded. This inactivation response is observed, for example, when chemical disinfectants are applied to oocysts and endospores. In the Rennecker–Mariñas model, this observation is addressed by incorporation of a lag term coefficient, b into Eq. 13-31 (Rennecker et al., 1997). The Rennecker–Mariñas model can be summarized as follows:

$$\ln\left(\frac{N}{N_0}\right) = \begin{cases} 0 & \text{for } Ct < b \quad (13\text{-}32) \\ -\Lambda_{CW}\,(Ct - b) & \text{for } Ct \geq b \quad (13\text{-}33) \end{cases}$$

where b = lag coefficient (in mg · min/L).

The lag coefficient b is the maximum value of Ct at which $\ln(N/N_0) = \ln(S_0) = 0$ (i.e., no inactivation has occurred). When b is zero, Eq. 13-33 corresponds to Eq. 13-31. It should be noted that the presentation of the mathematics used in the analysis of Eqs. 13-32 and 13-33 is consistent with but not identical to the approach used by Rennecker et al. (1997). Application of the Rennecker–Mariñas model is illustrated in Example 13-5.

Example 13-5 Application of the Rennecker–Mariñas model

Apply the Rennecker–Mariñas model to evaluate the coefficient of specific lethality and the lag coefficient for the inactivation of *C. parvum* using chlorine dioxide (ClO_2) based on the data given below. As shown in the data table, inactivation was measured at three concentrations of ClO_2 and at several time intervals. In analyzing the data, do not assume that it was possible to measure N_0 accurately (i.e., $N/N_0 \neq 0$ for $Ct < b$; instead, require $N/N_0 = \text{constant}$ for $Ct < b$). Analyze the data by developing a spreadsheet solution and use the Solver function in Excel to determine the model parameters. Also calculate the coefficient of determination (r^2). Data for the inactivation of *C. parvum* by ClO_2 (Corona-Vasquez et al., 2002) are provided in the following table:

C, mg/L	t, min	log(N/N_0)	C, mg/L	t, min	log(N/N_0)
0.96	0.0	−0.21	0.48	122.0	−1.08
0.96	15.5	−0.25	0.48	152.0	−1.68
0.96	30.8	−0.38	4.64	0.0	−0.15
0.96	46.1	−0.55	4.64	2.1	0.02
0.96	61.2	−1.04	4.64	4.2	−0.11
0.96	76.2	−1.66	4.64	6.2	−0.19
0.96	91.1	−2.03	4.64	8.2	−0.29
0.48	0.0	−0.17	4.64	10.0	−0.56
0.48	32.0	−0.12	4.64	12.0	−0.79
0.48	61.6	−0.31	4.64	13.9	−1.19
0.48	92.0	−0.60	4.64	15.8	−1.47

Solution

Construct a spreadsheet, as shown below, for analysis of the data and determination of the model parameters. Because the Solver function will be used, some of the calculations will need to be automated using advanced features of Excel, including the IF function, as described below.

1. Compute the value of Ct for each experiment and enter the corresponding inactivation value into the spreadsheet.
2. The measured log survival ratios are entered into the column labeled Data in the table below.
3. The value of the model parameters can be determined using an IF statement of the form "IF [$Ct < b$, log(S_0), log(S_0) + slope($b - Ct$)]."
4. Solver is used to minimize the sum of the [Data–model]^2 column by varying b, slope, and log(S_0). The results are displayed in the following table and figure:

Spreadsheet setup for model evaluation:

Ct,	$\log(N/N_0)$			
mg · min/L	Data	Model	$[\text{Data-Model}]^2$	$[\text{Data-Data}_{\text{avg}}]^2$
0.0	−0.21	−0.2	0.001	0.216
14.9	−0.25	−0.2	0.004	0.182
29.6	−0.38	−0.2	0.036	0.089
44.3	−0.55	−0.5	0.001	0.015
58.8	−1.04	−1.0	0.000	0.138
73.2	−1.66	−1.6	0.010	0.972
87.5	−2.03	−2.1	0.001	1.852
0.0	−0.17	−0.2	0.000	0.259
15.4	−0.12	−0.2	0.005	0.310
29.6	−0.31	−0.2	0.016	0.130
44.2	−0.60	−0.5	0.007	0.005
58.6	−1.08	−1.0	0.002	0.167
73.0	−1.68	−1.6	0.017	1.016
0.0	−0.15	−0.2	0.001	0.277
9.7	0.02	−0.2	0.043	0.483
19.3	−0.11	−0.2	0.006	0.319
28.8	−0.19	−0.2	0.000	0.230
38.2	−0.29	−0.3	0.000	0.147
46.6	−0.56	−0.6	0.002	0.012
55.6	−0.79	−0.9	0.020	0.013
64.6	−1.19	−1.3	0.003	0.271
73.5	−1.47	−1.6	0.009	0.642
Average:	−0.67	Sum:	0.184	7.745

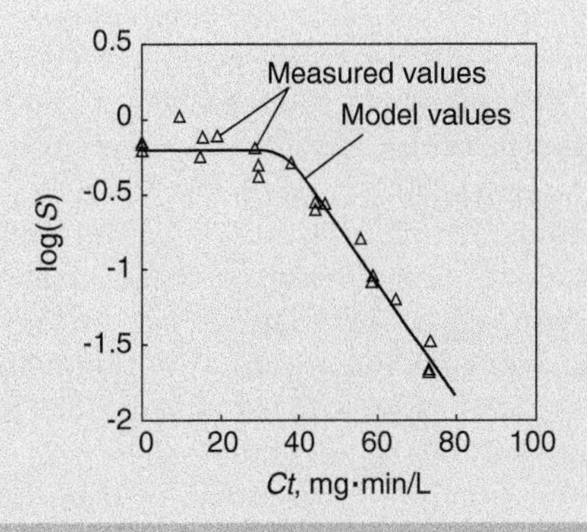

5. Solver minimizes the value of the sum of the [Data-model]^2 when the following values are used:

$$b = 34.9 \text{ mg} \cdot \text{min/L}$$

$$\text{Slope} = 0.036$$

$$\log(N/N_0) = -0.19$$

6. Since the data was plotted on a log (base 10) scale, the coefficient of specific lethality is calculated by multiplying the slope by ln(10):

$$\Lambda_{CW} = (\text{slope}) \ln(10) = 0.036(2.303) = 0.084 \text{ L/mg} \cdot \text{min}$$

7. The coefficient of determination (r^2) is calculated using the data in the spreadsheet as follows:

$$r^2 = 1 - \frac{\sum[\text{Data - Model}]^2}{\sum[\text{Data - Data}_{ave}]^2} = 1 - \frac{0.184}{7.745} = 0.98$$

The *Ct* Approach to Disinfection

In the approaches discussed in the previous section, disinfection effectiveness was related to the product *Ct*. In fact, the product *Ct* has long been used as a basis for disinfection requirements. It is equally practical when the Chick–Watson and Rennecker–Mariñas models are used. The *Ct* product required for achieving a given level of disinfection for a specific microorganism under defined conditions is a useful way of comparing alternate disinfectants and for comparing the resistance of a variety of pathogens. Indeed, the product *Ct* can be thought of as the dose of disinfectant.

The *Ct* concept also allows for the development of a broad overview of the relative effectiveness of different disinfectants and the resistance of different organisms. The *Ct* required to produce a 99 percent (2 log) inactivation of several microorganisms using the five disinfection techniques most often used in water treatment is illustrated on Fig. 13-18. Because of the difference in the behavior from one organism and one disinfectant combination to the next, *Ct* and *It* products range over seven orders of magnitude. For example, the *Ct* product required to inactivate *C. parvum* must be three orders of magnitude higher with combined chlorine than with ozone. Comparing UV disinfection to disinfection with chemical oxidants, little similarity exists between the *It* values and *Ct* values for a single organism. While the required UV doses vary over a range of two orders of magnitude, their variation is much less than that for other disinfectants. The reduced variation may be a result of the fact that UV disinfection of all microorganisms results from a similar protein dimerization mechanism.

The U.S. EPA began the practice of specifying *Ct* products that must be met as a way of regulating the control of pathogens in water treatment

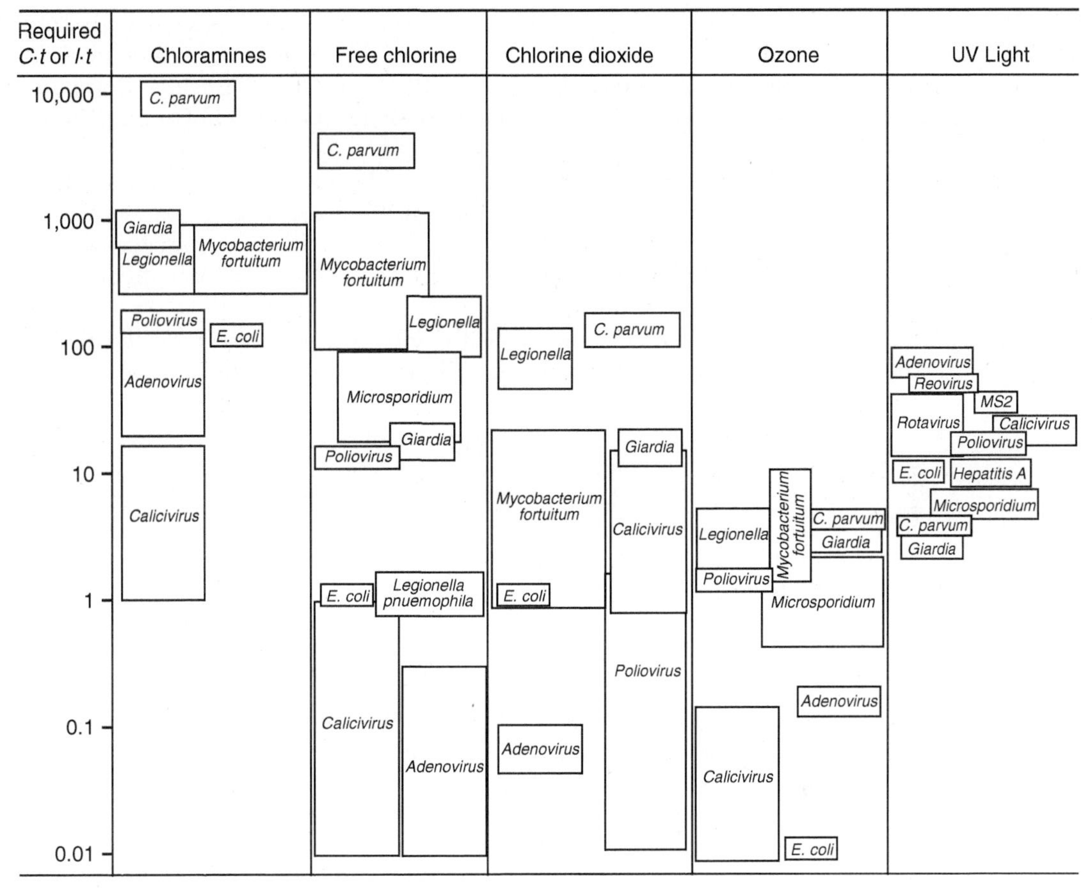

Figure 13-18
Overview of disinfection requirements for 99 percent inactivation. [Adapted from Jacangelo et al. (1997).]

with the promulgation of the Surface Water Treatment Rule. Tables of *Ct* and *It* values required to meet the primary disinfection requirements are available in the *Surface Water Treatment Rule Guidance Manual* available on the EPA website.

13-7 Disinfection Kinetics in Real Flow Reactors

The disinfection kinetics described in Sec. 13-6 were based on studies conducted in CMBRs. While the insight obtained from batch reactors is useful, full-scale continuous-flow systems exhibit more complex nonideal behavior. Of particular importance is the impact of dispersion on the progress of the reaction.

Modeling Approaches

The tanks-in-series (TIS) model, introduced in Sec. 4-11, can be used to model the performance of real (nonideal) reactors. Other methods, including the dispersed-flow model (DFM), and the segregated-flow model (SFM), are available. The DFM and SFM are described in Crittenden et al. (2012).

The TIS model is used to simulate the effects of dispersion on the residence time distribution (RTD) curve by an analogy between a real reactor and a series of CMFRs. The parameter that describes dispersion in the TIS model is the number of reactors in series, n. A high value of n corresponds to low dispersion.

The DFM is used to simulate the effects of dispersion on the RTD by including mass transport by axial dispersion in addition to advection into the mass balance of a plug flow reactor (PFR). In the DFM, it is assumed that all reactants are mixed completely in the lateral direction, but axial transport occurs by advection and dispersion. When dispersion is low, the TIS model and the DFM produce similar results. In the DFM, dispersion is described using the Peclet number (Pe) or the dispersion number (d, $\text{Pe} = 1/d$). A high value of Pe or a low value of d corresponds to low dispersion.

The SFM is used to simulate the effects of nonideal mixing by an analogy between a real reactor and a series of parallel PFRs having residence times that, in sum, match the RTD of the real reactor. In the SFM, it is assumed that the reactants are never completely blended in the reactor; rather the target reactant travels through the reactor in small cells or discrete elements that react with the bulk solution.While the TIS model and the DFM incorporate assumptions about the nature of the RTD curve, an RTD curve must be provided to use the SFM.

When Dispersion Is Important in Disinfection

Minimizing dispersion and short circuiting in disinfection contactors is accepted widely. The U.S. EPA limits the credit for disinfection contact time to the time it takes for the first 10 percent of a tracer to pass through a disinfection contactor (t_{10}), that is, the value of t in Ct is t_{10} instead of τ. California requires the minimum time to the peak concentration on the tracer curve (t_{modal}) to be 90 min and a minimum length-to-width ratio of 40:1 for baffled chlorine contactors in its regulations for reclaiming wastewater for nonrestricted reuse (Cal DHS, 1999).

As a general rule, reducing dispersion is more important when disinfection goals are substantial. For example, dispersion is more important in the design of a contactor that must achieve 4 log of inactivation than in the design of a contactor that must achieve 1 log of reduction. This effect is true regardless of the organism under consideration or its specific reaction kinetics.

A thought experiment can be used to illustrate this effect. Assume a disinfection process is designed to achieve a 4-log reduction of a particular virus and a 1 log reduction of a certain protozoa. Further assume the reactor

Prior to modifications:

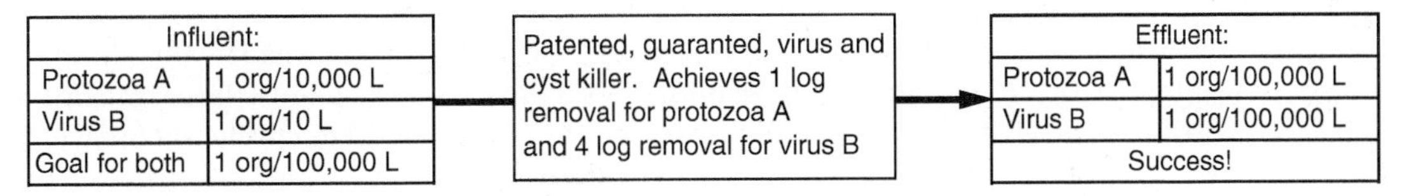

After modifications:

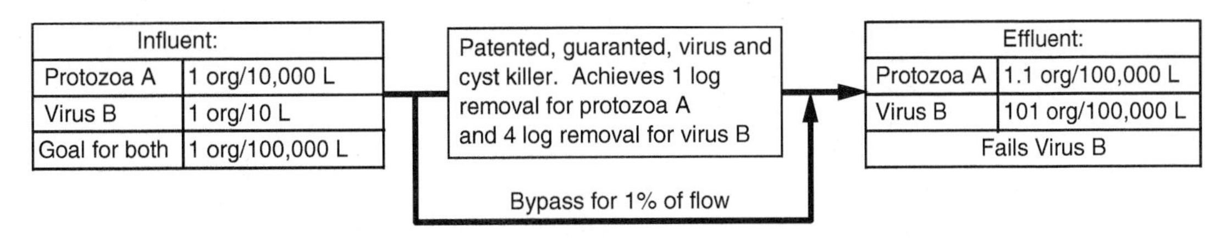

Calculations:
Effluent protozoa A = (1)(0.99) + (10)(0.01) = 1.09/100,000 L
Log removal = 0.96, somewhat below goal

Effluent virus B = (1)(0.99) + (10,000)(0.01) = 101/100,000 L
Log removal = 2.0, far below goal

Figure 13-19
Thought experiment: Dispersion and short circuiting are more important when removal goals are high.

operates as designed and achieves exactly those objectives. A small bypass pipe is installed, and 1 percent of the flow coming into the reactor is diverted so that it flows around the reactor and blends, without disinfection, with the treated water from the reactor. The result of the experiment is illustrated on Fig. 13-19. As illustrated, the small diversion has almost no impact on the removal of protozoa (only 9 percent increase in effluent concentration) but severely compromises the removal of the virus, exposing the consumer to virus levels over 100 times higher than the goal that was being sought.

13-8 Design of Disinfection Contactors with Low Dispersion

Because dispersion is so important in disinfection effectiveness, disinfectant contactors are now typically designed as a separate unit process. Engineered disinfectant contactors are typically of three types: (1) pipelines, (2) serpentine basins, and (3) over–under baffled contactors. Chlorine, combined chlorine, and chlorine dioxide contactors are typically pipelines or serpentine basins. Ozone contactors can be any of the three common types, and additionally deep U-tube contactors have also been used.

Pipeline Contactors

A long channel or pipeline with plug flow characteristics can be an ideal disinfectant contactor. Occasionally, a long pipeline leaving the plant has sufficient contact time to make it an attractive alternative for chlorine or chloramines disinfection. In small treatment plants, large pipelines arranged in a serpentine pattern have been used effectively as contactors with considerable savings in cost as compared to the construction of concrete basins.

Serpentine Basin Contactors

A pipeline is convenient if it is already necessary for some other purpose, but long, baffled, serpentine basins are generally more cost-effective means of achieving low dispersion. An optimal basin would be long and narrow, similar to the pipeline contactor. The approach used most commonly is to design serpentine basins to achieve a specified level of dispersion (Crittenden et al., 2012). However, because of U.S. regulatory requirements, these same facilities must be designed to meet a specified t_{10} value. Computational fluid dynamics are now commonly used to optimize the design of large disinfection contactors.

DESIGNING FOR A SPECIFIED DISPERSION NUMBER

The following equation, based on Taylor's dispersion equation, can be used to design serpentine disinfection contactors:

$$d = \frac{22.7 n R_h^{5/6}}{L} \tag{13-34}$$

where d = dispersion coefficient, m^2/s
n = Manning coefficient, unitless
R_h = hydraulic radius of channel, m
L = length of the contactor, m

Equation 13-34 may be rewritten to describe dispersion using the channel volume and height and length aspect ratios (Trussell and Chao, 1977):

$$d = 22.7 \frac{n}{\beta_L} \left(\frac{\beta_H}{2\beta_H + 1} \right)^{5/6} \left(\frac{\beta_H \beta_L}{V_{\text{ch}}} \right)^{1/18} \tag{13-35}$$

where β_H = height aspect ratio or H/W (channel height/channel width)
β_L = length aspect ratio or L/W (channel length/channel width)
V_{ch} = channel volume, m^3

The dispersion values computed using Eq. 13-35 are not sensitive to the range of β_H values typical for concrete contact chambers (1 to 3). As a result, the following abbreviated form of Eq. 13-35 can be used satisfactorily (Trussell and Chao, 1977):

$$d = \frac{0.14}{\beta_L} \tag{13-36}$$

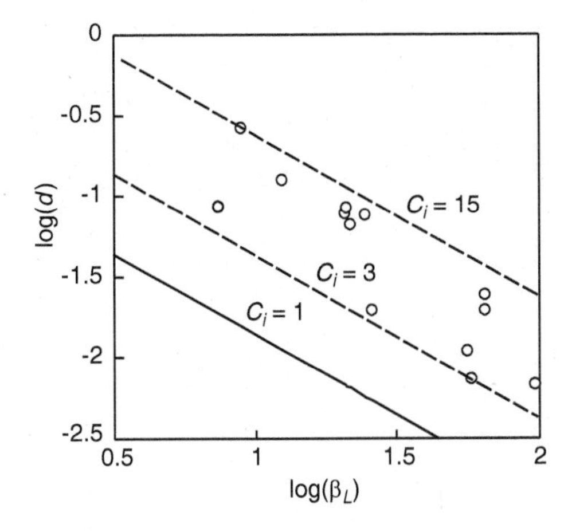

Figure 13-20
Impact of contactor aspect ratio on dispersion (adapted from Trussell and Pollock, 1983).

A plot of dispersion coefficients derived from tracer studies conducted on 17 different field-scale basins is illustrated on Fig. 13-20. Because the field tests were conducted in baffled, serpentine contactors, not long straight channels, none of the studies resulted in the performance predicted using Eq. 13-36. These basins include entrance effects, exit effects, 180° turns, and other nonidealities that would be expected to increase dispersion. Nevertheless, the results shown on Fig. 13-20 are encouraging for two reasons: (1) confirmation of the implication of Eq. 13-36 that dispersion is inversely proportional to the length aspect ratio and (2) the basins fall short of ideal performance, as expected. Recognizing this situation, a coefficient of ideality C_i was proposed (Trussell and Chao, 1977) such that

$$d = \frac{0.14C_i}{\beta_L} \tag{13-37}$$

where C_i = coefficient of ideality

Lines corresponding to C_i values between 3 and 15 are also displayed on Fig. 13-20 and all the data lie on or between them. Based on the data presented on Fig. 13-20, it appears that a good design should be able to equal or exceed the performance estimated by Eq. 13-37 with a C_i value of 3. A best-fit line corresponding to a C_i of 7.1 approximates the performance of a typical older reactor design.

Designing for a Specified t_{10}/τ

As noted in the previous section, the disinfection regulations in the United States account for dispersion by limiting the credit for disinfection contact time to the time it takes for the first 10 percent of a tracer to pass through the disinfection contactor, a value known as t_{10}. The size of a contactor is related to the hydraulic residence time (i.e., $\tau = V/Q$), so to determine the proper size of a contactor a means of estimating t_{10}/τ must be used to be

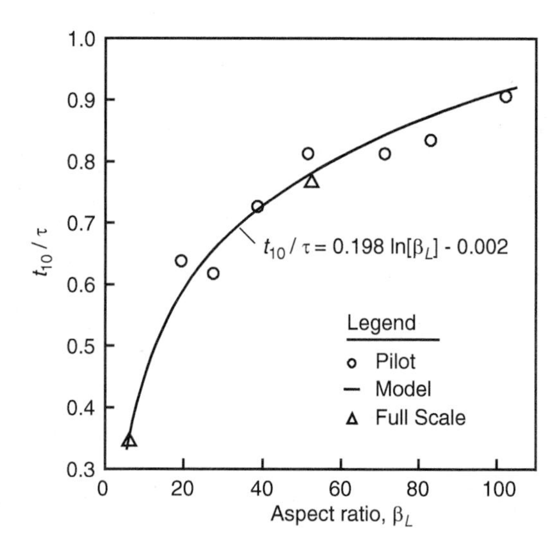

Figure 13-21
Impact of contactor aspect ratio on t_{10}. (Data from Crozes et al. (1999) and Ducoste et al. (2001).]

sure that the design will meet regulations (U.S. EPA, 1989). The impact of baffling rectangular contact tanks to improve hydraulic performance was evaluated by Crozes et al. (1999). A pilot contactor was baffled with nine different configurations having length aspect ratios ranging from 4.8 to 98. In addition, tracer tests were conducted on a full-scale, 34-ML/d (9-mgd) contactor before ($\beta_L = 6.1$) and after ($\beta_L = 52$) modifications. Finally, an empirical correlation between t_{10}/τ and β_L was developed and confirmed (Ducoste et al., 2001):

$$\left(\frac{t_{10}}{\tau}\right) = 0.198 \ln(\beta_L) - 0.002 \tag{13-38}$$

The data and correlation from the study are shown on Fig. 13-21. Note the results from full-scale tests lie close to model predictions.

Although the design of an effective disinfection contact basin requires attention to the length aspect ratio, other design details are also important. Any design detail that causes disturbances in flow is undesirable. Unnecessary gates, ports, or objects that constrict the flow lines are examples. In addition to minimizing the presence of these features, however, special attention should be given to three elements of design in every contactor: (1) inlet configuration, (2) outlet configuration, and (3) turns. Without proper attention, each of these is a likely cause of poor basin performance.

Over–Under Baffled Contactors

Over–under baffled contactors were the most common type of ozone contactor for many years, but are less common now because of increased use of pipe contactors or serpentine basins for ozone contact systems. Pipeline and serpentine basins have better hydraulic characteristics that improve disinfection and minimize bromate formation.

Multichamber over–under baffled contactors often have several chambers where the water alternately flows up over a baffle and down under the next baffle (Rakness, 2005). Schematics of such a contactor are shown on Fig. 13-22. Ozone is typically added to the first one or two chambers via porous stone diffusers situated at the bottom of the chambers. Water enters the first chamber from the top and exists from the bottom. This countercurrent flow configuration (between the water and the air) helps increase the overall ozone transfer efficiency. The water depth in the contactor is typically between 4.6 and 6 m (15 and 20 ft) to achieve high transfer efficiency of the added ozone.

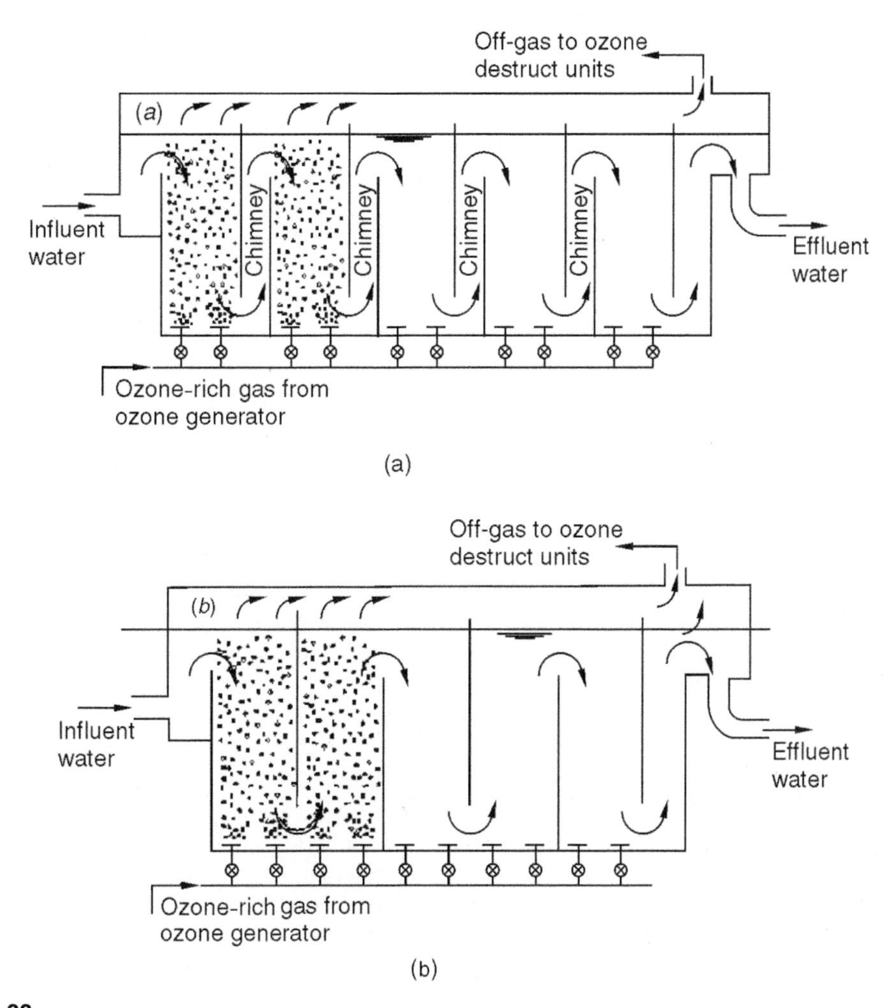

Figure 13-22
Schematics cross-sectional views of two alternate designs for five-chamber, over–under ozone contact chamber: (a) with chimneys and (b) without chimneys.

13-9 Disinfection By-products

The use of chemicals for disinfection and oxidation, as noted in Sec. 13-1, is common, occurring at nearly every drinking water treatment plant in the industrialized world. Familiarity with the by-products of the chemicals used for disinfection and oxidation is important because these by-products can impact consumer health. Some health effects are not fully understood, and some health effects have been identified only recently, making disinfection/oxidation by-products an ever-changing part of the drinking water treatment situation. A brief overview of disinfection by-products, including a historical perspective, some known by-products, and regulatory requirements is presented in this section. Additional details may be found on Chap. 19 of Crittenden et al. (2012).

Historical Perspective

In addition to use as a microbial disinfectant, chlorine—as well as other chemical disinfectants such as ozone and chlorine dioxide—has other benefits, including the ability to eliminate color and destroy many naturally occurring chemicals that cause objectionable taste and odor in the water. Consequently, water treatment plant operators commonly added as much disinfectant as necessary to achieve the desired aesthetic and microbial water quality.

In the early 1970s, researchers in the Netherlands and the United States were able to identify and quantify the formation of chloroform ($CHCl_3$) and other trihalomethanes (THMs) in drinking water and relate this formation to the use of chlorine during treatment. These early findings led to a large number of studies in the United States on the formation of these "by-products" of chlorination.

As additional studies were conducted, it was also found that chloroform was not the only chemical formed as a result of the reaction of chlorine with natural organic matter (NOM) present in water and that in the presence of the bromide ion (Br^-), the reaction between chlorine, bromide, and NOM resulted in the formation of a mix of chlorinated and brominated chemical by-products.

The formation of disinfection by-products (DBPs) is not limited to chlorine disinfection. It has also been found that the bromate ion (BrO_3^-) can be formed when ozone is added to waters containing bromide. Ozone addition to natural water was also implicated in the formation of numerous organic by-products, such as aldehydes (NAS, 1980). Bromate is now regulated, but thus far the presence of organic ozone by-products in drinking water has not been determined to be a public health concern at the typical levels at which they are formed.

Another disinfectant/oxidant used in water treatment is chloramines (combined chlorine). Initially, it was thought that chloramines did not form THMs. Later it was shown that monochloramine, the principal component of combined chlorine, is less reactive than free chlorine, but it also reacts to form DBPs but at much lower concentrations than are formed with free chlorine. The major category of DBPs formed during chloramination of potable water is haloacetic acids (HAAs), mostly the dihalogenated species. It has been found that the chloramination of drinking can result in the formation of NDMA (N-nitrosodimethylamine), which is believed to be of significant public health concern.

Chlorine dioxide forms by-products such as chlorite (ClO_2^-) and chlorate (ClO_3^-) ions, both of which have been suspected to cause health effects. While the health effects of chlorate and chlorite continue to be a topic of debate, there was sufficient concern that an MCL for chlorite was adopted by the U.S. EPA (1998).

Some Known By-products

Some of the known DBPs that form from the use of disinfectants during drinking water treatment are summarized in Table 13-4. While many of the DBPs listed in this table have been detected in some treated waters, they are typically present at very low concentrations.

REGULATORY REQUIREMENTS

Because of concerns about the health effects of chloroform, the U.S. EPA issued the THM Rule in 1979. The four regulated THMs were chloroform ($CHCl_3$), bromodichloromethane ($CHBrCl_2$), dibromochloromethane ($CHBr_2Cl$), and bromoform ($CHBr_3$). The THM Rule set an MCL of 0.10 mg/L (100 μg/L) for the total sum of these four THMs (on a mass basis) in the distribution system. The THM Rule required water systems to monitor a minimum of four locations throughout the distribution system on a quarterly basis. The running annual average (RAA) of all distribution system samples collected every quarter was not to exceed the MCL.

In 1998, the U.S. EPA expanded the number of regulated DBPs by issuing Stage 1 of the Disinfectants/Disinfection Byproducts (D/DBP) Rule (U.S. EPA, 1998). This rule reduced the MCL for total THMs from 0.10 to 0.080 mg/L and added MCLs for additional DBPs, as listed in Table 13-5.

In January 2006, the U.S. EPA promulgated the Stage 2 D/DBP Rule (U.S. EPA, 2006). The Stage 2 rule was designed to further reduce exposure to DBPs without undermining the control of microbial pathogens. The Stage 2 D/DBP Rule requires water utilities to conduct an initial distribution system evaluation (IDSE) to identify the locations in the system with the highest DBP concentrations. Once suitable sampling locations were identified, compliance was evaluated via a locational running annual average (LRAA). The LRAA requires a running annual average at each sampling site, whereas the former RAA required only a running annual average over all sampling sites in the entire system. The IDSE and the LRAA provide increased assurance that customers are receiving more consistent protection against exposure to DBPs, even in areas of a distribution system that had typically had higher DBP concentrations.

Table 13-4
Some known by-products of chlorine, combined chlorine (chloramines), chlorine dioxide, and ozone application during drinking water treatment

Class of Compound	By-product Name	Chemical Formula	By-product of
Trihalomethanes	Chloroform	$CHCl_3$	Chlorine
	Bromodichloromethane	$CHBrCl_2$	Chlorine
	Dibromochloromethane	$CHBr_2Cl$	Chlorine
	Bromoform	$CHBr_3$	Chlorine, ozone
	Dichloroiodomethane	$CHICl_2$	Chlorine
	Chlorodiiodomethane	CHI_2Cl	Chlorine
	Bromochloroiodomethane	$CHBrICl$	Chlorine
	Dibromoiodomethane	$CHBr_2I$	Chlorine
	Bromodiiodomethane	$CHBrI_2$	Chlorine
	Triiodomethane	CHI_3	Chlorine
Haloacetic acids	Monochloroacetic acid	$CH_2ClCOOH$	Chlorine
	Dichloroacetic acid	$CHCl_2COOH$	Chlorine
	Trichloroacetic acid	CCl3COOH	Chlorine
	Bromochloroacetic acid	$CHBrClCOOH$	Chlorine
	Bromodichloroacetic acid	$CBrCl_2COOH$	Chlorine
	Dibromochloroacetic acid	$CBr_2ClCOOH$	Chlorine
	Monobromoacetic acid	$CH_2BrCOOH$	Chlorine
	Dibromoacetic acid	$CHBr_2COOH$	Chlorine
	Tribromoacetic acid	CBr_3COOH	Chlorine
Haloacetonitriles	Trichloroacetonitrile	$CCl_3C{\equiv}N$	Chlorine
	Dichloroacetonitrile	$CHCl_2C{\equiv}N$	Chlorine
	Bromochloroacetonitrile	$CHBrClC{\equiv}N$	Chlorine
	Dibromoacetonitrile	$CHBr_2C{\equiv}N$	Chlorine
Haloketones	1,1-Dichloroacetone	$CHCl_2COCH_3$	Chlorine
	1,1,1-Trichloroacetone	CCl_3COCH_3	Chlorine
Aldehydes	Formaldehyde	$HCHO$	Ozone, chlorine
	Acetaldehyde	CH_3CHO	Ozone, chlorine
	Glyoxal	$OHCCHO$	Ozone, chlorine
	Methyl glyoxal	CH_3COCHO	Ozone, chlorine
Carboxylic acids	Formate	$HCOO^-$	Ozone
	Acetate	CH_3COO^-	Ozone
	Oxalate	$OOCCOO_2{}^-$	Ozone
Ketoacids	Glyoxylic acid	$OHCCOOH$	Ozone
	Pyruvic acid	$CH_3COCOOH$	Ozone
	Ketomalonic acid	$HOOCCOCOOH$	Ozone
Oxyhalides	Chlorite	$ClO_2{}^-$	Chlorine dioxide
	Chlorate	$ClO_3{}^-$	Chlorine dioxide
	Bromate	$BrO_3{}^-$	Ozone
Nitrosamines	N-Nitrosodimethylamine	$(CH_3)_2NNO$	Chloramines

Table 13-4
(*Continued*)

Class of Compound	By-product Name	Chemical Formula	By-product of
Cyanogen halides	Cyanogen chloride	$ClCN$	Chloramines
	Cyanogen bromide	$BrCN$	Chloramines
Miscellaneous	Chloral hydrate	$CCl_3CH(OH)_2$	Chlorine
Trihalonitromethanes	Trichloronitromethane (Chloropicrin)	CCl_3NO_2	Chlorine
	Bromodichloronitromethane	$CBrCl_2NO_2$	Chlorine
	Dibromochloronitromethane	CBr_2ClNO_2	Chlorine
	Tribromonitromethane	CBr_3NO_2	Chlorine

Source: Crittenden et al. (2012).

Table 13-5
Disinfection by-products regulated under the U.S. EPA D/DBP Rule

By-product	Regulatory Limit, mg/L	By-product of
Total THMs[a]	0.080	Chlorine
Five haloacetic acids (HAA5)[b]	0.060	Chlorine
Bromate (BrO_3^-)	0.010	Ozone
Chlorite (ClO_2^-)	1.0	Chlorine dioxide

[a]Sum of four THMs: chloroform, bromodichloromethane, dibromochloromethane, and bromoform.
[b]Sum of five HAAs: monochloroacetic acid, dichloroacetic acid, trichloroacetic acid, monobromoacetic acid, and dibromoacetic acid.

13-10 Residual Maintenance

A critical aspect of a water supply system is the quality of the water delivered to an individual home. Thus, an important consideration in evaluating the use of disinfectants is their ability to control the microbial quality of the treated water in the distribution system, which in addition to health concerns can cause taste and odor problems. Of the disinfectants considered in Sec. 13-1, only the chlorine-based compounds have the ability to maintain a residual, although the effectiveness varies considerably. Unfortunately, UV does not produce a residual, and an ozone residual lasts only a few minutes. In the United States, when ozone and UV are used as primary disinfectants, a secondary disinfectant must be added to maintain germicidal residual in the distribution system. A common disinfectant for this purpose is combined chlorine, primarily in the form of monochloramine. It is important to note that a chlorine residual is not maintained in European water distribution systems.

Typical practice in the United States involves the maintenance of a combined-chlorine residual of at least 0.2 mg/L. Unfortunately, it has been found that biofilms can form even when chlorine is present. Thus, even when maintaining a chloramine residual, it must be recognized that it is impossible to maintain a water distribution system completely free of microorganisms. However, biofilms in pipes may not affect water quality in customers' homes if the biofilms remain attached. Additional details on the control of microbiological quality in distribution systems may be found in LeChevallier et al. (2011).

The problems with maintaining a chloramine residual are related to the configuration and extent of the distribution system and the formation of THMs as described in the previous section. In large water distribution systems, it has been found to be difficult to maintain a residual at the extremities of the system when adding chlorine at a single location because of the decay of the chloramine residual. If enough chlorine is added to maintain the required residual, the formation of THMs can become excessive. In some large water distribution systems, chlorine is added at intermediate points within the system. Another problem with maintaining a residual is related to the design of the distribution system. Modern distribution system design relies on multiple flow paths to a particular point in the distribution system, a practice known as "looping." This design helps maintain pressure throughout the distribution system and prevents water from sitting stagnant in pipes that have low or no flow. In distribution systems without this design, it is common for the dead-end lines at the extremities of the distribution system to have no chlorine residual. In fact, it has been found difficult if not impossible to maintain a residual in dead-end lines.

13-11 Energy and Sustainability Considerations

Energy and sustainability considerations related to disinfection are related to the energy consumed in the production, transport, and use of the different disinfectants. Chlorine is produced in large industrial facilities and shipped to treatment plants in the form of a compressed gas in steel cylinders of various size [typically 68 and 909 kg (150 and 1 ton)]. In very large treatment facilities, chlorine is often delivered by rail in tanker cars. Sodium hypochlorite can be purchased in bulk or produced onsite. Chlorine dioxide is unstable and must be produced onsite. Ozone is produced onsite either from processed air or pure oxygen.

Energy Requirements

Typical values of energy consumption for the production of chemical disinfectants are presented in Table 13-6.

A range of values is given in Table 13-6 because different processes can be used for the production of a given disinfectant. For example, chlorine can be produced with three different processes: the membrane cell, the diaphragm cell, and mercury cell process. Also, in many cases, one or more of the disinfecting chemicals is produced as a by-product of another manufacturing process. The energy requirements for transport

Table 13-6
Approximate energy requirements for the production of various chemical disinfectants

Disinfectant	Formula	kWh/kg
Chlorine	Cl_2	1.6–3
Sodium hypochlorite	NaOCl	3–5
Chlorine dioxide	ClO_2	9–12
Ozone with air feed	O_3	12–20
Ozone with pure oxygen feed	O_3	6–10

Sources: Adapted in part from NCASI (2009) and Black & Veatch Corporation (2010).

and utilization of a given disinfectant are site specific, depending on the location of the treatment facility and the design of the application system.

Sustainability Issues

Sustainability issues with respect to chemical disinfectants are related to their toxicity, chemical properties, and cost. Because chlorine is highly toxic, the public has expressed concern over the transport of compressed chlorine in steel cylinders by truck over city streets. Recognizing the potential risks, some small water agencies have switched from gaseous chlorine to the use of sodium hypochlorite, which can also be generated locally and is deemed to be safer. However, in assessing the risks and long-term sustainability of using gaseous chlorine, the advantages and disadvantages listed in Table 13-7 must be evaluated.

Table 13-7
Some important advantages and disadvantages of using gaseous chlorine for the disinfection of treated water

Advantages	Disadvantages
The use of gaseous chlorine is a well-established technology.	All forms of chlorine are highly corrosive and toxic. Thus, storage, shipping, and handling pose a risk, requiring increased safety regulations, especially in light of the Uniform Fire Code.
Chlorine disinfection is reliable and effective against a wide spectrum of pathogenic organisms.	Chlorine oxidizes certain types of organic matter in water, creating more hazardous compounds [e.g., trihalomethanes (THMs)].
Chlorine is effective in oxidizing certain organic and inorganic compounds.	Some parasitic species have shown resistance to low doses of chlorine, including oocysts of *C. parvum*, cysts, of *Endamoeba histolytica*, and *G. lamblia*, and eggs of parasitic worms.
Chlorination has flexible dosing control.	
Chlorine can eliminate certain noxious odors while disinfecting.	
Germicidal chlorine residual can be maintained in long transmission lines.	

Adapted in part from Solomon et al. (1998) and Asano et al. (2007)

A similar analysis must be made for each disinfectant in evaluating alternatives. Cost is another issue that has an impact on sustainability. However, the cost of the disinfecting chemicals must be balanced against the benefits associated with providing a safe water.

13-12 Summary and Study Guide

After studying this chapter, you should be able to:

1. Define the following terms and phrases and describe the significance of each in the context of disinfection:

absorbance	disinfection by-products	reactivation
action spectra, UV	dose–response curve	sodium hypchlorite
breakpoint chlorination	Inactivation	transmittance
chloramines	over–under contactor	UV dose
combined chlorine residual	pipeline contactor	UV light
Ct		

2. Describe the three strategies used to reduce microbial contaminants in water treatment.
3. Describe the characteristics of disinfectants commonly used in municipal drinking water treatment and the trends regarding their use.
4. Identify the specific chlorine species present in free chlorine and combined-chlorine residuals. Identify which disinfectant is most effective germicidally.
5. Describe the concept of breakpoint chlorination, the role of ammonia, and the type of chlorine residual present in various regions of the chlorine breakpoint curve.
6. Identify the principal types of UV lamps used for disinfection.
7. Calculate the average UV dose in a collimated beam experiment.
8. Calculate the coefficient of specific lethality and lag coefficient from batch reactor disinfection kinetic data.
9. Calculate the log inactivation of microorganisms using the Chick-Watson Law if given the rate constant, disinfectant concentration, and disinfectant contact time.
10. Describe the relative strength of various disinfectants (chlorine, chloramines, chlorine dioxide, ozone, and UV light) with respect to inactivation of various pathogenic organisms (Giardia, Crypto, bacteria, viruses) (note: see Fig. 13-18).
11. Explain how disinfection inactivation credit (log inactivation) is determined using *Ct* values.

12. Calculate the Ct value, hydraulic residence time, and disinfection contactor volume to achieve a specific log inactivation of a microorganism.
13. Explain when dispersion is important in disinfection.
14. Calculate the t_{10}/τ ratio for a length:width ratio of a disinfection contactor.
15. Explain how the production of disinfection by-products affects the ability to accomplish effective disinfection.
16. Describe issues associated with maintaining a combined-chlorine residual in large water distribution systems.

Homework Problems

13-1 A utility desires to achieve a combined chlorine residual of 1.5 mg/L at the optimal chlorine: ammonia ratio for monochloramine formation. Dissolved organic matter in the water exerts an immediate demand that consumes 1 mg/L of chlorine as soon as it is added. Determine the doses of chlorine and ammonia necessary to achieve the desired residual.

13-2 Determine the average UV dose delivered to a sample in a collimated beam experiment that was conducted with the following conditions: $E_s = 10\ \text{mW/cm}^2$, $t = 30$, $R = 0.025$, $P_f = 0.94$, $k_{254} = 0.065\ \text{cm}^{-1}$, and $L = 48.0$ cm. The depth of water in the Petri dish is 10, 14, 15, 16, or 22 mm (water depth to be selected by instructor).

13-3 Hypochlorous acid is a weak acid that dissociates according to Eq. 13-3. For a total free-chlorine ($HOCl + OCl^-$) concentration of 1 mg/L, calculate and plot the concentrations of HOCl and OCl^- for pH values from 6 to 8. (see Sec. 4-4 for acid–base chemistry).

13-4 Assume that the inactivation of a bacteria is due to the parallel effects of HOCl and OCl^- that can be described with the equation

$$\ln\left(\frac{N}{N_0}\right) = -\Lambda_{HOCl} C_{HOCl} t - \Lambda_{OCl^-} C_{OCl^-} t$$

For this bacteria, OCl^- is a weaker disinfectant than HOCl, with $\Lambda_{HOCl} = 0.25$ L/mg·min and $\Lambda_{OCl} = 0.018$ L/mg·min. For the free-chlorine dose and pH range given in Problem 13-3, calculate and plot the required disinfectant contact time t as a function of pH to achieve 2-log inactivation of the bacteria.

13-5 Data from Wattie and Butterfield (1944) on the inactivation of *E. coli* with free chlorine at 2°C and pH 8 are given below. Fit the data to the Chick–Watson and Rennecker–Mariñas models, calculate the coefficients of specific lethality and lag coefficient, and comment on the results.

C, mg/L	T, min	$\log(N/N_0)$
0.05	1.0	−0.02
0.05	3.0	−0.09
0.05	4.9	−0.15
0.05	9.6	−0.68
0.05	18	−2.52
0.07	1.0	−0.06
0.07	3.0	−0.22
0.07	4.9	−0.58
0.07	9.7	−2.28
0.14	1.0	−0.24
0.14	2.8	−0.95
0.14	4.5	−2.15

13-6 Fit the following disinfection data to the Rennecker–Mariñas model and determine the coefficient of lethality and the lag coefficient.

C, mg/L	T, min	$\log(N/N_0)$
1.0	5	0.0
1.1	10	0.0
1.05	25	−1.0
1.03	30	−1.5
1.05	35	−2.1
2.05	20	−2.55
2.0	23	−3.1
2.03	25	−3.45
5.02	11	−4.1

13-7 For the given system (to be selected by instructor), determine the disinfection contact time required to achieve the specified inactivation.

a. 2-log removal of viruses with 1 mg/L chlorine, Chick–Watson kinetics, and $\Lambda = 3.4$ L/mg·min.

b. 4-log removal of viruses with 1.5 mg/L chlorine, Chick–Watson kinetics, and $\Lambda = 3.4$ L/mg·min.

c. 0.5-log removal of *G. lamblia* with 1 mg/L chlorine, Chick–Watson kinetics, and $\Lambda = 0.046$ L/mg·min.

d. 2-log removal of *Cryptosporidium* with 1.2 mg/L chloramines, Rennecker–Mariñas kinetics, $\Lambda = 0.00077$ L/mg·min, and $b = 5500$ mg·min/L.

e. 1-log removal of *Cryptosporidium* with 0.8 mg/L chlorine dioxide, Rennecker–Mariñas kinetics, $\Lambda = 0.083$ L/mg·min, and $b = 35$ mg·min/L.

13-8 For the system in Problem 13-7 (to be selected by instructor), determine the hydraulic residence time and volume of the disinfection contactor required to achieve the specified disinfection for the flow rate and t_{10}/τ ratio given below.
a. $Q = 75$ ML/d, $t_{10}/\tau = 0.45$.
b. $Q = 350$ ML/d, $t_{10}/\tau = 0.68$.
c. $Q = 10$ ML/d, $t_{10}/\tau = 0.45$.
d. $Q = 25$ ML/d, $t_{10}/\tau = 0.55$.
e. $Q = 120$ ML/d, $t_{10}/\tau = 0.62$.

13-9 For the disinfection contactor in Problem 13-8 (to be selected by instructor), determine the length and width of the contactor if the depth is 5 m.

13-10 A treatment plant has been designed to achieve 99 percent inactivation of *C. parvum* using ozonation. The engineer used data on ozonation of *C. parvum* at 20°C for the design, but the plant operates in a northern climate and the water temperature will be 0.5°C in winter. Estimate how much inactivation the plant will actually achieve when the water is at that temperature, assuming the plant will achieve the same Ct value, inactivation of *C. parvum* follows the Chick–Watson model (Eq. 13-31), and that disinfection kinetics follow the Arrhenius' equation (Eq. 4-33). Use an E_a value of 76 kJ/mol.

13-11 A full-scale UV reactor was tested with MS2 and was rated to have an effective fluence of 100 mJ/cm^2. Using an analogy to the thought experiment shown on Fig. 13-6, how much flow could bypass around the reactor during the test without changing $\log(N/N_0)$ for MS2 by more than 10 percent? Assuming no short circuiting, what log inactivation should the reactor achieve with *C. parvum*? What log inactivation of *C. parvum* would the reactor achieve if the bypass were to occur? Given: for UV disinfection, $\Lambda_{CW} = 0.96$ m^2/J for MS2, $\Lambda_{CW} = 25$ m^2/J for *C. parvum*. Discuss the significance of these results.

References

Asano, T., Burton, F. L., Leverenz, H., Tsuchihashi, R., and Tchobanoglous, G. (2007) *Water Reuse: Issues, Technologies, and Applications*, McGraw-Hill, New York.

Black & Veatch Corporation (2010) *White's Handbook of Chlorination and Alternative Disinfectants*, Wiley, Hoboken, NJ.

Cal DHS (1999) *Proposed Regulations: Water Recycling Criteria*, California Department of Health Services, Drinking Water Technical Programs Branch, Sacramento, CA.

Chick, H. (1908) "An Investigation of the Laws of Disinfection," *J. Hygiene*, **8**, 92–158.

Corona-Vasquez, B., Rennecker, J., Driedger, A., and Mariñas, B. (2002) "Sequential Inactivation of Cryptosporidium parvum Oocysts with Chlorine Dioxide Followed by Free Chlorine or Monochloramine," *Water Res.*, **36**, 1, 178–188.

Crittenden, J. C., Trussell, R. R., Hand, D. W., Howe, K. J., and Tchobanoglous, G. (2012) *MWH's Water Treatment: Principles and Design*, 3rd ed., Wiley, Hoboken, NJ.

Crozes, G., Hagstrom, J., Clark, M., Ducoste, J., and Burns, C. (1999) *Improving Clearwell Design for Ct Compliance*, American Water Works Association Research Foundation, Denver, CO.

Ducoste, J., Carlson, K., and Bellamy, W. (2001) "The Integrated Disinfection Design Framework Approach to Reactor Hydraulics Characterization," *J. Water Supply Res. Technol.-Aqua*, **50**, 44, 245–261.

Floyd, R., Sharp, D., and Johnson, J. (1978) "Inactivation of Single Poliovirus Particulates in Water by Hypobromite Ion, Molecular Bromine, Dibromamine and Tribromamide," *Environ. Sci. Technol.*, **16**, 7, 377–383.

Hoigné, J., and Bader, H. (1976) "Role of Hydroxyl Radical Reactions in Ozonation Processes in Aqueous Solutions," *Water Res.*, **10**, 377–386.

Jacangelo, J., Patania, N., Haas, C., Gerba, C., and Trussell, R. (1997) *Inactivation of Waterborne Emerging Pathogens by Selected Disinfectants*, Report No. 442, American Water Works Research Foundation, Denver, CO.

LeChevallier, N. W., Besner, M-C, Friedman, M., and Speight, V. L. (2011) Microbiological Quality Control in Distribution systems, Chap. 21, in J. K. Edzwald (ed.) *Water Quality & Treatment: A Handbook of Drinking Water*, 6th ed., McGraw-Hill, New York.

Linden, K., Shin, G., and Sobsey, M. (2001) "Comparative Effectiveness of UV Wavelengths for the Inactivation of *Cryptosporidium parvum* oocysts in water," *Water Sci. Technol.*, **43**, 12, 171–174.

Morris, C. (1975) Aspects of the Quantitative Assessment of Germicidal Efficiency, pp. 1–10 in D. Johnson (ed.), *Disinfection: Water and Wastewater*, Ann Arbor Science, Ann Arbor, MI.

Morris, J. C. (1966) "The Acid Ionization Constant of HOCl from 5 to 35°," *J. Phys. Chem.*, **70**, 12, 3798–3806.

NAS (1980) The Chemistry of Disinfectants in Water: Reactions and Products, Chap. III, in *Drinking Water and Health*, Vol. 2, National Academy of Sciences, Washington, DC.

NCASI (2009) *Electricity to Make Bleaching Chemicals*, National Council for Air and Stream Improvement, Research Triangle Park, NC

Palin, A. (1975) Water Disinfection—Chemical Aspects and Analytical Control, pp. 71–93, in J. Johnson (ed.), *Disinfection—Water and Wastewater*, Ann Arbor Science, Ann Arbor, MI.

Rakness, K. L. (2005) *Ozone in Drinking Water Treatment: Process Design, Operation, and Optimization*, American Water Works Association, Denver, CO.

Rauth, A. (1965) "The Physical State of Viral Nucleic Acid and the Sensitivity of Viruses to Ultraviolet Light", *Biophys. J.*, **5**, 257–273.

Reckhow, D., Legube, B., and Singer, P. (1986) "The Ozonation of Organic Halide Precursors: Effect of Bicarbonate," *Water Res.*, **20**, 8, 987–998.

Rennecker, J., Mariñas, B., Rice, E., and owens, J. (1997) Kinetics of *Cryptosporidium parvum* Oocyst Inactivation with Ozone, pp. 299–316, in *Proc. 1997 Annual AWWA Conference, Water Research, Vol. C.*, Atlanta, GA.

Saunier, B., and Selleck, R. (1979) "The Kinetics of Breakpoint Chlorination in Continuous Flow Systems," *J. AWWA*, **71**, 3, 164–172.

Scarpino, P., Cronier, S., Zink, M., and Brigano, F. (1977) Effect of Particulates on Disinfection of Enteroviruses and Coliform Bacteria in Water by Chlorine Dioxide, paper 2B–3 *Proceedings of AWWA Water Quality Technology Conference*, Denver, CO.

Solomon, C., Casey, P., Mackne, C., and Lake, A. (1998) Chlorine Disinfection: Fact Sheet, *National Small Flows Clearinghouse*, Morgantown, WV.

Staehelin, J., and Hoigné, J. (1982) "Decomposition of Ozone in Water: Rate of Initiation by Hydroxide Ions and Hydrogen Peroxide," *Environ. Sci. Tech.*, **16**, 10, 676–681.

Trussell, R., and Chao, J. (1977) "Rational Design of Chlorine Contact Tanks," *J. WPCF*, **49**, 4, 659–667.

Trussell, R. R., and Pollock, T. (1983) Design of Chlorination Facilities for Wastewater Disinfection, paper presented at Wastewater Disinfection Alternatives—Design, Operation, Effectiveness," Preconference Workshop for 56th WPCF Conference, Atlanta, GA.

U.S. EPA (1989) "Filtration and Disinfection; Turbidity, *Giardia lamblia,* Viruses, *Legionella,* and Heterotrophic Plate Count Bacteria. Final Rule," *Fed. Reg.* **54**, 124, June 29, 27486–27541.

U.S. EPA (1998) "Disinfectants and Disinfection By-Products Rule: Final Rule," *Fed. Reg.*, **63**, 241, Dec. 16, 69390.

U.S. EPA (2006) "National Primary Drinking Water Regulations: Stage 2 Disinfectant and Disinfection Byproduct Rule, Final Rule," *Fed. Reg.*, **71** 2, 388–493.

Watson, H. (1908) "A Note on the Variation of the Rate of Disinfection with Change in Concentration of the Disinfectant," *J. Hygiene,* **8**, 536.

Wattie, E., and Butterfield, C. (1944) "Relative Resistance of *Escherichia coli* and *Eberthella typhosa* to Chlorine and Chloramines," *Public Health Rep.*, **59**, 52, 1661–1671.

14 Residuals Management

Residuals management is the term used to describe the planning, design, and operation of facilities to reuse or dispose of water treatment residuals consisting of the liquid, semisolid, solid, and gaseous-phase by-products removed during the water treatment process. The principal objective of residuals management is to minimize the amount of material that must ultimately be disposed of by recovering recyclable materials and reducing the water content. Other considerations include minimizing environmental impacts and meeting discharge requirements established by regulatory agencies. Although the residuals from a number of different water treatment processes are identified, a primary focus of this chapter is the residuals from the coagulation and filtration processes. Additional details on the management of residuals from the many other processes used for the treatment of water may be found in the companion reference book to this textbook (Crittenden et al., 2012) and other references (Cornwell and Roth, 2011; U.S. EPA et al., 1996).

The purpose of this chapter is to provide an introduction to (1) the nature of the residuals management problem, including the sources of residuals, (2) the physical and chemical properties used to characterize water treatment plant residuals, (3) the residuals resulting from the coagulation process, (4) the residuals from the filtration process, (5) the options available for the management of residual liquid streams, (6) the options available for the management of residual sludge, and (7) the options available for the ultimate reuse and/or disposal of residuals after processing.

14-1 Defining the Problem

Residuals management can have an important impact on the design and operation of water treatment plants. For existing plants, residuals management systems may limit overall plant capacity if not designed and operated properly. Frequently, residuals are stored temporarily in the treatment process train before removal for treatment, recycle, and/or disposal. The removal of residuals must be optimized for the treatment process train and coordinated with the residuals management systems to maintain water quality. The problem of residuals management can be quantified with respect to (1) the sources, (2) quantities, (3) constituents of concern, (4) environmental constraints, and (5) regulatory constraints.

Sources of Residuals

The principal residuals generated from the treatment of water can be classified as (1) solids and semisolid sludge, (2) liquids, (3) liquids resulting from sludge thickening processes, and (4) gaseous wastes from specialized water treatment processes. The sources of these residuals and a brief description are presented in Table 14-1. The specific types of sludge and liquid waste streams will depend on the type of treatment train as illustrated on Fig. 14-1.

Quantities of Residuals

Typically, 3 to 5 percent of the volume of the raw water entering a conventional water treatment plant may end up as solid, semisolid, and liquid residuals. The bulk of that volume will be the filter waste wash water, which, in a conventional treatment plant, will contain less than 10 percent of the removed solids. Underflow from sedimentation basins typically contains on the order of 0.1 to 0.3 percent of the plant flow but contains most of the removed solids. In a direct or in-line filtration plant, however, all solids removal is accomplished in the filters. Typical values for the quantities of residuals produced by various treatment processes are summarized in Table 14-2. In general, the major portion of the cost associated with residuals management is associated with transport and ultimate disposal. Thus, the most economical solution is to reduce the volume of material for ultimate disposal.

Table 14-1
Sources of semisolid, solid, liquid, and gaseous residuals from the treatment of water

Source of Residual	Description
	Treatment Process Solid and Semi-solid Residuals
Chemical precipitation with alum and iron	Sludge resulting from the chemical precipitation of surface waters that may contain clay, silt, colloidal material, and microorganisms with coagulant chemicals and polymers.
Coarse screens	Coarse screens prevent the entry of debris and fish into the intake structure. The coarse solids retained on the screens include rags, stringy material, and large wood pieces.
Flotation	Float, which in time thickens to sludge, resulting from the flotation process. Sludge that settles is removed periodically in small plants and continuously in large plants.
Presedimentation	Sludge resulting from presedimentation to remove gross amounts of sediment prior to conventional treatment.
Slow sand filter scrapings	Semisolid material resulting from the scraping of the surface of slow sand filters.
Spent sorbents	Solid material used to sorb constituents from solution such as hardness, arsenic, fluoride, phosphorus, and selected organic constituents, which have lost significant adsorptive capacity and/or cannot be reactivated effectively
Traveling screens	Traveling screens are used to prevent grit, sand, and small rocks that have come through the intake from continuing into the treatment facility. Screenings include grit, sand, and small rocks.
Water softening	Lime sludge resulting from the removal of calcium and magnesium from hard waters during precipitation softening.
	Treatment Process Liquid Wastes
Brines and waste wash water from solid sorbents	Brine and rinse water from the reactivation of sorbents along with waste wash water used to clean the beds.
Concentrate	The water that contains the dissolved constituents removed by nanofiltration and reverse osmosis membranes.
Filter waste wash water	Waste wash water from backwashing filters to remove residual solids. Waste wash water is high in turbidity and may contain pathogenic organisms such as *Giardia* and *Cryptosporidium*.
Filter-to-waste water	Water used to condition filters after backwashing that has particles and turbidity above regulatory action levels.
Ion exchange brines and wash water	Brine and rinse water from the reactivation of ion exchange resins whose exchange capacity has been exhausted. Brines from resins used for softening typically containing sodium, chloride, and hardness ions; they are high in TDS, but low in suspended solids.
Slow sand filter wash water	Wash water high in turbidity that may contain pathogenic organisms such as *Giardia* and *Cryptosporidium* resulting from the cleaning of slow sand filter scrapings. (Note in many facilities the scraped sand is not washed for reuse on the filter beds but is used for other purposes.)

(*continues*)

Table 14-1
(Continued)

Source of Residual	Description
	Thickening/Dewatering Process Liquid Wastes
Centrate	Liquid resulting from centrifugal thickening of sludge.
Drying bed decant and underflow	Decant (supernatant) liquid from the surface and underflow (percolate) from sand and other types of drying beds.
Filtrate	Liquid resulting from plate and frame thickening of sludge.
Filtrate (pressate)	Liquid resulting from belt press thickening of sludge.
Supernatant flow	Clear water decanted off residual solids resulting from the gravity and flotation thickening of sludge.
	Treatment Process Gaseous Wastes
Stripping towers	Off-gas from stripping operations contains contaminants that may need to be removed before gas can be discharged.

Table 14-2
Typical production of residuals in water treatment facilities as percent of plant flow

	Percent of Plant Flow	
Type of Residual	**Range**	**Typical**
Alum coagulation sludge	0.08–0.3	0.1
Direct filtration backwash water	4–8	
Filter backwash water	2–5	2[a], 3[b]
Flotation sludge (from reactor surface)	0.01–0.1	0.06
Flotation sludge (from reactor bottom)	0.001–0.04	
Ion exchange brine	1.5–10	5–8
Iron coagulation sludge	0.08–0.3	0.1
Lime-softening sludge	0.3–6	4
Microfiltration backwash water	2–8	6
Reverse osmosis concentrate	10–50	10–50

[a]During warm months.
[b]During cold months.

Constituents of Concern

Constituents of concern contained in the residuals from treatment processes and thickening operations may include the following:

- Pathogenic bacteria and viruses
- *Giardia* cysts and *Cryptosporidium* oocysts
- Turbidity/particles
- Disinfection by-products (DBPs)
- Precursors in the formation of DBPs (natural organic matter)
- Total organic carbon (TOC)

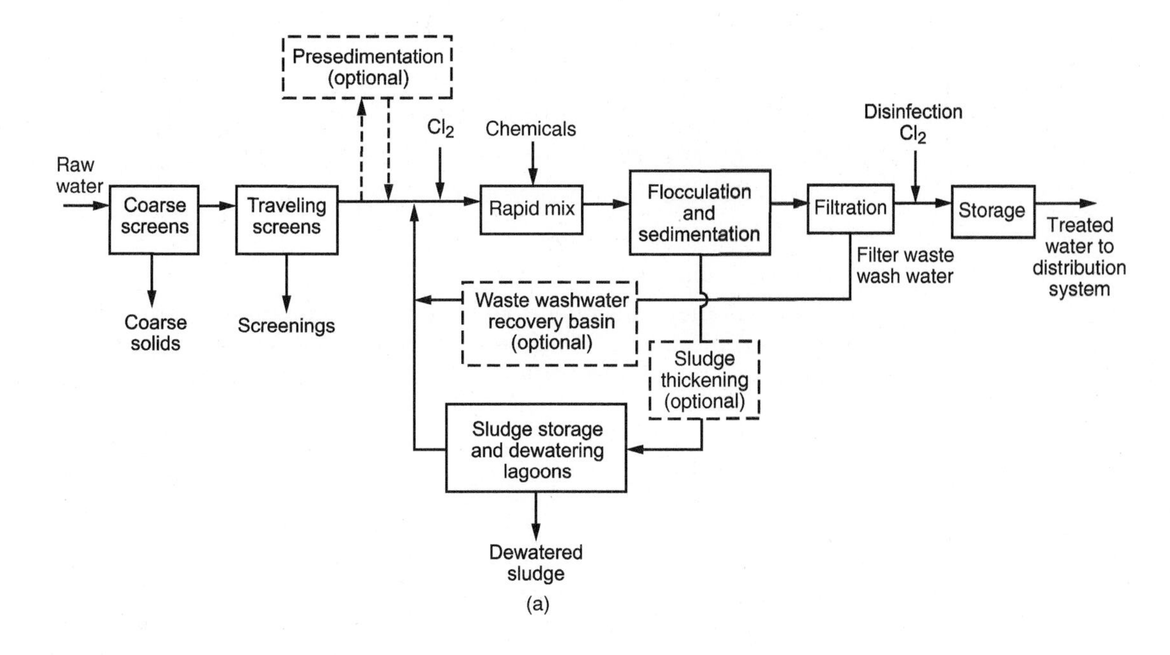

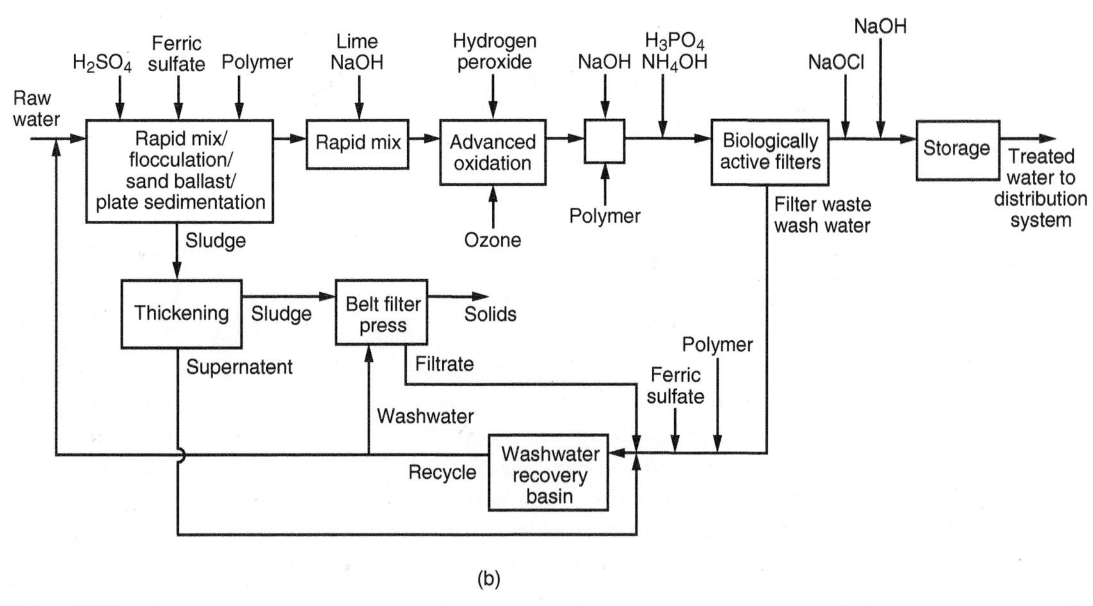

Figure 14-1
Typical water treatment process flow diagrams: (a) small plant with sludge storage lagoons. Future options include the addition of a waste wash-water recovery basin and sludge thickening before discharge to sludge lagoons. (b) Large plant with mechanically intensive sludge-processing facilities.

- ❑ Assimilative organic carbon (AOC)
- ❑ Taste- and odor-causing compounds
- ❑ Synthetic organic compounds (SOCs)
- ❑ Manganese and iron
- ❑ Arsenic or other toxic constituents
- ❑ Radioactive materials
- ❑ Dissolved solids/salt

A variety of other constituents and compounds may also be of concern depending on the source of the water. Where liquid wastes such as waste wash water and return flows from dewatering and thickening operations are recycled, these flows are typically returned to the headworks of the treatment process train. However, concern over the presence of one or more of the above constituents has led, in some cases, to the use of separate treatment facilities for these return flows.

Environmental Constraints

In the past, treatment plant residuals were often discharged to nearby streams, stored in lagoons, or spread on land with little or no processing, practices that created both negative aesthetic and environmental impacts. Aesthetic impacts include discoloration or increased turbidity in receiving waters, buildup of sludge deposits in waterways, and use of large land areas for lagoons. Impacts on the biota are, for sludge and waste wash water, related primarily to the impact(s) on fish from increased water turbidity, pH, and hardness. Redissolved iron and aluminum may also pose a problem. Brines may have toxic effects caused by the high salt concentrations, especially in localized areas around the discharge. Most sludge, if spread on land to any depth, will prevent or inhibit plant growth; however, if mixed adequately into the soil, sludge may have little or no impact on plant growth. Lime sludge may have beneficial impacts on soils containing clay, if used in appropriate amounts.

Regulatory Constraints

Regulatory constraints on residuals disposal have become increasingly severe in recent years. Prior to the late 1960s there was little concern for disposal of water treatment residuals. In most cases residuals were returned to the nearest receiving water, usually the source of the water supply. In the late 1960s some states began considering these residuals as pollutants and began establishing treatment or discharge standards for them. The 1972 Federal Water Pollution Control Act classified water treatment plant residuals as pollutants and categorized them as industrial waste. As such, they are now required to meet standards for best practicable control technology (BPT) currently available and best available technology (BAT) economically achievable. There has also been legislation, both federal and state, to control toxic and hazardous substances. Such regulations, while protecting public and environmental health, can severely limit the available residuals management options and add to the cost of disposal.

14-2 Physical, Chemical, and Biological Properties of Residuals

An understanding of the physical, chemical, and biological properties of the residuals produced by treatment processes is fundamental to determining appropriate management techniques and to design facilities to implement those techniques. The principal physical, chemical, and biological properties used to characterize water treatment plant residuals are reviewed in this section.

Physical Properties

The physical properties of water treatment plant residuals are important for sizing and design of residuals management facilities. The physical properties used most commonly to characterize residuals are summarized in Table 14-3.

Table 14-3
Physical, chemical, and biological properties used to characterize water treatment plant residuals

Parameter	Unit of Expression	Description
		Physical
Total solids	%	Measure of total mass of material that must be handled on dry basis as percent of combined mass of solute and material
Specific gravity	Unitless	Density relative to density of water at 4°C
Density, dry or wet	kg/m^3	Measure of mass per unit volume
Specific resistance	m/kg	Measure of rate at which sludge can be dewatered
Dynamic viscosity	$N \cdot s/m^2$	Measure of resistance to tangential or shear stress
Initial settling velocity	mm/s	Initial settling rate of a water–solids suspension
		Chemical
BOD	mg/L	Estimate of readily biodegradable organic content
COD	mg/L	Measure of oxygen equivalent of organic matter determined by chemical oxidation
pH	Unitless	Measure of effective acidity or alkalinity of solution
Alum content	% or mg/L	Derived from addition of coagulating chemical
Calcium, magnesium content	% or mg/L	Derived from addition of lime for water softening
Iron content	% or mg/L	Derived from addition of coagulating chemical
Silica and inert material	% or mg/L	Material present in surface water supplies
Trace constituents	μg/L or ng/L	Detection of specific constituents of concern
		Biological
Bacteria	no./100 mL	Variable depending on source of water and season
Protozoan cysts and oocysts	no./100 mL	Variable depending on source of water and season
Helminths	no./100 mL	Variable depending on source of water and season
Viruses	no./100 mL	Variable depending on source of water and season

TOTAL SOLIDS

Total solids, comprised of fixed and volatile solids, is the residue remaining after a sample has been evaporated and dried at a specified temperature (typically, 103 to 105°C). Fixed solids are what remains after a sample has been ignited at 500 ± 50°C. The mass of volatile solids corresponds to the difference between the mass of total and fixed solids.

SPECIFIC GRAVITY

The specific gravity of a sample depends on the mass of solids and the distribution between fixed and volatile solids. The specific gravity of the fixed and volatile solids found in water treatment plant residuals is typically 2.65 and about 1.0 to 1.025, respectively. The specific gravity of a sample containing fixed and volatile solids can be determined as follows:

$$\frac{W_s}{S_s \rho_w} = \frac{W_f}{S_f \rho_w} + \frac{W_v}{S_v \rho_w} \tag{14-1}$$

where W_s = weight of total dry solids, kg
S_s = specific gravity of total solids
ρ_w = density of water, kg/m^3
W_f = weight of fixed solids (mineral matter), kg
S_f = specific gravity of fixed solids
W_v = weight of volatile solids, kg
S_v = specific gravity of volatile solids

Thus, if 90 percent by weight of the solid matter in a sludge containing 95 percent water is composed of fixed mineral solids with a specific gravity of 2.5 and 10 percent is composed of volatile solids with a specific gravity of 1.0, then the specific gravity of all solids, S_s, would be equal to 2.17, computed using Eq. 14-1:

$$\frac{1}{S_s} = \frac{0.90}{2.5} + \frac{0.10}{1.0} = 0.46$$

$$S_s = \frac{1.0}{0.46} = 2.17$$

If the specific gravity of the water is taken to be 1.00, the specific gravity of the sludge, S_{sl}, is 1.03, as follows:

$$\frac{1}{S_{s1}} = \frac{0.05}{2.17} + \frac{0.95}{1.00} = 0.97$$

$$S_{s1} = \frac{1.0}{0.97} = 1.03$$

where S_{sl} = specific gravity of wet sludge, unitless

DENSITY OF SLUDGE

Sludge density is dependent on the solids and water content, varying from the density of water (1000 kg/m^3) for sludge below about 1 percent to

1100 kg/m^3 for a 14 percent sludge and higher for relatively dry sludge. The density of sludge, which is a mixture of solid matter and water, can be determined using the following expression:

$$\rho_{sl} = S_{sl}\rho_w \qquad (14\text{-}2)$$

where ρ_{sl} = density of wet sludge, kg/m^3

A reasonable estimate of the density of inorganic sludge, typical of alum or iron salts, can be made by assuming the dry density of the solids is about 2300 kg/m^3 (see Example 14-1).

VOLUME OF SLUDGE

The volume of sludge depends primarily on its water content and only slightly on the character of the solid matter. For example, a 5 percent sludge contains 95 percent water by weight. The volume of a wet sludge may be computed with the following expression:

$$V = \frac{W_s}{\rho_w S_{s1} P_s} \qquad (14\text{-}3)$$

where V = volume of sludge, m^3
P_s = percent solids expressed as a decimal

For approximate sludge volume calculations for a given solids content, it is simple to remember that the volume varies inversely with the percent of solid matter contained in the sludge, as given by

$$\frac{V_1}{V_2} = \frac{P_2}{P_1} \quad \text{(approximate)} \qquad (14\text{-}4)$$

where V_1, V_2 = sludge volumes, m^3
P_1, P_2 = percent solid matter

Application of the above relationships is illustrated in Examples 14-1 and 14-2.

Example 14-1 Estimating density and volume of alum sludge

Determine the density and liquid volume of 1000 kg of alum sludge with the following characteristics:

Item	Unit	Value
Solids	%	15
Volatile matter	%	6
Specific gravity of fixed solids	Unitless	2.65
Specific gravity of volatile solids	Unitless	1.0
Temperature	°C	20

Solution

1. Compute the specific gravity of all the solids in the sludge using Eq. 14-1:

$$\frac{1}{S_S} = \frac{0.94}{2.65} + \frac{0.06}{1.0} = 0.41$$

$$S_S = \frac{1}{0.41} = 2.44$$

2. Compute the specific gravity of the wet alum sludge:

$$\frac{1}{S_S} = \frac{0.15}{2.44} + \frac{0.85}{1.0} = 0.91$$

$$S_S = \frac{1}{0.91} = 1.10$$

3. Compute the density the wet alum sludge using Eq. 14-2. The density of water at 20°C from App. C = 998.2 kg/m^3.

$$\rho_{sl} = S_{sl}\rho_w = (998.2 \text{ kg/m}^3)(1.10) = 1{,}098 \text{ kg/m}^3$$

4. Compute the volume of wet sludge at 20°C using Eq. 14-3:

$$V = \frac{W_s}{\rho_w S_{sl} P_s} = \frac{1000 \text{ kg}}{(998.2 \text{ kg/m}^3)(1.10)(0.15)} = 6.07 \text{ m}^3$$

OTHER PHYSICAL PROPERTIES

Other important physical properties include specific resistance, dynamic viscosity, and initial settling velocity. All of these properties are dependent on the solids concentrations and the relative proportions of coagulant and other materials in the sludge. Specific resistance is a measure of the rate at which a sludge can be dewatered. Although developed for the vacuum filtration process, specific resistance has been found to be a useful parameter for assessing the dewaterability of sludge by gravity settling, centrifugation, belt filtration, plate and frame pressure filtration, and sand beds.

Chemical Properties

The chemical properties of residual sludge are related directly to the chemical content of the raw water and the coagulant chemicals. Important chemical characteristics are summarized in Table 14-3. The biochemical oxygen demand (BOD), chemical oxygen demand (COD), TOC, and related organic content are representative of the dissolved and suspended organic materials and algae removed from the water. The inorganic solids are derived from the coagulant chemicals and the clay and sediments removed from the raw water. The pH and dissolved solids in the liquid portion of the sludge are about the same as those in the water being treated.

Biological Properties

Water treatment plant residuals, as noted in Table 14-3, may contain a variety of microorganisms, depending on the source, the quality of the raw water, the treatment process employed (e.g., prechlorination), and the time of year. Coagulation sludge, filter waste wash water, and membrane concentrates will contain bacteria, protozoan cysts and oocysts, and viruses removed during treatment. It is not possible to generalize on the number of microorganisms that may be present per unit mass or volume for the reasons cited above.

14-3 Alum and Iron Coagulation Sludge

Coagulation sludge is produced by the coagulation and settling of natural turbidity by adding coagulant chemicals. In water treatment plants, coagulation sludge is collected in the sedimentation basins and on the filters. Residuals collected on the filters are removed from the filters during backwashing and, if the waste wash water is recovered, are removed from the waste wash water by settling.

Coagulation sludge is grouped according to the type of primary coagulant employed. The principal types of coagulant employed are (1) hydrolyzing metal salts of alum and iron, (2) prehydrolyzed metal salts such as polyaluminum chloride (PACl) and polyiron chloride (PICl), and (3) synthetic organic polymers. Estimating the volume and mass of sludge and representative physical properties of sludge is considered in this section.

Estimating Quantities of Coagulant Sludge

Typical overall values for the quantities of coagulant sludge produced were summarized previously in Table 14-2. For design purposes, the amount of sludge anticipated at a plant can be estimated based on the quality of the raw water and the type of chemical treatment. The suspended solids concentration of the sludge may be safely assumed to be equal to the suspended solids of the raw-water, or, if total suspended solids (TSS) data are not available, it may be estimated from turbidity data. It is important to note, however, that there is great variability in the TSS/turbidity ratio depending on the organic content of the water source. The ratio for most water sources will vary between 1 and 2, with a typical value being about 1.4. For raw-water turbidities less than 10 NTU the ratio is nearly equal to 1.

PRECIPITATION REACTIONS

Typical precipitation reactions for alum and iron when used as coagulants, as shown in Eqs. 5-6, 5-9, and 5-10 in Chap. 5, are as follows:

$$Al_2(SO_4)_3 \cdot 14H_2O \rightarrow 2Al(OH)_3 + 6H^+ + 3SO_4^{2-} + 8H_2O \qquad (14\text{-}5)$$

$$Fe_2(SO_4)_3{\cdot}9H_2O \rightarrow 2Fe(OH)_3 + 6H^+ + 3SO_4^{2-} + 3H_2O \quad (14\text{-}6)$$

$$FeCl_3{\cdot}6H_2O \rightarrow Fe(OH)_3 + 3H^+ + 3Cl^- + 3H_2O \quad (14\text{-}7)$$

For alum or iron sludge, the precipitates are largely aluminum and iron hydroxide, respectively, and the quantity precipitated can be calculated from the stoichiometry. Using Eq. 14-5, it can be calculated that a total of 0.26 g of sludge on a dry-solids basis will be produced for each gram of alum [$Al_2(SO_4)_3{\cdot}14H_2O$] added [e.g., 78 g/mol $Al(OH)_3$ × (2 mol/mol)/(594 g/mol alum)]. The corresponding values for ferric coagulants are 0.53 g sludge/g ferric sulfate [$Fe_2(SO_4)_3$]added, and 0.66 g sludge/g ferric chloride ($FeCl_3$) added. Typical values that can be used to estimate the quantity of alum and iron sludge are given in Table 14-4.

PACl [$Al_nCl_{(3n-m)}(OH)_m$] is supplied in solution form containing varying amounts of aluminum. The mass of sludge produced can be estimated using the relationship

$$\frac{\text{mg Al(OH)}_3}{\text{mg PACl}} = C_{Al}\frac{MW_{Al(OH)3}}{MW_{Al}} \quad (14\text{-}8)$$

where C_{Al} = concentration of aluminum in PACl, mg Al/mg PACl
$MW_{Al(OH)3}$ = molecular weight of $Al(OH)_3$, 78 g/mol
MW_{Al} = molecular weight of Al, 27 g/mol

Typical values that can be used to estimate the quantity of PACl sludge are given in Table 14-4.

For polymer sludge or sludge with polymer used as coagulant aid, the amount of polymer added should also be included in the calculation of the

Table 14-4
Typical values used to estimate quantities of sludge resulting from addition of coagulating chemicals and polymers, turbidity removal, and softening in water treatment processes

Process	Unit	Range	Typical Value
Coagulation			
Alum, $Al_2(SO_4)_3{\cdot}14H_2O$	kg dry sludge/kg coagulant	0.2–0.33[a]	0.26
Ferric sulfate, $Fe_2(SO_4)_3$	kg dry sludge/kg coagulant	0.5–0.53[a]	0.53
Ferric chloride, $FeCl_3$	kg dry sludge/kg coagulant	0.6–0.66[a]	0.66
PACl	kg dry sludge/kg PACl	(0.0372–0.0489) × Al (%)	(0.0489) × (Al, %)
Polymer addition	kg dry sludge/kg coagulant	1	1
Turbidity removal	mg TSS/NTU removed	1–2	1.4
Softening			
Ca^{2+}[b]	kg dry sludge/kg Ca^{2+} removed	2.0	2.0
Mg^{2+}[c]	kg dry sludge/kg Mg^{2+} removed	2.6	2.6

[a]Value without bound water.
[b]Sludge is expressed as $CaCO_3$.
[c]Sludge is expressed as $Mg(OH)_2$.

total amount of sludge produced. Other coagulant aids, such as bentonite or activated silica, should also be considered in the calculation as well as any other chemicals or materials, such as activated carbon, that may be collected in the basins or filters.

ESTIMATING SLUDGE MASS

The total sludge mass produced can be calculated as

$$C_{sl} = C_{coag}\,(\text{CR}) + \text{TSS} + C_{aid} \tag{14-9}$$

where C_{sl} = total concentration of sludge produced, kg/m^3
C_{coag} = coagulant dose, kg/m^3
CR = coagulant ratio of sludge produced to coagulant added, kg/kg
TSS = total suspended solids, kg/m^3
C_{aid} = dose of coagulant aids or other additives, kg/m^3 (or mg/L)

The ratio of sludge produced to coagulant added, CR, is given in Table 14-4. For the enhanced coagulation and precipitation process, which results in the production of additional sludge, U.S. EPA (1996) has recommended a value of CR = 0.36 to account for the additional sludge. Application of Eq. 14-9 for alum, ferric sulfate, and ferric chloride using the factors developed above and given in Table 14-4 for the expected quantities of sludge is as follows. Computation of the quantities of sludge using Eq. 14-9 is illustrated in Example 14-2.

Example 14-2 Determination of quantity of sludge from coagulant addition and the volume of sludge as percentage of the total treatment plant flow

Determine the mass and volume of sludge produced from alum precipitation for the removal of turbidity, given the following design criteria: (1) flow rate is 43.2 ML/d (0.5 m^3/s), (2) average raw-water turbidity is 25 NTU, (3) average alum dose is 30 mg/L, (4) average polymer dose is 1 mg/L, (5) the sludge solids concentration is 5 percent with a corresponding specific gravity of 1.05, and (6) the temperature is 15°C. Assume the ratio between TSS and turbidity expressed as NTU is 1.4. Also determine the volume of sludge as percentage of the total treatment plant flow.

Solution

1. Determine the concentration of dry sludge using Eq. 14-9:

$$C_{coag} = \frac{(30\ \text{mg/L})(10^3\ \text{L/m}^3)}{10^6\ \text{mg/kg}} = 0.030\ \text{kg/m}^3$$

$$\text{TSS} = (25\ \text{NTU})(1.4\ \text{g/m}^3\cdot\text{NTU})(10^{-3}\ \text{kg/g}) = 0.0035\ \text{kg/m}^3$$

$$C_{\text{aid}} = \frac{(1\ \text{mg/L})(10^3\ \text{L/m}^3)}{10^6\ \text{mg/kg}} = 0.001\ \text{kg/m}^3$$

$$C_{\text{sl}} = C_{\text{coag}}(\text{CR}) + \text{TSS} + C_{\text{aid}} = 0.030(0.26) + 0.035 + 0.001\ \text{kg/m}^3$$

$$= 0.0438\ \text{kg/m}^3$$

2. Determine the mass of dry sludge per day:

$$M_{\text{sl}} = QC_{\text{sl}} = (43.2\ \text{ML/d})(0.0438\ \text{kg/m}^3)(10^3\ \text{m}^3/\text{ML}) = 1890\ \text{kg/d}$$

3. Estimate the volume of the sludge using Eq. 14-3. The density of water at 15°C from App. C = 998.2 kg/m³:

$$V_{s1} = \frac{1890\ \text{kg/d}}{(998.2\ \text{kg/m}^3)(0.05)(1.05)} = 36.1\ \text{m}^3/\text{d}$$

4. Estimate the volume of the sludge as a percent of the total flow:

$$\text{Sludge volume, \% of total flow} = \frac{(36.1\ \text{m}^3/\text{d})(100)}{(43.2\ \text{ML/d})(10^3\ \text{m}^3/\text{ML})} = 0.084\%$$

Physical Properties of Coagulant Sludge

The solids concentrations and physical properties are the most important properties for sizing and design of residuals management facilities. Solids concentrations depend on the design and operation of the sedimentation basins in addition to the type of sludge and its composition. The physical properties of alum and iron sludge are summarized in Table 14-5. For example, alum sludge from an upflow clarifier would typically be drawn off at a concentration of 0.1 to 0.3 percent solids, compared to sludge from a horizontal-flow basin at 0.2 to 1.0 percent or more. However, because of the wide range of values that have been reported, the information given in Table 14-5 must be used with caution.

If sludge is allowed to accumulate for a month or longer in a horizontal-flow sedimentation basin, it may thicken to 4 to 6 percent solids. Sludge that has a relatively high proportion of alum or iron coagulant, resulting from the treatment of low-turbidity water, will have lower solids concentrations than will those with relatively higher proportions of turbidity or silt. Coagulation of waters having substantial algae concentrations will also result in light, low-solids-concentration sludge.

Chemical Properties of Coagulant Sludge

The chemical characteristics of coagulant sludge are directly related to the chemical content of the raw water and the coagulant chemicals. Typical data on the chemical characteristics of alum and iron coagulant sludge

Table 14-5
Typical physical properties and chemical constituents of alum and iron sludge from chemical precipitation

Item	Unit	Type of Sludge	
		Alum	Iron
	Physical Properties		
Volume	% water treated	0.05–0.15	0.06–0.15
Total solids	%	0.1–4	0.25–3.5
Dry density	kg/m^3	1200–1500	1200–1800
Wet density	kg/m^3	1025–1100	1050–1200
Specific resistance[a]	m/kg	$10-50 \times 10^{11}$	$40-150 \times 10^{11}$
Viscosity at 20°C	$N{\cdot}s/m^2$	$2-4 \times 10^{-3}$	$2-4 \times 10^{-3}$
Initial settling velocity	m/h	2.2–5.5	1–5
	Chemical Constituents		
BOD	mg/L	30–300	30–300
COD	mg/L	30–5000	30–5000
pH	Unitless	6–8	6–8
Solids			
$Al_2O_3{\cdot}5.5H_2O$	%	15–40	
Fe	%		4–21
Silicates and inert materials	%	35–70	35–70
Organics	%	10–25	5–15

[a]Values of secific resistance reported in literature in units of s^2/g must be multiplied by 9.81×10^3 [$(s^2/g)(9.81\ m/s^2)(10^3 g/kg) = m/kg$] to obtain units of m/kg.

are given in Table 14-5. Data for lime-softening sludge may be found in Crittenden et al. (2012).

14-4 Liquid Wastes from Granular Media Filters

Waste wash water from the cleaning of granular or membrane filters and water from the filter-to-waste operation are the most common types of liquid waste produced at water treatment plants. The corresponding quantities of water are difficult to estimate because they will depend on the raw-water quality, the degree and effectiveness of the treatment processes preceding the filtration step, the duration of the filter run, and the duration and type of back wash cycle employed. The physicochemical properties of the waste wash water will depend on the characteristics of the source water and the type of treatment.

Estimating Quantities of Filter Waste Wash Water

Based on the operating experience from a variety of water treatment plants, the quantity of waste wash water for both granular and membrane filters will typically comprise from 2 to 5 percent of the total amount of water processed. Some designers use 5 percent as a design value for the quantity of waste wash water as a factor of safety. Pilot plant studies, of sufficient scale to avoid wall effects, can be used to obtain information on backwashing rates and frequencies.

Filter design criteria that are relevant to estimating the quantity of filter waste wash water are the unit filter run volume (UFRV) and the unit backwash volume (UBWV). These concepts are introduced in Chap. 7 and are used to determine the recovery or production efficiency for a filter. The design criterion for production efficiency is typically 95 percent or greater. Typically, waste wash water quantities are 8 m^3/m^2 (200 gal/ft^2). To achieve a filter production efficiency of 95 percent, the UFRV would have to be at least 200 m^3/m^2 (5000 gal/ft^2) a run.

Filter-to-Waste Water

Another liquid waste stream from granular filters occurs when a granular media filter is initially brought online after backwashing and the initial filter effluent is wasted, called *filter-to-waste*. Filter-to-waste flow typically occurs for 15 min to 1 h after a filter is backwashed, but the specific time a filter operates in a filter-to-waste mode is based on the filter effluent quality. The filter-to-waste flow may be captured and recycled through the treatment plant headworks or, in some cases, directly upstream of the filters. Filter-to-waste water quality is different than both filter waste wash water and supernatant from dewatering processes; thus it may need to be separated from these other waste streams.

Physicochemical Properties of Waste Wash Water

The physicochemical properties of waste wash water are reported in Table 14-6. As reported in Table 14-6, the average total suspended solids

Table 14-6
Typical physical properties and chemical constituents of granular filter waste wash water

Item	Unit	Granular Filter
Physical Properties		
Volume	% water treated	1–5
Total solids	mg/L	100–1000
Specific gravity	sg	1.00–1.025
Specific resistance	m/kg	$11 - 120 \times 10^{10}$
Viscosity at 20°C	$N{\cdot}s/m^2$	$1 - 1.2 \times 10^{-3}$
Initial settling velocity	m/h	0.06–0.15
Chemical Constituents		
BOD	mg/L	2–10
COD	mg/L	20–200
pH	Unitless	7.2–7.8
Solids		
Al_2O_3 or Fe	%	20–50
Silicates and inert materials	%	30–40
Organics	%	15–22

concentration is typically on the order 100 to 1000 mg/L. Because of the low concentration of solids and their settleability, waste wash water has historically been (1) returned directly to the headworks of the treatment plant when comprising less than 10 percent of the plant flow; (2) discharged to a flow equalization basin and then returned to the headworks of the treatment plant; (3) discharged to waste wash water recovery ponds, basins, or lagoons where it is allowed to settle for 24 h or more before being decanted and returned to the headworks of the treatment plant; or (4) discharged to surface waters with the appropriate National Pollutant Discharge Elimination System (NPDES) permit in place.

14-5 Management of Residual Liquid Streams

The principal liquid waste streams in conventional water treatment plants utilizing granular media filtration are, as discussed above, filter waste wash water and filter-to-waste water. Other waste streams are comprised of recycle flows from sludge-processing operations, including centrate, filtrate, pressate, supernatant flow, and leachate. The combined volume of these waste streams may approach 4 to 5 percent of the total water treated, depending on the processes employed. In the past, as noted above, these streams were returned to the headworks, discharged to nearby water bodies, land applied, or discharged to wastewater collection systems. Because of new regulations, many of these past practices are no longer acceptable. As a result, the management of these liquid waste streams is a major issue in the design and operation of most water treatment plants. The management options for dealing with these constituents are considered below.

Flow Equalization Lagoons or Basins

Flow equalization is used to reduce the impact of the intermittent high-volume flows from backwashing operations (see Fig. 14-1b). By returning the waste wash water at a more constant rate, the impact on treatment process performance is minimized. Where the equalization basin also functions as a settling basin, the impact of the return flow is further mitigated (see Fig. 14-2). If the raw water has been coagulated effectively and flocculated prior to filtration, the solids in the filter waste wash water generally settle rapidly. Coagulants and coagulant aids such as alum and cationic polymer may be added to improve the settling characteristics of the solids in the waste wash water.

Treatment of Recycle Waste Streams

Because of concern over the presence and recycling of microorganisms, potential increases in the concentration of disinfection by-products, as well as other concerns such as taste and odor, separate treatment facilities are being utilized for waste wash water, especially in larger plants. Treatment options that have been utilized include:

- Lagoons without or with chemical addition
- Batch sedimentation without or with chemical addition

Figure 14-2
Typical waste wash water basins used for flow equalization and as settling basins.

- High-rate sedimentation without or with chemical addition and preflocculation
- Dissolved air flotation

Because of the larger area required for waste wash water storage basins, the use of high-rate sedimentation (see Fig. 14-3) has become common in larger water treatment plants. The sludge resulting from the high-rate sedimentation process as well as from the treatment options identified above is typically combined with other plant sludge for further treatment.

Disposal of Liquid Streams

In some cases, residual liquid waste streams have been discharged to surface waters and/or to wastewater collection systems. The ability to use either of these options is site specific.

DISPOSAL TO SURFACE WATERS

Surface water discharges are regulated under the Clean Water Act through the NPDES. These laws consider water treatment and supply to be an industry, and, therefore, consider water treatment residuals, such as concentrate, an industrial waste. The NPDES permits can specify a variety of water quality requirements, depending on classification of the receiving water body (e.g., potable water source, trout stream). State and local governments may impose additional restrictions on surface water discharges.

DISCHARGE TO WASTEWATER COLLECTION SYSTEM

The same laws that govern surface water discharge apply to wastewater collection system disposal. Pretreatment of the residual prior to discharge to the wastewater plant may be required because of state regulations or conditions imposed by the wastewater plant. In general, local pretreatment

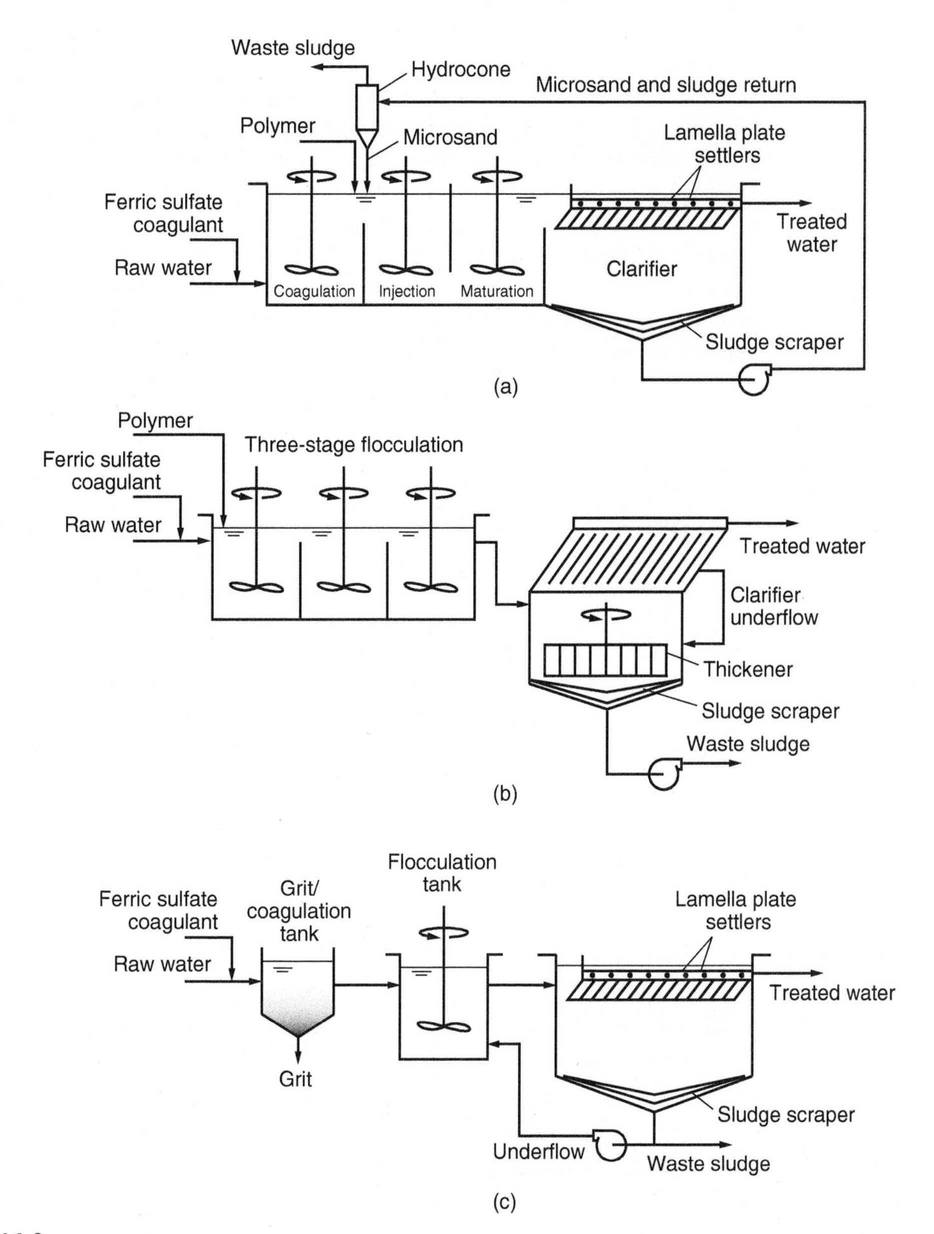

Figure 14-3
High-rate clarification processes: (a) ballasted flocculation, (b) lamella plate clarification, (c) dense sludge. (Adapted from Tchobanoglous et al., 2003.)

guidelines will cover the discharge from a water treatment plant to the wastewater collection system.

Direct discharge to a wastewater collection system has a low capital cost and may also have low operation and maintenance costs, depending on monitoring requirements and sewer use fees. An advantage of this method is simpler permitting requirements. Discharge to a collection system is the easiest disposal method if a local wastewater treatment plant is willing to accept the waste, an issue that is often facilitated when a municipality operates both the water and wastewater systems. In most cases a condition of discharge may be continuous monitoring of the organic strength and solids content of the residual flow.

14-6 Management of Residual Sludge

The major unit operations and processes that are employed for the management of residuals are reviewed in this section. A generalized process diagram showing the various unit processes that may be used in residuals management and the sequence in which they may be assembled to form complete treatment systems are shown on Fig. 14-4.

A complete residuals management system is made up of one unit process from one or more of the process steps shown (e.g., thickening/dewatering, conditioning) and must include one of the unit processes from the final reuse and/or disposal step. Some typical residuals management processes are as follows:

- For alum sludge: gravity thickening, chemical conditioning, centrifugation, and final disposal to sanitary or monofill landfill
- For alum sludge: sludge lagoons, decant recovery and recycle, and final disposal to a sanitary or monofill landfill or wastewater collection system
- For lime sludge: gravity thickening, filter press dewatering, heat drying, lime calcining, and reuse
- For lime sludge: sludge lagoons, drying beds, cropland application, or monofill landfill
- For reverse osmosis concentrate: final disposal directly to brackish surface water, the ocean, deep-well injection, or wastewater collection system
- For ion exchange brines: membrane concentration, thermal brine concentration, and evaporation ponds

Unit processes that have proven to be the most successful and to have significant capabilities for dewatering sludge from water treatment plants include drying beds, vacuum filtration, pressure filter press, belt filter press, centrifugation, and alum and lime recovery.

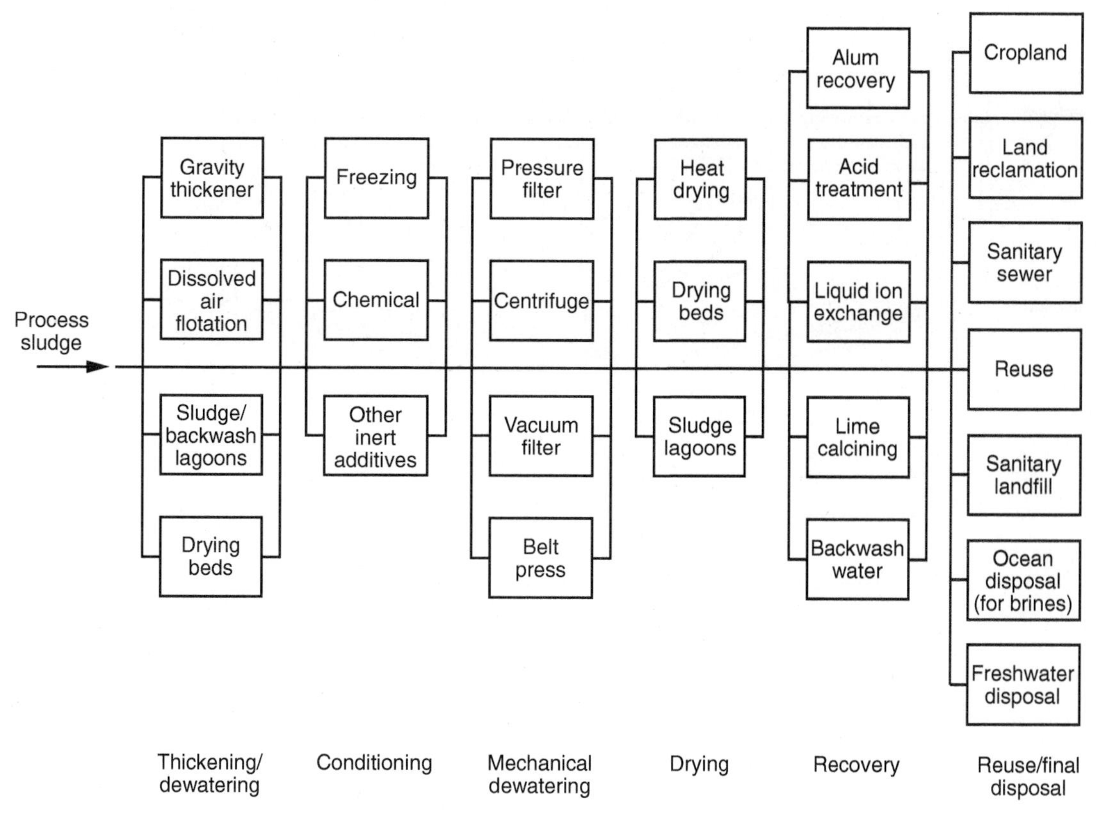

Figure 14-4
Unit operations and processes for management of water treatment plant sludge.

Thickening/ Dewatering

Thickening to increase the solids content of sludge involves concentration of the solids by settling and the removal of excess water by decanting. Decanted water is usually recovered unless the water contains objectionable tastes or odors or large numbers of algae or other microorganisms, while the solids are processed further or disposed of. The principal thickening/dewatering options are (1) mechanical gravity thickening, (2) flotation thickening, (3) sludge lagoons, and (4) sand drying beds.

MECHANICAL GRAVITY THICKENING

Mechanical gravity thickening is used most commonly as the first step in the residuals management process in larger plants; sludge lagoons, discussed below, are used most commonly for small plants. Mechanical gravity thickening is accomplished in specially designed reactors similar to a solids-contact clarifier or sedimentation tank. Sludge is introduced into the tank and allowed to settle and compact. Gentle agitation of the sludge prior to settling creates channels in the sludge matrix for water to escape

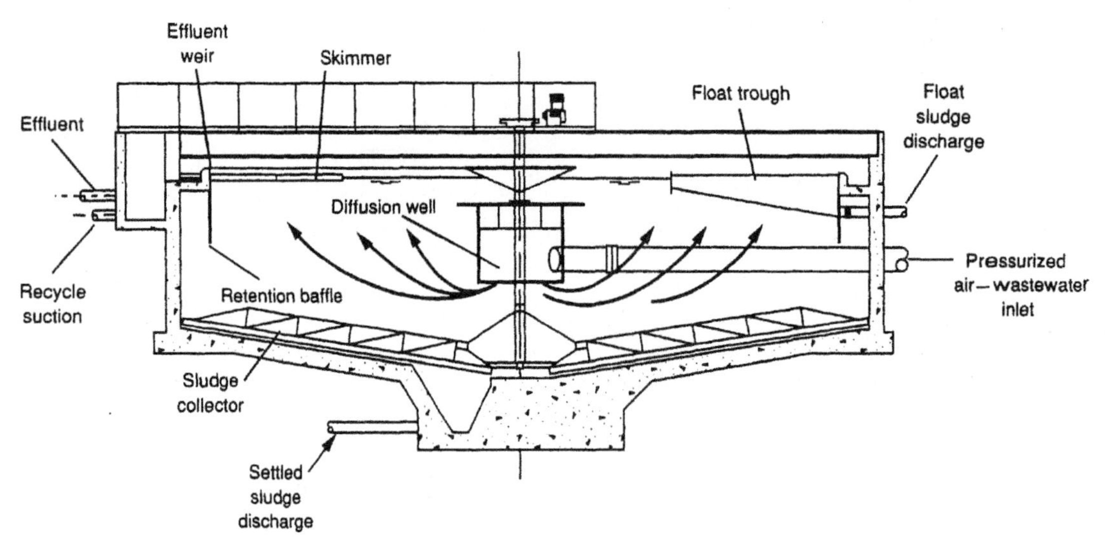

Figure 14-5
Section through typical flotation thickener for water treatment plant sludge.

and promote densification of the solids. The thickened sludge is collected and withdrawn at the bottom of the tank.

FLOTATION THICKENING

Dissolved air flotation (DAF) thickening (see Fig. 14-5) involving both sedimentation and flotation has been used successfully for dewatering. Typically, DAF thickening has been most successful with hydroxide sludge and at larger plants.

SLUDGE LAGOONS

If land is available, the use of sludge lagoons is a cost-effective nonmechanical means of storing and thickening residuals. Lagoons are commonly lined earthen basins equipped with inlet control devices and overflow structures (see Fig. 14-6). Wastes with settleable solids are discharged into the lagoons where the solids are separated by gravity sedimentation. Sludge lagoons can be classified by their mode of operation: permanent lagoons and dewatering lagoons. Permanent lagoons act as a final disposal site for settled water solids, whereas dewatering lagoons are cleaned periodically.

A common approach used at many water treatment plants in the United States is to use lagoons not only as thickeners (with continuous decanting) but also as drying beds after a predetermined filling period. Three months of filling and an average drying cycle of 3 months are the most common design parameters used. The required lagoon area can be determined using a sludge loading rate of 40 and 80 kg dry solids/m^2 of lagoon area (8.2 to 16.4 lb/ft^2) for wet and dry regions, respectively.

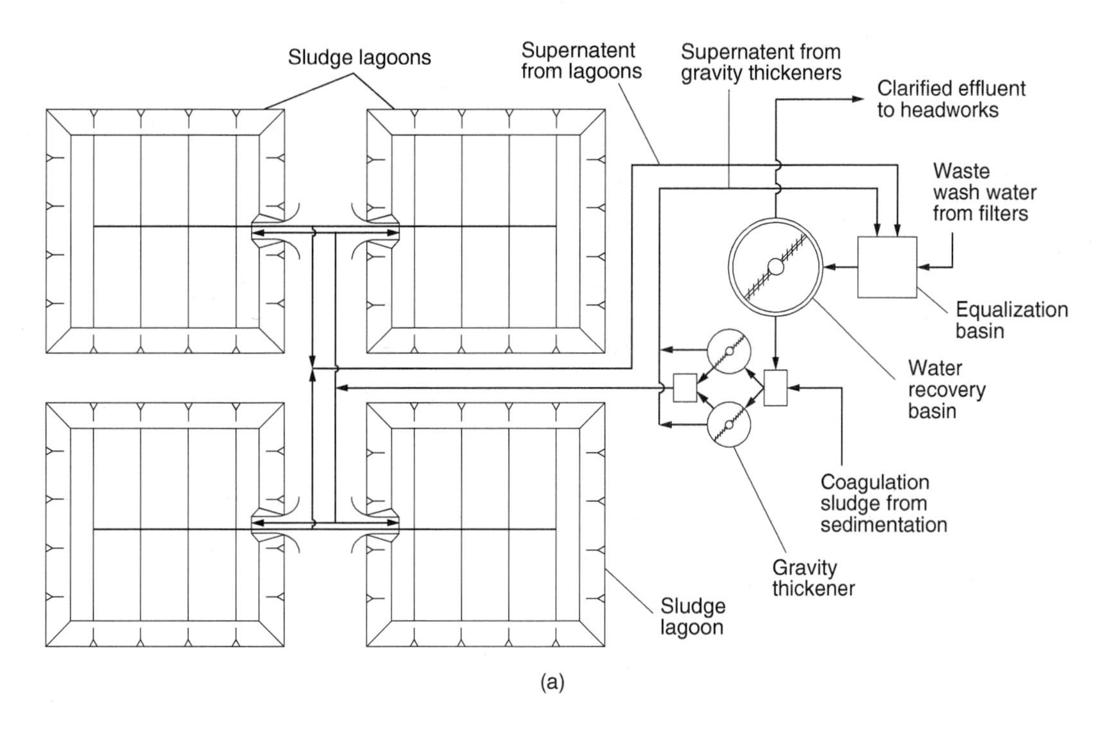

(a)

(b)

Figure 14-6
Typical sludge storage lagoons: (a) schematic (adapted from Qasim et al., 2000) and (b) view of large sludge lagoon.

For example, based on a loading rate of 80 kg/m^2, the effective area of lagoons required to handle alum sludge from the 43.2-ML/d (11.4-mgd) water treatment plant of Example 14-2 can be approximated as follows. Assuming two lagoons will be used with two loading and drying cycles per year in a dry region (e.g., ~92 days filling and ~91 days drying) the required effective area of each lagoon is 0.22 ha:

$$\text{Effective lagoon area} = \frac{(1892\ \text{kg/d})(365\ \text{d/yr})}{(80\ \text{kg/m}^2\text{·cycle·lagoon})(2\ \text{lagoon})(2\ \text{cycle/yr·lagoon})}$$

$$= 2158\ \text{m}^2/\text{lagoon} = 0.22\ \text{ha/lagoon}$$

The actual area required for a lagoon would be at least 1.5 times the area computed because of the additional area required for berms and access roads.

GRAVITY DEWATERING ON DRYING BEDS

Gravity dewatering involves placement of the sludge to be dewatered on a sand (see Fig. 14-7) or wedge wire filter surface and the subsequent drainage of water from the sludge through the filter material. A relatively dry, solid

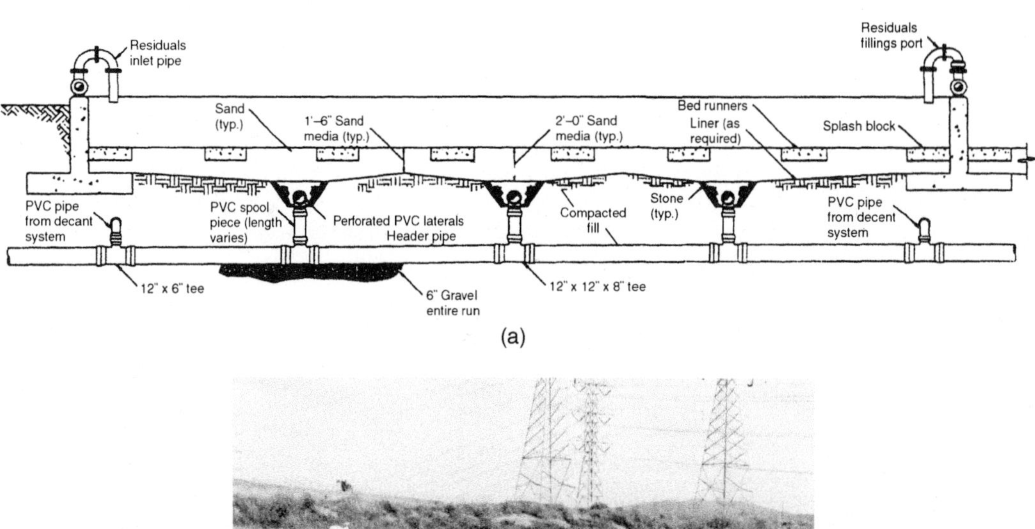

(b)

Figure 14-7
Typical sludge-drying beds for water treatment plant sludge: (a) section through sludge-drying bed and (b) view of sand-drying beds.

sludge for further treatment or disposal is produced from this process. Gravity dewatering may be combined with other drying and dewatering operations to produce a sludge of any desired dryness. Gravity dewatering is applicable to dewatering of sludge discharged directly from sedimentation basins or following thickening.

Conditioning

As in thickening, successful dewatering often depends on proper conditioning of the sludge in advance. The objectives of conditioning are to improve the physical properties of the sludge so that water will be released easily from the sludge matrix, improve the structural properties of the sludge to allow free draining of the released water, improve the solids recovery of the process (i.e., to reduce the fraction of solids lost in the removed water), and minimize dewatering process cycle times. The principal conditioning options are: chemical addition, freezing, and heat treatment.

CHEMICAL ADDITION

Polymers are the most commonly used conditioners for dewatering water treatment sludge. Based on full-scale operating experience, most types of polymers have been found to improve the dewatering characteristics of sludge. The selection of a polymer for a given application is based on bench tests or, preferably, pilot- or full-scale tests. For metal hydroxide sludge, polymer doses required are typically in the range of 10×10^{-4} to 100×10^{-4} kg polymer/kg sludge solids.

FREEZING

Freezing is very effective for metal hydroxide sludge such as alum and iron sludge (see Fig. 14-8). The effect is to destroy the gelatinous structure, leaving the sludge (after thawing) in the form of a fairly coarse granular material like sand or coffee grounds. The process is irreversible. Unfortunately, the mechanical efficiencies of equipment for freezing and thawing sludge are low, so this process is usually applied only where natural freezing will occur in a lagoon. Thus, the lagoon must have sufficient capacity to allow the sludge to sit over the winter.

Heat Treatment

Although heat treatment has been investigated as a sludge-conditioning process, results are not as dramatic as with freezing. Heat treatment of storage is not being employed at a full scale. With rising energy costs, heat treatment is not an attractive alternative for sludge conditioning.

MECHANICAL DEWATERING

Dewatering includes all those processes intended to remove free water from sludge beyond that which can be removed by decanting from a thickener. The objective is to reduce the sludge volume and produce a sludge that can be handled easily for further processing. As the use of open storage lagoons and drying beds becomes less feasible for dewatering due to the

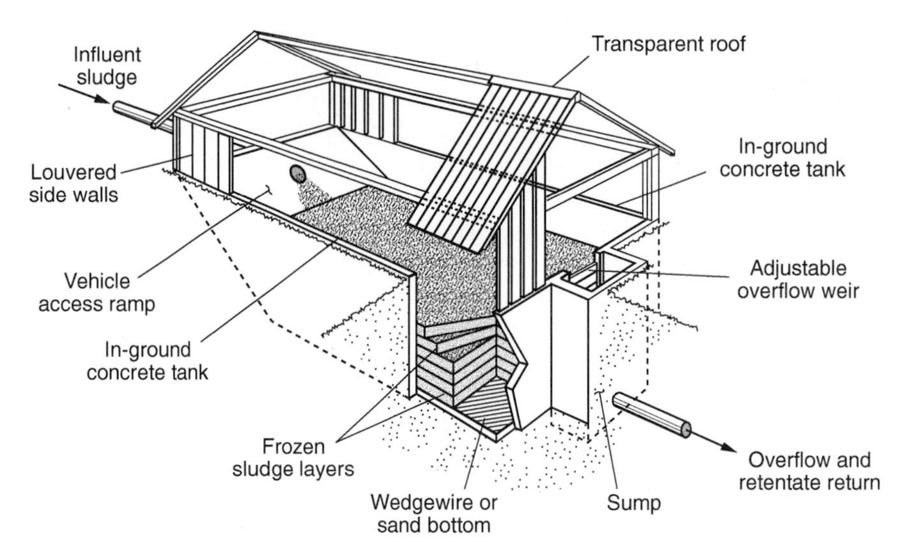

Figure 14-8
Typical installation for conditioning of sludge by freezing.

space required and potential for the formation of odors, some form of mechanical dewatering is now used at most large treatment plants. The principal types of mechanical dewatering devices now used are (1) belt filter presses, (2) centrifuges, and (2) plate-and-frame filter presses.

Gravity and Pressure Belt Filters

Thickening with a gravity belt filter involves two operational steps: (1) chemical conditioning of the sludge and (2) gravity drainage using a single belt as illustrated on Fig. 14-9a. In some designs a vacuum is applied to the underside of the belt to enhance dewatering. A belt filter press employing two belts for dewatering is illustrated schematically on Fig. 14-9b, and pictorially on Fig. 14-9c. Sludge dewatering with a two-belt filter press involves three operational steps: (1) chemical conditioning of the sludge, (2) gravity drainage, and (3) mechanical application of pressure. To accomplish the application of pressure two or more belts are used, depending on the manufacturer. For both types of belt filters (single and dual belt), the key to successful performance is the sludge chemical conditioning step.

In thickening dilute sludge, both coagulant and polymer addition is employed. Coagulant addition is used to concentrate the solids. Polymer addition is used to coagulate and flocculate the sludge before it is applied to the gravity belt thickener. Once applied to the belt thickener, the sludge is distributed uniformly across the width of the belt and moves with the belt. Fixed guide veins or plows located just above the surface of the moving belt create clear zones for free water released from the sludge to drain through the belt. Typically, from 70 to 80 percent of the free water is drained within

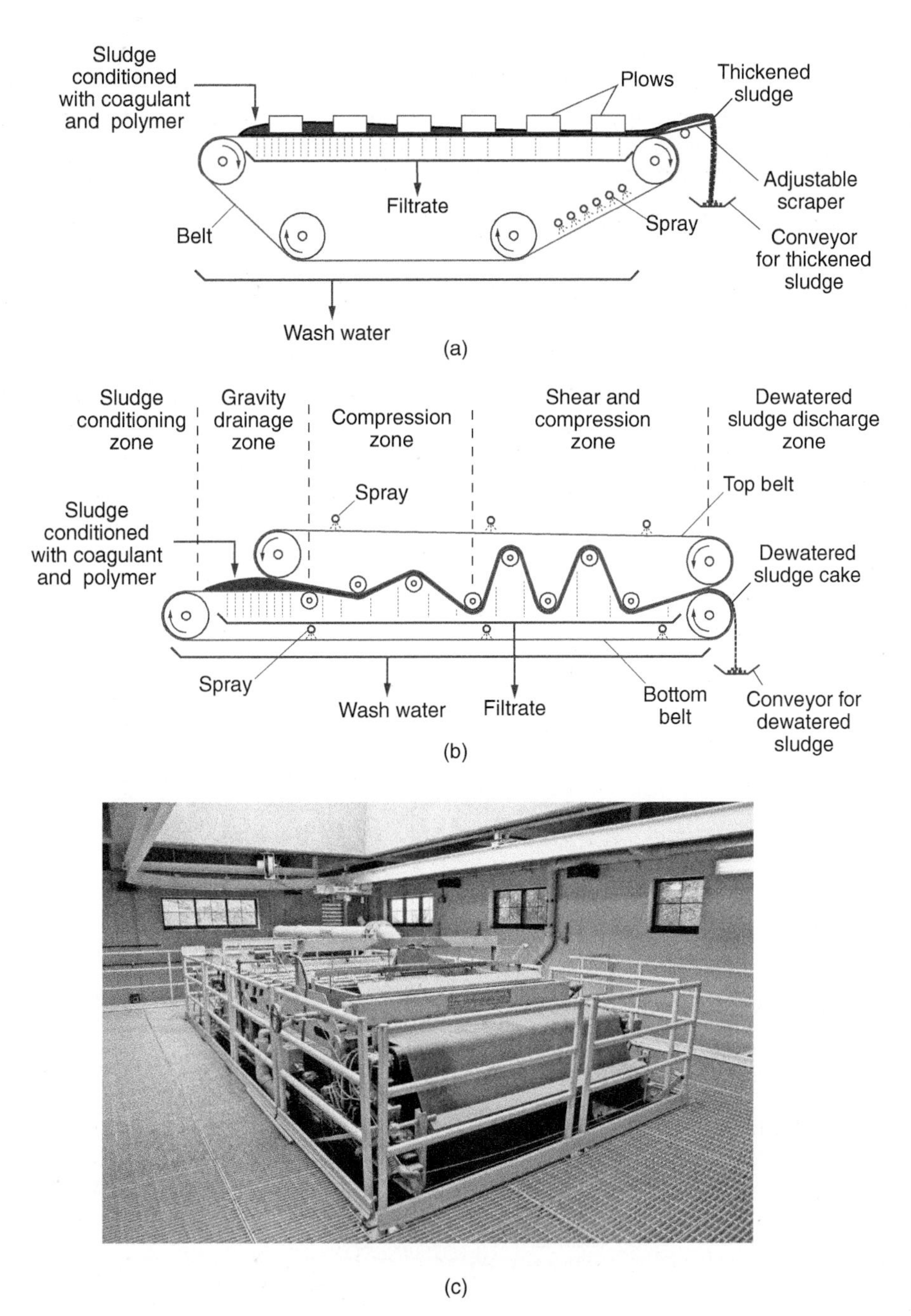

Figure 14-9
Belt filters for sludge thickening and dewatering: (a) schematic gravity belt thickener, (b) schematic belt filter press for dewatering sludge, and (c) view of typical belt filter press used for dewatering alum sludge.

the first meter. Thickened solids, scraped from the belt, are collected in a hopper for further processing, transport, or disposal. Thickened sludge cake with up to 20 percent solids is possible with proper conditioning.

CENTRIFUGES

Centrifuges are used to both thicken and dewater sludge. Twenty to 25 percent solids may be obtained from 3 to 4 percent solids alum sludge. The centrifuge is basically a sedimentation device in which the solids/liquid separation is improved by rotating the liquid at high speeds to increase the gravitational forces applied on the sludge. There are two basic types of centrifuges: (1) solid-bowl and (2) basket centrifuges. The two principal elements of centrifuges are the rotating bowl, which is the settling vessel, and the conveyor discharge of the settled solids (see Fig. 14-10).

Effective dewatering of alum sludge by centrifugation requires conditioning of the sludge with polymers and lime. Polymer doses of approximately 1 to 2 g/kg (2 to 4 lb/ton) of feed solids are typical. Feed solids concentration for alum sludge centrifugation is in the range of 1 to 6 percent, and 10 to 25 percent for lime sludge.

PLATE-AND-FRAME FILTER PRESSES

Filter press dewatering is achieved by forcing the water from the sludge under high pressure. Although the filter press produces high solids

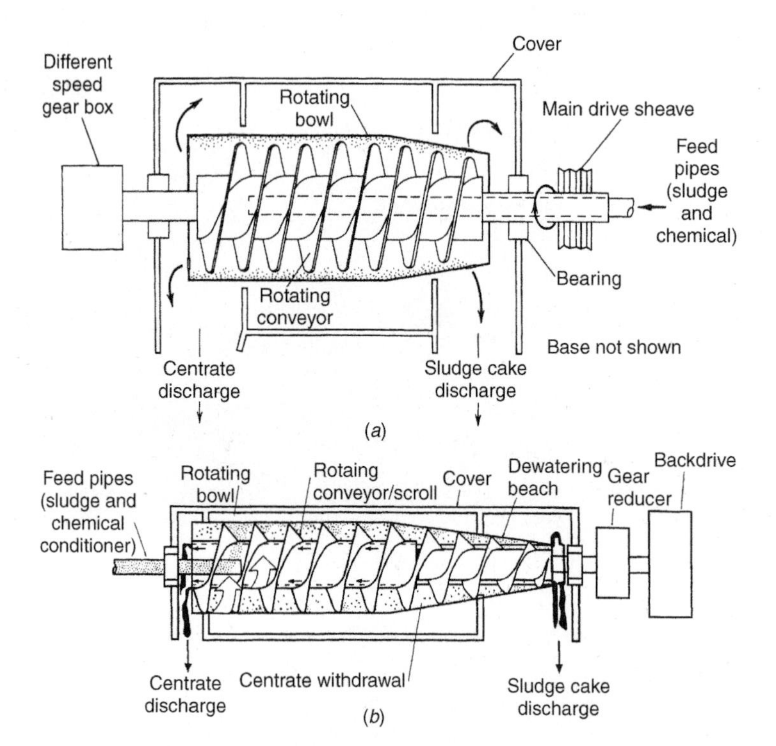

Figure 14-10
Typical centrifuges used for dewatering of water treatment plant sludge: (a) continuous countercurrent solid bowl and (b) continuous concurrent solid bowl.

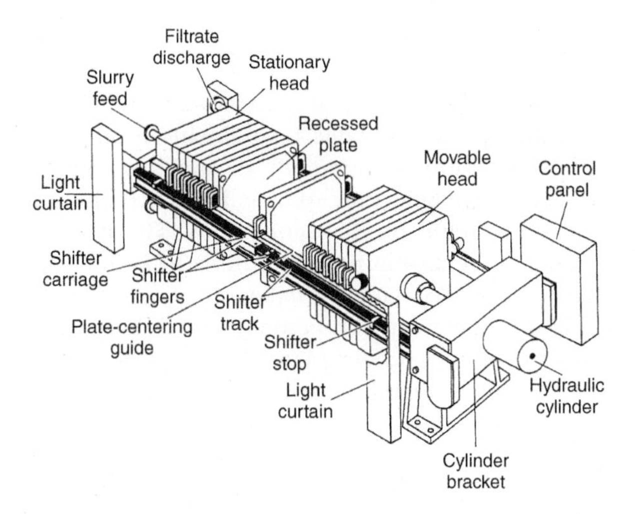

Figure 14-11
Schematic view of plate-and-frame filter press. (Adapted from Tchobanoglous et al. (2003).)

concentration and low chemical consumption, high labor costs and limitations on filter cloth life are disadvantages. A filter press consists of a number of plates or trays supported in a common frame (see Fig. 14-11). During sludge dewatering, these frames are pressed together either electromechanically or hydraulically between a fixed and moving end. A filter cloth is mounted on the face of each plate. Sludge is pumped into the press until the cavities or chambers between the trays are completely filled. Pressure is then applied, forcing the liquid through the filter cloth and plate outlet. The plates are then separated, which allows the thickened sludge cake to drop out under the force of gravity onto a collection belt below the press. Conditioning of the sludge prior to filtration is required and the degree of conditioning dictates the performance. Both capital and operation and maintenance (O&M) costs for this process are high.

Recovery of Coagulant

Alum and iron recovery can be accomplished by acidification with sulfuric acid. In simplified form, the reactions involved are

$$2Al(OH)_3 + 3H_2SO_4 \rightarrow Al_2(SO_4)_3 + 6H_2O \tag{14-10}$$

$$2Fe(OH)_3 + 3H_2SO_4 \rightarrow Fe_2(SO_4)_3 + 6H_2O \tag{14-11}$$

Normally, over 80 percent of alum and iron recovery is achieved at a pH of about 2.5. Unfortunately, heavy metals, manganese, and other organic compounds are often found in the recovered alum and iron. The presence of potential contaminants, as well as rising costs, has limited the recovery of alum and iron as a viable processing alternative. However, the recovered coagulant can be used for the treatment of wastewater. Lime recovery from lime-softeneing sludge using the recalcination process has been practiced for some time.

14-7 Ultimate Reuse and Disposal of Semisolid Residuals

Several alternatives are available for the disposal or reuse of water treatment plant residuals. In practice, the options available for ultimate disposal or reuse of water treatment plant residuals frequently dictate the type of in-plant handling system necessary. Selection of an alternative should be based on economic as well as regulatory considerations. The type of sludge and sludge characteristics are also important criteria to be used in developing disposal or reuse alternatives. It is critical that the ultimate solids disposal or reuse program be a reliable, environmentally sound practice to ensure that the primary goal of the treatment plant—the production of potable water—is not affected. Alternatives available for disposal or reuse of water treatment plant residuals include

- Landfilling
- Disposal on land (reuse as a soil amendment)
- Discharge to a wastewater collection system
- Codisposal with wastewater biosolids and
- Reuse in building or fill materials

Landfilling, land spreading, and lagoon storage followed by landfilling or spreading are typical land disposal options. Residuals disposed of in a wastewater collection system end up in the wastewater treatment plant, where they are removed and disposed of with wastewater sludge. Codisposal involves the mixing of water treatment plant residuals with wastewater treatment plant sludge followed by disposal or reuse. Reuse as building or fill material is site specific. However, before discussing the various disposal methods, it is appropriate to consider the impact of an element such as arsenic in residuals. Once arenic is present in treatment plant residuals, changes in pH or changes that result in a reducing environment may cause arsenic to resolubalize, which could lead to contamination of surface or groudwaters.

Landfilling

The most common disposal method for water treatment plant sludge in the United States is landfilling (see Fig. 14-12) in a commercial nonhazardous landfill, a monofill that receives only drinking water treatment plant residuals, or a hazardous waste landfill, which is regulated by the U.S. government.

Water treatment plant sludge is tested, using the U.S. EPA toxicity characteristic leaching procedure (TCLP), to determine if it is a hazardous (Resource Conservation and Recovery Act, RCRA, subtitle C) or nonhazardous (RCRA subtitle D) waste to determine which type of landfill is appropriate for final disposal. The TCLP test exposes a waste to a mildly acidic solution similar to what might be found in a municipal landfill (U.S. EPA, 1992). If the waste leachate contains regulated compounds at or

Figure 14-12
Dewatered sludge is placed in large storage containers that are hauled to a landfill, emptied, and returned.

above the minimum concentration in leachate for toxicity characteristics, it is considered to be toxic and, therefore, a hazardous waste.

California has more stringent regulations than the U.S. EPA and requires solid waste to be tested according to the California waste extraction test (WET) (State of California, 1991). The WET uses a slightly more aggressive leaching procedure than is used by the TCLP test, as shown in Table 14-7. Both the TCLP test and the WET are designed to simulate landfill leaching.

Table 14-7
Comparison of toxicity characteristic leaching procedure (TCLP) and the waste extraction test (WET)

	Test Procedure	
Parameter	**TCLP**[a]	**WET**[b]
Extraction fluid	Acetic acid	Citric acid
Extraction fluid pH	4.93	5.00
Extraction duration, h	18	48
Dilution of waste to extraction fluid of solid portion of waste	20-fold	10-fold
Anaerobic conditions	No	Yes, by purging with N_2 gas prior to agitation
Inorganic constituents measured	8	19
Organic constituents measured	23	18
Aggressiveness for inorganic constituents	Less	More

[a]U.S. EPA (1992).
[b]State of California (1991).

If the leachate contains any of the regulated compounds on the *List of Inorganic Persistent and Bioaccumulative Toxic Substances and Their Soluble Threshold Limit Concentration* (U.S. EPA, 1992) and the concentration of the compound is equal to or exceeds its listed soluble threshold limit concentration (STLC) or total threshold limit concentration (TTLC), the waste is considered toxic and therefore a hazardous waste.

A study of water treatment plant sludge leachate from plants that use either alum or iron as the primary coagulant was done by the American Water Works Research Foundation (Cornwell et al., 1992). The sludge was analyzed using the TCLP test and was found to be nonhazardous. Thus, landfilling of coagulant sludge in nonhazardous waste landfills is, in general, an appropriate disposal method.

Land Application

Land application of water treatment plant residuals is a disposal method that is regulated in the United States by the federal government under the Resource Conservation and Recovery Act (RCRA) as well as state and local governments. Sludge to be spread on land must be tested to determine if it is a hazardous (RCRA subtitle C) or nonhazardous (RCRA subtitle D) waste by either the TCLP or WET, which are compared in Table 14-7.

Residuals that have been land applied include coagulant sludge, lime-softening sludge, nanofiltration concentrate, and slow sand filter washings (Novak, 1993). Benefits from land application of coagulant sludge have not been clearly demonstrated (Gendebien et al., 2001). Specific concerns raised include aluminum having a negative impact on barley growth in soils where the pH is below 5.5; high levels of aluminum, reducing the availability of phosphorus and increasing soil compaction; and iron becoming concentrated in grazing land, resulting in a negative effect on copper metabolism, especially in sheep (Gendebien et al., 2001; Marshall, 2002).

14-8 Summary and Study Guide

After studying this chapter, you should be able to:

1. Define the following terms and phrases and describe the significance of each in the context of residuals management:

chemical conditioning	leachate	storage lagoons
conditioning	return flows	sludge
dewatering	residuals	supernatant flow
freezing	residuals management	underflows
heat treatment	sterilization	waste washwater

2. Describe the principal sources of semisolid, liquid, and gaseous waste residuals resulting from the treatment of water.

3. Describe the principal constituents of concern in water treatment plant residuals.
4. Calculate the specific gravity and density of various treatment plant sludge.
5. Calculate the mass and volume of sludge resulting from coagulant addition.
6. Describe the principal methods for the management of residual liquid streams.
7. Describe the principal methods for the treatment of recycle wate streams.
8. Describe the principal methods for the management of residual sludge.
9. Describe the principal methods for the reuse and diposal of semisolid resiuals.
10. Define and describe the toxicity characteristic leaching procedure (TCLP) and waste extraction test (WET) for sludge.

Homework Problems

14-1 An alum sludge contains 10 percent solids. If the density of alum is 2400 kg/m^3, estimate the density of the wet sludge at 25°C.

14-2 A ferric iron sludge contains 15 percent solids. If the density of iron is 2500 kg/m^3, estimate the density of the wet sludge at 11°C.

14-3 Determine the total that of sludge on a dry–solids basis that will be produced for each kilogram of alum [$Al_2(SO_4)_3 \cdot 6$ or 14 or $18H_2O$] added. Assume the TSS is equal to 12 mg/L. Value of bound water to be selected by instructor.

14-4 Determine the total mass of sludge on a dry–solids basis that will be produced for each kilogram of ferric sulfate [$Fe_2(SO_4)_3 \cdot 9H_2O$] and ferric chloide [$FeCl_3 \cdot 6H_2O$] added. How do the computed values compare to the values given in Table 14-4?

14-5 Determine the total mass of sludge on a dry–solids basis that will be produced for each kilogram of ferrous sulfate [$FeSO_4 \cdot 7H_2O$] added.

14-6 Determine the mass and volume of sludge produced and the volume of sludge as a percentage of the total flow from the use of alum [$Al_2(SO_4)_3 \cdot 6$ or 14 or $18H_2O$] for the removal of turbidity. Assume the following conditions apply: (1) flow rate is 0.05 m^3/s, (2) average raw-water turbidity is 45 NTU, and (3) average alum dose is 40 mg/L, sludge solids concentration is 5 percent with a corresponding specific gravity of 1.05, and temperature is 10°C. Assume the ratio between total suspended solids and turbidity expressed as NTU is 1.33 and

0.3 kg of alum sludge is produced per kilogram of alum added. Assume 1 mg/L of a coagulant aid will also be used. Value of bound water to be selected by instructor.

14-7 Determine the mass and volume of sludge produced and the volume of sludge as percentage of the total flow from the use of PACl for the removal of turbidity. Assume the following conditions apply: (1) flow rate is 0.1 m^3/s, (2) average raw-water turbidity is 20 NTU, and (3) average PACl dose is 45 mg/L, sludge solids concentration is 5 percent with a corresponding specific gravity of 1.05, and temperature is 15°C. Assume the ratio between total suspended solids and turbidity expressed as NTU is 1.5, the PACl contains 13 percent Al by weight and that 1.25 mg/L of a coagulant aid will be used.

14-8 Using a loading rate of 50 kg/m^2, estimate the effective area of lagoons required to handle alum sludge from a water treatment plant with a flow rate of 0.35 m^3/s. Assume that solids and alum dose are as described in Problem 14-6, that at least two lagoons will be used, and that two cycles (filling and drying) will be used per year. Allow an additional area of 40 percent times the lagoon area for berms and access roads.

References

Cornwell, D. A. and Roth, D. K. (2011) Water Treatment Plant Residuals Management, Chap. 22, in J.K. Edzwald (ed.)., *Water Quality and Treatment, A Handbook on Drinking Water,* 6th ed., American Water Works Association, Denver, CO.

Cornwell, D. A., Vandermeyden, C., Dillow, G., and Wang, M. (1992) *Landfilling of Water Treatment Plant Coagulant Sludges,* Environmental Engineering & Technology, American Water Works Association Research Foundation, Denver, CO.

Crittenden. J. C., Trussell, R. R., Hand, D. W., Howe, K. J., and Tchobanoglous, G. (2012) *MWH's Water Treatment: Principles and Design,* 3rd ed., Wiley, Hoboken, NJ.

Gendebien, A., Ferguson, R., Brink, J., Horth, H., Sullivan, M., Davis, R., Brunet, H., Dalimier, F., Landrea, B., Krack, D., Perot, J., and Orsi, C. (2001, July) *Survey of Wastes Spread on Land—Final Report,* WRC Study Contract B4–3040/99/110194/MAR/E3, European Commission-Directorate-General for Environment, European Communities, Luxembourg.

Marshall, T. (2002) "Sweeter Soil with Substantial Savings", *Ohio's Country J.* **11**, 5, 21–22.

Novak, J. T. (1993) *Demonstration of Cropland Application of Alum Sludges,* American Water Works Association Research Foundation, Denver, CO.

Qasim, S. R., Morley, E. M., and Zhu, G. (2000) *Water Works Engineering: Planning, Design and Operation,* Prentice-Hall PTR, Upper Saddle River, NJ.

State of California (ca. 1991) Waste Extraction Test (WET) Procedures, California Code of Regulations, Title 22, Division 4.5, Chapter 11, Article 5, Section 66261.126, Appendix II, Sacramento, CA.

Tchobanoglous, G., Burton, F. L., and Stensel, H. D. (2003) *Wastewater Engineering: Treatment and Reuse,* 4th ed., Metcalf and Eddy, McGraw-Hill, New York.

U.S. EPA (1992) *Test Methods for Evaluating Solid Waste, Physical/Chemical Methods,* EPA Publication SW-846 [Third Edition (September, 1986), as amended by Update I (July 1992)], U.S. Environmental Protection Agency, Washington, DC.

U.S. EPA, AWWA, and ASCE (1996) *Technology Transfer Handbook Management of Water Treatment Plant Residuals,* EPA/635/R-95/006, U.S. Environmental Protection Agency, Office of Research and Development, Washington DC.

APPENDIX A

Conversion Factors

Table A-1
Unit conversion factors, SI units to U.S. customary units and U.S. customary units to SI units

To convert, multiply in direction shown by arrows					
SI Unit Name	**Symbol**	→	←	**Symbol**	**U.S. Customary Unit Name**
Acceleration					
Meters per second squared	m/s^2	3.2808	0.3048	ft/s^2	Feet per second squared
Meters per second squared	m/s^2	39.3701	0.0254	$in./s^2$	Inches per second squared
Area					
Hectare (10,000 m^2)	ha	2.4711	0.4047	ac	Acre
Square centimeter	cm^2	0.1550	6.4516	$in.^2$	Square inch
Square kilometer	km^2	0.3861	2.5900	mi^2	Square mile
Square kilometer	km^2	247.1054	4.047×10^{-2}	ac	Acre
Square meter	m^2	10.7639	9.2903×10^{-2}	ft^2	Square foot
Square meter	m^2	1.1960	0.8361	yd^2	Square yard
Energy					
Kilojoule	kJ	0.9478	1.0551	Btu	British thermal unit
Joule	J	2.7778×10^{-7}	3.6×10^{6}	kWh	Kilowatt-hour
Joule	J	0.7376	1.356	$ft \cdot lb_f$	Foot-pound (force)
Joule	J	1.0000	1.0000	$W \cdot s$	Watt-second
Joule	J	0.2388	4.1876	cal	Calorie
Kilojoule	kJ	2.7778×10^{-4}	3600	kWh	Kilowatt-hour
Kilojoule	kJ	0.2778	3.600	$W \cdot h$	Watt-hour
Megajoule	kJ	0.3725	2.6845	$hp \cdot h$	Horsepower-hour
Force					
Newton	N	0.2248	4.4482	lb_f	Pound force
Flow rate					
Cubic meters per day	m^3/d	264.1720	3.785×10^{-3}	gal/d	Gallons per day
Cubic meters per day	m^3/d	2.6417×10^{-4}	3.7854×10^{3}	Mgal/d (mgd)	Million gallons per day
Cubic meters per second	m^3/s	35.3147	2.8317×10^{-2}	ft^3/s (cfs)	Cubic feet per second
Cubic meters per second	m^3/s	22.8245	4.3813×10^{-2}	Mgal/d (mgd)	Million gallons per day
Cubic meters per second	m^3/s	15,850.3	6.3090×10^{-5}	gal/min (gpm)	Gallons per minute
Million liters per day	ML/d	0.26417	3.7854	Mgal/d (mgd)	Million gallons per day
Liters per second	L/s	22,824.5	4.3813×10^{-2}	gal/d	Gallons per day

Liters per second	L/s	2.2825×10^{-2}	43.8126	Mgal/d (mgd)	Million gallons per day
Liters per second	L/s	15.8508	6.3090×10^{-2}	gal/min (gpm)	Gallons per minute
Length					
Centimeter	cm	0.3937	2.540	in.	Inch
Kilometer	km	0.6214	1.6093	mi	Mile
Meter	m	39.3701	2.54×10^{-2}	in.	Inch
Meter	m	3.2808	0.3048	ft	Foot
Meter	m	1.0936	0.9144	yd	Yard
Millimeter	mm	0.03937	25.4	in.	Inch
Mass					
Gram	g	0.0353	28.3495	oz	Ounce
Gram	g	0.0022	4.5359×10^{2}	lb	Pound
Kilogram	kg	2.2046	0.45359	lb	Pound
Megagram (10^3 kg)	Mg	1.1023	0.9072	ton	Ton (short: 2000 lb)
Megagram (10^3 kg)	Mg	0.9842	1.0160	ton	Ton (long: 2240)
Power					
Kilowatt	kW	0.9478	1.0551	Btu/s	British thermal units per second
Kilowatt	kW	1.3410	0.7457	hp	Horsepower
Watt	W	0.7376	1.3558	ft-lb_f/s	Foot-pounds (force) per second
Pressure (force/area)					
Pascal (newtons per square meter)	Pa (N/m^2)	1.4504×10^{-4}	6.8948×10^{3}	lb_f/ $in.^2$ (psi)	Pounds (force) per square inch
Pascal (newtons per square meter)	Pa (N/m^2)	2.0885×10^{-2}	47.8803	lb_f/ $in.^2$ (psi)	Pounds (force) per square foot
Pascal (newtons per square meter)	Pa (N/m^2)	2.9613×10^{-4}	3.3768×10^{3}	in. Hg	Inches of mercury
Pascal (newtons per square meter)	Pa (N/m^2)	4.0187×10^{-3}	2.4884×10^{2}	in. H_2O	Inches of water
Kilopascal (kilonewtons per square meter)	kPa (kN/m^2)	0.1450	6.8948	lb_f/ $in.^2$ (psi)	Pounds (force) per square inch
Kilopascal (kilonewtons per square meter)	kPa (kN/m^2)	0.0099	1.0133×10^{2}	atm	Atmosphere (standard)

(continued)

Table A-1 *(Continued)*

	To convert, multiply in direction shown by arrows				
SI Unit Name	**Symbol**	→	←	**Symbol**	**U.S. Customary Unit Name**
Temperature					
Degree Celsius (centigrade)	°C	1.8(°C) + 32	0.0555(°F) − 32	°F	Degree Fahrenheit
Degree kelvin	K	1.8(K) − 459.67	0.0555(°F) + 459.67	°F	Degree Fahrenheit
Velocity					
Kilometers per second	km/s	2.2369	0.44704	mi/h	Miles per hour
Meters per second	m/s	3.2808	0.3048	ft/s	Feet per second
Volume					
Cubic centimeter	cm^3	0.0610	16.3781	$in.^3$	Cubic inch
Cubic meter	m^3	35.3147	2.8317×10^{-2}	$in.^3$	Cubic foot
Cubic meter	m^3	1.3079	0.7646	yd^3	Cubic yard
Cubic meter	m^3	264.1720	3.7854×10^{-3}	gal	Gallon
Cubic meter	m^3	8.1071×10^{-4}	1.2335×10^3	ac · ft	Acre · foot
Liter	L	0.2642	3.7854	gal	Gallon
Liter	L	0.0353	28.3168	ft^3	Cubic foot
Liter	L	33.8150	2.9573×10^{-2}	oz	Ounce (U.S. fluid)

Table A-2
Conversion factors for commonly used water treatment plant design parameters

To convert, multiply in direction shown by arrows			
SI Units	→	←	**U.S. Units**
g/m^3	8.3454	0.1198	lb/Mgal
kg	2.2046	0.4536	lb
kg/ha	0.8922	1.1209	lb/acre
$kg/kW \cdot h$	1.6440	0.6083	$lb/hp \cdot h$
kg/m^2	0.2048	4.8824	lb/ft^2
kg/m^3	8345.4	1.1983×10^{-4}	lb/Mgal
$kg/m^3 \cdot d$	62.4280	0.0160	$lb/ft^3 \cdot d$
$kg/m^3 \cdot h$	0.0624	16.0185	$lb/ft^3 \cdot h$
kJ	0.9478	1.0551	Btu
kJ/kg	0.4303	2.3241	Btu/lb
kPa (gage)	0.1450	6.8948	$lb_f/in.^2$ (gage)
kPa Hg	0.2961	3.3768	in. Hg
kW/m^3	5.0763	0.197	$hp/10^3 gal$
$kW/10^3\ m^3$	0.0380	26.3342	$hp/10^3\ ft^3$
L	0.2642	3.7854	gal
L	0.0353	28.3168	ft^3
$L/m^2 \cdot d$	2.4542×10^{-2}	40.7458	$gal/ft^2 \cdot d$ (gfd)
$L/m^2 \cdot min$	0.0245	40.7458	$gal/ft^2 \cdot min$
$L/m^2 \cdot min$	35.3420	0.0283	$gal/ft^2 \cdot d$ (gfd)
m	3.2808	0.3048	ft
m/h	3.2808	0.3048	ft/h
m/h	0.0547	18.2880	ft/min
m/h	0.4090	2.4448	$gal/ft^2 \cdot min$
$m^2/10^3\ m^3 \cdot d$	0.0025	407.4611	$ft^2/Mgal \cdot d$
m^3	1.3079	0.7646	yd^3
m^3/capita	35.3147	0.0283	ft^3/capita
m^3/d	264.1720	3.785×10^{-3}	gal/d (gpd)
m^3/d	2.6417×10^{-4}	3.7854×10^3	Mgal/d (mgd)
m^3/h	0.5886	1.6990	ft^3/min
$m^3/ha \cdot d$	106.9064	0.0094	$gal/ac \cdot d$
m^3/kg	16.0185	0.0624	ft^3/lb
$m^3/m \cdot d$	80.5196	0.0124	$gal/ft \cdot d$
$m^3/m \cdot min$	10.7639	0.0929	$ft^3/ft \cdot min$
$m^3/m^2 \cdot d$	24.5424	0.0407	$gal/ft^2 \cdot d$ (gfd)
$m^3/m^2 \cdot d$	0.0170	58.6740	$gal/ft^2 \cdot min$
$m^3/m^2 \cdot d$	1.0691	0.9354	$Mgal/ac \cdot d$
$m^3/m^2 \cdot h$	3.2808	0.3048	$ft^3/ft^2 \cdot h$
$m^3/m^2 \cdot h$	589.0173	0.0017	$gal/ft^2 \cdot d$
m^3/m^3	0.1337	7.4805	ft^3/gal

(*continued*)

Table A-2 *(Continued)*

To convert, multiply in direction shown by arrows			
SI Units	→	←	**U.S. Units**
$m^3/10^3 m^3$	133.6805	7.04805×10^{-3}	ft^3/Mgal
$m^3/m^3 \cdot min$	133.6805	7.04805×10^{-3}	$ft^3/10^3\ gal \cdot min$
$m^3/m^3 \cdot min$	1000.0	0.001	$ft^3/10^3\ ft^3 \cdot min$
Mg/ha	0.4461	2.2417	ton/ac
mm	3.9370×10^{-2}	25.4	in.
ML/d	0.2642	3.785	Mgal/d (mgd)
ML/d	0.4087	2.4466	ft^3/s

APPENDIX B

Physical Properties of Selected Gases and Composition of Air

Table B-1
Molecular weight and density of gases at standard conditions (0°C, 1 atm)

Gas	Formula	Molecular Weight, g/mol	Density, g/L
Air	—	28.97[a]	1.2928
Ammonia	NH_3	17.03	0.7708
Carbon dioxide	CO_2	44.00	1.9768
Carbon monoxide	CO	28.00	1.2501
Hydrogen	H_2	2.016	0.0898
Hydrogen sulfide	H_2S	34.08	1.5392
Methane	CH_4	16.03	0.7167
Nitrogen	N_2	28.02	1.2507
Oxygen	O_2	32.00	1.4289

Source: Adapted from R. H. Perry, D. W. Green, and J. O. Maloney (1984) *Chemical Engineers' Handbook*, 6th ed., McGraw-Hill, New York.
[a]Values reported in the literature vary depending on the standard conditions. *Note:* $(0.7803 \times 28.02) + (0.2099 \times 32.00) + (0.0094 \times 39.95) + (0.0003 \times 44.00) = 28.97$ (see percents of gases in Table B-2).

Table B-2
Composition of dry air at 0°C and 1.0 atm

Gas	Formula	Percent by Volume	Percent by Weight
Nitrogen	N_2	78.03	75.47
Oxygen	O_2	20.99	23.18
Argon	Ar	0.94	1.30
Carbon dioxide	CO_2	0.03	0.05
Other[a]	—	0.01	—

Source: Adapted from *North American Combustion Handbook,* 2nd ed., North American Mfg., Cleveland, OH.
[a]Hydrogen, neon, helium, krypton, and xenon.

Table B-3
Density and viscosity of air at 1 atm (SI units)

Temperature T (°C)	Density ρ (kg/m^3)	Dynamic Viscosity[a,b] μ ($\times 10^{-5}$ kg/m · s)	Kinematic Viscosity ν ($\times 10^{-5}$ m^2/s)
0	1.293	1.736	1.343
5	1.269	1.762	1.388
10	1.247	1.787	1.433
15	1.225	1.812	1.479
20	1.204	1.837	1.525
25	1.184	1.862	1.572
30	1.165	1.886	1.619
35	1.146	1.910	1.667
40	1.127	1.934	1.716
45	1.110	1.958	1.765
50	1.093	1.982	1.814
60	1.060	2.029	1.915
70	1.029	2.075	2.017
80	1.000	2.121	2.121
90	0.972	2.166	2.228
100	0.946	2.210	2.336

[a]Dynamic viscosity can also be expressed in units of N · s/m^2.
[b]Dynamic viscosity calculated at <http://www.lmnoeng.com/Flow/GasViscosity.htm>.

B-1 Density of Air at Other Temperatures

The following relationship can be used to compute the density of air, ρ_a, at other temperatures at atmospheric pressure:

$$\rho_a = \frac{PM}{RT}$$

where ρ_a = density of air, g/m^3
P = pressure, N/m^2
M = molecular weight of air (see Table B-1), 28.97 g/mol
R = universal gas constant, $8.314 \, N \cdot m/mol \cdot K$
T = temperature, K (273.15 + °C)

For example, at 20°C and 1 atm ($1.01325 \times 10^5 \, N/m^2$), the density of air is

$$\rho_{a,20^\circ C} = \frac{(1.01325 \times 10^5 \, N/m^2)(28.97 \, g/mol)}{(8.314 \, N \cdot m/mol \cdot K)(293.15 \, K)}$$

$$= 1204 \, g/m^3 = 1.204 \, kg/m^3$$

B-2 Change in Atmospheric Pressure with Elevation

The following relationship can be used to compute the change in atmospheric pressure with elevation:

$$\frac{P_b}{P_a} = \exp\left[-\frac{gM(z_b - z_a)}{RT}\right]$$

where P_b = pressure at elevation z_b, N/m^2
P_a = atmospheric pressure at sea level, $1.01325 \times 10^5 \, N/m^2$
g = acceleration due to gravity, $9.81 \, m/s^2$
M = molecular weight of air (see Table B-1), 28.97 g/mol
z = elevation, m
R = universal gas constant, $8.314 \, N \cdot m/mol \cdot K$
T = temperature, K (273.15 + °C)

APPENDIX C

Physical Properties of Water

Table C-1
Physical properties of water (SI units)

Temperature T (°C)	Specific Weight γ (kN/m^3)	Density[a] ρ (kg/m^3)	Dynamic Viscosity[b] μ ($\times 10^{-3}$ kg/m·s)	Kinematic Viscosity ν ($\times 10^{-6}$ m^2/s)	Surface Tension[c] σ (N/m)	Modulus of Elasticity[a] E ($\times 10^9$ N/m^2)	Vapor Pressure P_v (kN/m^2)
0	9.805	999.8	1.781	1.785	0.0765	1.98	0.61
5	9.807	1000.0	1.518	1.519	0.0749	2.05	0.87
10	9.804	999.7	1.307	1.306	0.0742	2.10	1.23
15	9.798	999.1	1.139	1.139	0.0735	2.15	1.70
20	9.789	998.2	1.002	1.003	0.0728	2.17	2.34
25	9.777	997.0	0.890	0.893	0.0720	2.22	3.17
30	9.764	995.7	0.798	0.800	0.0712	2.25	4.24
40	9.730	992.2	0.653	0.658	0.0696	2.28	7.38
50	9.689	988.0	0.547	0.553	0.0679	2.29	12.33
60	9.642	983.2	0.466	0.474	0.0662	2.28	19.92
70	9.589	977.8	0.404	0.413	0.0644	2.25	31.16
80	9.530	971.8	0.354	0.364	0.0626	2.20	47.34
90	9.466	965.3	0.315	0.326	0.0608	2.14	70.10
100	9.399	958.4	0.282	0.294	0.0589	2.07	101.33

Source: Adapted from J. K. Venard and R. L. Street (1975). *Elementary Fluid Mechanics*, 5th ed., Wiley, New York.
[a]At atmospheric pressure.
[b]Dynamic viscosity can also be expressed in units of N·s/m^2.
[c]In contact with air.

The following equations can be used to compute the density ρ_w (kg/m^3) and dynamic viscosity μ_w (kg/m·s) at other temperatures:

$$\rho_w = \frac{\left[\begin{array}{c} 999.83952 + 16.945176(T) - 7.9870401 \times 10^{-3}(T)^2 \\ -46.170461 \times 10^{-6}(T)^3 + 105.56302 \times 10^{-9}(T)^4 - 280.54253 \times 10^{-12}(T)^5 \end{array}\right]}{1 + 16.879850 \times 10^{-3}(T)}$$

For $0 < T < 20°C$, $\mu_w = 10^{-3}(10^A)$

$$\text{where} \quad A = \frac{1301}{998.333 + 8.1855(T - 20) + 0.00585(T - 20)^2} - 1.30223$$

For $20 < T < 100°C$,

$$\mu_w = 1.002 \times 10^{-3}(10^B)$$

$$\text{where} \quad B = \frac{[1.3272(20 - T) - 0.001053(T - 20)^2]}{T + 105}$$

(see R. C. Weast, (1983) *CRC Handbook of Chemistry and Physics*, 64th ed. CRC Press, Boca Raton, FL.)

Table C-2
Physical properties of water (U.S. customary units)

Temperature T (°F)	Specific Weight γ (lb/ft^3)	Density[a] ρ (slug/ft^3)	Dynamic Viscosity μ (× 10^{-5} lb·s/ft^2)	Kinematic Viscosity ν (× 10^{-5} ft^2/s)	Surface Tension[b] σ (lb/ft)	Modulus of Elasticity[a] E (10^3 lb$_f$/in.^{2})	Vapor Pressure p_v (lb$_f$/in.^{2})
32	62.42	1.940	3.746	1.931	0.00518	287	0.09
49	62.43	1.940	3.229	1.664	0.00614	296	0.12
50	62.41	1.940	2.735	1.410	0.00509	305	0.18
60	62.37	1.938	2.359	1.217	0.00504	313	0.26
70	62.30	1.936	2.050	1.059	0.00498	319	0.36
80	62.22	1.934	1.799	0.930	0.00492	324	0.51
90	62.11	1.931	1.595	0.826	0.00486	328	0.70
100	62.00	1.927	1.424	0.739	0.00480	331	0.95
110	61.86	1.923	1.284	0.667	0.00473	332	1.27
120	61.71	1.918	1.168	0.609	0.00467	332	1.69
130	61.55	1.913	1.069	0.558	0.00460	331	2.22
140	61.38	1.908	0.981	0.514	0.00454	330	2.89
150	61.20	1.902	0.905	0.476	0.00447	328	3.72
160	61.00	1.896	0.838	0.442	0.00441	326	4.74
170	60.80	1.890	0.780	0.413	0.00434	322	5.99
180	60.58	1.883	0.726	0.385	0.00427	318	7.51
190	60.36	1.876	0.678	0.362	0.00420	313	9.34
200	60.12	1.868	0.637	0.341	0.00413	308	11.52
212	59.83	1.860	0.593	0.319	0.00404	300	14.70

Source: Adapted from J. K. Venard and R. L. Street (1975) *Elementary Fluid Mechanics*, 5th ed., Wiley, New York.
[a]At atmospheric pressure.
[b]In contact with the air.

APPENDIX D

Periodic Table

PERIODIC TABLE

Atomic Properties of the Elements

NIST — National Institute of Standards and Technology — U.S. Department of Commerce

Physics Laboratory — physics.nist.gov

Standard Reference Data — www.nist.gov/srd

Frequently used fundamental physical constants

For the most accurate values of these and other constants, visit physics.nist.gov/constants

1 second = 9 192 631 770 periods of radiation corresponding to the transition between the two hyperfine levels of the ground state of ^{133}Cs

Constant	Symbol	Value	
speed of light in vacuum	c	299 792 458 m s^{-1}	(exact)
Planck constant	h	6.6261×10^{-34} J s	($\hbar = h/2\pi$)
elementary charge	e	1.6022×10^{-19} C	
electron mass	m_e	9.1094×10^{-31} kg	
	$m_e c^2$	0.5110 MeV	
proton mass	m_p	1.6726×10^{-27} kg	
fine-structure constant	α	1/137.036	
Rydberg constant	R_∞	10 973 732 m^{-1}	
	$R_\infty c$	$3.289\ 842 \times 10^{15}$ Hz	
	$R_\infty hc$	13.6057 eV	
Boltzmann constant	k	1.3807×10^{-23} J K^{-1}	

Key: Solids / Liquids / Gases / Artificially Prepared

Groups: 1 IA, 2 IIA, 3 IIIB, 4 IVB, 5 VB, 6 VIB, 7 VIIB, 8–10 VIII, 11 IB, 12 IIB, 13 IIIA, 14 IVA, 15 VA, 16 VIA, 17 VIIA, 18 VIIIA. Periods 1–7. Lanthanides (57–71) and Actinides (89–103) shown separately.

Legend (example): Atomic Number 58; Ground-state Level $^1G^o_4$; Symbol Ce; Name Cerium; Atomic Weight† 140.116; Ground-state Configuration [Xe]$4f5d6s^2$; Ionization Energy (eV) 5.5387

Atomic Number	Symbol	Name	Ground-state Level	Atomic Weight†	Ground-state Configuration	Ionization Energy (eV)
1	H	Hydrogen	$^2S_{1/2}$	1.00794	$1s$	13.5984
2	He	Helium	$^1S_0$	4.002602	$1s^2$	24.5874
3	Li	Lithium	$^2S_{1/2}$	6.941	$1s^22s$	5.3917
4	Be	Beryllium	$^1S_0$	9.012182	$1s^22s^2$	9.3227
5	B	Boron	$^2P^o_{1/2}$	10.811	$1s^22s^22p$	8.2980
6	C	Carbon	$^3P_0$	12.0107	$1s^22s^22p^2$	11.2603
7	N	Nitrogen	$^4S^o_{3/2}$	14.0067	$1s^22s^22p^3$	14.5341
8	O	Oxygen	$^3P_2$	15.9994	$1s^22s^22p^4$	13.6181
9	F	Fluorine	$^2P^o_{3/2}$	18.9984032	$1s^22s^22p^5$	17.4228
10	Ne	Neon	$^1S_0$	20.1797	$1s^22s^22p^6$	21.5645
11	Na	Sodium	$^2S_{1/2}$	22.98976928	[Ne]$3s$	5.1391
12	Mg	Magnesium	$^1S_0$	24.3050	[Ne]$3s^2$	7.6462
13	Al	Aluminum	$^2P^o_{1/2}$	26.9815386	[Ne]$3s^23p$	5.9858
14	Si	Silicon	$^3P_0$	28.0855	[Ne]$3s^23p^2$	8.1517
15	P	Phosphorus	$^4S^o_{3/2}$	30.973762	[Ne]$3s^23p^3$	10.4867
16	S	Sulfur	$^3P_2$	32.065	[Ne]$3s^23p^4$	10.3600
17	Cl	Chlorine	$^2P^o_{3/2}$	35.453	[Ne]$3s^23p^5$	12.9676
18	Ar	Argon	$^1S_0$	39.948	[Ne]$3s^23p^6$	15.7596
19	K	Potassium	$^2S_{1/2}$	39.0983	[Ar]$4s$	4.3407
20	Ca	Calcium	$^1S_0$	40.078	[Ar]$4s^2$	6.1132
21	Sc	Scandium	$^2D_{3/2}$	44.955912	[Ar]$3d4s^2$	6.5615
22	Ti	Titanium	$^3F_2$	47.867	[Ar]$3d^24s^2$	6.8281
23	V	Vanadium	$^4F_{3/2}$	50.9415	[Ar]$3d^34s^2$	6.7462
24	Cr	Chromium	$^7S_3$	51.9961	[Ar]$3d^54s$	6.7665
25	Mn	Manganese	$^6S_{5/2}$	54.938045	[Ar]$3d^54s^2$	7.4340
26	Fe	Iron	$^5D_4$	55.845	[Ar]$3d^64s^2$	7.9024
27	Co	Cobalt	$^4F_{9/2}$	58.933195	[Ar]$3d^74s^2$	7.8810
28	Ni	Nickel	$^3F_4$	58.6934	[Ar]$3d^84s^2$	7.6399
29	Cu	Copper	$^2S_{1/2}$	63.546	[Ar]$3d^{10}4s$	7.7264
30	Zn	Zinc	$^1S_0$	65.38	[Ar]$3d^{10}4s^2$	9.3942
31	Ga	Gallium	$^2P^o_{1/2}$	69.723	[Ar]$3d^{10}4s^24p$	5.9993
32	Ge	Germanium	$^3P_0$	72.64	[Ar]$3d^{10}4s^24p^2$	7.8994
33	As	Arsenic	$^4S^o_{3/2}$	74.92160	[Ar]$3d^{10}4s^24p^3$	9.7886
34	Se	Selenium	$^3P_2$	78.96	[Ar]$3d^{10}4s^24p^4$	9.7524
35	Br	Bromine	$^2P^o_{3/2}$	79.904	[Ar]$3d^{10}4s^24p^5$	11.8138
36	Kr	Krypton	$^1S_0$	83.798	[Ar]$3d^{10}4s^24p^6$	13.9996
37	Rb	Rubidium	$^2S_{1/2}$	85.4678	[Kr]$5s$	4.1771
38	Sr	Strontium	$^1S_0$	87.62	[Kr]$5s^2$	5.6949
39	Y	Yttrium	$^2D_{3/2}$	88.90585	[Kr]$4d5s^2$	6.2173
40	Zr	Zirconium	$^3F_2$	91.224	[Kr]$4d^25s^2$	6.6339
41	Nb	Niobium	$^6D_{1/2}$	92.90638	[Kr]$4d^45s$	6.7589
42	Mo	Molybdenum	$^7S_3$	95.96	[Kr]$4d^55s$	7.0924
43	Tc	Technetium	$^6S_{5/2}$	(98)	[Kr]$4d^55s^2$	7.28
44	Ru	Ruthenium	$^5F_5$	101.07	[Kr]$4d^75s$	7.3605
45	Rh	Rhodium	$^4F_{9/2}$	102.90550	[Kr]$4d^85s$	7.4589
46	Pd	Palladium	$^1S_0$	106.42	[Kr]$4d^{10}$	8.3369
47	Ag	Silver	$^2S_{1/2}$	107.8682	[Kr]$4d^{10}5s$	7.5762
48	Cd	Cadmium	$^1S_0$	112.411	[Kr]$4d^{10}5s^2$	8.9938
49	In	Indium	$^2P^o_{1/2}$	114.818	[Kr]$4d^{10}5s^25p$	5.7864
50	Sn	Tin	$^3P_0$	118.710	[Kr]$4d^{10}5s^25p^2$	7.3439
51	Sb	Antimony	$^4S^o_{3/2}$	121.760	[Kr]$4d^{10}5s^25p^3$	8.6084
52	Te	Tellurium	$^3P_2$	127.60	[Kr]$4d^{10}5s^25p^4$	9.0096
53	I	Iodine	$^2P^o_{3/2}$	126.90447	[Kr]$4d^{10}5s^25p^5$	10.4513
54	Xe	Xenon	$^1S_0$	131.293	[Kr]$4d^{10}5s^25p^6$	12.1298
55	Cs	Cesium	$^2S_{1/2}$	132.9054519	[Xe]$6s$	3.8939
56	Ba	Barium	$^1S_0$	137.327	[Xe]$6s^2$	5.2117
57	La	Lanthanum	$^2D_{3/2}$	138.90547	[Xe]$5d6s^2$	5.5769
58	Ce	Cerium	$^1G^o_4$	140.116	[Xe]$4f5d6s^2$	5.5387
59	Pr	Praseodymium	$^4I^o_{9/2}$	140.90765	[Xe]$4f^36s^2$	5.473
60	Nd	Neodymium	$^5I_4$	144.242	[Xe]$4f^46s^2$	5.5250
61	Pm	Promethium	$^6H^o_{5/2}$	(145)	[Xe]$4f^56s^2$	5.582
62	Sm	Samarium	$^7F_0$	150.36	[Xe]$4f^66s^2$	5.6437
63	Eu	Europium	$^8S^o_{7/2}$	151.964	[Xe]$4f^76s^2$	5.6704
64	Gd	Gadolinium	$^9D^o_2$	157.25	[Xe]$4f^75d6s^2$	6.1498
65	Tb	Terbium	$^6H^o_{15/2}$	158.92535	[Xe]$4f^96s^2$	5.8638
66	Dy	Dysprosium	$^5I_8$	162.500	[Xe]$4f^{10}6s^2$	5.9389
67	Ho	Holmium	$^4I^o_{15/2}$	164.93032	[Xe]$4f^{11}6s^2$	6.0215
68	Er	Erbium	$^3H_6$	167.259	[Xe]$4f^{12}6s^2$	6.1077
69	Tm	Thulium	$^2F^o_{7/2}$	168.93421	[Xe]$4f^{13}6s^2$	6.1843
70	Yb	Ytterbium	$^1S_0$	173.054	[Xe]$4f^{14}6s^2$	6.2542
71	Lu	Lutetium	$^2D_{3/2}$	174.9668	[Xe]$4f^{14}5d6s^2$	5.4259
72	Hf	Hafnium	$^3F_2$	178.49	[Xe]$4f^{14}5d^26s^2$	6.8251
73	Ta	Tantalum	$^4F_{3/2}$	180.94788	[Xe]$4f^{14}5d^36s^2$	7.5496
74	W	Tungsten	$^5D_0$	183.84	[Xe]$4f^{14}5d^46s^2$	7.8640
75	Re	Rhenium	$^6S_{5/2}$	186.207	[Xe]$4f^{14}5d^56s^2$	7.8335
76	Os	Osmium	$^5D_4$	190.23	[Xe]$4f^{14}5d^66s^2$	8.4382
77	Ir	Iridium	$^4F_{9/2}$	192.217	[Xe]$4f^{14}5d^76s^2$	8.9670
78	Pt	Platinum	$^3D_3$	195.084	[Xe]$4f^{14}5d^96s$	8.9588
79	Au	Gold	$^2S_{1/2}$	196.966569	[Xe]$4f^{14}5d^{10}6s$	9.2255
80	Hg	Mercury	$^1S_0$	200.59	[Xe]$4f^{14}5d^{10}6s^2$	10.4375
81	Tl	Thallium	$^2P^o_{1/2}$	204.3833	[Hg]$6p$	6.1082
82	Pb	Lead	$^3P_0$	207.2	[Hg]$6p^2$	7.4167
83	Bi	Bismuth	$^4S^o_{3/2}$	208.98040	[Hg]$6p^3$	7.2855
84	Po	Polonium	$^3P_2$	(209)	[Hg]$6p^4$	8.414
85	At	Astatine	$^2P^o_{3/2}$	(210)	[Hg]$6p^5$	
86	Rn	Radon	$^1S_0$	(222)	[Hg]$6p^6$	10.7485
87	Fr	Francium	$^2S_{1/2}$	(223)	[Rn]$7s$	4.0727
88	Ra	Radium	$^1S_0$	(226)	[Rn]$7s^2$	5.2784
89	Ac	Actinium	$^2D_{3/2}$	(227)	[Rn]$6d7s^2$	5.3807
90	Th	Thorium	$^3F_2$	232.03806	[Rn]$6d^27s^2$	6.3067
91	Pa	Protactinium	$^4K_{11/2}$	231.03588	[Rn]$5f^26d7s^2$	5.89
92	U	Uranium	$^5L^o_6$	238.02891	[Rn]$5f^36d7s^2$	6.1939
93	Np	Neptunium	$^6L_{11/2}$	(237)	[Rn]$5f^46d7s^2$	6.2657
94	Pu	Plutonium	$^7F_0$	(244)	[Rn]$5f^67s^2$	6.0260
95	Am	Americium	$^8S^o_{7/2}$	(243)	[Rn]$5f^77s^2$	5.9738
96	Cm	Curium	$^9D^o_2$	(247)	[Rn]$5f^76d7s^2$	5.9914
97	Bk	Berkelium	$^6H^o_{15/2}$	(247)	[Rn]$5f^97s^2$	6.1979
98	Cf	Californium	$^5I_8$	(251)	[Rn]$5f^{10}7s^2$	6.2817
99	Es	Einsteinium	$^4I^o_{15/2}$	(252)	[Rn]$5f^{11}7s^2$	6.3676
100	Fm	Fermium	$^3H_6$	(257)	[Rn]$5f^{12}7s^2$	6.50
101	Md	Mendelevium	$^2F^o_{7/2}$	(258)	[Rn]$5f^{13}7s^2$	6.58
102	No	Nobelium	$^1S_0$	(259)	[Rn]$5f^{14}7s^2$	6.65
103	Lr	Lawrencium	$^2P^o_{1/2}$?	(262)	[Rn]$5f^{14}7s^27p$?	4.9 ?
104	Rf	Rutherfordium	$^3F_2$?	(265)	[Rn]$5f^{14}6d^27s^2$?	6.0?
105	Db	Dubnium		(268)		
106	Sg	Seaborgium		(271)		
107	Bh	Bohrium		(272)		
108	Hs	Hassium		(277)		
109	Mt	Meitnerium		(276)		
110	Ds	Darmstadtium		(281)		
111	Rg	Roentgenium		(280)		
112	Cn	Copernicium		(285)		
113	Uut	Ununtrium		(284)		
114	Uuq	Ununquadium		(289)		
115	Uup	Ununpentium		(288)		
116	Uuh	Ununhexium		(293)		
117	Uus	Ununseptium		(294)		
118	Uuo	Ununoctium		(294)		

†Based upon ^{12}C. () indicates the mass number of the longest-lived isotope.

For a description of the data, visit physics.nist.gov/data

NIST SP 966 (September 2010)

APPENDIX E

Electronic Resources Available on the John Wiley & Sons Website for This Textbook

Website URL: http://www.wiley.com/go/mwh

Table or Resource	Filename	Description
E1	Standard_Reduction_Potentials.pdf	Selected standard reduction potentials for inorganic compounds at 25°C.
E2	Ozone_Reactions.pdf	Reactions of ozone with inorganic and organic compounds.
E3	Extinction_Coefficients.pdf	Extinction coefficients for common inorganic chemicals.
E4	Hydroxyl_Rate_Constants.pdf	Second-order rate constants between hydroxyl radical and various species in water.
E5	Freundlich_Isotherm_Parameters.xlsx	Freundlich isotherm parameters K and $1/n$ for various organic compounds in aqueous and gaseous phases.
E10	AirStripCalc.xlsx	Spreadsheet designed to facilitate the calculations for countercurrent packed tower design.

Index